REINFORCED CONCRETE DESIGN

Fifth Edition

Chu-Kia Wang
Charles G. Salmon
University of Wisconsin — Madison

HarperCollins*Publishers*

Project Editor: Kristin Syverson
Art Direction: Julie Anderson
Text and Cover Design: Lucy Lesiak Design, Inc.: Cynthia Crampton
Cover Photo: Chicago Photographic Company
Production: Jeanie Berke, Beth Maglione, and Kathy Donnelly
Compositor: Syntax International
Printer and Binder: R. R. Donnelley & Sons Company
Cover Printer: Lehigh Press Lithographers

Reinforced Concrete Design, Fifth Edition
Copyright © 1992 by HarperCollins Publishers Inc.

Library of Congress Cataloging-in-Publication Data

Wang, Chu-Kia
 Reinforced concrete design/Chu-Kia Wang, Charles G. Salmon.—5th ed.
  p. cm.
 Includes bibliographical references and index.
 ISBN 0-06-046887-4
 1. Reinforced concrete construction. I. Salmon, Charles G.
 II. Title.
TA683.2.W3 1992
624.1′8341—dc20

91-29497
CIP

95 96 97 9 8 7 6 5 4

Preface

The content of the fifth edition of this text reflects the continuing change occurring in design procedures for reinforced concrete structures. The strength design philosophy is now well established. Current design recognizes that the "limit state" is reached when the *strength* of the member is reached, or when the member is not *serviceable* (such as when it deflects too much). Either *strength* or *serviceability* may control design; thus, both limit states must be considered.

The fifth edition incorporates the changes in design rules arising from the publication of the 1989 American Concrete Institute (ACI) Building Code and Commentary. The 1989 ACI Code includes those changes from the 1986 Supplement as well as those approved by ACI Committee 318 (Standard Building Code) since 1986.

This edition continues the modest treatment of SI units used in previous editions. The 1989 ACI Code has an SI version (known as ACI 318-89M), and the SI version of the ACI Code equations appear in this book as footnote equations with the same equation number. According to the ACI Code, the designer must use in its entirety either the Inch-Pound units version (ACI 318-89) or the SI version (ACI 318-89M). The authors believe that sufficient metrication should be included in a text on reinforced concrete so that some familiarity may be gained with SI units, but not so much as to interfere with learning the basic concepts of concrete design; constant conversion back and forth between Inch-Pound and SI units is more confusing than using *either one* exclusively. The text provides data on reinforcing bars in accordance with ASTM (American Society for Testing and Materials) Inch-Pound units, and also ASTM SI units (the bar sizes and strengths used in Canada but not yet actually used in the United States). Some design tables are provided for bars and material strengths in SI units, a few numerical examples are given in SI units, and many problems at the ends of chapters are given with an SI alternate in parentheses at the end of the problem statement.

Regarding the choice between the Standard Metric unit of force (kilogram force, kgf) or the SI unit of force (newton = kilogram meter per second squared), the authors have concluded that use of the newton in accordance with ASTM Standard E380 has a reasonable chance of becoming the accepted metric approach in the United States. Thus, in all parts of the book where metric versions are used, the newton (N) or kilonewton (kN) is used to measure force. The SI unit of stress is the pascal (Pa), or newton per meter squared, which (because of its typically large numerical value) is usually expressed in megapascals (MPa); that is, 10^6 pascals. A few diagrams have along the stress axis the kilogram force per centimeter squared (kgf/cm^2) in addition to Inch-Pound and SI units. In accordance with ASTM E380, the units traditionally used in the United States continue to be referred to as *Inch-Pound*, rather than the *US Customary* designation used in this text prior to the fourth edition. For the convenience of the reader, some conversion factors for forces, stresses, uniform loading, and moments are provided on a separate page immediately following the preface.

This new edition follows the same philosophical approach that has gained wide acceptance of users since the first edition was published in 1965. Herein, as previously, strength and behavior of concrete elements are treated with the primary objective of explaining and justifying the ACI Code rules and formulas.* Then numerous examples are presented illustrating the general approach to design and analysis. Considering the limited scope of most examples, attempts to reach practical results are made insofar as possible.

Considerable emphasis is placed on presenting for the student, as well as the practicing engineer, the basic concepts deemed essential to understand and apply properly the ACI Code rules and formulas. The treatment is incorporated into the chapters in such a way that the reader may either study in detail the concepts in logical sequence, or merely accept a qualitative explanation and proceed directly to the design process using the ACI Code.

Depending on the proficiency required of the student, this book may provide material for two courses of three or four semester-hours each. It is suggested that the beginning course in concrete structures for undergraduate students might contain the material of Chapters 1 through 9, 13, and the spread footing portion of Chapter 20, excepting Sections 5.14 through 5.17 and 13.18 through 13.21. In addition, the first portion of Chapter 21, "Introduction to Prestressed Concrete," is recommended for the first course. The second course may start with the continuous beam in Chapter 10, using that topic to review many of the subjects in Chapters 1 through 9. The remaining chapters—particularly Chapter 14 on deflections; Chapter 15 on length effects on compression members; Chapter 16 on two-way slab systems; the remaining sections of Chapter 5 on shear strength affected by axial force, deep beams, and brackets; Chapter 19 on torsion; and Chapter 21 on prestressed concrete—are suggested for inclusion.

Special features of the fifth edition are (a) completely rewritten Chapter 6 on development of reinforcement, in accordance with the new and more elaborate 1989 ACI Code rules; (b) completely revised portions of examples relating to the new development rules in Chapter 8 on one-way slabs, Chapter 10 on continuous beams and joists, Chapter 12 on retaining walls, Chapter 16 on two-way slab systems, and Chapter 20 on footings; (c) revised treatment of slab-column moment transfer provisions in Chapters 16 and 20; (d) revised treatment of shear strength requirements for deep beams in Chapter 5; and (e) completely revised Chapter 11 on monolithic beam-to-column joints.

— — — —

* Since nearly continuous reference is made to the 1989 ACI Code and Commentary, the reader will find it desirable to secure a copy of it from the American Concrete Institute, P.O. Box 19150, Detroit, MI 48219. Note that for the first time in 1989 the Code and Commentary are contained in a single document.

This complete revision has retained (striving to improve clarity and wording) the important features of the fourth edition, including (a) detailed treatment of beam deflections, in Chapter 14; (b) comprehensive treatment of design for torsion, in Chapter 19; (c) introductory treatment of yield line theory for slabs, in Chapter 18; (d) in addition to the new development length provisions, the important extensive treatment of the development length concepts and the use of the moment capacity diagram, in Chapter 6; (e) special analysis techniques for using equivalent frame in the analysis of slab-frame systems, in Chapter 17; (f) treatment of shear strength requirements for members subject to axial forces, and brackets and corbels, in Chapter 5; (g) prestressed concrete treatment necessary for understanding the basic concept, in Chapter 21; and (h) introductory treatment of composite concrete-on-concrete construction, in Chapter 22.

The authors continue to be indebted to students, colleagues, and other users of the first four editions, who have suggested improvements of wording, identified errors, and recommended items for inclusion or omission. These suggestions have been carefully considered and the result is reflected in this complete revision. The comments and suggestions of the reviewers of the fifth edition manuscript are also appreciated.

Users of this fifth edition are urged to communicate with the authors regarding all aspects of this book, particularly identification of errors and suggestions for improvement.

The authors again gratefully acknowledge the continuing patience and encouragement of their wives. To Vera Wang and in memory of the late Bette Salmon we affectionately dedicate this book.

Chu-Kia Wang
Charles G. Salmon

Conversion Factors

Some Conversion Factors, Between Inch-Pound Units and SI Units, Useful in Reinforced Concrete Design

	TO CONVERT	TO	MULTIPLY BY
Forces	kip force	kN	4.448
	N	lb	0.2248
Stresses	ksi	MPa	6.895
	MPa	ksi	0.1450
	psf	N/m^2	47.88
	N/m^2	psf	0.02089
Uniform Loading	kip/ft	kN/m	14.59
	kN/m	lb/ft	68.52
Moments	ft-kip	kN·m	1.356
	kN·m	ft-kip	0.7376

Basis (ASTM E380): 1 in. = 25.4 mm; 1 lb force = 4.448 221 615 260 5 newtons; 1 kg force = 9.80665 newtons.

Contents

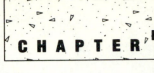

CHAPTER 1

Introduction, Materials, and Properties

1.1 REINFORCED CONCRETE STRUCTURES

The three most common materials from which most structures are built are wood, steel, and reinforced (including prestressed) concrete. Lightweight materials such as aluminum and plastics are also becoming more common in use. Reinforced concrete is unique in that two materials, reinforcing steel and concrete, are used together; thus the principles governing the structural design in reinforced concrete differ in many ways from those involving design in one material.

Many structures are built of reinforced concrete: bridges, viaducts, buildings, retaining walls, tunnels, tanks, conduits, and others. This text deals primarily with fundamental principles in the design and investigation of reinforced concrete members subjected to axial force, bending moment, shear, torsion, or combinations of these. These principles are basically applicable to the design of any structure, so long as information is known about the variation of axial force, shear, moment, etc., along the length of each member. Although *analysis* and *design* may be treated separately, they are inseparable in practice, especially in the case of reinforced concrete structures, which are usually statically indeterminate. In such cases, relative sizes of members are needed in the preliminary analysis that must precede the final design; so the final conciliation between analysis and design is largely a matter of trial, judgment, and experience.

Reinforced concrete is a logical union of two materials: plain concrete, which possesses high compressive strength but little tensile strength, and steel bars embedded in the concrete, which can provide the needed strength in tension. For instance, the strength of the beam shown in Fig. 1.1.1 is greatly increased by placing steel bars in the tension zone. However, since reinforcement steel is capable of resisting compression as well as tension, it is also used to provide part of the carrying capacity in reinforced concrete columns, and frequently in the compression zone of beams.

1

311 South Wacker Drive, Chicago, 967 ft high, world's tallest concrete building, shown in front of the Sears Tower (1454 ft world's tallest steel building). Completed January 1990. (Photo courtesy Steven M. Baldridge.)

Figure 1.1.1 Position of bars in a reinforced concrete beam.

Steel and concrete work readily in combination for several reasons: (1) bond (interaction between bars and surrounding hardened concrete) prevents slip of the bars relative to the concrete; (2) proper concrete mixes provide adequate impermeability of the concrete against water intrusion and bar corrosion; and (3) sufficiently similar rates of thermal expansion—that is, 0.0000055 to 0.0000075 for concrete and 0.0000065 for steel per degree Fahrenheit (°F), or 0.000010 to 0.000013 for concrete and 0.000012 for steel per degree Celsius (°C)—introduce negligible forces between steel and concrete under atmospheric changes of temperature.

Transverse cracks of small width may appear near the steel bars placed in the tension regions of ordinary reinforced concrete (unless prestressed); such cracks are expected and do not interfere with the performance of the member.

1.2 HISTORICAL BACKGROUND

Joseph Monier, the owner of an important nursery in Paris, generally deserves the credit for making the first practical use of reinforced concrete in 1867. He recognized many of its potential uses, and successfully undertook to expand the application of the new method [1.1].* Prior to Monier's work, however, the method of reinforcing concrete with iron was known and in some cases even protected by patents. Ancient Grecian structures have been found which show that builders knew something about the reinforcing of stonework for added strength [1.2].

In the mid-1800s, Lambot in France constructed a small boat which he exhibited at the ·Paris Exposition of 1854 and on which he received a patent in 1855. In Lambot's patent was shown a reinforced concrete beam and a column reinforced with four round iron bars. Another Frenchman, François Coignet, published a book in 1861 describing many applications and uses of reinforced concrete. In 1854, W. B. Wilkinson of England took out a patent for a reinforced concrete floor.

Monier acquired his first French patent in 1867 for iron reinforced concrete tubs. This was followed by his many other patents, such as for pipes and tanks in 1868,

---

* Numbers in brackets refer to the Selected References at the end of the chapter.

flat plates in 1869, bridges in 1873, and stairways in 1875. In 1880–1881, Monier received German patents for railroad ties, water feeding troughs, circular flower pots, flat plates, and irrigation channels, among others. Monier's iron reinforcement was made mainly to conform to the contour of the structural element and generally strengthen it. He apparently had no quantitative knowledge regarding its behavior or any method of making design calculations [1.1].

In the United States, the pioneering efforts were made by Thaddeus Hyatt, originally a lawyer, who conducted experiments on reinforced concrete beams in the 1850s. In a perfectly correct manner, the iron bars in Hyatt's beams were located in the tension zone, bent up near the supports, and anchored in the compression zone. Additionally, transverse reinforcement (known as vertical stirrups) was used near the supports. However, Hyatt's experiments were unknown until 1877 when he published his work privately.

The first cast-in-place reinforced concrete structure in the United States is generally credited to the William Ward house in Port Chester, New York, built in 1870 [1.3]. E. L. Ransome, head of the Concrete-Steel Company of San Francisco, apparently used some form of reinforced concrete in the early 1870s. He continued to increase the application of wire rope and hoop iron to many structures and was the first to use and patent in 1884 the deformed (twisted) bar.

In 1890, Ransome built the Leland Stanford Jr. Museum in San Francisco, a reinforced concrete building two stories high and 312 ft (95 m) long. Since that time, development of reinforced concrete in the United States has been rapid. During the period 1891–1894, various investigators in Europe published theories and test results; among them were Moeller (Germany), Wunsch (Hungary), Melan (Austria), Hennebique (France), and Emperger (Hungary), but practical use was less extensive than in the United States.

Throughout the entire period 1850–1900, relatively little was published, as the engineers working in the reinforced concrete field considered construction and computational methods as trade secrets. One of the first publications that might be classified as a textbook was that of Considère in 1899. By the turn of the century, there was a multiplicity of systems and methods with little uniformity in design procedures, allowable stresses, and systems of reinforcing. In 1903, with the formation in the United States of a joint committee of representatives of all organizations interested in reinforced concrete, uniform application of knowledge to design was initiated. The development of standard specifications is discussed in Chapter 2.

In the first decade of the twentieth century, progress in reinforced concrete was rapid. Extensive testing to determine beam behavior, compressive strength of concrete, and modulus of elasticity was conducted by Arthur N. Talbot at the University of Illinois, by Frederick E. Turneaure and Morton O. Withey at the University of Wisconsin, and by Bach in Germany, among others. From about 1916 to the mid-1930s, research centered on axially loaded column behavior. In the late 1930s and 1940s, eccentrically loaded columns, footings, and the ultimate strength of beams received special attention.

Since the mid-1950s, reinforced concrete design practice has made the transition from that based on elastic methods to one based on strength. Prestressed concrete (Chapter 21), wherein the steel reinforcement is stressed in tension and the concrete is in compression even before external loads are applied, has advanced from an experimental technique to a major structural material. There has been a transition from cast-in-place reinforced concrete to elements precast at a manufacturer's plant and shipped to the job site for assembly.

Recent summaries of concrete building construction in the United States are given by Cohen and Heun [1.3] and by the American Concrete Institute [1.4].

Understanding of reinforced concrete behavior is still far from complete; building codes and specifications that give design procedures are continually changing to reflect latest knowledge.

1.3 CONCRETE

Plain concrete is made by mixing cement, fine aggregate, coarse aggregate, water, and frequently admixtures. When reinforcing steel is placed in the forms and wet concrete mix is placed around it, the final solidified mass becomes reinforced concrete (see Fig. 1.3.1). The strength of concrete depends on many factors: notably the proportion of the ingredients and the conditions of temperature and moisture under which it is placed and cured.

Subsequent sections contain brief discussions of the materials in and the properties of plain concrete. The treatment is intended to be only introductory; an interested reader should consult standard references devoted entirely to the subject of plain concrete [1.5, 1.6, 1.7].

Figure 1.3.1 Cross-section of concrete. Cement and water paste completely coats each aggregate particle and fills all of the space between particles. (Courtesy of Portland Cement Association.)

1.4 CEMENT

Cement is a material that has adhesive and cohesive properties enabling it to bond mineral fragments into a solid mass. Although this definition can apply to many materials, the cements of interest for reinforced concrete construction are those that can set and harden in the presence of water—the so-called *hydraulic cements.* These consist primarily of silicates and aluminates of lime made from limestone and clay (or shale) which is ground, blended, fused in a kiln, and crushed to a powder. Such cements chemically combine with water (hydrate) to form a hardened mass. The usual hydraulic cement used for reinforced concrete is known as *portland cement,* because of its resemblance when hardened to Portland stone found near Dorset, England. The name was originated in a patent obtained by Joseph Aspdin of Leeds, England, in 1824.

Concrete made with portland cement ordinarily requires about 14 days to attain adequate strength so that forms can be removed and construction and dead loads carried. The design strength of such concrete is reached at about 28 days. This ordinary portland cement is identified by ASTM (American Society for Testing and Materials) C150 [1.8] as Type I. Other types of portland cement and their intended uses are given in Table 1.4.1. ACI Committee 225 provides a guide for selection and use of hydraulic cements [1.9].

There are also several categories of blended hydraulic cements (ASTM C595 [1.10]), such as *portland blast-furnace slag cement, portland-pozzolan cement, slag cement, pozzolan-modified portland cement,* and *slag-modified portland cement. Air-entraining portland cement* may be referred to for Types I, II, and III given in Table 1.4.1 by using the designation IA, IIA, or IIIA.

Air-entraining portland cement contains a chemical admixture finely ground with the cement to produce intentionally air bubbles on the order of 0.02 in. (0.5 mm) diameter uniformly distributed throughout the concrete. Such air entrainment will give the concrete improved durability against frost action, as well as better workability. Air-entraining agents may be added to the first three types of cement in ASTM C150 [1.8] or to the blended hydraulic cements in ASTM C595 [1.10] at the time the concrete is mixed.

Portland blast-furnace-slag cement has lower heat of hydration than ordinary Type I cement and is useful for mass concrete structures such as dams; and because

Table 1.4.1 Types of Portland Cement[a]

TYPE	USES
I	Ordinary construction where special properties are not required
II	Ordinary construction when moderate sulfate resistance or moderate heat of hydration is desired
III	When high early strength is desired
IV	When low heat of hydration is desired
V	When high sulfate resistance is desired

[a] According to ASTM C150 [1.8].

of its high sulfate resistance, it is used in seawater construction. *Portland-pozzolan cement* is a blended mixture of ordinary Type I cement with pozzolan. *Pozzolan* is a finely divided siliceous or siliceous and aluminous material which possesses little or no inherent cementitious property, but in the powdery form and in the presence of moisture, will chemically react with calcium hydroxide at ordinary temperatures to form compounds possessing cementitious properties. Blended cements with pozzolan gain strength more slowly than cements without pozzolan; hence they produce less heat during hydration, and thus are widely used in mass concrete construction.

1.5 AGGREGATES

Since aggregate usually occupies about 75% of the total volume of concrete, its properties have a definite influence on the behavior of hardened concrete. Not only does the strength of the aggregate affect the strength of the concrete, its properties also greatly affect durability (resistance to deterioration under freeze–thaw cycles). Since aggregate is less expensive than cement, it is logical to use the largest percentage feasible. In general, for maximum strength, durability, and best economy, the aggregate should be packed and cemented as densely as possible. Hence aggregates are usually graded by size and a proper mix specifies percentages of both *fine* and *coarse* aggregates.

Fine aggregate (sand) is any material passing through a No. 4 sieve* [i.e., less than about $\frac{3}{16}$-in. (5 mm) diameter]. Coarse aggregate (gravel) is any material of larger size. The nominal maximum size of coarse aggregate permitted (ACI-3.3.3)[†] is governed by the clearances between sides of forms and between adjacent bars and may not exceed "(a) $\frac{1}{5}$ the narrowest dimension between sides of forms, nor (b) $\frac{1}{3}$ the depth of slabs, nor (c) $\frac{3}{4}$ the minimum clear spacing between individual reinforcing bars. . . ." Additional information concerning aggregate selection and use is to be found in a report of ACI Committee 221 [1.12].

Natural stone aggregates conforming to ASTM C33 [1.13] are used in the majority of concrete construction, giving a unit weight for such concrete of about 145 pcf (pounds per cubic foot) or 2320 kg/m^3 (kilograms per cubic meter). When steel reinforcement is added, the unit weight of *normal-weight* reinforced concrete is taken for calculation purposes as 150 pcf, or 2400 kg/m^3. Actual weights for concrete and steel are rarely, if ever, computed separately. For special purposes, lightweight or extra heavy aggregates are used.

Structural lightweight concretes are usually made from aggregates conforming to ASTM C330 [1.14] which are produced artificially in a kiln, such as expanded clays and shales. The unit weight of such concretes typically ranges from 70 to

---

* 4.75 mm according to ASTM Standard E11.

† Numbers refer to sections in the "ACI Code," officially ACI 318–89, *Building Code Requirements for Reinforced Concrete* [1.11].

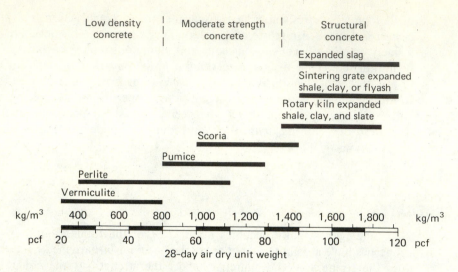

Figure 1.5.1 Approximate unit weight and use classification of lightweight aggregate concretes (from Ref. 1.15).

115 pcf (see Fig. 1.5.1). Lightweight concretes ranging down to 30 pcf (often known as cellular concretes) are also used for insulating purposes and for masonry units. When lightweight materials are used for both coarse and fine aggregates in structural concrete, it is termed *all-lightweight* concrete. When only the coarse aggregate is of lightweight material but normal weight sand is used for the fine aggregate, it is said to be *sand-lightweight* concrete. Often the term "sand replacement" is used in connection with lightweight concrete. This refers to replacing all or part of the lightweight aggregate fines with natural sand.

Steiger [1.15] provides historical background for the use of lightweight aggregate concrete, Mackie [1.16] has discussed recent uses, and ACI Committees 213 and 523 have guides for the use of structural lightweight aggregate concrete [1.17], cast-in-place low density concrete [1.18], and cellular concretes [1.19].

Heavyweight, high-density concrete is used for shielding against gamma and x radiation in nuclear reactor containers and other structures [1.20]. Naturally occurring iron ores, titaniferous iron ores, "hydrous iron ores" (i.e., containing bound and adsorbed water), and barites are crushed to suitable size for use as aggregates. Heavyweight concretes typically weigh from 200 to 350 pcf (3200 to 5600 kg/m³).

1.6 ADMIXTURES

In addition to cement, coarse and fine aggregates, and water, other materials known as *admixtures* may be added to the concrete mix immediately before or during the mixing. Admixtures may be used to modify the properties of the concrete to make it better serve its intended use or for better economy.

Some of the important purposes of admixtures are as follows:

1. To increase resistance to deterioration resulting from freeze–thaw cycles and the use of ice-removal salts (air-entraining admixtures, under ASTM C260).
2. To increase workability without increasing water content, or to decrease the water content at the same workability (finely divided materials including pozzolans, such as fly ash, are generally used for this purpose, under ASTM C618).
3. To accelerate the rate of strength development at early ages (calcium chloride is the best known and most widely used accelerator).
4. To retard the setting and thereby reduce heat evolution (ASTM C494 admixtures).
5. To increase the strength (water-reducing and set-controlling admixtures, ASTM C494, Chemical Admixtures for Concrete).

Air-entraining admixtures are probably the most widely used type. Air entrainment provides a high degree of resistance to the disruptive action of freezing and thawing and of deicing chemicals. Plasticity and workability are also improved, permitting a reduction in water content. Uniformity of placement with little segregation can be achieved. In addition, air-entrained concrete is more watertight and has greater resistance to sulfate action. For exposed concrete, the possible reduced strength (less than 15%) is far less important than the improved resistance to frost action.

ACI Committee 212 [1.21, 1.22] provides the essential guidance for use of admixtures. Mielenz [1.23] has provided a history of admixture use, and practical information on admixtures is available in *Concrete Construction* [1.24].

1.7 COMPRESSIVE STRENGTH

The strength of concrete is controlled by the proportioning of cement, coarse and fine aggregates, water, and various admixtures. The ratio of water to cement is the chief factor for determining concrete strength, as shown in Fig. 1.7.1. The lower the water–cement ratio, the higher the compressive strength. A recent review of the water–cement ratio relationship has been made by Popovics [1.81]. A certain minimum amount of water is necessary for the proper chemical action in the hardening of concrete; extra water increases the workability (how easily the concrete will flow) but reduces strength. A measure of the workability is obtained by a *slump test*. A truncated cone-shaped metal mold 12 in. (300 mm) high is filled with fresh concrete, the mold is then lifted off, and a measurement is made of the distance the top of the wet mass "slumps" from its original position before the mold was removed. The smaller the slump, the stiffer and less workable the mix. In building construction, a 3 to 4 in. (75 to 100 mm) slump is common. Vibration of the concrete mix will greatly improve workability and even very stiff no-slump concrete can be placed [1.25].

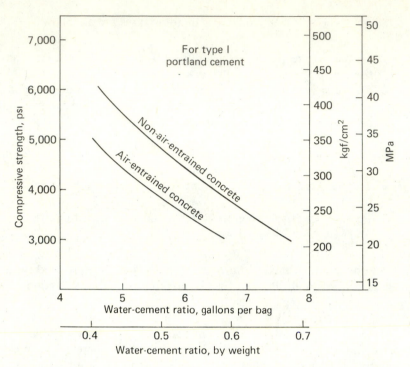

Figure 1.7.1 Effect of water–cement ratio on 28-day compressive strength. Average values for concrete containing 1.5 to 2% trapped air for non-air-entrained concrete, and no more than 5 to 6% air for air-entrained concrete (curves drawn from data in Ref. 1.27, Table 6.3.4a).

Information regarding proportioning of concrete mixes is available in *Design and Control of Concrete Mixtures* [1.26], as well as in ACI Standard 211.1 for normal-weight, heavyweight, and mass concrete [1.27], ACI Standard 211.2 for structural lightweight concrete [1.28], and ACI Standard 211.3 for no-slump concrete [1.25]. Actual strength of concrete in place in the structure is also greatly affected by quality control procedures for placement and inspection. Details regarding good practice are to be found in ACI Standard 304 [1.29], and in the *ACI Manual of Concrete Inspection* [1.30].

The strength of concrete is denoted in the United States by f'_c, which is the compressive strength in psi of test cylinders 6 in. (150 mm) in diameter by 12 in. (300 mm) high measured on the 28th day after they are made. In many other parts of the world, the standard test unit is the cube, frequently of 200 mm to a side.

Since nearly all reinforced concrete behavior is related to the standard 28-day compressive strength f'_c, it is important to note that such strength depends on the size and shape of the test specimen and the manner of testing. Factors such as tensile strength of concrete and size of the contact area of the testing machine have more effect on cube strength than on cylinder strength. As an average, the 6 × 12 in. (150 × 300 mm) cylinder strength is 80% of the 150-mm cube strength and 83% of the 200-mm cube strength [1.31]. For lightweight concrete, cylinder strength and cube strength are nearly equal.

In the United States, when taking test cores of concrete in existing structures, specimens less than the standard 6 in. diameter but having a 2:1 height-to-diameter ratio are usually cut from the concrete.

An interesting discussion of the cylinder test is given by Shilstone [1.32]. ACI Committee 214 has a recommended practice [1.34] for evaluating strength test results, and Tait [1.33] has discussed the use of test results. When an assessment of the strength of in-place concrete is desired, procedures ranging from tests of cylindrical cores cut from the structure to the use of nondestructive tests [1.35–1.41] are available.

The stress–strain behavior of concrete is dependent on its strength, age at loading, rate of loading, aggregates and cement properties, and type and size of specimens [1.42, 1.43]. Typical curves for specimens loaded in compression at 28 days using normal testing speeds are shown in Fig. 1.7.2.

Figure 1.7.2 Typical stress–strain curves for concrete in compression under short-time loading. (Curves represent a compromise adapted from curves and results given by Wang, Shah, and Naaman [1.43], Bertero [1.44], Naaman [1.45], and Nilson [1.46].)

Figure 1.7.3 Stress–strain curves for various strain rates of concentric loading (from Rüsch [1.47]).

The rate of applying strain during testing has a great influence on the result as shown by Fig. 1.7.3. Because of this strain rate, there is no agreement on what constitutes "typical curves." Particularly the portion of the curve after the maximum stress is reached varies considerably in test results.

One may note in Fig. 1.7.2 that lower-strength concrete has greater deformability (ductility) than higher-strength concrete, and the maximum stress is reached at a compressive strain between 0.002 and 0.0025. Ultimate strain at crushing of concrete varies from 0.003 to as high as 0.008; however, the maximum usable strain for practical cases is 0.003–0.004. The ACI Building Code [1.11] states (ACI-10.2.3): "Maximum usable strain at extreme concrete compression fiber shall be assumed equal to 0.003." The ACI maximum strain of 0.003 may not be conservative for high-strength concrete having f'_c in the range of 8000 to 12,000 psi (55 to 83 MPa).

In usual reinforced concrete design, specified concrete strengths f'_c of 3000 to 4000 psi (roughly 20 to 30 MPa) are used for nonprestressed structures, and 5000 to 6000 psi (roughly 35 to 42 MPa) for prestressed concrete. For special situations, particularly in the columns of tall buildings, concretes ranging from 6000 to 14,000 psi (42 to 97 MPa) have been used [1.48–1.50]. On the Pacific First Center in Seattle, the specified strength was 14,000 psi at 56 days [1.48]. The average strength obtained throughout the project was about 18,000 psi. Research continues on high-strength concrete [1.51, 1.52].

1.8 TENSILE STRENGTH

The strength of concrete in tension is an important property that greatly affects the extent and size of cracking in structures. Tensile strength is usually determined by using the *split-cylinder* test in accordance with ASTM C496 [1.53] in which the same size cylinder used for the compression test is placed on its side in the testing machine so that the compression load P is applied uniformly along the length of the cylinder in the direction of the diameter. The cylinder will split in half when the tensile strength is reached. The stress is computed by $2P/[\pi(\text{diameter})(\text{length})]$ based on theory of elasticity for a homogeneous material in a biaxial state of stress.* Tensile strength is a more variable property than compressive strength, and is about 10 to 15% of it. The split-cylinder tensile strength f_{ct} has been found to be proportional to $\sqrt{f'_c}$,[†] such that

$$f_{ct} = 6\sqrt{f'_c} \quad \text{to} \quad 7\sqrt{f'_c} \text{ psi} \quad \text{for normal-weight concrete}^{\ddagger}$$

$$f_{ct} = 5\sqrt{f'_c} \quad \text{to} \quad 6\sqrt{f'_c} \text{ psi} \quad \text{for lightweight concrete}^{\ddagger}$$

The ACI Code has indirectly used $f_{ct} = 6.7\sqrt{f'_c}$ psi for normal-weight concrete, $f_{ct} = 5.7\sqrt{f'_c}$ for sand-lightweight concrete, and $f_{ct} = 5\sqrt{f'_c}$ for all-lightweight concrete (ACI-11.2).

    Tensile strength in flexure, known as *modulus of rupture*, measured in accordance with ASTM C78 [1.54] is also important when considering cracking and deflection of beams. The modulus of rupture f_r, computed from the flexure formula $f = Mc/I$, gives higher values for tensile strength than the split-cylinder test, primarily because the concrete compressive stress distribution is not linear when tensile failure is imminent as is assumed in the computation of the nominal Mc/I stress. It is generally accepted (ACI-9.5.2.3) that an average value for the modulus of rupture f_r may be taken as $7.5\sqrt{f'_c}$ $(0.62\sqrt{f'_c}\text{ MPa})$ for normal-weight concrete and 75% of that value for all-lightweight concrete. Because of the large variability in modulus of rupture, as shown in Fig. 1.8.1, the selection of the coefficient 7.5, or even the entire expression $7.5\sqrt{f'_c}$, should be viewed as a practical choice for design purposes.

    One may note that neither the split-cylinder nor the modulus of rupture tensile strength is correctly a measure of the strength under axial tension. However, axial tension is difficult to measure accurately and when compared with the modulus of rupture or split-cylinder strength it does *not* give better correlation with tension-related failure behavior such as flexural cracking in beams, inclined cracking from shear and torsion, and splitting from interaction of reinforcing bars with surrounding concrete.

— — — —

* See, for example, S. Timoshenko and J. N. Goodier, *Theory of Elasticity*, 2nd ed., McGraw-Hill, 1951, pp. 85, 107.

† $\sqrt{f'_c}$ is in psi units; thus $f'_c = 3000$ psi, $\sqrt{f'_c} = 54.8$ psi. When f'_c is in kilogram-force per square centimeter (kgf/cm²), the constant in front of $\sqrt{f'_c}$ is to be multiplied by 0.265; when f'_c is in newtons per square millimeter, that is, megapascals (MPa), the constant in front of $\sqrt{f'_c}$ is to be multiplied by 0.083.

‡ In SI, with f'_c and f_{ct} in MPa,

$$f_{ct} = 0.5\sqrt{f'_c} \quad \text{to} \quad 0.6\sqrt{f'_c} \quad \text{for normal-weight concrete}$$

$$f_{ct} = 0.4\sqrt{f'_c} \quad \text{to} \quad 0.5\sqrt{f'_c} \quad \text{for lightweight concrete}$$

Figure 1.8.1 Comparison of test results for modulus of rupture with ACI Code expression (adapted from Mirza, Hatzinikolas, and MacGregor [1.55]).

1.9 MODULUS OF ELASTICITY

The modulus of elasticity of concrete varies, unlike that of steel, with strength. It also depends, though to a much lesser extent, on the age of concrete, properties of the aggregates and cement, rate of loading, and the type and size of specimen. Furthermore, since concrete exhibits some permanent set even under small loads, there are various definitions of the modulus of elasticity.

Referring to Fig. 1.9.1, representing a typical stress–strain curve for concrete in compression, the initial modulus, the tangent modulus, and the secant modulus are noted. Usually the secant modulus at from 25 to 50% of the compressive strength f'_c is considered to be the modulus of elasticity. For many years the modulus was approximated adequately as $1000f'_c$ by the ACI Code; but with the rapidly increasing use of lightweight concrete, the variable of density needed to be included. As a result of a statistical analysis of available data, the empirical formula given by ACI-8.5.1

Figure 1.9.1 Stress–strain curve for concrete in compression.

$$E_c = 33w_c^{1.5}\sqrt{f'_c} \qquad (1.9.1)*$$

was developed [1.56] for values of w_c between 90 and 155 lb/cu ft. Equation (1.9.1) may be considered as the secant modulus for a compressive stress at service load level. (A recent review of the applicability of Eq. (1.9.1) has been made by Shih, Lee, and Chang [1.82].) For normal-weight concrete weighing 145 pcf, the formula gives $E_c = 57{,}600\sqrt{f'_c}$. For normal-weight concrete ACI-8.5.1 suggests

$$E_c = 57{,}000\sqrt{f'_c} \qquad (1.9.2)^\dagger$$

Values of modulus of elasticity for various concrete strengths appear in Table 1.9.1.

— — — —

* For SI, with w_c in kg/m³ and E_c and f'_c in MPa,

$$E_c = 0.043w_c^{1.5}\sqrt{f'_c} \quad \text{(ACI 318–89M)} \qquad (1.9.1)$$

† For SI, with E_c and f'_c in MPa,

$$E_c = 5000\sqrt{f'_c} \quad \text{(ACI 318–89M)} \qquad (1.9.2)$$

and with E_c and f'_c in kgf/cm²,

$$E_c = 15{,}000\sqrt{f'_c} \quad \text{(approximate)} \qquad (1.9.2)$$

Table 1.9.1 Values of Modulus of Elasticity (Using $E_c = 33w_c^{1.5}\sqrt{f'_c}$ for Normal-Weight Concrete Weighing 145 pcf)

INCH-POUND UNITS		SI UNITS[b]	
f'_c (psi)	E_c (psi)	f'_c (MPa)	E_c[†] (MPa)
3000	3,150,000	21[a]	22,900
3500	3,400,000	24	24,500
4000	3,640,000	28	26,500
4500	3,860,000	31	27,800
5000	4,070,000	35	29,600

[a] These metric values are rounded values approximating concrete strengths in Inch-Pound units; actual equivalents for 3000, 3500, 4000, 4500, and 5000 psi are 20.7, 24.1, 27.6, 31.0, and 34.5 MPa, respectively.

[b] Multiply MPa values by 10.2 to obtain kgf/cm².

[†] Using $E_c = 5000\sqrt{f'_c}$ as per ACI 318-83M.

1.10 CREEP AND SHRINKAGE

Creep and shrinkage are time-dependent deformations, that, along with cracking, provide the greatest concern for the designer because of the inaccuracies and unknowns that surround them. Concrete is elastic only under loads of short duration; and, because of additional deformation with time, the effective behavior is that of an inelastic material. Deflection after a long period of time is therefore difficult to predict, but its control is needed to assure serviceability during the life of the structure.

Creep

Creep is the property of concrete (and other material) by which it continues to deform with time under sustained loads at unit stresses within the accepted elastic range (say, below $0.5f'_c$). This inelastic deformation increases at a decreasing rate during the time of loading, and its total magnitude may be several times as large as the short-time elastic deformation. Frequently creep is associated with shrinkage, since both are occurring simultaneously and often provide the same net effect: increased deformation with time. As may be noted by the general relationship of deformation versus time in Fig. 1.10.1, the "true elastic strain" decreases since the modulus of elasticity E_c is a function of concrete strength f'_c which increases with time.

Although creep is separate from shrinkage, it is related to it. Detailed information is available for predicting creep [1.57, 1.58]. The internal mechanism of creep, or "plastic flow" as it is sometimes called, may be due to any one or a combination of the following: (1) crystalline flow in the aggregate and hardened cement paste;

Figure 1.10.1 Change in strain of a loaded and drying specimen; t_0 is the time of application of load.

(2) plastic flow of the cement paste surrounding the aggregate; (3) closing of internal voids; and (4) the flow of water out of the cement gel due to external load and drying.

Factors affecting the magnitude of creep are (1) the constituents—such as the composition and fineness of the cement, the admixtures, and the size, grading, and mineral content of the aggregates; (2) proportions such as water content and water-cement ratio; (3) curing temperature and humidity; (4) relative humidity during period of use; (5) age at loading; (6) duration of loading; (7) magnitude of stress; (8) surface-volume ratio of the member; and (9) slump.

Accurate prediction of creep is complicated because of the variables involved; however, a general prediction method developed by Branson [1.58] gives a standard creep coefficient equation (4 in. or less slump, 40% relative humidity, moist cured, and loading age of 7 days),

$$C_t = \frac{\text{creep strain}}{\text{initial elastic strain}}$$

$$= \frac{t^{0.60}}{10 + t^{0.60}} C_u \qquad\qquad (1.10.1)$$

shown in Fig. 1.10.2, where t is the duration of loading (days) and C_u is the ultimate

Figure 1.10.2 Standard creep coefficient variation with duration of loading (for 4 in. or less slump, 40% relative humidity, moist cured, and loading age of 7 days).

Figure 1.10.3 Typical creep and recovery relationship.

creep coefficient. (Branson [1.58] suggests using an average of 2.35 for standard conditions but the range is shown to be from 1.3 to 4.15.) Correction factors are given for relative humidity, loading age, minimum thickness of member, slump, percent fines, and air content. For practical purposes, the only factors significant enough to require correction are humidity and age at loading.

The effect of unloading may be seen from Fig. 1.10.3 where at a certain time t_1 the load is removed. There is an immediate elastic recovery and a long-time creep recovery, but a residual deformation remains.

Shrinkage

Shrinkage, broadly defined, is volume change that is unrelated to load application. It is possible for concrete cured continuously under water to increase in volume; however, the usual concern is with a decrease in volume. A discussion of the mechanisms of shrinkage may be found in Ref. 1.6 (pp. 481–500). In general, the same factors have been found to influence shrinkage strain as those that influence creep—primarily those factors related to moisture loss.

The Branson general prediction method [1.58] gives a standard shrinkage strain equation (for 4 in. or less slump, 40% ambient relative humidity and minimum thickness of member 6 in. or less, after 7 days moist cured)

$$\epsilon_{sh} = \left(\frac{t}{35 + t}\right)(\epsilon_{sh})_u \tag{1.10.2}$$

shown in Fig. 1.10.4, where t is time (days) after moist curing, and $(\epsilon_{sh})_u$ is ultimate shrinkage strain. (Branson [1.58] suggests using 800×10^{-6} in./in. for average conditions but the range is from 415 to over 1000×10^{-6}.) Correction factors are given with the primary one relating to humidity H,

correction factor = $1.40 - 0.01H$ for $40\% \leq H \leq 80\%$

correction factor = $3.00 - 0.03H$ for $80\% \leq H \leq 100\%$

Shrinkage, particularly when restrained unsymmetrically by reinforcement, causes deformations generally additive to those of creep. For proper serviceability, it is desirable to predict or compensate for shrinkage in the structure.

Figure 1.10.4 Standard shrinkage strain variation with time after moist curing (for 4 in. or less slump, 40% ambient relative humidity and minimum thickness of member 6 in. or less, after 7 days moist cured).

1.11 CONCRETE QUALITY CONTROL

In reinforced concrete design the concrete sections are proportioned and reinforced using a *specified compressive strength* f'_c (28-day compressive strength). The strength f'_c for which each part of a structure has been designed should be clearly indicated on the design drawings. In the United States, as indicated in Section 1.7, f'_c is based on cylinder strength (6 × 12 in. cylinders), evaluated in accordance with ACI Standard 214 [1.34].

Concrete is a material whose strength and other properties are not precisely predictable, so that test cylinders from a mix designed to provide, say, 3000 psi (roughly 20 MPa) concrete will show considerable variability. Because of this, mixes must be designed to provide an average compressive strength greater than the specified value f'_c.

ACI-5.6.2.3 indicates that adequate control of strength occurs when both of the following requirements are met:

1. Average of all sets of three consecutive strength tests* equal or exceed f'_c.
2. No individual strength test (average of two cylinders) falls below f'_c by more than 500 psi (3.4 MPa).

Statistical variations are to be expected and strength tests failing to meet the above criteria will occur perhaps once in 100 tests even though all proper procedures have been followed.

When the ready-mix plant or other concrete production facility has a record based on at least 30 consecutive strength tests for materials and conditions similar

---

* According to ACI-5.6.1.4, a strength test is the "average of the strengths of two cylinders made from the same sample of concrete and tested at 28 days or at test age designated for determination of f'_c."

to those expected, the *standard deviation s* can be computed based on those tests to establish how variable the concrete strength is.

Standard deviation is calculated by first computing the simple average of the tests results, taking the absolute value of the difference (deviation) between each test value and the average, then obtaining the square root of the average of the squares of the deviations. The smaller the standard deviation, the more consistent the results.

ACI-5.3.2 indicates that the *required average compressive strength* f'_{cr} used for proportioning the mix must exceed the specified f'_c by increasing amounts for increasing values of the standard deviation s, using the *larger* of the following:

$$f'_{cr} = f'_c + 1.34s \qquad\qquad \textbf{(1.11.1)}$$

or

$$f'_{cr} = f'_c + 2.33s - 500 \qquad\qquad \textbf{(1.11.2)}$$

when at least 30 consecutive strength tests are the basis for computing the standard deviation s. When fewer than 30 tests are available a modification in accordance with ACI-5.3.1.2 will require using a higher value of f'_{cr}. Thus, if the designer has used a specified strength f'_c of 4000 psi, and the concrete producer has shown a standard deviation of 450 psi, the mix should be designed for an average strength of 4600 psi [i.e., the larger of 4000 + 1.34(450) or 4000 + 2.33(450) − 500]. Further, when data are not available to establish a standard deviation, the mix must be designed to produce an average strength 1200 psi (8.3 MPa) above the specified strength f'_c for f'_c of 3000 to 5000 psi (21 to 35 MPa).

With regard to evaluation and acceptance of concrete, ACI-5.6 provides requirements for the numbers of samples that must be taken. A discussion on the risks inherent in the consideration of limited test data is provided by Tait [1.59].

It should be noted that the term "quality control" entails much more than designing the concrete mix and evaluating the cylinder strength tests. The foregoing discussion of concrete strength variation should merely give an awareness of the fact that concrete having a specified compressive strength f'_c cannot be expected to provide precisely known actual strength and other properties.

Quality control in the broader sense for reinforced concrete construction is a subject of great importance, but generally lies outside the scope of this text. The ACI Committee 121 Report [1.60] and the papers by Tuthill [1.61], Mather [1.62], Newman [1.63], and Scanlon [1.64] provide an excellent overall treatment of this subject.

1.12 STEEL REINFORCEMENT

Steel reinforcement may consist of bars, welded wire fabric, or wires. For usual construction, bars (called *deformed bars*) having lugs or protrusions (*deformations*) are used (Fig. 1.12.1). Such deformations inhibit longitudinal movement of the bar relative to the concrete that surrounds it. These deformed bars are available in the United States in sizes $\frac{3}{8}$ to $2\frac{1}{4}$ in. (9.5 to 57 mm) nominal diameter.

Sizes of ASTM bars (in Inch-Pound units) are indicated by numbers (see Table 1.12.1). For sizes #3 through #8, they are based on the number of eighths of an

Figure 1.12.1 Deformed reinforcing bars. (Courtesy of Concrete Reinforcing Steel Institute.)

inch included in the nominal diameter of the bars. Bars of designation #9 through #11 are round bars corresponding to the former 1 in. square, $1\frac{1}{8}$ in. square, and $1\frac{1}{4}$ in. square sizes, and bars designated #14 and #18 are round bars having cross-sectional areas equal to those of $1\frac{1}{2}$ and 2 in. square sizes, respectively. The nominal

Table 1.12.1 Standard Reinforcing Bar Dimensions and Weights (Bars in Inch-Pound Units According to ASTM A615 [1.65], A616 [1.66], A617 [1.67], and A706 [1.68])

BAR NUMBER	NOMINAL DIMENSIONS					
	DIAMETER		AREA		WEIGHT	
	(in.)	(mm)	(sq in.)	(cm²)	(lb/ft)	(kg/m)
3	0.375	9.5	0.11	0.71	0.376	0.559
4	0.500	12.7	0.20	1.29	0.668	0.994
5	0.625	15.9	0.31	2.00	1.043	1.552
6	0.750	19.1	0.44	2.84	1.502	2.235
7	0.875	22.2	0.60	3.87	2.044	3.041
8	1.000	25.4	0.79	5.10	2.670	3.973
9	1.128	28.7	1.00	6.45	3.400	5.059
10	1.270	32.3	1.27	8.19	4.303	6.403
11	1.410	35.8	1.56	10.06	5.313	7.906
14	1.693	43.0	2.25	14.52	7.65	11.38
18	2.257	57.3	4.00	25.81	13.60	20.24

Table 1.12.2 ASTM (and Canadian Standard) Metric Reinforcing Bar Dimensions and Weights in SI Units

BAR NUMBER	DIAMETER (mm)	AREA (cm^2)	WEIGHT (kg/m)	COMPARISON TO U.S. BARS BAR NUMBER (AREA)
10M	11.3	1.0	0.784	— 3 (0.71 cm^2)
				— 4 (1.29 cm^2)
15M	16.0	2.0	1.568	— 5 (2.00 cm^2)
20M	19.5	3.0	2.352	— 6 (2.84 cm^2)
				— 7 (3.87 cm^2)
25M	25.2	5.0	3.920	— 8 (5.10 cm^2)
				— 9 (6.45 cm^2)
30M	29.9	7.0	5.488	—10 (8.19 cm^2)
35M	35.7	10.0	7.840	—11 (10.06 cm^2)
45M	43.7	15.0	11.760	—14 (14.52 cm^2)
55M	56.4	25.0	19.600	—18 (25.81 cm^2)

Table 1.12.3 Reinforcing Bar Steels

ASTM DESIGNATION	GRADE	BAR SIZES	MINIMUM YIELD STRESS,[a] f_y		MINIMUM TENSILE STRENGTH	
			ksi	MPa	ksi	MPa
A615	40	#3–#6	40		70	
(Billet steel)	60	#3–#18	60		90	
	75	#11, #14, #18	75		100	
A615[b]	300	10–20		300		500
(Billet steel	400	10–55		400		600
	500	35, 45, 55		500		700
A616/A616M	50	#3–#11	50		80	
(Rail steel)	60	#3–#11	60		90	
	350	10–35		350		550
	400	10–35		400		600
A617/A617M[c]	40	#3–#11	40		70	
(Axle steel)	60	#3–#11	60		90	
	300	10–35		300		500
	400	10–35		400		600
A706/A706M	60	#3–#18	60		80	
(Low-alloy steel)	400	10–55		400		550

[a] The term "yield stress" refers to either *yield point*, the well-defined deviation from perfect elasticity, or *yield strength*, the value obtained by a specified offset strain for material having no well-defined yield point.

[b] Metric (SI) specification applies to bars designated numbers 10 through 55, as given in Table 1.12.2.

[c] Although rail steel is no longer considered a practical source of bar reinforcement, its use is still permitted provided the bend test requirements for axle steel, ASTM 617/617M, Grade 60, can be satisfied.

Figure 1.12.2 Typical stress–strain curves for reinforcing bar steels in tension.

diameter of a deformed bar is equivalent to the diameter of a plain bar having the same weight per foot as the deformed bar.

The ASTM standard sizes for metric reinforcing bars agree with the Canadian standard sizes for such bars, the designations and dimensions of which are given in Table 1.12.2.

Reinforcing bar steel in the United States is covered under ASTM designations as shown in Table 1.12.3. The "Grade" of steel is the minimum specified yield stress* expressed in ksi for Inch-Pound reinforcing bar Grades 40, 50, 60, and 75, and in MPa for SI reinforcing bar Grades 300, 350, 400, and 500.

Billet steel (ASTM A615 and A615M) is newly made steel having its chemical content sufficiently controlled to provide necessary ductility. Axle and rail steel bars, both of which are rarely used now, are rerolled from old axles and rails, and are generally less ductile than bars of billet steel.

Grade 60 steel is the primary reinforcement material. Formerly, Grade 40 was more economical and readily available; today (1992), Grade 60 is economical and Grade 40 billet steel is available only in the smaller bars #3 through #6. Both Grades 40 and 60 exhibit the well-defined yield point and elastic–plastic strain behavior shown in Fig. 1.12.2(a). The overall relationships are shown in Fig. 1.12.2(b).

Welded wire fabric is used in thin slabs, thin shells, and other locations where available space would not permit the placement of deformed bars with proper cover and clearance. Welded wire fabric (WWF) covered under ASTM A185 [1.70] and

— — — —

* The term "yield stress" refers to either *yield point*, the well-defined deviation from perfect elasticity, or *yield strength*, the value obtained by a specified offset strain for material having no well-defined yield point.

A497 [1.71], consists of cold-drawn wire in orthogonal patterns, square or rectangular, resistance welded at all intersections. The wires may be smooth (ASTM A82 [1.72]) or deformed (ASTM A496 [1.73]). The wire is specified by the symbol W (for smooth wires) or D (for deformed wires) followed by a number representing the cross-sectional area in hundredths of a square inch, varying from 1.5 to 31. On design drawings it is usually indicated by the symbol WWF followed by spacings of the wires in the two 90° directions. Thus WWF6 × 8 − W7 × W7 indicates welded wire fabric with 6-in. longitudinal wire spacing, 8-in. transverse wire spacing, and both sets of wires smooth and having a cross-sectional area of 0.07 sq in. Properties for welded wire fabric steels are given in Table 1.12.4. Additional information about welded wire fabric may be obtained from the Wire Reinforcement Institute [1.74].

Wires in the form of individual wires or groups of wires forming *strands* are used for prestressed concrete. There is a great variety of wire and strands of different strengths and properties, the most prevalent being the 7-wire strands, either stress-relieved or low-relaxation, conforming to ASTM A416 [1.75]. These strands have a center wire enclosed by six helically wound outside wires. Usual nominal diameters for 7-wire strands are $\frac{1}{4}$, $\frac{3}{8}$, and $\frac{1}{2}$ in. The minimum tensile strength for Grade 250

Table 1.12.4   Wire and Welded Wire Fabric Steels

ASTM DESIGNATION	WIRE SIZE DESIGNATION	MINIMUM YIELD STRESS,[a] f_y		MINIMUM TENSILE STRENGTH	
		ksi	MPa	ksi	MPa
A82 (cold-drawn wire) (properties apply when material is to be used for fabric)	W1.2 and larger[b]	65	450	75	515
	Smaller than W1.2	56	385	70	485
A185 (welded wire fabric)	Same as A82; this is A82 material fabricated into sheet (so-called "mesh") by the process of electric welding.				
A496 (deformed steel wire) (properties apply when material is to be used for fabric)	D-1 through D-31[c]	70	485	80	550
A497	Same as A82 or A496; this specification applies for fabric made from A496, or from a combination of A496 and A82 wires.				

[a] The term "yield stress" refers to either *yield point*, the well-defined deviation from perfect elasticity, or *yield strength*, the value obtained by a specified offset strain for material having no well-defined yield point.

[b] The W number represents the nominal cross-sectional area in square inches multiplied by 100, for smooth wires.

[c] The D number represents the nominal cross-sectional area in square inches multiplied by 100, for deformed wires; as D-31 has cross-sectional area 0.31 sq in.

strands is 250,000 psi (1720 MPa), and for Grade 270 is 270,000 psi (1860 MPa); there is no well-defined yield point. The *yield strength* measured at 1% extension under load is specified as a minimum of 85% of tensile strength for stress-relieved strand and 90% of tensile strength for low-relaxation strand. Typically under service conditions these prestressed strands have a stress of 150,000 to 160,000 psi (1030 to 1100 MPa).

The modulus of elasticity for all nonprestressed steel may be taken (ACI-8.5.2) as 29,000,000 psi (200,000 MPa). For prestressing steel it is lower and somewhat variable; therefore it must be obtained from the manufacturer. A value of 27,000,000 psi (186,000 MPa) is often used for 7-wire strands conforming to ASTM A416.

ACI Committee 439 [1.76] provides an excellent review of steel reinforcement properties and availability.

1.13 SI UNITS

The use of SI units has made little advance in the construction industry in the United States. In Canada, formal conversion to SI was accomplished with the adoption of the 1977 Canadian Concrete Code [1.77]. The American Concrete Institute adopted an SI version of the 1983 ACI Code (see Section 2.1) and continues with a 1989 SI version. It is not anticipated, however, that wide use of SI units will occur in the United States for some time.

The metric system, although not SI, is used in nearly all western hemisphere countries (other than the USA and Canada). In these countries the MKS (metre-kilogram-second) system is used, where instead of the kilogram (kg) as a unit of mass, as in SI, the kilogram (kgf) is used as a unit of force. Thus, stresses are given in kilograms per square centimeter (kgf/cm^2). Throughout this text, curves involving stresses have three parallel axes: pounds per square inch (psi) or kips per square inch (ksi) for Inch-Pound units, megapascals (MPa) for SI units, and kilogram-force per square centimeter (kgf/cm^2) for the other metric (MKS) system. The authors believe that some familiarity with metric units is essential; on the other hand, overemphasis on units in design calculations and procedures detracts from concepts. In general, throughout the remaining chapters, design equations that involve units will have an SI version (usually ACI 318–89M) given in a footnote. The reader interested in proper use of SI units should consult the ASTM *Standard for Metric Practice* [1.78] and the *Standard Practice for the Use of Metric (SI) Units in Building Design and Construction* [1.79].

SELECTED REFERENCES

1.1.   Hans Straub. *A History of Civil Engineering*. Cambridge, Massachusetts: M.I.T. Press, 1964 (pp. 205–215); *also* London: Leonard Hill Ltd, 1952.

1.2.   R. S. Kirby, Sidney Withington, A. B. Darling, and F. G. Kilgour. *Engineering in History*. New York: McGraw-Hill, 1956.

1.3. Edward Cohen and Raymond C. Heun. "100 Years of Concrete Building Construction in the United States," *Concrete International*, **1**, March 1979, 38–46.

1.4. "Memorable Miscellany," *Concrete International*, **1**, October 1979, 46–62.

1.5. George E. Troxell, Harmer E. Davis, and Joe W. Kelly. *Composition and Properties of Concrete* (2nd ed.). New York: McGraw-Hill, 1968.

1.6. Sidney Mindness and J. Francis Young. *Concrete*. Englewood Cliffs, New Jersey: Prentice-Hall, Inc., 1981.

1.7. P. Kumar Mehta. *Concrete Structure, Properties, and Materials*. Englewood Cliffs, New Jersey: Prentice-Hall, Inc., 1986.

1.8. ASTM. *Standard Specification for Portland Cement* (C150–89). Philadelphia: American Society for Testing and Materials, 1989.

1.9. ACI Committee 225. "Guide to the Selection and Use of Hydraulic Cements," *ACI Journal, Proceedings*, **82**, November–December 1985, 901–928. (Abstract: *Concrete International*, **9**, January 1987, 45–50.)

1.10. ASTM. *Standard Specification for Blended Hydraulic Cements* (C595–89). Philadelphia: American Society for Testing and Materials, 1989.

1.11. ACI. *Building Code Requirements for Reinforced Concrete* (ACI 318–89). Detroit: American Concrete Institute, 1989.

1.12. ACI Committee 221. "Guide for Use of Normal Weight Aggregates in Concrete," *ACI Journal, Proceedings*, **81**, March–April 1984, 115–139. (Abstract: *Concrete International*, **8**, November 1986, 56–62.)

1.13. ASTM. *Standard Specification for Concrete Aggregates* (C33–90). Philadelphia: American Society for Testing and Materials, 1990.

1.14. ASTM. *Standard Specification for Lightweight Aggregates for Structural Concrete* (C330–89). Philadelphia: American Society for Testing and Materials, 1989.

1.15. Richard W. Steiger. "Development of Lightweight Aggregate Concrete," *Concrete Construction*, **30**, June 1985, 519–525.

1.16. George K. Mackie II. "Recent Uses of Structural Lightweight Concrete," *Concrete Construction*, **30**, June 1985, 497–502.

1.17. ACI Committee 213. "Guide for Structural Lightweight Aggregate Concrete," *Concrete International*, **1**, February 1979, 33–62. Disc. August 1979, 66.

1.18. ACI Committee 523. "Guide for Cast-in-Place Low-Density Concrete," *ACI Journal, Proceedings*, **83**, September–October 1986, 830–837. (Abstract: *Concrete International*, **12**, December 1987, 57–60.)

1.19. ACI Committee 523. "Guide for Cellular Concretes Above 50 pcf, and for Aggregate Concretes Above 50 pcf with Compressive Strengths Less Than 2500 psi," *ACI Journal, Proceedings*, **72**, February 1975, 50–66.

1.20. ACI Committee 304. "High Density Concrete: Measuring, Mixing, Transporting, and Placing," *ACI Journal, Proceedings*, **72**, August 1975, 407–414. (See also *ACI Manual of Standard Practice*.)

1.21. ACI Committee 212. "Chemical Admixtures for Concrete," *ACI Materials Journal*, **86**, May–June 1989, 297–327.

1.22. ACI Committee 212. Guide for Use of Admixtures in Concrete. *Concrete International*, **3**, May 1981, 53–65.

1.23. Richard C. Mielenz. "History of Chemical Admixtures for Concrete," *Concrete International*, **6**, April 1984, 40–53.

1.24. ⸺. *Concrete Construction*, **30**, April 1985 (Entire issue devoted to admixtures.)

1.25. ACI. *Standard Practice for Selecting Proportions for No-Slump Concrete* (ACI 211.3–75, Revised 1980). Detroit: American Concrete Institute, 1980.

1.26. PCA. *Design and Control of Concrete Mixtures*. Skokie, Illinois: Portland Cement Association, 1985.

1.27. ACI. *Standard Practice for Selecting Proportions for Normal, Heavyweight, and Mass Concrete* (ACI 211.1–89). Detroit: American Concrete Institute, 1989 [Published as "Proposed Revisions to: Standard Practice for Selecting Proportions for Normal, Heavyweight, and Mass Concrete (ACI 211.1–81) (Revised 1985)," *ACI Materials Journal*, **86**, January–February 1989, 82–100.]

1.28. ACI. *Standard Practice for Selecting Proportions for Structural Lightweight Concrete* (ACI 211.2–91). Detroit: American Concrete Institute, 1991. [Published as "Standard Practice for Selecting Proportions for Structural Lightweight Concrete (ACI 211.2)," *ACI Materials Journal*, **87**, November–December 1990, 638–651.]

1.29. ACI. *Recommended Practice for Measuring, Mixing, Transporting, and Placing Concrete* (ACI 304–73). Detroit: American Concrete Institute, 1973. (Reaffirmed 1983.)

1.30. ACI. *ACI Manual of Concrete Inspection*, (SP-2, 7th ed.). Detroit: American Concrete Institute, 1981.

1.31. UNESCO. *Reinforced Concrete: An International Manual*. London: Butterworth, 1971 (pp. 19–22).

1.32. James M. Shilstone, Jr. "The Cylinder Test—Reliable Informer or False Prophet," *Concrete International*, **2**, July 1980, 63–68.

1.33. J. Bruce Tait. "Making the Most of Concrete Strength Test Results," *ACI Journal, Proceedings*, **83**, May–June 1986, 383–388.

1.34. ACI. *Recommended Practice for Evaluation of Strength Test Results of Concrete* (ACI 214–77). Detroit: American Concrete Institute, 1977.

1.35. John R. Smith. "Estimating Later Age Strengths of Concrete," *ACI Journal, Proceedings*, **81**, November–December 1984, 609–612.

1.36. A. P. Keiller. "Assessing the Strength of In Situ Concrete," *Concrete International*, **7**, February 1985, 15–21.

1.37. Robert S. Jenkins. "Nondestructive Testing—An Evaluation Tool," *Concrete International*, **7**, February 1985, 22–26.

1.38. William C. Stone and Bruce J. Giza. "The Effect of Geometry and Aggregate on the Reliability of the Pullout Test," *Concrete International*, **7**, February 1985, 27–36.

1.39.   Thomas J. Parsons and Tarun R. Naik. "Early Age Concrete Strength Determination by Maturity," *Concrete International*, **7**, February 1985, 37–43.

1.40.   Kal R. Hindo and Wayne R. Bergstrom. "Statistical Evaluation of the In-Place Compressive Strength of Concrete," *Concrete International*, **7**, February 1985, 44–48.

1.41.   V. M. Malhotra. "Testing Early-Age Strength of Concrete In-Place," *Concrete International*, **7**, April 1985, 39–41.

1.42.   Hjalmar Granholm. *A General Flexural Theory of Reinforced Concrete*. New York: Wiley, 1965 (pp. 23–36).

1.43.   P. T. Wang, S. P. Shah, and A. E. Naaman. "Stress-Strain Curves of Normal and Lightweight Concrete in Compression," *ACI Journal, Proceedings*, **75**, November 1975, 603–611.

1.44.   Vitelmo V. Bertero. "Inelastic Behavior of Structural Elements and Structures," *High Strength Concrete*, Proceedings of a Workshop, University of Illinois at Chicago Circle, December 2–4, 1979, pp. 102–103, 153.

1.45.   Antoine E. Naaman. *Prestressed Concrete Analysis and Design Fundamentals*. New York: McGraw-Hill, 1982 (p. 54).

1.46.   Arthur H. Nilson. *Design of Prestressed Concrete*. New York: Wiley, 1978 (p. 45).

1.47.   Hubert Rüsch. "Researches Toward a General Flexural Theory for Structural Concrete," *ACI Journal, Proceedings*, **57**, July 1960, 1–28.

1.48.   Vaughan Randall and Kenneth Foot. "High-Strength Concrete for Pacific First Center," *Concrete International*, **11**, April 1989, 14–16.

1.49.   Nathan L. Howard and David M. Leatham. "The Production and Delivery of High-Strength Concrete," *Concrete International*, **11**, April 1989, 26–30.

1.50.   Gregory J. Smith and Franz N. Rad. "Economic Advantages of High-Strength Concretes in Columns," *Concrete International*, **11**, April 1989, 37–43.

1.51.   S. P. Shah, editor. *High Strength Concrete*, Proceedings of a Workshop Held at the University of Illinois at Chicago Circle, December 2–4, 1979 (sponsored by National Science Foundation), 226 pp.

1.52.   Anthony E. Fiorato. "PCA Research on High-Strength Concrete," *Concrete International*, **11**, April 1989, 44–50.

1.53.   ASTM. *Standard Method of Test for Splitting Tensile Strength of Cylindrical Concrete Specimens* (C496–86). Philadelphia: American Society for Testing and Materials, 1986.

1.54.   ASTM. *Standard Test Method for Flexural Strength of Concrete* (*Using Simple Beam with Third-Point Loading*) (*C*78–75) (Reapproved 1982). Philadelphia: American Society for Testing and Materials, 1982.

1.55.   Sher Ali Mirza, Michael Hatzinikolas, and James G. MacGregor. "Statistical Descriptions of Strength of Concrete," *Journal of the Structural Division*, ASCE, **105**, ST6 (June 1979), 1021–1037.

1.56.   Adrian Pauw. "Static Modulus of Elasticity of Concrete as Affected by Density," *ACI Journal, Proceedings*, **57**, December 1960, 679–687.

1.57. ACI Committee 209. "Prediction of Creep, Shrinkage and Temperature Effects in Concrete Structures," *Designing for Creep and Shrinkage in Concrete Structures* (SP–76). Detroit: American Concrete Institute, 1982 (pp. 193–300).

1.58. Dan E. Branson. *Deformations of Concrete Structures.* New York: McGraw-Hill, 1977 (pp. 11–27, 44–55).

1.59. J. Bruce Tait. "Concrete Quality Assurance Based on Strength Tests," *Concrete International*, **3**, September 1981, 79–87.

1.60. ACI Committee 121. "Quality Assurance Systems for Concrete Construction," *ACI Journal, Proceedings*, **82**, July–August 1985, 537–543.

1.61. Lewis H. Tuthill. "Obtaining Quality in Concrete Construction," *Concrete International*, **8**, March 1986, 24–29.

1.62. Bryant Mather. "Selecting Relevant Levels of Quality," *Concrete International*, **8**, March 1986, 30–36.

1.63. Ken Newman. "Common Quality in Concrete Construction," *Concrete International*, **8**, March 1986, 37–49.

1.64. John M. Scanlon. "Quality Control During Hot and Cold Weather," *Concrete International*, **1**, September 1979, 58–65.

1.65. ASTM. *Standard Specification for Deformed and Plain Billet-Steel Bars for Concrete Reinforcement* (A615–90). Philadelphia: American Society for Testing and Materials, 1990.

1.66. ASTM. *Standard Specification for Rail-Steel Deformed and Plain Bars for Concrete Reinforcement* (A616/A616M–90). Philadelphia: American Society for Testing and Materials, 1990.

1.67. ASTM. *Standard Specification for Axle-Steel Deformed and Plain Bars for Concrete Reinforcement* (A617/A617M–90). Philadelphia: American Society for Testing and Materials, 1990.

1.68. ASTM. *Standard Specification for Low-Alloy Steel Deformed Bars for Concrete Reinforcement* (A706/A706M–90). Philadelphia: American Society for Testing and Materials, 1990.

1.69. ASTM. *Standard Specification for Deformed and Plain Billet-Steel Bars for Concrete Reinforcement* [Metric] (A615M–88). Philadelphia: American Society for Testing and Materials, 1988.

1.70. ASTM. *Standard Specification for Steel Welded Wire Fabric, Plain, for Concrete Reinforcement* (A185–90a). Philadelphia: American Society for Testing and Materials, 1990.

1.71. ASTM. *Standard Specification for Steel Welded Wire Fabric, Deformed, for Concrete Reinforcement* (A497–90a). Philadelphia: American Society for Testing and Materials, 1990.

1.72. ASTM. *Standard Specification for Steel Wire, Plain, for Concrete Reinforcement* (A82–90a). Philadelphia: American Society for Testing and Materials, 1990.

1.73. ASTM. *Standard Specification for Steel Wire, Deformed, for Concrete Reinforcement* (A496–90a). Philadelphia: American Society for Testing and Materials, 1990.

1.74. WRI. *Welded Wire Fabric—Manual of Standard Practice* (3rd ed.). McLean, Virginia: Wire Reinforcement Institute, 1979 (8361A Greensboro Drive, McLean, Virginia 22102).

1.75. ASTM. *Standard Specification for Steel Strand, Uncoated Seven-Wire Stress-Relieved for Prestressed Concrete* (A416–90a) (including Supplement I. Low-Relaxation Strand). Philadelphia: American Society for Testing and Materials, 1990.

1.76. ACI Committee 439. "Steel Reinforcement—Physical Properties and U.S. Availability," *ACI Materials Journal*, **86**, January–February 1989, 63–76.

1.77. CSA. *Code for the Design of Concrete Structures for Buildings* (CAN3–A23.3–M84). Rexdale, Ontario, Canada: Canadian Standards Association, 1984.

1.78. ASTM. *Standard Practice for Use of the International System of Units (SI) (The Modernized Metric System)* (E380–89a). Philadelphia: American Society for Testing and Materials, 1989.

1.79. ASTM. *Standard Practice for the Use of Metric (SI) Units in Building Design and Construction* (Committee E6 Supplement to E–380) (E621–78). Philadelphia: American Society for Testing and Materials, 1978.

1.80. Tom Cosgrove. "Tall Concrete Buildings Come of Age." *Engineering News Record*, November 30, 1989, 25–27.

1.81. Sandor Popovics. "Analysis of the Concrete Strength versus Water–Cement Ratio Relationship," *ACI Materials Journal*, **87**, September–October 1990, 517–529.

1.82. T. S. Shih, G. C. Lee, and K. C. Chang. "On Static Modulus of Elasticity of Normal-Weight Concrete," *Journal of Structural Engineering*, ASCE, **115**, 10 (October 1989), 2579–2587.

2

Design Methods
and Requirements

2.1 ACI BUILDING CODE

When two different materials, such as steel and concrete, act together, it is understandable that the analysis for strength of a reinforced concrete member has to be partly empirical, although mostly rational. These semirational principles and methods are being constantly revised and improved as results of theoretical and experimental research accumulate. The American Concrete Institute, serving as a clearinghouse for these changes, issues building code requirements, the most recent of which is the *Building Code Requirements for Reinforced Concrete* (ACI 318–89), hereafter referred to as the ACI Code [1.11].*

The ACI Code is a Standard of the American Concrete Institute. In order to achieve legal status, it must be adopted by a governing body as a part of its general building code. The ACI Code is partly a specification-type code, which gives acceptable design and construction methods in detail, and partly a performance code, which states desired results rather than details of how such results are to be obtained. A building code, legally adopted, is intended to prevent people from being harmed; therefore, it specifies minimum requirements consistent with good safety. It is important to realize that a building code is not a recommended practice, nor is it a design handbook, nor is it intended to replace engineering knowledge, judgment, or experience. It does *not* relieve the designer of the responsibility for having a safe economical structure.

— — — —

* The reader is advised to have the ACI Code as a ready reference while using this text.

Lake Point Tower, Chicago, apartment building, 70 stories high. (Photo by C. G. Salmon.)

2.2 STRENGTH DESIGN AND WORKING STRESS DESIGN METHODS

Two philosophies of design have long been prevalent. The *working stress method*, focusing on conditions at service load (that is, when the structure is being used), was the principal method used from the early 1900s until the early 1960s. Today (1992), with few exceptions, the *strength design method* is used, focusing on conditions at loads greater than service loads when failure may be imminent. The strength design method is deemed conceptually more realistic to establish structural safety.

2.3 WORKING STRESS METHOD

In the working stress method (referred to by the present ACI Code as the "alternate design method"), a structural element is so designed that the stresses resulting from the action of *service loads* (also called *working loads*) and computed by the mechanics of elastic members do not exceed some predesignated allowable values. Service load is the load, such as dead, live, snow, wind, and earthquake, which is assumed actually to occur when the structure is in service.

The working stress method may be expressed by the following:

$$f \le [\text{allowable stress}, f_{\text{allow}}] \qquad (2.3.1)$$

where

f = an elastically computed stress, such as by using the
flexure formula $f = Mc/I$ for a beam

and

f_{allow} = a limiting stress prescribed by a building code as a
percentage of the compressive strength f'_c for concrete,
or of the yield stress f_y for the steel reinforcing bars

Some of the obstacles to the working stress method are as follows.

1. Since the limitation is on the total stress under service load, there is no simple way to account for different degrees of uncertainty of various kinds of load. Generally the dead load (gravity load due to weight of structural elements and permanent attachments) is known more accurately than the live load, which is code prescribed and may have unknown and variable distribution.
2. Creep and shrinkage, which contribute major time-dependent effects on a structure, are not easily accounted for by elastic calculation of stresses.
3. Concrete stress is not proportional to strain up to its crushing strength, so that the inherent safety provided is unknown when a percentage of f'_c is used as the allowable stress.

2.4 STRENGTH DESIGN METHOD

In the *strength design method* (formerly called *ultimate strength method*), the service loads are increased sufficiently by factors to obtain the load at which failure is considered to be "imminent." This load is called the *factored load* or *factored service load*.

The structure or structural element is then proportioned such that the strength is reached when the factored load is acting. The computation of this strength takes into account the nonlinear stress-strain behavior of concrete.

The strength design method may be expressed by the following,

$$\text{strength provided} \geq \begin{bmatrix} \text{strength required to} \\ \text{carry factored loads} \end{bmatrix} \tag{2.4.1}$$

where the "strength provided" (such as moment strength) is computed in accordance with rules and assumptions of behavior prescribed by a building code, and the "strength required" is that obtained by performing a structural analysis using factored loads.

The "strength provided" has been commonly referred to by practitioners as "ultimate strength." However, it is a code-defined value for strength, and is not necessarily "ultimate" in the sense of being a value above which it is impossible to reach. The ACI Code uses a conservative definition of strength; thus the modifier "ultimate" is not appropriate.

When the strength design method is used, the comparison of provided strength with required strength (that is, axial force, shear, or bending moment, caused by factored loads) *does not* imply that any material "yields" or "fails" under service load conditions. In fact, at service loads, the behavior of the structure is essentially elastic. The use of the term "imminent failure" under factored loads is only a device for establishing adequate safety parameters.

2.5 COMMENTS ON DESIGN METHODS

Historically, "ultimate" strength was the earliest approach to design, since the failure load could be measured by test without a knowledge of the magnitude or distribution of internal stresses. With the interest in and understanding of the elastic methods of analysis in the early 1900s, the elastic working stress method was adopted almost universally by codes as the best for design. As more detailed understanding of the actual behavior of reinforced concrete structures subjected to loads in excess of the service loads developed, adjustments in the theory and in the design procedures were made.

The first modification of the elastic working stress method resulted from the study of axially loaded columns in the early 1930s. By 1940, the design of axially loaded columns was based on ultimate strength. Next, the working stress method was modified to account for creep of concrete in beams with compression steel and in eccentrically loaded columns. The early history of the ACI Code has been summarized by Kerekes and Reid [2.1].

The 1956 ACI Code was the first that officially recognized and permitted the strength design method, the result of work by ACI–ASCE Committee 327 [2.2]. The 1963 ACI Code treated the working stress method and the strength design method on an equal basis; but, actually, the major portion of the working stress method was based on strength. With the relegation of the working stress method to a small section referred to as the "alternate method," the 1971 ACI Code entirely accepted the strength design method. Codes since 1971 have had the "alternate design method" in an appendix (currently Appendix A). It is likely that in the near future the "alternate design method" will disappear and the only use for the elastic concepts of the working stress method will be in the computation of deflections, which are of interest under service loads rather than at factored loads.

No matter which of the above philosophies is employed in a design, *serviceability* must also be considered. Serviceability factors that may be as important as strength are excessive deflection, detrimental cracking, excessive amplitude or undesirable frequency of vibration, and excessive noise transmission. The designer must consider both *strength* and *serviceability*. Any one, or a combination, of the strength and serviceability factors may provide a criterion for the limit of structural usefulness.

The term "limit state" has come into use in recent years. Strength and a serviceability consideration such as a limit on deflection may be thought of as "limit states." MacGregor [2.3] has provided an excellent treatment of limit states design as applied to reinforced concrete.

2.6 SAFETY PROVISIONS

Structures and structural members must always be designed to carry some reserve load above what is expected under normal use. Such reserve capacity is provided to account for a variety of factors, which may be grouped in two general categories: factors relating to overload and factors relating to understrength (that is, less strength than computed by acceptable calculation procedures). Overloads may arise from changing the use for which the structure was designed, from underestimation of the effect of loads by oversimplification in calculation procedures, and from effects of construction sequence and methods. Understrength may result from adverse variations in material strengths, workmanship, dimensions, control, and degree of supervision, even though individually these items may be within required tolerances.

Conventionally, the term "safety factor" has been used in working stress design to designate nominally the ratio between the yield stress (real, as for steel; nominally defined, as for concrete) and the allowable working stress. Such use has resulted in structures and structural elements with the same "safety factor" but considerably variant in their strength to service load ratio. Thus the term "safety factor" as conventionally applied has little meaning as far as the prediction of strength is concerned.

The variability in the ratio of the strength to service load under the working stress method has been a major factor in the transition to the use of the strength design method. To distinguish between the conventional term "safety factor" and the strength to service load ratio, the term "load factor" has traditionally been adopted for the latter.

The ACI Code has separated the safety provision into factors U for overload and factors ϕ for understrength. The basic overload equation (ACI-9.2.1) for structures in such locations and of such proportions that the effects of wind and earthquake may be neglected is

$$U = 1.4D + 1.7L \qquad \text{(2.6.1)}$$

where

U = required strength (based on possible overload)

D = dead load under service conditions

L = live load under service conditions

The factors ϕ for understrength, called *strength reduction factors* according to ACI-9.3, are prescribed as follows:

		ϕ Factors
1.	Flexure, with or without axial tension	0.90
2.	Axial tension	0.90
3.	Shear and torsion	0.85
4.	Compression members, spirally reinforced	0.75*
5.	Compression members, other	0.70*
6.	Bearing on concrete	0.70
7.	Bending in plain concrete	0.65[†]

Even though the overload factors U and the strength reduction factors ϕ are itemized separately in the ACI Code, they are in fact complementary parts of the provision to ensure adequate safety. The total required nominal (theoretical) strength to be provided in design is U/ϕ. Because the ϕ factors vary for different types of member action, the designer generally will use them at various stages in the calculations. The overload provision is best accounted for at the beginning of the design, where the service loads are multiplied by their respective factors in U, giving what are called *factored loads*. The term "design load" is to be avoided since in common use it is unclear whether service load or factored load is referred to.

The purpose of a safety provision is to limit the probability of failure and yet to permit economical structures. Obviously, if cost is no object, it is easy to design a structure whose probability of failure is nil. To arrive properly at a suitable degree of safety, the relative importance of various items must be established. Some of those items are

1. Seriousness of a failure, either to humans or goods.

2. Reliability of workmanship and inspection.

3. Expectation of overload and to what magnitude.

4. Importance of the member in the structure.

5. Chance of warning prior to a failure.

By assigning percentages to the above items and evaluating the circumstances for any given situation, proper values for U and ϕ may be determined for each case.

— — — —

* For combined compression and flexure, *both* axial load and bending moment are subject to the same ϕ factor, which may be variable, and increase to 0.90 as the axial compression decreases to zero.

[†] From *Building Code Requirements for Structural Plain Concrete* (ACI 318–89) *and Commentary*. Detroit: American Concrete Institute, 1989 (Sec. 6.2.2).

The ACI Code committee has combined experience with historical precedent to arrive at the prescribed overload and strength reduction factors. The normal range for U/ϕ is from 1.55 to 2.4 for reinforced concrete structural elements. The subject of safety provisions for reinforced concrete has been summarized by Winter [2.4].

In recent years, considerable attention has been focused on using a theory of probability as a basis for a design code, thus providing a more rational basis for the various components comprising the U and ϕ factors. A series of ACI Committee 348 (structural safety) papers [2.5–2.9] gives an excellent overview of this approach, and an interesting collection of discussion and opinion is also available [2.8]. Benjamin and Lind [2.6] have stated that the following five safety conditions form the basis for the current practice in structural analysis and design.

1. The probability of a real loading in excess of the nominal service load $D + L$ must be satisfactorily small.
2. The probability of a real loading in excess of the factored loading, say $U = 1.4D + 1.7L$, must be very small or near to zero during the life of the structure.
3. The probability of unsatisfactory performance at the factored load U must be satisfactorily small.
4. The probability of unsatisfactory performance under a load test, say $0.85U$ as given in ACI-20.4.3, must be very small or near to zero.
5. The probability of unsatisfactory performance at the service load $D + L$ must be practically zero.

It is noted that conditions 1 and 2 relate to the overload provision U along with analysis methods, whereas conditions 3, 4, and 5 relate to the factor ϕ for understrength as well as the methods for computing strength.

Currently under review by the ACI Code committee is a proposal to adopt the probability-based factors for overload and the combinations of factored loads given by the 1988 ASCE Standard 7-88 (formerly ANSI A58.1) [2.10]. These factors and loading combinations are based on the report, *Development of a Probability Based Load Criterion for American National Standard A58* [2.11], which recommends using lower factors U, as well as different combinations than discussed in Section 2.7, to account for overload. Since the overall objective is to use the *same* factored loads and factored load combinations no matter what material is used, the use of lower factors for overload in reinforced concrete design must be accompanied by the use of lower strength reduction factors ϕ (so-called *resistance factors* according to ASCE 7-88), because the overall safety in present concrete structures design is considered appropriate.

Detailed discussions of the implications of using design reliability and probability-based load criteria are given by Ellingwood [2.12, 2.13], MacGregor, Mirza, and Ellingwood [2.14], and MacGregor [2.15], and the subject is treated in general by Galambos, Ellingwood, MacGregor, and Cornell [2.16, 2.17].

Since the 1986 adoption by the American Institute of Steel Construction (AISC) of Load and Resistance Factor Design (LRFD) [2.18] using the ASCE 7-88 load factors [2.10], there has been further impetus to use the same load factors in the ACI Code. Corotis [2.19] and Israel, Ellingwood, and Corotis [2.20] reflect the more recent thinking on this.

2.7 OVERLOAD PROVISIONS FOR VARIOUS LOAD COMBINATIONS

In addition to the basic provision for load factors on dead load plus live load given by Eq. (2.6.1), other service loads may also act.

For wind load W acting in combination with other loads, ACI-9.2.2 provides

$$U = 0.75(1.4D + 1.7L + 1.7W) \qquad (2.7.1)$$

Because wind is of a transient nature and acts with its maximum magnitude for a short duration, it has been traditional to allow an overstress of $33\frac{1}{3}\%$ under the working stress method. The same effect is accomplished by using three-quarters of the factored load when wind effect is included.

Sometimes a more severe situation arises with wind loading when live load is absent; this possibility must be considered. When live load is absent, Eq. (2.7.1) becomes

$$U = 1.05D + 1.275W \qquad (2.7.2)$$

Furthermore, for situations in which dead load is a gravity stabilizing effect in combination with wind (such as a tower or wall), the possibility of a reduced dead load must be considered rather than overload; thus ACI-9.2.2 also gives

$$U = 0.9D + 1.3W \qquad (2.7.3)$$

When lateral earth pressure H is involved, it is treated as live load; thus

$$U = 1.4D + 1.7L + 1.7H \qquad (2.7.4)$$

but when dead or live load (or both) reduces the effect of earth pressure, then

$$U = 0.9D + 1.7H \qquad (2.7.5)$$

For liquid pressure F having a controllable maximum height, Eqs. (2.7.4) and (2.7.5) are used, except $1.7H$ is replaced with $1.4F$. Since liquid density is generally accurately known, its pressure is treated as dead load using the 1.4 factor. On the other hand, soil properties are more variable; thus earth pressure is treated as live load using the 1.7 factor.

Where the structural effects T of differential settlement, creep, shrinkage, or temperature change may be significant, they are to be included with dead load (ACI-9.2.7),

$$U = 0.75(1.4D + 1.4T + 1.7L) \qquad (2.7.6)$$

but not less than

$$U = 1.4(D + T) \qquad (2.7.7)$$

The 0.75 factor is used to recognize the low probability of having these effects occur simultaneously with full dead and live load.

Any structure or structural element *must be designed for the most severe of any of the load combinations* given by Eq. (2.6.1) and Eqs. (2.7.1) to (2.7.7). Although not described here, overload provisions for earthquake and live load impact effects are in ACI-9.2.3 and 9.2.6.

2.8 HANDBOOKS AND COMPUTER SOFTWARE

It may be pointed out that this textbook does not appreciably consider office practices or the use of design aids. Once concepts and principles are thoroughly understood, the use of curves and tables can greatly speed up design. There are several handbooks in common use; the *Strength Design Handbook*, Vols. 1 and 2 [2.21, 2.22], published by ACI, the *CRSI Handbook* [2.23], published by the Concrete Reinforcing Steel Institute, and the *PCI Design Handbook* [2.24], published by the Prestressed Concrete Institute. These publications contain useful tables and charts that can speed up design for the experienced designer.

The importance of correct and clear detailing work cannot be overemphasized. For this the reader is referred to the *ACI Detailing Manual* (SP-66) [2.25] which contains typical detailing of steel reinforcement, engineering and placing drawings, and other reference data regarding materials and sizes.

In modern design offices, computer software is used for a considerable portion of the design computations. Many commercial organizations offer such software for sale. The American Concrete Institute serves as a clearinghouse to expedite the sale of concrete-related software. The designer, especially the inexperienced engineer, should be wary of using purchased software to do computations that the designer/engineer does not thoroughly understand and would be unable to formulate in the absence of a computer program. Though ACI serves as a clearinghouse for dissemination of software, *ACI does not review programs in detail nor does it assume any responsibility for program correctness or for compliance with the ACI Code.* Currently, there is no commercial software that has the same general acceptance as the handbooks mentioned previously.

2.9 DIMENSIONS AND TOLERANCES

Although the designer may tend to think of dimensions, clearances, and bar locations as exact, practical considerations require that there be accepted tolerances. These tolerances are the permissible variations from dimensions given on drawings.

Overall dimensions of reinforced concrete members are usually specified by the engineer in whole inches for beams, columns, and walls, sometimes half inches for thin slabs, and often 3-in. increments for more massive elements such as plan dimensions for footings. Formwork for the placing of these members must be built carefully so that it does not deform excessively under the action of workmen, construction machinery loads, and wet concrete. Accepted tolerances for variation in cross-sectional dimensions of columns and beams and in the thickness of slabs and walls are $+\frac{1}{2}$ in. and $-\frac{1}{4}$ in. when the specified dimension is greater than 12 in. but not exceeding 3 ft [2.27]. For concrete footings, accepted variations in plan dimensions are $+2$ in. and $-\frac{1}{2}$ in. [2.27], whereas the thickness has an accepted tolerance of -5% of the specified thickness [2.27]. The strength reduction factor ϕ is intended to account for a situation in which several acceptable tolerances may combine to reduce the strength from that computed using specified dimensions.

Reinforcing bars are normally specified in 3-in. length increments and the placement tolerances are given in the ACI Code (ACI-7.5.2). For minimum clear concrete protection and for the effective depth d (distance from compression face of concrete to center of tension steel) in flexural members, walls, and compression members, the specified tolerances (ACI-7.5.2.1) are as follows:

| EFFECTIVE DEPTH, d | | TOLERANCES | | | |
| | | ON EFFECTIVE DEPTH | | ON MINIMUM CONCRETE COVER | |
(in.)	(mm)	(in.)	(mm)	(in.)	(mm)
$d \leq 8$	200^a	$\pm 3/8$	± 10	$-3/8$	-10
$d > 8$	200	$\pm 1/2$	± 12	$-1/2$	-12

^{a} These values are from ACI 318–89M, and are not strictly conversions from ACI 318–89.

Notwithstanding the stated tolerances on cover, the resulting cover shall not be less than two-thirds of the minimum cover specified on structural drawings or in specifications. Since the effective depth and the clear concrete cover are both components of total depth, the tolerances on those dimensions are directly related. When the tolerances on bar placement and cover accumulate, the overall dimension tolerance may be exceeded; thus field adjustment may have to be made. This may be particularly important for very thin sections such as in precast and shell structures.

For location of bars along the longitudinal dimension, and of bar bends, the tolerance is ± 2 in. (± 50 mm*) except at discontinuous ends where the tolerance shall be $\pm \frac{1}{2}$ in. (± 12 mm*) (ACI-7.5.2.2).

A summary of all tolerances for concrete construction and materials has been provided by ACI Committee 117 [2.27].

2.10 ACCURACY OF COMPUTATIONS

When one understands that variations exist in material strength for both steel and concrete and that variations in dimensions are inevitable (and acceptable), it becomes clear that design calculations for reinforced concrete structures do not require a high degree of precision.

The designer should place highest priority on determining proper location and length of steel reinforcement to carry the tension forces, thus making up for that capacity which is deficient in the concrete. Failures, when they occur, generally result from gross underestimating of tensile forces or lack of identification of how the structure or element will behave under loads. Failures are rarely the result of carrying too few significant figures in the design computations. However, significant figures may be lost in arithmetic operations, and gross errors may sometimes result from sloppiness.

— — — —

* According to ACI 318–89M, Section 7.5.2.

Results indicated on the display of an electronic calculator or computer, no matter how many digits, rarely indicate a precision of more than two or three significant figures. Recording of results on computation sheets should not exceed four significant figures, primarily for systematic control and checking of computations. While the extra digits recorded may not at first seem harmful, the resulting many-digit values make difficult the scanning of computations to detect gross errors (in addition to being a waste of time).

SELECTED REFERENCES

2.1. Frank Kerekes and Harold B. Reid, Jr. "Fifty Years of Development in Building Code Requirements for Reinforced Concrete," *ACI Journal, Proceedings*, **50**, February 1954, 441.

2.2 ACI-ASCE Committee 327. "Ultimate Strength Design," *ACI Journal, Proceedings*, **52**, January 1956, 505–524.

2.3. J. G. MacGregor. "Safety and Limit States Design for Reinforced Concrete," *Canadian Journal of Civil Engineering*, **3**, December 1976, 484–513.

2.4. George Winter. "Safety and Serviceability Provisions in the ACI Building Code," *Concrete Design: U.S. and European Practices* (SP–59). Detroit: American Concrete Institute, 1979.

2.5. Robert G. Sexsmith and Mark F. Nelson. "Limitations in Application of Probabilistic Concepts," *ACI Journal, Proceedings*, **66**, October 1969, 823–828.

2.6. Jack R. Benjamin and N. C. Lind. "A Probabilistic Basis for a Deterministic Code," *ACI Journal, Proceedings*, **66**, November 1969, 857–865.

2.7. C. Allin Cornell. "A Probability-Based Structural Code," *ACI Journal, Proceedings*, **66**, December 1969, 974–985.

2.8. R. C. Reese, D. E. Allen, C. A. Cornell, Luis Esteva, R. N. White, R. G. Sexsmith, and George Winter. "Probabilistic Approaches to Structural Safety," *ACI Journal, Proceedings*, **73**, January 1976, 37–49.

2.9. Haresh C. Shah and Robert G. Sexsmith. "A Probabilistic Basis for the ACI Code," *ACI Journal, Proceedings*, **74**, December 1977, 610–611.

2.10. ASCE. *American Society of Civil Engineers, Minimum Design Loads for Buildings and Other Structures* (ASCE 7-88; formerly ANSI A58.1). New York: American Society of Civil Engineers, 1988.

2.11. Bruce Ellingwood, Theodore V. Galambos, James G. MacGregor, and C. Allin Cornell. *Development of a Probability Based Load Criterion for American National Standard A58* (NBS Special Publication 577). Washington, D.C.: U.S. Department of Commerce, National Bureau of Standards, June 1980.

2.12. Bruce Ellingwood. "Reliability of Current Reinforced Concrete Designs," *Journal of the Structural Division*, ASCE, **105**, ST4 (April 1979), 699–712.

2.13. Bruce Ellingwood. "Reliability Based Criteria for Reinforced Concrete Design," *Journal of the Structural Division*, ASCE, **105**, ST4 (April 1979), 713–727.

2.14. J. G. MacGregor, S. A. Mirza, and B. Ellingwood. "Statistical Analysis of Resistance of Reinforced and Prestressed Concrete Members," *ACI Journal, Proceedings*, **80**, May–June 1983, 167–176.

2.15. James G. MacGregor. "Load and Resistance Factors for Concrete Design," *ACI Journal, Proceedings*, **80**, July–August 1983, 279–287.

2.16. Theodore V. Galambos, Bruce Ellingwood, James G. MacGregor, and C. Allin Cornell. "Probability Based Load Criteria: Assessment of Current Design Practice," *Journal of the Structural Division*, ASCE, **108**, ST5 (May 1982), 959–977.

2.17. Bruce Ellingwood, James G. MacGregor, Theodore V. Galambos, and C. Allin Cornell. "Probability Based Load Criteria: Load Factors and Load Combinations," *Journal of the Structural Division*, ASCE, **108**, ST5 (May 1982), 978–997.

2.18. AISC. *Load and Resistance Factor Design Specification for Structural Steel Buildings.* Chicago: American Institute of Steel Construction, 1986.

2.19. Ross B. Corotis. "Probability-based Design Codes," *Concrete International*, **7**, April 1985, 42–49.

2.20. Morris Israel, Bruce Ellingwood, and Ross Corotis. "Reliability-Based Code Formulations for Reinforced Concrete Buildings," *Journal of Structural Engineering*, ASCE, **113**, 10 (October 1987), 2235–2252.

2.21. ACI Committee 340. *Design Handbook In Accordance with the Strength Design Method of ACI 318–89. Vol. 1—Beams, One-Way Slabs, Brackets, Footings, and Pile Caps* [SP–17(91)] (5th ed.). Detroit: American Concrete Institute, 1991.

2.22. ACI Committee 340. *Design Handbook In Accordance with the Strength Design Method of ACI 318–89. Vol. 2—Columns* [SP–17A(90)] (3rd ed.). Detroit: American Concrete Institute, 1990.

2.23. CRSI. *CRSI Handbook 1984.* Schaumberg, Illinois: Concrete Reinforcing Steel Institute (933 North Plum Grove Road, Schaumberg, IL 60195), 1984.

2.24. PCI. PCI *Design Handbook—Precast and Prestressed Concrete* (3rd ed.). Chicago: Prestressed Concrete Institute (201 North Wells Street, Chicago, Illinois 60606), 1985.

2.25. ACI Committee 315. *ACI Detailing Manual–1988* [SP–66(88)]. Detroit: American Concrete Institute, 1988.

2.26. ACI Committee 347. *Guide to Formwork for Concrete.* Detroit: American Concrete Institute, 1989. (33 pp.)

2.27. ACI Committee 117. "Standard Specifications for Tolerances for Concrete Construction and Materials (ACI 117)", *ACI Materials Journal*, **85**, November–December 1988, 560–571.

Strength of Rectangular Sections in Bending

3.1 GENERAL INTRODUCTION

Until 1956, the ACI Code method for the determination of whether a reinforced concrete member had adequate strength was made solely on the basis of allowable working stresses, service (or working) loads, and the straightline theory of flexure. Since 1971, the concepts involved in the working stress method have been used only for establishing serviceability at service loads. The working stress method and serviceability considerations are therefore dealt with in Chapter 4.

The strength design method, in which factored loads and the computed strength of sections at imminent failure are used, was first permitted as an alternate method of design in the 1956 ACI Code and later became a substantial part of the 1963 Code. The 1971 ACI Code relegated the working stress method to a small section of the Code and utilized primarily the strength method of design. The adjective "ultimate" was dropped since there was little chance of confusion over the meaning of the word "strength." The 1977 ACI Code removed the alternate design (working stress) method from the body of the Code and placed it in an appendix, where it still remains in the 1989 ACI Code.

"Strength" is used in the general sense to include various modes of failure. The deletion of the term "ultimate" recognizes that the strength used for design is a code-defined strength, and may not in fact be "ultimate" in its literal sense.

In the strength method, the *factored loads* (including moments, shears, axial forces, etc.) are obtained by multiplying the service loads by factors U to cover possible overloads and variations in design assumptions. The *design strength* of a section is obtained by multiplying the *nominal strength* (based on statical equilibrium and compatibility of stress and strain) by a strength reduction factor ϕ to account for adverse variations in material strengths, workmanship, dimensions, control, and degree of supervision, even though all are within accepted tolerances.

Rectangular tapered beams, cantilevers, and rigid frame, University of Wisconsin Stadium, Madison, Wis. (Photo by C. G. Salmon.)

In discussing the ACI strength design method for reinforced concrete structures, attention must be called to the difference between loads on the structure as a whole and loads on the cross sections of individual members. The elastic methods of structural analysis are used first to compute the service loads in the individual members due to the action of service loads on the entire structure. Only then are the overload factors applied to the service loads acting on the individual cross sections. This emphasis is more conceptual than practical. In practice, one may apply the factored loads on the entire structure and still make an elastic analysis (which theoretically is no longer valid) to obtain the factored loads (such as shears and bending moments) in the individual members.

Research is continuing on the feasibility of using the inelastic, or limit, method of structural analysis in which factored loads in the individual members are determined correctly from the factored loads acting on the whole structure. This is called "limit design" and is introduced briefly in Section 10.12.

3.2 BASIS OF NOMINAL FLEXURAL STRENGTH

The modern analytical approach to reinforced concrete beam strength was originated by F. Stüssi in 1932 [3.1]. The general stress–strain behavior for concrete as presented in Section 1.7 shows the nonlinearity of stress and strain at stress levels

d = effective depth; distance from compression face to centroid of
   tension steel

(a) Singly   (b) Strain condition   (c) Stress condition
    reinforced   when M_n is   at strain condition
    beam   reached   when M_n is reached

Figure 3.2.1 Conditions when nominal strength M_n in flexure is reached.

above about $0.5f_c'$. Since the compression zone of a beam should be expected to have the same general variation of stress and strain as the test specimen (standard cylinder or cube), the compressive stress distribution for a beam that has achieved its theoretical (nominal) strength [3.2, 3.3, 3.4] should be as shown in Fig. 3.2.1(c).

The nominal strength is assumed to be reached when the strain in the extreme compression fiber is equal to the crushing strain ϵ_{cu} of concrete (see Fig. 3.2.2). When crushing occurs (usually a somewhat sudden occurrence), the strain in the tension

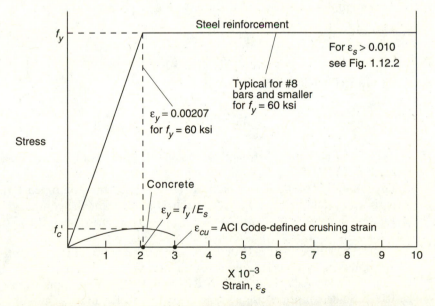

Figure 3.2.2 Stress–strain curves for concrete and steel reinforcement, and definitions of ϵ_y and ϵ_{cu}.

steel A_s *could* be either larger or smaller than the strain $\epsilon_y = f_y/E_s$ at first yield, depending on the relative proportion of steel to concrete. If the steel amount were low enough, it would yield prior to crushing of the concrete, resulting in a ductile failure mode in which there is large deformation. On the other hand, a large quantity of steel would allow the steel to remain elastic at the time of crushing of the concrete, causing a brittle or sudden mode of failure. The ACI Code has provisions which, by limiting the amount of tension steel, are intended to ensure the ductile mode of failure when the nominal strength is reached.

Although the compressive stress distribution in a beam has the same general shape as a test cylinder, the maximum stress is less than f'_c, say $k_3 f'_c$ [see Fig. 3.2.1(c)]. The average stress over a beam of constant width is $k_1 k_3 f'_c$, and the centroid location of the roughly parabolic distribution is $k_2 x$ measured from the compression face, where x is the neutral axis distance. Thus the compressive force C is the summation of the compressive stresses acting on the compression concrete area, which may be considered a "stress solid" volume, as

$$C = k_1 k_3 f'_c x b \qquad (3.2.1)$$

and for the ductile failure condition, the tensile force T is

$$T = A_s f_y \qquad (3.2.2)$$

Equilibrium requires $C = T$, from which

$$x = \frac{A_s f_y}{k_1 k_3 f'_c b} \qquad (3.2.3)$$

The nominal flexural strength then may be expressed as

$$M_n = T(\text{arm}) = T(d - k_2 x)$$
$$= A_s f_y (d - k_2 x) \qquad (3.2.4)$$

Substituting Eq. (3.2.3) for x into Eq. (3.2.4) gives

$$M_n = A_s f_y \left(d - \frac{k_2}{k_1 k_3} \frac{A_s f_y}{f'_c b} \right) \qquad (3.2.5)$$

One may note that if strength is the quantity that is of interest, it is readily obtainable from Eq. (3.2.5) *if the quantity $k_2/(k_1 k_3)$ is known.* It is not necessary to have values for k_1, k_2, or k_3 individually if the value for the combined term is known. Experimental results [3.2, 3.3] have established values for the combined term, as well as the individual k values, with some of the results shown in Fig. 3.2.3. From that figure, $k_2/(k_1 k_3)$ ranges from about 0.55 to 0.63.

The values experimentally determined when crushing of the concrete occurred at the compression face necessarily involved variation in the crushing strain ϵ_{cu} for the various tests. The ACI Code (ACI-10.2.3) considers the maximum usable strain ϵ_{cu} to be 0.003. Some countries use a value of 0.0035, which makes little difference in the computed flexural strength.

Figure 3.2.3 "Stress solid" parameters (adapted from Ref. 3.2).

3.3 WHITNEY RECTANGULAR STRESS DISTRIBUTION

The computation of flexural strength M_n based on the approximately parabolic stress distribution of Fig. 3.2.1(c) may be done using Eq. (3.2.5) with given values of $k_2/(k_1 k_3)$. However, it is desirable for the designer to have a simple method in which basic static equilibrium is used.

In the 1930s, Whitney [3.5, 3.6] proposed the use of a rectangular compressive stress distribution to replace that of Fig. 3.2.1(c). As shown in Fig. 3.3.1(c), an average stress of $0.85f'_c$ is used with a rectangle of depth $a = \beta_1 x$, determined so that $a/2 = k_2 x$. Whitney determined that β_1 should be 0.85 for concrete with $f'_c \leq 4000$ psi, and 0.05 less for each 1000 psi of f'_c in excess of 4000 psi.* The value of β_1 may not be taken less than 0.65 (ACI-10.2.7.3).

The flexural strength M_n, using the equivalent rectangle, is obtained from Fig. 3.3.1 as follows:

$$C = 0.85f'_c ba \tag{3.3.1}$$

$$T = A_s f_y \tag{3.3.2}$$

---

* For SI, ACI 318–89M gives $\beta_1 = 0.85$ for $f'_c \leq 30$ MPa and reduces by 0.08 for each 10 MPa of f'_c in excess of 30 MPa, but not less than 0.65.

d = effective depth; distance from compression face of concrete
   to centroid of tension steel

(a) Beam   (b) Actual stress   (c) Whitney
                            distribution   rectangular
                                                        stress block

Figure 3.3.1 Definition of Whitney rectangular stress distribution.

where the use of f_y assumes that the steel yields prior to crushing of the concrete. Equating $C = T$ gives

$$a = \frac{A_s f_y}{0.85 f_c' b} \tag{3.3.3}$$

$$M_n = A_s f_y (d - a/2) \tag{3.3.4}$$

which on substituting Eq. (3.3.3) into Eq. (3.3.4), gives

$$M_n = A_s f_y \left(d - 0.59 \frac{A_s f_y}{f_c' b} \right) \tag{3.3.5}$$

Note that 0.59 corresponds to $k_2/(k_1 k_3)$ of Eq. (3.2.5). The ACI Code explicitly accepts the Whitney rectangle (ACI-10.2.7). One may also note that β_1 is needed only to establish the neutral axis location for determining the steel strain. So long as the steel strain exceeds $\epsilon_y = f_y/E_s$, as shown in Fig. 3.2.2, the flexural strength M_n is not affected by the value of β_1.

3.4 INVESTIGATION OF RECTANGULAR SECTIONS IN BENDING WITH TENSION REINFORCEMENT ONLY

The quantities defining a rectangular section with tension reinforcement only are b, d, and A_s [Fig. 3.3.1(a)]. Such a section is said to be *singly reinforced*. The steel area A_s is, of course, furnished by the combined area of an actual number of reinforcing bars. Protective concrete cover is necessary around the bars in order to make the steel and concrete act together and, also very importantly, to provide fire protection. At high temperatures (say above 600°F), the yield strength and modulus of elasticity begin to reduce markedly, so that concrete cover is needed for insulation. Minimum cover requirements are generally prescribed by code (see ACI-7.7).

Since the tensile strength of concrete is neglected in flexure calculations (ACI-10.2.5), the cross-sectional shape of the beam on the tension side of the neutral axis and the amount of concrete cover do not affect the flexural strength. Thus the important depth dimension for computing strength is the *effective depth d* rather than the overall depth *h*. The effective depth is defined as the distance from the extreme fiber in compression to the centroid of the tension steel area. When the tension steel is comprised of bars in several layers satisfying the minimum spacing requirement between layers, the centroid of the combined area is usually used, with all bars assumed to have the same strain.

Investigation of a rectangular section in bending is made to establish whether or not the section has adequate *strength* for given applied service loads (or service load moments). The system must also be serviceable; that is, it must perform satisfactorily under service loads without detrimental effects, such as excessive deflection, cracking, or vibration. *Serviceability* in general is treated in Chapter 4, and the detailed discussion of deflection appears in Chapter 14.

ACI Code Strength Method.   According to the ACI Code, the required strength U for a section in bending is obtained by applying load factors (i.e., overload provision) to the moment due to dead load D, live load L, wind load W, earthquake load E, lateral earth pressure H, fluid pressure F, and structural effects T, as discussed in Section 2.6 (ACI-9.2). The design flexural strength ϕM_n is the product of the strength reduction factor ϕ and the nominal flexural strength M_n.

The basic load factors U in the overload provision are 1.4 and 1.7, used as follows:

$$U = 1.4D + 1.7L$$

and the ϕ factor for flexure is 0.90. The terms U, D, and L of the overload provision may also represent quantities that are functions of load, such as moment, shear, and axial force. If M_u is defined as the moment under factored load and M_D and M_L are the service dead load and live load moments, respectively, the basic overload provision for flexure may be expressed as

$$M_u = 1.4M_D + 1.7M_L \tag{3.4.1}$$

The *strength requirement* for flexure may then be expressed as

$$\phi M_n \geq M_u \tag{3.4.2}$$

For computation of the nominal flexural strength M_n, the following assumptions (ACI-10.2) are made:

1. The strength of members shall be based on satisfying the applicable conditions of equilibrium and compatibility of strains.
2. Strain in the steel reinforcement and in the concrete shall be assumed directly proportional to the distance from the neutral axis (except for deep members covered under ACI-10.7).
3. The maximum usable strain ϵ_{cu} at the extreme concrete compression fiber shall be assumed equal to 0.003.

4. The tensile strength of the concrete is to be neglected (except for certain prestressed concrete conditions).

5. The modulus of elasticity of nonprestressed steel reinforcement may be taken as 29,000,000 psi (200,000 MPa or 2,040,000 kgf/cm²).

6. For practical purposes the relationship between the concrete compressive stress distribution and the concrete strain when nominal strength is reached may be taken as an equivalent rectangular stress distribution (ACI-10.2.7), wherein a concrete stress intensity of $0.85f'_c$ is assumed to be uniformly distributed over an equivalent compression zone bounded by the edges of the cross-section and a straight line located parallel to the neutral axis at a distance $a = \beta_1 x$ from the fiber of maximum compressive strain. The distance x from the fiber of maximum strain to the neutral axis is measured in a direction perpendicular to that axis. The value of β_1 is given by the following equations:

For $f'_c \leq 4000$ psi,

$$\beta_1 = 0.85 \tag{3.4.3}$$

For $f'_c > 4000$ psi,

$$\beta_1 = 0.85 - 0.05\left(\frac{f'_c - 4000}{1000}\right) \geq 0.65 \tag{3.4.4}$$

It should be noted that assumption (6) describes the Whitney rectangular compressive stress distribution (see Section 3.3), but other shapes of stress solids, such as the trapezoid and the parabola, have been used [3.7] and are acceptable for use according to ACI-10.2.6.

■ **EXAMPLE 3.4.1** Determine the nominal flexural strength M_n of the rectangular section shown in Fig. 3.4.1, given $f'_c = 5000$ psi, $f_y = 50,000$ psi, $b = 14$ in., $d = 21.5$ in., and $A_s = 4$-#10 bars.

(a) Cross section   (b) Actual condition

Figure 3.4.1 Singly reinforced beam of Example 3.4.1.

Solution: Assume the steel has already yielded when the strength is reached. From Fig. 3.4.1 the internal forces are

$$C = 0.85f'_c ba = 0.85(5)(14)a = 59.5a$$

$$T = A_s f_y = 5.08(50) = 254 \text{ kips}$$

For equilibrium, $C = T$; therefore

$$a = \frac{254}{59.5} = 4.27 \text{ in.}$$

$$\beta_1 = 0.80 \quad \text{for } f'_c = 5000 \text{ psi}$$

The neutral axis position is

$$x = \frac{a}{\beta_1} = \frac{4.27}{0.80} = 5.34 \text{ in.}$$

The strain in the tension steel when the compressive strain 0.003 is reached at the extreme concrete fiber is, by straightline proportion,

$$\epsilon_s = \frac{d - x}{x}(0.003) = \frac{16.16}{5.34}(0.003) = 0.0091$$

$$\epsilon_y = \frac{f_y}{E_s} = \frac{50}{29,000} = 0.00172$$

When the nominal flexural strength is reached, ϵ_s is 5.3 times ϵ_y, which means that large and gradual deflection has occurred before the crushing of concrete. The assumption that the steel yields has been shown valid.

The nominal flexural strength is

$$M_n = C\left(d - \frac{a}{2}\right) \quad \text{or} \quad T\left(d - \frac{a}{2}\right)$$
$$= 254(21.5 - 2.13)\tfrac{1}{12} = 410 \text{ ft-kips} \quad \blacksquare$$

■ **EXAMPLE 3.4.2**   For the beam of Example 3.4.1, determine the safe service moment M_w that may be applied according to the ACI Code, if 60% of the total moment is dead load and 40% is live load.

Solution: From Example 3.4.1,

$$M_n = \text{nominal strength} = 410 \text{ ft-kips}$$

$$M_u = \text{factored moment} = 1.4M_D + 1.7M_L$$

$$\phi = 0.90 \text{ for flexure}$$

For safety, Eq. (3.4.2) requires that

$$\phi M_n \geq M_u$$
$$\geq [1.4M_D + 1.7M_L]$$

which, if $M_D = 0.60M_w$ and $M_L = 0.40\ M_w$, gives

$$\text{required } M_n = \frac{M_u}{\phi} = \frac{1.4(0.60)M_w + 1.7(0.40)M_w}{0.90} = 1.69M_w$$

Thus the safe service moment is

$$M_w = \frac{M_n}{1.69} = \frac{410}{1.69} = 243 \text{ ft-kips}$$

The safety factor between nominal strength and service load is 1.69 for this case. ∎

3.5 DEFINITION OF BALANCED STRAIN CONDITION

At the *balanced strain* condition [Fig. 3.5.1(b)] the maximum strain ϵ_{cu} at the extreme concrete compression fiber just reaches 0.003 simultaneously with the tension steel reaching a strain $\epsilon_y = f_y/E_s$ (Fig. 3.2.2 shows the definition of ϵ_{cu} and ϵ_y). An amount of tension steel A_{sb} would provide the neutral axis distance x_b for this balanced strain condition. If the actual A_s were greater than A_{sb}, equilibrium of internal forces ($C = T$) would mean an increase in the depth a of the compression stress block (and thereby also make x exceed x_b), so that the strain ϵ_s would be less than ϵ_y when $\epsilon_{cu} = 0.003$. The failure of this beam would be sudden when the concrete reaches the strain 0.003 although the beam would exhibit little deformation (steel does not yield) to warn of impending failure.

Figure 3.5.1 Balanced strain condition.

On the other hand, when the actual A_s is less than A_{sb}, the tensile force reduces so that internal force equilibrium reduces the depth a of the compression stress block (and thereby makes x less than x_b), giving a strain ϵ_s greater than ϵ_y. In this case, with the steel having yielded, the beam will have noticeable deflection prior to the concrete reaching the crushing strain of 0.003.

Therefore, the relative amount of tension steel compared to that in the balanced strain condition will determine whether the failure is ductile (gradual, giving warning) or brittle (sudden, without warning).

Reinforcement Ratio at Balanced Strain Condition for Rectangular Beam Having Tension Reinforcement Only.   The symbol ρ, the *reinforcement ratio* (often called *reinforcement percentage*), may be conveniently used to represent the relative amount of tension reinforcement in a beam. Thus using the dimensions of Fig. 3.5.1,

$$\rho = \frac{A_s}{bd} \tag{3.5.1}$$

The reinforcement ratio ρ_b in the balanced strain condition may be obtained by applying the equilibrium and compatibility conditions. From the linear strain condition, Fig. 3.5.1(b),

$$\frac{x_b}{d} = \frac{\epsilon_{cu}}{\epsilon_{cu} + \epsilon_y} = \frac{0.003}{0.003 + f_y/29,000,000} = \frac{87,000}{87,000 + f_y} \tag{3.5.2}$$

The compressive force C_b is

$$C_b = 0.85 f_c' b \beta_1 x_b$$

The tensile force T_b is

$$T_b = f_y A_{sb} = \rho_b b \, d f_y$$

Equating C_b to T_b gives

$$0.85 f_c' b \beta_1 x_b = \rho_b b \, d f_y$$

$$\rho_b = \frac{0.85 f_c'}{f_y} \beta_1 \left(\frac{x_b}{d}\right) \tag{3.5.3}$$

which on substitution of Eq. (3.5.2) gives

$$\rho_b = \frac{0.85 f_c'}{f_y} \beta_1 \left(\frac{87,000}{87,000 + f_y}\right) \tag{3.5.4}*$$

where the stresses f_y and f_c' are in psi.

— — — —

* For SI, ACI 318–89M gives

$$\rho_b = \frac{0.85 f_c'}{f_y} \beta_1 \left(\frac{600}{600 + f_y}\right) \tag{3.5.4}$$

with f_c' and f_y in MPa, and β_1 as given in footnote on p. 47.

3.6 MAXIMUM REINFORCEMENT RATIO ρ

In order to have reasonable assurance for a ductile mode of failure in flexure, ACI-10.3.3 limits the amount of tension steel to not more than 75% of the amount in the balanced strain condition, that is,

$$\max \rho = 0.75\rho_b \tag{3.6.1}$$

The limitation on reinforcement ratio ρ is an indirect way of controlling the strain diagram when "failure" (i.e., crushing of the concrete at the extreme compression fiber) is imminent.

A more direct way of controlling ductility is to prescribe a maximum value for the neutral axis distance x at the " failure" imminent condition. It can be shown that for singly reinforced rectangular sections, ACI-10.3.3 is equivalent to (see Fig. 3.6.1)

$$\max x = 0.75x_b \tag{3.6.2}$$

Thus the direct and indirect concepts lead to identical results of ductile failure for singly reinforced rectangular sections.

For beams containing steel in the compression zone or having a nonrectangular compression zone, the literal wording of ACI-10.3.3 as expressed by Eq. (3.6.1) may give somewhat different results than limiting x to $0.75x_b$, as expressed by Eq. (3.6.2). In the majority of cases, however, the two concepts lead to identical results. This subject is further treated in Section 3.10.

In actual structures the provided design strength ϕM_n, being equal to or slightly greater than the factored moment M_u, is always much larger than the actual service moment acting on the beam. Thus the actual attainment of M_n in a beam is an imaginary situation or, put in another way, a philosophy of design. In this philosophy, it is considered appropriate to restrict the "failure" imminent condition to that of Fig. 3.6.1(b), wherein the ductile failure mode can be expected.

Figure 3.6.1 Maximum reinforcement ratio (ACI-10.3.3); max $\rho \le 0.75\rho_b$ or max $x \le 0.75x_b$.

■ **EXAMPLE 3.6.1** Determine whether or not the amount of steel used in the beam ($b = 14$ in., $d = 21.5$ in., $A_s = 5.08$ sq in., $f'_c = 5000$ psi, and $f_y = 50,000$ psi) of Example 3.4.1 (Fig. 3.4.1) is acceptable according to the ACI Code.

Solution: (a) Determine A_{sb} in the balanced strain condition using basic principles, according to the ACI Code. From Fig. 3.6.2,

$$x_b = \frac{0.003(21.5)}{0.003 + 50/29,000} = 13.65 \text{ in.}$$

$$\beta_1 = 0.80 \ [\text{for } f'_c = 5000 \text{ psi}]$$

$$a_b = 0.80(13.65) = 10.92 \text{ in.}$$

$$C_b = 0.85(5)(14)(10.92) = 650 \text{ kips}$$

$$A_{sb} = \frac{650}{50} = 13.0 \text{ sq in.}$$

Note that the balanced strain condition is a *reference* condition that is not permitted by the ACI Code to actually occur in a beam.

(b) Compare actual steel used with the maximum permitted by the ACI Code.

$$\max A_s = 0.75A_{sb} = 0.75(13.0) = 9.75 \text{ sq in.}$$

$$\text{actual } A_s = 5.08 \text{ sq in.} < 9.75 \text{ sq in.} \qquad \textbf{OK}$$

ACI-10.3.3 is satisfied. Ordinarily this check should be made *before* an investigation of strength is made in cases where the ACI Code is to be satisfied.

If the maximum steel permitted is expressed in terms of the reinforcement ratio ρ,

$$\max \rho = 0.75\rho_b = \frac{9.75}{14(21.5)} = 0.0324$$

Values of maximum ρ for beams having a rectangular compression zone and various material strengths are given in Table 3.6.1. ■

(a) Cross section   (b) Balanced condition

Figure 3.6.2 Balanced strain condition for Example 3.6.1 (actual condition shown in Fig. 3.4.1).

Table 3.6.1   Maximum Reinforcement Ratio ρ for Singly Reinforced Rectangular Beams (Corresponding to $0.75\rho_b$)

f_y	$f'_c = 3000$ psi $\beta_1 = 0.85$	$f'_c = 3500$ psi $\beta_1 = 0.85$	$f'_c = 4000$ psi $\beta_1 = 0.85$	$f'_c = 5000$ psi $\beta_1 = 0.80$	$f'_c = 6000$ psi $\beta_1 = 0.75$
40,000 psi	0.0278	0.0325	0.0371	0.0437	0.0491
50,000 psi	0.0206	0.0241	0.0275	0.0324	0.0364
60,000 psi	0.0160	0.0187	0.0214	0.0252	0.0283

f_y	$f'_c = 20$ MPa $\beta_1 = 0.85$	$f'_c = 25$ MPa $\beta_1 = 0.85$	$f'_c = 30$ MPa $\beta_1 = 0.85$	$f'_c = 35$ MPa $\beta_1 = 0.81$	$f'_c = 40$ MPa $\beta_1 = 0.77$
300 MPa	0.0241	0.0301	0.0361	0.0402	0.0436
350 MPa	0.0196	0.0244	0.0293	0.0326	0.0354
400 MPa	0.0163	0.0203	0.0244	0.0271	0.0295

f_y	$f'_c = 200$ kgf/cm^2 $\beta_1 = 0.85$	$f'_c = 240$ kgf/cm^2 $\beta_1 = 0.85$	$f'_c = 280$ kgf/cm^2 $\beta_1 = 0.85$	$f'_c = 320$ kgf/cm^2 $\beta_1 = 0.82$	$f'_c = 360$ kgf/cm^2 $\beta_1 = 0.79$
2800 kgf/cm^2	0.0266	0.0319	0.0372	0.0410	0.0444
3500 kgf/cm^2	0.0197	0.0236	0.0276	0.0304	0.0330
4200 kgf/cm^2	0.0153	0.0184	0.0214	0.0236	0.0256

3.7 MINIMUM REINFORCEMENT RATIO ρ

When the steel reinforcement in a flexural member is only a small amount because the factored moment M_u is low, the beam may perform in service in its uncracked state. The method used for computing flexural strength, however, is based on the assumption that the tension concrete is cracked. Thus, it would be possible for the computed nominal strength M_n for a section having a small amount of reinforcement, based on cracked tension concrete, to be less than the strength M_n (called M_{cr}) of the same section as a plain concrete beam (i.e., having no steel reinforcement). Since a ductile failure mode is desired, the lowest amount of steel permitted should be the amount that would equal the strength of an unreinforced beam. The desired relationship may be developed as follows:

$$\begin{bmatrix} \text{strength of reinforced} \\ \text{concrete beam, } \phi M_n \end{bmatrix} \geq \begin{bmatrix} \text{strength of plain} \\ \text{concrete beam, } M_{cr} \end{bmatrix} \qquad \textbf{(3.7.1)}$$

The strength of a plain concrete beam is referred to as the *cracking moment* M_{cr} and is achieved when the extreme fiber in tension reaches the modulus of rupture f_r (see Section 1.8). For normal weight concrete, the ACI Code uses

$$f_r = 7.5\sqrt{f'_c} \qquad \textbf{(3.7.2)}$$

as given by ACI-9.5.2.3. Considering plain (nonreinforced) concrete as an elastic homogeneous material, the flexure formula gives M_{cr} as

$$M_{cr} = f_r \frac{I_g}{y_t} \qquad (3.7.3)$$

where

     I_g = moment of inertia of the gross concrete cross section

     y_t = distance from the neutral axis to the extreme fiber in tension

Thus, for a rectangular section,

$$M_{cr} = 7.5\sqrt{f_c'} \frac{bh^3/12}{h/2} = \frac{7.5\sqrt{f_c'}\, bh^2}{6} \qquad (3.7.4)$$

For a reinforced concrete rectangular beam, using Eq. (3.3.4),

$$\phi M_n = \phi A_s f_y (d - a/2) \qquad [3.3.4]$$

Substituting Eqs. (3.7.4) and (3.3.4) into Eq. (3.7.1) gives

$$\phi A_s f_y \left(d - \frac{a}{2} \right) \geq \frac{7.5\sqrt{f_c'}\, bh^2}{6} \qquad (3.7.5)$$

Letting $A_s = \rho bd$, estimating $a/2$ as $\approx 0.05d$ for small values of ρ, and using $\phi = 0.9$ for flexure, Eq. (3.7.5) gives

$$(0.9)\rho b\, df_y (0.95d) \geq 1.25\sqrt{f_c'}\, bh^2$$

or

$$\rho \geq \frac{1.25\sqrt{f_c'}}{0.855 f_y} \left(\frac{h}{d} \right)^2 \qquad (3.7.6)$$

Then, if $d \approx 0.9h$,

$$\min \rho \geq \frac{1.80\sqrt{f_c'}}{f_y} \qquad (3.7.7)$$

For various values of f_c', the min ρ becomes

$$f_c' = 3000 \text{ psi}; \qquad \min \rho = \frac{99}{f_y}$$

$$f_c' = 4000 \text{ psi}; \qquad \min \rho = \frac{114}{f_y}$$

$$f_c' = 5000 \text{ psi}; \qquad \min \rho = \frac{128}{f_y}$$

The requirement of ACI-10.5.1 states

$$\min \rho \geq \frac{200}{f_y} \qquad (3.7.8)$$

Equation (3.7.8) is conservative for rectangular sections. Even if h/d becomes 1.25, as it might for a thin slab, the coefficient would only go as high as 145. Because it is recognized that Eq. (3.7.8) is extremely conservative for slabs, ACI-10.5.3 directs the designer to ACI-7.12, because the reinforcement indicated for temperature and shrinkage is approximately the amount indicated by Eq. (3.7.7). Temperature and shrinkage behavior is not involved but the amounts in ACI-7.12 are the approximate minimums.

Since beams in monolithic concrete construction are commonly T-shaped (see Chapter 9) rather than rectangular, the same analysis used to obtain Eq. (3.7.7) should be made for, say, the T-section with the top (the flange or slab) of the T in tension, and also with the bottom (the stem of width b_w) of the T in tension. Such an analysis will show that with the *slab in compression*, Eq. (3.7.7) using $\rho = \rho_w = A_s/b_w d$ will give

$$\min \rho_w \geq \frac{(150 \text{ to } 250)^*}{f_y} \tag{3.7.9}$$

For the *slab in tension* on the T-section,

$$\min \rho_w \geq \frac{(300 \text{ to } 600)^\dagger}{f_y} \tag{3.7.10}$$

Thus, ACI-10.5.1 is a conservative requirement for isolated rectangular beams and reasonable for the usual positive moment orientation for T-sections. It is unconservative for T-sections with the slab in tension. The authors believe that minimum reinforcement should be used in both positive and negative moment regions, and particularly at the support on cantilevers.

Note that ACI-10.5.2 permits a waiver of Eq. (3.7.8) whenever ϕM_n equals or exceeds $1.33 M_u$. The philosophy is that extra safety is a satisfactory substitute for having a ductile failure mode.

3.8 DESIGN OF RECTANGULAR SECTIONS IN BENDING WITH TENSION REINFORCEMENT ONLY

In the design of rectangular sections in bending with tension reinforcement only, the problem is to determine b, d, and A_s from the required value of $M_n = M_u/\phi$, and the given material properties f'_c and f_y.

The two conditions of equilibrium are

$$C = T \tag{3.8.1a}$$

and

$$M_n = (C \text{ or } T)\left(d - \frac{a}{2}\right) \tag{3.8.1b}$$

– – – –

* 150 for flange width equal to four times web width, flange thickness equal to 0.2 of gross depth, for $f'_c = 4000$ psi, and assuming h/d equal to 1.10.

$\dagger$ 280 for the proportions and data stated above.

Since there are three unknowns but only two conditions, there are many possible solutions. If the reinforcement ratio ρ is preset, then, from Eq. (3.8.1a),

$$0.85f'_c ba = \rho bdf_y$$

$$a = \rho\left(\frac{f_y}{0.85f'_c}\right)d \tag{3.8.2}$$

Substituting Eq. (3.8.2) into Eq. (3.8.1b),

$$M_n = \rho bdf_y\left[d - \frac{\rho}{2}\left(\frac{f_y}{0.85f'_c}\right)d\right] \tag{3.8.3}$$

A strength *coefficient of resistance* R_n may be obtained by dividing Eq. (3.8.3) by bd^2 and letting

$$m = \frac{f_y}{0.85f'_c} \tag{3.8.4a}$$

Thus

$$R_n = \frac{M_n}{bd^2} = \rho f_y(1 - \tfrac{1}{2}\rho m) \tag{3.8.4b}$$

The relationship between ρ and R_n for various values of f'_c and f_y is shown in Fig. 3.8.1.

In some situations the values of b and d may be preset, which is equivalent to having R_n preset; then ρ may be determined by solving the quadratic equation (3.8.4). Thus

$$R_n = \rho f_y(1 - \tfrac{1}{2}\rho m)$$

from which

$$\rho = \frac{1}{m}\left(1 - \sqrt{1 - \frac{2mR_n}{f_y}}\right) \tag{3.8.5}$$

The procedure (without being concerned with certain practical decisions) to be used in the strength design of rectangular sections with tension reinforcement only involves the following steps:

1. Assume a value of ρ equal to or less than $0.75\rho_b$, but greater than the minimum ρ of $200/f_y$ (ACI-10.5). The balanced value ρ_b may be obtained from basic principles or from Eq. (3.5.4),

$$\rho_b = \frac{0.85\beta_1 f'_c}{f_y}\left(\frac{87,000}{87,000 + f_y}\right)$$

and

$$\text{for } f'_c \leq 4000 \text{ psi} \quad \beta_1 = 0.85$$

$$\text{for } f'_c > 4000 \text{ psi} \quad \beta_1 = 0.85 - 0.05\left(\frac{f'_c - 4000}{1000}\right) \geq 0.65$$

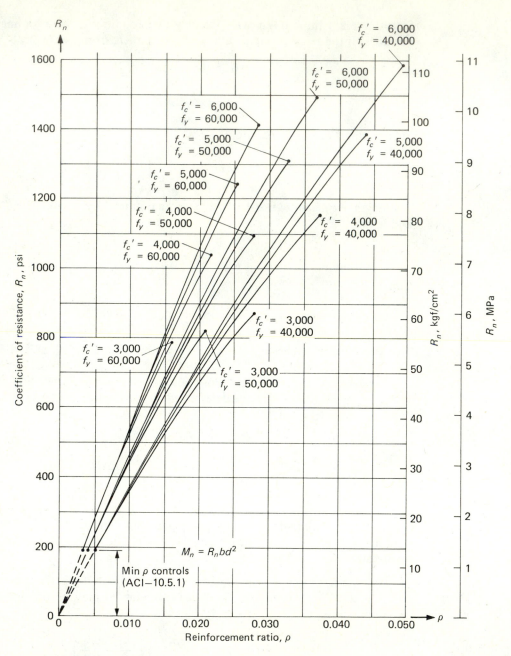

Figure 3.8.1 Strength curves (R_n vs ρ) for singly reinforced rectangular sections. Upper limit of curves is at $0.75\rho_b$.

2. Determine the required bd^2 from

$$\text{required } bd^2 = \frac{\text{required } M_n}{R_n}$$

in which

$$R_n = \rho f_y(1 - \tfrac{1}{2}\rho m)$$

with

$$m = \frac{f_y}{0.85 f'_c}$$

3. Choose a suitable set of values of b and d so that the provided bd^2 is approximately equal to the required bd^2. (Note: Actually, d is not chosen; rather the overall depth h is chosen and d is computed from h while maintaining the desired minimum protective cover.)

4. Determine the revised value of ρ after computing $R_n = M_n/bd^2$ for the selected section using one of the following methods:

   a. By formula (most exact),

   $$\rho = \frac{1}{m}\left(1 - \sqrt{1 - \frac{2mR_n}{f_y}}\right)$$

   b. By curves (Fig. 3.8.1).

   c. By approximate proportion (on the safe side if revised R_n is smaller than original R_n), the revised ρ is

   $$\rho \approx (\text{original } \rho)\frac{(\text{revised } R_n)}{(\text{original } R_n)}$$

   It may be noted from Fig. 3.8.1 that the relationship between R_n and ρ is approximately linear over short distances, even though the actual equation is a quadratic one.

5. Compute A_s from

$$A_s = (\text{revised } \rho)(\text{actual } bd)$$

6. Select reinforcement and check the strength of the section to be certain that

$$\phi M_n \geq M_u$$

■ **EXAMPLE 3.8.1**   Determine a set of values of b, d, and A_s that will carry a factored bending moment M_u of 400 ft-kips. Use $f'_c = 4000$ psi and $f_y = 40,000$ psi.

Solution: (a) Establish the limits within which the reinforcement ratio ρ can be chosen. From Eq. (3.5.4),

$$\rho_b = \frac{0.85\beta_1 f'_c}{f_y}\left(\frac{87,000}{87,000 + f_y}\right) = \frac{0.85(0.85)(4)}{40}\left(\frac{87}{87 + 40}\right) = 0.0495$$

$$\max \rho = 0.75\,\rho_b = 0.0371$$

From ACI-10.5.1,

$$\min \rho = \frac{200}{f_y} = \frac{200}{40,000} = 0.005$$

(b) Choose reinforcement ratio ρ, determine corresponding required beam size, and select actual beam size. Arbitrarily assume that $\rho = 0.03$ (any choice within prescribed limits is acceptable here),

$$m = \frac{f_y}{0.85f'_c} = \frac{40}{0.85(4)} = 11.76$$

$$R_n = \rho f_y(1 - \tfrac{1}{2}\rho m) = 0.03(40,000)\left[1 - \tfrac{1}{2}(0.03)(11.76)\right] = 988 \text{ psi}$$

$$\text{required } M_n = \frac{M_u}{\phi} = \frac{400}{0.90} = 444 \text{ ft-kips}$$

$$\text{required } bd^2 = \frac{\text{required } M_n}{R_n} = \frac{444(12,000)}{988} = 5400 \text{ in.}^3$$

Try $b = 14$ in. and $d = \sqrt{5400/14} = 19.64$ in. Use $b = 14$ in. and compute d to be 21.5 in. after selecting overall depth (d is usually $2\frac{3}{8}$ to $2\frac{5}{8}$ in. less than overall depth if one layer of steel can be used; this is illustrated in Example 3.9.2).

(c) Compute required A_s and select actual steel bars, using actual beam size. The effective depth d is now greater than required for $\rho = 0.03$ in order to get a convenient depth, or perhaps to reduce deflection.

$$\text{required } R_n = \frac{\text{required } M_n}{\text{provided } bd^2} = \frac{444(12,000)}{14(21.5)^2} = 824 \text{ psi}$$

$$\rho = \frac{1}{m}\left(1 - \sqrt{1 - \frac{2mR_n}{f_y}}\right)$$

$$= \frac{1}{11.76}\left[1 - \sqrt{1 - \frac{2(11.76)(824)}{40,000}}\right] = 0.0240$$

$$A_s = \rho bd = 0.0240(14)(21.5) = 7.22 \text{ sq in.}$$

Two layers of reinforcement will be required. Select 3-#11 and 2-#10, $A_s = 7.22$ sq in.

(d) Check design. A review of the correctness of the above computations in which formulas are used should be made using the basic statics shown in Fig. 3.8.2.

$$T = A_s f_y = 7.22(40) = 289 \text{ kips}$$

$$a = \frac{C \text{ or } T}{0.85f'_c b} = \frac{289}{0.85(4)(14)} = 6.07 \text{ in.}$$

$$M_n = (C \text{ or } T)\left(d - \frac{a}{2}\right) = 289[21.5 - 0.5(6.07)]\tfrac{1}{12}$$

$$= 444 \text{ ft-kips} \qquad\qquad \textbf{OK}$$

Figure 3.8.2 Section for Example 3.8.1.

(e) Final decision and design sketch. Every design requires a clear decision as its conclusion, along with an appropriate design sketch.

Use 3-#11 and 2-#10 in the 14 by 25 in. beam shown in Fig. 3.8.2. (The *d* was computed from the overall 25-in. depth as discussed in succeeding examples.) ∎

It should be noted that the beam of Example 3.8.1 with 2.4% reinforcement, which requires placement in two layers, is significantly smaller in concrete cross-sectional area than usually used. The higher the value of ρ, the smaller the beam cross-section. For practical bar placement in one layer and the need for larger cross-section to reduce deflection, ρ does not usually exceed $0.5\rho_{\max}$. Deflection thus becomes an essential consideration in designing beams by the strength method. The subject of deflection is treated in Chapter 14.

3.9 PRACTICAL SELECTION FOR BEAM SIZES, BAR SIZES, AND BAR PLACEMENT

In the previous section, the procedure and example for the design of rectangular sections in bending with tension reinforcement only have been treated on the assumption that the factored moment M_u is known. This is rarely the case, however, because the factored moment must include the effect of the weight of the beam itself, which has not yet been designed. In reality, then, the dead weight of the beam has to be assumed at the outset; a trial beam size is then obtained and may be readjusted if its effect on the factored moment significantly differs from the assumed value.

The choice of the steel reinforcement ratio ρ is very much dependent on the limitation on the deflection of the beam. Years of experience with the working stress method showed that deflection problems were rarely encountered with beams having a steel reinforcement ratio ρ not more than one-half the maximum permissible value. The use of this amount (one-half of $0.75\rho_b = 0.375\rho_b$) may provide a suitable *guide* for the preliminary choice of the reinforcement ratio. Abendroth and Salmon [3.8]

Table 3.9.1 Total Areas for Various Numbers of Reinforcing Bars

BAR SIZE	NOMINAL DIAMETER (in.)	WEIGHT (lb/ft)	NUMBER OF BARS									
			1	2	3	4	5	6	7	8	9	10
#3	0.375	0.376	0.11	0.22	0.33	0.44	0.55	0.66	0.77	0.88	0.99	1.10
#4	0.500	0.668	0.20	0.40	0.60	0.80	1.00	1.20	1.40	1.60	1.80	2.00
#5	0.625	1.043	0.31	0.62	0.93	1.24	1.55	1.86	2.17	2.48	2.79	3.10
#6	0.750	1.502	0.44	0.88	1.32	1.76	2.20	2.64	3.08	3.52	3.96	4.40
#7	0.875	2.044	0.60	1.20	1.80	2.40	3.00	3.60	4.20	4.80	5.40	6.00
#8	1.000	2.670	0.79	1.58	2.37	3.16	3.95	4.74	5.53	6.32	7.11	7.90
#9	1.128	3.400	1.00	2.00	3.00	4.00	5.00	6.00	7.00	8.00	9.00	10.00
#10	1.270	4.303	1.27	2.54	3.81	5.08	6.35	7.62	8.89	10.16	11.43	12.70
#11	1.410	5.313	1.56	3.12	4.68	6.24	7.80	9.36	10.92	12.48	14.04	15.60
#14[a]	1.693	7.65	2.25	4.50	6.75	9.00	11.25	13.50	15.75	18.00	20.25	22.50
#18[a]	2.257	13.60	4.00	8.00	12.00	16.00	20.00	24.00	28.00	32.00	36.00	40.00

[a] #14 and #18 bars are used primarily as column reinforcement and are rarely used in beams.

have presented a sensitivity study to show how the variables in simply supported beam design affect the economics of the design.

For the selection of an actual number of bars to meet a total steel area requirement, it is desirable to tabulate the combined area of several bars at a time. Table 3.9.1 gives bar areas for up to 10 bars of different sizes. Note that once the area of one bar, say of #8, is set at 0.79 sq in., the area of 10 bars becomes 7.90, not 7.85; this practice has been carried on by tradition.

For the placement of bars within the beam width, ACI-7.6.1 specifies the clearance needed between bars to permit proper concrete placement around them. This clearance for bars in a layer is 1 in. or the nominal diameter of the bar, whichever is greater. When two or more layers of bars are required, the minimum clearance between layers is 1 in. (ACI-7.6.2). The bars in different layers must be located directly above one another and not staggered (ACI-7.6.2). Table 3.9.2 gives minimum beam widths for various numbers of equal-sized bars, computed in the manner described above.

Since the tensile force in reinforcing bars requires an interaction of the bars with the surrounding concrete, larger than minimum spacing and clear cover of bars will make the concrete confine the bars more efficiently. This benefit is explicitly recognized for the first time in the 1989 ACI Code. As will be discussed in Chapter 6 on development of the tension force in the bars, the use of at least $2d_b$ clear lateral spacing of bars (a category of improved performance) or $3d_b$ clear lateral spacing of bars (the best category of performance) will accrue benefit when determining the lengths of reinforcing bars to be used. Since the selection of cross-section and bars is all that has been treated thus far in this chapter, suffice it to say that shorter length bars can be used when $2d_b$ clear lateral spacing can be used (minimum width in Table 3.9.3), and when $3d_b$ clear lateral spacing can be used (minimum width in Table 3.9.4) the minimum lengths of reinforcement will be required.

Table 3.9.2   Minimum Beam Width (Inches) According to the ACI Code[a]

SIZE OF BARS	NUMBER OF BARS IN SINGLE LAYER OF REINFORCEMENT							ADD FOR EACH ADDED BAR
	2	3	4	5	6	7	8	
#4	6.8	8.3	9.8	11.3	12.8	14.3	15.8	1.50
#5	6.9	8.5	10.2	11.8	13.4	15.0	16.7	1.63
#6	7.0	8.8	10.5	12.3	14.0	15.8	17.5	1.75
#7	7.2	9.0	10.9	12.8	14.7	16.5	18.4	1.88
#8	7.3	9.3	11.3	13.3	15.3	17.3	19.3	2.00
#9	7.6	9.8	12.2	14.3	16.6	18.8	21.1	2.26
#10	7.8	10.4	12.9	15.5	18.0	20.5	23.1	2.54
#11	8.1	10.9	13.8	16.6	19.4	22.2	25.0	2.82
#14	8.9	12.3	15.7	19.1	22.5	25.9	29.3	3.40
#18	10.6	15.1	19.6	24.1	28.6	33.1	37.6	4.51

Table shows minimum beam widths when #3 stirrups are used.

For additional bars, add dimension in last column for each added bar.

For bars of different size, determine from table the beam width for smaller size bars and then add last column figure for each larger bar used.

[a] Assumes maximum aggregate size does not exceed three-fourths of the clear space between bars (ACI–3.3.2). Table computation procedure is in agreement with the ACI Code interpretation of ACI Committee 340, as used in the *Strength Design Handbook* [2.21].

$A = 1\frac{1}{2}$-in. clear cover to stirrup

$B = \frac{3}{8}$-in. stirrup bar diameter

$C =$ For #11 and smaller bars, use twice the diameter of #3 stirrups (i.e., $C = 0.75$ in.). For #14 and #18 bars, use $C = 0.5d_b$

$D =$ clear distance between bars $= d_b$ or 1 in., whichever is greater (where d_b is the diameter of the larger adjacent longitudinal bar)

Diameter of corner bar is assumed to be located to intersect the horizontal tangent to stirrup bend

Table 3.9.3 Minimum Beam Width (inches) To Satisfy
2 Bar Diameters Clear Spacing

BAR SIZE	NUMBER OF BARS IN SINGLE LAYER						
	2	3	4	5	6	7	8
#4	6.8	8.3	9.8	11.3	12.8	14.3	15.8
#5	7.1	9.0	10.9	12.8	14.6	16.5	18.4
#6	7.5	9.8	12.0	14.3	16.5	18.8	21.0
#7	7.9	10.5	13.1	15.8	18.4	21.0	23.6
#8	8.3	11.3	14.3	17.3	20.3	23.3	26.3
#9	8.6	12.0	15.4	18.8	22.2	25.6	28.9
#10	9.1	12.9	16.7	20.5	24.3	28.1	31.9
#11	9.5	13.7	17.9	22.2	26.4	30.6	34.9
#14	12.2	15.9	20.9	26.0	31.1	36.2	41.2
#18	15.0	19.8	26.6	33.3	40.1	46.9	53.7

Table Assumptions:

a. Side cover 1.5 in. each side.
b. #3 stirrups for bars #11 and smaller.
c. #4 stirrups for bars #14 and #18.
d. Since stirrups are bent around 4 stirrup bar diameters, the distance from centroid of bar nearest side face of beam to inside face of #3 stirrup is taken as 0.75 in. for bars #11 and smaller; and equal to the bar radius for #14 and #18 bars.

Table 3.9.4 Minimum Beam Width (inches) To Satisfy
3 Bar Diameters Clear Spacing

BAR SIZE	NUMBER OF BARS IN SINGLE LAYER						
	2	3	4	5	6	7	8
#4	7.3	9.3	11.3	13.3	15.3	17.3	19.3
#5	7.8	10.3	12.8	15.3	17.8	20.3	22.8
#6	8.3	11.3	14.3	17.3	20.3	23.3	26.3
#7	8.8	12.3	15.8	19.3	22.8	26.3	29.8
#8	9.3	13.3	17.3	21.3	25.3	29.3	33.3
#9	9.8	14.3	18.8	23.3	27.8	32.3	36.8
#10	10.3	15.4	20.5	25.6	30.7	35.7	40.8
#11	10.9	16.5	22.2	27.8	33.5	39.1	44.7
#14	12.5	19.2	26.0	32.8	39.6	46.3	53.1
#18	15.3	24.3	33.3	42.4	51.4	60.4	69.5

* Table Assumptions:

a. Side cover 1.5 in. each side.
b. #3 stirrups for bars #11 and smaller.
c. #4 stirrups for bars #14 and #18.
d. Since stirrups are bent around 4 stirrup bar diameters, the distance from centroid of bar nearest side face of beam to inside face of #3 stirrup is taken as 0.75 in. for bars #11 and smaller; and equal to the bar radius for #14 and #18 bars.

Table 3.9.5 Standard Types and Sizes of Bar Supports (Adapted from Ref. 2.26, p. 194)

SYMBOL	BAR SUPPORT ILLUSTRATION	TYPE OF SUPPORT	STANDARD SIZES
SB	5″	Slab Bolster	$\frac{3}{4}$, 1, $1\frac{1}{2}$, and 2 in. heights in 5-ft and 10-ft lengths
SBU	5″	Slab Bolster Upper	Same as SB
BB	$2\frac{1}{2}″$	Beam Bolster	1, $1\frac{1}{2}$, 2, over 2 to 5 in. heights in increments of $\frac{1}{4}$ in. in lengths of 5 ft
BBU	$2\frac{1}{2}″$	Beam Bolster Upper	Same as BB
BC		Individual Bar Chair	$\frac{3}{4}$, 1, $1\frac{1}{2}$, and $1\frac{3}{4}$ in. heights
HC		Individual High Chair	2 to 15 in. heights in increments of $\frac{1}{4}$ in.
CHC	8″	Continuous High Chair	Same as HC in 5-ft and 10-ft lengths
CHCU	8″	Continuous High Chair Upper	Same as CHC

Bars are supported from the bottom of the forms, and layers of bars are separated by various types of bar supports, known as *bolsters* and *chairs*, some of which are shown in Table 3.9.5. Bar supports may be made of concrete, metal, or other approved materials. Usually they are standard factory-made wire bar supports. They remain in place after the concrete is cast and must have special rust protection on the portions nearest the face of the concrete.

In order to assist the designer further in making choices for beam sizes, bar sizes, and bar placement, the following guidelines are suggested. These may be regarded as accepted practice, and are *not* ACI Code requirements. Undoubtedly, situations will arise in which the experienced designer will, for good and proper reasons, make a selection not conforming to the guidelines.

For Beam Sizes

1. Use whole inches for overall dimensions; except slabs may be in $\frac{1}{2}$-in. increments.
2. Beam stem widths are most often in multiples of 2 or 3 in., such as 9, 10, 12, 14, 15, 16, and 18.
3. Minimum specified clear cover is measured from outside the stirrup or tie to the face of the concrete. (Thus beam effective depth d has rarely, if ever, a dimension to the whole inch.)
4. An economical rectangular beam proportion is one in which the overall depth-to-width ratio is between about 1.5 to 2.0.
5. For T-shaped beams, typically the flange thickness represents about 20% of overall depth (see Chapter 9 for treatment of T-shaped sections).

For Reinforcing Bars

6. Maintain bar symmetry about the centroidal axis which lies at right angles to the bending axis (i.e., symmetry about the vertical axis in usual situations).
7. Use at least two bars wherever flexural reinforcement is required.
8. Use bars #11 and smaller for usual sized beams.
9. Use no more than two bar sizes and no more than two standard sizes apart for steel in one face at a given location in the span (i.e., #7 and #9 bars may be acceptable, but #9 and #4 bars would not).
10. Place bars in one layer if practicable. Try to select bar size so that no less than two and no more than five or six bars are put in one layer.
11. Follow requirements of ACI-7.6.1 and 7.6.2 for clear distance between bars and between layers, and arrangement between layers.
12. When different sizes of bars are used in several layers at a location, place the largest bars in the layer nearest the face of beam.

■ **EXAMPLE 3.9.1**   Select an economical rectangular beam size and select bars using the ACI strength method. The beam is a simply supported span of 40 ft and it is to carry a live load of 1.3 kips/ft and a dead load of 0.8 kip/ft (not including beam weight). Without actually checking deflection, use a reinforcement ratio ρ such that excessive deflection is unlikely. Use $f_c' = 4000$ psi, and $f_y = 60,000$ psi.

Solution: (a) Decide on a reinforcement ratio ρ to use. To have reasonable expectation that deflection will not be excessive, choose ρ at about half the maximum permitted ($0.75 \rho_b = 0.0214$). Use $\rho = 0.011$.

(b) Determine the desired R_n (corresponding to the desired ρ) using Eq. (3.8.4).

$$m = \frac{f_y}{0.85 f_c'} = \frac{60}{0.85(4)} = 17.65$$

$$R_n = \rho f_y (1 - \tfrac{1}{2}\rho m)$$
$$= 0.011(60,000)[1 - 0.5(0.011)(17.65)] = 596 \text{ psi}$$

(c) Determine factored moments.

$$M_u = 1.4M_D + 1.7M_L$$

$$M_L = \frac{1.3(40)^2}{8} = 260 \text{ ft-kips}$$

$$M_D = \frac{(0.8 + 0.4)(40)^2}{8} = 240 \text{ ft-kips}$$

using a beam weight estimated at 0.4 kip/ft, based on unit weight of reinforced concrete at 150 pcf.

$$M_u = 1.4(240) + 1.7(260) = 336 + 442 = 778 \text{ ft-kips}$$

$$\text{required } M_n = \frac{M_u}{\phi} = \frac{778}{0.90} = 864 \text{ ft-kips}$$

(d) Determine required bd^2 from desired R_n.

$$\text{required } bd^2 = \frac{M_n}{R_n} = \frac{864(12,000)}{596} = 17,400 \text{ in.}^3$$

(e) Establish beam size. Select width b and determine the corresponding required value for effective depth d. Make a table of possibilities.

CHOSEN b	REQUIRED d
12	38.1
15	34.1
18	31.1 ← Try
20	29.5

Selecting the 18-in. width will give a beam whose overall depth is between $1\frac{1}{2}$ and 2 times its width (suggested guideline).

Determine overall depth in order to verify the assumed weight. Assuming that the bars to be selected will fit in one layer, the minimum overall depth may be computed (see Fig. 3.9.1)

$h = d + 1\frac{1}{2}$ in. cover $+ \frac{3}{8}$ diameter stirrup $+$ bar radius, say $\frac{1}{2}$ in.

$= d + (2\frac{3}{8} \text{ to } 2\frac{1}{2} \text{ in.})$

$= 31.1 + 2.5 = 33.6 \text{ in.}$

The overall depth would be in whole inches; so try 34 in. Since the guideline value of $R_n = 596$ psi for $\rho = 0.011$ is not a rigorous requirement, the overall depth selected could be somewhat less or somewhat more than the computed requirement in order to obtain a desired dimension. The stirrup is reinforcement to provide shear strength for the beam and should always be allowed for at this stage of the design. The actual

Figure 3.9.1 Quantities added to effective depth d to get overall depth h for beams with one layer of tension steel.

size and spacing of stirrups is determined after the cross section and tension bars have been selected, and the subject is treated in Chapter 5. *Try h = 34 in.*

(f) Check weight, revise M_u, and select reinforcement.

$$w = \frac{18(34)}{144}(0.15) = 0.638 \text{ kip/ft}$$

$$\text{revised } M_D = \frac{(0.8 + 0.638)}{8}(40)^2 = 288 \text{ ft-kips}$$

$$\text{revised } M_u = 1.4(288) + 442 = 404 + 442 = 846 \text{ ft-kips}$$

$$\text{revised required } M_n = \frac{846}{0.90} = 940 \text{ ft-kips}$$

Compute the value of d from the overall dimension h,

$$\text{actual } d = h - (\approx 2.5 \text{ in.}) \quad \text{for one layer of bars} = 34 - 2.5 = 31.5 \text{ in.}$$

When the overall depth is increased or decreased, the clear cover distance (ACI-7.7.1) is the one that is held constant; thus the effective depth will shift.

$$\text{required } R_n = \frac{\text{required } M_n}{bd^2} = \frac{940(12,000)}{18(31.5)^2} = 632 \text{ psi}$$

The steel requirement may be determined from Eq. (3.8.5), from the curves of Fig. 3.8.1, or approximated by straightline proportion. Using the latter and knowing that

$$R_n = 596 \text{ psi} \quad \text{for } \rho = 0.011$$

find approximate ρ for $R_n = 632$ psi,

$$\text{approximate } \rho = 0.011\left(\frac{632}{596}\right) = 0.0117$$

[using Eq. (3.8.5) would have given about the same value]

approximate $A_s = \rho bd = 0.0117(18)(31.5) = 6.63$ sq in.

Select 4-#10 and 2-#9 bars, $A_s = 7.08$ sq in. (Table 3.9.1). (Note that 3-#10 and 3-#9 having $A_s = 6.81$ sq in. provide a smaller A_s, but the bars cannot be placed in one layer and still have symmetry about the vertical beam axis, and 7-#9 will not fit in one layer.) Check whether 4-#10 and 2-#9 will fit into an 18-in. width in one layer.

$$\frac{\text{approx clear spacing}}{\text{between bars}} = \frac{18 - 2(1.5) - 2(0.375) - 4(1.27) - 2(1.128)}{5}$$

$$= 1.38 \text{ in.} > [1.27 \text{ in.} = \text{diameter of } \#10] \qquad \textbf{OK}$$

Subtracted from the overall width are the combined values of the minimum clear cover on both sides (3.0), one stirrup diameter on both sides (0.75), 4-#10 bar diameters (5.08), and 2-#9 bar diameters (2.26). The result is divided by the number of spaces between bars, and this is the *approximate* clearance that must exceed the diameter of the larger bar (ACI-7.6.1). Table 3.9.2 gives the minimum beam width for 6-#10 bars as 18.0 in.

Note that the above clearance computation is approximate because it assumes the #3 stirrup may be bent tightly around the corner longitudinal bar. ACI-7.2.2 requires the inside diameter of bends for stirrups to be not less than four stirrup bar diameters for #5 stirrups and smaller; thus for #3 stirrups the actual curve of the stirrup at the corner has a radius of $\frac{3}{4}$ in., which is larger than the longitudinal bar radius for #11 bars and smaller. Table 3.9.2 is based on the conservative assumption that the diameter of the corner longitudinal bar is located to intersect the horizontal tangent to the stirrup bend (see sketch under Table 3.9.2). Using Table 3.9.2, the minimum required width for 4-#10 and 2-#9 is

$$\text{min } b = 7.6 + 4(2.54) = 17.76 \text{ in.}$$

The 7.6 is for 2-#9 and the 2.54 is for each of the additional #10 bars.

(g) Check strength and provide design sketch. Using computed $d = 31.5$ in.,

$$C = 0.85f'_c ba = 0.85(4)18a = 61.2a$$

$$T = A_s f_y = 7.08(60) = 425 \text{ kips}$$

$$a = \frac{425}{61.2} = 6.95 \text{ in.}$$

$$M_n = A_s f_y \left(d - \frac{a}{2} \right) = 425(31.5 - 3.47)\tfrac{1}{12} = 994 \text{ ft-kips}$$

$$[\phi M_n = 0.90(994) = 895 \text{ ft-kips}] > [M_u = 846 \text{ ft-kips}] \qquad \textbf{OK}$$

Use 18 × 34 *beam with* 4-#10 *and* 2-#9 *bars, as shown in Fig. 3.9.2.* ∎

(a) Final design   (b) Internal forces

Figure 3.9.2 Design for Example 3.9.1.

■ **EXAMPLE 3.9.2** Design a floor slab to carry a uniform live load of 350 psf (pounds per square foot) on a simply supported span of 23 ft, given $f'_c = 3000$ psi and $f_y = 60,000$ psi. Use the ACI Code. Keep the reinforcement ratio at about 30% of the amount in the balanced strain condition so that deflection is unlikely to be excessive.

Solution: The design of a slab is usually made by taking a 1-ft wide typical strip for calculation purposes rather than the entire slab width. This is known as a one-way slab that acts as a wide beam. The common situation of the one-way slab continuous across several beams is treated in detail in Chapter 8.

(a) Determine factored moment.

$$U = 1.4D + 1.7L = 1.4(0.15 \text{ kip/ft estimated}) + 1.7(0.350)$$
$$= 0.210 + 0.595 = 0.805 \text{ kip/ft/ft of width}$$

$$M_u = \frac{0.805(23)^2}{8} = 53.2 \text{ ft-kips/ft of width}$$

$$\text{required } M_n = \frac{M_u}{\phi} = \frac{53.2}{0.90} = 59.2 \text{ ft-kips/ft}$$

(b) Since $\rho = 0.3\rho_b$ is desired, using $\rho = 0.0160(0.3/0.75) = 0.0064$ based on Table 3.6.1, determine the corresponding desired R_n,

$$R_n = \rho f_y(1 - \tfrac{1}{2}\rho m)$$

where

$$m = \frac{f_y}{0.85f'_c} = \frac{60}{0.85(3)} = 23.5$$

$$R_n = 0.0064(60,000)[1 - 0.5(0.0064)(23.5)] = 355 \text{ psi}$$

(c) Determine required bd^2 from desired R_n and select trial slab thickness.

$$\text{required } bd^2 = \frac{M_n}{R_n} = \frac{59.2(12,000)}{355} = 2000 \text{ in.}^3$$

Since a slab is designed by using a 1-ft strip, b is 12 in. Then required $d = 12.9$ in. The required total thickness is obtained by adding on the required clear cover ($\frac{3}{4}$-in. minimum as per ACI-7.7.1) and the bar radius. Stirrups are rarely used in slabs so the $\frac{3}{8}$-in. allowance used for beams (Example 3.9.1) is not included here.

$$\text{total thickness, } h = 12.9 + 0.75 + 0.50 = 14.15 \text{ in.} \quad \text{Try } 14\frac{1}{2} \text{ in.}$$

(d) Check weight and revise required M_n.

$$w = \frac{14.5}{12}(0.15) = 0.181 \text{ kip/ft}$$

This value exceeds the amount estimated, but a repeat of the preceding steps will show that the theoretical total thickness is still about 14.5 in. Correct the moment for use in selecting steel.

$$\text{revised } U = 1.4(0.181) + 1.7(0.350) = 0.849 \text{ kip/ft}$$

$$\text{revised required } M_n = 59.2\left(\frac{0.849}{0.805}\right) = 62.4 \text{ ft-kips/ft}$$

(e) Determine the steel to be used.

$$\text{actual } d = h - 0.75 - 0.5(\text{est}) = 14.5 - 1.25 = 13.25 \text{ in.}$$

$$\text{required } R_n = \frac{\text{required } M_n}{bd^2} = \frac{62.4(12,000)}{12(13.25)^2} = 355 \text{ psi}$$

The actual reinforcement ratio ρ may be determined from Eq. (3.8.5), from the curves of Fig. 3.8.1, or approximated by straightline proportion (see Example 3.9.1). Since the required R_u is equal to that corresponding to $\rho = 0.3\rho_b = 0.0064$,

$$\text{required } A_s = 0.0064(12)(13.25) = 1.02 \text{ sq in./ft}$$

Try #7 @ 7 in. spacing, $A_s = 1.03$ sq in./ft.

For slabs, bars are not selected by picking a total number; instead, they are selected on the basis of an average area provided per foot of width. One #7 gives 0.60 sq in., or 0.60 sq in./ft if a bar is spaced every 12 in. For #7 bars spaced at 7 in., the average area provided is 0.60(12/7) = 1.03 sq in./ft. Table 3.9.6 gives average areas per foot of width provided by various bar spacings. Typically, bar spacings from 4 in. to 9 in. should be used for main reinforcement.

(f) Check strength and provide design sketch. Make a check of strength by using statics (not illustrated here; see Example 3.9.1); select steel transverse to the main steel for temperature, shrinkage, and distribution of loading (not illustrated here, but treated in Chapter 8); and draw design sketch (see Fig. 3.9.3). *Use* $14\frac{1}{2}$-in. thick slab, with #7 @ 7 as main reinforcement. ∎

Table 3.9.6 Average Area per Foot of Width Provided by Various Bar Spacings

BAR SIZE NUMBER	NOMINAL DIAMETER (in.)	SPACING OF BARS IN INCHES													
		2	$2\frac{1}{2}$	3	$3\frac{1}{2}$	4	$4\frac{1}{2}$	5	$5\frac{1}{2}$	6	7	8	9	10	12
3	0.375	0.66	0.53	0.44	0.38	0.33	0.29	0.26	0.24	0.22	0.19	0.17	0.15	0.13	0.11
4	0.500	1.18	0.94	0.78	0.67	0.59	0.52	0.47	0.43	0.39	0.34	0.29	0.26	0.24	0.20
5	0.625	1.84	1.47	1.23	1.05	0.92	0.82	0.74	0.67	0.61	0.53	0.46	0.41	0.37	0.31
6	0.750	2.65	2.12	1.77	1.51	1.32	1.18	1.06	0.96	0.88	0.76	0.66	0.59	0.53	0.44
7	0.875	3.61	2.88	2.40	2.06	1.80	1.60	1.44	1.31	1.20	1.03	0.90	0.80	0.72	0.60
8	1.000		3.77	3.14	2.69	2.36	2.09	1.88	1.71	1.57	1.35	1.18	1.05	0.94	0.78
9	1.128		4.80	4.00	3.43	3.00	2.67	2.40	2.18	2.00	1.71	1.50	1.33	1.20	1.00
10	1.270			5.06	4.34	3.80	3.37	3.04	2.76	2.53	2.17	1.89	1.69	1.52	1.27
11	1.410			6.25	5.36	4.69	4.17	3.75	3.41	3.12	2.68	2.34	2.08	1.87	1.56

Figure 3.9.3 Slab design of Example 3.9.2.

■ **EXAMPLE 3.9.3** Compute, using SI dimensions, the nominal strength M_n of the beam of Fig. 3.9.4, 500 mm wide and 760 mm deep overall, containing 3-#35M bars in the outer layer and 2-#30M bars in the inner layer. The stirrup is #10M, and minimum clear cover is 40 mm with 25 mm between layers. Use $f'_c = 25$ MPa, $f_y = 400$ MPa, and $E_s = 200,000$ MPa.

Figure 3.9.4 Computation of beam strength with metric dimensions, Example 3.9.3.

Solution: **(a)** Determine that the steel is within the permissible limits of the ACI Code. Examine the balanced strain condition (see Section 3.5).

$$\epsilon_y = \frac{f_y}{E_s} = \frac{400}{200{,}000} = 0.0020$$

$$x_b = \frac{0.003}{0.003 + 0.002}\, d = 0.600d$$

$$d = 760 - 40 - 11 - \frac{3(10.0)18 + 2(7.0)76}{3(10.0) + 2(7.0)} = 673 \text{ mm}$$

$$f'_c = 25 \text{ MPa} < f'_c = 30 \text{ MPa}; \qquad \beta_1 = 0.85 \quad \text{(see Table 3.6.1)}$$

$$a_b = \beta_1 x_b = 0.85(0.600)673 = 343 \text{ mm}$$

$$C_b = 0.85f'_c b a_b = 0.85(0.025)(500)(343) = 3640 \text{ kN}$$

$$T_b = A_{sb} f_y$$

$$C_b = T_b$$

$$A_{sb} = \frac{3640(1000)}{400} = 9100 \text{ mm}^2 \; (91.0 \text{ cm}^2)$$

$$\max A_s = 0.75(91.0) = 68.3 \text{ cm}^2$$

Since the actual $A_s = 44.0$ cm^2 is less than the maximum permitted, the beam has an acceptable reinforcement ratio (deflection might be a problem, so it should be checked).

(b) Determine the nominal flexural strength M_n.

$$C = 0.85f'_c b a = 0.85(0.025)(500)a = 10.62a$$

$$T = A_s f_y = 4400(0.400) = 1760 \text{ kN}$$

$$a = \frac{1760}{10.62} = 166 \text{ mm}$$

$$M_n = T(d - 0.5a) = 1760[673 - 0.5(166)]\tfrac{1}{1000} = 1038 \text{ kN·m} \quad \blacksquare$$

3.10 INVESTIGATION OF RECTANGULAR SECTIONS IN BENDING WITH BOTH TENSION AND COMPRESSION REINFORCEMENT

Rectangular sections with both tension and compression reinforcement are also called "doubly reinforced" sections. Because the compressive strength of concrete is high, the need for compression reinforcement *to obtain adequate strength* is not great. In beams where compression reinforcement might be used in order to reduce the size of the cross section, deflection may be excessive, and there may be difficulty in placing all of the tension reinforcement within the width of the beam, even if two or more layers of bars are used. In addition, shear stress will become high so that a large amount of shear reinforcement might be required.

The use of compression reinforcement for deflection control (to reduce the creep and shrinkage deflection) is the usual reason for its use when the strength method is used.

The investigation of a doubly reinforced section such as that shown in Fig. 3.10.1 involves the determination of the nominal flexural strength M_n with b, d, d', A_s, A_s', f_c', and f_y as the given data. The investigation is similar to that for the singly reinforced beam except the compressive force C is composed of two parts, one in the concrete and the other in the steel. The actual stress in the compression steel when the nominal strength M_n is reached may be the yield stress, or it may be something less, depending on the position of the neutral axis. The stress used in the compression steel must be compatible with the strain in the compression steel when nominal strength is reached.

In the case of the beam with compression steel, the ACI Code states that ρ shall not exceed $0.75\rho_b$ (just as for the singly reinforced beam), but in addition states (ACI-10.3.3) that "the portion of ρ_b equalized by compression reinforcement need not be reduced by the 0.75 factor." The use of the above quoted statement was intended to make computation easier.

Philosophically, the ductility requirement, in terms of how large the strain in the tension steel should be when the extreme concrete compression fiber reaches a strain of 0.003, should be the same for a beam having compression steel as for one without it. Controlling the strain diagram such that the neutral axis distance x cannot exceed $0.75x_b$ is the easiest to understand and the simplest in computation as the way to determine the maximum reinforcement ratio ρ permitted in a beam having compression steel.

The literal application of ACI-10.3.3 containing the above-quoted phrase gives the same result (for rectangular sections) as limiting the actual x to a maximum of $0.75x_b$ *when the compression steel yields*; but raises the maximum value of ρ slightly *when the compression steel does not yield*. Since rarely, if ever, does a doubly reinforced beam have its maximum limit of tension reinforcement, and since the compression steel most often yields, the subtle distinction between the results of limiting the actual

Figure 3.10.1 Doubly reinforced beam cross-section.

x to a maximum of $0.75x_b$ and a literal use of the quoted clause in ACI-10.3.3 is of little consequence.

■ **EXAMPLE 3.10.1**   Determine the nominal moment strength M_n of the rectangular section shown in Fig. 3.10.2(a), given $f_c' = 5000$ psi, $f_y = 60,000$ psi, $b = 14$ in., $d = 26$ in., $d' = 3$ in., $A_s' = 2$-#8 bars and $A_s = 8$-#10 bars.

Solution: **(a)** Determine A_{sy} in the compression steel yield condition. Refer to Fig. 3.10.2(c) where the strain diagram is defined by points A and B. Point B is obtained by letting the strain in the compression steel exactly equal ϵ_y.

$$x_y = \frac{0.003(3)}{0.003 - 0.00207} = 9.67 \text{ in.}$$

$$a_y = \beta_1 x_y = 0.80(9.67) = 7.74 \text{ in.}$$

$$C_{cy} = 0.85f_c'ba_y = 4.25(14)(7.74) = 460 \text{ kips}$$

$$C_{sy} = (f_y - 0.85f_c')A_s' = (60 - 4.25)(1.58) = 88 \text{ kips*}$$

$$T_y = C_{cy} + C_{sy} = 460 + 88 = 548 \text{ kips}$$

$$A_{sy} = \frac{548}{60} = 9.13 \text{ sq in.}$$

Since the actual A_s is greater than A_{sy}, actual x is larger than x_y, $\epsilon_s' > \epsilon_y$, and compression steel yields. An alternative would be to assume compression steel yields and check it afterwards, as shown in part **(c)**.

(b) Determine the maximum permissible A_s. From Fig. 3.10.2(b), using the balanced strain condition defined by points A and C on the strain diagram,

$$x_b = \frac{0.003(26)}{0.003 + 0.00207} = 15.39 \text{ in.}$$

$$\text{max } x = 0.75x_b = 0.75(15.39) = 11.54 \text{ in.}$$

which defines the strain line through points A and D in Fig. 3.10.2(b).

$$\text{max } a = \beta_1 x = 0.80(11.54) = 9.23 \text{ in.}$$

$$\text{max } C_c = 0.85f_c'ba = 4.25(14)(9.23) = 549 \text{ kips}$$

$$\epsilon_s' = \frac{11.54 - 3.00}{11.54}(0.003)$$

$$= 0.00222 > \epsilon_y \quad \text{(compression steel yields)}$$

— — — —

* It is noted that the components of the compressive force nominally are the portions carried by the steel and by the concrete, respectively. For the convenience of knowing the point of action, however, the compressive force in the concrete is taken larger than the real amount by stressing $0.85f_c'$ over the steel area, whereas the compressive force in the steel is taken less by the same amount. The total compressive force is correct. The components C_c and C_s will hereinafter be computed as shown in this example.

(a) Actual condition

(b) max A_s condition

(c) Compression steel yield condition

Figure 3.10.2 Section for Example 3.10.1.

$$C_s = (f_y - 0.85f'_c)A'_s = (60 - 4.25)1.58 = 88 \text{ kips}$$

$$\max T = 549 + 88 = 637 \text{ kips}$$

One may note that in determining the maximum T the portion 88 kips "equalized by compression reinforcement" (ACI-10.3.3) is *not* reduced by the 0.75 factor. It is believed that the thought process illustrated here provides better understanding of the concept than does the wording in ACI-10.3.3.

$$\max \text{ permissible } A_s = \frac{637}{60} = 10.62 \text{ sq in.} > (\text{actual } A_s = 10.16 \text{ sq in.}) \qquad \textbf{OK}$$

One may note that the maximum permissible amount of tension steel in this case is 78% of the full balanced amount (i.e., if A_{sb} were computed using $x = x_b$), greater than the 75% acceptable for a beam without compression steel. This is reasonable since compression steel provides ductility; so that the greater the proportion of the compression force is carried by compression steel, the closer the maximum tension steel can approach the balanced amount.

(c) Determine M_n in the actual condition. From Fig. 3.10.2(a), using a strain condition defined by point A and the requirement of equilibrium ($C = T$),

$$T = 60(10.16) = 610 \text{ kips}$$

$$C_s = (60 - 4.25)(1.58) = 88 \text{ kips}$$

$$C_c = 0.85f'_c ba = 4.25(14)a$$

$$a = \frac{C_c}{4.25(14)} = \frac{610 - 88}{4.25(14)} = 8.77 \text{ in.}$$

$$x = \frac{8.77}{\beta_1} = \frac{8.77}{0.80} = 10.96 \text{ in.}$$

Instead of doing the work in part **(b)**, this value of $x = 10.96$ in. may be checked against $0.75x_b = 11.54$ in.

$$\epsilon'_s = \frac{10.96 - 3.00}{10.96}(0.003) = 0.00218 > \epsilon_y \qquad \textbf{OK}$$

The check on ϵ'_s verifies the conclusion in part **(a)**.

$$M_n = C_c\left(d - \frac{a}{2}\right) + C_s(d - d')$$

$$= 522[26 - 0.5(8.77)]\tfrac{1}{12} + 88(26 - 3)\tfrac{1}{12} = 941 + 169$$

$$= 1110 \text{ ft-kips}$$

Note that the work shown in parts **(a)** and **(b)** is to explain why the checks on (1) if actual $x < 0.75x_b$ and (2) if actual $\epsilon'_s < \epsilon_y$ are needed in part **(c)**. In a usual problem, the work shown in part **(c)** is all that is necessary (if both checks can be verified).

■ **EXAMPLE 3.10.2** Repeat the solution for Example 3.10.1, except that A_s is 4-#11 bars instead of 8-#10 bars. Also determine the percentage increase in the strength of the beam over the same beam without compression steel.

Solution: **(a)** Determine A_{sy} in the compression steel yield condition. From Example 3.10.1, $A_{sy} = 9.13$ sq in. Since actual A_s (6.24 sq in.) is less than A_{sy}, the compression steel does not yield.

(b) Determine M_n in the actual condition. Let the location of the neutral axis in Fig. 3.10.3 be the unknown. Equating $(C_c + C_s)$ to T,

$$4.25(14)(0.80x) + \left[\frac{x-3}{x}(0.003)(29,000) - 4.25\right](1.58) = 60(6.24)$$

Solving the quadratic equation for x,

$$x = 6.46 \text{ in.}$$

$$a = \beta_1 x = 0.80(6.46) = 5.17 \text{ in.}$$

Then

$$C_c = 4.25(14)(5.17) = 307.6 \text{ kips}$$

$$\epsilon'_s = \left(\frac{6.46 - 3.00}{6.46}\right)0.003 = 0.00161 < \epsilon_y \quad \text{(as assumed)}$$

$$C_s = [29(1.61) - 4.25](1.58) = 66.9 \text{ kips}$$

$$C_c + C_s = 307.6 + 66.9 = 374.5 \text{ kips}$$

$$T = A_s f_y = 6.24(60) = 374.4 \text{ kips} \quad \text{(Check)}$$

$$M_n = 307.6[26 - 0.5(5.17)]\tfrac{1}{12} + 66.9(26 - 3)\tfrac{1}{12} = 600 + 128$$

$$= 728 \text{ ft-kips}$$

Figure 3.10.3 Section for Example 3.10.2.

(c) Determine M_n if the compression steel had not existed. From Fig. 3.10.2(b) in Example 3.10.1(b), check whether or not $A_s = 6.24$ sq in. exceeds $0.75A_{sb}$ without compression steel. Using maximum C_c,

$$\text{max } A_s = \frac{549}{60} = 9.15 \text{ sq in.} > (\text{actual } A_s = 6.24 \text{ sq in.}) \qquad \textbf{OK}$$

$$C = 0.85f'_cba = 4.25(14)a$$

$$T = 6.24(60) = 374.4 \text{ kips}$$

$$a = \frac{374.4}{4.25(14)} = 6.29 \text{ in.}$$

$$M_n = 374.4[26 - 0.5(6.29)]\tfrac{1}{12} = 713 \text{ ft-kips}$$

Thus, even though in part **(b)** the 128 ft-kips representing the contribution of compression steel was 17.5% of the total, the addition of the compression steel to the singly reinforced beam actually added only 2.1% $[(728 - 713)/713]$ to the original capacity. This is a common situation; the compression steel was not used because the compression zone without it was inadequate for strength, but instead it was used for deflection control. ■

3.11 CRITERION FOR THE COMPRESSION STEEL YIELD CONDITION

It may be shown that the criterion to ensure that the compression steel in a doubly reinforced section yields when the nominal strength is reached is

$$\rho - \rho'\left(1 - \frac{0.85f'_c}{f_y}\right) \geq 0.85\beta_1\left(\frac{f'_cd'}{f_yd}\right)\left(\frac{87,000}{87,000 - f_y}\right) \qquad \textbf{(3.11.1)}$$

Referring to Fig. 3.11.1, the criterion for yielding of the compression steel is

$$\epsilon'_s \geq \epsilon_y \qquad \textbf{(3.11.2)}$$

Figure 3.11.1 Compression steel yield condition.

The internal forces in Fig. 3.11.1 are

$$T = \rho b d f_y$$

$$C_c = 0.85 f'_c \beta_1 x b$$

$$C_s = (f_y - 0.85 f'_c) \rho' b d$$

Equating T to $C_c + C_s$ and solving for x,

$$x = \frac{f_y d}{0.85 \beta_1 f'_c} \left[\rho - \rho' \left(1 - \frac{0.85 f'_c}{f_y} \right) \right] \qquad (3.11.3)$$

From Fig. 3.11.1(b)

$$\epsilon'_s = \frac{\epsilon_{cu}}{x} (x - d') \qquad (3.11.4)$$

Substituting Eq. (3.11.4) in Eq. (3.11.2) gives

$$\frac{\epsilon_{cu}}{x} (x - d') \geq \epsilon_y \quad \text{or} \quad \frac{87,000}{x} (x - d') \geq f_y$$

or

$$x \geq \frac{87,000 d'}{87,000 - f_y} \qquad (3.11.5)$$

Equation (3.11.1) may be obtained by substituting Eq. (3.11.3) in Eq. (3.11.5).

Use of Eq. (3.11.1) is the alternative to use of basic statics as done in Example 3.10.1, part (a), to determine if compression steel does yield when nominal strength M_n is reached. Although solution by basic statics is preferable in longhand computations, a formula like Eq. (3.11.1) may be useful in computer programming.

3.12 DESIGN OF RECTANGULAR SECTIONS IN BENDING WITH BOTH TENSION AND COMPRESSION REINFORCEMENT

When a rectangular section of given dimensions is required to have a strength greater than it may have when it is reinforced with the maximum permissible amount of tension reinforcement, compression reinforcement becomes necessary. However, such necessary use of compression steel for strength is rare. The principal reason for using compression reinforcement is to reduce long-time deflection due to creep and shrinkage.

The logical procedure for designing a doubly reinforced section is to determine first whether compression steel is needed for strength. This may be done by comparing the required moment strength with the moment strength of a singly reinforced section in which the maximum permissible amount of tension steel has been used.

Having decided that compression steel is to be used, be it required for strength or desirable for deflection control, the designer now needs to select the appropriate tension steel A_s and compression steel A'_s. For this purpose, the two equilibrium equations may be utilized; namely,

$$C_c + C_s = T \tag{3.12.1}$$

and

$$M_n = C_c\left(d - \frac{\beta_1 x}{2}\right) + C_s(d - d') \tag{3.12.2}$$

wherein the symbols are those defined in Fig. 3.10.1. Further, verifications are necessary to see that the tension steel as finally used does not make the neutral axis distance x exceed 75% of the value in the balanced strain condition and that the compression steel does in fact yield [Eq. (3.11.1)] when it has been so assumed in the equilibrium equations [Eqs. (3.12.1) and (3.12.2)]. If the compression steel does not yield, the equilibrium conditions must be reformulated using a stress f'_s in the compression steel proportional to the strain in that steel.

A unique solution is obtainable once the actual x is preset. This actual x has a maximum limit of $0.75x_b$. If deflection control is an important consideration, the actual x may be preset at a much smaller value than $0.75x_b$, perhaps $0.375x_b$ or smaller.

■ **EXAMPLE 3.12.1** Determine the A_s and A'_s required to carry a service live load moment of 390 ft-kips and a service dead load moment of 200 ft-kips, using $b = 14$ in., $d = 26$ in., $d' = 3$ in., $f'_c = 5000$ psi, $f_y = 60,000$ psi, and the ACI Code, as shown in Fig. 3.12.1(a).

Solution: (a) Determine the required strength.

$$M_u = 1.4(200) + 1.7(390) = 280 + 663 = 943 \text{ ft-kips}$$

$$\text{required } M_n = \frac{M_u}{\phi} = \frac{943}{0.90} = 1048 \text{ ft-kips}$$

(b) Determine the maximum strength and reinforcement allowed by ACI Code for a singly reinforced section. The location of the neutral axis at the balanced strain condition [Fig. 3.12.1(b)] may be obtained as

$$x_b = \frac{\epsilon_{cu}}{\epsilon_{cu} + \epsilon_y} d = \frac{3}{3 + 60/29} 26 = 15.39 \text{ in.}$$

$$\text{max } x = 0.75(15.39) = 11.54 \text{ in.} \quad [\text{Fig. 3.12.1(c)}]$$

$$\text{max } C = 0.85f'_c b \beta_1 x = 4.25(14)(0.80)(11.54) = 549 \text{ kips}$$

$$\text{max } A_s \text{ in singly reinforced section} = 549/60 = 9.15 \text{ sq in.}$$

$$\text{max } M_n \text{ of singly reinforced section} = (549)(26 - 4.62)\tfrac{1}{12} = 978 \text{ ft-kips}$$

The required M_n exceeds the maximum strength obtainable without compression steel; therefore compression steel is needed for strength.

Figure 3.12.1 Section for Examples 3.12.1 and 3.12.2.

(c) Determine the minimum compression reinforcement required. Maintain max $x = 0.75x_b = 11.54$ in. [Fig. 3.12.1(d)]. Let

$$M_{nc} = 978 \text{ ft-kips} \quad [\text{from part (b)}]$$

$$M_{ns} = M_n - M_{nc} = 1048 - 978 = 70 \text{ ft-kips}$$

$$\text{required } C_s = \frac{70(12)}{26 - 3} = 36.5 \text{ kips}$$

Will compression steel yield when $x = 11.54$ in.?

$$\epsilon'_s = \frac{11.54 - 3.00}{11.54}(0.003) = 0.0022 > \epsilon_y$$

$$C_s = A'_s(f_y - 0.85f'_c) = 36.5 \text{ kips}$$

$$\text{required } A'_s = \frac{36.5}{60 - 4.25} = 0.65 \text{ sq in.}$$

$$T = \max C_c + C_s = 549 + 36.5 = 585.5 \text{ kips}$$

$$\text{required } A_s = \frac{585.5}{60} = 9.76 \text{ sq in.}$$

This amount of tension steel is exactly the maximum permitted by the ACI Code.

In this example the remaining steps of selecting actual bars and making a final check of strength are not shown. ∎

∎ **EXAMPLE 3.12.2** Redesign the section of Example 3.12.1 so that the actual neutral axis location is at 0.375 of that in the balanced strain condition. Assume that the purpose is for deflection control. From Example 3.12.1, $M_u = 943$ ft-kips.

Solution: (a) Determine M_{nc} corresponding to locating the actual neutral axis at 0.375 of the value in the balanced strain condition. Use of this assumption seems a rational choice consistent with the guideline value for deflection control on a singly reinforced beam, as discussed in Section 3.9. Thus, referring to Fig. 3.12.1(e),

$$\text{actual } x = 0.375 x_b = 0.375(15.39) = 5.77 \text{ in.}$$

$$\text{actual } a = 0.80 x = 0.80(5.77) = 4.62 \text{ in.}$$

$$C_c = 0.85 f'_c ba = 0.85(5)(14)(4.62) = 275 \text{ kips}$$

$$T_1 = C_c = 275 \text{ kips}$$

$$M_{nc} = 275[26 - 0.5(4.62)]\tfrac{1}{12} = 543 \text{ ft-kips}$$

Let A_{sc} be the part of tension steel to match the concrete in compression; then

$$A_{sc} = \frac{T_1}{f_y} = \frac{275}{60} = 4.58 \text{ sq in.}$$

(b) Determine steel requirements for both faces of the beam. Using $M_u/\phi = 1048$ ft-kips as required M_n from Example 3.12.1,

$$M_{ns} = 1048 - 543 = 505 \text{ ft-kips}$$

$$C_s = T_2 = \frac{505(12)}{26 - 3} = 264 \text{ kips}$$

$$\epsilon'_s = \frac{5.77 - 3.0}{5.77}(0.003) = 0.00144 < \epsilon_y$$

Compression steel does not yield. If $x = 5.77$ in. is still to be maintained, then

$$f'_s = 0.00144(29,000 \text{ ksi}) = 41.8 \text{ ksi}$$

so that the stress used is consistent with the strain on the compression steel.

$$A'_s = \frac{264}{41.8 - 4.25} = 7.03 \text{ sq in.}$$

$$A_{ss} = \frac{264}{60} = 4.40 \text{ sq in.}$$

$$A_s = A_{sc} + A_{ss} = 4.58 + 4.40 = 8.98 \text{ sq in.}$$

Select 9-#9 in two layers for tension steel, and 4-#11 as compression reinforcement. The 4-#11 ($A'_s = 6.24$ sq in.) is the maximum steel that will fit into one layer.

(c) Check the strength of the section. The above solution of $A_s = 8.98$ and $A'_s = 7.03$ is the correct one for $x = 0.375x_b$. When $A'_s = 6.24$ is used instead, both x and A_s will have to change. A trial-and-error procedure is presented below to determine the stress f'_s acting in the compression steel, an alternative to that used in Example 3.10.2 wherein a quadratic equation was solved for the neutral axis location x.

Since a smaller amount of compression steel is used than that required to locate the neutral axis at $0.375x_b$, the actual neutral axis will be lower if the section is to carry the same bending moment. Also, the stress f'_s must exceed 41.8 ksi for $x = 0.375x_b$. Estimate the compression steel stress to be 45 ksi.

$$C_c = 0.85f'_c ba = 0.85(5)(14)a = 59.5a$$

$$C_s = A'_s(f'_s - 0.85f'_c) = 6.24(45 - 4.25) = 254 \text{ kips}$$

$$T = A_s f_y = 9.00(60) = 540 \text{ kips}$$

$$C_c + C_s = T$$

$$a = \frac{540 - 254}{59.5} = 4.81 \text{ in.}; \qquad x = \frac{4.81}{0.80} = 6.00 \text{ in.}$$

$$\epsilon'_s = \frac{6.00 - 3.00}{6.00}(0.003) = 0.0015$$

$$f'_s = 0.0015(29,000) = 43.5 \text{ ksi}$$

Since this does not agree with the 45 ksi assumed, make a new assumption, say $f'_s = 44$ ksi:

$$\text{revised } C_s = 6.24(44 - 4.25) = 248 \text{ kips}$$

$$a = \frac{540 - 248}{59.5} = 4.91 \text{ in.}; \qquad x = \frac{4.91}{0.80} = 6.14 \text{ in.}$$

$$\epsilon'_s = \frac{6.14 - 3.00}{6.14}(0.003) = 0.00153$$

$$f'_s = 0.00153(29,000) = 44.4 \text{ ksi}$$

Repeated trials may be made until computed f'_s agrees as closely as desired with the assumed value. In this case assume present agreement is close enough.

$$C_c = 59.5a = 59.5(4.91) = 292 \text{ kips}$$

$$C_s = 248 \text{ kips}$$

$$M_n = 292[26 - 0.5(4.91)]\tfrac{1}{12} + 248(26 - 3)\tfrac{1}{12} = 573 + 475 = 1048 \text{ ft-kips}$$

$$\phi M_n = 0.90(1048) = 943 \text{ ft·kips} = [M_u = 943 \text{ ft-kips}] \qquad \textbf{OK}$$

A comparison of three possible designs for this doubly reinforced section as worked out in detail in Examples 3.12.1 and 3.12.2 is tabulated below:

Example 3.12.1, part **(c)**,	$x = 11.54$ in.	$A_s = 9.76$ sq in.	$A_s' = 0.65$ sq in.
Example 3.12.2, part **(b)**,	$x = 5.77$ in.	$A_s = 8.98$ sq in.	$A_s' = 7.03$ sq in.
Example 3.12.2, part **(c)**,	$x = 6.14$ in.	$A_s = 9.00$ sq in.	$A_s' = 6.24$ sq in.

It may be observed that when there is a large amount of compression steel that does not yield due to the linear strain variation across the depth of the section, varying the amount of compression steel has essentially the effect of changing only the proportions in M_{nc} and M_{ns}. It has a negligible effect (say typically 3 to 4%) on total capacity. The choice of 9-#9 (tension steel) and 4-#10 (compression steel) is acceptable. Deflection may still need checking but it will probably be within acceptable limits. ■

3.13 NON-RECTANGULAR BEAMS

When non-rectangular beams are used, the shape of the compression zone dictates whether the formulas developed in this chapter are applicable. The concept of using the Whitney rectangular compressive stress distribution may be used in accordance with ACI-10.2.7 for any shape of the compression zone in the cross section. For ductility, it is appropriate to limit actual x so that it does not exceed $0.75x_b$, particularly for notched or other shaped beams having less compression concrete area at the extreme fiber than at the neutral axis. The maximum amount of tension reinforcement may be obtained by setting $x = 0.75x_b$ and evaluating the internal forces to be compatible with the strain in the maximum x condition. Then equating C to T, the maximum tension steel permitted can be obtained. Alternatively, of course, ACI-10.3.3 may be used by limiting A_s to a maximum of $0.75A_{sb}$. The treatment of compression steel in a non-rectangular compression zone would be identical to that in a rectangular beam, as discussed in Section 3.10.

For sections having a non-rectangular compression zone, the ductility (that is, the amount of tension steel strain achieved at maximum A_s condition) obtained by using the literal wording of ACI-10.3.3 may be either *more* or *less* than obtained by a consistent use of maximum $x = 0.75x_b$. When the section has the greater compression area near the extreme fiber, as for a T-section, the tension steel strain achieved will be greater than for the rectangular beam; that is, maximum x will be less than $0.75x_b$. When the section has the greater compression area near the neutral axis, the tension steel strain achieved will be less than for the rectangular beam; that is, maximum x will be greater than $0.75x_b$. Rational treatment would set the same maximum x limit for all beams. ACI-10.3.3 was not specifically worded with the intention of providing varying amounts of tension steel strain achieved when the nominal strength is reached; rather, it was intended to make computation easier.

Though no examples of non-rectangular sections are provided, there are some problems provided at the end of the chapter for practice.

SELECTED REFERENCES

3.1.   Jack R. Janney, Eivind Hognestad, and Douglas McHenry. "Ultimate Flexural Strength of Prestressed and Conventionally Reinforced Concrete Beams," *ACI Journal*, *Proceedings*, **52**, February 1956, 601–620.

3.2.   Eivind Hognestad, N. W. Hanson, and Douglas McHenry. "Concrete Stress Distribution in Ultimate Strength Design," *ACI Journal*, *Proceedings*, **52**, December 1955, 455–479.

3.3.   Hjalmar Granholm. *A General Flexural Theory of Reinforced Concrete*. New York: Wiley, 1965.

3.4.   Eivind Hognestad. "Confirmation of Inelastic Stress Distribution in Concrete," *Journal of the Structural Division*, ASCE, **83**, Paper No. 1189, ST2, March 1957.

3.5.   Charles. S. Whitney. "Plastic Theory of Reinforced Concrete Design," *Transactions ASCE*, **107**, 1942, 251–326.

3.6.   C. S. Whitney and Edward Cohen. "Guide for Ultimate Strength Design of Reinforced Concrete," *ACI Journal*, *Proceedings*, **53**, November 1956, 455–475.

3.7.   ACI-ASCE Joint Committee: "Report on ASCE-ACI Joint Committee on Ultimate-Strength Design," *ASCE, Proceedings*, **81**, Paper No. 809, October 1955. See also *ACI Journal*, *Proceedings*, **52**, January 1956, 505–524.

3.8.   Robert E. Abendroth and Charles G. Salmon. "Sensitivity Study of Reinforced Concrete Simple Beams," *ACI Journal*, *Proceedings*, **83**, September–October 1986, 764–771.

PROBLEMS

All problems* are to be done in accordance with the strength design method of the ACI Code (except Problems 3.1 and 3.2), and all loads given are *service* loads, unless otherwise indicated. Wherever possible basic principles are to be used for solutions, avoiding the direct use of formulas. All design problems require a clear statement of the final choice at the end of the calculations, along with a design sketch drawn to scale.

— — — —

* Most problems may be solved as problems stated in Inch-Pound units, or as problems in SI units using quantities in parenthesis at the end of the statement. The metric·conversions are approximate to avoid implying higher precision for the given information in metric units than that for the Inch-Pound units.

Review Problems in Mechanics of Materials

3.1   An elastic homogeneous beam of material capable of carrying both tension and compression is simply supported over a span of 20 ft (6 m) center to center of supports. It carries a uniformly distributed load of 2 klf (kips per linear ft) (30 kN/m) in addition to a concentrated load of 10 kips (44 kN) located at 5 ft (1.5 m) from the right end of the span.

(a) Compute *independently* the maximum flexural tensile and compressive stresses on the section shown in the accompanying figure (use the Case assigned by the instructor), using basic statics involving internal forces and internal couple, that is, using $C = T$ and $M = (C$ or $T)$ times (moment arm between points of action of C and T).

(b) Check the correctness of part **(a)** by computing both extreme fiber stresses using the flexure formula, $f = Mc/I$.

Problem 3.1a

CASE	h	b_1	t_1	t_w	b_2	t_2
			(DIMENSIONS IN INCHES)			
1	30	12	2	1	12	2
2	20	40	3	4	40	3
3	20	40	4	4	20	4
4	20	60	4	12	12	0
5	24	50	4	14	14	0
6	28	40	6	14	14	0
7	30	30	4	12	12	0
8	30	10	3	2	10	3
9	20	10	4	20	20	0
			(DIMENSIONS IN MILLIMETERS)			
10	750	300	50	25	300	50
11	500	1500	100	300	300	0

3.2   The cross section of Problem 3.1 (use the Case assigned by instructor) is used for an elastic homogeneous beam on a simply supported span of 25 ft (7.5 m), and carries a uniformly distributed load of 1 kip/ft (15 kN/m) plus a concentrated load of 9 kips (40 kN) at 8 ft (2.4 m) from the left end of the span. Compute the stresses as required by **(a)** and **(b)** of Problem 3.1.

Reinforced Concrete Problems

3.3   For the Case (or Cases) assigned, compute the nominal flexural strength M_n using basic statics (i.e., no formula) with the Whitney rectangular stress block and the internal couple. Also, compute the service moment capacity $(M_D + M_L)$, using the gravity dead load plus live load overload provision of

ACI-9.2.1, and assuming the total service load is 60% live load. Assume the bars given are in the tension face of the beam in one layer (unless two layers are indicated), and the amount of reinforcement is acceptable according to the ACI Code.

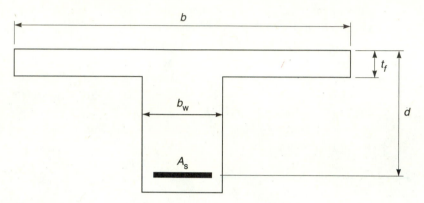

Problem 3.3

CASE	d	b	b_w	t_f	BARS	f'_c	f_y	SHAPE
	(DIMENSIONS IN INCHES)					(STRESSES, psi)		
1	19.5	12	12	0	3-#7	3500	60,000	Rectangular
2	19.5	12	12	0	3-#10	3500	60,000	Rectangular
3	19.5	10	10	0	6-#7	3500	40,000	Rectangular
					(Two layers of three)			
4	14.7	10	10	0	5-#9	3500	40,000	Rectangular
					(3-#9 Outside layer & 2-#9 inside layer)			
5	36.25	18	18	0	8-#11	4000	60,000	Rectangular
					(Two layers of four)			
6	15.75	36	6	4	2-#6	3000	60,000	T-section
7	24.6	50	13	2	8-#8	3000	50,000	T-section
					(Two layers of four)			
8	36	30	14	7	8-#11	3000	50,000	T-section
					(Two layers of four)			
	(DIMENSIONS IN MILLIMETERS)					(STRESSES, MPa)		
9	495	300	300	0	3-#20M	25	400	
10	495	300	300	0	3-#30M	25	4000	

3.4   Using basic principles, prove *by calculating strains in the tension steel* whether or not there is yield of the tension steel when the nominal strength M_n is reached (that is, when the extreme fiber in compression reaches a strain of 0.003). Basic principles mean the use of statics with the Whitney rectangular stress block and the internal couple (i.e., no formulas). Does the beam violate the ACI Code? If so, in what respect? Use $f'_c = 4000$ psi and $f_y = 58,000$ psi. ($f'_c = 30$ MPa and $f_y = 400$ MPa.)

(4 – # 30M)
4 – # 10

14"
(350 mm)

14"
(350 mm)

Problem 3.4

3.5   Determine the required effective size (width b rounded to whole inches and effective depth d) rectangular section and theoretical steel area A_s to carry uniformly distributed dead and live loads on a simply supported span as given. Do the Case assigned by the instructor. *Do not make the practical decisions of choosing actual overall beam dimensions and selecting bars.* Assume the dead load includes the beam weight. Use basic principles (i.e., no formulas) to obtain reinforcement ratio limits and verify by comparing with Table 3.6.1. Derive the formula for R_n in terms of ρ and then use it to get the limiting values.

(a) Obtain the *largest* effective size permitted by the ACI Code.

(b) Obtain the *smallest* effective size permitted by the ACI Code.

CASE	f'_c (psi)	f_y (psi)	w_D (kips/ft)	w_L (kips/ft)	SPAN (ft)	APPROX d/b
1	3000	60,000	1.3	1.0	24	1.75
2	4000	60,000	1.3	1.0	24	2.0
3	4000	60,000	1.3	1.0	24	1.25
4	3500	60,000	1.0	1.6	24	1.5
5	3500	60,000	0.8	1.2	18	1.5
6	3000	40,000	0.8	1.4	18	2.0

3.6   For the Case assigned in Problem 3.5, design the *largest* practical size beam. This time, assume the given dead load does *not* include the beam weight, and make the practical design decisions, including the choice of overall dimensions, selecting the bars, and checking the strength. The beam is to be designed using tension steel only (that is, no compression steel).

3.7   For the Case assigned in Problem 3.5, design the *smallest* practical size beam. All other requirements are as in Problem 3.6.

3.8   Design a rectangular cross-section having tension steel only *of such size that excessive deflection would not be expected under normal circumstances.* The beam is simply supported and is to carry given uniformly distributed dead and live loads, in addition to the weight of the beam. Make all the practical decisions for size of cross-section, select the reinforcement, and check the strength using statics and basic principles. Do the Case assigned by the instructor. (*Hint:* Use ρ = approximately one-half the maximum permitted by the ACI Code.)

CASE	f'_c (psi)	f_y (psi)	w_D (kips/ft)	w_L (kips/ft)	SPAN (ft)	APPROX d/b
1	3500	40,000	1.0	2.0	30	1.5–2.0
2	3000	60,000	0.8	1.8	30	1.5–2.0
3	3500	60,000	1.3	2.5	30	1.5–2.0
4	4000	60,000	1.2	2.5	30	1.5–2.0
5	4000	60,000	0.8	1.5	30	1.5–2.0
6	4000	60,000	1.2	2.2	32	1.5–2.0
7	4000	60,000	1.6	2.5	28	1.5–2.0
8	4000	60,000	1.8	2.7	26	1.5–2.0
9	3000	60,000	0.8	2.8	24	1.5–2.0

3.9   Design the *smallest practical size* rectangular cross-section having tension steel only, for the Case in Problem 3.8 assigned by the instructor. The beam is simply supported and is to carry given uniformly distributed dead and live loads, in addition to the weight of the beam. Make all the practical decisions for size of cross-section, select the reinforcement, and check the strength using statics and basic principles.

3.10 Design a rectangular beam to carry a live load of 1.5 kips/ft and a dead load of 1.5 kips/ft, in addition to the beam weight, for a simple span of 32 ft. Use an approximate steel ratio ρ of about 0.375 of the balanced amount (one-half of the maximum permitted), and also satisfy ACI-Table 9.5a. Use f'_c = 4000 psi and f_y = 60,000 psi. (Live load = 22 kN/m; dead load = 22 kN/m; span = 9.8 m; f'_c = 30 MPa; f_y = 400 MPa.)

3.11 Select the economical reinforcement for a beam 12 in. wide by 22 in. deep overall to carry dead load and live load moments of 15 ft-kips and 60 ft-kips, respectively. Use f'_c = 3000 psi, f_y = 40,000 psi. (Width = 300 mm; depth = 560 mm; dead load moment = 20 kN·m; live load moment = 80 kN·m; f'_c = 20 MPa; f_y = 300 MPa.)

3.12 Select economical reinforcement for a beam 20 in. wide by 40 in. overall depth to carry a live load moment of 500 ft-kips and a dead load moment (including beam weight) of 300 ft-kips. Use reinforcement in only one face. Though a check of deflection cannot be made with the given information, would you expect such a check to show that deflection is excessive? Explain your answer. Use $f'_c = 4000$ psi and $f_y = 60,000$ psi. (Beam size = 500 mm × 1000 mm; live load moment = 700 kN·m; dead load moment = 400 kN·m; $f'_c = 30$ MPa; $f_y = 400$ MPa.)

3.13   Assuming no deflection limitation, and without using compression steel, select the smallest practical size rectangular section permitted by the ACI Code, using the same size for the entire length (main span plus cantilever) of beam, as shown in the accompanying figure. The loads given are in addition to the beam weight. Select reinforcement for both the positive and negative moment zones. Use the Case assigned by the instructor. (*Note:* Live load is always to be applied in the locations and distributions to cause the most severe effects; spans may be fully loaded, partially loaded, or unloaded, as necessary to obtain maximum effects.)

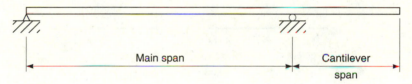

Main span   Cantilever span

Problem 3.13

CASE	f'_c (psi)	f_y (psi)	w_D (kips/ft)	w_L (kips/ft)	MAIN SPAN (ft)	CANTILEVER SPAN (ft)
1	3500	40,000	1.0	1.5	20	8
2	3500	40,000	1.25	2.0	24	8
3	3500	60,000	1.0	1.5	16	10
4	4000	60,000	1.0	2.0	18	8
5	4000	60,000	0.8	1.5	20	8
6	4000	60,000	1.2	2.2	20	8
7	4000	60,000	0.8	1.5	22	10
8	4000	60,000	1.2	2.2	22	10
9	4000	60,000	1.2	2.5	24	10
	(MPa)	(MPa)	(kN/m)	(kN/m)	(m)	(m)
10	25	300	15	20	6	2.5
11	25	300	17	30	7.3	2.4

3.14   For the double overhanging cantilever beam shown, design the *smallest* practical rectangular cross-section (without compression steel) to be used for the entire length (main span plus cantilevers) of beam. The given dead load does *not* include beam weight. Select reinforcement for both positive moment and negative moment regions. Do the Case assigned by the instructor. Assume there is no deflection limit, and refer to the note at the end of Problem 3.13.

Problem 3.14

CASE	f'_c (psi)	f_y (psi)	w_D (kips/ft)	w_L (kips/ft)	MAIN SPAN (ft)	CANTILEVER SPANS (ft)
1	3500	60,000	1.0	1.5	30	10
2	3500	60,000	1.0	1.8	28	10
3	3500	60,000	1.3	2.5	28	9
4	4000	60,000	1.2	2.5	26	9
5	4000	60,000	0.8	1.5	32	8
6	4000	60,000	1.2	2.2	32	12
7	4000	60,000	1.6	2.5	24	8
8	4000	60,000	1.8	2.7	24	6
9	3000	60,000	0.8	2.8	24	8
10	4000	60,000	1.2	1.9	27	10
	(STRESSES, MPa)		(LOADS, kN/m)		(SPANS, m)	
11	25	400	14	22	9.2	3
12	25	400	14	26	8.5	3

3.15 For the loading and beam span conditions of Problem 3.14, design the beam assuming it is part of a floor system supporting nonstructural elements likely to be damaged by large deflections. (*Hint:* Refer to ACI–Table 9.5a and Problem 3.8.) Do the Case assigned by the instructor.

3.16 Compute the nominal strength M_n for the assigned Case of beam having compression steel, using principles of statics with internal couple. As a normal part of the procedure, verify by using basic principles whether or not compression steel has yielded when nominal strength is reached; if it does not yield, use a compression steel stress proportional to the strain in the compression steel. Verify that tension steel does not exceed the maximum permitted by the ACI Code.

Compression face

b

d'

A'_s

d

A_s

Tension face

Problem 3.16

CASE	f'_c (psi)	f_y (psi)	b (in.)	d (in.)	d' (in.)	A_s BARS	A'_s BARS
1	4000	40,000	18	36.1	2.5	10-#11	6-#7
2	3000	60,000	16	27.3	3.0	4-#11	3-#11
3	4000	60,000	16	27.3	3.0	4-#11	3-#11
4	5000	60,000	16	27.3	3.0	4-#11	3-#11
5	3500	60,000	12	21.5	2.5	3-#10	2-#10
6	3500	40,000	12	20.5	2.44	3-#10, 3-#9	3-#9
7	3500	60,000	12	17.6	2.38	3-#9	3-#8
8	3500	60,000	14	19.5	2.5	2-#11, 2-#9	2-#10
9	4000	60,000	20	28.4	2.38	6-#10, 2-#9	2-#8
	(MPa)	(MPa)	(mm)	(mm)	(mm)	BARS	BARS
10	30	300	460	917	64	10–35M	5–25M
11	30	400	400	693	76	4–35M	3–35M

3.17 A rectangular section with $b = 14$ in. and effective depth $d = 21.5$ in. has 4-#10 as tension reinforcement and 2-#10 as compression reinforcement centered 2.5 in. from the face of the beam. Determine the strength M_n for this section. How much can the strength be increased by adding tension steel only? At the point where compression steel is needed to further increase capacity, what ratio of A'_s to A_s would be required to be added? Use $f'_c = 5000$ psi and $f_y = 60,000$ psi. ($b = 350$ mm; $d = 546$ mm; tension steel, 4-#30M; compression steel, 2-#30M centered 63.5 mm from face; $f'_c = 35$ MPa; $f_y = 400$ MPa.)

3.18   For the cross-section obtained for the assigned data Case of Problem 3.13, redesign as a beam with compression steel, making the net reinforcement ratio $(\rho - \rho')$ about one-half the maximum permitted for a singly reinforced beam. If Problem 3.13 has not been solved previously, use the smallest rectangular cross-section permitted by the ACI Code for a beam having tension reinforcement only.

3.19   Repeat the requirements of Problem 3.18 except provide for M_{nc} to be carried by the singly reinforced beam using $\rho = 0.375\rho_b$, instead of making $(\rho - \rho')$ one-half the maximum permitted for a singly reinforced beam.

3.20   For the conditions of Problem 3.14, Case 1, use a rectangular cross-section 14 by 22 in. overall, and design compression steel such that A'_s is about $0.5A_s$ in order to control creep and shrinkage deflection (section 350 by 560 mm).

3.21   For the beam of the accompanying figure, it is desired to utilize 3-#7 bars as compression reinforcement. The factored moment M_u to be carried is 210 ft-kips; $f'_c = 3000$ psi and $f_y = 60,000$ psi. Determine the tension steel required, select the bars, and check the section.

Compression face

3 – #7

#3

22"

$1\frac{1}{2}''$ clear cover all sides

$A_s = ?$

10"

Problem 3.21

Nonrectangular Section Problems

3.22 to 3.26   For the beam cross-section of the accompanying figure, assuming $f'_c = 3000$ psi, $f_y = 60,000$ psi, and the clear cover from the bottom (tension) face to the bars = 2 in., compute the following:

(a) Compare the given tension reinforcement with the maximum (max $A_s = 0.75A_{sb}$) permitted by the ACI Code, and with the maximum using a max $x = 0.75x_b$ requirement. Use basic principles starting with the balanced strain condition.

(b) Using basic principles with the Whitney rectangular stress distribution, compute the nominal strength M_n for the cross-section.

Problem 3.22

Problem 3.23

Problem 3.24

Problem 3.25

Problem 3.26

(c) Neglecting the reinforcement and assuming that the concrete is a homogeneous elastic material, compute the cracking moment M_{cr} when the extreme fiber in tension reaches the modulus of rupture value given by ACI (i.e., $7.5\sqrt{f'_c}$ for normal weight concrete).

(d) Using the result in part **(c)**, determine the reinforcement ratio $\rho = A_s/b_w d$ that makes ϕM_n for a reinforced concrete beam (ignore the given bars for this part) equal to M_{cr}. Give answer in terms of min $\rho = \text{coefficient}/f_y$ and compare with ACI-10.5.1. Use for b_w the narrowest width on the tension side of the neutral axis.

(e) Assuming that compression steel of an amount equal to the given tension steel were to be used and located with 2-in. clear cover to the top (compression) face of the beam, determine the maximum amount of tension steel the ACI Code would permit. Do the computation using both the literal wording of ACI-10.3.3 and the maximum $x = 0.75x_b$ approach.

CHAPTER 4

Rectangular Sections in Bending Under Service Load Conditions

4.1 GENERAL INTRODUCTION

As discussed in Chapter 2, there have been two generally accepted philosophies of design and investigation—working stress and strength design. In Chapter 3, the strength design method was treated using overload factors, strength reduction factors, and the flexural strength of sections when "failure" is imminent and stress is no longer proportional to strain. This chapter considers the working stress method in which service load stresses, allowable working stresses, and the linear relationship between stress and strain are used.

The working stress method is referred to by the present ACI Code as the *alternate design method*. With increasing emphasis on a clear distinction between the design criteria of *strength* and *serviceability*, the term "strength" is associated with behavior at imminent failure at a multiple of service load (say, 1.4 to 1.7 times service load), whereas "serviceability" means satisfactory performance under service load (i.e., working stress) conditions. Satisfactory performance may be defined in terms of (1) deflection within acceptable limits so that supported nonstructural elements such as walls, partitions, and ceilings are not damaged; and (2) cracking controlled to prevent large crack widths that are either unsightly or may permit water to enter, causing corrosion of steel and perhaps deterioration of concrete. Other serviceability requirements, such as vibration and noise control, may also be important but lie outside the scope of this book. Though its use is not encouraged, the ACI "alternate design method" does provide another way of design which gives about the same results as long as the tension steel used is less than about one-half of the maximum permitted in the strength design method.

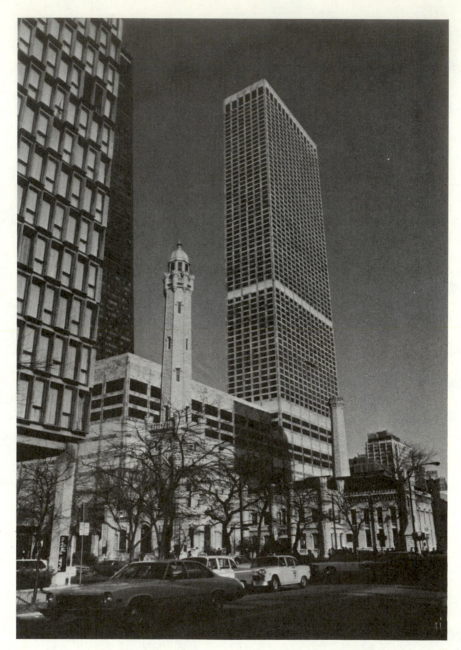

Water Tower Place, 74 stories, 859 ft high, hotel–condominium–shopping complex, Chicago, Ill, completed 1976. (Courtesy of Portland Cement Association.)

4.2 FUNDAMENTAL ASSUMPTIONS

Four basic assumptions are used in the working stress method for flexural members:

1. Plane sections remain plane after bending; that is, the variation in strain is linear across the depth of the member.
2. Stress is proportional to strain.
3. Concrete does not take tension (concrete cracks under tension).
4. No slip occurs between the steels bars and the surrounding concrete during the development of the tensile forces in the bars.

Of the four assumptions, only the second assumption, that stress is proportional to strain, is unique to the working stress method, because it is reasonably accurate for stresses below about one-half f_c', the 28-day compressive strength. Regarding the third assumption, it is true that more of the concrete close to the neutral axis on the tension side will not be cracked under service load than under factored load; however, the magnitude of tensile stress transmitted is generally so small that, for reason of simplicity, the steel is assumed to take all of the tensile stress. The subject of development of the forces in the bars by means of the longitudinal interaction between reinforcing bars and the surrounding concrete is treated in Chapter 6.

4.3 MODULUS OF ELASTICITY RATIO *n*

As discussed in Chapter 1, the stress–strain curve for reinforcement steel is linear below the yield stress, but that of concrete is only approximately linear, even at or below the service load stress. The modulus of elasticity of steel varies little with its strength, whereas that of concrete varies with its density and strength. In the ACI Code, E_s is taken to be 29,000,000 psi (200,000 MPa), and E_c is expressed as $w_c^{1.5}\,33\sqrt{f_c'}$, in which w_c is the density of concrete. Table 1.9.1 gives values for the modulus of elasticity of concrete, E_c.

In the working stress procedure , it will be shown that the ratio *n* of the modulus of elasticity of steel to that of concrete is needed for computing working (i.e., service load) stresses rather than the actual values of E_c or E_s.

Table 4.3.1 Practical Values for Modular Ratio *n*

INCH-POUND		SI	
f_c' (psi)	*n*	f_c' (MPa)[a]	*n*
3000	9	20	9
3500	8.5	25	8
4000	8	30	7.5
4500	7.5	35	7
5000	7	40	6.5
6000	6.5		

[a] For practical use, multiply MPa by 10 to obtain value in kgf/cm².

The modular ratio $n = E_s/E_c$ may be taken (ACI-Appendix A.5.4) as the nearest whole number, but not less than 6. Except in deflection calculations, the value of n for lightweight concrete shall be assumed to be the same as for normal weight concrete of the same strength. It is suggested that the values in Table 4.3.1 be used for normal weight concrete.

4.4 EQUILIBRIUM CONDITIONS

Two equilibrium conditions apply to a section subjected to bending only: (1) the resultant internal compressive force must be equal to the resultant internal tensile force; and (2) the moment of the internal couple, composed of the resultant compressive and tensile forces, must be equal to the applied bending moment. In fact, these two equilibrium conditions must hold true regardless of whether the service load is acting or failure is imminent under overload, the only difference being that the stress distribution across the depth is linear in one but not in the other.

The resultant compressive force may be entirely from concrete stresses or it may be from a combination of the stresses in concrete and those in the compression reinforcement. The resultant tensile force, of course, comes entirely from the tension reinforcement.

■ **EXAMPLE 4.4.1**   Using the equilibrium conditions, determine the working stresses in the steel and on the extreme compression fiber of concrete in the section of Fig. 4.4.1(a) due to an applied service load moment M_w of 2000 in.-kips. Use $n = 8$ for the ratio of the moduli of elasticity of steel to concrete.

Solution: The assumptions of linear strain and stress proportional to strain are shown in Fig. 4.4.1(b) and (c). The first step in the solution is to locate the neutral axis (NA). The internal compressive force is obtained by integration of the stress times the area on which it acts, equivalent to computing the volume of the stress solid. Thus

(a) Section   (b) Strains   (c) Stresses

Figure 4.4.1 Section for Example 4.4.1.

the internal compressive force in the concrete is the volume of a triangular wedge, as in Fig. 4.4.1(a),

$$C = \tfrac{1}{2}f_c bx = 6.0 f_c x$$

The internal tensile force is

$$T = f_s A_s = f_s(4)(1.27) = 5.08 f_s$$

Equating C to T gives

$$\frac{f_s}{f_c} = \frac{6.0x}{5.08}$$

The ratio of f_s to f_c may also be obtained using the linear strain relationship and taking stress proportional to strain, as shown in Fig. 4.4.1,

$$\frac{\epsilon_s}{\epsilon_c} = \frac{24 - x}{x}$$

$$\frac{f_s}{f_c} = \frac{E_s \epsilon_s}{E_c \epsilon_c} = n\frac{\epsilon_s}{\epsilon_c} = 8\frac{(24 - x)}{x}$$

Note that the actual values of E_s and E_c are not needed. Equating the two expressions for f_s/f_c,

$$\frac{6.0x}{5.08} = \frac{8(24 - x)}{x}$$

$$6x^2 = 40.64(24 - x)$$

$$x^2 + 6.77x = 162.6$$

$$x = 9.81 \text{ in.}$$

Note that, other than the *ratio* of the moduli of elasticity, only the properties of the section (depth, width, and steel area) affect the position of the neutral axis. The loading does *not* affect the neutral axis location.

The moment *arm* of the internal couple, or the distance from centroid of compressive solid to centroid of tension steel, equals for the case of the triangular wedge compressive solid,

$$\text{arm} = 24 - \frac{x}{3} = 24 - \frac{9.81}{3} = 20.73 \text{ in.}$$

Then, using the second equilibrium condition,

$$M_w = 2000 \text{ in.-kips} = (C \text{ or } T) \times \text{arm}$$

$$C = T = \frac{M_w}{\text{arm}} = \frac{2000}{20.73} = 96.5 \text{ kips}$$

The stresses are determined from the expressions for C and T,

$$f_c = \frac{C}{6x} = \frac{96,500}{6(9.81)} = 1640 \text{ psi}$$

$$f_s = \frac{T}{5.08} = \frac{96,500}{5.08} = 19,000 \text{ psi} \quad \blacksquare$$

4.5 METHOD OF TRANSFORMED SECTION

In the method of transformed section, the cross section containing steel and concrete is transformed into a homogeneous section of one material all having the modulus of elasticity of concrete. This requires the replacement of the actual steel area by an equivalent area (i.e., imaginary area) in concrete. In the transformation, two conditions must be satisfied. Let A_s and A_t be the areas of and f_s and f_t be the tensile stresses in the actual steel and the equivalent concrete, respectively. First, the equilibrium condition requires that the total tensile force be the same, or

$$A_s f_s = A_t f_t \tag{4.5.1}$$

Second, the compatibility of deformation condition requires that the unit elongation be the same, or

$$\frac{f_s}{E_s} = \frac{f_t}{E_c} \tag{4.5.2}$$

Utilizing the modular ratio $n = E_s/E_c$ and solving Eqs. (4.5.1) and (4.5.2),

$$A_t = nA_s \tag{4.5.3}$$

$$f_t = \frac{f_s}{n} \tag{4.5.4}$$

Thus the equivalent concrete area A_t is n times the actual steel area, and the equivalent tensile stress f_t (i.e., imaginary stress) is $1/n$ times the actual tensile stress.

Equations (4.5.3) and (4.5.4) are useful whenever elastic properties of the section are needed; a reinforced concrete section may then be treated as a section of one material, with the equivalent concrete on the tension side taking tension. Elastic properties are needed in working stress design computations, and more importantly, whenever deflections are needed.

4.6 INVESTIGATION OF RECTANGULAR SECTIONS IN BENDING WITH TENSION REINFORCEMENT ONLY

In investigation problems, the cross-sectional dimensions (including reinforcement area and location), the modular ratio, and the allowable working stresses are given. The problem may be (1) to compare the actual working stresses with the allowable working stresses for a given service load bending moment, or (2) to determine the allowable service load bending moment that the section may carry.

Two methods for investigation may be used. In the "internal-force" method, as used in Example 4.4.1, the external bending moment is equated to the internal resisting couple. This couple is composed of an internal compressive force C on one side of the neutral axis and an internal tensile force T on the other side of the neutral axis, at a distance apart equal to the moment arm. In the transformed section method, the well-known flexure formula Mc/I is used, in which I is the moment of inertia I_{cr} of the transformed cracked section about its centroidal axis. Note that the neutral axis coincides with the centroidal axis only in a section under pure bending (without axial load).

If the neutral-axis location, obtained during the investigation, happens to be situated so that the allowable stresses for both steel and concrete are reached simultaneously, it is said to be at the ideal location. If, however, the neutral (or centroidal) axis is nearer to the compression face than the ideal location, only the allowable steel stress, not the allowable concrete stress, can be reached under the allowable service loading. In this case, one may say *steel controls*. On the other hand, if the neutral axis is farther from the compressive face than the ideal location, only the allowable concrete stress can be reached. This situation may be referred to as *concrete controls*.

Generally, most sections having the elastic neutral axis much farther from the compression face than the ideal location from the viewpoint of the working stress method have less tension steel than 75% of the amount in the balanced strain condition as defined in the strength method. There is no *explicit* upper limit on ρ in the working stress method; there is an *implicit* upper limit when use of compression steel is the only practical design choice, as discussed in Example 4.7.2.

The detailed procedure for investigating a rectangular section in bending with tension reinforcement only is illustrated in the following two examples.

■ **EXAMPLE 4.6.1** Given $f'_c = 3000$ psi, allowable $f_s = 20{,}000$ psi, and the ACI Code, determine the allowable service load bending moment M_w that the rectangular section as shown in Fig. 4.6.1(a) may carry. Locate the elastic neutral axis by the transformed section method; then find M_w using both the internal-force and the transformed section methods.

Solution: (a) Locate the elastic neutral axis using transformed area [Fig. 4.6.1(b)]. For $f'_c = 3000$ psi, the modular ratio n may be taken as 9 (see Table 4.3.1). The area $11.50x$ above the neutral axis is in compression, whereas the equivalent tension area $nA_s = 16.20$ sq in. is assumed to be concentrated at a distance $(20.50 - x)$ below the neutral axis.

(a) Section   (b) Transformed section   (c) Applicable stress distribution

*The $11\frac{1}{2}$-in. beam width is an uncommon dimension, but is used here to provide a number that can be conveniently followed in calculations. For typical beam widths refer to Sec. 3.9.

Figure 4.6.1 Section for Example 4.6.1.

Equating the first moments of the compression and tension areas about the neutral or centroidal axis,

$$\tfrac{1}{2}(11.50)x^2 = 16.20(20.50 - x)$$

$$x = 6.32 \text{ in.}$$

(b) Determine applicable stress distribution. If the allowable compressive stress (see ACI-Appendix A.3.1) in concrete of $f_c = 0.45f'_c = 1350$ psi is realized at the extreme compressive face, the corresponding actual tensile stress f_t in the equivalent concrete would be

$$\text{actual } f_t = (1350)\frac{14.18}{6.32} = 3029 \text{ psi}$$

which cannot be since it exceeds the allowable f_t of $20,000/9 = 2222$ psi. The applicable stress variation is shown in Fig. 4.6.1(c), in which the actual f_t is made equal to the allowable f_t of 2222 psi. The actual f_c is, by proportion,

$$\text{actual } f_c = 2222(6.32/14.18) = 990 \text{ psi}$$

(c) Use the stress-solid internal-couple method to find allowable M_w.

$$\text{actual } C = \tfrac{1}{2}(0.990)(11.50)(6.32) = 36.0 \text{ kips}$$

$$\text{actual } T = 2.222(16.20) \qquad \text{or} \qquad 20(1.80) = 36.0 \text{ kips}$$

Note that the two internal forces are equal. This is always a good check on the correctness of x.

$$\text{actual arm} = 20.50 - \frac{6.32}{3} = 18.39 \text{ in.}$$

$$\text{allowable } M_w = 36.0(18.39)\tfrac{1}{12} = 55.2 \text{ ft-kips}$$

(d) Use the transformed section method to find M_w. The transformed cracked section moment of inertia is

$$I_{cr} = \tfrac{1}{3}(11.5)(6.32)^3 + 16.20(14.18)^2$$

$$= 968 + 3260 = 4230 \text{ in.}^4$$

By the flexure formula,

$$\text{allowable } M_w = \frac{0.990(4230)}{6.32(12)} \qquad \text{or} \qquad \frac{2.222(4230)}{14.18(12)} = 55.2 \text{ ft-kips} \quad \blacksquare$$

■ **EXAMPLE 4.6.2** For the rectangular section in Example 4.6.1, determine the allowable service load bending moment M_w if the tension reinforcement is increased from 3-#7, first to 3-#9, and then to 3-#11.

Solution: **(a)** Summary of results. The solutions for the three amounts of steel area are shown in Fig. 4.6.2. The solution in Fig. 4.6.2(a) is from Example 4.6.1. The solutions in Fig. 4.6.2(b) and (c) are obtained by following the same procedure described in Example 4.6.1.

(a) Elastic NA above ideal location; $M_w = 55.2$ ft-kips

(b) Elastic NA at ideal location; $M_w = 89.6$ ft-kips

(c) Elastic NA below ideal location; $M_w = 103.1$ ft-kips

Figure 4.6.2 Section for Example 4.6.2.

(b) Discussion. The reinforcement ratio for the section in Fig. 4.6.2(c), where the steel stress is limited to 15,130 psi by the assumption of linear stress variation (i.e., concrete controls), is 0.0198, while $0.75\rho_b$ for $f'_c = 3000$ psi and $f_y = 40,000$ psi is 0.0278 in strength design. In working stress design, even when the steel reinforcement seems large compared to the concrete compression area, it is not as large as the amount permitted in strength design. ∎

4.7 DESIGN OF RECTANGULAR SECTIONS IN BENDING WITH TENSION REINFORCEMENT ONLY

In problems of design, the bending moment, the modular ratio, and the allowable working stresses are given. The designer is to select the values of b, d, and A_s. The attempt to locate the neutral axis at the ideal location can rarely be accomplished with exactness, because b and h (d is derived from h) are adjusted to a desirable whole inch (occasionally $\frac{1}{2}$ in.) and A_s must be provided by an actual number of bars. In building construction, it is common to use the same b and h for a group of beams having approximately equal loadings and span lengths.

The procedure for determining the theoretical values of b, d, and A_s to place the neutral axis at the ideal location is described in what follows, without reference for the time being to the adjustment to convenient values of b and h or to the choice of bars. Consider the section of Fig. 4.7.1 where the neutral axis is at the ideal location. From similar triangles obc and dbf, $bc/bf = oc/df$ or

$$k = \frac{\text{ideal } x}{d} = \frac{\text{allowable } f_c}{\text{allowable } f_s/n + \text{allowable } f_c} \tag{4.7.1}$$

The term k is defined as the ratio of the ideal x to the effective depth d. Then

$$\text{ideal moment arm } jd = d - \frac{x}{3} = d - \frac{kd}{3}$$

or

$$j = 1 - \frac{k}{3} \tag{4.7.2}$$

Figure 4.7.1 Elastic neutral axis at the ideal location.

The term j is defined as the ratio of the ideal moment arm jd to the effective depth d. Equating the internal forces C and T and letting $A_s = \rho bd$,

$$\tfrac{1}{2}(\text{allowable } f_c)(bkd) = (\text{allowable } f_s)(\rho bd)$$

from which

$$\rho = \frac{k\,(\text{allowable } f_c)}{2\,(\text{allowable } f_s)} \tag{4.7.3}$$

This ρ is the reinforcement ratio to place the neutral axis at the ideal location. Equating the bending moment M_w to Cjd,

$$M_w = \tfrac{1}{2}(\text{allowable } f_c)(bkd)(jd) = \tfrac{1}{2}(\text{allowable } f_c)jkbd^2 = Rbd^2$$

in which

$$R = \tfrac{1}{2}(\text{allowable } f_c)jk \tag{4.7.4}$$

Equating the bending moment M_w to Tjd, $M_w = A_s(\text{allowable } f_s)jd$, from which

$$A_s = \frac{M_w}{(\text{allowable } f_s)jd} \tag{4.7.5}$$

The constants k, j, ρ, and R as expressed in Eqs. (4.7.1) to (4.7.4) will be called the four *design constants* aimed to place the elastic neutral axis at the ideal location. Their values depend on the allowable working stresses f_c and f_s and on the value of n. It is to be noted that k, j, and ρ are dimensionless constants, but R is usually in psi.

In general, the reinforcement ratio ρ of Eq. (4.7.3) is approximately one-half of the maximum permitted by the strength method. Other values for this reinforcement ratio are given in Table 4.7.1.

The procedure for determining the theoretical values of b, d, and A_s for placing the elastic neutral axis at the ideal location may be summarized as follows:

1. Find the required value of bd^2 from M_w/R.
2. Assume a value of b and determine d (select b and h and check weight). See Section 3.9 for practical selection of beam sizes.
3. Determine A_s from ρbd and check its value from $M_w/[(\text{allowable } f_s)(jd)]$.

Table 4.7.1 Ratio ρ for placing the Elastic Neutral Axis at the Ideal Location According to Working Stress Method[a]

f_y (psi)	(MPa)	$f'_c = 3000$ psi (21 MPa) $n = 9$	$f'_c = 3500$ psi (24 MPa) $n = 8.5$	$f'_c = 4000$ psi (28 MPa) $n = 8$	$f'_c = 5000$ psi (35 MPa) $n = 7$
40,000	280	0.0128	0.0158	0.0188	0.0248
50,000	350	0.0128	0.0158	0.0188	0.0248
60,000	410	0.0095	0.0117	0.0141	0.0186

[a] *Note:* SI equivalents are approximate, given to two significant figures.

It is to be noted that there may be many usable solutions in accordance with the above procedure.

■ **EXAMPLE 4.7.1** Design the cross section for a rectangular beam to carry a uniform live load of 1.9 kips/ft and a uniform dead load of 1.0 kip/ft (not including beam weight) on a simply supported span of 32 ft. Use the ACI Code with $f'_c = 4000$ psi and $f_y = 60,000$ psi (Grade 60 steel).

Solution: **(a)** Determine design constants relating to the ideal location of the elastic neutral axis. The allowable stresses (ACI-Appendix A.3) are $f_c = 0.45f'_c = 1800$ psi, $f_s = 24,000$ psi for Grade 60 steel, and $n = 8$. Referring to Fig. 4.7.2(b) and using similar triangles,

$$k = \frac{1800}{1800 + 24,000/8} = 0.375$$

$$j = 1 - \frac{k}{3} = 0.875$$

$$R = \tfrac{1}{2}f_c kj = \tfrac{1}{2}(1800)(0.375)(0.875) = 295 \text{ psi}$$

$$\rho = \frac{1}{2}\frac{kf_c}{f_s} = \frac{1}{2}(0.375)\frac{1800}{24,000} = 0.01405$$

(b) Determine preliminary size. Estimating beam weight per foot at 0.5 kip/ft,

$$M_w \text{ (live load)} = \tfrac{1}{8}(1.9)(32)^2 = 243 \text{ ft-kips}$$

$$M_w \text{ (dead load)} = \tfrac{1}{8}(1.0 + 0.5)(32)^2 = 192 \text{ ft-kips}$$

$$M_w = 243 + 192 = 435 \text{ ft-kips}$$

(a) Selected cross section

(b) Ideally reinforced stresses

(c) Actual beam stresses

Figure 4.7.2 Section for Example 4.7.1.

Since there is no size limitation, it is desired to make the beam have its elastic neutral axis located at the ideal location. Thus

$$\text{required } bd^2 = \frac{M_w}{R} = \frac{435(12,000)}{295} = 17,700 \text{ in.}^3$$

TRIAL b	REQUIRED d	
12	38.4	
15	34.4	
18	31.4	←Try

Overall depth required (assuming two layers of equal sized bars) = 31.4 + clear cover + stirrup diameter + bar diameter + 0.5 of clear distance between layers, where clear cover is 1.5 in. (ACI-7.7.1) and stirrup diameter is $\frac{3}{8}$ or $\frac{1}{2}$ in. usually. [*Note:* Stirrups enclosing the main steel, as shown in Fig. 4.7.2, provide additional shear strength to that provided by concrete alone. Since nearly all beams contain stirrups, whose size is generally not definitely established until later in the design process (see Chapter 5), provision for them should be made.] Bar diameter is, say, 1 in. One-half clear between layers is 0.5 in. (ACI-7.6.2). Thus for two layers of bars the overall depth is about $3\frac{1}{2}$ to 4 in. greater than the required d.

Overall depth $h = 31.4 + 4.0 = 35.4$ in. Try a beam 18×36.

(c) Check weight, and revise step (b) if necessary.

$$\text{weight} = \frac{18(36)}{144}(0.15) = 0.675 \text{ kip/ft}$$

$$\text{corrected } M_w = 243 + \tfrac{1}{8}(1.675)(32)^2 = 457 \text{ ft-kips}$$

For $b = 18$ in.,

$$\text{revised required } d = 32.1 \text{ in.}$$

$$\text{required } h = 32.1 + \approx 4 = 36.1 \approx 36 \text{ in.} \qquad \textbf{OK}$$

A beam 18×36 appears acceptable.

(d) Determine steel area required. If #8 bars may possibly be used, 3.5 in. instead of 4.0 might be subtracted from h to obtain d.

$$\text{actual } d \approx 36 - 3.5 = 32.5 \text{ in.}$$

$$\text{required } A_s = \frac{M_w}{f_s jd} = \frac{457(12)}{24(0.875)(32.5)} = 8.03 \text{ sq in.}$$

Try 8-#9 in two layers; $A_s = 8.00$ sq in.

(e) Check the section:

$$d = 36 - 1.5 - 0.5 - 1.128 - 0.5 = 32.37 \text{ in.}$$
$$\text{↑— (stirrup diameter estimate)}$$

Locate neutral axis:

$$18x\left(\frac{x}{2}\right) = 8.0(8)(32.37 - x)$$

$$x^2 + 7.12x = 230$$

$$x = 12.02 \text{ in.}$$

$$\text{arm} = 32.37 - 12.02/3 = 28.36 \text{ in.}$$

$$C = T = \frac{M_w}{\text{arm}} = \frac{457(12)}{28.36} = 193.5 \text{ kips}$$

$$\text{actual } f_s = \frac{T}{A_s} = \frac{193.5}{8.0} = 24.2 \text{ ksi} \approx 24 \text{ ksi} \qquad \textbf{OK}$$

$$\text{actual } f_c = \frac{C}{(\frac{1}{2})bx} = \frac{193.5}{0.5(18)12.02} = 1.79 \text{ ksi} < 1.80 \text{ ksi} \qquad \textbf{OK}$$

Even though the steel stress is slightly high, usual practice might accept a slight overstress.

Selection of beam size and of bars has been done according to the general guidelines of Section 3.9. The designer must also be certain that the bars fit into the beam width (see Table 3.9.2) without violating code clearance requirements (ACI-7.6.1). The same cover requirements apply for the side of a beam as for the top or bottom. The selected cross section is shown in Fig. 4.7.2. ■

DESIGN OF A SECTION WHERE CONCRETE CONTROLS

■ **EXAMPLE 4.7.2**   For the service load moment $M_w = 457$ ft-kips (final M_w for Example 4.7.1), determine the required A_s if a section 18×34 in. were to be used. Use $f'_c = 4000$ psi, $f_y = 60,000$ psi, and the ACI Code. (The value of M_w could be revised due to reduction in weight of beam, but the purpose here is to illustrate the increase in required A_s for the same M_w.)

Solution: (a) Design constants are $k = 0.375$, $j = 0.875$, and $R = 295$ psi (see Example 4.7.1 for computation).

(b) Determine the *required* position of the elastic neutral axis relative to the ideal location for this beam. In this case, based on the solution to Example 4.7.1, the neutral axis must fall below the ideal location to increase the internal compressive force and thus counteract the effect of the shortened moment arm. Estimating two layers of steel, $d \approx 30$ in.

$$\text{required } R = \frac{M_w}{bd^2} = \frac{457(12,000)}{18(30)^2} = 339 \text{ psi}$$

The fact that the required R of 339 psi exceeds the ideal R value of 295 psi indicates that the neutral axis distance from the compression face must be larger than the ideal value for that distance, even had Example 4.7.1 not been done. One way this can be accomplished is by adding extra tension steel for the sole purpose of lowering the neutral axis. Thus the extra steel is there to give the necessary neutral axis location and will be acting under very low actual stress. This is the main drawback of the working stress method; it does not recognize the "strength" of a section where the elastic neutral axis measured from the compression face of the beam is larger than the distance to its ideal location.

(c) Determine the required stress distribution. Referring to Fig. 4.7.3,

$$M_w = (\text{required } C)(d - x/3)$$

$$457(12) = \tfrac{1}{2}(1.80)(18x)(30 - x/3)$$

Solving the quadratic equation for the required neutral axis distance x,

$$\text{required } x = 13.23 \text{ in.}$$

$$C = \tfrac{1}{2}(1.80)(18)(13.23) = 214 \text{ kips}$$

$$\text{arm} = d - x/3 = 25.59 \text{ in.}$$

$$M_w = C \text{ times (arm)} = 214(25.59)\tfrac{1}{12} = 457 \text{ ft-kips} \quad \text{(Check)}$$

In order for the neutral axis to be at $x = 13.23$ in., the stress used to compute the required steel area must be the value defined by the straightline stress relationship; a much smaller value than the code-specified allowable f_s of 24 ksi. Thus the reduced allowable value is

$$\text{allowable } f_s = nf_c\left(\frac{d-x}{x}\right)$$

$$= 8(1.80)\left(\frac{30-13.23}{13.23}\right) = 18.3 \text{ ksi}$$

Figure 4.7.3 Section for Example 4.7.2.

(d) Determine required steel area.

$$\text{required } A_s = \frac{T}{\text{allowable } f_s} = \frac{214}{18.3} = 11.7 \text{ sq in.}$$

The selection of bars, recomputation of d, and a final check of stresses are steps omitted here.

(e) Discussion. As a section requires more and more steel to increase the distance from the compression face to the neutral axis because an increased concrete compression zone is required, the reduced allowable steel stress may become so small as to make the design impractical or impossible. The alternative is to use compression steel to increase the compressive force.

Computation by the ACI strength method shows that the use of a maximum reinforcement ratio of $0.75\rho_b$ (a steel area of 11.6 sq in.) would give a nominal strength M_n of 1410 ft-kips, which is 3.1 times the working moment M_w of 457 ft-kips. Even if almost all of the working moment is from live load, the U/ϕ value can only be as large as $1.7/0.90 = 1.89$. This generally means that the designer will use compression steel as an alternative, thus greatly reducing long-time deflection due to creep and skrinkage. The ACI strength method, as dealt with in Chapter 3, provides a clear separation between the strength of a beam and its serviceability requirements for deflection control. ■

DESIGN OF A SECTION WHERE STEEL CONTROLS

■ **EXAMPLE 4.7.3**   For the service load moment $M_w = 457$ ft-kips (final M_w for Example 4.7.1), determine the required A_s if a section 18×40 in. were to be used. Use $f'_c = 4000$ psi, $f_y = 60,000$ psi, and the ACI Code. (The value of M_w should be revised due to increase in weight of the beam, but the purpose here is to illustrate the decrease in required A_s for the same M_w.)

Solution: **(a)** Design constants are $k = 0.375$, $j = 0.875$, and $R = 295$ psi (see Example 4.7.1 for computation).

(b) Determine the *required* position of the elastic neutral axis relative to the ideal location for this beam. Estimating two layers of steel, $d \approx 36$ in.,

$$\text{required } R = \frac{M_w}{bd^2} = \frac{457(12,000)}{18(36)^2} = 235 \text{ psi}$$

Since this is less than the ideal R, less steel than the ideal amount will be needed, and the neutral axis distance from the compression face of the beam should be less than the distance to the ideal location. This is a situation where the *steel controls*.

(c) Determine the required stress distribution. Referring to Fig. 4.7.4,

$$M_w = (\text{actual } C)(d - x/3)$$

$$457(12) = \frac{1}{2}\left(\frac{3.00x}{36 - x}\right)(18x)\left(36 - \frac{x}{3}\right)$$

$$x^3 - 108x^2 - 609x + 22,000 = 0$$

Figure 4.7.4 Section for Example 4.7.3.

Solving the cubic equation for the required neutral axis position,

$$\text{required } x = 12.29 \text{ in.}$$

$$f_c = \frac{3000x}{36 - x} = \frac{3000(12.29)}{36 - 12.29} = 1560 \text{ psi}$$

$$C = \tfrac{1}{2}f_c bx = \tfrac{1}{2}(1.56)(18)(12.29) = 172 \text{ kips}$$

$$\text{available arm} = 36 - 12.29/3 = 31.9 \text{ in.}$$

$$M_w = 172(31.9)\tfrac{1}{12} = 457 \text{ ft-kips} \qquad \text{(Check)}$$

$$\text{required } A_s = \frac{T \text{ or } C}{\text{allowable } f_s} = \frac{172}{24} = 7.16 \text{ sq in.}$$

(d) Discussion. The effort of solving the cubic equation only provides the slightly larger moment arm (31.9 in.) compared with the ideal jd (31.5 in.); thus an approximate value of the A_s required may be determined from the following formula:

$$A_s \left(\begin{array}{c} \text{in } underreinforced \\ \text{sections only} \end{array} \right) = \frac{M_w}{(\text{allowable } f_s)(\text{design constant } j)(\text{actual } d)} \qquad \textbf{(4.7.6)}$$

For the present problem,

$$A_s = \frac{457(12)}{24(0.875)(36)} = 7.25 \text{ sq in.}$$

This value of A_s is very close to its correct value using the cubic equation, and the error is on the safe side.

When Code-specified allowable f_s is reached before allowable f_c as in this example, Eq. (4.7.6) would usually be used. However, when allowable f_c is reached before allowable f_s, Eq. (4.7.6) should *never* be used. For instance, had it been used in Example 4.7.2, the A_s required would have been $(457)(12)/[24(0.875)(30)] = 8.7$ sq in., which is far from the correct value of 11.7 sq in. ■

4.8 DESIGN OF RECTANGULAR SECTIONS IN BENDING WITH BOTH TENSION AND COMPRESSION REINFORCEMENT

Often the size of a rectangular section is restricted such that the bending moment is decidedly larger than the resisting moment of the singly reinforced section having its neutral axis in the ideal position; that is, wherein there is simultaneous attainment of the Code-specified allowable concrete and steel stresses. When this situation occurs, the working stress method indicates the use of both tension and compression reinforcement. This would, in effect, limit the amount of tension steel in a singly reinforced section to about 40 to 60% of the maximum permitted in the strength method.

The only unknowns in such a case are the values of the tension steel A_s and the compression steel A'_s. In practice there may be complications as to whether all the required A_s can be placed in one layer within the width of the section, because the effective depth d will have to be further reduced from the available h if two layers of steel become necessary. In this discussion, it is assumed that the effective depth d is given, and this will be measured from the extreme compression face of concrete to the centroid of the tension steel, regardless of the number of layers used.

The "two-couple" method is commonly used to determine the required values of A_s and A'_s in a doubly reinforced section. In this method, the neutral axis is first placed at the ideal location, and then the bending moment is taken as the sum of two resisting couples. One of the resisting couples is provided by the concrete stress solid along with a portion A_{s1} of the tension steel, and the other couple is from the compression steel and the remainder A_{s2} of the tension steel. Referring to Fig. 4.8.1, the procedure for determining A_s and A'_s is as follows:

1. Compute $M_{w1} = Rbd^2$.
2. Compute $A_{s1} = \rho bd$ and check with $A_{s1} = M_{w1}/[(\text{allowable } f_s)(jd)]$.
3. Compute $M_{w2} = M_w - M_{w1}$.
4. Compute $C_2 = T_2 = M_{w2}/(d - d')$.

Figure 4.8.1 Two-couple method for doubly reinforced sections.

5. Compute $A_{s2} = T_2/$allowable f_s.
6. Compute $A_s = A_{s1} + A_{s2}$.
7. Compute $f_{c1} = $ (allowable f_c)$(kd - d')/kd$.
8. Compare $2nf_{c1}$ with allowable f_s.
9. Compute $A'_s = C_2/[(2n - 1)f_{c1}]$ if $2nf_{c1} \leq$ allowable f_s;
   compute $A'_s = C_2/$(allowable $f_s - f_{c1}$) if $2nf_{c1} >$ allowable f_s.

Steps 1 to 6 are self-explanatory. Steps 7 to 9 are explained in the following.

In a rectangular section with tension reinforcement only, concrete alone takes the compression. In a doubly reinforced section, concrete and steel act together to take the compression. If concrete and steel were both elastic, that is, if all deformation occurred at the instant when load was applied and disappeared when the load was removed, the method of transformed section should still apply; that is, the compression steel would be replaced by an equivalent concrete area equal to n times its actual area. However, as discussed in Section 1.10, concrete under stress deforms (creeps) with time and it is also subject to shrinkage over a period of time. These time-dependent effects do not occur in the steel. Hence as concrete deforms, even at or below its allowable working stress, there is a continuous transfer of load from concrete to steel. In order that the straightline variation of stress across the depth of the section may be applied to doubly reinforced sections, one way of approximating the effect of this transfer of load is to increase the equivalent concrete area of compression steel to more than n times the actual steel area.

An effective modular ratio of $2n$ is permitted by ACI-Appendix A.5.5 when transforming the compression reinforcement into equivalent concrete area for *stress computations*. The ACI provision means actually that the stress in the compression reinforcement in an elastic analysis is $2nf_{c1}$ (see Fig. 4.8.1), but it may not exceed the allowable f_s. This accounts approximately for the transfer of load to the compression steel, but also prevents one from assuming a compression capacity of steel that exceeds its tensile capacity. Thus steps 7 and 8 are explained.

The compression steel actually displaces some concrete area which has already been counted on in taking the internal force C_1 in the first resisting couple. Thus the internal force C_2 represents the excess from the compression reinforcement over that required to replenish the little force $f_{c1}A'_s$ already included in the internal force C_1; or $C_2 = [(2nf_{c1}$ or allowable f_s, whichever is smaller) $A'_s - f_{c1}A'_s]$. The expressions in step 9 are obtained by solving the above equation for A'_s.

4.9 INVESTIGATION OF RECTANGULAR SECTIONS IN BENDING WITH BOTH TENSION AND COMPRESSION REINFORCEMENT

The investigation of a rectangular section in bending with both tension and compression reinforcement is complicated by the provision that the stress in the latter can be either $2nf_{c1}$ or allowable f_s, whichever is smaller. On the other hand when computing the moment of inertia of the transformed section in deflection calculations, the equivalent area in concrete of the compression steel is definitely to be taken at n times, not $2n$ times, the actual area of the compression steel, because creep and shrinkage (time-dependent effects) are accounted for in a different way (see Chapter 14 on deflections).

Returning to the problem of finding the allowable service load moment for a doubly reinforced section, the important step is to determine the neutral axis distance x. If f_{c1} is the value on the linear stress diagram at the level of the compression steel, then $f_{c1} = f_c(x - d')/x$. At the outset it is unknown whether the additional compressive force C_s provided by the compression steel is (allowable $f_s - f_{c1}$) times A'_s or $(2nf_{c1} - f_{c1})$ times A'_s, with the latter prevailing if $2nf_{c1}$ is less than allowable f_s. Thus, it is necessary to select one possibility and make the confirmation after x is solved from the equilibrium condition $C_c + C_s = T$; or

$$\frac{1}{2} f_c bx + \left(\begin{array}{c} 2nf_{c1} - f_{c1} \\ \text{or} \\ \text{allowable } f_s - f_{c1} \end{array} \right) A'_s = f_c \frac{(x - d)}{x} A_s \qquad \textbf{(4.9.1)}$$

Note that the term f_c, which is the actual extreme fiber concrete stress, cancels out in the above equation so that x is the only unknown.

Numerical examples for design and investigation of doubly reinforced sections by the working stress method appeared in earlier editions of this book.

4.10 SERVICEABILITY—DEFLECTIONS

Whether safety of a beam or floor system is established by the working stress method (ACI "alternate design method") or by the strength design method, excessive deflection may make the system unserviceable. When flexural members support or are attached to partitions and other construction likely to be damaged by large deflection, deflection computations *under service load conditions* will usually be necessary. Even in situations in which excessive deflection may not crack or damage anything, large noticeable deflection is psychologically disturbing to humans.

For situations of "members not supporting or attached to partitions or other construction likely to be damaged by large deflections," ACI-Table 9.5(a) provides minimum thickness for beams and one-way slabs unless deflections are computed. When the minimum thickness requirement of ACI-Table 9.5(a) is not satisfied *or* if excessive deflection may cause a problem, deflection must be computed and must satisfy the limits of ACI-Table 9.5(b).

Since the deflection that concerns the designer is a service load phenomenon, the elastic beam properties (such as moment of inertia of the transformed cracked section) are a necessary part of the computations. Properties such as neutral axis location and moment of inertia involve the basic assumptions of the working stress method, and hence are dealt with in this chapter.

Since the applied service load moment usually far exceeds the moment that causes the tension concrete to crack, the properties of the so-called "cracked section" are needed. In earlier sections of this chapter, the neutral axis of the cracked section has been obtained from either (a) equating the compressive force C to the tensile force T, or (b) locating the centroid of the tranformed section (see Sections 4.4 and 4.6). Deflection calculations involve using a formula of the type,

$$\Delta = \beta_a \frac{M_w L^2}{E_c I_e} \qquad \textbf{(4.10.1)}$$

where

β_a = constant that depends on the loading and support conditions

L = span length

M_w = maximum service load moment

E_c = concrete modulus of elasticity

I_e = effective moment of inertia

The effective moment of inertia I_e used by the ACI Code (ACI-9.5) involves both the gross section moment of inertia I_g (commonly without steel) and the transformed cracked section moment of inertia I_{cr}. The following example shows the computation of the moment of inertia of the transformed cracked section for a section having both tension and compression steel.

Since deflection computations are based on *elastic* conditions and service loads, it is the *elastic* section properties such as centroid and moment of inertia that are needed. Thus, in obtaining the transformed section all steel, both tension and compression, is transformed into equivalent concrete using the elastic modular ratio n. According to ACI-9.5, the long time deflection under sustained loads due to creep and shrinkage is obtained by multiplying the *elastic* "instantaneous" deflection by a time-dependent multiplier. Thus, even for computation of the time-dependent deflection, the elastic I_{cr} is used.

■ **EXAMPLE 4.10.1**   Compute the moment of inertia I_{cr} of the transformed cracked section to be used in deflection calculation for the doubly reinforced section shown in Fig. 4.10.1, using $f'_c = 4000$ psi ($n = 8$).

Solution: (a) Locate the neutral axis. The steel is transformed into equivalent concrete by using $A_t = nA_s$ for both compression and tension steel. Referring to Fig. 4.10.1,

$$\frac{13x^2}{2} + 10.5(x - 2.5) = 56.64(22 - x)$$

$(n - 1)A_s' = 10.5$ sq in.

13"

2.5"

$x = 9.76"$

22"

$nA_s = 56.64$ sq in.

2 – # 6 ⎱
2 – # 5 ⎰
$A_s' = 1.50$

4 – # 10 ⎱
2 – # 9 ⎰
$A_s = 7.08$

Figure 4.10.1 Doubly reinforced section of Example 4.10.1.

Solving for x,

$$x = 9.76 \text{ in.}$$

(b) Determine the transformed cracked section moment of inertia.

$$I_{cr} = \tfrac{1}{3}(13)(9.76)^3 + 10.5(9.76 - 2.5)^2 + 56.64(22 - 9.76)^2$$
$$= 4030 + 550 + 8480 = 13,100 \text{ in.}^4$$

For examples of computing deflection according to the ACI Code (ACI-9.5), the reader is referred to Section 14.12.   ∎

4.11 SERVICEABILITY—FLEXURAL CRACK CONTROL FOR BEAMS AND ONE-WAY SLABS

Cracking in concrete is generally the result of the following actions [4.1–4.3]: (1) volumetric change, including that due to drying shrinkage, creep under sustained load, thermal stresses, and chemical incompatibility of concrete components; (2) internal or external direct stress due to continuity, reversible load, long-time deflection, camber in prestressed concrete, or differential movement in structures; and (3) flexural stress due to bending.

Visible cracking is generally initiated by either internal microcracking (volumetric change would usually induce this type) or flexural microcracks. Flexural microcracks are surface cracks that are not visible except by careful close investigation and are generally initiated by flexural stress. Once flexural microcracks have formed, a slight increase in flexural load causes these cracks to open up suddenly to measurable widths. An excellent source for information on control of cracking resulting from flexural and other causes is contained in an ACI Committee 224 report [4.1, 4.2].

Strength designs using Grade 60 reinforcement can lead to service load *strains* in the tension steel which are 50% higher than those in existing structures designed by the working stress method. Higher strains mean wider cracks. For example, using a typical overload factor of, say, 1.55 divided by strength reduction factor $\phi = 0.90$ for flexure, gives 1.72 as the total factor for safety. This corresponds approximately to $f_y/1.72 = 0.58f_y$, say $0.60f_y$, as the service load steel stress. For Grade 60 steel, $0.60f_y = 36$ ksi, which is 50% more than the 24 ksi allowed for Grade 60 in the working stress method.

Because of the compatibility in strain between concrete and steel, the 50% or more increase in steel strain would seem to indicate the same percentage increase in crack width. Furthermore, even at low steel stress levels, say 9000 psi (63 MPa), flexural microcracking has been found to occur [4.3]. As larger cracks form, the exposure to a corrosive environment may be detrimental to the steel. Factors such as humidity, salt air, alternate wetting and drying, or freezing and thawing may accelerate corrosion and contribute to concrete deterioration in the vicinity of large width cracks. Increased cover provides thicker protection but may result in wider cracks at the beam face, influencing corrosion [4.4, 4.5]. Wide cracks may also be unsightly and contribute to doubt about structural safety. Although cracking cannot be eliminated, it is generally more desirable to have many fine hairline cracks than a few wide cracks. Thus crack control is a matter of controlling the distribution and size of cracks rather than eliminating them.

To control cracking, it is better to use several smaller bars at moderate spacing than larger bars of equivalent area. The objective is, therefore, one of distributing the reinforcement in the concrete tension zone; hence the words in the title of ACI-10.6, "Distribution of flexural reinforcement," which contains the crack control provisions for beams and one-way slabs. Control of cracking is particularly important when reinforcement with a yield stress in excess of 40,000 psi (280 MPa) is used, or when reinforcement percentages exceed the amount of reinforcement traditionally used in the working stress method. For the strength method, the comparable percentage is about $0.375\rho_b$ (one-half the maximum permitted amount).

Good bar arrangement in the cross section will usually lead to adequate crack control even when Grade 60 bars are used. Entirely satisfactory structures have been built, particularly in Europe, using design yield stresses exceeding 80,000 psi (560 MPa), which is the current specified limit in ACI-9.4.

Many studies [4.6–4.12] have verified the generally accepted belief that carck width is proportional to steel stress. Two other significant variables have been found to be the thickness of concrete cover and the area of concrete surrounding each individual reinforcing bar in the zone of maximum tension.

The ACI Code provisions (ACI-10.6.4) are based on the Gergely-Lutz [4.7] expression for crack width, as follows:

$$w = C\beta_h f_s \sqrt[3]{d_c A} \qquad (4.11.1)$$

where, referring to Fig. 4.11.1,

w = crack width at the tension face of the beam (in. or mm)

$\beta_h = h_2/h_1$, the ratio of the distances to the working stress neutral axis from the extreme tension fiber and from the centroid of the main tension reinforcement

f_s = service load stress in the steel (ksi or MPa)

d_c = thickness of concrete cover measured from the extreme tension fiber to the center of the bar located closest to that fiber (in. or mm)

Figure 4.11.1 Dimensional notation for Gergely-Lutz equation (4.11.1).

$A = A_e/m$, effective tension area of concrete surrounding the main tension rein-forcing bars and having the same centroid as that reinforcement, divided by the number of bars (sq in. or mm^2)

m = number of bars; for different sizes use $m = A_s/(A_b$ for largest bar)

C = an experimental constant (76×10^{-6} sq in./kip for Inch-Pound units or 11.0×10^{-6} mm^2/N for SI units)

Equation (4.11.1) represents the most probable maximum crack width on the bottom face of a beam. Because of considerable scatter expected, even under care-fully controlled laboratory conditions, crack width calculations should serve only as a guide to good detailing of bars and not as values to use for comparison with mea-sured cracks in a building that is in service. Consequently, Eq. (4.11.1) has been con-verted into a form in which a simple calculation could be made to arrive at reasonable reinforcing details as indicated by experience and laboratory tests, without actually emphasizing crack width. Thus dividing Eq. (4.11.1) by $C\beta_h$, the quantity z is defined,

$$z = f_s \sqrt[3]{d_c A} = \frac{w}{C\beta_h} \tag{4.11.2}$$

For simplification β_h may be taken at an approximate value of 1.2. In order to have numerical values for the quantity $w/(C\beta_h)$ in the ACI Code, crack widths have been limited to 0.016 and 0.013 in. (0.41 and 0.33 mm) for interior and exterior exposure, respectively. Substitution of these values in Eq. (4.11.2) gives

$$z_{\text{limit}} = \frac{16}{0.076(1.2)} = 175 \text{ kips/in. (interior) (30.6 MN/m)}$$

$$z_{\text{limit}} = \frac{13}{0.076(1.2)} = 142.5 \approx 145 \text{ kips/in. (exterior) (25.4 MN/m)}$$

which are limits specified in ACI-10.6.4* when the design yield stress f_y for tension reinforcement exceeds 40,000 psi (280 MPa).

The Gergely-Lutz expression has been found [4.12, 4.13] to apply also to one-way slabs (wide beams). However, the average value for $\beta_h = h_2/h_1$ (see Fig. 4.11.1) is about 1.35 for floor slabs, rather than 1.2 which applies to beams. Accordingly, for the same crack control on one-way slabs, as for beams, the limits of z should be re-duced by the factor 1.20/1.35 [i.e., z should be limited to 155 kips/in. (27.1 MN/m) for interior exposure and about 130 kips/in. (22.9 MN/m) for exterior exposure]. Oh and Kang [4.15] have recently reviewed the subject of cracking in flexural members.

--- --- --- ---

* For SI, ACI 318–89M, Section 10.6.4, gives

$$z_{\text{limit}} = 30 \text{ MN/m (interior)}$$

$$z_{\text{limit}} = 25 \text{ MN/m (exterior)}$$

For two-way slabs, the above crack control procedure and limiting z values do not apply; other recommendations are given in the ACI Committee 224 Report [4.1]. Chapter 16 on two-way slabs also contains brief treatment.

When structures are "subject to very aggressive exposure or designed to be watertight," the requirements of ACI-10.6.5 relating to flexural members and one-way slabs are not sufficient. The limitations for z may not be low enough, and possibly the use of ACI Eq. (10-4) may not even be appropriate. According to ACI Commentary-R10.6.5, "clear experimental evidence is not available regarding crack width beyond which a corrosion danger exists." For guidance in the design of sanitary structures, ACI Committee 350 [4.14, Table 2.6.4b] recommends z not to exceed 115 kips/in. (20.1 MN/m) for "normal sanitary exposure" nor 95 kips/in. (16.6 MN/m) for "severe exposure."

In recognition of the undesirability of performing an additional analysis using working stress method to determine f_s for use in Eq. (4.11.2) when the strength method is otherwise being used, ACI-10.6.4 permits taking f_s as 60% of the specified yield stress f_y in lieu of using the working stress computation. This will be a conservative approach because it will frequently overestimate the stress f_s. When a steel percentage approximately equal to that giving rise to an ideally located elastic neutral axis in the working stress method is used (corresponds in strength method to $\rho = 0.375\rho_b$), the value of f_s for Grade 60 steel would be about 24,000 psi which is only 40% of the yield stress.

■ **EXAMPLE 4.11.1**   Check the crack control provisions of the ACI Code against the cross section selected in Example 4.7.1 [Fig. 4.7.2(a)]. The selected beam has $b = 18$ in., $h = 36$ in., 8-#9 bars in two layers, #3 stirrups, 1.5 in. clear cover at bottom, and 1 in. clear between layers; $f'_c = 4000$ psi and Grade 60 steel are used.

Solution: The crack control provision of the ACI Code is given under the heading "Distribution of flexural reinforcement in beams and one-way slabs." Compute z of ACI-10.6.4:

$$z = f_s \sqrt[3]{d_c A}$$

$f_s = 24.2$ ksi (computed in Example 4.7.1)

d_c = distance to centroid of bottom layer

   = 1.5 (cover) + 0.375 (stirrup) + 0.564 (bar radius)

   = 2.44 in.

$A = A_e/m$

$A_e = 2d_s b$

d_s = distance to centroid of tension bar group

   = 1.5 (cover) + 0.375 (stirrup) + 1.128 (bar diameter) + 0.5

   = 3.50 in.

$$A_e = 2(3.50)18 = 126 \text{ sq in.}$$

$$m = \text{number of tension bars} = 8$$

$$A = 126/8 = 15.75 \text{ sq in./bar}$$

$$z = 24.2 \sqrt[3]{2.44(15.73)} = 81.5 \text{ kips/in.}$$

Since this does not exceed the 175 and 145 kips/in. (30.6 and 25.4 MN/m) allowed (ACI-10.6.4) for interior and exterior exposure, respectively, the selected cross section is acceptable. ■

■ **EXAMPLE 4.11.2**   Examine the crack control situation of Example 4.11.1 if 2-#18 bars had been used instead of 8-#9.

Solution: Referring to Fig. 4.11.2 for this cross section

$$d_c = d_s = 1.5 + 0.375 + \frac{2.257}{2} = 3.00 \text{ in.}$$

The steel stress f_s should be computed using elastic properties of the cross section, as follows:

$$18x\left(\frac{x}{2}\right) = 8.0(8)(33.0 - x)$$

$$x = 12.17 \text{ in.}$$

$$f_s = \frac{M_w}{A_s(\text{arm})} = \frac{457(12)}{8.0(33 - 12.17/3)} = 23.7 \text{ ksi}$$

If the elastic analysis is not made, f_s is to be taken at $0.60f_y$, which is 36.0 ksi (ACI-10.6.4).

Figure 4.11.2 Cross-section for Example 4.11.2 (an unsatisfactory bar selection.

$$A_e = 2(3.0)18 = 108 \text{ sq in.}$$

$$m = 2$$

$$A = \frac{A_e}{m} = \frac{108}{2} = 54 \text{ sq in./bar}$$

$$z = f_s \sqrt[3]{d_c A} = 23.7 \sqrt[3]{3.0(54)}$$
$$= 129 \text{ kips/in.} < 145 \text{ kips/in.} \qquad \textbf{OK}$$

Even this generally undesirable bar arrangement (see guidelines of Section 3.9) seems to satisfy the crack control provision. If $0.60f_y$ had been used instead of the computed value for f_s,

$$z = 36.0 \sqrt[3]{3.0(54)} = 196 \text{ kips/in.} > 175 \text{ kips/in.} \qquad \textbf{NG}$$

For the use of bundled bars, as might be necessary for very large beams, Lutz [4.16] suggests taking parameter $m = 1.4$ for each bundle used, regardless of the number of bars in the bundle.

Because of the low allowable stresses in the working stress method, it does not seem likely that crack control will be a problem when that method is used. When the ACI "strength design method" is used, the computed flexural stress at service load may be significantly higher, because the maximum percentage of reinforcement permitted in that method is nearly twice that for the ideal amount that would be used in the working stress method. ■

4.12 SERVICEABILITY—SIDE FACE CRACK CONTROL FOR LARGE BEAMS

On deep beams, the maximum crack width may occur along the side faces of beams between the neutral axis and the main tension reinforcement [4.17]. In such cases, the main reinforcement provides a restraining effect on the opening of cracks. Thus, the maximum width occurs somewhere between the neutral axis, where there should be no flexural cracking, and the main tension reinforcement where the crack opening is restrained. Unacceptably wide side face cracks have been observed that are three times as wide as the crack width at the main tension reinforcement [4.17, 4.18].

Therefore, ACI-10.6.7 has new (1989) provisions requiring side face (skin) reinforcement. When the "depth of a web exceeds 3 ft, longitudinal skin reinforcement" must be provided. The skin reinforcement A_{sk} sq in. per ft of height required on each side face (see Fig. 4.12.1) is

$$A_{sk} \geq 0.012(d - 30) \qquad (4.12.1)$$

The maximum spacing of the skin reinforcement may not exceed the lesser of $d/6$ and 12 in., and the total area used in both faces as skin reinforcement need not exceed one-half of the required flexural tension reinforcement.

Figure 4.12.1 Skin reinforcement for deep beams.

■ **EXAMPLE 4.12.1**   Design the "skin" reinforcement according to the ACI Code for a rectangular beam 18 in. wide by 48 in. deep, having 10-#9 as tension reinforcement (Fig. 4.12.2); $f'_c = 4000$ psi and $f_y = 60,000$ psi.

Solution: **(a)** Establish whether or not skin reinforcement is needed. Since the "web", in this case the overall depth, exceeds 3 ft longitudinal skin reinforcement is needed according to ACI-10.6.7.

(b) Determine the required A_{sk}.

$$A_{sk} = 0.012(d - 30)$$
$$= 0.012(44.9 - 30) = 0.179 \text{ sq in./ft} \qquad \textbf{[4.12.1]}$$

(c) Determine the spacing to be used for skin reinforcement. The maximum spacing is the lesser of $d/6$ and 12 in. (ACI-10.6.7).

$$d/6 = 44.9/6 = \underline{7.5 \text{ in.}} \text{ or } 12 \text{ in.}$$

Use #3 @ 7, $A_{sk} = 0.19$ sq in./ft. Place within the distance $d/2$ nearest the main tension steel. The total amount of A_{sk} used is 6-#3, giving an area of 0.66 sq in., which is less than $A_s/2$ (i.e., 5.00 sq in.), and thus is appropriate. Details are shown in Figure 4.12.2. ■

Figure 4.12.2 Side face reinforcement, Example 4.12.1.

SELECTED REFERENCES

4.1.   ACI Committee 224. "Control of Cracking in Concrete Structures," *Concrete International*, **2**, October 1980, 35–76. Disc. **3**, 101–110.

4.2.   ACI Committee 224. "Causes, Evaluation, and Repair of Cracks in Concrete Structures," *ACI Journal, Proceedings*, **81**, May–June 1984, 211–230.

4.3.   Edward G. Nawy. "Crack Control in Reinforced Concrete Structures," *ACI Journal, Proceedings*, **65**, October 1968, 825–836. Disc. **66**, 308–311.

4.4.   David Darwin, David G. Manning, Eivind Hognestad, Andrew W. Beeby, Paul F. Rice, and Abdul Q. Ghowrwal. "Debate: Crack Width, Cover, and Corrosion," *Concrete International*, **7**, May 1985, 20–35.

4.5.   A. W. Beeby. "Cracking, Cover, and Corrosion of Reinforcement," *Concrete International*, **5**, February 1983, 35–40.

4.6.   Bengt B. Broms and LeRoy A. Lutz. "Effects of Arrangement of Reinforcement on Crack Width and Spacing of Reinforced Concrete Members," *ACI Journal, Proceedings*, **62**, November 1965, 1395–1410. Disc. **62**, 1807–1812.

4.7.   Peter Gergely and LeRoy A. Lutz. "Maximum Crack Width in Reinforced Concrete Flexural Members," *Causes, Mechanism, and Control of Cracking in Concrete* (SP-20). Detroit: American Concrete Institute, 1968 (pp. 87–117).

4.8.   Paul H. Kaar. "High Strength Bars as Concrete Reinforcement, Part 8: Similitude in Flexural Cracking of T-Beams Flanges," *Journal PCA Research and Development Laboratories*, **8**(2), May 1966, 2–12.

4.9.   G. D. Base, J. B. Reed, A. W. Beeby, and H. P. J. Taylor. *An Investigation of the Crack Control Characteristics of Various Types of Bar in Reinforced Concrete Beams.* Research Report No. 18. London: Cement and Concrete Association, December 1966, (44 pp.).

4.10.   LeRoy A. Lutz, Nand K. Sharma, and Peter Gergely. "Increase in Crack Width in Reinforced Concrete Beams Under Sustained Loading," *ACI Journal, Proceedings,* **64**, September 1967, 538–546.

4.11.   A. W. Beeby. "The Prediction and Control of Flexural Cracking in Reinforced Concrete Members," *Cracking, Deflection, and Ultimate Load of Concrete Slab Systems* (SP-30). Detroit: American Concrete Institute, 1971, (pp. 55–75).

4.12.   John P. Lloyd, Hassen M. Rejali, and Clyde E. Kesler. "Crack Control in One-Way Slabs Reinforced With Deformed Wire Fabric," *ACI Journal, Proceedings,* **66**, May 1969, 366–376.

4.13.   A. W. Beeby. "An Investigation of Cracking in Slabs Spanning One Way," *Technical Report No. TRA 433,* London: Cement and Concrete Association, April 1970, (32 pp).

4.14.   ACI Committee 350. "Concrete Sanitary Engineering Structures," *ACI Journal, Proceedings,* **80**, November–December 1983, 467–486.

4.15.   Byung Hwan Oh and Young-Jin Kang. "New Formulas for Maximum Crack Width and Crack Spacing in Reinforced Concrete Flexural Members," *ACI Structural Journal,* **84**, March–April 1987, 103–112. Disc. **85**, January–February 1988, 109–110.

4.16.   LeRoy A. Lutz. "Crack Control Factor for Bundled Bars and for Bars of Different Sizes," *ACI Journal, Proceedings,* **71**, January 1974, 9–10.

4.17.   Gregory C. Frantz and John E. Breen. "Cracking on the Side Faces of Large Reinforced Concrete Beams," *ACI Journal, Proceedings,* **77**, September–October 1980, 307–313.

4.18.   Gregory C. Frantz and John E. Breen. "Design Proposal for Side Face Crack Control Reinforcement for Large Reinforced Concrete Beams," *Concrete International,* **2**, October 1980, 29–34.

PROBLEMS

All problems are to be done in accordance with the ACI Code "alternate design method" (i.e., working stress method) unless otherwise indicated. When selecting bars be certain that they fit into the beam width (see Table 3.9.2) and allow for #3 stirrups (for SI use #10M) around longitudinal bars. For all designs a check of stresses is required. In addition, a design sketch (to scale) is required, showing cross-sectional dimensions, number, size, and location of bars, and amount of protective clear cover. Use the modulus of elasticity ratio *n* as given in Table 4.3.1.

The SI units in parenthesis provide a problem approximately the same as that in Inch-Pound units; the conversions have been adjusted so that given information indicates comparable precision to that of the original given data. For all metric unit problems, consider the ACI allowable stresses for steel to be $f_s = 150$ MPa for $f_y = 300$ MPa and for $f_y = 350$ MPa; and $f_s = 160$ MPa for $f_y = 400$ MPa.

4.1   For the Case assigned by the instructor, calculate the stress f_s at the centroid of the steel and $f_c(\max)$ at the extreme compression fiber of the concrete due to application of the given service load bending moment M_w.

(a) Use the stress solid internal couple method; and

(b) Use the flexure formula method with transformed section.

CASE	E_s/E_c	M_w (ft-kips)	d (in.)	b (in.)	TENSION BARS
1	10	85	19.5	12	3-#9
2	9	52	19.5	12	3-#7
3	9	92	19.5	12	3-#10
4	8	400	32.7	16	7-#9 (2 layers; 5 and 2 bars)
5	8	400	27.5	22	7-#10
6	8	110	18.5	12	2-#8, 2-#10
		(kN·m)	(mm)	(mm)	
7	10	115	500	300	3-#30M
8	9	70	500	300	3-#20M

4.2   For the Case assigned by the instructor for Problem 4.1, calculate the maximum allowable bending moment M_w using the allowable stress Case assigned from the table below. The most appropriate allowable stresses in relation to the modulus of elasticity ratio used in Problem 4.1 are given in column 4.

CASE	ALLOWABLE f_s (psi)	ALLOWABLE f_c (psi)	MOST APPROPRIATE FOR PROBLEM 4.1 CASES GIVEN BELOW
1	22,000	1200	1
2	19,000	1350	1
3	20,000	1350	2, 3
4	24,000	1350	2, 3
5	20,000	1800	4, 5, 6
6	24,000	1800	4, 5, 6
	(MPa)	(MPa)	
7	150	8.3	7
8	130	9.5	7

4.3 For the Case assigned by the instructor, calculate the allowable bending moment the beam will be permitted to carry according to the working stress method (Alternate Method) of the ACI Code. Is the service load moment capacity controlled by concrete or by steel, according to the working stress method?

CASE	f'_c (psi)	f_y (psi)	d (in.)	b (in.)	TENSION BARS
1	3000	40,000	19.5	12	3-#7
2	3000	40,000	19.5	12	3-#10
3	3500	60,000	19.5	12	3-#8
4	4000	60,000	19.5	14	3-#8
5	4000	60,000	19.5	14	4-#8
6	4000	60,000	21.5	14	4-#6
7	4000	60,000	21.5	14	4-#9
8	4000	60,000	21.5	14	4-#10
9	3000	60,000	23.5	14	4-#7
	(MPa)	(MPa)	(mm)	(mm)	
10	20	300	500	300	3-#20M
11	20	300	500	300	3-#30M

4.4 Design for a given service load moment M_w a rectangular beam where *either* the width *or* the overall depth is predetermined (see table below). The beam is to have tension reinforcement only. Assume the given M_w includes the moment due to the beam weight. Do the Case assigned by the instructor. For practical proportions and other guidelines, refer to Section 3.9.

CASE	f'_c (psi)	f_y (psi)	M_w (ft-kips)	WIDTH b_w RESTRICTED (in.)	DEPTH h RESTRICTED (in.)
1	4000	60,000	400	16	NO
2	4000	60,000	400	NO	30
3	4000	50,000	110	12	NO
4	3500	60,000	450	20	NO
5	3500	60,000	320	18	NO
	(MPa)	(MPa)	(kN·m)	(mm)	(mm)
6	30	400	540	400	NO
7	30	400	540	NO	760

4.5   Design a simply supported rectangular beam having one restricted dimension (width or overall depth) to carry a service *live load* moment M_L as given. Assume the only dead load is the self-weight of the beam. The beam is to have tension reinforcement only. Do the Case assigned by the instructor. For practical proportions and other guidelines, refer to Section 3.9.

CASE	f'_c (psi)	f_y (psi)	M_L (ft-kips)	SPAN (ft)	WIDTH b_w RESTRICTED (in.)	DEPTH h RESTRICTED (in.)
1	4000	60,000	225	20	14	NO
2	4000	60,000	225	20	NO	30
3	4000	50,000	225	22	16	NO
4	3500	60,000	150	22	14	NO
5	3500	60,000	150	24	14	NO
	(MPa)	(MPa)	(kN·m)	(m)	(mm)	(mm)
6	30	400	305	6.1	350	NO
7	30	400	305	6.7	400	NO

4.6   In order to reuse forms and avoid multiple framing details, the same size beam is to be used in many places. For the Case assigned by the instructor, design a simply supported rectangular beam of predetermined overall dimensions. The service live load moment is given. The beam is to have tension reinforcement only. As a simplifying assumption, assume the only dead load is the beam weight.

CASE	f'_c (psi)	f_y (psi)	M_L (ft-kips)	WIDTH b (in.)	DEPTH h (in.)	SPAN (ft)
1	3000	40,000	100	15	30	24
2	3000	40,000	200	15	30	26
3	3500	60,000	275	18	32	28
4	4000	60,000	130	18	32	26
5	4000	60,000	190	14	26	22
	(MPa)	(MPa)	(kN·m)	(mm)	(mm)	(m)
6	20	300	136	380	760	7.3
7	20	300	270	380	760	8.0

4.7 Design a simply supported rectangular beam whose cross-sectional dimensions are to be selected by the designer. The beam is to have tension reinforcement only. As a simplifying assumption, assume the only dead load is the beam weight. Do the Case assigned by the instructor. For practical proportions and other guidelines, refer to Section 3.9.

CASE	f'_c (psi)	f_y (psi)	w_L (kips/ft)	SPAN (ft)
1	3000	60,000	5.0	28
2	4000	60,000	4.0	32
3	3500	60,000	3.0	32
4	4000	60,000	2.0	26
5	4000	60,000	2.0	22
	(MPa)	(MPa)	(kN/m)	(m)
6	20	400	73	8.5
7	30	400	60	8.0

4.8 Design a simply supported one-way slab to carry the given uniformly distributed live load w_L. The slab is to have tension reinforcement only. As a simplifying assumption, assume the only dead load is the slab weight. Do the Case assigned by the instructor. Include the temperature and shrinkage requirements of ACI-7.12 and refer to Example 3.9.2. For practical proportions and other guidelines, refer to Section 3.9.

CASE	f'_c (psi)	f_y (psi)	w_L (psf)	Span (ft)
1	3000	40,000	175	18
2	3500	60,000	175	18
3	4000	60,000	175	14
4	4000	60,000	175	16
5	4000	60,000	150	18
	(MPa)	(MPa)	(kN/m²)	(m)
6	20	300	8.4	5.5
7	30	400	8.4	5.5

4.9 Design a rectangular beam having compression and tension reinforcement, and having the given overall dimensions. The total service load bending M_w is given. Assume minimum size stirrups will be used around both tension and compression steel. Do the Case assigned by the instructor.

CASE	f'_c (psi)	f_y (psi)	M_w (ft-kips)	b (in.)	h (in.)
1	4000	60,000	475	20	32
2	3500	60,000	600	20	38
3	4000	60,000	250	14	26
4	4000	60,000	340	16	30
5	4000	60,000	300	16	28
6	4500	60,000	480	18	32
	(MPa)	(MPa)	(kN·m)	(mm)	(mm)
6	30	400	645	500	800
7	25	400	815	500	960

4.10 For the beam of the accompanying figure, determine the allowable moment capacity M_w. Use $f'_c = 4000$ psi and Grade 60 steel. ($f'_c = 30$ MPa; $f_y = 400$ MPa.)

Problems 4.10 and 4.11

4.11 For the beam of Problem 4.10, compute the transformed cracked section moment of inertia I_{cr} that would be needed for a deflection computation [i.e., used in ACI Formula (9-7)].

4.12   For the given beam of the accompanying figure, compute the transformed cracked section moment of inertia I_{cr} that would be needed for a deflection calculation. Use $f'_c = 4000$ psi and $f_y = 60,000$ psi. ($f'_c = 30$ MPa; $f_y = 400$ MPa.)

Problem 4.12

4.13   Design an irregular-shaped beam for a building whose floor system is composed of precast slabs. The 3-in. ledges are required for the support of the precast sections. The total moment to be carried is 50 ft-kips, and the cross section to be used is given in the accompanying figure. Because of forming costs, a large number of beams are being made with the same cross section. Use $f'_c = 3000$ psi and $f_y = 40,000$ psi.

Problem 4.13

4.14 **(a)** Derive an equation for the neutral axis location in nondimensional form using x/d in terms of ρ and n for investigating rectangular sections with tension reinforcement only. Plot x/d vs ρ for $n = 9$ from $\rho = 0.005$ to 0.035 in increments of 0.005.

(b) Derive an equation for the moment arm of the internal couple, using arm/d as the nondimensional quantity. Plot as in part **(a)**.

4.15 Check the ACI crack control provisions for the tension steel detail of Problem 4.10.

4.16 Check the ACI crack control provisions for the tension steel detail of Problem 4.12. Assume 1.5-in. (38-mm) clear cover and that #3 stirrups are used.

4.17 Check the ACI crack control provisions for the tension steel details of the beams designed for whichever of Problems 4.4 through 4.8 have already been worked.

CHAPTER 5

Shear Strength and Shear Reinforcement

5.1 INTRODUCTION

In this chapter, the shear strength of nonprestressed flexural members is treated. Effects of axial compression and tension are included in Section 5.14. Consideration is also given to the special requirements for deep beams (Section 5.15), the shear-friction concept (Section 5.16), and brackets and corbels (Section 5.17). Shear strength of prestressed concrete members is treated in Chapter 21. Shear effects arising from torsion and from its combination with bending or shear, or both, are treated in Chapter 19.

For the simple beam shown in Fig. 5.1.1, the bending moment M at section A-A causes compressive stresses in the concrete above the neutral axis, and tensile stresses in the reinforcement and in the concrete below the neutral axis had it not yet been cracked. To satisfy the vertical force equilibrium, the summation of the vertical shear stresses across the section must be equal to the shear force V. Below the neutral axis there is nearly a state of pure shear as shown in Fig. 5.1.2, which gives rise to a tensile stress equal to the shear stress, and acting on a 45° plane. This diagonal tension constitutes the main cause of inclined cracking. Thus the failures in beams commonly

Figure 5.1.1 Shear force and bending moment in a simply supported beam.

Inclined shear crack; test by D. R. Buettner at the University of Wisconsin, Madison.

referred to as "shear failures" are actually tension failures at the inclined cracks. One of the earliest to recognize this was E. Mörsch in Germany in the early 1900s [5.1].

Bresler and MacGregor [5.2] have presented an excellent systematic treatment of the various situations in which shear-related cracks develop, and their work has been expanded by ACI-ASCE Committee 426 [5.3]. More recently, Collins [5.4], Collins and Mitchell [5.5], Marti [5.6], Vecchio and Collins [5.7, 5.8], and Schlaich, Schäfer, and Jennewein [5.9] have explained the truss model (also known as the strut-and-tie model and compression-field theory) as the most rational approach to shear-related behavior of beams. This approach is briefly presented in Section 5.7.

In this chapter, the concepts of horizontal and vertical shear, with the resulting principal tensile stress, are presented first so that the reader may gain a feeling for

Figure 5.1.2 A state of pure shear (that is, no tensile or compressive stresses on the faces of the element).

the potential direction of inclined cracks. Then follows a presentation of recent knowledge in regard to the variables affecting shear strength, leading up to the ACI Code provisions.

5.2 THE SHEAR STRESS FORMULA BASED ON LINEAR STRESS DISTRIBUTION

A uniformly loaded simple beam is shown in Fig. 5.2.1. Consider the free body of the elemental block $abcd$, as shown in Fig. 5.2.1(b), where kd is the elastic neutral axis distance* (not necessarily the ideal value used in Chapter 4). Horizontal force equilibrium requires

$$v_y b\, dz = C_2 - C_1 \tag{5.2.1}$$

in which v_y is the unit horizontal shear stress on a plane at a distance y from the neutral axis. In the working stress range,

$$C_1 = \tfrac{1}{2}(f_{c1} + f_{c1y})b(kd - y)$$

$$f_{c1y} = \frac{y}{kd}\, f_{c1}$$

$$C_1 = \frac{1}{2} f_{c1}\left(1 + \frac{y}{kd}\right)(kd - y)b = \frac{1}{2} f_{c1} b k d \left[1 - \left(\frac{y}{kd}\right)^2\right] \tag{5.2.2}$$

Dividing the bending moment M_1 on section 1-1 by the arm of the internal couple gives the full compressive force on section 1-1, or

$$\frac{M_1}{\text{arm}} = \frac{1}{2} f_{c1} b(kd)$$

Substitution of f_{c1} from the above expression into Eq. (5.2.2) gives

$$C_1 = \frac{M_1}{\text{arm}}\left[1 - \frac{y^2}{(kd)^2}\right] \tag{5.2.3}$$

Similarly,

$$C_2 = \frac{M_2}{\text{arm}}\left[1 - \frac{y^2}{(kd)^2}\right] \tag{5.2.4}$$

Substituting Eqs. (5.2.3) and (5.2.4) into Eq. (5.2.1) gives

$$v_y = \left(\frac{M_2 - M_1}{dz}\right)\frac{1}{b(\text{arm})}\left[1 - \frac{y^2}{(kd)^2}\right]$$

$$= \frac{V}{b(\text{arm})}\left[1 - \frac{y^2}{(kd)^2}\right] \tag{5.2.5}$$

- - - - -

* In most places in this book, the symbol x is used for the neutral axis distance from the compression face of a beam in the strength design method as well as in the working stress method. Here y is used to measure the distance from the neutral axis to any point in the compression area.

Figure 5.2.1 Horizontal shear stress in a beam.

Equation (5.2.5) is valid from $y = 0$ at the neutral axis to the extreme concrete compression fiber where $y = kd$. Proceeding from the extreme compression fiber to the neutral axis, the differential horizontal force $C_2 - C_1$ increases to a maximum. Thus at the neutral axis $y = 0$, Eq. (5.2.5) gives the maximum shear stress

$$v = \frac{V}{b(\text{arm})} \tag{5.2.6}$$

The horizontal shear stress v of Eq. (5.2.6) is also the vertical shear stress, because shear stresses on two perpendicular planes must be equal. In regions where bending stress is low or where flexural cracks exist, the stress condition is close to a state of pure shear. In such a case, the maximum principal stress (see Fig. 5.1.2), a tensile stress acting at 45°, equals the shear stress.

For strength design, the shear force V is the factored shear force and the distribution of flexural stress is no longer linear in the compression region. The tensile stress v of Eq. (5.2.6) is then *not* the actual tensile stress; however, the relative magnitude obtained from this equation is still a measure of the potential for inclined cracking. In the ACI Code, the denominator in Eq. (5.2.6) is thus replaced by b times the full value of the effective depth d.

5.3 THE COMBINED STRESS FORMULA

If at a certain point below the neutral axis in a homogeneous beam the tensile stress is f_t and the shear stress is v, the principal tensile stress $f_t(\text{max})$ is given by

$$f_t(\text{max}) = \tfrac{1}{2}f_t + \sqrt{(\tfrac{1}{2}f_t)^2 + v^2} \tag{5.3.1}$$

The derivation of Eq. (5.3.1) is available in most textbooks on mechanics of materials. But because of its importance here, it will be shown again.

Figure 5.3.1 Stress condition on an elemental block.

Using equilibrium of the forces acting on the free body in the directions of f_t' and v' shown in Fig. 5.3.1(b) and calling the width of the beam b,

$$f_t'\left(\frac{b\,dz}{\sin\alpha}\right) = f_t\left(\frac{b\,dz}{\tan\alpha}\right)\cos\alpha + v(b\,dz)\cos\alpha + v\left(\frac{b\,dz}{\tan\alpha}\right)\sin\alpha$$

$$v'\left(\frac{b\,dz}{\sin\alpha}\right) = f_t\left(\frac{b\,dz}{\tan\alpha}\right)\sin\alpha + v(b\,dz)\sin\alpha - v\left(\frac{b\,dz}{\tan\alpha}\right)\cos\alpha$$

from which

$$f_t' = \tfrac{1}{2}f_t(1 + \cos 2\alpha) + v\sin 2\alpha$$

$$v' = \tfrac{1}{2}f_t\sin 2\alpha - v\cos 2\alpha$$

The value of α_{max}, that is, the α which makes f_t' maximum and at the same time makes v' zero, may be found either by differentiating the expression for f_t' with respect to α, or by setting the expression for v' to zero. Equation (5.3.1) may be obtained by substituting

$$\tan 2\alpha_{max} = \frac{v}{\tfrac{1}{2}f_t} \tag{5.3.2}$$

into the expression for f_t'.

The principal tensile stress $f_t(max)$ in the diagonal direction, which is at an angle α_{max} with the beam axis, is at least as large as either f_t or v. It is nearly equal to the longitudinal tensile stress f_t if the shear stress v is small and its direction is nearly horizontal. It is nearly equal to the shear stress v if the longitudinal tensile stress f_t is small and its direction is nearly at 45° with the beam axis. Since concrete is weak in tension, these principal tensile stresses are undoubtedly correlated to inclined cracking as shown in Fig. 5.3.2.

Figure 5.3.2 Directions of potential cracks in a simply supported beam.

5.4 BEHAVIOR OF BEAMS WITHOUT SHEAR REINFORCEMENT

As shown in Section 5.3, high shear stress on a beam results in the formation of inclined cracks. In order to prevent the formation of inclined cracks, transverse reinforcement (known as "shear reinforcement") in the form of closed or U-shaped stirrups is used, normally in the vertical direction to enclose the main longitudinal reinforcement along the faces of the beam (see Fig. 5.6.1 in Section 5.6).

Inclined cracking in the webs of reinforced or prestressed concrete beams may develop either in the absence of flexural cracks in the vicinity or as an extension of a previously developed flexural crack. An inclined crack occurring in a beam that was previously uncracked due to flexure is known as a *web-shear crack* as shown in Fig. 5.4.1(a). As inclined crack originating at the top of and becoming an extension to a previously existing flexural crack is known as a *flexure-shear crack*, as shown in Fig. 5.4.1(b). The critical flexural crack is referred to as the "initiating crack."

Flexure-shear cracks are the usual type found in both reinforced and prestressed concrete. Web-shear cracks are relatively rare in nonprestressed beams. These cracks occur in thin-webbed I-shaped beams having relatively large flanges, common only in prestressed concrete construction. This is discussed further in Chapter 21 which is devoted entirely to prestressed concrete. Web-shear cracks may also occur near the inflection points or bar cutoff points on continuous reinforced concrete beams subjected to axial tension [5.10, 5.11].

The transfer of shear in reinforced concrete members occurs by a combination of the following mechanisms [5.3], as shown in Fig. 5.4.2:

1. Shear resistance of the uncracked concrete, V_{cz}.
2. Aggregate interlock (or interface shear transfer) force V_a, tangentially along a crack [5.12–5.14], and similar to a frictional force due to irregular interlocking of the aggregates along the rough concrete surfaces on each side of the crack.

(a) Web-shear crack

A–A

Flexure-shear crack

Initiating crack
Secondary crack

(b) Flexure-shear crack

Figure 5.4.1 Types of inclined cracks (from Reference 5.2).

Figure 5.4.2 Redistribution of shear resistance after formation of inclined crack.

3. Dowel action, V_d, the resistance of the longitudinal reinforcement to a transverse force [5.15–5.19].
4. Arch action [see Fig. 5.4.5(a)] on relatively deep beams.
5. Shear reinforcement resistance, V_s, from stirrups (not available in beams without shear reinforcement).

The ability of a beam to carry additional load after an inclined crack has formed depends on whether or not the portion of shear formerly carried by uncracked concrete can be redistributed across the inclined crack. Mechanisms 1 through 4 mentioned above all participate in the redistribution, the success of which determines the shear strength and the degree of seriousness of the crack formation.

For rectangular beams without shear reinforcement, it is reported [5.3] that after an inclined crack has formed, the proportion of the shear transferred by the various mechanisms is as follows: 15 to 25% by dowel action; 20 to 40% by the uncracked concrete compression zone; and 33 to 50% by aggregate interlock or interface shear transfer.

Inclined cracks begin and grow depending on the relative magnitudes of shear stress v and flexural stress f_t, as used in Section 5.3. These controlling stresses may be expressed

$$v = k_1 \frac{V}{bd} \tag{5.4.1a}$$

$$f_t = k_2 \frac{M}{bd^2} \tag{5.4.1b}$$

where k_1 and k_2 are proportionality constants. The discussion in Section 5.2 may serve to justify the shear stress as proportional to V/bd, and in the presentation of Chapters 3 and 4 the flexural capacity M has been related to bd^2 through the use of a coefficient of resistance R which has the same units as the flexural stress.

From Section 5.3 one may note that the principal tensile stress is a function of the ratio f_t/v. Another expression for f_t/v may be obtained from Eqs. (5.4.1); thus

$$\frac{f_t}{v} = \frac{k_2}{k_1}\frac{M}{Vd} = k_3\frac{M}{Vd} \tag{5.4.2}$$

For a simply supported beam symmetrically loaded with two equal concentrated loads (see Fig. 5.4.3), the ratio M/V may be thought of as the distance a over which the shear is constant. This distance a is known as the *shear span*. For the general case where the shear is continually varying, the "shear span" may be expressed as

$$a = \frac{M}{V} \tag{5.4.3}$$

which has a value at every point along a beam.

Using Eq. (5.4.3) in Eq. (5.4.2), the ratio f_t/v becomes

$$\frac{f_t}{v} = k_3\left(\frac{a}{d}\right) \tag{5.4.4}$$

The shear span to depth ratio a/d has also been shown experimentally to be a highly influential factor in establishing shear strength [5.1–5.3, 5.21]. When factors other than a/d are kept constant, the variation in shear capacity may be illustrated by Fig. 5.4.4 using the results for rectangular beams.

From Fig. 5.4.4, four general categories of failure may be established: (1) deep beams with $a/d < 1$; (2) short beams with a/d ratios from 1 to about $2\frac{1}{2}$, in which the shear strength exceeds the inclined cracking capacity; (3) usual beams of intermediate length having a/d ratios from about $2\frac{1}{2}$ to 6, in which the shear strength equals the inclined cracking strength; and (4) long beams with a/d greater than 6, whose flexural strength is less than their shear strength.

Deep Beams. $a/d \leq 1$. For a deep beam, shear stress has the predominant effect. After inclined cracking occurs, this simply supported beam tends to behave like a

Figure 5.4.3 Basic definition of shear span a.

Figure 5.4.4 Variation in shear strength with a/d for rectangular beams (adapted from Reference 5.2).

tied-arch wherein the load is carried by direct compression extending around the shaded area of Fig. 5.4.5(a) and by the tension in the longitudinal steel. Once the shear-related crack develops, the beam transforms quickly into a tied-arch which exhibits considerable reserve capacity. Several modes of failure are possible [5.2] for the tied-arch system, as shown in Fig. 5.4.5(b).

Possible modes of failure are indicated in Fig. 5.4.5(b); they are (1) an anchorage failure; that is, pullout of the tension reinforcement at the support; (2) a crushing failure at the reactions; (3) a "flexural failure" arising from either a crushing of concrete near the top of the arch or a yielding of the tension reinforcement; or (4) failure of the

Figure 5.4.5 Modes of failure in deep beams, $a/d \le 1.0$ (adapted from Reference 5.2).

(a) Shear-tension failure

(b) Shear-compression failure

Figure 5.4.6 Typical shear failures in short beams, $a/d = 1$ to $2\frac{1}{2}$ (adapted from Reference 5.2).

arch rib due to an eccentricity of the arch thrust, resulting in either a tension crack over the support at point 4 of Fig. 5.4.5(b) or a crushing of concrete on the underside of the rib at point 5.

Short Beams. $1 < a/d \leq 2\frac{1}{2}$. Just as for deep beams, short beams have a shear strength that exceeds the inclined cracking strength. After the flexure-shear crack develops, the crack extends further into the compression zone as the load increases. It also propagates as a secondary crack toward the tension reinforcement and then progresses horizontally along that reinforcement. Failure eventually results, either (1) by an anchorage failure at the tension reinforcement, called a "shear-tension" failure [Fig. 5.4.6(a)]; or (2) by a crushing failure in the concrete near the compression face, called a "shear-compression" failure [Fig. 5.4.6(b)].

Usual Beams of Intermediate Length. $2\frac{1}{2} < a/d \leq 6$. For intermediate length beams, vertical flexural cracks are the first to form, followed by the inclined flexure-shear cracks. At the beginning several flexural cracks tend to bend over creating beam segments between cracks (the "teeth" shown in Fig. 5.4.7). When the root of the "tooth," as a result of the increasing number of flexural cracks, is so reduced [5.22] in size that it becomes unable to carry the moment arising from ΔT, it breaks to form the

Figure 5.4.7 "Diagonal tension failure" or "tooth cracking failure" on intermediate length beams; a/d about $2\frac{1}{2}$ to 6.

inclined flexure-shear crack. At the sudden occurrence of the inclined crack, the beam is not able to redistribute the load, as in the situation of smaller a/d ratio. In other words, the formation of the inclined crack represents the shear strength of beams in this category, for which the term "diagonal tension failure" has been given [5.2]. This is the usual category for beam design.

Long Beams. $a/d > 6$. The failure of long beams starts with yielding of the tension reinforcement and ends by crushing of the concrete at the section of maximum bending moment. In addition to the nearly vertical flexural cracks at the section of maximum bending moment, prior to failure slightly inclined (from the vertical) cracks may be present between the support and the section of maximum bending moment. Nevertheless, the strength of the beam is entirely dependent on the magnitude of the maximum bending moment and is not affected by the size of the shear force.

In summary, shear tends to cause inclined cracks. When no such cracks form before the nominal flexural strength is reached, the effect of shear is negligible. Beams may fail upon formation of an inclined crack, as for the so-called "diagonal tension" failures when a/d is between about $2\frac{1}{2}$ and 6, or there may be considerable reserve strength. In order for the beams with reserve capacity to maintain a state of equilibrium, stresses must be redistributed after the formation of the inclined crack. Present knowledge of how the redistribution actually takes place is limited; thus for the design of all but deep beams the shear strength is assumed to be reached when the inclined crack forms.

5.5 SHEAR STRENGTH OF BEAMS WITHOUT SHEAR REINFORCEMENT—ACI CODE

The strength at which an inclined crack [usually a flexure-shear crack as in Fig. 5.4.1(b)] forms is taken to be the shear strength of a beam without shear reinforcement, according to the intent of the ACI Code. After establishing on a rational basis those variables that are involved, the relationship between them was statistically determined from test results [5.1].

It is assumed that the strength is reached when the principal tensile stress in Eq. (5.3.1) reaches the tensile strength of concrete, which is proportional to $\sqrt{f'_c}$. Although the exact distributions of the flexural and shear stresses in a cross section are not known, it may be assumed that flexural tensile stress f_t varies as E_c/E_s times the tensile stress in the reinforcement and that v varies as the average shear stress. Assume that E_c is also proportional to $\sqrt{f'_c}$ and let V_n and M_n be the nominal shear force and nominal bending moment at a section when the inclined crack forms.

As already shown by Eq. (5.4.1a), the shear stress may be written

$$v = k_1 \frac{V_n}{bd} \tag{5.5.1}$$

The stress in the steel is proportional to $M_n/A_s d$, and the tensile stress f_t in the concrete then becomes

$$f_t \propto \frac{E_c f_s}{E_s} \propto \frac{E_c M_n}{E_s d A_s} \propto \frac{M_n \sqrt{f'_c}}{E_s d A_s} \propto \frac{M_n}{bd^2}\left(\frac{\sqrt{f'_c}}{\rho E_s}\right)$$

The above expression for f_t may be written as

$$f_t = \frac{k_4}{E_s}\left(\frac{\sqrt{f_c'}}{\rho}\right)\frac{M_n}{bd^2} \tag{5.5.2}$$

in which k_4 is a dimensionless constant and E_s has a definitely known value. Also, the tensile strength of concrete may be represented by

$$f_t(\text{max}) = k_5\sqrt{f_c'} \tag{5.5.3}$$

Substituting Eqs. (5.5.1) to (5.5.3) into the principal stress equation, Eq. (5.3.1),

$$k_5\sqrt{f_c'} = \frac{V_n}{bd}\left[\frac{1}{2}\frac{k_4}{E_s}\frac{M_n}{V_nd}\frac{\sqrt{f_c'}}{\rho} + \sqrt{\left(\frac{1}{2}\frac{k_4}{E_s}\frac{M_n}{V_nd}\frac{\sqrt{f_c'}}{\rho}\right)^2 + k_1^2}\right]$$

or

$$\frac{V_n}{bd\sqrt{f_c'}} = k_5\Bigg/\left[\frac{1}{2}\frac{k_4}{E_s}\frac{M_n}{V_nd}\frac{\sqrt{f_c'}}{\rho} + \sqrt{\left(\frac{1}{2}\frac{k_4}{E_s}\frac{M_n}{V_nd}\frac{\sqrt{f_c'}}{\rho}\right)^2 + k_1^2}\right] \tag{5.5.4}$$

In Eq. (5.5.4) the variables are observed to be $V_n/(bd\sqrt{f_c'})$ and $M_n\sqrt{f_c'}/(E_s\rho V_nd)$. It may be noted that these two variables are nondimensional quantities, because $\sqrt{f_c'}$ has units of force per unit area. In the statistical study, the nominal shear force V_n was defined as that causing the critical inclined crack and M_n as the corresponding moment at the top of the initiating crack (Fig. 5.4.1). On the basis of 440 tests [5.1], as shown in Fig. 5.5.1, the relationship between these two variables, on using the correct value of E_s, was obtained as follows:

$$\frac{V_n}{bd\sqrt{f_c'}} = 1.9 + 2500\frac{\rho V_nd}{M_n\sqrt{f_c'}} \leq 3.5 \tag{5.5.5}$$

Equation (5.5.5) is generally considered [5.3, 5.22–5.24] to be acceptable for predicting the flexure-shear cracking load, particularly for $M_n/(V_nd)$ (i.e., shear span/depth) ratios of about $2\frac{1}{2}$ to 6, with considerable conservatism for lower $M_n/(V_nd)$ values. Thus the favorable factors for the shear strength of beams without shear reinforcement are a high percentage ρ of longitudinal reinforcement and a high ratio of V_nd to M_n, that is, a low a/d ratio.

Since 1963, the ACI Code has accepted the relationship of Eq. (5.5.5) as the shear (inclined cracking) strength of beams without shear reinforcement. Thus defining V_c as the nominal strength of such beams, Eq. (5.5.5) using the web width* b_w for b becomes

$$V_c = \left(1.9\sqrt{f_c'} + 2500\frac{\rho_w V_ud}{M_u}\right)b_wd \leq 3.5\sqrt{f_c'}\,b_wd \tag{5.5.6}\dagger$$

— — — —

* The term "web" refers to the width dimension of a beam at its neutral axis. For T-shaped beams the flange width b is distinguished from the stem, or web, width b_w.

† For SI, ACI 318–89M, with f_c' in MPa, gives

$$V_c = \frac{1}{6}\left(\sqrt{f_c'} + 100\frac{\rho_w V_ud}{M_u}\right)b_wd \leq 0.3\sqrt{f_c'}\,b_wd \tag{5.5.6}$$

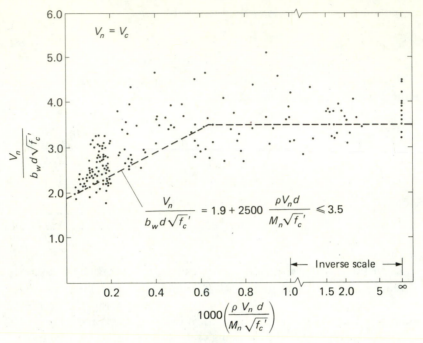

Figure 5.5.1 Derivation of ACI shear strength equation, Eq. (5.5.6), for beams without shear reinforcement, that is, $V_n = V_c$ (adapted from Reference 5.1).

which is ACI Formula (11-6). The use of the factored shear force V_u and the factored moment M_u instead of the values $V_n = V_u/\phi$ and $M_n = M_u/\phi$ makes little difference because the ratio V_u/M_u remains approximately equal to the ratio V_n/M_n, in spite of some difference in the strength reduction factors ϕ for shear and for moment or combined axial compression with bending. Note also that the reinforcement ratio $\rho_w = A_s/(b_w d)$ is used in the ACI Code formula, where b_w is the web width for a T-section rather than the flange width. The ACI Code defines b_w as "web width." For tapered webs such a definition is unclear. In general, when the web is subject to flexural tension the "average web width" should be used as b_w. However, when the web is subject to flexural compression (as for negative moment regions) the use of average web width may be unsafe [5.25]. For such negative moment regions the use of a value lower than the average, perhaps the minimum, web width is more appropriate.

The value of $V_u d/M_u$ shall not be taken greater than 1.0 except when axial compression is present [see Eq. (5.14.3) where M_m replaces M_u] which has the effect of limiting V_c at and near the points of inflection.

Continuous Beams.   The application of Eq. (5.5.6) for continuous beams has recently been subject to question [5.3]. The compression-strut action on a continuous beam is shown in Fig. 5.5.2. The analogy in Fig. 5.4.3 that M/V equals the shear span a implies that at the point of zero moment there is a support to accomodate a com-

Figure 5.5.2 Crack pattern, compression strut, and bending moment diagram near support on continuous beam.

pression-strut [see Fig. 5.4.5(a)]. For a continuous beam as in Fig. 5.5.2, the distances a_1 and a_2 are analogous to a in Fig. 5.4.3; however, there is no support to take a compression-strut reaction at the inflection point. ACI-ASCE Committee 426 [5.10, 5.11] recommends that ACI Formula (11-6) no longer be used; instead V_c should be taken as $2\sqrt{f'_c}\,b_w d$.

Lightweight Concrete. It has been shown [5.26–5.28] that the same general relationships as those for normal-weight concrete are valid for lightweight concrete, with a slight modification. In lightweight concrete, the tensile strength f_{ct} based on the split cylinder (see Section 1.8) provides a better correlation with inclined cracking strength than does the compressive strength f'_c. Since for normal-weight concrete the splitting tensile strength f_{ct} is approximately $6.7\sqrt{f'_c}$, this value may be substituted into Eq. (5.5.6); thus for lightweight concrete

$$V_c = \left[1.9\left(\frac{f_{ct}}{6.7}\right) + 2500\,\frac{\rho_w V_u d}{M_u}\right]b_w d \le 3.5\left(\frac{f_{ct}}{6.7}\right)b_w d \qquad (5.5.7)$$

as prescribed in accordance with ACI-11.2.1.1. In order that shear strength for lightweight concrete beams not exceed that for normal-weight concrete, $f_{ct}/6.7$ may not be taken to exceed $\sqrt{f'_c}$.

Tests have shown [5.26] that inclined cracking strength for lightweight concrete varies from about 60 to 100% of the values for normal-weight concrete of the same nominal compressive strength, depending on the particular aggregates used. More recent studies [5.28] have shown that using 0.75 to 0.85 of the tensile strength related term $\sqrt{f'_c}$ for normal-weight concrete in shear-strength equations is a reasonable and generally conservative approach.

Thus ACI-11.2.1.2 permits multiplying all values of $\sqrt{f'_c}$ appearing in shear (or torsion) equations by 0.75 for "all-lightweight" concrete and by 0.85 for "sand-lightweight" concrete when the split-cylinder strength f_{ct} is not specified. For "all-lightweight" concrete,

$$V_c = \left[0.75(1.9\sqrt{f'_c}) + 2500 \frac{\rho_w V_u d}{M_u} \right] b_w d \leq 0.75(3.5)\sqrt{f'_c} b_w d \qquad (5.5.8)$$

For "sand-lightweight" concrete,

$$V_c = \left[0.85(1.9\sqrt{f'_c}) + 2500 \frac{\rho_w V_u d}{M_u} \right] b_w d \leq 0.85(3.5)\sqrt{f'_c} b_w d \qquad (5.5.9)$$

Linear interpolation is permitted when partial sand replacement is used.

High Strength Concrete. In response to the increasing use of concrete having f'_c greater than 8000 psi, studies [5.29–5.36] have indicated that shear-related strength seems not to continue to increase in proportion to $\sqrt{f'_c}$. Though a definitive relationship for the shear-related strength of beams using high strength concrete has not been established, for design using concrete strength f'_c above 10,000 psi ACI Committee 318 (ACI-11.1.2) decided "it was prudent to limit $\sqrt{f'_c}$ to 100 psi in calculations of shear strength and development length." One exception to this rule (ACI-11.1.2.1, is that values of $\sqrt{f'_c}$ greater than 100 psi may be used in computing V_c when A_v is at least $f'_c/500$ times ($50\, b_w d/f_y$) but need not be more than ($150\, b_w s/f_y$).

5.6 FUNCTION OF SHEAR REINFORCEMENT

The types of shear reinforcement recognized for reinforced nonprestressed concrete by the ACI Code (ACI-11.5.1) are (1) vertical stirrups, that is, stirrups perpendicular to the longitudinal reinforcement, as shown in Fig. 5.6.1; (2) welded wire fabric [see Section 1.12] with wires located perpendicular to the axis of the member; (3) inclined stirrups, where the plane of the stirrup makes an angle of 45° or more with the longitudinal reinforcement; (4) longitudinal tension bars bent toward the compression zone of the member so that the axis of the bent portion makes an angle of 30° or more with the axis of the longitudinal portion; and (5) combinations of (1) or (3) with (4). Spirals are also permitted. Vertical stirrups (as opposed to inclined stirrups and bent bars) are used almost exclusively for shear reinforcement in current (1991) design practice.

The most generally accepted model for the behavior of reinforced concrete beams containing shear reinforcement is the truss model, originated by Ritter and Mörsch [5.9]. The current thinking related to the truss model is well presented by Schlaich, Schäfer, and Jennewein [5.9]. In a simply supported steel truss as shown in Fig. 5.6.2(a), the upper and lower chords are in compression and tension, respectively, and the diagonal and vertical members, called web members, are alternately in tension and compression. In the reinforced concrete beam of Fig. 5.6.2(b) the concrete performs the task of carrying the compressive forces while steel reinforcement is used

Figure 5.6.1 Vertical stirrups as shear reinforcement.

(a) A steel truss

(b) Truss action in a reinforced concrete beam

(c) Reinforced concrete beam with vertical shear reinforcement

Figure 5.6.2 Truss analogies.

Figure 5.6.3 Distribution of internal shears in beams with shear reinforcement (adapted from Reference 5.3).

for the tensile forces. Since the vertical transverse reinforcement (stirrups) acts in a reinforced concrete beam like the tension web members of a steel truss, the term "web reinforcement" is often used for shear reinforcement. The shear reinforcement wraps around the longitudinal tension reinforcement and must be anchored in the compression zone, usually by hooking it around longitudinal bars. (These bars are either compression reinforcement or are provided solely to hold in place and anchor the shear reinforcement.)

While shear reinforcement provides shear *strength*, its contribution to the strength occurs only after inclined cracks form. Prior to the formation of inclined cracks, the concrete performs the task of carrying the shear. However, shear reinforcement is necessary in order to allow a redistribution of internal forces across any inclined crack that may form, and thus prevent a sudden failure upon formation of the crack.

Therefore, shear reinforcement has several functions that combine to allow a redistribution of the internal forces upon the formation of an inclined crack. The primary functions [5.3] (see Fig. 5.6.3) are (1) to carry part of the shear V_s; (2) to restrict the growth of the inclined crack and thus help maintain *aggregate interlock* (also known as *interface shear transfer*) strength V_a; and (3) to tie the longitudinal bars in place and thereby increase their strength V_d to resist transverse forces (known as *dowel action*). In addition, dowel action on the stirrups may transfer a small force across a crack, and the confining action of the stirrups on the compression concrete may slightly increase its strength.

If the amount of shear reinforcement is too little, it will yield immediately at the formation of an inclined crack, and the beam then fails. If the amount of shear reinforcement is too much, there will be a shear-compression failure without the yielding of the shear reinforcement. The optimum amount should be such that both the shear reinforcement and the compression zone of the beam each continue to carry increasing shear after the formation of the inclined crack, until the shear reinforcement yields, resulting in a ductile failure.

5.7 TRUSS MODEL FOR REINFORCED CONCRETE BEAM

Referring to Fig. 5.7.1(a), a reinforced concrete beam is shown containing vertical stirrups, with equal spacing s for simplicity in explanation. According to the truss model, the reinforced concrete beam would behave as shown in Fig. 5.7.1(b). The diagonals, such as fc, would carry compressive forces in a band width (known as a field). The top chord, such as cd, would carry the internal compressive force, M/arm, as treated in Chapters 3 and 4, either with or without compression reinforcement. The bottom chord, i.e., the main tension reinforcement, would carry the internal tensile force, M/arm.

Even though compression member design is not treated until Chapter 13, from mechanics of materials it is known that compression member strength is related to (a) material (i.e., concrete) compressive strength and (b) member length (slenderness ratio). Thus, for example, the member fc must be capable of resisting the diagonal force resulting from the loading in order to transfer that force to the stirrup bf. When the diagonal force is large the diagonal compression-field would have to intercept more than one stirrup.

Refer to Fig. 5.7.1(c) where additional stirrups are used to reduce the spacing to $s/2$. In the vicinity of concentrated loads, there is no way of knowing how the load will "fan out"; in the model there are now shown four compression struts (gd, jd, kd, and id) to the midspan concentrated load P. Likewise at the supports, the reaction force $P/2$ will "fan out," probably taking more than one angle for transfer of the compressive force into the beam.

In the terminology of the truss model, also called "compression-field theory" and "strut-and-tie-model", the zone in the vicinity of a concentrated load, or at an abrupt discontinuity in the member (such as an abrupt change in member depth) is called a "D-region" [5.9]. The term D stands for discontinuity, disturbance, or detail [5.9]. In such regions, the plane section theory is not applicable, and the true forces are not those obtained by shear and moment diagrams and first-order elastic beam theory, particularly when the sections are cracked. In general, the reinforcement of the D-region has relied upon "past experience" or "good practice." However, the advocates for using the truss model believe shear-related design in D-regions can be simpler and more logical [5.9]. In the regions of a beam where the use of plane strain (the Bernoulli hypothesis) is appropriate, such as panel bc of Fig. 5.7.1(c), and therefore design for the forces at each section along the member is appropriate, the term "B-region" is used (where B stands for beam or Bernoulli).

The truss model also shows that in beams containing flexural cracks (that is, most beams) and also containing shear reinforcement (again, most beams) the use of forces from the homogeneous beam bending moment diagram, such as mno in Fig. 5.7.1(d), to design the main tension reinforcement is not correct. Rather, because

(a) Reinforced concrete beam

(b) Truss model for concrete beam

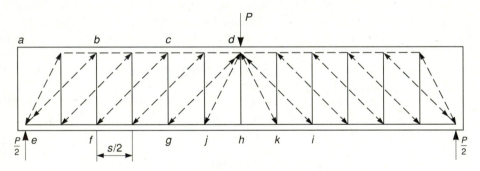

(c) Concrete beam showing possible compression fields

(d) Internal tensile or compressive force from bending moment, M/arm

Figure 5.7.1 Truss model for reinforced concrete beam.

154

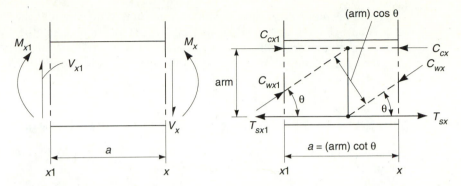

Figure 5.7.2 Internal forces for equilibrium of beam element.

of the truss behavior, the bending moment (and thus the tensile force T in the main reinforcement) should correspond to the "jumped" diagram of Fig. 5.7.1(d). The tensile force in the truss will be constant over the panel distance; thus, it will change only at a stirrup.

Since the real situation is not a pin-jointed truss, and there can be many angles comprising the compression-field, the requirement for the main tension steel would actually be somewhere between the M/arm (based on ordinary bending moment diagram) and the stepped diagram for a truss, as shown in Fig. 5.7.1(d). Shown in Fig. 5.7.2 is a segment along a beam, along with the internal forces acting on that segment; at the section x, the forces C_{cx} (concrete compression chord), C_{wx} (the diagonal compression-field "strut"), and T_{sx} (the tensile force carried by the main tension steel). These forces can be expressed,

$$C_{cx} = \frac{M_x}{\mathrm{arm}} - \frac{V_x}{2} \cot \theta \qquad (5.7.1)$$

$$T_{sx} = \frac{M_x}{\mathrm{arm}} + \frac{V_x}{2} \cot \theta \qquad (5.7.2)$$

$$C_{wx} = \frac{V_x}{\sin \theta} \qquad (5.7.3)$$

The important observation is that the truss model recognizes that the main tension steel requirement *is everywhere larger than* that obtained from the bending moment diagram by the amount $(V_x/2) \cot \theta$.

In general, this effect has been accounted for in ACI Code design by extending the bars a distance d beyond the point where the calculated requirement is made. However, the truss model shows that the shear reinforcement behavior and the forces in longitudinal bars cannot be entirely separated. This is discussed further in Chapter 6, which is devoted to determining bar lengths and the distance along the span that bars must extend.

The truss model illustrates the importance of reinforcement at the "joints" of the truss. The behavior of the stirrups, their proper detailing, and their effectiveness, have been studied recently by Hsiung and Frantz [5.37], Johnson and Ramirez [5.38], Anderson and Ramirez [5.39], Mphonde [5.40], Mphonde and Frantz [5.41], and Belarbi and Hsu [5.42].

5.8 SHEAR STRENGTH OF BEAMS WITH SHEAR REINFORCEMENT—ACI CODE

The traditional ACI approach to design for shear strength is to consider the total nominal shear strength V_n as the sum of two parts,

$$V_n = V_c + V_s \tag{5.8.1}$$

in which V_n is the nominal shear strength; V_c is the shear strength of the beam attributable to the concrete (see Fig. 5.6.3 and Section 5.5), and V_s is the shear strength attributable to the shear reinforcement.

An expression for V_s may be developed from the truss analogy. Assume that an inclined crack in the 45° direction extends all the way from the longitudinal reinforcement to the compression surface and that it intersects an average of N shear reinforcing bars, as shown in Fig. 5.8.1. The portion V_s carried across the crack by the shear reinforcement equals the sum of the vertical components of the tensile forces developed in the shear reinforcement. Thus

$$V_s = N A_v f_y \sin \alpha \tag{5.8.2}$$

in which A_v is the area of shear reinforcement within a distance s, and f_y is the tensile yield stress for the shear reinforcement. From the trigonometry,

$$Ns = d(\cot 45° + \cot \alpha) = d(1 + \cot \alpha)$$

Thus

$$V_s = \frac{d(1 + \cot \alpha)}{s} A_v f_y \sin \alpha = \frac{A_v f_y (\sin \alpha + \cos \alpha) d}{s} \tag{5.8.3}$$

Figure 5.8.1 Shear strength V_s attributable to shear reinforcement.

or, if $\alpha = 90°$,

$$V_s = \frac{A_v f_y d}{s} \tag{5.8.4}$$

The use of the 45° inclined crack assumption has been traditional since the early twentieth century. The variable angle truss using the general truss analogy [5.5–5.9] has been accepted by the Canadian Standard CAN3-A23.3-M84 [5.43], *Design of Concrete Structures for Buildings* since 1984, when it was adopted as the "General Method" and described in detail in Appendix D of that document. The "Simplified Method", essentially identical to the 1989 ACI Code provisions, is still permitted.

5.9 LOWER AND UPPER LIMITS FOR AMOUNT OF SHEAR REINFORCEMENT

It has been noted in Section 5.6 that the amount of shear reinforcement should be neither too low nor too high in order to ensure the yielding of steel when the failure strength in shear is reached. The ACI Code [ACI Formula (11-14)] requires a minimum shear reinforcement area A_v equal to

$$\min A_v = 50 \frac{b_w s}{f_y} \tag{5.9.1}*$$

in which b_w is the beam web width.

From Eq. (5.8.4) this minimum amount corresponds to

$$V_s = \frac{A_v f_y d}{s} = \frac{f_y d}{s}\left(50\frac{b_w s}{f_y}\right) = 50 b_w d \tag{5.9.2}*$$

or in terms of nominal unit stress on area $b_w d$,

$$v_s = \frac{V_s}{b_w d} = \frac{50 b_w d}{b_w d} = 50 \text{ psi} \tag{5.9.3}$$

To ensure that the amount of shear reinforcement is not too high, the designer will usually keep V_s below the following range,

$$V_s \le 6\sqrt{f'_c}\, b_w d \quad \text{to} \quad 8\sqrt{f'_c}\, b_w d$$

ACI-11.5.6.8 gives the upper limit for V_s as $8\sqrt{f'_c}\, b_w d$.

— — — —

* For SI, ACI 318–89M, for A_v in mm^2, b_w and s in mm, f_y in MPa, and V_s in meganewtons (MN), gives

$$\min A_v = \frac{b_u s}{3 f_y} \tag{5.9.1}$$

$$V_s = \left(\frac{1}{3}\text{ MPa}\right) b_w d \tag{5.9.2}$$

5.10 CRITICAL SECTION FOR NOMINAL SHEAR STRENGTH CALCULATION

In experimental work the *critical section* for computing the nominal shear strength was the location of the first inclined crack. Since most testing was made on simply supported beams under simple loading arrangements, it was difficult to extend such results to generalized loadings on continuous structures.

In order to plot the test points for the development of Eq. (5.5.6), two assumptions based on observations were used: (1) For shear span to depth ratio (a/d) greater than 2, the critical inclined crack is expected at d from the section of maximum moment: and (2) for shear span to depth ratio (a/d) less than 2, an inclined crack is expected at the center of the shear span.

Thus for gradually varying shear force (such as for uniform loading), ACI-11.1.3 *permits* taking the critical section at a distance d from the face of support, in recognition of the fact that the support reaction being in the direction of the applied shear introduces compression into the end region of the member. This compression in the end region would occur when the beam is gravity loaded and supported by columns or walls. *Shear reinforcement must be provided, however, between the face of support and the distance d therefrom,* using the same requirements as at the critical section.

The critical section must be taken at the face of support when one of the following occurs:

1. Factored shear V_u does gradually decrease from the face of support but the support is itself a beam or girder and therefore does *not* introduce compression into the end region of the member (see Fereig and Smith [5.44]).
2. When a concentrated load occurs between the face of support and the distance d therefrom.
3. When any loading may cause a potential inclined crack to occur *at* the face of support or *extend into* instead of away from the support (see Fig. 5.10.1).

Figure 5.10.1 Inclined crack when the reaction induces tension in the member. Critical crack is at the face of support.

5.11 ACI CODE PROVISIONS FOR SHEAR STRENGTH OF BEAMS

In the ACI strength design method for shear, it is required that

$$\phi V_n \geq V_u \qquad\qquad (5.11.1)$$

where V_u is the factored shear force and ϕV_n is the design strength in shear. The strength reduction factor ϕ is 0.85 for shear. The nominal shear strength V_n is

$$V_n = V_c + V_s \tag{5.11.2}$$

where V_c and V_s are the portions of the shear strength attributable to the concrete (see Section 5.5) and to the shear reinforcement (see Section 5.8), respectively.

Strength V_c Attributable to Concrete. The development of the detailed equation, Eq. (5.5.6), for V_c has been shown in Section 5.5 (see also Fig. 5.5.1). Since that equation is not easy to use as a design equation, and because of the wide scatter of test results, ACI-11.3.1 and 11.3.2 permit using either of the following:

   1. For the *simplified method*,

$$V_c = 2\sqrt{f'_c}\,b_w d \tag{5.11.3}*$$

From Fig. 5.5.1 this value appears to be conservative; however, recent studies [5.3, 5.45–5.47] have shown otherwise when ρ_w is below about 0.012. For values of ρ_w (that is, $A_s/b_w d$) lower than 0.012, the following is suggested [5.45]

$$V_c = (0.8 + 100\rho_w)\sqrt{f'_c}\,b_w d \tag{5.11.4}†$$

   2. For the *more detailed method*,

$$V_c = \left(1.9\sqrt{f'_c} + 2500\rho_w\,\frac{V_u d}{M_u}\right)b_w d \le 3.5\sqrt{f'_c}\,b_w d \tag{5.11.5}‡$$

Equation (5.11.5) is identical to Eq. (5.5.6). The value of $V_u d/M_u$ may not exceed 1.0 (ACI-11.3.2.1), and M_u is the factored moment occurring simultaneously with the V_u for which shear strength is being provided.

   ACI-ASCE Committee 426 has recommended [5.10] against further use of Eq. (5.11.5). The primary practical use of that equation is and has been to justify slightly larger stirrup spacings in high shear regions where spacings using Eq. (5.11.3) are small (say less than 3 in.).

——— ———

* For SI, ACI 318–89M, with f'_c in MPa, gives

$$V_c = \frac{1}{6}\sqrt{f'_c}\,b_w d \tag{5.11.3}$$

† For SI, an approximate conversion with f'_c in MPa is

$$V_c = (0.07 + 8.3\rho_w)\sqrt{f'_c}\,b_w d \tag{5.11.4}$$

‡ For SI, ACI 318–89M, with f'_c in MPa, gives

$$V_c = \frac{1}{6}\left(\sqrt{f'_c} + 100\,\frac{\rho_w V_u d}{M_u}\right)b_w d \le 0.3\sqrt{f'_c}\,b_w d \tag{5.11.5}$$

Strength V_s Attributable to Shear Reinforcement.   The contribution of shear reinforcement, as developed in Section 5.8, is (ACI-11.5.6)

$$V_s = \frac{A_v f_y d}{s} (\sin \alpha + \cos \alpha) \qquad (5.11.6)$$

and when vertical stirrups are used ($\alpha = 90°$),

$$V_s = \frac{A_v f_y d}{s} \qquad (5.11.7)$$

Design Categories and Requirements.   For design, the shear envelope for V_u is the starting point. From a practical point of view it is better to plot the V_u diagram rather than the "required V_n" diagram (equal to V_u/ϕ). The basic diagrams used should be those of M_u and V_u due to factored load; the ϕ factor is different for moment than for shear and should not be included in the load-related diagrams.

The design for shear may be separated into the following categories:

1. $$V_u \le 0.5\phi V_c \qquad (5.11.8)$$

In this category, no shear reinforcement is required (ACI-11.5.5.1).

2. $$0.5\phi V_c < V_u \le \phi V_c \qquad (5.11.9)$$

Minimum shear reinforcement is required except for thin slablike flexural members which experience has shown may perform satisfactorily without shear reinforcement. The thin slablike member exceptions include (a) slabs, (b) footings, (c) floor joist construction, and (d) beams where the total depth does not exceed 10 in., $2\frac{1}{2}$ times the flange thickness for T-shaped sections, or one-half of the web width. Though the largest of these three limits could indicate a very large beam, the authors believe that only wide shallow beams (giving emphasis to the 10-in. depth) should be excepted from minimum shear reinforcement when V_u exceeds $0.5\phi V_c$. The requirement of minimum shear reinforcement may be waived if tests are conducted which show that the required flexural and shear strengths can be developed.

For this category, the shear reinforcement must satisfy ACI-11.5.5.3 and 11.5.4.1, as follows:

$$\text{required } \phi V_s = \min \phi V_s = \phi(50) b_w d \qquad (5.11.10)^*$$

and

$$\text{maximum spacing } s \le \frac{d}{2} \le 24 \text{ in.} \qquad (5.11.11)$$

3. $$\phi V_c < V_u \le [\phi V_c + \min \phi V_s] \qquad (5.11.12)$$

— — — —

* For SI, ACI 318–89M, for b_w and d in mm and V_s in MN, gives

$$\min \phi V_s = \phi \left(\frac{1}{3} \text{ MPa} \right) b_w d \qquad (5.11.10)$$

For all flexural members, including those exempted from shear reinforcement in Category 2, shear reinforcement must be provided satisfying Eqs. (5.11.10) and (5.11.11).

4.
$$[\phi V_c + \min \phi V_s] < V_u \leq [\phi V_c + \phi(4\sqrt{f'_c})b_w d] \qquad (5.11.13)*$$

For this category, the computed shear reinforcement requirement will exceed the min ϕV_s requirement, and the shear reinforcement must satisfy ACI Formula (11-2), ACI-11.5.6, 11.5.4.1, and 11.5.4.3, as follows:

$$\text{required } \phi V_s = V_u - \phi V_c \qquad (5.11.14)$$

$$\text{provided } \phi V_s = \frac{\phi A_v f_y d}{s} \quad \text{(for } \alpha = 90°) \qquad (5.11.15)$$

$$\text{maximum } s = \frac{d}{2} \leq 24 \text{ in.} \qquad (5.11.16)$$

Note that in terms of nominal stress $v_s = V_s/b_w d = 4\sqrt{f'_c}$ psi is the maximum v_s for which the $d/2$ maximum spacing limit applies.

5.
$$[\phi V_c + \phi(4\sqrt{f'_c})b_w d] < V_u \leq [\phi V_c + \phi(8\sqrt{f'_c})b_w d] \qquad (5.11.17)†$$

The difference between categories 4 and 5 is that for all regions of a beam where the nominal stress v_s to be taken by shear reinforcement is between $4\sqrt{f'_c}$ and $8\sqrt{f'_c}$ the maximum shear reinforcement spacing s may not exceed $d/4$. The shear reinforcement provided in this category must satisfy ACI Formula (11-2), ACI-11.5.6, 11.5.4.3, 11.5.6.8, as follows:

$$\text{required } \phi V_S = V_u - \phi V_c \qquad (5.11.18)$$

$$\text{provided } \phi V_s = \frac{\phi A_v f_y d}{s} \quad \text{(for } \alpha = 90°) \qquad (5.11.19)$$

$$\text{maximum spacing } s \leq \frac{d}{4} \leq 12 \text{ in.} \qquad (5.11.20)$$

The factored shear V_u may not exceed the upper limit of Eq. (5.11.17) according to ACI-11.5.6.8.

The maximum factored shear that must be provided for on any beam is that occurring at the *critical section*, as defined in Section 5.10. The V_u requirement between the face of support and the critical section is to be taken as constant, equal to the value at the critical section.

The ACI Code provisions for shear strength of beams described in this section are summarized in Table 5.11.1.

––––

* For SI, ACI 318–89M gives in place of $4\sqrt{f'_c}$ psi, $\sqrt{f'_c}/3$ when f'_c is in MPa.

† For SI, ACI 318–89M gives in place of $4\sqrt{f'_c}$ and $8\sqrt{f'_c}$ psi, $\sqrt{f'_c}/3$ and $2\sqrt{f'_c}/3$, respectively, when f'_c is in MPa.

Table 5.11.1 Shear Strength of Members Under Bending Only—ACI Code

ITEM	STRENGTH DESIGN ($\phi = 0.85$)	CODE
1	$\phi V_n \geq V_u$	Formula (11-1), 11.1.1
	Maximum V_u at a distance d from face of support in usual situations (three exceptions)	11.1.2.1
2	$V_n = V_c + V_s$	Formula (11-2), 11.1.1
3	$\sqrt{f_c'} \leq 100$ psi	11.1.2
	unless $A_v \geq \dfrac{f_c'}{5000}\left(50\,\dfrac{b_w s}{f_y}\right)$	
	but need not exceed $150\,\dfrac{b_w s}{f_y}$	
4	Simplified method:	
	$V_c = 2\sqrt{f_c'}\,b_w d$	Formula (11-3), 11.3.1.1
	$V_c = (0.8 + 100\rho_w)\sqrt{f_c'}\,b_w d$ for $\rho_w < 0.012$	Refs. 5.45
	More detailed method:	
	$V_c = \left(1.9\sqrt{f_c'} + 2500\rho_w \dfrac{V_u d}{M_u}\right)b_w d \leq 3.5\sqrt{f_c'}\,b_w d$	Formula (11-6), 11.3.2.1
	$V_u d/M_u$ not to exceed unity	
	Allow 10% increase for joists	8.11.8
	Lightweight concrete when f_{ct} is specified: Use smaller of $f_{ct}/6.7$ or $\sqrt{f_c'}$ for $\sqrt{f_c'}$	11.2.1.1
	Lightweight concrete when f_{ct} is not specified: Use $0.75\sqrt{f_c'}$ to $0.85\sqrt{f_c'}$ in cases of all lightweight to sand lightweight concrete	11.2.1.2
5	$V_s = \dfrac{A_v f_y d}{s}(\sin\alpha + \cos\alpha)$	Formula (11-18), 11.5.6.3
	$V_s = A_v f_y \sin\alpha \leq 3\sqrt{f_c'}\,b_w d$ (single bar)	Formula (11-19), 11.5.6.4
	f_y not to exceed 60,000 psi	11.5.2
	V_s not to exceed $8\sqrt{f_c'}\,b_w d$	11.5.6.8

Table 5.11.1 (Continued)

ITEM	STRENGTH DESIGN ($\phi = 0.85$)	CODE
6	For $0.5\phi V_c < V_u \le \phi V_c$ Use $s = \dfrac{A_v f_y}{50 b_w}$, max $s = \dfrac{d}{2} \le 24$ in. except for slabs, footings, joists, and small beams shallower than 10 in., $2\frac{1}{2}$ times flange thickness, or $b_w/2$; for these cases, no shear reinforcement required unless $V_u > \phi V_c$.	Formula (11-14), 11.5.5.3 11.5.5.1, 11.5.4.1
7	For $\phi V_c < V_u \le [\phi V_c + \min \phi V_s]$ $\min \phi V_s = \phi 50 b_w d$ Use $s = \dfrac{A_v f_y}{50 b_w}$, max $s = \dfrac{d}{2} \le 24$ in.	11.5.5.3 11.5.4.1
8	For $[\phi V_c + \min \phi V_s] < V_u \le [\phi V_c + \phi(4\sqrt{f'_c}\, b_w d)]$ $\min \phi V_s = \phi 50 b_w d$ Design shear reinforcement max $s = \dfrac{d}{2} \le 24$ in.	11.5.5.3 11.5.4.1
9	For $[\phi V_c + \phi(4\sqrt{f'_c}\, b_w d)] < V_u \le [\phi V_c + \phi(8\sqrt{f'_c}\, b_w d)]$ Design shear reinforcement max $s = \dfrac{d}{4} \le 12$ in.	11.5.6.8 11.5.4.3

CANNOT DESIGN FOR STEEL CAPACITY $> 5\phi V_c$

5.12 WORKING STRESS METHOD—ACI CODE, APPENDIX A

In strength design, the preference is to use shear force rather than shear stress, because stresses are less definable when the loading is above the service load range. It may still be noted, however, that the difference between computing shear force and shear stress (however nominal the stress may be) is only in the multiplier $b_w d$, which is the effective area of the web cross section.

The nominal shear stress in the "alternate design method," as prescribed in Appendix A.7 of the ACI Code is to be based on the service load shears without applying the overload factors U or the strength reduction factors ϕ; thus

$$v = \frac{V_D + V_L}{b_w d} \qquad (5.12.1)$$

The allowable concrete stresses and the limiting maximum stresses for shear are taken as approximately 55% for beams, joists, walls, and one-way slabs, and 50% for two-way slabs and footings, respectively, of the stresses used for strength design.

Thus for beams the allowable stresses for concrete, using approximately 55% of Eqs. (5.11.3) and (5.11.5) are (ACI-Appendix A.3.1)

$$v_c = 1.1\sqrt{f_c'} \qquad (5.12.2)$$

or (ACI-Appendix A.7.4.4)

$$v_c = \sqrt{f_c'} + 1300\rho_w \frac{Vd}{M} \le 1.9\sqrt{f_c'} \qquad (5.12.3)$$

Note that reducing the allowable stresses to 55% is equivalent to using a U/ϕ value of $1/0.55 = 1.82$. Assuming equal dead and live loads gives an average overload factor of $\frac{1}{2}(1.4 + 1.7) = 1.55$, which with $\phi = 0.85$ results in a U/ϕ value of 1.82.

The procedure for the working stress method is identical to that for the strength method. The minimum area of shear reinforcement is the same as for the strength method, as given by Eq. (5.9.1). The limiting maximum stresses, as discussed in Section 5.9 and as related to maximum web reinforcement spacing (ACI-Appendix A.7.5.4), are also reduced to 55 or 50% of their corresponding values in the strength method.

Since the procedures are the same as for the strength method, the use of the alternate method for designing shear reinforcement offers little advantage, except the results could be more or less conservative depending on whether the dead load is larger or smaller than the live load.

5.13 SHEAR STRENGTH OF BEAMS—DESIGN EXAMPLES

Four examples are presented to illustrate the basic procedure of designing vertical stirrups. In the first example, the complete design procedure for flexure and shear is shown for a simply supported beam using the simplified procedure. The second example illustrates the calculations for using the more detailed method of obtaining V_c. Some of the more practical aspects are discussed in the third example. Use of metric units is shown for the fourth example. Design of shear reinforcement in the continuous spans of a slab-beam-girder floor system is treated in Chapter 10.

■ **EXAMPLE 5.13.1** For the given beam shown in Fig. 5.13.1(a), first determine the maximum uniform dead and live loads under service condition permitted by the ACI strength method. Then using those maximum service loads design the shear reinforcement using vertical stirrups and the strength method with the simplified proce-

Figure 5.13.1 Beam for Example 5.13.1.

dure using constant V_c. Assume that service live load to dead load ratio is 1.5, $f'_c = 4000$ psi, and $f_y = 60,000$ psi.

Solution: (a) Check whether tension steel exceeds the maximum amount permitted [Fig. 5.13.1(b)].

$$x_b = \frac{0.003(22.5)}{0.003 + 0.00207} = 13.31 \text{ in.}$$

$$\text{max } x = 0.75x_b = 0.75(13.31) = 9.98 \text{ in.}$$

For $x = 9.98$ in.,

$$\epsilon'_s = \frac{9.98 - 2.5}{9.98}(0.003) = 0.00225 > \epsilon_y; \qquad f'_s = f_y$$

$$\text{max } C_c = 0.85f'_c b\beta_1(\text{max } x)$$
$$= 0.85(4)(14)(0.85)(9.98) = 404 \text{ kips}$$

$$C_s = A'_s(f_y - 0.85f'_c) = 2.40(60 - 3.4) = 136 \text{ kips}$$

$$\text{max } T = \text{max } C = 404 + 136 = 540 \text{ kips}$$

$$\text{max } A_s = \frac{\text{max } T}{f_y} = \frac{540}{60} = 9.00 \text{ sq in.} > (8\text{-}\#9 = 8.00) \qquad \textbf{OK}$$

(b) Find the nominal flexural strength M_n [Fig. 5.13.1(c)] and allowable service load. Assuming compression steel yields (this is not certain even though compression steel does yield in the max x condition),

$$0.85f'_c b\beta_1 x + A'_s(f_y - 0.85f'_c) = A_s f_y$$

$$0.85(4)(14)(0.85x) + 2.40(60 - 3.4) = 8.00(60)$$

$$x = 8.51 \text{ in.}$$

$$\epsilon'_s = 0.003 \frac{8.51 - 2.5}{8.51} = 0.00212 > \epsilon_y$$

Assumption is confirmed; compression steel yields ($f'_s = f_y$) in actual condition.

$$C_c = 0.85f'_c b\beta_1 x = 0.85(4)(14)(0.85)(8.51) = 344 \text{ kips}$$

$$C_s = A'_s(f_y - 0.85f'_c) = 2.40(60 - 3.4) = 136 \text{ kips}$$

$$T = A_s f_y = 8.00(60) = 480 \text{ kips}$$

$$d - \frac{a}{2} = 22.5 - \tfrac{1}{2}(0.85)(8.51) = 18.88 \text{ in.}$$

$$M_n = 344(18.88)\tfrac{1}{12} + 136(20)\tfrac{1}{12} = 541 + 227 = 768 \text{ ft-kips}$$

$$M_u = \tfrac{1}{8}(w_u)(24)^2 = \phi M_n = 0.90(768) = 691 \text{ ft-kips}$$

$$w_u = 9.60 \text{ kips/ft}$$

$$w_L = 1.5w_D$$

$$w_u = 1.4w_D + 1.7(1.5w_D)$$

$$\text{service dead load } w_D = \frac{9.60}{1.4 + 2.55} = 2.43 \text{ kips/ft}$$

$$\text{service live load } w_L = 3.64 \text{ kips/ft}$$

(c) Design of shear reinforcement, using the simplified method with a constant value for V_c. The factored shear to be designed for must be the maximum that may possibly act at each point along the span; that is, an envelope of maximum shear is needed. A knowledge of influence lines tells the designer that for bending moment on a simply supported span the maximum value occurs at every point along the span when the full span is loaded with live load. However, for shear, the maximum shear occurs with partial span loading for every point along the span except at the supports. Unless the designer can justify other treatment, the live load should always be treated as variable position loading acting wherever it may cause the greatest effect, whereas the dead load would be fixed position loading. For most ordinary situations an approximate shear envelope may be acceptable, using a straightline relationship between the maximum shear at the support and the maximum shear at midspan. Such a procedure will always be conservative.

Figure 5.13.2 The factored shear V_u diagram for Example 5.13.1.

For this beam (see Fig. 5.13.2), the maximum factored shear at the centerline of support is

$$V_u = \frac{w_u L}{2} = \frac{[2.43(1.4) + 3.64(1.7)]24}{2} = 115.1 \text{ kips}$$

The maximum factored shear possible at midspan occurs with live load on half the span,

$$V_u = \frac{w_L L}{8} = \frac{3.64(1.7)(24)}{8} = 18.6 \text{ kips}$$

The dead load shear (with full span loaded) is zero at midspan.

The critical section for determining the closest stirrup spacing may be taken, according to ACI-11.1.2.1, at a distance d from *face* of support; in this case, $d = 22.5$ in. and the support width is 12 in., making the critical section $(22.5 + 6 = 28.5 \text{ in.})$ 2.38 ft from the center of the support. By linear interpolation, V_u at d from the face is

$$V_u = 115.1 - \left(\frac{115.1 - 18.6}{12}\right)2.38 = 96.0 \text{ kips}$$

The design requirement between the face of support and the critical section is considered to be constant (in this case, 96.0 kips).

The factored shear V_u diagram for the left half of this symmetrical structure is shown enlarged in Fig. 5.13.3. The design may be done primarily *on* the factored shear V_u diagram (Fig. 5.13.3).

The design shear strength attributable to the concrete is

$$\phi V_c = \phi(2\sqrt{f_c'} b_w d)$$
$$= 0.85(2\sqrt{4000})(14)(22.5)\tfrac{1}{1000} = 34 \text{ kips}$$

Figure 5.13.3 Design of shear reinforcement in Example 5.13.1 using constant ϕV_c.

The difference (the portion crosshatched in Fig. 5.13.3) between V_u and ϕV_c must be provided for by shear reinforcement. For most beams (ACI-11.5.5.1) shear reinforcement may *not* be terminated when V_u equals ϕV_c, but rather must be continued until $V_u = 0.5\phi V_c$ (Fig. 5.13.3). The distance over which shear reinforcement is required *from the face of support* is in this case the entire span, because $0.5\phi V_c$ never exceeds V_u.

Examination of Fig. 5.13.3 shows that V_u exceeds ϕV_c over most of the span, putting most of the design in categories 3 through 5 of Section 5.11. At the critical section,

$$\text{required } \phi V_s = V_u - \phi V_c = 96.0 - 34.0 = 62.0 \text{ kips}$$

Check limits on ϕV_s,

$$\min \phi V_s \text{ (ACI-11.5.5.3)} = 0.85(50)(14)(22.5)\tfrac{1}{1000} = 13.4 \text{ kips}$$

This value of ϕV_s corresponds to the location where the factored shear V_u would equal 47.4 kips (that is, $\min \phi V_s + \phi V_c$), in this case 95 in. from face of support (see Table 5.13.1).

Table 5.13.1 Spacing and Strength Relationships for Vertical Stirrups in Example 5.13.1

s	$\phi V_s = \dfrac{\phi A_v f_y d}{s(1000)} = \dfrac{252}{s}$	$z =$ DISTANCE FROM FACE OF SUPPORT TO INTERSECTION OF ϕV_n (i.e., $\phi V_c + \phi V_s$) WITH V_u $= 22.5 + \dfrac{62.0 - \phi V_s}{96.0 - 18.6}(144 - 28.5)$
4.1 in.	62.0 kips (max)	0 to 22.5 in.
5	50.4	$22.5 + \dfrac{115.5}{77.4}(11.6) = 22.5 + 17.3 = 40$ in.
6	42.0	$(20.0) = 22.5 + 29.8 = 52$
8	31.5	$(30.5) = 22.5 + 45.5 = 68$
10	25.2	$(36.8) = 22.5 + 54.9 = 77$
11.25 max	22.4	$(39.6) = 22.5 + 59.1 = 82$
18.9 (not usable)	13.4 min	$(48.6) = 22.5 + 72.5 = 95$

$$\text{max } \phi V_s \left(\text{for } s = \frac{d}{2}\right) = 0.85(4\sqrt{4000})(14)(22.5)\tfrac{1}{1000} = 67.7 \text{ kips}$$

(ACI-11.5.4.1 and 11.5.4.3)

$$\text{max } \phi V_s \left(\text{for } s = \frac{d}{4}\right) = 0.85(8\sqrt{4000})(14)(22.5)\tfrac{1}{1000} = 135.5 \text{ kips}$$

(ACI-11.5.6.8 and 11.5.4.3)

Since the required ϕV_s at the critical section (which is the maximum for the beam) lies between 13.4 and 67.7 kips, design for the portion of the beam between the face of support and the critical section is in category 4 (max $s = d/2$). The strength requirement using ϕV_s computed from V_u will control. Using #3 vertical U stirrups, in which two stirrup bar areas comprise A_v, in Eq. (5.11.15),

$$s \text{ (at critical section)} = \frac{\phi A_v f_y d}{\phi V_s} = \frac{(0.85)(2)(0.11)(60)22.5}{62.0} = 4.1 \text{ in.}$$

This means that #3 stirrups may not be spaced farther apart than 4.1 in. for the region from the face of support to the critical section at a distance d from the face of support.

The determination of max s at the critical section usually controls what size stirrups are to be used. If the spacing for #3 stirrups were required to be too small, say less than about 3 in., the stirrup bar size would be increased to #4, or occasionally #5. In this case, a spacing of 4 in. is acceptable.

For the region where minimum stirrups are required, the spacing permitted will be the lesser of $d/2$ (11.25 in. for this beam) and

$$s = \frac{A_v f_y}{50 b_w} = \frac{0.22(60,000)}{50(14)} = 18.9 \text{ in.}$$

The 11.25-in. restriction controls.

As the V_u requirement decreases toward midspan, the stirrup spacing can be increased. A table of potentially acceptable spacings (s vs ϕV_s) should be made, as in the first two columns of Table 5.13.1. Spacings are chosen until the spacing reaches $d/2$ (in this example, 11.25 in.) or until ϕV_s gets down to min ϕV_s (in this example, 13.4 kips). In the table, the $d/2$ spacing limit, not the 18.9-in. limit, happens to control out to $z = 95$ in. In fact, between $z = 95$ in. and midspan, $d/2$ still controls in this example because V_u at midspan is larger than $0.5\ \phi V_c$.

To determine the set of spacings to actually be used, the designer may prefer to draw on the V_u diagram the lines representing the ϕV_n provided by different spacings, as shown on Fig. 5.13.3, and then scale from the diagram the locations z where the various spacings are permissible (marked along the base line of Fig. 5.13.3); or, alternatively, the locations z may be computed as in the third column of Table 5.13.1. Note that in this column, the quantity $(144 - 28.5)/(96.0 - 18.6)$ is the distance in which the value of V_u drops 1 kip. Whether location z is computed or scaled, a table of ϕV_s versus s should be used. The limiting spacings of 4.1 and 11.25 have already been explained. The intermediate spacings of 5, 6, 8, and 10 in. are those chosen by the designer as practical possibilities.

Since spacings of stirrups cannot be varied continuously, they must change by "jumps." One conservative policy is to use a spacing, say of 6 in., only beyond the theoretical point at which a 6-in. spacing may be used (in this case, 52 in. or more from the face of support). In a less conservative manner, the next larger spacing may be used somewhat before the point at which this spacing may be used. With this in mind, the set of spacings to be used is

$$3 \text{ in.} \mid 9 @ 4 \text{ in.} \mid 3 @ 5 \text{ in.} \mid 4 @ 6 \text{ in.} \mid 6 @ 10 \text{ in.} \mid = 138 \text{ in.}$$

$$\qquad\qquad\qquad\qquad\qquad\qquad\qquad\qquad\qquad\qquad\qquad\qquad\text{required}$$

$$\quad 3 \qquad\quad 39 \qquad\quad 54 \qquad\quad 78 \qquad\quad 138 \;\; \text{(midspan)}$$

The first stirrup is placed at 3 in. from the support, more than the usual half-space in order to make uniform spacing at midspan. Many designers prefer to place the first stirrup a full space from the face of support for closely spaced stirrups. Note that in the adopted set of spacings, 5-in. spacing is used after $z = 39$ in. ($z = 40$ in. theoretically required); 6-in. spacing is used after $z = 54$ in. ($z = 52$ in. theoretically required); and 10-in. spacing is used after $z = 78$ in. ($z = 77$ in. theoretically required). ◼

■ **EXAMPLE 5.13.2**   For the beam designed in Example 5.13.1, investigate whether or not the designer could justify increasing the 4-in. spacing used near the support to 5 in. by applying the more detailed procedure for V_c involving $\rho_w V_u d / M_u$ (ACI-11.3.2.1).

Solution: (a) Materials used and factored loads. From Example 5.13.1, the factored shear V_u envelope is given in Fig. 5.13.2. Also $f_c' = 4000$ psi and $f_y = 60{,}000$ psi.

(b) Compute ϕV_c. In general, this is not constant along the span because $\rho_w V_u d / M_u$ varies along the span. Since only the region near the support is of interest because of the problem statement, the value at the critical section is computed.

At the critical section,

$$V_u = 96.0 \text{ kips} \quad (\text{Fig. 5.13.3})$$

$$V_c = \left(1.9\sqrt{f_c'} + 2500\rho_w \frac{V_u d}{M_u}\right)b_w d \le 3.5\sqrt{f_c'}b_w d$$

$$\rho_w = \frac{A_s}{b_w d} = \frac{8(1.0)}{14(22.5)} = 0.0254$$

$$M_u = \frac{w_u}{2}(2.38)(24 - 2.38) = \frac{9.60}{2}(2.38)(21.62) = 247 \text{ ft-kips}$$

$$\frac{V_u d}{M_u} = \frac{96.0(22.5)}{247(12)} = 0.729 < 1.0 \text{ max} \qquad \textbf{OK}$$

$$V_c = [1.9\sqrt{4000} + 2500(0.0254)(0.729)](14)(22.5)\tfrac{1}{1000}$$
$$= 52.4 \text{ kips} < (3.5\sqrt{f_c'}b_w d = 69.7 \text{ kips}) \qquad \textbf{OK}$$

$$\phi V_c = 0.85(52.4) = 44.4 \text{ kips}$$

(c) Stirrup spacing for #3 U stirrups. The maximum requirement for ϕV_s is at the critical section,

$$\text{max required } \phi V_s = V_u - \phi V_c = 96.0 - 44.4 = 51.6 \text{ kips}$$

The stirrup spacing based on the strength requirement (ACI-11.5.6.2) at the critical section is

$$\text{max } s \text{ (at critical section)} = \frac{\phi A_v f_y d}{\phi V_s} = \frac{0.85(2)(0.11)(60)22.5}{51.6} = 4.9 \text{ in.}$$

Because the 4.9 in. above is close to the desired 5 in., a reduced value of V_u at the critical section may be obtained by using the shear envelope computed from partial span loading of the live load, rather than the linear approximation in Fig. 5.13.3. This value is found to be 93.0 kips, with a corresponding M_u of 230.5 ft-kips. Then, ϕV_c becomes 48.0 kips, and max s (at critical section) = 5.3 in. ∎

■ **EXAMPLE 5.13.3** Design the locations of #3 vertical U stirrups to be used in the beam of Fig. 5.13.4. The beam is to carry service dead and live loads of 5.2 and 6.0 kips/ft, respectively. Use $f_c' = 3500$ psi and $f_y = 60,000$ psi. Use the alternate approach of plotting v_n (v_n is the nominal shear stress which is the required nominal shear strength V_u/ϕ divided by $b_w d$) diagram and then providing the v_c attributable to the concrete and v_s attributable to shear reinforcement.

Solution: **(a)** The factored shear force V_u and nominal shear stress v_n diagrams (envelopes of maximum values for different loading conditions) for design are as shown in Fig. 5.13.4. The effect of partial span live load on shear is approximated

(a) The beam

12'−0" center to center of supports

Symmetrical about $\mathbb{C}$ of beam

$d = 21.5"$

4 − # 9

3 U stirrups

2" 5 @ 5" 2 @ 6" 3 @ 9" 12"

(b) Shear force diagram

V_u shear force, kips

105

Critical section
Straightline approximation
True maximum shear envelope for uniform live load

$d = 21.5"$

15.3

12" Face of support

(c) Shear stress v_n diagram

$\dfrac{V_u}{\phi b_w d}$ shear stress, psi

478

$d = 21.5"$

Shaded portion indicates stress to be carried by shear reinforcement

323 $s = 6"$

$s = 8"$

$s = 10"$

Required v_s

$v_c/2$

69.8

$v_c = 118$ psi carried by concrete

(d) Stirrup requirement diagram

s stirrup spacing, inches

12

8

4

2"

6" 8" 10"

Max. s $(d/2) = 10.75"$

Face of support

Permitted

Provided

5 @ 5 = 2'−1" 2 @ 6" 3 @ 9"

3 stirrups

Figure 5.13.4 Stirrup placement for Example 5.13.3

by passing a straight line through the maximum shear values at the support and at midspan.

$$1.4w_D = 1.4(5.2) = 7.28 \text{ kips/ft}$$

$$1.7w_L = 1.7(6.0) = 10.20 \text{ kips/ft}$$

At the support,

$$V_u = \tfrac{1}{2}(7.28 + 10.20)(12) = 105 \text{ kips}$$

$$v_n = \frac{V_u}{\phi bd} = \frac{105,000}{0.85(12)21.5} = 478 \text{ psi}$$

At midspan,

$$V_u = \tfrac{1}{8}(10.20)(12) = 15.3 \text{ kips}$$

$$v_n = \frac{V_u}{\phi bd} = \frac{15,300}{0.85(12)(21.5)} = 69.8 \text{ psi}$$

Note that the shear at midspan due to dead load is zero; but positive live load shear at midspan is largest when only the right half of the span is loaded.

(b) Simplified method.

$$v_c = 2\sqrt{f'_c} = 2\sqrt{3500} = 118 \text{ psi}$$

The nominal shear stress at the distance d from the face of the support is

$$v_n \text{ (at } d) = 478 - \frac{(6 + 21.5)}{72}(478 - 69.8) = 323 \text{ psi}$$

$$\text{required } v_s = v_n - v_c = 323 - 118 = 205 \text{ psi}$$

$$4\sqrt{f'_c} = 237 \text{ psi} \quad (\text{ACI-11.5.4.3})$$

Where $v_s > 4\sqrt{f'_c}$, the maximum spacing of vertical stirrups may not exceed $d/4$, otherwise $d/2$. In this case, the spacing limitation $d/2$ applies wherever stirrups are needed. Note that this simple check on whether or not required v_s exceeds $4\sqrt{f'_c}$ is often advantageous over using the criterion shown in item 9 of Table 5.11.1.

Try #3 vertical U stirrups.

$$A_v f_y = 2(0.11)(60,000) = v_s b_w s$$

$$v_s s = \frac{13,200}{12} = 1100$$

which for the v_s of 205 psi at the critical section permits

$$\text{max } s \text{ (at critical section)} = \frac{1100}{205} = 5.4 \text{ in.}$$

The maximum spacing for shear reinforcement to provide for a minimum v_s of 50 psi is

$$\text{max } s \text{ (for min } A_v) = \frac{1100}{50} = 22 \text{ in.} > \left[\frac{d}{2} = 10.75 \text{ in.}\right]$$

Table 5.13.2   Spacing of Vertical Stirrups (Example 5.13.3)

s	$v_s = \dfrac{1100}{s}$	$v_n = v_c + v_s$
5.4	205	323
6	183	301
8	138	256
10	110	228
max 10.75	102	220
(22)	(50)	(168)

Thus the spacing cannot be more than 10.75 in. even when v_s gets small. If the above computation had given a value less than $d/2$ instead of 22 in., then that lesser value would be the maximum s for the region wherever stirrups *are* required ($v_n > v_c/2$).

The placement of stirrups will be done by scaling from the diagram of Fig. 5.13.4(c). To facilitate this, Table 5.13.2 is computed using spacings considered desirable by the designer. The shear stress v_s values are plotted as horizontal lines marked $s = 6$, 8, and 10 in. on Fig. 5.13.4(c); then their intersections with the maximum shear stress v_n line are projected downward to the base line. Stirrups are then laid out with a scale beginning a distance $s/2$ from the face of support. (Some designers place the first stirrup a full space from the face of support for small spacings.) The same spacing necessary at the critical section is specified by ACI-11.1.2.1 to be used between the face of support and the critical section. Thus one may start at $s/2$ from support with a 5-in. spacing until within $s/2$ of the capacity line for the next desired spacing, which in this case is slightly beyond the vertical line projected from $s = 6$ in. The ACI Code requires shear reinforcement until the stress $v_n \leq v_c/2$. In this case, the entire beam must be provided with stirrups, because the smallest v_n (69.8 psi) is larger than $v_c/2$ (59 psi).

To illustrate clearly what has been done, a diagram showing permitted (actual spacings must be below permitted line) and provided stirrup spacings is given in Fig. 5.13.4(d). In this case, some shifting of spacing has been made to avoid leaving a small fragmental space at midspan. One more space at 5 in. was used than necessary and the three spaces adjacent to midspan were reduced to 9 in. from the permitted 10 in. in order to eliminate a small space near midspan.

The final design details are shown in Fig. 5.13.4(a). ■

■ **EXAMPLE 5.13.4**   Determine the vertical stirrup requirement for the beam of Fig. 5.13.5. Use #10M bars (see Table 1.12.2) for U stirrups, $f'_c = 25$ MPa, and $f_y = 300$ MPa. The service live load is 30 kN/m and the service dead load is 43 kN/m (including beam weight).

Solution: (a) Determine maximum factored shear V_u envelope.

$$w_u = 1.4(43) + 1.7(30) = 111.2 \text{ kN/m}$$

Figure 5.13.5 Beam and stirrup design for Example 5.13.4.

For maximum shear at d (480 mm) from face of support, place live load on remainder ($3.2 - 0.15 - 0.48 = 2.57$ m) of the span.

$$V_u \text{ at critical section} = 1.4(43)(1.6 - 0.15 - 0.48) + 1.7(30)\frac{(2.57)^2}{2(3.2)}$$

$$= 58.4 + 52.6 = 111 \text{ kN}$$

$$V_u \text{ at midspan} = \tfrac{1}{8}(51.0)(3.2) = 20.4 \text{ kN}$$

The straightline approximation maximum shear envelope is given as Fig. 5.13.5(c). (Note that, in the previous example, the straightline approximation was made between centerline of support and midspan. The present approximation is closer to the "exact" maximum shear curve.)

(b) Determine stirrup spacing. If the factored shear V_u diagram is used,

$$v_c = \sqrt{f'_c}/6 \quad \text{(from ACI 318–89M, with } f'_c \text{ in MPa)}$$
$$= \sqrt{25}/6 = 0.833 \text{ MPa}$$

$$\phi V_c = \phi v_c b_w d = 0.85(0.833)(280)(480)\tfrac{1}{1000} = 95.2 \text{ kN}$$

At the critical section,

$$\text{required } \phi V_s = V_u - \phi V_c = 111 - 95.2 = 15.8 \text{ kN}$$

The limiting $\phi V_s = \phi(\sqrt{f'_c}/12)b_w d$ for $d/2$ stirrup spacing limitation can be conveniently obtained from $2\phi V_c$; thus

$$\text{limiting } \phi V_s = 2\phi V_c = 2(95.2) = 190 \text{ kN} > [\text{required } \phi V_s = 15.8 \text{ kN}]$$

Thus maximum spacing cannot exceed $d/2$.

For #10M U stirrups, $A_v = 2(100) = 200 \text{ mm}^2$. The spacing requirement for strength is

$$s = \frac{\phi A_v f_y d}{\phi V_s} = \frac{(0.85)(200)(0.300)480}{\phi V_s} = \frac{24{,}480}{\phi V_s \text{ (in kN)}}$$

The stirrup spacing requirements can be summarized as follows:

1. Maximum spacing for strength requirement at the critical section,

$$\text{max } s = \frac{24{,}480}{15.8} = 1550 \text{ mm}$$

2. Maximum spacing $d/2$,

$$\text{max } s = \frac{d}{2} = \frac{480}{2} = 240 \text{ mm} \qquad \text{(Controls)}$$

3. Maximum spacing for minimum shear reinforcement (ACI-11.5.5.3),

$$\text{min } \phi V_s = \phi(\tfrac{1}{3} \text{ MPa})b_w d \quad \text{(ACI 318–89M)}$$
$$= 0.85(0.333)(280)(480)\tfrac{1}{1000} = 38.0 \text{ kN}$$

$$\text{max } s = \frac{24{,}480}{38.0} = 644 \text{ mm} > \frac{d}{2}$$

4. Conclusion. Use 240-mm spacing for the portion where stirrups are required (i.e., from face of support to location where $V_u = 0.5\phi V_c$).

The final stirrup arrangement is shown in Fig. 5.13.5(c). ■

5.14 SHEAR STRENGTH OF MEMBERS UNDER COMBINED BENDING AND AXIAL LOAD

The presence of an axial compressive load on a reinforced concrete flexural member decreases the longitudinal tensile stress and the resulting tendency for inclined cracking. Conversely, the addition of an axial tensile load increases the longitudinal tensile stress and the tendency for inclined cracking. Thus, for the same bending moment, the shear strength of a member is increased by the addition of an axial compressive load and decreased by an axial tensile load. Some experimental work is available on shear strength in the presence of axial load [5.3, 5.48–5.52].

Axial Compression. Since inclined cracking is dependent on the combination of tensile or compressive stress due to flexure and shear stress as discussed in Sections 5.3 through 5.5, the addition of axial compression tends to delay the opening of the shear crack and prevent its extending as far into the beam.

When the simplified method is used with axial compression, a linear increase with axial compression is given by ACI-11.3.1.2:

$$V_c = 2\left(1 + \frac{N_u}{2000A_g}\right)\sqrt{f'_c}\, b_w d \qquad \textbf{(5.14.1)*}$$

which is ACI Formula (11-4). Note that N_u is the factored axial compressive load and A_g is the gross area of the concrete cross-section.

The more detailed equation for V_c is also permitted by ACI-11.3.2 for use with flexure and axial compression,

$$V_c = \left(1.9\sqrt{f'_c} + 2500\rho_w\frac{V_u d}{M_u}\right)b_w d \leq 3.5\sqrt{f'_c}\, b_w d \qquad \textbf{(5.14.2)}$$

which is ACI Formula (11-6). In the use of this equation for axial compression, M_m replaces M_u, where

$$M_m = M_u - N_u\left(\frac{4h - d}{8}\right) \qquad \textbf{(5.14.3)}$$

which is ACI Formula (11-7).

The rationale for M_m may be seen by referring to Fig. 5.14.1, which shows a free body of a short length of beam dz. Moment equilibrium about point A, the line of action of the internal compressive force C, gives

$$T\left(d - \frac{a}{2}\right) = M_u - N_u\left(\frac{h}{2} - \frac{a}{2}\right) \qquad \textbf{(5.14.4)}$$

When the moment arm $(d - a/2)$ is approximated as $7d/8$, thereby making $a/2 = d/8$, the right side of Eq. (5.14.4) becomes identical to Eq. (5.14.3), and the left side represents an *equivalent moment*, now called M_m. Note that N_u is positive for compression and that M_m must never be used as a negative value. Also, $V_u d/M_m$ is permitted to have values greater than unity. This equivalent M procedure is to be used in the more detailed method for combined axial *compression* and bending only. Experience [5.48] has shown the method to be unsafe for axial *tension*.

— — — —

* For SI, ACI 318–89M, for N_u/A_g and f'_c in MPa, gives

$$v_c = \left(1 + \frac{N_u}{14A_g}\right)\left(\frac{\sqrt{f'_c}}{6}\right)b_w d \qquad \textbf{(5.14.1)}$$

Figure 5.14.1 Member under combined axial compression and bending moment.

The upper limit for the nominal shear strength V_c of members without shear reinforcement subject to bending only is given in Eq. (5.14.2) as $3.5\sqrt{f_c'}\,b_w d$. This upper limit should be adjusted upward in the presence of axial compression. As explained below, this adjustment factor may be rationally put into the form

$$\text{adjustment factor} = \sqrt{1 + \frac{N_u}{500 A_g}} \qquad (5.14.5)$$

The formula for the principal tensile stress $f_t\,(\text{max})$ in terms of the tensile stress f_t and the shear stress v has been derived in Section 5.3 to be

$$f_t\,(\text{max}) = \tfrac{1}{2}f_t + \sqrt{(\tfrac{1}{2}f_t)^2 + v^2}$$

Solving this equation for v,

$$v = f_t\,(\text{max})\sqrt{1 - \frac{f_t}{f_t\,(\text{max})}} \qquad (5.14.6)$$

It may be seen from Eq. (5.14.6) that the shear strength of a member under bending only becomes a constant if f_t is zero, that is, if the bending moment approaches zero. Empirically this constant is the upper limit $3.5\sqrt{f_c'}$. For a member under an axial compressive load N_u without bending moment, f_t is a constant that is equal to $-N_u/A_g$. Substituting this value of f_t in Eq. (5.14.6) and using an average value of 500 psi for $f_t\,(\text{max})$, the upper limit of strength V_c in members under combined bending and axial compression becomes

$$\cdot \quad V_c\,(\text{upper limit}) = v b_w d = 3.5\sqrt{f_c'}\sqrt{1 + \frac{N_u}{500 A_g}}\,b_w d \qquad (5.14.7)^*$$

which is ACI Formula (11-8).

— — — —

* For SI, ACI 318–89M, for N_u/A_g and f_c' in MPa, gives

$$V_c\,(\text{upper limit}) = 0.3\sqrt{f_c'}\sqrt{1 + \frac{0.3 N_u}{A_g}}\,b_w d \qquad (5.14.7)$$

Figure 5.14.2 Effect of axial load on inclined cracking shear stress (dots indicate test results) (adapted from Reference 5.48).

Axial Tension. When a flexural member is subject to axial tension, ACI in the simplified method states in ACI-11.3.1.3 that V_c is to be zero. In the more detailed method, a simple linear reduction for V_c has been specified in ACI-11.3.2.3. Thus

$$V_c = 2\left(1 + \frac{N_u}{500A_g}\right)\sqrt{f'_c}\, b_w d \qquad (5.14.8)*$$

which is ACI Formula (11-9). Note that N_u is negative for tension.

In order to show the relationship between ACI formulas and experimental results, Fig. 5.14.2 is presented. The crosshatched portion on the compression side represents the reasonable range when using the $\rho V d/M$ procedure.

Parts of the ACI Code are summarized in Table 5.14.1 so that the shear strength of members under bending only may be compared with that of members under combined bending and axial load.

---

* For SI, ACI 318–89M, for N_u/A_g and f'_c in MPa, gives

$$V_c = \left(1 + \frac{0.3N_u}{A_g}\right)\left(\frac{\sqrt{f'_c}}{6}\right) b_w d \qquad (5.14.8)$$

Table 5.14.1   Effect of Axial Load on the Shear Strength of Members Without Shear Reinforcement—ACI Code

	SIMPLIFIED METHOD	MORE DETAILED METHOD
Bending only	Formula (11-3), 11.3.1.1 $$V_c = 2\sqrt{f'_c}\, b_w d$$	Formula (11-6), 11.3.2.1 $$V_c = \left(1.9\sqrt{f'_c} + 2500\rho_w \frac{V_u d}{M_u}\right) b_w d$$ $$\leq 3.5\sqrt{f'_c}\, b_w d$$ $V_u d / M_u$ not to exceed unity
Bending and axial compression	Formula (11-4), 11.3.1.2 $$V_c = 2\left(1 + \frac{N_u}{2000 A_g}\right)\sqrt{f'_c}\, b_w d$$ Formula (11-8), 11.3.2.2 $$V_c \leq 3.5\sqrt{f'_c}\, b_w d \sqrt{1 + \frac{N_u}{500 A_g}}$$ N_u is positive for compression and N_u/A_g is in psi	Formula (11-7), 11.3.2.2 $$M_m = M_u - N_u\left(\frac{4h - d}{8}\right)$$ Use M_m for M_u in Formula (11-6) $V_u d / M_u$ has no limitation Formula (11-8), 11.3.2.2 $$V_c \leq 3.5\sqrt{f'_c}\, b_w d \sqrt{1 + \frac{N_u}{500 A_g}}$$ N_u is positive for compression and N_u/A_g is in psi
Bending and axial tension	11.3.1.3 $$V_c = 0$$ Design shear reinforcement for total shear	Formula (11-9), 11.3.2.3 $$V_c = 2\left(1 + \frac{N_u}{500 A_g}\right)\sqrt{f'_c}\, b_w d$$ N_u is negative for tension and N_u/A_g is in psi

■ **EXAMPLE 5.14.1**   Show the effect of axial load on the ACI shear strength V_c for the beam of Example 5.13.1 (Fig. 5.13.1) when it contains no shear reinforcement. Compute V_c for the critical section at d from the face of support, where $V_u = 96.0$ kips and $M_u = 247$ ft-kips (computed in Example 5.13.2). As in Example 5.13.1 use $f'_c = 4000$ psi, $b = 14$ in., $d = 22.5$ in., $h = 26$ in., and 8-#9 bars for the tension steel.

Solution: **(a)** Axial compression. Using the simplified method, Eq. (5.14.1) or ACI Formula (11-4),

$$V_c = 2\left(1 + \frac{N_u}{2000 A_g}\right)\sqrt{f'_c}\, b_w d$$

$$\frac{V_c}{\sqrt{f'_c}\, b_w d} = 2\left(1 + \frac{N_u}{2000 A_g}\right)$$

Figure 5.14.3 Shear strength variation with axial load: Example 5.14.1, with $f'_c = 4000$ psi, $\rho_w = 0.0254$, $M_u = 247$ ft-kips, $V_u = 96.0$ kips, $d = 22.5$ in., $h = 26$ in., and $A_g = 364$ sq in. (Note that $v_c = V_c/b_w d$.)

The right side of the above equation is 2 for $N_u/A_g = 0$; it varies linearly to 2.8 for $N_u/A_g = 800$ psi. This linear expression is plotted on the left side of the vertical axis in Fig. 5.14.3.

Using the more detailed equation, Eq. (5.14.2) or ACI Formula (11-6),

$$\rho_w = \frac{A_s}{bd} = \frac{8(1.00)}{14(22.5)} = 0.0254$$

$$V_c = \left(1.9\sqrt{f'_c} + 2500\rho_w \frac{V_u d}{M_u}\right)b_w d$$

$$\frac{V_c}{\sqrt{f'_c}\,b_w d} = 1.9 + \frac{2500(0.0254)(1.88)}{\sqrt{4000}}\left(\frac{V_u}{M_u}\right)$$

$$= 1.9 + 1.89\frac{V_u}{M_u}$$

Equation (5.14.3) [ACI Formula (11-7)] gives M_m to replace M_u in the above equation,

$$M_m = M_u - N_u\left(\frac{4h - d}{8}\right)$$

$$= M_u - N_u\left[\frac{4(26) - 22.5}{8(12)}\right] = M_u - 0.849N_u$$

where M_u is in ft-kips and N_u is in kips.

Table 5.14.2   Shear Strength with Axial Compression—Example 5.14.1

N_u/A_g (psi)	M_m (ft-kips)	$1.89V_u/M_m$	$V_c/(\sqrt{f'_c}\,b_w d)$	$3.5\sqrt{1+0.002N_u/A_g}$
50	232	0.78	2.68	$3.5(1.05) = 3.67$
100	216	0.84	2.74	$3.5(1.095) = 3.83$
200	185	0.98	2.88	$3.5(1.183) = 4.14$
400	123	1.48	3.38	$3.5(1.342) = 4.70$
600	62	2.93	4.83	$3.5(1.484) = 5.19$
800	0	—	—	$3.5(1.612) = 5.64$

At the critical section, $M_u = 247$ ft-kips and $V_u = 96.0$ kips,

$$M_m = 247 - 0.849N_u = 247 - \frac{N_u}{A_g}(0.849)(14)(26)$$

$$= 247 - 309\frac{N_u}{A_g}$$

Values for the $\rho_w V_u d/M_u$ formula as well as the upper limit equation, Eq. (5.14.7) [ACI Formula (11-8)], are tabulated in Table 5.14.2. The results are shown in Fig. 5.14.3.

(b) Axial tension. The simple linear expression, Eq. (5.14.10) [ACI Formula (11-9)], is plotted on the right side of Fig. 5.14.3. ■

5.15 DEEP BEAMS

When the shear span ($a = M/V$) to depth ratio of beams is lower than about 2 as shown in Fig. 5.4.4, the shear strength exceeds that when a shear-related crack occurs. However, such a crack may be a large undesirable one unless proper reinforcement is provided. The behavior of a simply supported deep beam will be as shown in Fig. 5.15.1(a); that is, two compressive struts will develop in the concrete to carry the load to the supports. If unreinforced, large cracks may open at lower midspan, as in Fig. 5.15.1(b). The use of both horizontal and vertical reinforcement (Fig. 5.15.1(c)) may prevent the formation of such cracks. As the concrete here is subject to a multiaxial state of stress, transverse compressive stress is favorable but transverse tensile stress is unfavorable. Thus, anchorage of the main tension reinforcement is of great importance.

As discussed in Section 5.7, behavior near a concentrated load, that is, in a *D*-region ("disturbed region") [5.9], is complex since ordinary beam theory using shear and bending moment diagrams does not apply. Deep beams are entirely *D*-region situations; thus, special procedures must be used for design. Three approaches may be taken [5.53]: (1) use empirical design procedures, as in the 1989 ACI Code;

(a) Mechanism   (b) Struts not   (c) Struts
                                   reinforced   reinforced

Figure 5.15.1 Deep beam struts (from Schlaich et al. [5.9]).

(2) use two- or three-dimensional analyses, linear or nonlinear; or (3) use the truss analogy comprised of concrete struts for compression and steel ties for tension.

The ACI Code classifies a deep beam in ACI-11.8 as one having a ratio L_n/d of clear span L_n to effective depth d less than 5. For a concentrated load at mid-span, an L_n/d ratio of 5 would be equivalent to a shear span to depth ratio a/d of about 2.5. Preferably, the shear span a to depth ratio d should be used to distinguish between a deep beam for shear purposes and a "shallow" or ordinary beam. A concentrated load may be located near a support, creating a deep beam situation between the load and the near support, but an ordinary beam situation still exists between the load and the more distant support.

Prior to the 1986 ACI Code Supplement, the provisions of ACI-11.8 seemingly applied to all beams having L_n/d less than 5, *whether simply supported or continuous and no matter how the load was applied.* The provisions of ACI-11.8 were based on simply supported beams [5.54–5.56] loaded at the top face and supported at the bottom face. Tests on continuous beams by Rogowsky, MacGregor, and Ong [5.57] indicated that while ACI-11.8 is conservative for simply supported beams, it may significantly underestimate the strength of continuous beams. Particularly, the strength of continuous beams having horizontal shear reinforcement or without shear reinforcement was underestimated by the ACI-11.8 procedure of the 1983 ACI Code.

Though there have been several recent studies on deep beams [5.58–5.62], a truly adequate procedure has not been adopted by the ACI Code. There is a clear trend toward acceptance of the truss model, the so-called "strut-and-tie" approach, presented in general by Schlaich et al. [5.9], and in detail for deep beam situations by Rogowsky and MacGregor [5.53], Adebar, Kuchma, and Collins [5.63], Schlaich and Anagnostou [5.64], and Cook and Mitchell [5.65]. An excellent detailed example is presented by MacGregor [5.66].

The 1989 ACI Code, ACI-11.8.1, applies to members that are "loaded on one face and supported on the opposite face so that compressive struts can develop between the loads and the supports." ACI-11.8.2 applies for *simply supported* beams; that is, use the long-standing deep beam provisions which are conservative. ACI-11.8.3 indicates that *continuous* beams are to be designed using the regular beam procedure of ACI-11.1 to 11.5 used for long shallow beams, *unless "methods satisfying equilibrium and strength requirements" are used.* The latter would permit the "strut-and-tie" approach. Continuous beam design must satisfy the limit on V_n of ACI-11.8.4, and the minimum horizontal and vertical shear reinforcement requirements of ACI-11.8.9 and 11.8.10. Furthermore, though it is not clearly stated, *the critical section for continuous beams should be considered to be at the face of support.* For simply supported beams, ACI-11.8.5 (not ACI-11.1.3) should be used for locating the critical section.

Basic Shear Strength.   In accordance with the general requirement of ACI-11.1.1, it is required that

$$\phi V_n \geq V_u \tag{5.15.1}$$

where

$$V_n = V_c + V_s \tag{5.15.2}$$

where V_c and V_s in Eq. (5.15.2) are for deep beams treated differently for simply supported beams than for continuous beams. Shear reinforcement is required whenever V_u exceeds ϕV_c (ACI-11.8.8) for simply supported beams; however, it is required for continuous beams wherever V_u exceeds $0.5\phi V_c$ (ACI-11.5.5.1).

Location for Computing Factored Shear.   On deep beams, shear reinforcement usually will be the same throughout a span; on a simply supported deep beam, the shear reinforcement is *required* to be the same throughout the span. The location for computing V_u is different for simply supported than for continuous beams.

(a) Simply Supported Beams. The *critical section* is defined as located at a distance (say z) from the face of support; thus, ACI-11.8.5 defines

1. $z = 0.15L_n \geq d$ for uniform loading

2. $z = 0.50a \geq d$ for concentrated loading

where the shear span a is the distance from the face of support to the concentrated load. ACI-ASCE Committee 426 recommends [5.11, 5.56] limiting a to a maximum of 1.15 times the clear distance from the face of the load to the face of the support.

(b) Continuous Beams. The critical section is to be taken at the face of support (*not the distance d from the face*). There is no requirement for the shear reinforcement to be the same throughout the span; however, it should have uniform vertical and horizontal spacing within any diagonal compressive strut region.

Strength V_c—Simply Supported Beams.   ACI-11.8.6 permits using the *simplified method* with the same V_c as for ordinary beams; thus

$$V_c = 2\sqrt{f_c'}\,b_w d \tag{5.15.3}$$

The *more detailed procedure* using ACI Formula (11-30) puts a multiplier on the shear strength V_c used in the more detailed procedure for ordinary beams; thus,

$$V_c = \left(3.5 - 2.5\,\frac{M_u}{V_u d}\right)\left(1.9\sqrt{f_c'} + 2500\rho_w\,\frac{V_u d}{M_u}\right)b_w d \qquad \text{(5.15.4)}$$

where the upper limit on V_c is

$$\text{limit } V_c \le 6\sqrt{f_c'}\,b_w d \qquad \text{(5.15.5)}$$

In addition, the multiplier has an upper limit,

$$\left(3.5 - 2.5\,\frac{M_u}{V_u d}\right) \le 2.5 \qquad \text{(5.15.6)}$$

which is reached at $M_u/(V_u d) = 0.4$.

The value of M_u is the factored bending moment occurring at the critical section simultaneously with maximum factored shear V_u.

Strength V_s—Simply Supported Beams. The equation requiring both horizontal and vertical reinforcement is given in ACI-11.8.8 as ACI Formula (11-31),

$$V_s = \left[\frac{A_v}{s}\left(\frac{1 + L_n/d}{12}\right) + \frac{A_{vh}}{s_2}\left(\frac{11 - L_n/d}{12}\right)\right]f_y d \qquad \text{(5.15.7)}$$

where

$\quad L_n$ = clear span

$\quad A_v$ = vertical stirrup area

$\quad A_{vh}$ = longitudinal shear reinforcement area

$\quad\quad s$ = spacing of vertical stirrups

$\quad\quad s_2$ = vertical spacing of longitudinal shear reinforcement

ACI-11.8.9 and 11.8.10 provide for minimum amounts of shear reinforcement, A_v and A_{vh}.

$$\text{min } A_v = 0.0015 b_w s \qquad \text{(5.15.8)}$$

$$s \le \frac{d}{5} \le 18 \text{ in.*}$$

$$\text{min } A_{vh} = 0.0025 b_w s_2 \qquad \text{(5.15.9)}$$

$$s_2 \le \frac{d}{3} \le 18 \text{ in.*}$$

where b_w is the web width. The minimum amount of horizontal and vertical shear reinforcement is required throughout the span (ACI-11.8.11) whenever $V_u > \phi V_c$ (ACI-11.8.8).

$\quad$ * For SI, ACI 318–89M gives the upper limit as 500 mm.

Strength V_c—Continuous Beams.   ACI-11.8.3 indicates that ordinary beam expressions for V_c given in ACI-11.3 shall apply to continuous deep beams. Thus, ACI-11.3.1 gives

$$V_c = 2\sqrt{f'_c}\, b_w d \qquad\qquad (5.15.10)$$

or, alternatively, ACI-11.3.2 may be used

$$V_c = \left[1.9\sqrt{f'_c} + 2500\rho_w \frac{V_u d}{M_u} \right] b_w d \le 3.5\sqrt{f'_c}\, b_w d \qquad\qquad (5.15.11)$$

Strength V_s—Continuous Beams.   ACI-11.8.3 refers to ACI-11.5, where

$$V_s = \frac{A_v f_y d}{s} \qquad\qquad (5.15.12)$$

where the minimum amounts of shear reinforcement, A_v and A_{vh}, required are the same for continuous beams as for simply supported beams; the amounts are given (ACI-11.8.9 and 11.8.10) by Eqs. (5.15.8) and (5.15.9). This minimum reinforcement is required wherever V_u exceeds $0.5\phi V_c$ (ACI-11.5.5.1).

Limitations on Nominal Shear Strength.   The limitations on strength are best itemized in terms of the nominal stress $v_n = V_u/(\phi b_w d)$ since they will then be independent of $b_w d$.

The maximum shear stress v_n allowed by ACI-11.8.4 is

$$\max v_n \le 8\sqrt{f'_c} \qquad\qquad \text{for } \frac{L_n}{d} < 2 \qquad\qquad (5.15.13)$$

$$\max v_n = \frac{2}{3}\left(10 + \frac{L_n}{d} \right)\sqrt{f'_c} \quad \text{for } \frac{L_n}{d} \ge 2 \qquad\qquad (5.15.14)$$

This gives a maximum value of v_n varying from $8\sqrt{f'_c}$ to $10\sqrt{f'_c}$. The upper value $10\sqrt{f'_c}$ that is permitted when $L_n/d = 5$ is approximately the same as the limit of $8\sqrt{f'_c}$ on v_s for longer beams (ACI-11.5.6.8).

■ **EXAMPLE 5.15.1**   Design the shear reinforcement for a simply supported beam that carries two concentrated service live loads of 126 kips each on a clear span of 12 ft, as shown in Fig. 5.15.2. The beam has a width of 14 in. and an effective depth d of 36 in. Use $f'_c = 3500$ psi and $f_y = 60,000$ psi.

Solution: (a) Determine whether or not deep beam provisions of ACI-11.8 apply. For this beam, $L_n/d = 144/36 = 4$; thus it is a deep beam.

(b) Critical section (ACI-11.8.5). For concentrated loads, using the shear span of $a = 4$ ft,

$$0.50a = 0.50(4.0) = 2.0 \text{ ft} < [d = 3 \text{ ft}]$$

Critical section is taken at 2.0 ft from face of support.

Figure 5.15.2 Beam for Example 5.15.1.

(c) Shear strength of beam without shear reinforcement at critical section (ACI-11.8.5). The factored concentrated load is

$$126(1.7) = 214 \text{ kips}$$

Neglect the uniform dead load which is small compared to the concentrated load. Using the more detailed procedure, at the critical section,

$$\frac{M_u}{V_u d} = \frac{214(2)}{214(3)} = 0.67$$

The multiplier for deep beam action is then

$$3.5 - 2.5 \frac{M_u}{V_u d} = 3.5 - 2.5(0.67) = 1.83 < 2.5 \qquad \textbf{OK}$$

$$v_c = 1.83 \left[1.9\sqrt{f_c'} + 2500\rho_w \frac{V_u d}{M_u} \right]$$

$$\rho_w = \frac{4(1.56)}{14(36)} = 0.0124$$

$$v_c = 1.83 \left[1.9\sqrt{3500} + \frac{2500(0.0124)}{0.67} \right]$$

$$= 1.83[112 + 46.3] = 290 \text{ psi} \qquad \text{(controls)}$$

$$\text{upper limit } v_c = 6\sqrt{f_c'} = 6(59.2) = 355 \text{ psi}$$

$$V_c = v_c b_w d = 290(14)(36)\tfrac{1}{1000} = 146 \text{ kips}$$

(d) Nominal shear strength. At critical section,

$$\text{required } V_n = \frac{V_u}{\phi} = \frac{214}{0.85} = 252 \text{ kips}$$

$$v_n \text{ (max allowed)} = \frac{2}{3}\left(10 + \frac{L_n}{d}\right)\sqrt{f'_c}$$

$$= \frac{2}{3}(10 + 4)\sqrt{f'_c} = 9.33\sqrt{f'_c} = 552 \text{ psi}$$

$$\text{max } V_n = (0.552)(14)(36) = 278 \text{ kips} > 252 \text{ kips} \qquad \textbf{OK}$$

Since required $V_n > V_c$ (that is, 252 > 146), vertical and horizontal shear reinforcement is required for strength.

(e) Shear reinforcement. According to ACI-11.8.8,

$$\frac{A_v}{s}\left(\frac{1 + L_n/d}{12}\right) + \frac{A_{vh}}{s_2}\left(\frac{11 - L_n/d}{12}\right) = \frac{V_s}{f_y d}$$

or for $L_n/d = 4$, $V_s = 252 - 146 = 106$ kips, $b = 14$ in., and $f_y = 60,000$ psi,

$$\frac{A_v}{s}\left(\frac{5}{12}\right) + \frac{A_{vh}}{s_2}\left(\frac{7}{12}\right) = \frac{106}{60(36)} = 0.049$$

$$\min A_v = 0.0015 b_w s, \qquad \max s = \frac{d}{5} = 7.2 \text{ in.}$$

$$\min A_{vh} = 0.0025 b_w s_2, \qquad \max s_2 = \frac{d}{3} = 12 \text{ in.}$$

Try using #4 bars horizontally in each face at about 11 in.

$$\min \text{ required } A_{vh} = 0.0025(14)11 = 0.38 \text{ sq in.}$$

$$\text{provided } A_{vh} = (0.20)2 = 0.40 \text{ sq in.} > 0.38 \text{ sq in.} \qquad \textbf{OK}$$

$$\frac{A_v}{s}\left(\frac{5}{12}\right) + \frac{0.40}{11}\left(\frac{7}{12}\right) = 0.049$$

$$\frac{A_v}{s} = [0.049 - 0.021]\frac{12}{5} = 0.067$$

For the #4 U stirrups, $A_v = 2(0.20) = 0.40$ sq in.,

$$\text{required } s = \frac{0.40}{0.067} = 6.0 \text{ in.} < \frac{d}{5} \qquad \textbf{OK}$$

Use #4 vertical stirrups @ 6 in. *throughout span.* Note that the main flexural steel is extended into the supports as far as practical and bent upward to obtain maximum embedment. Anchorage of bars is the subject of Chapter 6. ∎

■ **EXAMPLE 5.15.2** Design the shear reinforcement for the continuous beam of Fig. 5.15.3, which is to support heavy machinery. The factored moment and shear diagrams are given in Fig. 5.15.3. Use $f'_c = 3500$ psi and $f_y = 60,000$ psi.

Figure 5.15.3 Continuous deep beam, Example 5.15.2.

Solution: Since the beam is of short span, it is likely a deep beam according to ACI-11.8.

(a) Determine whether or not deep beam provisions of ACI-11.8 apply.

$$\frac{L_n}{d} = \frac{8(12)}{23.5} = 4.1 < 5 \quad \text{ACI-11.8 applies!}$$

(b) Critical section. Since this is a continuous beam, ACI-11.8.5 does *not* apply. The critical section should be taken *at the face of the support.*

(c) Shear strength V_c at critical section. For continuous beams, ACI-11.3.1.1 or 11.3.2.1 may be used to compute V_c. Using the *simplified method*, and in terms of nominal stress v,

$$v_c = 2\sqrt{f'_c} = 2\sqrt{3500} = 118 \text{ psi}$$

Using the *more detailed procedure,*

$$v_c = \left[1.9\sqrt{f'_c} + 2500\rho_w \frac{V_u d}{M_u} \right] \le 3.5\sqrt{f'_c} \qquad \text{[5.15.11]}$$

$$V_u = 92 \text{ kips} \quad \text{(given from Fig. 5.15.3)}$$

$$M_u = 145 \text{ ft-kips} \quad \text{(given from Fig. 5.15.3)}$$

$$\frac{V_u d}{M_u} = \frac{92(23.5)}{145(12)} = 1.24 > 1.0 \text{ limit}; \quad \text{use } 1.0 \text{ [ACI-11.3.2.1]}$$

$$v_c = 1.9\sqrt{3500} + 2500\left(\frac{3(0.79)}{14(23.5)} \right)(1.0) = 130 \text{ psi}$$

$$\text{upper limit } v_c = 3.5\sqrt{f'_c} = 3.5(59.2) = 207 \text{ psi} \qquad \textbf{OK}$$

(d) Nominal shear stress. At face of support,

$$v_n = \frac{V_u}{\phi b_w d} = \frac{92(1000)}{0.85(14)23.5} = 329 \text{ psi}$$

Since $v_n > 0.5v_c$, that is, 329 psi > 130/2 psi, shear reinforcement is required. Check upper limit on v_n as given by ACI-11.8.4,

$$\max v_n = \frac{2}{3}\left(10 + \frac{L_n}{d} \right)\sqrt{f'_c} = \frac{2}{3}(10 + 4.1)\sqrt{f'_c}$$

$$= 9.4\sqrt{f'_c} = 556 \text{ psi} > 329 \text{ psi} \qquad \textbf{OK}$$

(e) Vertical shear reinforcement (ACI-11.5, 11.8.9). Using #3 stirrups ($A_v = 0.22$ sq in.) if practicable,

$$\phi V_s = V_u - \phi V_c$$

$$\phi V_c = 0.85(130)(14)(23.5)/1000 = 36.5 \text{ kips}$$

$$\text{required } \phi V_s = 92 - 36.5 = 55.5 \text{ kips}$$

The maximum spacing for #3 U stirrups to satisfy the strength requirement is

$$\max s = \frac{\phi A_v f_y d}{\text{reqd}(\phi V_s)} = \frac{0.85(0.22)(60)23.5}{55.5} = 4.75 \text{ in.}$$

The maximum spacing to satisfy the minimum amount of vertical stirrups requirement of ACI-11.8.9 is

$$\max s = \frac{A_v}{0.0015 b_w} = \frac{0.22}{0.0015(14)} = 10.5 \text{ in.}$$

but the spacing shall not exceed

$$\max s = \frac{d}{5} = \frac{23.5}{5} = \underline{4.70 \text{ in.}} \quad \text{or} \quad 18 \text{ in.}$$

(f) Horizontal shear reinforcement (ACI-11.8.10). Trying #3 bars in pairs (that is, one in each face), giving $A_{vh} = 2(0.11) = 0.22$ sq in.,

$$\text{min } A_{vh} = 0.0025 b_w s_2$$

$$\text{max } s_2 = \frac{2(0.11)}{0.0025(14)} = 6.3 \text{ in. (controls)}$$

but the spacing shall not exceed

$$\text{max } s_2 = \frac{d}{3} = \frac{23.5}{3} = 7.8 \text{ in. } \quad \text{or} \quad 18 \text{ in.}$$

Use #3 vertical U stirrups spaced at $4\frac{1}{2}$ in. throughout the beam, and #3 bars horizontally in each face spaced at about 5 in. to give proper cover (4-#3 bars horizontally in each face). The results are shown in Fig. 5.15.3(a). ■

5.16 SHEAR-FRICTION

Even though uncracked concrete is relatively strong in shear, and shear-related cracks in usual beams are inclined cracks (diagonal tension cracks), such shear-related cracks become more vertical as the member becomes deeper compared to the shear span, as discussed in Section 5.4. The ACI Code design procedures for beams as discussed in Sections 5.5 through 5.14 are intended to prevent *inclined* cracking (diagonal tension cracking).

For situations in which a crack may form and slippage *along* that crack interface might occur if no steel reinforcement crosses the crack, *and* the usual design procedures for shear reinforcement to resist inclined cracking are inappropriate (such as for a/d less than about 1.0), the shear-friction concept of shear transfer should be applied. The shear-friction concept *is* appropriate for providing a shear transfer mechanism in such cases as:

1. At the interface between concretes cast at different times.
2. At the junction of a corbel (bracket) with a column, such as shown in Fig. 5.16.1(a).
3. At the junction of elements in precast concrete construction, such as the bearing detail shown in Fig. 5.16.1(b).
4. At an interface between steel and concrete, such as the steel bracket attachment to a concrete column shown in Fig. 5.16.1(c).

The crack for which shear-friction reinforcement is required *may* not have been caused by shear. It could, for instance, have been caused by shrinkage. However, once the crack has occurred (from whatever source) a shear transfer mechanism must be provided. The design approach is to assume that a crack will occur and then provide reinforcement across the assumed crack to resist relative displacement along the crack.

Consider a block of concrete as in Fig. 5.16.2(a) acted on by collinear shear forces V such that a failure plane would form along plane *a-a*. Since the crack along *a-a*

(a) Corbel (see Sec. 5.17 for details) (b) Bearing region of beam

(c) Column face plate

Figure 5.16.1 Uses for the shear-friction concept.

(a) (b)

Figure 5.16.2 Idealization of the shear-friction concept.

will tend to be rough, the sliding motion will produce a separation, as in Fig. 5.16.2(b). One might imagine slippage along a sawtooth that forces the crack to open; as it opens the reinforcement is put in tension, with a resulting compression or clamping force on the concrete. A frictional force is then developed equal to the compression in the concrete (or the tension in the bars) times the coefficient of friction. If one may assume that the separation is sufficient to load the steel reinforcement to its yield stress, the shearing resistance then equals the frictional force; thus (ACI-11.7.4) states

$$V_n = A_{vf} f_y \mu \qquad (5.16.1)$$

where A_{vf} is the area of reinforcement extending across the potential crack at 90° to it, and μ is the coefficient of friction between materials along the potential crack. This concept of shear-friction has been verified experimentally [5.67–5.79].

ACI-11.7.1 states that shear-friction provisions apply where "it is appropriate to consider shear transfer across a given plane, such as: an existing or potential crack, an interface between dissimilar materials, or an interface between two concretes cast at different times."

If the shear-friction reinforcement is inclined at an angle to the assumed crack, such that the shear force produces tension in the shear-friction reinforcement, as shown in Fig. 5.16.3, the shear strength V_n becomes

$$V_n = A_{vf} f_y (\mu \sin \alpha_f + \cos \alpha_f) \qquad (5.16.2)$$

where α_f is the angle between the shear-friction reinforcement and the shear plane.

The logic of Eq. (5.16.2) may be observed from Fig. 5.16.3, where the shear strength provided along the potential crack consists of two parts: the vertical component $T \cos \alpha_f$ of the force in the reinforcement A_{vf} and the frictional force μC, which is the same as $\mu T \sin \alpha_f$. Thus,

$$V_n = T \cos \alpha_f + \mu C \qquad (5.16.3)$$

Figure 5.16.3 Action of shear-friction reinforcement when inclined to shear plane.

One should note that the component of the tensile force in the shear-friction reinforcement normal to the potential crack causes a compression at the crack interface, giving rise to the friction force μC.

The required nominal shear-friction strength V_n is V_u / ϕ, in which case Eq. (5.16.1) becomes

$$\text{required } A_{vf} = \frac{V_u}{\phi f_y \mu} \qquad (5.16.4)$$

and when α_f is less than 90°, the required A_{vf} from Eq. (5.16.2) is

$$\text{required } A_{vf} = \frac{V_u}{\phi f_y (\mu \sin \alpha_f + \cos \alpha_f)} \qquad (5.16.5)$$

Note that Eq. (5.16.5) becomes Eq. (5.16.4) when $\alpha_f = 90°$. Just as for regular stirrups, f_y may not be taken greater than 60,000 psi. The strength reduction factor ϕ is taken to be 0.85 for shear in Eqs. (5.16.4) and (5.16.5).

The coefficient of friction μ is to be taken (ACI-11.7.4.3) as follows:

1. Concrete cast monolithically. $\qquad \mu = 1.4\,\lambda$
2. Concrete placed against hardened concrete with surface intentionally roughened to an "amplitude of approximately $\frac{1}{4}$ in." and the surface must be "clean and free of laitance." $\qquad \mu = 1.0\,\lambda$
3. Concrete placed against hardened concrete not intentionally roughened, but "clean and free of laitance." $\qquad \mu = 0.6\,\lambda$
4. Concrete anchored to as-rolled structural steel by headed studs or by reinforcing bars, with as-rolled steel "clean and free of paint." $\qquad \mu = 0.7\,\lambda$

In the above expressions for μ, the multiplier λ shall be 1.0 for normal-weight concrete, 0.85 for "sand-lightweight" concrete, and 0.75 for "all-lightweight" concrete. Linear interpolation is permitted when partial sand replacement is used.

The maximum nominal shear stress may not exceed $0.2f'_c$ nor 800 psi, which means according to ACI-11.7.5 that

$$\max V_n = v_n A_c = 0.2f'_c A_c \leq 800 A_c \qquad (5.16.6)$$

where A_c is the area of concrete section resisting shear transfer (sq in.).

Since the steel A_{vf} across the potential crack as determined by Eqs. (5.16.4) and (5.16.5) is only that necessary to provide the clamping action that produces friction, any *external* direct tension on the assumed crack must be provided for by additional reinforcement (ACI-11.7.7).

To ensure attainment of a uniform frictional force along the assumed crack, the "shear-friction reinforcement shall be appropriately placed along the shear plane and shall be anchored to develop the specified yield strength on both sides by embedment, hooks, or welding to special devices" (ACI-11.7.8).

The application of the shear-friction provisions to brackets and corbels appears in Section 5.17, which is devoted entirely to that topic.

For guidance in the application of shear-friction to bearings and other special situations commonly encountered in precast concrete construction, the designer should consult the PCI publication, *Design and Typical Details of Connections for Precast and Prestressed Concrete* [5.80], as well as the *PCI Design Handbook* [2.24]. The following example presents an application other than for brackets and corbels.

■ **EXAMPLE 5.16.1**   Design the reinforcement needed at the bearing region of a precast beam 14 in. wide by 28 in. deep supported on a 4-in. bearing pad. The factored shear V_u is 105 kips. The horizontal force resulting from restraint of volume change movements due to creep, shrinkage, and temperature effects, is 0.3 of the factored shear V_u. Grade 60 steel is to be used for reinforcement.

Solution: **(a)** Identify the potential crack location. One should assume that a crack will form in the most undesirable manner. According to the *PCI Design Handbook* [2.24], an appropriate assumption for the crack angle θ, as shown in Fig. 5.16.4(a), is approximately 20°. Recent information [5.80, p. 4–13] indicates that θ is likely to be closer to zero. The crack may intersect the bottom of the beam immediately adjacent to the bearing pad, which in this example is taken as 4 in. A rolled structural steel angle is used for confinement across the width of the beam at the bearing.

(b) Determine the shear-friction reinforcement A_{vf} required. Presumably, it would be appropriate to resolve the V_u and N_{uc} in Fig. 5.16.4(a) into components parallel and perpendicular to the potential crack when θ is assumed other than zero. Even in such case, it will be simpler and more practical to assume that all of V_u will act parallel to the crack. Thus, using Eq. (5.16.4),

$$\text{required } A_{vf} = \frac{V_u}{\phi f_y \mu} = \frac{105}{0.85(60)1.4} = 1.47 \text{ sq in.}$$

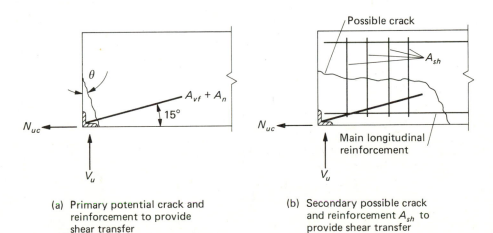

(a) Primary potential crack and reinforcement to provide shear transfer

(b) Secondary possible crack and reinforcement A_{sh} to provide shear transfer

Figure 5.16.4 Shear-friction concept applied to bearing region of a beam.

(c) Determine the additional reinforcement A_n to provide for the net tension across the potential crack. It will be conservative not to use the sum of components V_u and N_{uc} perpendicular to any assumed non-vertical crack, but rather to use N_{uc} as if for a vertical crack. According to ACI-11.7.7,

$$\text{required } A_n = \frac{N_{uc}}{\phi f_y} = \frac{0.3(105)}{0.85(60)} = 0.62 \text{ sq in.}$$

Note that N_{uc} was given as 0.3 of V_u, presumably as the result of an analysis for volume change effects. It is recommended [2.24, 5.81] (and *required* by ACI-11.9.3.4 in the case of brackets and corbels) that unless all tensile N_{uc} can be eliminated by appropriate design, the value of N_{uc} should not be taken less than $0.2V_u$. The ϕ factor of 0.85 for shear is considered appropriate for the above calculation, even though 0.90 for axial tension is indicated by ACI-9.3.2. For brackets and corbels, in the similar situation of reinforcement for the tensile N_{uc}, ACI-11.9.3.1 specifies taking ϕ as 0.85.

(d) Total reinforcement to restrain primary crack [Fig. 5.16.4(a)]. The total reinforcement required is

$$A_s = A_{vf} + A_n = 1.47 + 0.62 = 2.09 \text{ sq in.}$$

Use 5-#6, $A_s = 2.20$ sq in. Distribute as shown in Fig. 5.16.5, place at the recommended [2.24] 15° with the horizontal, weld to the steel angle on one end, and embed the other end into the beam to develop the tensile strength of the #6 bars beyond the potential crack. (Development length requirements are treated in Chapter 6.)

(e) Reinforcement for the potential secondary horizontal crack that may form as shown in Fig. 5.16.4(b). If a vertical crack begins near the corner region where the main shear-friction reinforcement terminates, then either with or without the tensile force N_{uc} acting, there would be a potential horizontal crack due to the tensile force

Figure 5.16.5 Final design for Example 5.16.1.

developed in the *main* shear-friction reinforcement. The maximum shear that could act along such a failure plane would be the horizontal shear-friction force arising from the tensile capacity of the main shear-friction reinforcement. Thus the required vertical stirrup shear-friction reinforcement A_{sh} [Fig. 5.16.4(b)] is

$$\text{reqd } A_{sh} = \frac{\text{tensile capacity of shear-friction reinf.}}{\mu f_y}$$

$$= \frac{2.09(60)}{1.4(60)} = 1.49 \text{ sq in.}$$

<u>Use 4-#4 U stirrups.</u> $A_{sh} = 4(0.4) = 1.60$ sq in.

(f) Additional confinement reinforcement. A conservative approach, recommended by Mast [5.68], is to provide reinforcement to prevent splitting in the vertical plane of the beam equal to 25% of the support reaction. This confinement reinforcement is divided equally into horizontal A_{ch} and vertical A_{cv} portions. Thus,

$$A_{ch} = A_{cv} = \frac{V_u}{8 f_y} = \frac{105}{8(60)} = 0.22 \text{ sq in.}$$

<u>Use 2-#4 vertical and 2-#3 U stirrups horizontal.</u> The final design is shown in Fig. 5.16.5. ∎

5.17 BRACKETS AND CORBELS

Brackets and corbels projecting from the faces of columns are widely used in precast concrete construction to support beams and girders, as shown in Fig. 5.17.1. It is inappropriate to design brackets and corbels as cantilever beams using the usual beam provisions for shear as described in Sections 5.5 through 5.14. As discussed in Section 5.4 and shown in Fig. 5.4.4, when a/d is less than about 1.0, deep beam theory, rather than simple flexural theory, should apply. Brackets and corbels, furthermore, differ from the deep beams discussed in Section 5.15 in that design calculations for horizontal forces must also be made. Because the beams are attached to the bracket, the restraint on the beams due to creep, shrinkage, and temperature deformations give rise to horizontal forces (N_{uc} in Fig. 5.17.1).

Figure 5.16.1 Bracket or corbel.

(a) As shown on
     drawing

(b) As bent

Potential
failure
surface

(c) Cracking in corbel
     with too shallow
     an outer face

(d) Cracking in corbel
     with outer face of
     sufficient depth

Figure 5.17.2 Corbel details and possible failure modes (from Kriz and Raths [5.81]).

Typically, reinforcement for brackets or corbels has in the past consisted of several bars across the width of the bracket bent as shown in Fig. 5.17.2(a). When minimum bend radii are considered, the actual arrangement is as in Fig. 5.17.2(b), in which case a potential failure surface is indicated by the dashed line. When the outer face is too shallow, the critical inclined crack will form in the location shown in Fig. 5.17.2(c). When the bracket is deep enough, the crack will tend to extend back into the column [Fig. 5.17.2(d)], with the portion between the crack and the sloping face acting as a compression element. If the compression strut can be developed, the bracket will have reserve capacity after the crack forms; if the strut cannot develop [as in Fig. 5.17.2(c)], failure will be instantaneous upon formation of the crack.

Research by Mattock et al. [5.74, 5.82] has shown that the shear-friction concept can be applied to bracket (corbel) design for a/d as high as 1.0. The design recommendations of Mattock [5.83], the suggestions of ACI-ASCE Committee 426 [5.11], and the further discussions of MacGregor and Hawkins [5.56] are the basis for the present ACI-11.9. Additional discussions of the subject are provided by Shaikh [5.84] and Solanki and Sabnis [5.85].

Basic Equilibrium Equations.   Referring to Fig. 5.17.3, the strengths in shear V_n and in tension N_{uc} are related to the internal forces such that statics is satisfied. From vertical force equilibrium,

$$V_n = \mu C \qquad\qquad (5.17.1)$$

Figure 5.17.3. Equilibrium of forces acting on a bracket or corbel.

From horizontal force equilibrium,

$$N_{nc} = T - C \tag{5.17.2}$$

and from moment equilibrium, taken about point A,

$$V_a a + N_{uc}\left(h - \frac{a_1}{2}\right) = T\left(d - \frac{a_1}{2}\right) \tag{5.17.3}$$

Substituting Eq. (5.17.1) for C into Eq. (5.17.2), and taking $T = A_s f_y$,

$$N_{nc} = A_s f_y - \frac{V_n}{\mu}$$

or

$$A_s = \frac{V_n}{f_y \mu} + \frac{N_{nc}}{f_y} \tag{5.17.4}$$

Substitution of $A_s f_y$ for T in Eq. (5.17.3) gives

$$V_n a + N_{nc}(h - d) + N_{nc}\left(d - \frac{a_1}{2}\right) = A_s f_y\left(d - \frac{a_1}{2}\right)$$

and solving for A_s gives

$$A_s = \frac{V_n a + N_{nc}(h - d)}{f_y(d - a_1/2)} + \frac{N_{nc}}{f_y} \tag{5.17.5}$$

Equations (5.17.4) and (5.17.5) give the formulas for A_s to provide the required strengths V_n and N_{nc}.

For design, the factored loads V_u and N_{uc} divided by ϕ are the required strengths V_n and N_{nc}, respectively. Thus, Eq. (5.17.4) becomes

$$\text{required } A_s = \frac{V_u}{\phi f_y \mu} + \frac{N_{uc}}{\phi f_y} \tag{5.17.6}$$

and Eq. (5.17.5) becomes

$$\text{required } A_s = \frac{V_u a + N_{uc}(h-d)}{\phi f_y(d - a_1/2)} + \frac{N_{uc}}{\phi f_y} \tag{5.17.7}$$

Note that $N_{uc}/\phi f_y$ is the reinforcement A_n required for axial tension (using the symbol A_n used in Example 5.16.1) and $V_u/(\phi f_y \mu)$ is the shear-friction reinforcement A_{vf} given by Eq. (5.16.4). Further, observe that if the numerator of the first term in Eq. (5.17.7) is treated as an "equivalent" moment, it would represent the reinforcement A_f required for a beam, corresponding to A_s of Eq. (3.3.4).

To summarize the steel area requirements for brackets and corbels, the following may be stated:

$$\text{required } A_s = A_{vf} + A_n \tag{5.17.8}$$

or

$$\text{required } A_s = A_f + A_n \tag{5.17.9}$$

in which

$$A_{vf} = \frac{V_u}{\phi f_y \mu} \tag{5.17.10}$$

$$A_n = \frac{N_{uc}}{\phi f_y} \tag{5.17.11}$$

$$A_f = \frac{\text{equivalent } M_u}{\phi f_y(d - a_1/2)} \tag{5.17.12}$$

and

$$\text{equivalent } M_u = V_u a + N_{uc}(h-d) \tag{5.17.13}$$

Minimum Horizontal Stirrups. In addition to the steel A_s in Fig. 5.17.3, stirrups in the horizontal plane are needed across the vertical potential crack in order to prevent premature diagonal tension failure. These stirrups or ties must be closed hoops, having a total area A_h. It was conservative to neglect this steel A_h in the development of Eqs. (5.17.8) and (5.17.9) since it could have been deducted from the right side of those equations. Thus Eq. (5.17.8) could become

$$\text{required } A_s = A_{vf} + A_n - A_h \tag{5.17.14}$$

Tests [5.75] on brackets (corbels) indicate that minimum A_h for the horizontal stirrups must be

$$\min A_h \geq \frac{1}{2} A_f \tag{5.17.15}$$

and

$$\text{min } A_h \geq \frac{1}{3}A_{vf} \qquad (5.17.16)$$

ACI Code Provisions. ACI-11.9.3.5 requires the area of primary tension reinforcement A_s to be the greater of the following:

$$\text{required } A_s = \frac{2}{3}A_{vf} + A_n \qquad (5.17.17)$$

$$\text{required } A_s = A_f + A_n \qquad (5.17.18)$$

and, according to ACI-11.9.4, closed stirrups or ties parallel to A_s must be used, having a total area A_h not less than the following:

$$\text{required } A_h \geq 0.5(A_s - A_n) \qquad (5.17.19)$$

The area A_h is to be uniformly distributed within two-thirds of the effective depth from A_s.

Note that the three ACI requirements, expressed by Eqs. (5.17.17), (5.17.18), and (5.17.19) automatically satisfy the basic equilibrium requirements of Eqs. (5.17.14) and (5.17.9), as well as the minimum A_h requirements of Eqs. (5.17.15) and (5.17.16). That Eq. (5.17.16) is always satisfied can be proved by substituting Eq. (5.17.17) into Eq. (5.17.19), or

$$\text{required } A_h \geq 0.5\left(\frac{2}{3}A_{vf} + A_n - A_n\right)$$

$$\text{required } A_h \geq \frac{1}{3}A_{vf}$$

Equation (5.17.17) satisfies Eq. (5.17.14) in recognition that A_h is at least $\frac{1}{3}A_{vf}$.

In addition to Eqs. (5.17.17) to (5.17.19), ACI Code limitations and requirements in the design of brackets and corbels, including those previously described in the introductory material, are:

1. Shear span to depth ratio a/d may not exceed 1.0 (ACI-11.9.1).
2. Factored tensile force N_{uc} may not exceed factored shear V_u (ACI-11.9.1).
3. Factored tensile force N_{uc} may not be taken less than $0.2V_u$ unless special precautions are taken to avoid tensile forces (ACI-11.9.3.4).
4. Critical section is *at* face of support, where the effective depth d is to be measured (ACI-11.9.1), as shown in Fig. 5.17.4. The effective depth d may not be more than twice the depth d_1 at the outer edge of the bearing area (ACI-11.9.2).
5. The strength reduction factor ϕ is to be taken as 0.85 for all calculations relating to the design of brackets and corbels (ACI-11.9.3.1).
6. The maximum strength V_n for which brackets and corbels may be designed using normal weight concrete is

$$\text{max } V_n \leq 0.2f'_c b_w d \leq (800 \text{ psi})b_w d \qquad (5.17.20)$$

Figure 5.17.4 Effective depth of bracket or corbel.

according to ACI-11.9.3.2.1. For "all-lightweight" or "sand-lightweight" concrete, the maximum (ACI-11.9.3.2.2) is

$$\max V_n \leq \left(0.2 - 0.07\frac{a}{d}\right) f'_c b_w d \leq \left(800 \text{ psi} - \frac{(280 \text{ psi})a}{d}\right) b_w d \qquad \textbf{(5.17.21)}$$

7. The minimum reinforcement ratio ρ for the main tension steel A_s is

$$\min \rho = 0.04 \frac{f'_c}{f_y} \qquad \textbf{(5.17.22)}$$

8. Primary reinforcement A_s at front face must be anchored by (a) a structural weld to transverse bar of at least equal size to develop a force of $f_y A_s$, or (b) bending A_s bars back to form a horizontal loop, or (c) some means of positive anchorage (ACI-11.9.6).
9. Bearing area must not project beyond straight portion of A_s bars, nor beyond the interior face of the transverse anchor bar if one is provided (ACI-11.9.7).

Additional Recommendations for Detailing.   Kriz and Raths [5.81] provide several recommendations for detailing, as shown in Fig. 5.17.5. Detail A in that figure is essentially that of ACI-11.9.6. Alternatively, a confinement angle as in Fig. 5.17.6 can be used, to which the main tension bars are welded at the underside. The use of the confinement angle is one of the recommendations of the *PCA Design Handbook* [2.24]. Kriz and Raths also recommended that the outer edge of a bearing plate resting on a corbel should be placed not closer than 2 in. from the outer edge of the corbel. The *PCA Design Handbook* recommends a 1-in. minimum setback when *no* confinement angle is used but does not indicate a setback when a confinement angle is used. Further, for good practice, when corbels are designed to resist horizontal forces, steel bearing plates welded to the tension reinforcement should be used to transfer the horizontal forces directly to the tension reinforcement. Details and design aids for brackets and corbels are given in the *PCA Design Handbook* [2.24].

Note:
    Distance x should be great
    enough to prevent contact
    between outer corbel edge
    and beam due to possible
    rotations

$\frac{1}{2}''$ min A_s

Same
bar
size   Weld must develop the
    yield strength of A_s

Detail A

Unrestrained
precast beam

Elastomeric pad

x 2" min $\frac{5}{8}$ min

$\frac{5}{8}''$ min →   see A s

$\frac{3}{4}''$ min → $h/2$ min s

h d

4" min

(a) Corbels subject to vertical load only

Restrained
precast beam

Beam
reinforcement
not shown

Steel ⌶

x 2" min

$\frac{5}{8}''$ min →   see A s

$\frac{3}{4}''$ min → $h/2$ min s

h d

4" min

(b) Corbels subject to vertical load and
    restrained creep and shrinkage force.
    Steel ⌶'s welded or not welded.

Figure 5.17.5 Recommended corbel details (from Kriz and Raths [5.81]).

Steel angle
say 6 × 4 × $\frac{1}{2}$

V_u

Face of column

A_s

Main tension steel
welded to angle
along contact between
bars and inside of angle

Figure 5.17.6 Anchorage of main steel provided by
welding to a confinement angle.

■ EXAMPLE 5.17.1   Design a typical interior bracket that projects from a 14-in. square tied column. It must support a dead load reaction of 30 kips and a live load reaction of 50 kips, resulting from gravity loads. Assume that suitable bearings are provided for the supported prestressed concrete girder so that horizontal restraint forces are eliminated. The tolerance gap between the beam end and column face is 1 in. Use $f'_c = 5000$ psi, $f_y = 60,000$ psi, and the ACI Code.

Solution: **(a)** Factored loads:

$$V_u = 1.4V_D + 1.7V_L = 1.4(30) + 1.7(50) = 127 \text{ kips}$$

(b) Preliminary bracket size. The shear span a is dependent on the bearing length required to support the reaction on the concrete. ACI-10.15 gives nominal bearing strength as $0.85f'_c A_1$, so that using $\phi = 0.70$ (ACI-9.3.2),

$$V_u = \phi(0.85f'_c)A_1$$

$$\text{bearing plate width} = \frac{127,000}{0.70(0.85)(5000)14} = 3.1 \text{ in.}$$

Use $3\frac{1}{2}$ in. for bearing plate width. Allowing the tolerance gap of 1-in. clear between face of column and beam for possible overrun in beam length and also because the beam might be 1 in. too short, the shear span is

$$a = 2 + \tfrac{1}{2}(\text{bearing plate width}) = 2 + 1.75 = 3.75 \text{ in.}$$

(c) Determine depth of bracket. Based on the maximum strength V_n (Eq. 5.17.20) permitted by ACI-11.9.3.2.1,

$$\max V_n = 0.2f'_c b_w d \le (800 \text{ psi})b_w d \qquad \textbf{[5.17.20]}$$

Since $0.2f'_c = 1000$ psi, max $v_n = 800$ psi; then

$$\min d = \frac{V_u}{\phi b(\max v_n)} = \frac{127,000}{0.85(14)800} = 13.3 \text{ in.}$$

If overall $h = 15$ in., $d \approx 13.5$ in. (allowing 1-in. cover), check

$$\frac{a}{d} = \frac{3.75}{13.5} = 0.28 < 1.0 \quad (\text{ACI-11.9.1})$$

(d) Determine the shear-friction reinforcement A_{vf}. Using Eq. (5.16.4) or Eq. (5.17.10) according to ACI-11.7.4.1,

$$A_{vf} = \frac{V_u}{\phi f_y \mu} = \frac{127}{0.85(60)1.4} = 1.78 \text{ sq in.}$$

using $\mu = 1.4$ for monolithically cast concrete.

(e) Determine main tension reinforcement A_s. First calculate the requirement A_f for flexure.

$$M_u = V_u a + N_{uc}(h - d)$$
$$= V_u a = 127(3.75)\tfrac{1}{12} = 39.7 \text{ ft-kips}$$

$$\text{required } R_n = \frac{M_u}{\phi b d^2} = \frac{39.7(12{,}000)}{0.85(14)(13.5)^2} = 220 \text{ psi}$$

Note that $\phi = 0.85$, according to ACI-11.9.3.1. The use of $\phi = 0.90$ if it had been already included in design aids would make little difference. From Eq. (3.8.5) or from Fig. 3.8.1,

$$\text{required } \rho = 0.0035$$

From ACI-11.9.5,

$$\min \rho = 0.04 \frac{f_c'}{f_y} = 0.04 \left(\frac{5}{60}\right) = 0.0033$$

Then,

$$\text{required } A_f = 0.0035(14)(13.5) = 0.66 \text{ sq in.}$$

From ACI-11.9.3.5, the main steel A_s requirement is the larger of Eqs. (5.17.17) and (5.17.18), as follows:

$$A_s = \frac{2}{3} A_{vf} + A_n \text{ (zero in this example)} = \frac{2}{3}(1.78) = 1.19 \text{ sq in.}$$

or

$$A_s = A_f + A_n \text{ (zero in this example)} = 0.66 \text{ sq in.}$$

Use 4-#5 ($A_s = 1.24$ sq in.).

(f) Design closed stirrups or ties. In accordance with ACI-11.9.4, Eq. (5.17.19) requires

$$\text{required } A_h = 0.5(A_s - A_n) = 0.5(1.19) = 0.60 \text{ sq in.}$$

Note that in this design, with small $a/d = 0.28$, the flexure requirement A_f does not affect the design; the shear-friction requirement A_{vf} for V_u controls with $A_{vf} = A_s + A_h$ (Eq. 5.17.14).

Use 3-#3 closed hoops [$A_h = 2(3)0.11 = 0.66$ sq in.]. The spacing of the hoops must be within the upper two-thirds of the effective depth (ACI-11.9.4).

(g) Final design. Referring to Fig. 5.17.7 the overall depth at the face of column is

$$h = 13.5 + 1(\text{cover}) + 0.3125(\text{bar radius}) = 14.8 \text{ in.}$$

Use $h = 15$ in.

The 1-in. cover is used here but it could be as little as the $\frac{5}{8}$-in. bar diameter for precast columns (ACI-7.7.2), but would presumably have to be $1\frac{1}{2}$ in. for cast-in-place members.

Figure 5.17.7 Corbel designed for Example 5.17.1.

At the outer edge of the bearing area, the effective depth d_1 must be at least half of that used at the face of column. In this case, making the outer face 8 in. will just about provide the required d_1 of 13.7/2 (see Fig. 5.17.4).

Another important aspect of the bracket design is the provision of adequate anchorage into the column so that the tensile force $A_s f_y$ is available at the face of the column. Development of reinforcement is treated in Chapter 6. ■

■ **EXAMPLE 5.17.2**   Design a bracket that is to support gravity dead and live loads of 15 and 25 kips, respectively. The vertical reaction is 10 in. from the face of a 14-in. square column. Provide a horizontal reaction of 10 kips due to creep and shrinkage of a restrained beam. Use $f'_c = 5000$ psi, $f_y = 40,000$ psi, and the ACI Code.

Solution: **(a)** Factored loads:

$$V_u = 1.4(15) + 1.7(25) = 21 + 42.5 = 63.5 \text{ kips}$$

$$N_{uc} = 1.7(10) = 17 \text{ kips}$$

ACI-11.9.3.4 states that N_{uc} is to be regarded as live load when it results from creep, shrinkage, or temperature change.

$$\frac{N_{uc}}{V_u} = \frac{17}{63.5} = 0.268 > 0.20 \text{ min} \qquad\qquad \textbf{OK}$$

(b) Depth of bracket for shear.

$$\max V_n = 0.2 f'_c b_w d \le (800 \text{ psi}) b_w d$$

Since $0.2f'_c = 1000$ psi is larger than 800 psi, use 800 psi; then

$$\min d = \frac{V_u}{\phi b_w(800)} = \frac{63,500}{0.85(14)(800)} = 6.7 \text{ in.}$$

Since this is very small, perhaps the flexure requirement will require a larger d (this is a good possibility because the load on the bracket is 10 in. from the face of column).

(c) Depth of bracket for flexure.

$$M_u = V_u a + N_{uc}(h - d)$$
$$= 63.5(10) + 17(h - d)$$

Estimating $(h - d)$ at 2 in.,

$$M_u = 63.5(10) + 17(2) = 669 \text{ in.-kips}$$

Using the minimum reinforcement ratio,

$$\min \rho = 0.04 \frac{f'_c}{f_y} = 0.04 \left(\frac{5}{40} \right) = 0.005$$

which corresponds to minimum $R_n = 193$ psi,

$$\text{required } d = \sqrt{\frac{M_u}{\phi R_n b}} = \sqrt{\frac{669,000}{0.85(193)14}} = 17.1 \text{ in.}$$

For the maximum reinforcement ratio corresponding to 0.75 of that in the balanced strain condition, maximum $R_n = 1209$ psi, which gives

$$\text{required } d = \sqrt{\frac{669,000}{0.85(1209)14}} = 6.8 \text{ in.}$$

(d) Select bracket depth. Since the ACI rules for bracket and corbel design only apply when a/d does not exceed 1.0,

$$\min d = a = 10 \text{ in.}$$

Try a bracket somewhat deeper, say 15 in. overall. This would make $d \approx 13.5$ in.

(e) Determine the shear-friction reinforcement A_{vf}. Using Eqs. (5.16.4) or (5.17.10) according to ACI-11.7.4.1,

$$\text{required } A_{vf} = \frac{V_u}{\phi f_y \mu} = \frac{63.5}{0.85(40)1.4} = 1.33 \text{ sq in.}$$

where $\mu = 1.4$ for monolithic concrete.

(f) Determine the flexure reinforcement A_f. Following ACI-11.9.3.3,

$$\text{required } R_n = \frac{M_u}{\phi b d^2}$$

where

$$M_u = V_u a + N_{uc}(h - d) = 63.5(10) + 17(1.5) = 661 \text{ in.-kips}$$

$$\text{required } R_n = \frac{661,000}{0.85(14)(13.5)^2} = 305 \text{ psi}$$

required $\rho = 0.0081$ [from Eq. (3.8.5) or Fig. 3.8.1]

required $A_f = 0.0081(14)13.5 = 1.53 \text{ sq in.}$

(g) Determine additional reinforcement A_n for axial tension. In accordance with ACI-11.9.3.4, using Eq. (5.17.11),

$$\text{required } A_n = \frac{N_{uc}}{\phi f_y} = \frac{17}{0.85(40)} = 0.50 \text{ sq in.}$$

(h) Total main tension reinforcement A_s. According to ACI-11.9.3.5, the required A_s is the larger of the values from Eqs. (5.17.17) and (5.17.18), as follows:

$$\text{required } A_s = A_f + A_n = 1.53 + 0.50 = 2.03 \text{ sq in.}$$

or

$$\text{required } A_s = \frac{2}{3}A_{vf} + A_n = \frac{2}{3}(1.33) + 0.50 = 1.39 \text{ sq in.}$$

The required $A_s = 2.03 \text{ sq in.}$

Use 5-#6 for main tension steel, $A_s = 2.20 \text{ sq in.}$

(i) Determine stirrup (or tie) requirements. According to ACI-11.9.4,

$$\text{required } A_h = 0.5(A_s - A_n) = 0.5(2.03 - 0.50) = 0.77 \text{ sq in.}$$

Use 3-#4 closed stirrups, $A_h = 0.40(3) = 1.20 \text{ sq in.}$, the spacing of which should be $(\frac{2}{3})13.5/3 = 3.0 \text{ in.}$ (ACI-11.9.4). Use 3-in. spacing.

(j) Overall bracket dimensions. Assuming that a 1-in. thick bearing plate is to be welded to the main tension reinforcement, the overall depth is

$$h = \text{bearing plate} + \text{bar radius} + \text{effective depth, } d$$
$$= 1 + 0.44 + 13.5 = 14.94 \text{ in., say 15 in.}$$

$$\text{bearing plate length} = \frac{V_u}{\phi 0.85 f_c' (\text{column width})}$$

$$= \frac{63,500}{0.70(0.85)5000(14)} = 1.53 \text{ in.}$$

Use a 3-in. plate length as the practical minimum.

$$\text{length of bracket projection} = 2 \text{ in.} + \tfrac{1}{2} \text{ bearing plate} + \text{shear span, } a$$
$$= 2 + 1.5 + 10 = 13.5 \text{ in.}$$

$$\text{depth of outer face of bracket} = \tfrac{1}{2} \text{ overall depth} = 7\tfrac{1}{2} \text{ in.}$$

Final design is shown in Fig. 5.17.8.   ■

Figure 5.17.8 Final design for Example 5.17.2.

SELECTED REFERENCES

GENERAL

5.1. ACI–ASCE Committee 326. "Shear and Diagonal Tension," *ACI Journal, Proceedings*, **59**, January, February, and March 1962, 1–30, 277–344, and 352–396.

5.2. Boris Bresler and James G. MacGregor. "Review of Concrete Beams Failing in Shear," *Journal of the Structural Division*, ASCE, **93**, ST1 (February 1967), 343–372.

5.3. ACI-ASCE Committee 426. "The Shear Strength of Reinforced Concrete Members— Chapters 1 to 4," *Journal of the Structural Division*, ASCE, **99**, ST6 (June 1973), 1091–1187.

5.4. Michael P. Collins. "Towards a Rational Theory for RC Members in Shear," *Journal of the Structural Division*, ASCE, **104**, ST4 (April 1978), 649–666.

5.5. Michael P. Collins and Denis Mitchell. "Shear and Torsion Design of Prestressed and Non-Prestressed Concrete Beams," *PCI Journal*, **25**, September–October 1980, 32–100. Disc. **26**, November–December 1981, 96–118.

5.6. Peter Marti. "Basic Tools of Reinforced Concrete Beam Design," *ACI Journal, Proceedings*, **82**, January–February 1985, 46–56.

5.7. Frank J. Vecchio and Michael P. Collins. "The Modified Compression-Field Theory for Reinforced Concrete Elements Subjected to Shear," *ACI Journal, Proceedings*, **83**, March–April 1986, 219–231. Disc. 84, January–February 1987, 87–90.

5.8. Frank J. Vecchio and Michael P. Collins. "Predicting the Response of Reinforced Concrete Beams Subjected to Shear Using Modified Compression Field Theory," *ACI Structural Journal*, **85**, May–June 1988, 258–268.

5.9. Jörg Schlaich, Kurt Schäfer, and Mattias Jennewein. "Toward a Consistent Design of Structural Concrete," *PCI Journal*, **32**, May–June 1987, 74–150.

5.10. ACI–ASCE Committee 426. "Suggested Revisions to Shear Provisions for Building Codes," (Abstract), *ACI Journal*, *Proceedings*, **74**, September 1977, 458–469.

5.11. ACI–ASCE Committee 426. *Suggested Revisions to Shear Provisions for Building Codes.* Detroit, Michigan: American Concrete Institute, 1979. 82 pp.

AGGREGATE INTERLOCK

5.12. T. Paulay and P. J. Loeber. "Shear Transfer by Aggregate Interlock," *Shear in Reinforced Concrete*, Vol. 1 (SP–42). Detroit: American Concrete Institute, 1974 (pp. 503–537).

5.13. Theodossius P. Tassios and Elizabeth N. Vintzeleou, "Concrete-to-Concrete Friction," *Journal of Structural Engineering*, ASCE, **113**, 4 (April 1987), 832–849.

5.14. Sandro Dei Poli, Pietro G. Gambarova, and Cengiz Karakoc. "Aggregate Interlock Role in R.C. Thin-Webbed Beams in Shear," *Journal of Structural Engineering*, ASCE, **113**, 1 (January 1987), 1–19.

DOWEL ACTION

5.15. David W. Johnston and Paul Zia. "Analysis of Dowel Action," *Journal of the Structural Division*, ASCE, **97**, ST5 (May 1971), 1611–1630.

5.16. Rafael Jimenez, Richard N. White, and Peter Gergely. "Cyclic Shear and Dowel Action Models in R/C," *Journal of the Structural Division*, ASCE, **108**, ST5 (May 1982), 1106–1123.

5.17. Parviz Soroushian, Kienuwa Obaseki, Maximo C. Rojas, and Jongsung Sim. "Analysis of Dowel Bars Acting Against Concrete Core," *ACI Journal*, *Proceedings*, **83**, July–August 1986, 642–649.

5.18. E. N. Vintzeleou and T. P. Tassios. "Behavior of Dowels under Cyclic Deformations," *ACI Structural Journal*, **84**, January–February 1987, 18–30.

5.19. P. Soroushian, K. Obaseki, M. Rojas, and H. S. Najm. "Behavior of Bars in Dowel Action Against Concrete Cover," *ACI Structural Journal*, **84**, March–April 1987, 170–176.

5.20. Elizabeth N. Vintzeleou and Theodossius P. Tassios. "Eccentric Dowels Loaded Against Core of Concrete Sections," *Journal of Structural Engineering*, ASCE, **116**, 10 (October 1990), 2621–2633.

BEAMS WITHOUT SHEAR REINFORCEMENT

5.21. JoDean Morrow and I. M. Viest. "Shear Strength of Reinforced Concrete Frame Members Without Web Reinforcement," *ACI Journal*, *Proceedings*, **53**, March 1957, 833–869.

5.22. G. N. J. Kani. "The Riddle of Shear Failure and Its Solution," *ACI Journal, Proceedings*, **61**, April 1964, 441–467.

5.23. G. N. J. Kani. "Basic Facts Concerning Shear Failure," *ACI Journal, Proceedings*, **63**, June 1966, 675–692. Disc. pp. 1511–1528.

5.24. G. N. J. Kani. "How Safe Are Our Large Reinforced Concrete Beams," *ACI Journal, Proceedings*, **64**, March 1967, 128–141.

5.25. John M. Hanson. *Square Openings in Webs of Continuous Joists* (RD001.01D). Skokie, Illinois: Portland Cement Association, 1969.

LIGHTWEIGHT AGGREGATE CONCRETE

5.26. J. A. Hanson. "Tensile Strength and Diagonal Tension Resistance of Structural Lightweight Concrete," *ACI Journal, Proceedings*, **58**, July 1961, 1–40.

5.27. E. Hognestad, R. C. Elstner, and J. A. Hanson. "Shear Strength of Reinforced Structural Lightweight Aggregate Concrete Slabs," *ACI Journal, Proceedings*, **61**, June 1964, 643–656.

5.28. Don L. Ivey and Eugene Buth. "Shear Capacity of Lightweight Concrete Beams," *ACI Journal, Proceedings*, **64**, October 1967, 634–643.

HIGH-STRENGTH CONCRETE

5.29. Andrew G. Mphonde and Gregory C. Frantz. "Shear Tests of High- and Low-Strength Concrete Beams Without Stirrups," *ACI Journal, Proceedings*, **81**, July–August 1984, 350–357.

5.30. Ashraf H. Elzanaty, Arthur H. Nilson, and Floyd O. Slate. "Shear Capacity of Reinforced Concrete Beams Using High-Strength Concrete," *ACI Journal, Proceedings*, **83**, March–April 1986, 290–296.

5.31. Shuaib H. Ahmad, A. R. Khaloo, and A. Poveda. "Shear Capacity of Reinforced High-Strength Concrete Beams," *ACI Journal, Proceedings*, **83**, March–April 1986, 297–305.

5.32. Ashraf H. Elzanaty, Arthur H. Nilson, and Floyd O. Slate. "Shear Capacity of Prestressed Concrete Beams Using High-Strength Concrete," *ACI Journal, Proceedings*, **83**, May–June 1986, 359–368.

5.33. Shuaib H. Ahmad and D. M. Lue. "Flexure-Shear Interaction of Reinforced High-Strength Concrete Beams," *ACI Structural Journal*, **84**, July–August 1987, 330–341. Disc. **85**, May–June 1988, 354–358.

5.34. M. Keith Kaufman and Julio A. Ramirez. "Re-evaluation of the Ultimate Shear Behavior of High-Strength Concrete Prestressed I-Beams," *ACI Structural Journal*, **85**, May–June 1988, 295–303.

5.35. Miguel A. Salandra and Shuaib H. Ahmad. "Shear Capacity of Reinforced Lightweight High-Strength Concrete Beams," *ACI Structural Journal*, **86**, November–December 1989, 697–704.

5.36. John J. Roller and Henry G. Russell. "Shear Strength of High-Strength Concrete Beams with Web Reinforcement," *ACI Structural Journal*, **87**, March–April 1990, 191–198.

SHEAR REINFORCEMENT

5.37. Wayne Hsiung and Gregory C. Frantz. "Transverse Stirrup Spacing in R/C Beams," *Journal of Structural Engineering*, ASCE, **111**, 2 (February 1985), 353–362. Disc. **113**, 1 (January 1987), 174–177.

5.38. Mark K. Johnson and Julio A. Ramirez. "Minimum Shear Reinforcement in Beams with Higher Strength Concrete," *ACI Structural Journal*, **86**, July–August 1989, 376–382. Disc. **87**, May–June 1989, 362–364.

5.39. Neal S. Anderson and Julio A. Ramirez. "Detailing of Stirrup Reinforcement," *ACI Structural Journal*, **86**, September–October 1989, 507–515.

5.40. Andrew G. Mphonde. "Use of Stirrup Effectiveness in Shear Design of Concrete Beams," *ACI Structural Journal*, **86**, September–October 1989, 541–545.

5.41. Andrew G. Mphonde and Gregory C. Frantz. "Shear Tests of High and Low-Strength Concrete Beams with Stirrups, *High Strength Concrete* (SP-87). Detroit: American Concrete Institute, 1985 (pp. 179–196).

5.42. Abdeldjelil Belarbi and Thomas T. C. Hsu. "Stirrup Stresses in Reinforced Concrete Beams," *ACI Structural Journal*, **87**, September–October 1990, 530–538.

5.43. CSA. *Design of Concrete Structures for Buildings* (CAN3-A23.3-M84). Rexdale, Ontario: Canadian Standards Association (178 Rexdale Blvd, Rexdale, Ontario, Canada M9W 1R3), 1984 (281 pp).

5.44. S. M. Fereig and K. N. Smith. "Indirect Loading on Beams with Short Shear Span," *ACI Journal*, *Proceedings*, **74**, May 1977, 220–222.

5.45. K. S. Rajagopalan and P. M. Ferguson. "Exploratory Shear Tests Emphasizing Percentage of Longitudinal Steel," *ACI Journal*, *Proceedings*, **65**, August 1968, 634–638. Disc. **66**, 150–154.

5.46. Barrington deV. Batchelor and Mankit Kwun. "Shear in RC Beams Without Web Reinforcement," *Journal of the Structural Division*, ASCE, **107**, ST5 (May 1981), 907–921.

5.47. Michael N. Palaskas, Emmanuel K. Attiogbe, and David Darwin. "Shear Strength of Lightly Reinforced T-Beams," *ACI Journal*, *Proceedings*, **78**, November–December 1981, 447–455.

AXIAL LOAD EFFECT

5.48. James G. MacGregor and John M. Hanson. "Proposed Changes in Shear Provisions for Reinforced and Prestressed Concrete Beams," *ACI Journal, Proceedings*, **66**, April 1969, 276–288. Disc. 849–851.

5.49. Alan H. Mattock. "Diagonal Tension Cracking in Concrete Beams With Axial Forces," *Journal of the Structural Division*. ASCE, **95**, ST9 (September 1969), 1887–1900.

5.50. Munther J. Haddadin, Sheu-Tien Hong, and Alan M. Mattock. "Stirrup Effectiveness in Reinforced Concrete Beams With Axial Force," *Journal of the Structural Division*, ASCE, **97**, ST9 (September 1971), 2277–2297.

5.51.   Shrinivas B. Bhide and Michael P. Collins. "Influence of Axial Tension on the Shear Capacity of Reinforced Concrete Members," *ACI Structural Journal*, **86**, September–October 1989, 570–581.

5.52.   Alan H. Mattock and Zuhua Wang. "Shear Strength of Reinforced Concrete Members Subject to High Axial Compressive Stress," *ACI Journal, Proceedings*, **81**, May–June 1984, 287–298.

DEEP BEAMS

5.53.   D. M. Rogowsky and J. G. MacGregor. "Design of Reinforced Concrete Deep Beams," *Concrete International*, **8**, August 1986, 49–58.

5.54.   H. A. Rawdon de Paiva and Chester P. Siess. "Strength and Behavior of Deep Beams in Shear," *Journal of the Structural Division*, ASCE, **91**, ST5 (October 1965), 19–42.

5.55.   R. A. Crist. "Shear Behavior of Deep Reinforced Concrete Beams," *Proceedings*, Symposium on the Effects of Repeated Loading of Materials and Structural Elements (Mexico City, 1966), **4**, RILEM, Paris (31 pp). (Published by Instisto Mexicano del Cemento y del Concreto, Mexico D.F., Mexico)

5.56.   J. G. MacGregor and N. M. Hawkins. "Suggested Revisions to ACI Building Code Clauses Dealing with Shear Friction and Shear in Deep Beams and Corbels," *ACI Journal, Proceedings*, **74**, November 1977, 537–545. Disc., **75**, May 1978, 221–224.

5.57.   David M. Rogowsky, James G. MacGregor, and See Y. Ong. "Tests of Reinforced Concrete Deep Beams," *ACI Journal, Proceedings*, **83**, July–August 1986, 614–623.

5.58.   K. N. Smith and S. M. Fereig. "Effect of Loading and Supporting Conditions on the Shear Strength of Deep Beams," *Shear in Reinforced Concrete*, Vol. 2 (SP–42). Detroit: American Concrete Institute, 1974 (pp. 441–460).

5.59.   K. N. Smith and A. S. Vantsiotis. "Shear Strength of Deep Beams," *ACI Journal, Proceedings*, **79**, May–June 1982, 201–213.

5.60.   S. T. Mau and Thomas T. C. Hsu. "Shear Strength Prediction for Deep Beams with Web Reinforcement," *ACI Structural Journal*, **84**, November–December 1987, 513–523.

5.61.   S. T. Mau and Thomas T. C. Hsu. "Formula for the Shear Strength of Deep Beams," *ACI Structural Journal*, **86**, September–October 1989, 516–523.

5.62.   M. Haque, Rasheeduzzafar, and A. H. J. Al-Tayyib. "Stress Distribution in Deep Beams with Web Openings," *Journal of Structural Engineering*, ASCE, **112**, 5, (May 1986), 1147–1165.

5.63.   Perry Adebar, Daniel Kuchma, and Michael P. Collins. "Strut-and-Tie Models for the Design of Pile Caps: An Experimental Study," *ACI Structural Journal*, **87**, January–February 1990, 81–92.

5.64.   Michael Schlaich and Georg Anagnostou. "Stress Fields for Nodes of Strut-and-Tie Models," *Journal of Structural Engineering*, ASCE, **116**, 1 (January 1990), 13–23.

5.65.   William D. Cook and Denis Mitchell. "Studies of Disturbed Regions near Discontinuities in Reinforced Concrete Members," *ACI Structural Journal*, **85**, March–April 1988, 206–216.

5.66. James G. MacGregor. *Reinforced Concrete: Mechanics and Design.* Englewood Cliffs, New Jersey: Prentice-Hall, 1988 (pp. 687–708).

SHEAR TRANSFER; SHEAR-FRICTION; CORBELS

5.67. Philip W. Birkeland and Halvard W. Birkeland. "Connections in Precast Concrete Construction," *ACI Journal, Proceedings,* **63**, March 1966, 345–368.

5.68. R. F. Mast. "Auxiliary Reinforcement in Precast Concrete Connections," *Journal of the Structural Division*, ASCE, **94**, ST6 (June 1968), 1485–1504.

5.69. J. A. Hofbeck, I. O. Ibrahim, and Alan H. Mattock. "Shear Transfer in Reinforced Concrete," *ACI Journal, Proceedings,* **66**, February 1969, 119–128. Disc. 678–680.

5.70. A. H. Mattock and N. M. Hawkins. "Research on Shear Transfer in Reinforced Concrete," *PCI Journal*, **17**, March–April 1972, 55–75.

5.71. Bjorn R. Hermansen and John Cowan. "Modified Shear-Friction Theory for Bracket Design," *ACI Journal, Proceedings,* **71**, February 1974, 55–60.

5.72. A. H. Mattock. Disc. of "Modified Shear-Friction Theory for Bracket Design," by B. R. Hermansen and J. Cowan, *ACI Journal, Proceedings,* **71**, August 1974, 421–423.

5.73. A. H. Mattock. "Shear Transfer in Concrete Having Reinforcement at an Angle to the Shear Plane," *Shear in Reinforced Concrete*, Vol. 1 (SP–42). Detroit: American Concrete Institute, 1974 (pp. 17–42).

5.74. Alan H. Mattock, L. Johal, and H. C. Chow. "Shear Transfer in Reinforced Concrete with Moment or Tension Acting Across the Shear Plane," *PCI Journal*, **20**, July–August 1975, 76–93.

5.75. Alan H. Mattock, W. K. Li, and T. C. Wang. "Shear Transfer in Lightweight Reinforced Concrete," *PCI Journal*, **21**, January–February 1976, 20–39. *PCI Design Handbook* (2nd ed.). Chicago: Prestressed Concrete Institute, 1978.

5.76. Joost Walraven, Jerome Frenay, and Arjan Pruijssers. "Influence of Concrete Strength and Load History on the Shear Friction Capacity of Concrete Members," *PCI Journal*, **32**, January–February 1987, 66–84.

5.77. Thomas T. C. Hsu, S. T. Mau, and Bin Chen. "Theory of Shear Transfer Strength of Reinforced Concrete," *ACI Structural Journal*, **84**, March–April 1987, 149–160.

5.78. Robert A. Bass, Ramon L. Carrasquillo, and James O. Jirsa. "Shear Transfer Across New and Existing Concrete Interfaces," *ACI Structural Journal*, **86**, July–August 1989, 383–393.

5.79. G. Annamalal and Robert C. Brown, Jr. "Shear-Transfer Behavior of Post-Tensioned Grouted Shear-Key Connections in Precast Concrete-Framed Structures," *ACI Structural Journal*, **87**, January–February 1990, 53–59.

5.80. PCI Committee on Connection Details. *Design and Typical Details of Connections for Precast and Prestressed Concrete*, 2nd ed. Chicago: Prestressed Concrete Institute, 1988.

5.81. L. B. Kriz and C. H. Raths. "Connections in Precast Concrete Structures—Strength of Corbels," *PCI Journal*, **10**, 1 (February 1965), 16–47.

5.82. Alan H. Mattock, K. C. Chen, and K. Soongswang. "The Behavior of Reinforced Concrete Corbels," *PCI Journal*, **21**, March–April 1976, 52–77.

5.83. Alan H. Mattock. "Design Proposals for Reinforced Concrete Corbels," *PCI Journal*, **21**, May–June 1976, 18–42. Disc., **22**, March–April 1977, 90–109.

5.84. A. Fattah Shaikh. "Proposed Revisions to Shear-Friction Provisions," *PCI Journal*, **23**, March–April 1978, 12–21.

5.85. Himat Solanki and Gajanan M. Sabnis. "Reinforced Concrete Corbels—Simplified," *ACI Structural Journal*, **84**, September–October 1987, 428–432.

PROBLEMS

All problems* are to be worked in accordance with the strength method of the ACI Code unless otherwise indicated, and all stated loads are service loads. The "correct" shear envelope is to be used (i.e., live load must be appropriately applied such as to cause maximum effect). All shear diagrams are to be drawn to scale in terms of V_u directly below the diagram of the beam, also drawn to scale. Use the V_u diagram for design by scaling values from it; also scale from it the locations where stirrup spacings are permitted. Computation of V_u values and locations of stirrup spacings may be made to confirm scaled information.

5.1 The simply supported beam of 16-ft span is to carry a uniform dead load of 1.5 kips/ft (including beam weight) and a uniform live load of 2.4 kips/ft. Use $f'_c = 3500$ psi and $f_y = 40,000$ psi.

(a) Determine the adequacy of the #3 U stirrups that are spaced at 8 in. Use the simplified method with constant V_c.

(b) Draw superimposed on the factored shear V_u diagram the diagram of strength provided ϕV_n. From the comparison of ϕV_n with V_u, determine if the stirrups are adequate for the entire beam. If not, make recommendations to obtain a beam having adequate shear strength.

Problem 5.1

— — — —

* Some problems may be solved as problems stated in Inch-Pound units, or as problems in SI units using quantities in parenthesis at the end of the statement. The metric conversions are approximate to avoid implying higher precision for the given information in SI units than that given for the Inch-Pound units.

5.2   The beam of the accompanying figure carries a uniform live load of 3.0 kips/ft in addition to its own weight. Assume a support width of 12 in., and use $f'_c = 3000$ psi and $f_y = 40,000$ psi.

(a) Draw the maximum factored shear V_u envelope. Calculate the value at the critical section, at the $\frac{1}{4}$ point, and at midspan; then draw a smooth curve through these points for design use.

(b) Draw the curve of required stirrup spacing using the simplified procedure permitted by ACI-11.3.1.1; show on the same diagram the spacings provided.

(c) Evaluate whether or not the spacings used satisfy the ACI Code using the simplified method of constant V_c.

(d) Repeat (b) using the more detailed procedure of ACI-11.3.2.1; show also the spacings provided.

(e) Evaluate whether or not the spacings used are satisfactory according to the analysis of (d).

Problem 5.2

5.3   The beam of the accompanying figure is to carry 1.6 kips/ft live load and 0.90 kip/ft dead load (including beam weight). Using $f'_c = 3000$ psi and $f_y = 40,000$ psi, investigate the beam for stirrup adequacy according to the simplified method using constant V_c. If design is not adequate, indicate what revision is necessary.

Problems 5.3 and 5.15

5.4   For the portion of the continuous beam shown, with the given portion of the factored shear V_u envelope, determine the spacings to be used for #3 U stirrups. Dimension and show the stirrups on the given portion of beam. Use the simplified method of constant V_c, with $f'_c = 3500$ psi and $f_y = 60,000$ psi. (Beam: $b = 300$ mm; $d = 530$ mm; support width $= 300$ mm; half-span $= 2.7$ m; V_u at support $= 260$ kN; V_u at midspan $= 80$ kN; use #10M stirrups; $f'_c = 25$ MPa; $f_y = 400$ MPa.)

Problem 5.4

5.5   For the Case assigned by the instructor, use the simplified method of ACI-11.3.1.1 to design vertical U stirrups (that is, specify their size, dimension their locations, and show them on a side view of the beam). Use whole inches for spacings. The beam is simply supported, has a support width of 18 in., and the given dead load does *not* include the beam weight. Use a stirrup size such that the spacing will not be closer than 3 in. Explicitly state the length of the beam over which stirrups are required.

CASE	SPAN (ft)	b_w (in.)	d (in.)	A_s (sq in.)	DEAD LOAD (kips/ft)	LIVE LOAD (kips/ft)	f'_c (psi)	f_y (psi)
1	20	18	36.3	5-#10 5-#11	6	10	4000	40,000
2	26	18	29.6	6-#9	1.6	2.7	4000	60,000
3	30	16	24.6	9-#9	1.2	2.5	4000	60,000
4	28	18	29.6	6-#9	1.6	2.5	4000	60,000

Problem 5.5

5.6   For the simply supported beam of the accompanying figure, having a span of 32 ft, with uniform dead load of 2.3 kips/ft (including beam weight) and live load of 3.7 kips/ft, design #4 vertical U stirrups using whole inches for spacings. Show the stirrups in a side view of the beam and dimension their locations. Use $f'_c = 3750$ psi and $f_y = 50,000$ psi, and the simplified method with constant V_c. The support width is 12 in.

Problem 5.6

5.7   A reinforced concrete simply supported beam ($b = 250$ mm, $d = 410$ mm) must carry on a span of 5.5 m the single concentrated moving load of 45 kN plus a uniform dead load of 30 kN/m (including beam weight). Design and detail the stirrup spacing (use 20 mm increments) for #10M vertical U stirrups; support width is 300 mm; $f'_c = 25$ MPa; $f_y = 400$ MPa. Apply the simplified method using a constant value for V_c. See SI footnote to Eq. (5.11.3).

5.8   The beam in the accompanying figure is to carry dead load of 1.4 kips/ft (including beam weight) and live load of 1.6 kips/ft. Use $f'_c = 3500$ psi, $f_y = 40,000$ psi, and neglect any compression steel effect. Use a "correct" factored shear V_u envelope. Using the simplified procedure of constant V_c, design and detail the spacings for #3 U stirrups for the beam. Use whole inches for spacings.

5.9   Repeat Problem 5.8, except use an 18-ft main span and a 7-ft cantilever, with dead load of 1.5 kips/ft (including beam weight) and live load of 1.7 kips/ft. All other dimensions and reinforcing bars are the same as in Problem 5.8.

Problems 5.8 and 5.9

5.10   The beam in the accompanying figure is to carry a dead load of 55 kN/m (including beam weight) and a live load of 72 kN/m. Use $f'_c = 25$ MPa and $f_y = 400$ MPa. Using the simplified procedure of constant V_c; design and detail on the beam 20 mm increment spacings of U stirrups for the beam. Use stirrups of a size such that the 20-mm increment spacing will not be closer than 80 mm.

Problem 5.10

5.11   For a simply supported beam of 16-ft span, having support widths of 12 in., design and detail on the beam U stirrups (use no spacing less than 3 in.). The dead load is 1.6 kips/ft (including weight of beam) and the live load is 3.0 kips/ft. Use $f'_c = 3000$ psi and $f_y = 60,000$ psi. Use the ACI Code simplified procedure with constant V_c. Assume $b_w = 14$ in. and $d = 21.5$ in.

5.12   The beam of the accompanying figure carries a live load of 2.7 kips/ft in addition to the weight of the slab and beam. Design and detail on the beam the vertical U stirrup spacing for #3 stirrups. Use $f'_c = 3000$ psi and $f_y = 40,000$ psi. Use the ACI Code simplified procedure with constant V_c.

Problem 5.12

5.13 A reinforced concrete simply supported beam of span 5.5 m carries a concentrated dead load of 23 kN at 1.8 m from the left support, and a uniform dead load of 90 kN/m. The width of support is 300 mm. The rectangular beam has a 300-mm width and a 650 mm effective depth d. Use $f'_c = 30$ MPa and $f_y = 400$ MPa. Design and specify by dimensioning, the spacings to be used for #10M U stirrups. Use the simplified procedure of ACI-11.3.1.1.

5.14 For a rectangular beam of 14 in. width and effective depth 22.5 in. with $f'_c = 4000$ psi and $f_y = 60,000$ psi, determine the maximum factored shear V_u for this beam for the following conditions:

(a) When no stirrups are to be used.

(b) When minimum percentage of shear reinforcement (#3 U stirrups) is used according to ACI-11.5.5.3; specify the spacing to be used.

(c) When maximum percentage of stirrups is used (#4 U stirrups); specify the spacing to be used.

5.15 Completely design and detail the stirrups for the beam of Problem 5.3 (ignore the spacings given), using the more detailed $\rho Vd/M$ method of ACI-11.3.2.1.

5.16 For the Case assigned by the instructor, repeat the requirements of Problem 5.5, except use the more detailed $\rho Vd/M$ method of ACI-11.3.2.1.

5.17 Repeat the requirements of Problem 5.8, except use the more detailed $\rho Vd/M$ method of ACI-11.3.2.1.

5.18 The beam of the accompanying figure is to carry a uniform live load of 0.6 kips/ft in addition to the concentrated load shown and the beam weight. Use $f'_c = 4000$ psi and $f_y = 50,000$ psi.

(a) If the bent-up longitudinal bars are counted on to resist shear, is the beam adequate as shown?

(b) If the beam is not adequate, prescribe the number and location of #3 U stirrups required to act along with the bent-up bars.

Problem 5.18

5.19   If a factored axial compression N_{uc} of 140 kips is acting additionally on the beam of Example 5.13.1, determine the number of stirrups that may be eliminated by taking the compression into account when computing V_c by the more detailed procedure involving $\rho Vd/M$. (Note: This compressive force is approximately $0.1f'_cA_g$ and might reasonably be neglected when designing the section for flexure.)

5.20   Reinvestigate the shear reinforcement for the beam of Problem 5.1 if an axial tensile force of 35 kips live load is acting. Redesign stirrups for the beam using the more detailed procedure of ACI-11.3.2.1.

5.21   Redesign the stirrups for the beam case of Problem 5.5 assigned by the instructor if an axial compressive force of 70 kips dead load and 120 kips live load is acting.

(a) Use simplified method using ACI Formula (11-4).

(b) Use more detailed $\rho Vd/M$ method.

5.22   Design the shear reinforcement for the beam shown. The rectangular continuous beam is to support heavy machinery, and the service loading is 12 kips/ft dead load and 40 kips/ft live load (including impact). The factored moment M_u diagram is given in the accompanying figure. Use $f'_c = 4000$ psi and $f_y = 60,000$ psi.

Problem 5.22

5.23   Design the flexural and shear reinforcement for a beam 16 in. wide × 52 in. deep overall to carry two concentrated live loads of 200 kips each symmetrically placed at 4 ft from the ends of a 20-ft span having support widths of 2 ft. Assume simple support similar to Fig. 5.15.2. Use $f'_c = 4000$ psi and $f_y = 60,000$ psi. (Beam: $b = 400$ mm; $h = 1.3$ m; concentrated loads $= 890$ kN; distance from ends $= 1.2$ m; span $= 6$ m; support width $= 0.6$ m; $f'_c = 30$ MPa; $f_y = 400$ MPa.)

5.24   Design the details of the bearing shoe on a prestressed girder of 12 in. width. Assume an angle will be used across the width for bearing as in Fig. 5.16.1(b). The reaction is 35 kips dead load and 40 kips live load. The girder concrete has $f'_c = 6000$ psi. Assume no horizontal restraint is developed.

5.25   Design a bracket (corbel) that projects from one side of a 16 × 16 column to support a vertical load of 35 kips dead load and 65 kips live load. Assume that suitable bearings are provided so that horizontal restraint is eliminated. The reaction is located 5 in. from the column face. Use $f'_c = 5000$ psi and $f_y = 60,000$ psi. (Column size $= 400 × 400$ mm; dead load $= 160$ kN; live load $= 290$ kN; reaction 130 mm from column face; $f'_c = 35$ MPa; $f_y = 400$ MPa.)

5.26   Redesign the bracket (corbel) of Problem 5.25 if the reaction is 3.5 in. (90 mm) from the column face.

5.27   Design for the conditions of Problem 5.25 except take the reaction location 9 in. (230 mm) from the column face.

5.28   Repeat Problem 5.25 if the reaction is from a restrained beam that induces a horizontal tension equal to 50% of the total gravity reaction.

5.29   Redesign the bracket (corbel) of Problem 5.28 if the reaction is 3.5 in. (90 mm) from the column face.

5.30   Repeat Problem 5.27 if the reaction is from a restrained beam that induces a horizontal tension equal to 40% of the total gravity reaction.

5.31   Redesign the bracket (corbel) of Example 5.17.1 considering that the supported prestressed girder is welded to the bracket. Creep, shrinkage, and temperature effects on the restrained girder induce a horizontal force of 50 kips (unfactored) on the bracket.

CHAPTER 6

Development of Reinforcement

6.1 GENERAL

A basic requirement in reinforced concrete construction is that there is adequate means for transfer of the force in the reinforcement to the surrounding concrete. This transfer may arise from the adhesion at the surface area of the reinforcing bar, as well as from bearing of the raised ribs, or "lugs", of the deformed bars against the concrete. The interacting force preventing slip of the longitudinal bars relative to the surrounding concrete has traditionally been referred to as *bond*.

In recent years, the term *development length* has been used because it better identifies a certain length of embedment, through which the force in a bar may increase from zero at its end to a point where its full strength is available.

When the failure mode is slippage of a reinforcing bar relative to its surrounding concrete, the development length is roughly proportional to the bar diameter. This would be the case for plain bars (that is, bars circular in cross-section but without surface deformations), bars having small deformations (commonly called lugs), and bars of small diameter. This mode is often called a "pullout" failure.

The more common failure mode is "splitting" of the surrounding concrete, resulting from excessive bearing of the reinforcing bar deformations against that concrete. In this case, the development length is roughly proportional to the bar area. Since splitting depends on the ability of the concrete to resist tension, favorable conditions are increased concrete cover and larger spacing between bars, as well as the confinement capabilities of any ties and stirrups.

The concepts relating to the transfer of force between reinforcement and the surrounding concrete by a development length, with or without additional mechanical end anchorage, are presented in the next several sections. The mechanics of the behavior has been explained by ACI Committee 408 [6.1], Lutz and Gergely [6.2], Orangun, Jirsa, and Breen [6.3], Losberg and Olsson [6.4], Jirsa, Lutz, and Gergely [6.5], Yankelevsky [6.6], and Kemp [6.7].

Concrete cantilever beams, Denver, Colorado. (Photo by C. G. Salmon.)

6.2 DEVELOPMENT LENGTH

Design of longitudinal and shear reinforcement to accommodate the moment and shear at sections along a beam has been treated in Chapters 3, 4, and 5. To resist bending moment, an area of longitudinal steel is provided to carry the tensile force. However, the bars must be embedded in the concrete sufficiently so that the tensile force can *develop*. If there is inadequate development length, either the bars will pull out or split the surrounding concrete.

The moment strength (capacity) of a beam is, therefore, a three-dimensional relationship involving not only the cross-sectional properties at a location along the span, but also the embedment lengths of the steel bars in both directions therefrom.

Consider a uniformly loaded cantilever beam as shown in Fig. 6.2.1(a), which has been properly proportioned so that at nominal strength M_n the steel force is $A_s f_y$. To illustrate the concept of development length, assume the tension reinforcement consists of a single bar of diameter d_b. When the bar segment AB is considered as a free body as shown in Fig. 6.2.1(b), the tensile force at B, which is $f_y(\pi d_b^2/4)$, must be transmitted to the concrete by the interaction between the bar and the surrounding

Figure 6.2.1 Development of reinforcement.

concrete in the development length $L_1 = AB$. If u_s is the failure stress against slippage acting over the nominal surface area $\pi d_b L_1$, then

$$u_s \pi d_b L_1 = f_y \pi \frac{d_b^2}{4}$$

or

$$L_1 = \frac{f_y}{4u_s} d_b \tag{6.2.1}$$

On the other hand, if u_b is the failure stress against splitting and A_{br} is the average bearing area per unit length, then

$$u_b A_{br} L_1 = f_y \pi \frac{d_b^2}{4}$$

or

$$L_1 = \frac{f_y}{A_{br} u_b} \pi \frac{d_b^2}{4} \tag{6.2.2}$$

The same situation exists in free body BC, as shown in Fig. 6.2.1(c). Thus the maximum tensile force at B has to develop by embedment in both directions from B; that is, both the AB and BC distances. Where space limitations prevent providing the proper amount of straight embedment, such bars may be terminated by standard hooks (as defined in ACI-7.1). A standard hook is permitted to be considered as contributing to an equivalent development length by mechanical action (ACI-12.5), thus reducing the total embedment dimension required. Section 6.11 provides treatment of development length with standard hooks.

Adequate development length must be provided for a reinforcing bar in compression as well as in tension.

6.3 FLEXURAL BOND

As bending moment varies along a span, the tensile force in the steel also varies; this induces a longitudinal interaction between the bars and the surrounding concrete, known as *flexural bond*. High localized stress, either at the bar surface or at the bearing of the steel reinforcement lugs against the concrete, exists at locations along the span where the rate of change of tensile force in the bars is high.

Consider a segment DD' of the reinforcing bar in the cantilever beam used in Section 6.2. As shown by the free body of DD' in Fig. 6.3.1(b), T_D is slightly greater than $T_{D'}$. The bending moment M_D equals the internal force C or T times the moment arm between them; thus,

$$T_D = \frac{M_D}{\text{arm}} \quad \text{and} \quad T_{D'} = \frac{M_{D'}}{\text{arm}} \tag{6.3.1}$$

Also, from horizontal force equilibrium,

$$u_s \pi d_b \, dz + u_b \frac{\pi d_b^2}{4} A_{br} \, dz = T_D - T_{D'} \tag{6.3.2}$$

in which u_s is the localized surface stress over the nominal contact area between the steel bar and the concrete, d_b is the diameter of the single bar, and u_b is the localized bearing stress over the area A_{br} per unit length between the lugs and the concrete.

Whether the action represented by the left side of Eq. (6.3.2) is more "pullout" or "splitting", the magnitude depends on $(T_D - T_{D'})/dz$, which from Eq. (6.3.1) is

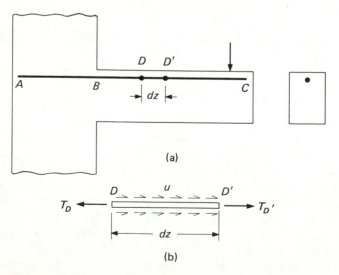

(a)

(b)

Figure 6.3.1 Flexural bond in a tension bar.

$dM/(\text{arm})\,dz = V/(\text{arm})$. Thus, this *localized* interaction (i.e., flexural bond) between bar and surrounding concrete is proportional in magnitude to the shear.

Even though high localized surface stress on bars resulting from the rate of change of moment may seem to be important for proper design, research and experience in practice have shown that prevention of bar "pullout" failure or "splitting" failure by having adequate development length of embedment is a sufficient criterion. Therefore, high localized stress need not be of concern.

6.4 FAILURE MODES FROM INADEQUATE DEVELOPMENT OF REINFORCEMENT

The term "bond failures" has been given to the mechanisms by which failures occur when inadequate development length is provided. Years ago, when plain bars (relatively smooth bars without lug deformations) were used, slip resistance ("bond") was thought of as adhesion between concrete paste and the surface of the bar. Yet even with low tensile stress in the reinforcement, there was sufficient slip immediately adjacent to a flexural crack in the concrete to break the adhesion, leaving only friction to resist bar movement relative to the surrounding concrete over the slip length.

Shrinkage can also cause frictional drag against the bars. Typically, a hot-rolled *plain* bar may pull loose by longitudinal splitting when the adhesion and friction resistances are high, or just pull out leaving a cylindrical hole when adhesion and friction resistances are low.

Deformed bars were created to change the behavior pattern so that there would be less reliance on friction and adhesion (though they still exist) and more reliance on the bearing of the lugs against the concrete. The bearing forces act at an angle to the axis of the bar, causing radial outward components against the concrete, as shown in Fig. 6.4.1. When inadequate development length is provided, deformed bars in normal-weight concrete give rise to a splitting mode of failure (i.e., "bond failure") [6.1, 6.5, 6.7]. A splitting failure occurs when the wedging action of the steel lugs on a deformed bar causes cracks in the surrounding concrete parallel to the bar. These cracks occur between the bar and the nearest concrete face, as shown in Fig. 6.4.2(a, b), or over the short distance between bars when bars are closely spaced, as in Fig. 6.4.2(c).

When small size bars are used with large cover, the lugs may crush the concrete by bearing and result in a pullout failure without splitting the concrete. This nonsplitting failure has also been reported for larger bars in structural lightweight concrete [6.1].

(a) On bar   (b) On concrete   (c) Components on concrete

Figure 6.4.1 Forces between bar and surrounding concrete.

Figure 6.4.2 Splitting cracks and ultimate splitting failure modes (from Reference 6.1).

Although splitting is the usual failure mode, an initial splitting crack on one face of a beam is *not* considered failure. The distress sign indicating failure is *progressive splitting.* Confinement of tension steel by stirrups, ties, or spirals usually will delay collapse (commonly defined as an increase in loading that results in no increase in resistance) until several splitting cracks have formed.

Originally, development length requirements were based on pullout tests [6.8] of plain bars, followed by pullout tests [6.9–6.15] of deformed bars, including the related load-slip data. Since confinement exists in pullout tests, the early work did not give sufficient emphasis to the splitting mode of failure. Splitting has been emphasized in the more recent studies by Orangun, Jirsa, and Breen [6.3], Untrauer and Warren [6.16], Losberg and Olsson [6.4], Kemp and Wilhelm [6.17], Morita and Kaku [6.18], Jimenez, White, and Gergely [6.19], and Kemp [6.7].

The studies of Orangun, Jirsa, and Breen [6.3] and Untrauer and Warren [6.16] have hypothesized that the action of splitting arises from a stress condition analogous to a concrete cylinder surrounding a reinforcing bar and acted upon by the outward radial components [Fig. 6.4.1(c)] of the bearing forces from the bar. The cylinder would have an inner diameter equal to the bar diameter d_b and a thickness C equal to the smaller of C_b, the clear bottom cover, or C_s, half of the clear spacing to the next adjacent bar (see Fig. 6.4.3). The tensile strength of this concrete cylinder determines the resistance against splitting. If $C_s < C_b$, a side-split type of failure occurs [Fig. 6.4.2(c)]. When $C_s > C_b$, longitudinal cracks through the bottom cover form

Figure 6.4.3 Concrete cylinder hypothesis for splitting failure (from Reference 6.3).

first [first splitting cracks in Fig. 6.4.2(a),(b)]. If C_s is only nominally greater than C_b, the secondary splitting will be side splitting along the plane of the bars. If C_s is significantly greater than C_b, the secondary splitting will also be through the bottom cover to create a V-notch failure [Fig. 6.4.2(b)].

The proposal of ACI Committee 408 [6.5, 6.20] recognizes the cylinder hypothesis for splitting failure. The portion of the proposal relating to hooks (see Section 6.11) was adopted for the 1983 ACI Code, and the portion relating to straight bar development length forms the basis for the development length rules in the 1989 ACI Code.

6.5 REASONS FOR NOT USING LOCALIZED SURFACE STRESS IN STRENGTH DESIGN

The localized resistance (usually known as *flexural bond stress*) between the bar and its surrounding concrete over a short length, say dz, as discussed in Section 6.3, does not provide an adequate measure of the margin of safety against a splitting failure.

Several situations may be identified where the local situation has little significance regarding *strength* against a splitting failure:

1. The tension concrete is uncracked in a region of low bending moment; thus the change in force, say $(T_D - T_{D'})$ in Fig. 6.3.1, in the bars is overestimated because concrete actually carries part of the tensile force.
2. At a point where high bending moment exists and therefore where a flexural crack will likely occur, the low rate of change of moment (i.e., shear) would indicate low $(T_D - T_{D'})$. *Adjacent* to the flexural crack (see Fig. 6.5.1), however, the surface stress is likely to be high, because the concrete shares in carrying the tension, whereas at the crack the concrete flexural tension is zero [6.12].
3. In the vicinity of a shear-related inclined crack, such as that of Fig. 5.4.2, not only does high surface stress exist adjacent to the crack but, in addition, the tensile force T computed at z from the support actually acts much closer to the support (refer to the truss model discussion in Section 5.7 and Fig. 5.7.1).

Flexure cracks

Actual f_s

Calculated f_s (at crack)

Near max ultimate surface stress

0 —— 0

Figure 6.5.1 Probable surface stress between cracks when beam shear is zero.

4. At locations where some bars are terminated in a tension zone, the abrupt change in the distribution of the total tensile force among the bars gives rise to very high surface stresses.

Thus, the localized situation, relating to rate of change in moment, does not directly correlate with the development-length-related strength of the member. When the bars are properly anchored, that is, they have adequate development length provided and continue to carry their required tensile force, the localized stress condition is not of concern.

6.6 MOMENT CAPACITY DIAGRAM—BAR BENDS AND CUTOFFS

As stated in Section 6.2, the moment capacity of a beam at any section along its length is a function of its cross section and the actual embedment length of its reinforcement. The concept of a diagram showing this three-dimensional relationship can be a valuable aid in determining cutoff or bend points of longitudinal reinforcement. It may be recalled from Chapter 3 that in terms of the cross section, the moment capacity (i.e., strength) for a singly reinforced rectangular beam may be expressed

$$M_n = A_s f_y (d - a/2) \qquad \textbf{[3.8.1]}$$

Equation (3.8.1) assumes that the steel reinforcement comprising A_s is adequately embedded *in each direction* by the required development length L_d from the section where M_n is computed such that the stress f_y is reached.

■ **EXAMPLE 6.6.1** Compute and draw the moment capacity diagram qualitatively for the beam of Fig. 6.6.1.

Solution: The procedure is basically the same whether strength (M_n or ϕM_n) or working stress moment capacity is desired.

The maximum capacity in each region is represented by the horizontal portions of the diagram in Fig. 6.6.1. In this example, there are five bars of one size in section *C-C*; thus the maximum moment capacity represented by each bar is in this case approximately one-fifth of the total capacity. Actually, the sections with four and two bars will have a little more than four-fifths and two-fifths, respectively, of the total capacity of the section containing five bars, due to the slight increase in moment arm when the number of bars in the section decreases.

At point *a*, the location where the fifth bar terminates, this bar has zero embedment length to the left and thus has zero capacity. Proceeding to the right from point *a*, the bar may be counted on to carry a tensile force proportional to its embedment from point *a* up to the development length L_d. Thus, in Fig. 6.6.1, point *b* represents the point where the fifth bar is fully developed through the distance L_d and can therefore carry its full tensile capacity. The other cutoff points are treated in the same way. ■

Figure 6.6.1 Moment capacity diagram.

■ **EXAMPLE 6.6.2** Demonstrate qualitatively the practical use of the moment capacity ϕM_n diagram for verification of the locations of cutoff or bend points in a design. Assume that the main cross section with five equal-sized bars provides exactly the required strength at midspan for this simply supported beam with uniform load, as shown in Fig. 6.6.2.

Solution: **(a)** Compute the actual ϕM_n for each potential bar grouping that may be used; in the present case, for five bars, four bars, and two bars.
(b) Decide which bars must extend entirely across the span and into the support. ACI-12.11.1 states that "At least one-third the positive moment reinforcement in

Figure 6.6.2 Verification of bar cutoffs with the moment capacity diagram.

simple members . . . shall extend along the same face of member into the support." In beams, the reinforcement must extend into the support at least 6 in. In this case, two bars should extend into the support.

(c) Decide on the order of cutting or bending the remaining bars. The least amount of longitudinal reinforcement will be obtained when the resulting moment capacity ϕM_n diagram is closest to the factored moment M_u diagram. With that thought in mind, and proceeding from maximum moment region to the support, cut off one bar as soon as permissible.

(d) Cutoff restrictions. Point A of Fig. 6.6.2 is the theoretical location to the left of which the capacity represented by the remaining four bars is adequate. To provide for a safety factor against shifting of the moment M_u diagram (especially in continuous spans) and to provide partially for the complexity arising from a potential diagonal crack, the ACI Code provides that there must be an extension beyond the point where a bar theoretically may be terminated, or it may be bent into the compression face. In ACI-12.10.3 is the statement, "Reinforcement shall extend beyond the point at which it is no longer required to resist flexure for a distance equal to the effective depth of the member or 12 bar diameters, whichever is greater, except at supports of simple spans and at free end of cantilevers."

(e) Once cutoff or bend points are located, a check is made by drawing the moment capacity ϕM_n diagram to ensure no encroachment on the factored moment M_u diagram.

(f) Other restrictions. Since points B and C of Fig. 6.6.2 are bar terminations in a tension zone, the stress concentrations described in Section 6.5 are present, effectively reducing the shear strength of the beam [6.21, 6.22]. Thus, *one* of the three special conditions of ACI-12.10.5 must be satisfied for cutoffs to be acceptable. However, if these bars were bent up and anchored in the compression zone, no further investigation would be necessary. ■

6.7 DEVELOPMENT LENGTH FOR TENSION REINFORCEMENT—ACI CODE

The term "development length" has been defined in Section 6.2 as the length of embedment needed to develop the yield stress in the reinforcement. As described in Section 6.4, the development length requirement is primarily a function of the splitting resistance of the concrete surrounding the bars rather than a frictional–adhesional pullout resistance. The splitting resistance is roughly proportional to the bar area, indicated by Eq. (6.2.2), whereas the pullout resistance is roughly proportional to the bar diameter, indicated by Eq. (6.2.1).

In the 1989 ACI Code, completely new bar development provisions were adopted (ACI-12.2), recognizing the effects of (a) lateral spacing of bars being developed, (b) clear cover over bars being developed, and (c) confinement, if any, by stirrups, ties, or spirals around the bars being developed.

Before explaining the specific provisions in detail, a general outline of the procedure for determining the development length L_d is presented. The steps are:

1. Apply formulas of ACI-12.2.2 to compute *basic development length L_{db}*.
2. Determine the category for conditions relating to lateral spacing, cover, and confinement according to Category A (ACI-12.2.3.1), Category B (ACI-12.2.3.3), and Category C (ACI-12.2.3.2), thereby establishing the appropriate multiplier α_{abc} of 1.0, 1.4, or 2.0, respectively.
3. Establish applicability of favorable conditions Category D (ACI-12.2.3.4) for wide lateral spacing of bars or Category E (ACI-12.2.3.5) for confinement by closely spaced ties or spirals, giving a multiplier α_{de} of 0.8 for Category D, or 0.75 for Category E.
4. Apply the appropriate multiplier α_{abc} from Step 2 to L_{db} from Step 1; then, if applicable, apply the multiplier α_{de} from Step 3 to the result.
5. Since the Step 4 result ($\alpha_{abc}\alpha_{de}L_{db}$) cannot be taken less than a minimum from ACI-12.2.3.6, the larger of $\alpha_{abc}\alpha_{de}L_{db}$ or Min L_d is used. This result is referred to as the "first-level L_d."

6. The "first-level L_d" from Step 5 is then multiplied cumulatively by the appropriate additional factors α_{top} for top reinforcement (ACI-12.2.4.1), $\alpha_{\ell w}$ for lightweight aggregate concrete (ACI-12.2.4.2), α_{epx} for epoxy-coated reinforcement (ACI-12.2.4.3), and then optionally by α_{exs} modify for excess reinforcement (ACI-12.2.5).

7. The final result $[(\text{first-level } L_d)\alpha_{top}\alpha_{epx}\alpha_{\ell w}]$ from Step 6 cannot be taken less than 12 in. (300 mm) (ACI-12.2.1).

This sequence is significantly more complicated than that in previous ACI Codes. A recent report [6.23] of a Task Group has presented justification for a simpler procedure; such a simplification has a reasonable chance of being adopted in a 1992 Code Supplement, or at least in the 1995 ACI Code.

In summary, the following seven items are needed to obtain the ACI Code development length L_d: these are L_{db}, α_{abc}, α_{de}, Min L_d, α_{top}, α_{epx}, and $\alpha_{\ell w}$. The eighth item, α_{exs}, is optional.

Basic Development Length L_{db}. ACI-12.2.2 states that the basic development length L_{db} shall be:

1. For #11 bar and smaller and deformed wire,

$$L_{db} = 0.04A_b \frac{f_y}{\sqrt{f_c'}}$$

(6.7.1)*

2. For #14 bars, (with $0.04A_b = 0.09$ taken as 0.085 instead)

$$L_{db} = 0.085 \frac{f_y}{\sqrt{f_c'}}$$

(6.7.2)*

3. For #18 bars, (with $0.04A_b = 0.16$ taken as 0.125 instead)

$$L_{db} = 0.125 \frac{f_y}{\sqrt{f_c'}}$$

(6.7.3)*

---

* For SI, ACI 318–89M, with L_{db} and d_b in mm, A_b in mm², and f_y and f_c' in MPa, gives for #35M bars or smaller,

$$L_{db} = 0.02 \frac{A_b f_y}{\sqrt{f_c'}}$$

(6.7.1)

for #45M,

$$L_{db} = \frac{25 f_y}{\sqrt{f_c'}}$$

(6.7.2)

for #55M,

$$L_{db} = \frac{40 f_y}{\sqrt{f_c'}}$$

(6.7.3)

Further, the value used for $\sqrt{f'_c}$ in the computation for L_d shall not exceed 100 psi* (ACI-12.1.2). As discussed in the shear chapter (Chapter 5), research data is meager on concrete having f'_c exceeding 10,000 psi (70 MPa); thus, the limitation. Values for basic development length L_{db} are given in Table 6.7.1.

Table 6.7.1 Basic Development Length L_{db}

INCH-POUND BARS WITH L_{db} IN INCHES[a]

BAR	$f_y = 40,000$ psi f'_c (psi)			$f_y = 60,000$ psi f'_c (psi)		
	3000	4000	5000	3000	4000	5000
#3	3.2	2.8	2.5	4.8	4.2	3.7
#4	5.8	5.1	4.5	8.8	7.6	6.8
#5	9.1	7.8	7.0	13.6	11.8	10.5
#6	12.9	11.1	10.0	19.3	16.7	14.9
#7	17.5	15.2	13.6	26.3	22.8	20.4
#8	23.1	20.0	17.9	34.6	30.0	26.8
#9	29.2	25.3	22.6	43.8	37.9	33.9
#10	37.1	32.1	28.7	55.6	48.2	43.1
#11	45.6	39.5	35.3	68.4	59.2	52.9
#14	62.1	53.8	48.1	93.1	80.6	72.1
#18	91.3	79.1	70.7	136	119	106

METRIC BARS[b] WITH L_{db} IN CENTIMETERS[c]

BAR	$f_y = 300$ MPa f'_c (MPa)			$f_y = 400$ MPa f'_c (MPa)		
	25	30	35	25	30	35
#10M	12.0	11.0	10.1	16.0	14.6	13.5
#15M	24.0	21.9	20.3	32.0	29.2	27.0
#20M	36.0	32.9	30.4	48.0	43.8	40.6
#25M	60.0	54.8	50.7	80.0	73.0	67.6
#30M	84.0	76.7	71.0	112	102	94.7
#35M	120	110	101	160	146	135
#45M	150	137	127	200	183	169
#55M	240	219	203	320	292	270

[a] Using equations of ACI–12.2.2.

[b] Metric bars are from Table 1.12.2.

[c] Using equations of ACI–12.2.2 from ACI 318–89M.

* For SI, ACI–89M gives 0.7 MPa.

Adjustment α_{abc} for Bar Spacing, Cover, and Stirrup Confinement. After determining the basic development length L_{db}, the adjustment based on bar spacing, cover, and confinement must be obtained according to whether the design situation is in Category A, B, or C, giving multipliers α_{abc} on L_{db} of 1.0, 1.4, or 2.0, respectively. These modifications essentially impose a penalty (i.e., require longer development length) for unfavorable conditions of (1) small clear lateral spacing between bars, (2) small clear cover on the bars, and (3) insufficient, or lack of, transverse reinforcement to enclose the bars. The conditions for Categories A, B, and C are as follows:

Category A (ACI-12.2.3.1) (Multiplier $\alpha_{abc} = 1.0$). Any *one* of the following conditions (1, 2, 3, or 4) will satisfy this most favorable situation:

1. **(a)** clear lateral spacing between bars at least $3d_b$, *and*
   (b) cover at least minimum as per ACI-7.7.1, *and*
   (c) minimum stirrups or ties along the development length (as per ACI-7.10.5 for ties, or ACI-11.5.4 combined with ACI-11.5.5.3 for stirrups).
2. **(a)** clear lateral spacing between bars of at least $3d_b$, *and*
   (b) cover not less than $2d_b$.
3. **(a)** cover at least minimum as per ACI-7.7.1, *and*
   (b) bars enclosed within transverse reinforcement A_{tr} along the development length, satisfying

$$A_{tr} \geq \frac{d_b s N}{40} \tag{6.7.4}$$

where s is the spacing of stirrups or ties along the development length, and N is the number of bars being developed.
4. **(a)** inner layer* of *slab or wall reinforcement*, *and*
   (b) clear lateral spacing of bars at least $3d_b$.

Category B (ACI-12.2.3.3) (Multiplier $\alpha_{abc} = 1.4$).

1. **(a)** clear lateral spacing between bars greater than $2d_b$ but less than $3d_b$, *and*
   (b) cover in excess of d_b.

Category C (ACI-12.2.3.2) (Multiplier $\alpha_{abc} = 2.0$).

1. clear lateral spacing between bars $2d_b$ or less; *OR*
2. cover d_b or less.

Actually, the ACI Code specifically defines the requirements for using multiplier $\alpha_{abc} = 1.0$ (ACI-12.2.3.1) and for using $\alpha_{abc} = 2.0$ (ACI-12.2.3.2), and then states that 1.4 is to be used when conditions are not applicable for using 1.0 or 2.0. Category B is defined above as it would usually apply for beams in practical design.

----

* Note that an *inner* layer of bars is one that has *transverse* bars located between the face of the slab or wall and the bars being developed, thereby retarding formation of a split directly from the developing bar to the face of the slab or wall.

The practicality for applying the rules in ordinary reinforced concrete construction is that most beams will contain at least minimum stirrups (thereby satisfying Category A, item 1c), and cover must satisfy ACI-7.7.1 in any case (thereby satisfying Category A, item 1b). Thus, the crucial item for satisfying Category A is whether or not the bars being developed satisfy lateral clear spacing of at least $3d_b$. This can easily be checked by using the minimum width for $3d_b$ in Table 3.9.4 of Chapter 3. Even for beams without stirrups, the 1.5-in. minimum cover required for cast-in-place construction will equal at least $2d_b$ for #6 bars and smaller (thereby satisfying Category A, item 2b), in which case the important issue is still the $3d_b$ lateral clear spacing.

Thus, in general for cast-in-place construction having at least minimum stirrups, one need only determine whether the lateral clear spacing of the bars being developed is at least $3d_b$ (qualifying for Category A), greater than $2d_b$ but less than $3d_b$ (qualifying for Category B), or at least the minimum of ACI-7.6.1 but $2d_b$ or less (meaning Category C). The minimum width tables for satisfying $3d_b$, $2d_b$, and minimum of ACI-7.6.1 are given in Section 3.9.

Adjustment α_{de} for Wide Lateral Spacing or Enclosure Within Spirals or Ties.

The multiplier α_{de} on ($\alpha_{abc} L_{db}$) is to recognize the favorable conditions described in Category D or Category E as follows:

Category D (ACI-12.2.3.4) ($\alpha_{de} = 0.80$). This adjustment recognizes a favorable condition of wide lateral clear spacing common in slabs not less than $5d_b$, and with side clear cover from the edge bar at least $2.5d_b$. This adjustment is to be applied (see Step 3 above) *after* using the α_{abc} multiplier of Category A, B, or C. Seemingly, this wide spacing would satisfy Category A; however, since stirrups are not usually used on slabs, the cover measured perpendicular to the plane of the bars will be less than in beams. In slabs, ACI-7.7.1 permits as little as $\frac{3}{4}$-in. cover. Since item 2 under Category A above is the only one that applies when no stirrups are used, cover must be $2d_b$ perpendicular to the plane of the bars in order for Category A to apply. Thus, even with 1-in. cover, slab situations will rarely satisfy the $2d_b$ Category A requirement.

With the common situation of $\frac{3}{4}$-in. or 1-in. cover measured perpendicular to the plane of bars, such cover would equal d_b for #6 (for $\frac{3}{4}$-in. cover) or #8 (for 1-in. cover) bars; thus, giving Category B (multiplier $\alpha_{abc} = 1.4$). For larger bars, Category C (multiplier $\alpha_{abc} = 2.0$) would apply.

After establishing the α_{abc} multiplier of Category A, B, or C; *then, if Category D applies*, the $\alpha_{de} = 0.80$ multiplier is used. For example, widely spaced #8 bars having minimum $\frac{3}{4}$-in. cover would be Category C ($\alpha_{abc} = 2.0$), as well as Category D ($\alpha_{de} = 0.80$); thus, α_{abc} times α_{de} gives 1.6 as the multiplier at Step 4.

Category E: (ACI-12.2.3.5) ($\alpha_{de} = 0.75$). This adjustment applies primarily for tied or spirally reinforced compression members. Special confinement permits reduced development length when: (1) #4 or larger ties are spaced at not more than 4 in. and alternate bars have support provided by a corner of a tie having an included angle of 135° or less; or (2) $\frac{1}{4}$-in. or larger spirals, or #4 or larger circular ties, having a center-to-center spacing not exceeding 4 in.

In the same manner as for Category D, the α_{abc} multiplier of Category A, B, or C is established. Then, *if Category E applies*, the $\alpha_{de} = 0.75$ multiplier is used. Thus, at Step 4 in the procedure, the multiplier is α_{abc} times α_{de}.

Though theoretically possible, in actual practice one will not have both Category D and Category E. Therefore, α_{de} will be either 0.80 for slab situations in Category D, or 0.75 for columns in Category E, or 1.0 if neither of these apply. In practical design, Category E will occur only for spirally reinforced columns; seldom would ties be spaced as close as 4 in. Tied and spirally reinforced columns are treated in Chapter 13.

Minimum Development Length. In Step 5, after computing L_{db} and multiplying by the appropriate α_{abc} and then by α_{de}, if applicable, the result may be referred to as the

Table 6.7.2   Minimum Development Length—ACI 12.2.3.6

INCH-POUND BARS WITH L_{db} IN INCHES[a]

BAR	$f_y = 40,000$ psi f'_c (psi)			$f_y = 60,000$ psi f'_c (psi)		
	3000	4000	5000	3000	4000	5000
#3	8.2	7.1	6.4	12.3	10.7	9.5
#4	11.0	9.5	8.5	16.4	14.2	12.7
#5	13.7	11.9	10.6	20.5	17.8	15.9
#6	16.4	14.2	12.7	24.6	21.3	19.1
#7	19.2	16.6	14.8	28.8	24.9	22.3
#8	21.9	19.0	17.0	32.9	28.5	25.5
#9	24.7	21.4	19.1	37.1	32.1	28.7
#10	27.8	24.1	21.6	41.7	36.1	32.3
#11	30.9	26.8	23.9	46.3	40.1	35.9
#14	37.1	32.1	28.7	55.6	48.2	43.1
#18	49.4	42.8	38.3	74.2	64.2	57.5

METRIC BARS[b] WITH L_{db} IN CENTIMETERS[c]

BAR	$f_y = 300$ MPa f'_c (MPa)			$f_y = 400$ MPa f'_c (MPa)		
	25	30	35	25	30	35
#10M	25.4	23.2	21.5	33.9	30.9	28.7
#15M	36.0	32.9	30.4	48.0	43.8	40.6
#20M	43.9	40.1	37.1	58.5	53.4	49.4
#25M	56.7	51.8	47.9	75.6	69.0	63.9
#30M	67.3	61.4	56.9	89.7	81.9	75.8
#35M	80.3	73.3	67.9	107	98	91
#45M	98.3	89.8	83.1	131	120	111
#55M	127	116	107	169	154	143

[a] ACI–12.2.3.6, Min $L_d = 0.03d_b f_y/\sqrt{f'_c}$
[b] Metric bars are from Table 1.12.2.
[c] ACI–12.2.3.6, ACI 318–89M, Min $L_d = 3d_b f_y/(8\sqrt{f'_c})$.

"first-level L_d." This "first-level L_d" cannot exceed a minimum prescribed by ACI-12.2.3.6, as follows:

$$\text{Min } L_d = 0.03 d_b \frac{f_y}{\sqrt{f_c'}} \tag{6.7.5}*$$

This minimum is based on the possibility of a pullout type of failure, rather than splitting. The more concrete surrounding the bars and/or the more confinement, the more likely a pullout mode of failure will occur. Values for Eq. (6.7.5) are given in Table 6.7.2. This brings the calculation of L_d to the end of Step 5 of the sequence listed at the beginning of this section.

6.8 SECOND-LEVEL MODIFICATIONS TO FIRST-LEVEL DEVELOPMENT LENGTH

Further increases (see Step 6) to the "first-level L_d" are required by ACI-12.2.4 for top reinforcing bars, bars in lightweight aggregate concrete, and epoxy-coated bars. Reduction is permitted by ACI-12.2.5 when excess reinforcement is used.

Top Reinforcing Bars, α_{top}. When horizontal bars are placed so that more than 12 in. (300 mm) of concrete is cast in the member below the bars, they are referred to as *top bars*. This is usually taken to mean that if the depth of concrete mix placed in one lift under the horizontal bar exceeds 12 in., it is a top bar. The modification factor $\alpha_{\text{top}} = 1.3$ is a recognition by the ACI Code (ACI 12.2.4.1) that top cast bars exhibit reduced strength, apparently due to the settling away of concrete from the bar on its underside [6.1, 6.11, 6.24]. The value α_{top} was reduced from its long-time ACI Code value of 1.4 to its current value of 1.3 in recognition of the 1989 inclusion of α_{abc} factors relating to cover, bar spacing, and stirrup confinement.

Epoxy-Coated Bars, α_{epx}. The rapidly increasing use of epoxy-coated bars is reflected in this new Code provision in ACI-12.2.4.3 specifying two different modification factors α_{epx}, depending on the cover and clear spacing of the bars. The values are:

1. Bars having cover less than $3d_b$, *or*
   having clear spacing less than $6d_b$ $\quad\quad\quad\quad\quad\quad\quad\quad \alpha_{\text{epx}} = 1.5$
2. Others $\quad\quad\quad\quad\quad\quad\quad\quad\quad\quad\quad\quad\quad\quad\quad\quad\quad\quad\quad \alpha_{\text{epx}} = 1.2$

When the epoxy-coated bars are in the unfavorable category of $\alpha_{\text{epx}} = 1.5$, *and* the bars are top bars ($\alpha_{\text{top}} = 1.3$), ACI-12.2.4 indicates that the product $\alpha_{\text{top}}\alpha_{\text{epx}}$ need not exceed 1.7 (it would otherwise be 1.95).

---- -

* For SI, ACI 318–89M, with L_d and d_b in mm, and f_y and f_c' in MPa,

$$\text{Min } L_d = 3 d_b \frac{f_y}{8\sqrt{f_c'}} \tag{6.7.5}$$

Lightweight Aggregate Concrete, $\alpha_{\ell w}$. There is increased potential for a pullout mode of development failure in lightweight aggregate concrete. However, the $\alpha_{\ell w}$ of 1.3 (or $6.7\sqrt{f'_c}/f_{ct}$ but not less than 1.0) specified by ACI-12.2.4.2 is a compromise value reflecting meager research results [6.5]. The 1989 ACI Code eliminates the different values for "sand-lightweight" and "all-lightweight" aggregate concrete, accepting the higher value of 1.3.

Excess Reinforcement, α_{exs}. When more steel reinforcement has been provided than is required by the loading, the designer is *permitted* to reduce the development length in proportion that the A_s required is less than A_s provided. When anchorage to develop f_y is specifically required, the development length is not permitted to be reduced. Such situations include (a) when the flexural member is part of the lateral load resisting system (ACI-12.11.2), (b) temperature and shrinkage reinforcement (ACI-7.12.2.3), and (c) the integrity reinforcement required by ACI-7.13 and 13.4.8.5.

Summary. The "second-level L_d" modification factors are summarized in Table 6.8.1.

Table 6.8.1 Second-Level L_d Modification Factors

SITUATION	SECOND-LEVEL MODIFICATION FACTOR
Top reinforcement[a]	$\alpha_{top} = 1.3$
Epoxy-coated bars[a]	$\alpha_{epx} = 1.5$ for bars with cover $< 3d_b$
	or bars with spacing $< 6d_b$
	$\alpha_{epx} = 1.2$ otherwise
Lightweight aggregate concrete	$\alpha_{\ell w} = 1.3$
	or $6.7\sqrt{f'_c}/f_{ct} \geq 1.0$
Excess reinforcement	$\alpha_{exs} = \dfrac{\text{required } A_s}{\text{provided } A_s} \leq 1.0$

[a] The product $\alpha_{top}\alpha_{exp}$ need not exceed 1.7 (ACI–12.2.4).

Thus, the development length L_d can be formulated as:

$$L_d = (\alpha_{abc}\alpha_{de}L_{db} \geq \text{Min } L_d)(\alpha_{top}\alpha_{epx} \leq 1.7)\alpha_{\ell w}\alpha_{exs} \geq 12 \text{ in.} \qquad \textbf{(6.8.1)}$$

Note that after all modifications the resulting L_d cannot be taken less than 12 in. (300 mm).

■ **EXAMPLE 6.8.1** Determine the development length L_d required for the #9 bars A on the top of a 15-in. slab, as shown in Fig. 6.8.1. Use $f_y = 70,000$ psi and $f'_c = 4000$ psi with normal-weight aggregate.

Solution: (a) Determine the basic development length L_{db}. For a #9 bar, according to Eq. (6.7.1),

$$L_{db} = 0.04A_b f_y/\sqrt{f'_c} = 0.04(1.00)(70,000)/\sqrt{4000} = 44.3 \text{ in.}$$

Figure 6.8.1 Top bars for Example 6.8.1.

(b) Determine whether Category A, B, or C applies for α_{abc}. Since there are no stirrups, Category A, item 2, is the only possibility to be satisfied for $\alpha_{abc} = 1.0$. Check clear cover and clear spacing. Referring to Fig. 6.8.1, bars A are developed over distance 1–2, while bars B are developed over the distance 2–3. The spacing to be used for bars A is the spacing of the closest bars that terminate at the same point. In other words, the spacing for both bars A and B is 8 in.

$$\text{Clear spacing} = \frac{8}{d_b} - 1.0 = \frac{8}{1.128} - 1.0 = 6.1d_b > 3d_b \qquad \textbf{OK}$$

$$\text{Clear cover} = \frac{1.5}{d_b} = \frac{1.5}{1.128} = 1.3d_b < 2d_b \text{ for Category A} \qquad \textbf{NG}$$

With cover in excess of d_b though less than $2d_b$, Category B applies; $\alpha_{abc} = 1.4$. For slabs, cover is usually the determining factor to obtain α_{abc}.

(c) Determine whether Category D applies to obtain α_{de}. Since spacing of bars exceeds $5d_b$ and side cover exceeds $2.5d_b$, Category D applies and $\alpha_{de} = 0.80$. Multiplying L_{db} by α_{abc} and then by α_{de},

$$\alpha_{abc}\alpha_{de}L_{db} = 1.4(0.80)(44.3) = 49.6 \text{ in.}$$

(d) Compare with Min L_d and obtain "first-level L_d."

$$\text{Min } L_d = 0.03d_b f_y/\sqrt{f'_c}$$
$$= 0.03(1.128)(70,000)/\sqrt{4000} = 37.5 \text{ in.}$$

Since Min L_d is less than $\alpha_{abc}\alpha_{de}L_{db}$, the "first-level L_d" is 49.6 in.

(e) Modification for top bars. Since the negative moment region bars are cast with more than 12 in. of concrete below them, they are top bars according to ACI-12.2.4.1; thus $\alpha_{top} = 1.3$. Since the bars are not epoxy-coated and normal-weight concrete is used, the other multipliers can be thought of as 1.0 (i.e., not used).

(f) Final L_d. Applying α_{top};

$$\text{Final } L_d = \alpha_{top} \times (\text{first-level } L_d) = 1.3(49.6) = 64.5 \text{ in.}$$

Use $L_d = 64.5$ in. since it exceeds the 12-in. minimum. ■

6.9 DEVELOPMENT LENGTH FOR COMPRESSION REINFORCEMENT

Relatively less is known about the development length for compression bars than for tension bars, except that the weakening effect of flexural tension cracks is not present and there is beneficial effect of the end bearing of the bars on the concrete. ACI-12.3 gives as the basic development length L_{db},

$$L_{db} = 0.02 d_b \frac{f_y}{\sqrt{f'_c}} \tag{6.9.1}*$$

which is basically two-thirds of the minimum development length (Min L_d) for tension reinforcement to prevent a "pullout" mode of failure. ACI-12.3 also states that L_{db} must not be less than

$$L_{db} \geq 0.0003 d_b f_y \tag{6.9.2}*$$

which really means that only f'_c up to about 4400 psi may be counted upon. Thus the basic development length L_{db} is to be taken as the larger of Eqs. (6.9.1) and (6.9.2).

When excess bar area is provided such that provided A_s exceeds required A_s, Eqs. (6.9.1) or (6.9.2), whichever controls, may be reduced by applying the multiplier (required A_s/provided A_s).

Reduction in development length is permitted when reinforcement is enclosed by spirals or closely spaced ties (typically in columns; see Chapter 13) which are not less than $\frac{1}{4}$-in. diameter for spirals (ACI-7.10.4), or #4 bars for ties (ACI-7.10.5), and having a pitch (for spirals) or center-to-center spacing (for ties) not exceeding 4 in. Under these confinement conditions, L_{db} may be reduced 25%. After all modifications, the development length L_d is not permitted to be less than 8 in. (200 mm). Thus, in general, for compression reinforcement

$$L_d = \begin{bmatrix} \text{Eqs. (6.9.1)} \\ \text{or (6.9.2)} \end{bmatrix} \begin{bmatrix} \text{required } A_s \\ \overline{\text{provided } A_s} \end{bmatrix} \begin{bmatrix} 0.75 \text{ for enclosure} \\ \text{by spirals or ties} \end{bmatrix} \geq 8 \text{ in.} \tag{6.9.3}$$

— — — —

* For SI, ACI 318–89M, for L_{db} and d_b in mm, and f'_c and f_y in MPa, gives

$$L_{db} = 24 d_b \frac{f_y}{\sqrt{f'_c}} \tag{6.9.1}$$

$$L_{db} = 0.044 f_y d_b \tag{6.9.2}$$

6.10 DEVELOPMENT LENGTH FOR BUNDLED BARS

When space for proper clearance is restricted and large steel areas are required, groups of parallel bars are sometimes bundled. No more than four bars bundled in contact (ACI-7.6.6), enclosed by stirrups or ties, may be arranged with no more than two bars in the same plane into typical bundle shapes, such as triangular, square, or L-shaped for three- and four-bar bundles (see Fig. 6.10.1). Bars larger than #11 shall not be bundled in beams or girders, primarily to ensure proper control of cracking (see ACI Commentary-R7.6.6). In flexural members, termination of individual bars within a bundle at points along the span must be at different points offset by at least 40 bar diameters. Where spacing requirements and minimum clear cover are based on bar size, one bundle of bars shall be treated as a single bar of an equivalent diameter derived from the total area of the bars in the bundle. In applying the crack control provisions of ACI-10.6.4 (see Chapter 4, Section 4.12) to bundled bars, this assumption of treating a bundle of bars as a single large bar may be overly conservative. An alternative is suggested by Lutz [6.25].

Figure 6.10.1 Bundled bar arrangements (bars are in contact with each other).

When considering the development length requirement for a bundle of bars, the three-bar triangular pattern and the four-bar arrangement will have $16\frac{2}{3}$ and 25% reduction, respectively, in the total nominal surface contact (considering all bars cylindrical having the nominal diameter) of the bars with surrounding concrete. In order to allow further for the difficulty of attaining good resistance to splitting or pullout at the reentrant corner where the bars in the bundle touch each other, additional reduction in resistance appears logical.

The requirements of ACI-12.4.1 specify that the development length for a bundle shall be based on that for the individual bar in the bundle, increased by 20% for a three-bar bundle and 33% for a four-bar bundle. For determination of all needed modification factors, ACI-12.4.2 states that "a unit of bundled bars shall be treated as a single bar of a diameter derived from the equivalent total area."

Experimental results for beams and columns having bundled reinforcement are reported by Hanson and Reiffenstuhl [6.26], and some practical applications are given by Steiner [6.27].

6.11 DEVELOPMENT LENGTH FOR A TENSION BAR TERMINATING IN A STANDARD HOOK

When straight embedment is inadequate to provide for the necessary development length of a tension bar, or when it is desired to have the full capacity of a bar available in the shortest distance of embedment, a standard hook as defined in ACI-7.1 and 7.2 and shown in Fig. 6.11.1 may be used. Hooks are not considered effective in adding to the compressive resistance of reinforcement.

Figure 6.11.1 Standard hooks for development of main tension reinforcement.

In general, the resistance of a hook *in mass concrete* is about the same as that of a straight bar with the same total length of embedment [6.28, 6.29]. Quite commonly, hooks in *structural members* are located close to a free surface where splitting forces proportional to the tensile capacity of the bar will govern the hook capacity. Because of the tendency for splitting failures, hooks may have less capacity than provided by an equal length of straight embedment.

ACI Code Procedure. Based on the works of Hribar and Vasko [6.28], Orangun, Jirsa, and Breen [6.3], Minor and Jirsa [6.29], Marques and Jirsa [6.30], and Pinc, Watkins, and Jirsa [6.31], the 1989 ACI Code prescribes in a direct manner the required embedment (development length L_{dh} as in Fig. 6.11.2) to develop f_y in a hooked bar, uncoupled from the embedment required (development length L_d) to develop f_y in a straight bar. A summary of the rationale for this procedure is given by Jirsa, Lutz, and Gergely [6.5].

The procedure as prescribed by ACI-12.5 is to compute a *basic development length L_{hb}* required to develop the yield strength f_y of a hooked bar, measured from

Figure 6.11.2 Development length L_{dh} for hooked bar.

Table 6.11.1   Factors to Modify Basic Development Length L_{hb} for a Hooked Bar (ACI-12.5.3)

CONDITION	MULTIPLIER
1. *Yield stress:* For f_y other than 60,000 psi	$\dfrac{f_y}{60,000}$
2. *Concrete cover:* 180° hook, #11 bars and smaller having side cover (normal to plane of hook) $\geq 2\frac{1}{2}$ in. 90° hook, same as for 180° hook *plus* cover on bar extension beyond hook ≥ 2 in.	0.7
3. *Ties or stirrups:* For bars #11 and smaller, hook enclosed vertically or horizontally within ties or stirrup-ties spaced along the full development length L_{dh} not greater than $3d_b$ (d_b = diameter of hooked bar). [Note: Multiplier cannot be used if the hook occurs at the discontinuous end of a member when clear cover to hooked bar is less than $2\frac{1}{2}$ in. (ACI-12.5.4).]	0.8
4. *Excess reinforcement:* Where anchorage or development for f_y is not specifically required, and there is reinforcement in excess of that required	$\dfrac{\text{required } A_s}{\text{provided } A_s} \leq 1.0$
5. *Lightweight aggregate concrete*	1.3

the location where the yield strength is needed to the extreme outside face of the hook, as follows:

$$L_{hb} = \frac{1200d_b}{\sqrt{f'_c}}$$ **(6.11.1)***

for $f_y = 60,000$ psi.

Modification factors are then applied as given in Table 6.11.1.

Once the basic value L_{hb} has been multiplied by the appropriate modification factors, the resulting required development length L_{dh} for the hooked bar is the largest of (1) the modified L_{hb}, (2) 8 hooked bar diameters, and (3) 6 in.

Most hooked bars are embedded in joints where other concrete members frame in at the sides and therefore provide some lateral confinement, in addition to the vertical confinement provided by the force in the column. When these other members are not present, such as at the discontinuous end of a cantilever as shown in Fig. 6.11.3, ACI-12.5.4 requires either $2\frac{1}{2}$ in. or more of cover to the hooked bar or that the hooked bar be enclosed in stirrups or ties over the development length L_{dh}.

— — — —

* For SI, ACI 318–89M, for L_{hb} and d_b in mm, and f'_c in MPa, gives

$$L_{hb} = \frac{100d_b}{\sqrt{f'_c}}$$ **(6.11.1)**

for $f_y = 400$ MPa.

Figure 6.11.3 Special requirements for standard hooks developed at the discontinuous end of a member (ACI-12.5.4).

Table 6.11.2 Development Length L_{dh} (in.) for Hooked Bars[a]

HOOKED BAR	$f_y = 40,000$ psi			$f_y = 60,000$ psi		
	f_c' (psi)			f_c' (psi)		
	3000	4000	5000	3000	4000	5000
#3	5.5[b]	4.7[b]	4.3[b]	8.2	7.1	6.4
#4	7.3	6.3	5.7[b]	11.0	9.5	8.5
#5	9.1	7.9	7.1	13.7	11.9	10.6
#6	10.9	9.5	8.5	16.4	14.2	12.7
#7	12.8	11.1	9.9	19.2	16.6	14.8
#8	14.6	12.7	11.3	21.9	19.0	17.0
#9	16.5	14.3	12.7	24.7	21.4	19.1
#10	18.5	16.1	14.4	27.8	24.1	21.6
#11	20.6	17.9	15.9	30.9	26.8	23.9
#14	24.7	21.4	19.1	37.1	32.1	28.7
#18	32.9	28.5	25.5	49.4	42.8	38.3

[a] Assumes normal-weight concrete and side cover to hooked bar less than $2\frac{1}{2}$ in., computed in accordance with Eq. (6.11.1). For $f_y = 40,000$ psi, modification 1 from Table 6.11.1 has been applied to Eq. (6.11.1).

[b] If no modification greater than 1.0 is used, value is 6 in.

Values of the development length L_{dh}, assuming that L_{dh} equals the basic development length L_{hb}, according to Eq. (6.11.1), for $f_y = 60,000$ psi, and L_{hb} modified by $(f_y/60,000)$ for $f_y = 40,000$ psi are shown in Table 6.11.2.

6.12 BAR CUTOFFS IN NEGATIVE MOMENT REGION OF CONTINUOUS BEAMS

The general concept of drawing the moment capacity diagram to investigate the adequacy of a given beam has been presented in Section 6.6. In this and the next two sections, several situations are discussed in which the ACI Code provisions are applied to establish cutoff points, aided by the drawing of the moment capacity diagram. Three factors are involved: (1) adequate horizontal offset from theoretical cutoff points in recognition of the truss action described in Section 5.7 where the tensile force is shown to act at a distance offset from the section where M_u is computed, as well as to provide safety against the possible shifting of the moment diagram due to unusual loading arrangements; (2) adequate development lengths so that full bar capacity is available where needed; and (3) sufficient relief from stress concentrations when bars are terminated in a tension zone.

Cutoffs in the negative moment region of continuous beams can best be explained by means of a specific case. (For an example of a complete factored moment M_u envelope for a continuous beam, the reader is referred to Fig. 10.6.1.) Referring to the partial envelope for M_u in Fig. 6.12.1, assume that it is desired to cut two out

Figure 6.12.1 Bar cutoff in negative moment region of continuous beams.

of four bars as close as possible (point C) to the support, and then terminate the remaining two as soon as feasible (point E). In accordance with ACI-12.12.3, the area provided by bars $R2$ must exceed one-third the total reinforcement area provided for negative moment. In a general situation, the horizontal distance between points C and E is greater than the development length L_d for bars $R2$; the special case wherein this is not so will be discussed later.

Point A represents the maximum factored moment M_u at the face of support, the critical location for bending moment. The higher factored moment (shown dashed in Fig. 6.12.1) within the support is resisted by a larger cross section than the one at the face of support; thus, the factored moment M_u at point A is considered the largest value for which strength must be provided. The full moment capacity ϕM_n provided by bars $R1$ and $R2$ is somewhat greater than required. The theoretical cutoff for the two $R1$ bars is at point B. Point C is then located (ACI-12.10.3) horizontally to the right from point B a distance equal to the effective depth of the member or 12 diameters of bars $R1$. Point D is located horizontally to the left of point C a distance equal to the development length L_d for bars $R1$. Since points A and D will generally be approximately horizontal with one another, point D must lie at or to the right of point A in order to provide adequate capacity at the face of support. When extra capacity is provided, the safety may be considered adequate when the line CD intersects the face of support line at or below point A (satisfying ACI-12.12.2).

Point E is next established by extension, according to ACI-12.12.3, beyond the point of inflection (point of zero moment) "not less than effective depth of the member, $12d_b$, or one-sixteenth the clear span, whichever is greater." Point F, where the bars $R2$ will become capable of carrying their full capacity, is located L_d horizontally to the left of point E. In order that the moment capacity diagram does not encroach closer than $12d_b$ or d to point B, point F must lie to the right of point C.

The special case wherein the horizontal distance between points C and E is smaller than the development length L_d for bars $R2$ is shown in Fig. 6.12.2. The moment capacity ϕM_n diagram is indicated by $EC'F'D$, because the variation between the moment capacities at C' and F' has to be linear since development of capacity is assumed to be proportional to the amount of embedment up to a maximum length of L_d. The requirement to allow for the shifting of the factored moment M_u curve is to be satisfied by making sure that the distance BM is larger than the effective depth d or 12 diameters of bars $R1$, whichever is greater.

Cutting Bars in the Tension Zone. Since point C in Fig. 6.12.1 is still in the tension zone (though only slightly), the provisions of ACI-12.10.5 to minimize stress concentrations must be checked. *One of the following three conditions must be satisfied at point C:*

1. The factored shear V_u at the cutoff point does not exceed two-thirds of the shear strength ϕV_n, or

$$V_u \leq [\tfrac{2}{3}\phi V_n = \tfrac{2}{3}\phi(V_c + V_s) \tag{6.12.1}$$

2. Excess stirrup area is provided to give a strength $\phi V_s = \phi(60 \text{ psi})b_w d$ in excess of the required $\phi V_s = V_u - \phi V_c$. The excess stirrups are to be used along the terminated bar over a distance from the termination point equal to three-

Figure 6.12.2 Special case of Fig. 6.12.1 wherein the distance CE is less than L_d for bars $R2$.

fourths of the effective depth of the member. The spacing of such stirrups must not exceed

$$\max s = \frac{d}{8\beta_b} \qquad (6.12.2)$$

where β_b is the ratio of area of longitudinal bars cut off to the total area of bars at the section.

3. For #11 and smaller bars, the following may be satisfied at the cutoff point:

$$V_u \le [\tfrac{3}{4}\phi V_n = \tfrac{3}{4}\phi(V_c + V_s)] \qquad (6.12.3)$$

and

$$\phi M_n \ge 2M_u \qquad (6.12.4)$$

When it may be impractical or undesirable to satisfy the above described provisions of ACI-12.10.5, the cutoff point may be extended until it is in the compression zone, or the bars may be bent across into the opposite face of the beam and then continued or terminated.

6.13 BAR CUTOFFS IN POSITIVE MOMENT REGION OF CONTINUOUS BEAMS

Bar cutoffs in the positive moment region of continuous beams are to be explained by reference to Fig. 6.13.1. Assume that it is desired to cut two of the four bars that are used for resisting the maximum positive factored moment M_u. The area provided by bars $R4$ must exceed one-fourth (one-third for simple spans) of the total reinforcement area provided for positive moment, in accordance with ACI-12.11.1. In a general situation, the horizontal distance between points J and H is greater than the development length L_d for bars $R4$; the special case wherein this is not so will be discussed later in the section.

Point G is first located by computing the moment capacity ϕM_n of the continuing bars $R4$. The cutoff point H must lie to the left of point G at least 12 diameters of bars $R3$ or the effective depth d, whichever is larger. Point I is then located horizontally to the right from point H a distance equal to the development length L_d for the bars $R3$. Point J is located at the end of the bars $R4$, and point K is located horizontally to the right of point J, a distance equal to the development length L_d for the bars $R4$. The cutoff at point H must satisfy ACI-12.10.3 by giving a moment capacity ϕM_n diagram that is offset horizontally from the factored moment diagram at every point (except at the support) by 12 bar diameters or the effective depth d, whichever is greater.

It is to be noted that whereas ACI-12.11.1 requires $R4$ bars to extend into the support a distance of 6 in. minimum, ACI-12.11.2 requires full development of the

Figure 6.13.1 Bar cutoff in positive moment region of continuous beams.

Figure 6.13.2 Special case of Fig. 6.13.1 wherein the distance JH is less than L_d for bars $R4$.

tensile yield strength of such bars at the face of support when the flexural member is part of the primary lateral load resisting system. Such would be the case if significant moments are developed in the member as a result of wind or earthquake loading.

The special case wherein the horizontal distance between points J and H is smaller than the development length L_d for bars $R4$ is shown in Fig. 6.13.2. The moment capacity ϕM_n diagram is indicated by $JH'K'I$. The requirement to allow for the shifting of the factored moment curve is satisfied by making sure that the distance GN is larger than the effective depth d or 12 diameters of bars $R3$, whichever is greater.

In addition, since the cutoff at point H lies in the tension zone, one of the conditions of ACI-12.10.5 must be satisfied [see Eqs. (6.12.1) through (6.12.4)].

6.14 BAR CUTOFFS IN UNIFORMLY LOADED CANTILEVER BEAMS

Bar cutoffs in uniformly loaded cantilever beams can best be discussed by examining Fig. 6.14.1. Assume that of the six bars provided for the maximum factored moment M_u, it is desired to cut two bars $R5$ as soon as possible, then cut two more bars $R6$, and run the remaining two bars $R7$ out to the end of the cantilever.

Figure 6.14.1 Bar cutoff in a uniformly loaded cantilever beam.

The following steps illustrate the cutoff determination and check:

1. Locate the theoretical cutoff point A where the moment capacity ϕM_n of four bars (bars $R6$ and $R7$) is adequate; extend 12 diameters of bars $R5$ or the effective depth of the member, whichever is greater, to arrive at point B.
2. Determine the development length L_d for the bars $R5$ that are intended to be cut. Full capacity from the $R5$ bars is available at the distance L_d to the left of the cutoff point.
3. Since the horizontal distance CB is less than L_d, less than full capacity of the $R5$ bars will be available at the support had they been cut at point B; therefore, extend cutoff location to point D so that the horizontal distance CD equals L_d. (ACI-12.10.5 must also be satisfied, since the proposed cut location lies in a tension zone.)

4. Locate point E, the theoretical location where only the two $R7$ bars are required for moment; extend 12 diameters of bars $R6$ or the effective depth of the member, whichever is greater, to arrive at point F. (ACI-12.10.5 must also be satisfied for point F, since it lies in a tension zone.)
5. Determine the development length L_d required for the $R6$ bars being cut; these two bars will have their full capacity available at point G, a distance L_d to the left of point F.
6. In the region from G to D the moment capacity consists of partial contributions from the bars $R5$ and $R6$ plus the full contribution of bars $R7$. The combined moment capacity (represented by line IH) is the sum of the linear contributions. The intent of ACI-12.10.3 is satisfied if the moment capacity diagram encloses the factored moment diagram by an adequate amount of horizontal offset equal to the effective depth of the member or 12 bar diameters (of the cutoff bars), whichever is greater. At a support or the end of a cantilever, this horizontal offset need not be provided because the factored moment curve cannot shift left or right at such locations. In this case, the line IH passes by chance through point B, satisfying the offset requirement. Point H should also have the necessary horizontal offset.
7. When any part of the line IH encroaches on the factored moment M_u diagram closer than $12d_b$ of the $R5$ bars or d, whichever is greater, then either or both of the proposed cut locations (points D and F) must be extended toward the end of the cantilever. A convenient and practical procedure is to extend point F to the right until point G coincides with point D.

■ **EXAMPLE 6.14.1** For the cantilever beam shown in Fig. 6.14.2 determine the distance L_1 from the support to the point where 2- #8 bars may be cut off. Assume the #4 stirrups shown (solid, not the dashed ones) have been preliminarily designed. Assume there will be at least L_d embedment of the bars into the support. Draw the resulting moment capacity ϕM_n diagram for the entire beam. Use $f'_c = 3000$ psi and $f_y = 60,000$ psi.

Solution: (a) Compute the maximum moment capacity ϕM_n of the section.

$$0.75\rho_b \text{ (Table 3.6.1)} = 0.0160$$

$$\rho = \frac{3(1.27) + 2(0.79)}{16(28)} = 0.0120 < 0.75\rho_b \qquad \textbf{OK}$$

$$C = 0.85(3)16a = 40.8a$$

$$T = [3(1.27) + 2(0.79)]60 = (3.81 + 1.58)60 = 323 \text{ kips}$$

$$a = \frac{323}{40.8} = 7.92 \text{ in.}$$

$$M_n = 323[28 - 0.5(7.92)]\tfrac{1}{12} = 647 \text{ ft-kips}$$

$$\phi M_n = 0.90(647) = 582 \text{ ft-kips} \approx M_u = 590 \text{ ft-kips} \qquad \textbf{OK}$$

Figure 6.14.2 Beam of Example 6.14.1.

(b) Determine the theoretical cutoff point for 2-#8 bars. The moment capacity ϕM_n remaining with 3-#10 bars is

$$C = 40.8a$$

$$T = 3.81(60) = 229 \text{ kips}$$

$$a = \frac{229}{40.8} = 5.61 \text{ in.}$$

$$\phi M_n = 0.90(229)[28 - 0.5(5.61)]\tfrac{1}{12} = 433 \text{ ft-kips}$$

Plot on the factored moment M_u diagram and locate the theoretical cutoff point A. Extend to the right 12 bar diameters (of the #8 bars that are to be cut) or the effective depth of the member, whichever is greater, to arrive at point B.

$$d = 28 \text{ in. } (2.33 \text{ ft}) > [12d_b = 12(1.0) = 12 \text{ in.}]$$

(c) Determine development length for #8 bars. The basic development length is obtained by using Eq. (6.7.1); thus

$$L_{db} = 0.04 \frac{f_y A_b}{\sqrt{f'_c}} = 0.04 \left(\frac{0.79(60,000)}{\sqrt{3000}} \right) = 34.6 \text{ in.}$$

Next, it is necessary to determine Category A, B, or C for the #8 bars. Categories D and E are usually eliminated because this is neither a column nor is it a slab; thus, α_{de} likely is 1.0. Assuming the bars, though unequal in size, are uniformly spaced, the clear spacing between them is

$$\text{clear spacing} = \frac{16 - 2(1.5) - 2(0.5) - 3(1.27) - 2(1.0)}{4} = 1.55 \text{ in.}$$

Since only the 2-#8 bars are being developed, and the 3-#10 are presumed to continue beyond the #8 cutoff location, it is the spacing between the 2-#8 that determines the Category. The failure mode would have splitting from a #8 bar to the side or top face of the member or between the two #8 bars. It is the intention of the 1989 ACI Code development length rules to treat a bar (or bars) as inert when it is not being developed within the development length region of other bars. Thus, when the #10 bars of this example have a development length from their termination near the end of the cantilever that is less than the distance to the #8 bar cut, the #10 bars are considered to have no influence on L_d for the #8 bars. It is a matter of opinion whether or not the #10 itself should be treated as concrete. That is, in this case whether to use the full spacing between the #8 bars, $2(1.55) + 1.27$ diam. of #10 = 4.37 in., or to use only the concrete width between the #8 bars, $2(1.55)$ = 3.10 in. The authors believe it appropriate in this case to consider the spacing of the #8 to be 4.37 in. for the purpose of satisfying a Category A requirement, *assuming L_d for the #10 bars does not overlap the L_d for the #8 bars.*

Even if the concrete width between the #8 bars, $2(1.55) = 3.10$ in., it still exceeds the $3d_b$ for the #8 bar to satisfy Category A, item 2(a), given in Section 6.7 (ACI-12.2.3.1a), as well as item 2(b) (ACI-12.2.3.1d), because cover to the top face is 2 in., which equals the $2d_b$ needed for that item.

Thus, Category A applies: $\alpha_{abc} = 1.0$. Since the 4.37 in. spacing does not satisfy the $5d_b$ requirement for Category D, and the member is not a column with ties or spirals satisfying Category E, $\alpha_{de} = 1.0$. At this point, $\alpha_{abc}\alpha_{de}L_{db}$ must be compared with Min L_d of ACI-12.2.3.6.

The minimum development length, relating to a pullout failure mode, from Eq. (6.7.5) is

$$\text{Min } L_d = 0.03d_b \frac{f_y}{\sqrt{f'_c}} = \frac{0.03(1.0)60,000}{\sqrt{3000}} = 32.9 \text{ in.}$$

The first-level L_d is

$$\alpha_{abc}\alpha_{de}L_{db} = 1.0(1.0)34.6 = 34.6 \text{ in.} > (\text{Min } L_d = 32.9 \text{ in.})$$

The first-level L_d is 34.6 in. The only second-level adjustment applicable is that for top bars (ACI-12.2.4.1); thus, α_{top} is 1.3. Therefore,

$$L_d \text{ (for } \#8) = \alpha_{top}(34.6) = 1.3(34.6) = 45.0 \text{ in. } (3.75 \text{ ft})$$

Since point B, the proposed cutoff point, lies only about 3.5 ft from the support, the $\#8$ bars would not have full capacity at the support. Therefore, extend the proposed cutoff to point C, which is located at L_d (for $\#8) = 3.75$ ft from the support.

(d) Check ACI-12.10.5 for cutting bars at point C in the tension zone. The shear strength, including contribution of stirrups, is first computed. Using the simplified method of constant V_c,

$$V_c = 2\sqrt{f_c'}\,b_w d = 2\sqrt{3000}(16)(28)\tfrac{1}{1000} = 49.1 \text{ kips}$$

For the 14-in. spaced $\#4$ stirrups in the vicinity of the potential cut point C,

$$V_s = \frac{A_v f_y d}{s} = \frac{2(0.20)(60)28}{14} = 48.0 \text{ kips}$$

The shear strength ϕV_n at point C is

$$\phi V_n = \phi(V_c + V_s) = 0.85(49.1 + 48.0) = 82.5 \text{ kips}$$

$$\text{percent stressed in shear} = \frac{V_u}{\phi V_n} = \frac{78.4}{82.5} = 95\% > 75\% \qquad \textbf{NG}$$

Even when only 50% of the moment strength ϕM_n is used by M_u, the percent stressed in shear cannot exceed 75% (see Condition 3, Eqs. 6.12.3 and 6.12.4). Try using one more 8-in. stirrup spacing to cover the potential cut at point C, and see whether or not Condition 1, Eq. (6.12.1), is satisfied.

$$V_s = 48.0\left(\frac{14}{8}\right) = 84.0 \text{ kips}$$

$$\text{percent stressed in shear} = \frac{78.4}{0.85(49.1 + 84)} = 69\%$$

This is borderline to satisfy the two-thirds limit of ACI-12.10.5.1 (Eq. 6.12.1). Extend the $\#8$ bars to point C' 4 ft from face of support.

(e) Check whether the continuing $\#10$ bars have adequate development length to the right of point C. The basic development length for a $\#10$ bar is obtained from Eq. (6.7.1); thus

$$L_{db} = 0.04A_b \frac{f_y}{\sqrt{f_c'}} = \frac{0.04(1.27)60,000}{\sqrt{3000}} = 55.6 \text{ in.}$$

The clear spacing between the continuing three #10 bars is

$$\frac{\text{clear}}{\text{spacing}} = \frac{16 - 2(1.5) - 2(0.5) - 3(1.27)}{2} = 4.1 \text{ in.}$$

which exceeds the $3d_b$ of 3.81 in. required for Category A. Since the stirrups have previously been designed for shear, they presumably satisfy the minimum requirement of ACI-11.5.5.3. Verify the spacing for the #4 stirrups to satisfy

$$s = \frac{A_v f_y}{50 b_w} = \frac{2(0.20)60,000}{50(14)} = 34.3 \text{ in.} > \left(\frac{d}{2} = 14 \text{ in.} \right)$$

The 14 in. originally used satisfied the $d/2$ controlling maximum spacing of 14 in. Thus, Category A, item 1 (ACI-12.2.3.1a), is satisfied; $\alpha_{abc} = 1.0$. Categories D and E do not apply; $\alpha_{de} = 1.0$. The minimum L_d must next be compared.

The minimum development length, relating to a pullout failure mode, from Eq. (6.7.5) is

$$\text{Min } L_d = 0.03 d_b \frac{f_y}{\sqrt{f_c'}} = \frac{0.03(1.27)(60,000)}{\sqrt{3000}} = 41.7 \text{ in.}$$

The first-level L_d is

$$\alpha_{abc} \alpha_{de} L_{db} = 1.0(1.0)55.6 = 55.6 \text{ in.} > (\text{Min } L_d = 41.7 \text{ in.})$$

The first-level L_d is 55.6 in. The only second-level adjustment applicable is that for top bars (ACI-12.2.4.1); thus, α_{top} is 1.3. Therefore,

$$L_d \text{ (for #8)} = \alpha_{top}(55.6) = 1.3(55.6) = 72.3 \text{ in. (6.0 ft)}$$

Furthermore, embedment of 6 ft measured from the end of straight #10 bars would overlap the development length region of the #8 bars, placing the #8 bars in Category C (i.e., doubling L_d for the #8 bars). The #10 bars would satisfy literally the statement of ACI-12.10.4, which requires "Continuing reinforcement shall have an embedment length not less than the development length L_d beyond the point where bent or terminated tension reinforcement is no longer required to resist flexure." In other words, the distance from point A to the end of the cantilever must be at least L_d (for #10). The authors believe in a somewhat more conservative approach, requiring the moment capacity ϕM_n diagram to have an offset from the factored moment M_u diagram, except at or near a simple support or the free end of a cantilever, equal to 12 bar diameters or the effective depth d, whichever is greater.

In this case, try standard 90° hooks (see Fig. 6.11.1) on the ends of the #10 bars. Since the beam has the usual 1.5-in. clear cover and #4 stirrups, the cover to the hooked bars is 2 in., which is less than the $2\frac{1}{2}$ in. required by ACI-12.5.4; thus, the special provisions of that Code section must be satisfied.

The development length L_{dh} for the #10 hooked bar is the basic value L_{hb} (i.e., no modification to L_{hb} applies) given by Eq. (6.11.1) and Table 6.11.2. Thus, for #10 hooked bar,

$$L_{dh} = L_{hb} = \frac{1200d_b}{\sqrt{f_c'}} = \frac{1200(1.27)}{\sqrt{3000}} = 27.8 \text{ in.}$$

which exceeds the minimum $8d_b$ or 6 in., whichever is greater (ACI-12.5.1). The L_{dh} of 27.8 in. is dimensioned from the outside face of the tail of the hook, as shown in Fig. 6.14.2. Of course, here stirrups spaced at not more than $3d_b$ (4.23 in.) must be provided along the 27.8 in. of development distance, in accordance with ACI-12.5.4.

(f) Moment capacity ϕM_n diagram. The full strength ϕM_n for the beam with 3-#10 hooked bars will be available at 27.8 in. from the outside of the hook on the end of the beam. Assuming 1.5 in. cover, full capacity is available at 29.3 in. (2.44 ft) from end of beam (point D). The dashed line in Fig. 6.14.2 has been drawn from zero strength at the end of the hook to full strength $\phi M_n = 433$ ft-kips at 2.44 ft from end of beam; however, it is *not* intended to imply that the hooked bar develops its strength linearly since that is highly improbable.

(g) Final decision. Cut 2-#8 bars at 4 ft–0 in. from the support; use 90° standard hooks on the 3-#10 bars; use 8-#4 U stirrups as confinement over the L_{dh} distance, a shown in Fig. 6.14.2 by the dashed stirrups.

The use of #10 bars in this cantilever beam is not a practical design but serves to illustrate the need for extending the cut location from B to C and then to C', and the need for and treatment of hooked bars. ■

6.15 DEVELOPMENT OF REINFORCEMENT AT SIMPLE SUPPORTS AND AT POINTS OF INFLECTION

The concept of requiring the development of reinforcement on both sides of a section where the bars are to be fully stressed may also be applied to the continuation of positive moment tension reinforcement beyond either the centerline of a simple support or a point of inflection.

Simple Supports. Referring to Fig. 6.15.1, consider the point A on the factored moment M_u curve near a simple support, where the factored moment M_u equals the moment capacity ϕM_n of the bars continuing into the support. The distance from point A to the end of the bars must be at least equal to the required development length L_d as computed from ACI-12.2. This requirement given in ACI-12.11.3 is that the available embedment length must equal or exceed L_d, or

$$\text{available embedment length} = L_a + 1.3 \frac{M_n}{V_u} \geq L_d \qquad \textbf{(6.15.1)}$$

where

M_n = nominal flexural strength of the remaining bars

$\qquad = A_s f_y \left(d - \dfrac{a}{2} \right)$, or

$\qquad$ = moment capacity in the working stress method (ACI-Appendix A.4 and A.5),

Figure 6.15.1 Development of reinforcement at a simple support.

V_u = factored shear at the support, or

= the service load shear in the working stress method (ACI-Appendix A.4),

L_a = the straight embedment length beyond the centerline of support to the end of the bars [When the bars are hooked, Eq. (6.15.1) is considered automatically satisfied.]

The distance x between the point A and the centerline of support in Fig. 6.15.1 is approximately equal to

$$x = \frac{M_n}{V_u} \qquad (6.15.2)$$

because the area of the shaded portion of the shear diagram equals the *change* in moment between the center of support and point A; thus the ordinate on the factored moment diagram at A is

$$V_{avg} \text{ (say, } 0.9\ V_u)x = \phi M_n$$

Comparing Eq. (6.15.1) and Eq. (6.15.2), it is seen that Eq. (6.15.1) is identical to

$$L_a + 1.30x \geq L_d$$

wherein the 1.30 factor is in recognition of the fact that the bars extending into a simple support have less tendency to cause splitting when confined by a compressive reaction [6.32].

The 1989 ACI Code exempts the requirement of ACI-12.11.3 "for reinforcement terminating beyond centerline of simple supports by a standard hook, or a mechanical anchorage at least equivalent to a standard hook."

Inflection Points. Since an inflection point is a point of zero moment located away from a support (refer to Fig. 6.15.2), bars in that region are not confined by a compressive reaction; therefore the 1.30 factor is interpreted as not to apply. In this case, the embedment length that must exceed the required development length L_d (ACI-12.11.3) may be stated as

$$\text{available embedment length} = \begin{bmatrix} \text{actual } L_a\text{, but} \\ \text{not exceeding the} \\ \text{larger of } 12d_b \text{ or d} \end{bmatrix} + \frac{M_n}{V_u} \geq L_d \quad \textbf{(6.15.3)}$$

wherein M_n and V_u refer to the nominal flexural strength and the factored shear at the point of inflection. The limitation of the usable L_a to 12 bar diameters or the effective depth has been applied because, according to the ACI Commentary-R12.11.3, there is no experimental evidence to show that long anchorage length will be fully effective in developing a bar in a short length between the point of inflection and a point of maximum stress.

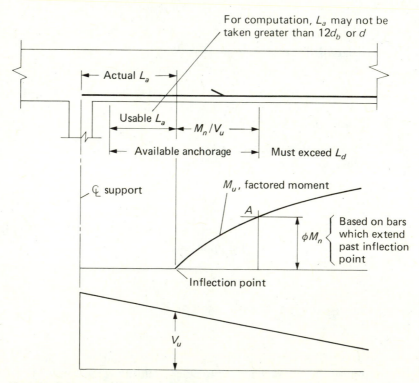

Figure 6.15.2 Development of reinforcement at an inflection point.

Additional development of reinforcement at the face of support is required by ACI-12.11.2 when the flexural member is part of the primary lateral load resisting system.

6.16 DEVELOPMENT OF SHEAR REINFORCEMENT

Reinforcement in the web of a beam, whether it be for shear or for torsion (see Chapter 19), must be properly anchored so that its full tensile capacity is available at or near the middepth of a beam. For proper function, the shear reinforcement must be "carried as close to compression and tension surfaces of member as cover requirements and proximity of other reinforcement will permit" (ACI-12.13.1). It is especially important to extend the stirrups as close to the compression face as possible because the flexural tension cracks may extend deeply into the compression zone when the nominal strength of the member is approached.

The ends of single leg, simple U, or multiple U stirrups of deformed bars or wire must be anchored by means of a standard stirrup hook around longitudinal reinforcement as shown in Fig. 6.16.1. Such stirrups may be inclined, but the angle between the stirrups and the longitudinal bars must be at least 45° in accordance with ACI-11.5.1.2.

The standard hooks for stirrups and ties have somewhat relaxed requirements as compared with the standard hooks used for main bars as shown in Fig. 6.11.1. Also, hooked stirrups and hooked ties are not permitted for bars larger than #8. As noted in Chapter 5, stirrups and ties are usually #3 or #4 bars, occasionally #5 or #6, and rarely, if ever, larger than #8, so the ACI Code limit on size of hooked stirrups should rarely apply. Note also that the 180° hook is not used for stirrups and ties. The requirements for standard hooks at ends of ties and stirrups are shown in Fig. 6.16.2.

(a) For #5 bar and D31 wire, and smaller, or For #6, #7, and #8 bar with $f_y \leq 40,000$ psi (ACI – 12.13.2.1)

(b) For #6, #7, #8 stirrups with $f_y > 40,000$ psi (ACI – 12.13.2.2)

Figure 6.16.1 Development of deformed bar or deformed wire stirrups.

(a) 90° hook
(for # 5 bars and smaller)

(b) 90° hook
(for # 6, # 7, and # 8 bars)

(c) 135° hook
(for # 8 bars and smaller)

Figure 6.16.2 Standard hooks for stirrups and ties (ACI-7.1).

When *closed* stirrups are desired, one practical procedure is to use a pair of U stirrups without hooks, placed to form a closed unit. If this is done, ACI-12.13.5 requires laps of $1.3L_d$ for proper splicing. When members are at least 18 in. deep and the tensile capacity $A_b f_y$ of the stirrup does not exceed 9 kips per leg, splices are adequate if the legs extend the full available depth of the member.

For U stirrups of welded smooth wire fabric, anchorage may be accomplished (ACI-12.13.2.3) using either "(a) Two longitudinal wires spaced at a 2-in. spacing along the member at the top of the U," or "(b) One longitudinal wire located not more than $d/4$ from the compression face and a second wire closer to the compression face and spaced not less than 2 in. from the first wire. The second wire may be located on the stirrup leg beyond a bend, or on a bend with an inside diameter of bend not less than $8d_b$." These provisions are illustrated in Fig. 6.16.3.

For *single-leg* stirrups of welded smooth or deformed wire fabric, ACI-12.13.2.4 includes the recommendations [6.33] of the PCI/WRI Ad Hoc Committee on Welded Wire Fabric for Shear Reinforcement. Single-leg stirrups are practical and common in precast prestressed concrete T-sections. The provisions of ACI-12.13.2.4 are detailed in Fig. 6.16.4.

For guidance in placing stirrups and ties to obtain proper anchorage, reference should be made to the *ACI Detailing Manual—1988* [p. 33 of Ref. 2.25].

(a)

(b)

Figure 6.16.3 Development of welded smooth wire fabric stirrups (ACI-12.13.2.3).

See ACI—12.13.1

Two horizontal wires top and bottom

2″ minimum

Smooth or deformed vertical wires

Greater of $d/4$ or 2″

Mid-depth

Greater of $d/4$ or 2″

Main reinforcement

2″ minimum

"as close to tension surface as cover requirements and proximity of other reinforcement will permit" (ACI—12.13.1)

Outer wire not above lowest main reinforcement

Figure 6.16.4 Development of single-leg smooth or deformed welded wire fabric shear reinforcement (ACI-12.13.2.4) (adapted from Reference 6.33).

6.17 TENSION LAP SPLICES

Wherever bar lengths required in a structure exceed the length available or the length that may be economically shipped, splices are necessary. Splicing may be accomplished by simple lapping of bars either in contact or separated. As an alternative, butt connections may be made by welding.

In general, splices should be located away from points of maximum tensile stress, and splicing should be staggered along the length of the bars. In other words all of the bars should not be spliced at one location.

The beam with splices should be as ductile as one without splices. The ACI Code provisions are intended to assure that no splice failure will occur when the full nominal strength in flexure is reached at the spliced location. Requirements for minimum clear spacing of *contact* splices (ACI-7.6.4) are to ensure an adequate amount of surrounding concrete to resist splitting; but, in *noncontact* lap splices the individual bars should *not* be spaced transversely too far apart (ACI-12.14.2.3).

The overlap distance required in tension lap splices should be equal to or larger than the development length L_d of the bar because stress concentrations near the splice ends tend to produce splitting at early stages of loading unless special precautions are taken [6.3, 6.34–6.37]. Classes A and B tension lap splices are defined by ACI-12.15.1 to have overlap distances of $1.0L_d$ and $1.3L_d$, respectively, but a total lap of not less than 12 in. The class of tension lap splice to be used depends on (1) the percentage of bars being spliced of the total number within the required lap distance, and (2) the stress level in the unspliced bars at the splice location.

A summary of the requirements for the two classes of tension lap splices appears in Table 6.17.1. The provisions apply equally to deformed bar or deformed wire splices. Tension lap splices may not be used for bars larger than #11 (ACI-12.14.2.1).

Table 6.17.1 Tension Lap Splices (ACI-12.15.2)

$\left(\dfrac{A_s \text{ REQUIRED}}{A_s \text{ PROVIDED}}\right)$ AT THE SPLICE	MAXIMUM PERCENT OF A_s SPLICED	SPLICE CLASS	REQUIRED LAP	NOTES
≤ 0.5	50	A	L_d	Desirable
	100	B	$1.3L_d$	OK
> 0.5	100	B	$1.3L_d$	Avoid more than 0.5

As stated previously in Section 6.10, the lap lengths prescribed in Table 6.17.1 shall be increased 20% for a three-bar bundle and 33% for a four-bar bundle (ACI-12.14.2.2). Bars spliced by noncontact lap splices in flexural members shall not be spaced transversely farther apart than one-fifth of the required lap length nor 6 in. (ACI-12.14.2.3).

The required overlap distances shown in Table 6.17.1 are determined by the development length requirements of ACI-12.2 as described in Sections 6.7 and 6.8. For determining the appropriate modification factors, a unit of bundled bars is to be treated as a single bar of a diameter derived from the equivalent total area.

The ratio (A_s required)/(A_s provided) column in Table 6.17.1 refers to the percentage of available capacity that is utilized. The ratio may also be considered as the percent of f_y to which the bars are stressed. When the factored moment M_u is only 50% of the moment capacity ϕM_n, the ratio would be considered 0.5. In general, temperature, shrinkage, and load distribution reinforcement should be considered as fully stressed for the purpose of designing splices.

6.18 WELDED TENSION SPLICES AND MECHANICAL CONNECTIONS

A welded tension butt splice or mechanical connection is used in situations where large tensile forces are to be transmitted across the splice or large bars need to be spliced and the lap splice may be impractical or prohibited. Bars larger than #11 may not be lap spliced to carry tension (ACI-12.14.2.1). Tension tie members also may not be lap spliced (ACI-12.15.5). A tension tie is a member (a) carrying a tensile force large enough to cause tension over the entire section, (b) having a stress level in the reinforcement high enough to require every bar to be fully effective, and (c) having limited concrete cover on its sides (ACI-R12.15.5).

These tension splices, referred to as *full welded splices* or *full mechanical connections* (ACI-12.14.3), are required to develop in tension at least 125% of the specified yield strength of the bar when used in regions of high stress.

The full welded splice is intended primarily for relatively large bars (#6 and larger) in main members (ACI-R12.15.5). The tensile capacity required is intended to ensure sound full penetration welds—that is, to produce splices capable of developing the strength $A_s f_y$ of the bars spliced. According to ACI-R12.14.3.3, "the 25 percent increase above the specified yield strength was selected as both an adequate minimum for safety and a practical maximum for economy."

In regions of low stress (i.e., where less than 50% of available capacity is being used), welded splices or mechanical connections of less capacity than 125% are permitted (ACI-12.14.3.5 and 12.15.4). In these situations welded lap joints of reinforcing bars, either with or without backup material, or other welded bar arrangements, may be allowed. However, such splices must be "staggered at least 24 in. and in such manner as to develop at every section at least twice the calculated tensile force at that section but not less than 20,000 psi for total area of reinforcement provided."

In computing the capacity, spliced bars may be rated at the specified splice strength (assuming it is less than the strength of the bars), whereas any unspliced bars are to be rated in proportion to their length of bar development to the splice point (but not to exceed the maximum bar capacity at embedment L_d). For example, a welded splice may be specified to provide 90% of the full capacity of the bars being spliced, in which case that bar area is taken as 90% of its actual area for computing capacity. For unspliced bars that have been embedded, say only $0.5L_d$, the bar area used for strength calculation should be taken as one-half the actual area.

6.19 COMPRESSION LAP SPLICES

Whereas bars larger than #11 acting in tension are not permitted to be lap spliced, #14 and #18 bars in *compression* may be lap spliced to #11 and smaller bars (ACI-12.16.2).

The minimum overlap in compression lap splices when f'_c is not less than 3000 psi must be at least equal to the following (ACI-12.16.1):

For $f_y \leq 60,000$ psi,

$$\text{lap} = 0.0005 f_y d_b \qquad \text{or 12 in.} \quad \text{(whichever is larger)} \qquad \textbf{(6.19.1)*}$$

For $f_y > 60,000$ psi,

$$\text{lap} = (0.0009 f_y - 24) d_b \quad \text{or 12 in.} \quad \text{(whichever is larger)} \qquad \textbf{(6.19.2)*}$$

When f'_c is less than 3000 psi, the lap length is to be increased by one-third. When lapping bars of two different sizes, the lap splice length is to be the larger of (1) the compression splice length of the smaller bar, or (2) the compression development length L_d of the larger bar (see Section 6.9).

For compression members whose main steel is surrounded by closed ties throughout the lap length, the required lap may be taken at 0.83 of that otherwise required, but not less than 12 in. A minimum percentage of column tie area is also required by ACI-12.17.2.4. Column ties are discussed in Chapter 13 (Section 13.8).

— — — —

* For SI, ACI 318–89M, for lap, L_d, and d_b in mm, f'_c and f_y in MPa, gives

For $f_y \leq 400$ MPa,

$$\text{lap} = 0.073 f_y d_b \qquad \text{or 300 mm} \qquad \textbf{(6.19.1)}$$

For $f_y > 400$ MPa,

$$\text{lap} = (0.13 f_y - 24) d_b \quad \text{or 300 mm} \qquad \textbf{(6.19.2)}$$

For members whose main steel is surrounded by a closely wound spiral, the required lap may be taken at 0.75 of that otherwise required, but not less than 12 in. (ACI-12.17.2.5). Spiral reinforcement is discussed in Chapter 13 (Section 13.9).

The number of bar diameters required for the overlap in compression lap splices is summarized in Table 6.19.1.

Table 6.19.1 Bar Diameters Required for Compression Lap Splices for $f'_c \geq 3000$ psi (ACI-12.16.1)

YIELD STRESS f_y (ksi)	BAR DIAMETERS[a]		
	SPIRAL COLUMN	TIED COLUMN	OTHERS
40	15	16.6	20
50	18.75	20.75	25
60	22.5	24.9	30
75	32.6	36.2	43.5
80	36.0	39.9	48.0

[a] When computing splice length, the minimum to be used is 12 in.

6.20 COMPRESSION END BEARING CONNECTIONS, WELDED SPLICES, AND MECHANICAL CONNECTIONS

End bearing connections are allowed for compression only, wherein the load in the bars is transmitted by bearing of square cut ends held in concentric contact by a suitable device. According to ACI-12.16.6.2, bar ends must terminate in flat surfaces within $1\frac{1}{2}°$ of right angles to the axis of the bars and be fitted within $3°$ of full bearing after assembly. End bearing splices are only permitted when the member contains closed ties, closed stirrups, or spirals.

When welded splices or mechanical connections are used in compression, the requirements are the same as for tension splices—that is, the development of 125% of the yield strength of the bars, except where less than 50% of the full unspliced bar capacity is required by the design load (ACI-12.14.3 and 12.15.4).

6.21 SPLICES FOR MEMBERS UNDER COMPRESSION AND BENDING

For compression members there are three categories of special splice provisions (ACI-12.17):

1. Design in the upper part of the "compression controls" region (Fig. 13.6.2) such that the bar stress due to factored loads is *compressive*: Any type of splices may be used, including lap splices according to ACI-12.16.1 and 12.16.2, "full welded" splices according to ACI-12.14.3.3, "full mechanical" connections ac-

cording to ACI-12.14.3.4, and "end bearing splices" according to ACI-12.16.4 and 12.17.4. For lap splices, the lap lengths must satisfy Eqs. (6.19.1) or (6.19.2) and cannot be less than 12 in.

2. Design in the middle part of the "compression controls" region (Fig. 13.6.2) such that the bar stress resulting from factored loads is *tensile* and does not exceed $0.5f_y$ in tension. Lap splices may be used according to ACI-12.17.2.2; alternatively, "full welded" splices according to ACI-12.14.3.3 or "full mechanical" connections according to ACI-12.14.3.4 may be used. When lap splices are used (for tension not exceeding $0.5f_y$), Class B tension lap splices must be used when more than 50% of the bars are spliced at the section. When fewer than 50% of the bars are spliced at a section and the splices are staggered by an amount L_d, Class A splices may be used.

3. Design in the lower part of the "compression controls" region or in "tension controls" region (Fig. 13.6.2) such that the bar stress resulting from factored loads is greater than $0.5f_y$ in *tension*. When lap splices are used, they must be Class B splices according to ACI-12.17.2.3; alternatively, "full welded" splices according to ACI-12.14.3.3 or "full mechanical" connections according to ACI-12.14.3.4. Even if these *are* tension splices, ACI-12.17.3 apparently restricts from using ACI-12.14.3.5 when excess bars are used, since the former applies to splices or connectors in "columns".

Even in "end bearing splices" for bars in compression, ACI-12.17.4 requires the continuing bars in each face to have a tensile strength across the splice equal to $0.25f_y$ times the area of the longitudinal reinforcement in each face. Thus, whatever mode is used for making a splice in a column, this minimum tensile capacity is required.

Applicable to lap splices in all three categories, when the main longitudinal steel is surrounded by qualifying ties (with effective area not less than $0.0015hs$) or spiral, the required lap may be reduced using a multiplier of 0.83 or 0.75, respectively, on the lap otherwise required, but the lap must be at least 12 in. (ACI-12.17.2.4 and 12.17.2.5).

6.22 DESIGN EXAMPLES

Two complete examples in the design of reinforced concrete flexural members are presented here for the purpose of showing the design for flexure, shear, and development of reinforcement, all in the same beam.

■ **EXAMPLE 6.22.1** Design the simply supported beam shown in Fig. 6.22.1(a). The dead load is 0.9 kip/ft, not including the weight of the beam. The live load consists of a concentrated load of 13 kips at midspan. Use $f'_c = 3000$ psi, $f_y = 40,000$ psi, and the ACI strength method.

Solution: (a) Design for flexure. Assume that a rectangular section with tension reinforcement only will be used at a reinforcement ratio somewhat lower than the maximum permissible value. Using basic principles as illustrated in Section 3.6, or the value from Table 3.6.1,

$$\text{max } \rho = 0.75\rho_b = 0.0278$$

Figure 6.22.1 Simply supported beam of Example 6.22.1.

Arbitrarily selecting an approximate $\rho = 0.025$ and using Eq. (3.8.4) or the direct statics as illustrated in Section 3.8,

$$m = \frac{f_y}{0.85 f'_c} = 15.7$$

$$R_n = \rho f_y (1 - \tfrac{1}{2}\rho m) = 804 \text{ psi}$$

Assume weight of beam is 0.2 kip/ft,

$$w_u = 1.4(0.9 + 0.2) = 1.54 \text{ kips/ft (dead load)}$$

$$W_u = 1.7(13) = 22.1 \text{ kips (live load)}$$

$$M_u = \tfrac{1}{8}(1.54)(20)^2 + \tfrac{1}{4}(22.1)(20) = 77 + 110.5 = 188 \text{ ft-kips}$$

$$\text{required } M_n = \frac{M_u}{\phi} = \frac{188}{0.90} = 209 \text{ ft-kips}$$

$$\text{required } bd^2 = \frac{M_n}{R_n} = \frac{209(12{,}000)}{804} = 3120 \text{ in}^3.$$

If $b = 12$ in.

$$d = \sqrt{\frac{3120}{12}} = 16.1 \text{ in.}$$

$$\text{required } h = d + \text{approx } 2\tfrac{1}{2} \text{ in. for one layer of bars}$$

$$= 16.1 + 2.5 = 18.6 \text{ in.}$$

$$\text{weight of beam} = \frac{12(20)}{144}(0.15) = 0.25 \text{ kip/ft}$$

$$\text{revised } w_u = 1.4(0.9 + 0.25) = 1.61 \text{ kips/ft (dead load)}$$

$$\text{revised } M_u = \tfrac{1}{8}(1.61)(20)^2 + 110.5 = 80.5 + 110.5 = 191 \text{ ft-kips}$$

$$\text{revised required } M_n = \frac{191}{0.90} = 212 \text{ ft-kips}$$

$$\text{required } R_n = \frac{M_n}{bd^2} = \frac{212(12{,}000)}{12(17.5)^2} = 692 \text{ psi}$$

The steel reinforcement ratio may then be found from Eq. (3.8.5), Fig. 3.8.1, or approximately by straightline proportion,

$$A_s \approx 0.025(12)(17.5)\left(\frac{692}{804}\right) = 4.52 \text{ sq in.}$$

Note again that the approximation using straightline proportion is slightly on the safe side when revised R_n is smaller, as in this case.

Try 2-#11 and 1-#10 bars ($A_s = 4.39$ sq in.). Assuming #3 U stirrups, the minimum width of beam to accommodate these bars is 10.8 in. (see Table 3.9.2) which is less than the width of beam being used. Note that the entry of Table 3.9.2 must be with the two corner bars (#11) and then add for the one smaller bar (#10), and then add the difference in diameters between the larger and smaller bars, giving 10.64 in. from the table plus 0.14 in., making a total of 10.78 in. (say 10.8 in.).

Check capacity.

$$C = 0.85f'_c ba = 0.85(3)12a = 30.6a$$

$$T = A_s f_y = 4.39(40) = 176 \text{ kips}$$

$$a = \frac{176}{30.6} = 5.75 \text{ in.}$$

$$M_n = T\left(d - \frac{a}{2}\right) = 176[17.5 - 0.5(5.75)]\tfrac{1}{12}$$

$$= 214 \text{ ft-kips} > 212 \text{ ft-kips required} \qquad \qquad \textbf{OK}$$

(b) Make the preliminary selection of the cutoff point for 1-#10. The remaining moment capacity with 2-#11 bars is

$$C = 30.6a$$

$$T = 2(1.56)40 = 125 \text{ kips}$$

$$a = \frac{125}{30.6} = 4.08 \text{ in.}$$

$$\phi M_n = 0.90(125)[17.5 - 0.5(4.08)]\tfrac{1}{12} = 145 \text{ ft-kips}$$

The value of 145 ft-kips is plotted on the factored M_u diagram to locate the theoretical cutoff point A in Fig. 6.22.1. The actual potential cutoff location (point B) is found by extending from point A toward the support a distance of 12 bar diameters or the effective depth of the member, whichever is greater (ACI-12.10.3).

$$12d_b = 12\left(\frac{1.27}{12}\right) = 1.27 \text{ ft}$$

$$d = \frac{17.5}{12} = 1.46 \text{ ft} \qquad \qquad \textbf{(Controls)}$$

The cutoff at point B will be acceptable only if the shear does not exceed two-thirds of the shear strength at point B (ACI-12.10.5.1). An alternative is to provide extra stirrups in accordance with ACI-12.10.5.2.

(c) Determine development length for the #10 bar that is to be terminated at point B. The basic development length using Eq. (6.7.1) is

$$L_{db} (\#10) = 0.04A_b \frac{f_y}{\sqrt{f_c'}} = \frac{0.04(1.27)40,000}{\sqrt{3000}} = 37.1 \text{ in.}$$

Next, since the #11 bars are (presumably) not being developed over the portion of the beam where the #10 will be developed, the #11 bars do not influence the determination of Category A, B, or C. Since the #10 has nearly 5 in. cover to the side faces of the beam, will be confined within *at least* minimum stirrups, and has bottom cover of $1\tfrac{7}{8}$ (with #3 stirrups), Category A, item 1 (ACI-12.2.3.1a) applies; $\alpha_{abc} = 1.0$. Categories D and E do not apply; thus, $\alpha_{de} = 1.0$. Compare $\alpha_{abc}\alpha_{de}L_{db}$ with Min L_d,

$$\alpha_{abc}\alpha_{de}L_{db} = 1.0(1.0)37.1 = 37.1 \text{ in.} \qquad \qquad \textbf{(Controls)}$$

$$\text{Min } L_d = 0.03d_b \frac{f_y}{\sqrt{f_c'}} = \frac{0.03(1.27)(40,000)}{\sqrt{3000}} = 27.8 \text{ in.}$$

Thus, "first-level L_d" = 37.1 in. These are bottom bars and no second-level multipliers are applicable; thus,

$$L_d \, (\#10) = 37.1 \text{ in. } (3.1 \text{ ft})$$

(d) Determine the development length for the #11 bars that extend into the support. The basic development length using Eq. (6.7.1) is

$$L_{db} \, (\#11) = 0.04 A_b \frac{f_y}{\sqrt{f_c'}} = \frac{0.04(1.56)40,000}{\sqrt{3000}} = 45.6 \text{ in.}$$

A check of Table 3.9.4 giving minimum beam width such that clear spacing of 2-#11 will equal $3d_b$, shows a beam 10.9 in. wide or more will qualify. Since this is less than the 12 in. used, the #11 bars satisfy Category A, item 1; $\alpha_{abc} = 1.0$. In this case, Min L_d for #11 is 30.9 in. (see Table 6.7.2 or compute as for the #10 bar). Thus, the "first-level L_d" is 45.6 in. Since no second-level modifications apply,

$$L_d \, (\#11) = 45.6 \text{ in. } (3.8 \text{ ft})$$

(e) Design of shear reinforcement; simplified method with constant V_c.

$$V_u \text{ (at centerline of support)} = 1.61(10) + 11.1 = 27.2 \text{ kips}$$

$$V_u \text{ (at } d \text{ from face of support)} = 27.2 - 1.96(1.61) = 24.0 \text{ kips}$$

$$\phi V_c = \phi(2\sqrt{f_c'} \, b_w d) = 0.85(2\sqrt{3000})(12)(17.5)\tfrac{1}{1000} = 19.6 \text{ kips}$$

$$\text{required } \phi V_s = V_u - \phi V_c = 24.0 - 19.6 = 4.4 \text{ kips}$$

$$\text{min } \phi V_s = \phi(50 b_w d) = 0.85(50)(12)(17.5)\tfrac{1}{1000} = 8.9 \text{ kips}$$

Since required $\phi V_s = 4.4$ kips is less than min $\phi V_s = 8.9$ kips to satisfy the requirement of ACI-11.5.5.3 based on $v_s = 50$ psi, use min $\phi V_s = 8.9$ kips to determine the stirrup spacing at the critical section. For #3 U stirrups, using Eq. (5.11.7),

$$\phi V_s = \frac{\phi A_v f_y d}{s}$$

$$\text{max } s = \frac{\phi A_v f_y d}{\phi V_s} = \frac{0.85(0.22)(40)(17.5)}{8.9} = 14.7 \text{ in.}$$

However, the stirrup spacing may not exceed $d/2 = 8.75$ in.

Try #3 stirrups @ $8\tfrac{1}{2}$ in. spacing. Stirrups at this spacing must be used until $V_u \le 0.5\phi V_c$ which for this beam [see Fig. 6.22.1(c)] means the entire span.

(f) Check cutoff point for satisfying the shear requirement of ACI-12.10.5 for cutting bars in the tension zone. The shear strength provided by #3 stirrups at $8\tfrac{1}{2}$ in. spacing is

$$\phi V_n = \phi V_c + \phi V_s = 19.6 + \frac{0.85(0.22)(40)(17.5)}{8.5} = 35.0 \text{ kips}$$

$$V_u \text{ at proposed cutoff point } B = 19.2 \text{ kips}$$

$$\text{percent stressed in shear} = \frac{V_u}{\phi V_n} = \frac{19.2}{35.0} = 55\% < 66\tfrac{2}{3}\% \qquad \textbf{OK}$$

Use 1-#10, 10′–0″ long placed symmetrically about midspan.

(g) Check development length requirement at the support. According to ACI-12.11.3, it is required that

$$1.30 \frac{M_n}{V_u} + L_a \geq L_d$$

$$M_n \text{ for 2-#11 bars} = \frac{145}{0.9} = 161 \text{ ft-kips}$$

V_u = factored shear at centerline of support = 27.2 kips

L_a = embedment length beyond the center of support; assume zero here

$$1.30 \frac{161(12)}{27.2} = 92.5 \text{ in.} > [L_d = 45.6 \text{ in.}] \qquad \textbf{OK}$$

Actually, the moment capacity ϕM_n diagram provides this same check but more conservatively (i.e., without the 1.30 factor) since the horizontal distance from the center of support to point A exceeds L_d.

(h) Design sketch. The final conclusions are presented in Fig. 6.22.1(a). Since $f_y = 40,000$ psi, the crack control provisions of ACI-10.6.4 need not be checked. If deflection control is important to prevent damage to partitions or other construction, the deflection must be checked according to ACI-9.5. Computation for deflections is treated in Chapter 14. ■

■ **EXAMPLE 6.22.2**   Design the overhanging beam shown in Fig. 6.22.2. The superimposed service uniform dead load is 4 kips/ft. Use $f'_c = 4000$ psi, $f_y = 60,000$ psi, and the ACI strength method. (Note that live load is not used in this example because the purpose here is to show how flexure, shear, and development of reinforcement are to be considered together in one simple case.)

Solution: **(a)** Design for flexure. Since tension reinforcement will be required in the top of the overhang, it may be desirable to run some bars straight across the top of the entire beam. This would also help reduce creep and shrinkage deflection under sustained load.

For deflection control, a guideline value of ρ equal to one-half the maximum permissible value may serve to establish the beam size.

$$\max \rho = 0.75 \, \rho_b = 0.0214 \quad \text{(Table 3.6.1)}$$

$$\text{Arbitrarily choose } \rho = 0.011 \qquad m = f_y/0.85f'_c = 17.6$$

$$R_n = \frac{M_u}{\phi bd^2} = \rho f_y (1 - \tfrac{1}{2}\rho m)$$

$$= 0.011(60,000)[1 - \tfrac{1}{2}(0.011)(17.6)] = 596 \text{ psi}$$

Estimating the beam weight at 0.3 kip/ft, the factored uniform loading is

$$w_u = 1.4(4.3) = 6.02 \text{ kips/ft}$$

Figure 6.22.2 Overhanging beam for Example 6.22.2.

The factored shear at the left support is

$$V_u = \frac{6.02(18)}{2} - \frac{0.5(6.02)(6.5)^2}{18} = 47.1 \text{ kips}$$

$$\max (+)M_u = \frac{(47.1)^2}{2(6.02)} = 184 \text{ ft-kips}$$

$$\max (-)M_u = \tfrac{1}{2}(6.02)(6.5)^2 = 127 \text{ ft-kips}$$

$$\text{required } bd^2 = \frac{M_u}{\phi R_n} = \frac{184(12,000)}{0.90(596)} = 4120 \text{ in}^3.$$

b	d	h
12	18.5	21
14	17.2	20

In selecting a size, the designer need not adhere rigidly to the value of $R_n = 596$ psi but may go either higher or lower, since the value of ρ is about in the middle of the permissible range. Use a rectangular section with $b = 14$ in. and $h = 20$ in., which gives $d \approx 17.5$ in. The revised beam weight is $14(20)(0.15)/144 = 0.29$ kip/ft, giving $w_u = 1.4(0.29 + 4.00) = 6.01$ kips/ft.

The factored shear and bending moment diagrams (i.e., using factored service loads) are shown in Fig. 6.22.2. At section *A-A*,

$$\text{required } R_n = \frac{M_u}{\phi bd^2} = \frac{184(12,000)}{0.90(14)(17.5)^2} = 572 \text{ psi}$$

$$A_s \approx 0.011bd\left(\frac{\text{required } R_n}{596}\right)$$

$$A_s = 0.011(14)(17.5)\left(\frac{572}{596}\right) = 2.59 \text{ sq in.}$$

Use 2-#9 and 1-#7 ($A_s = 2.60$ sq in.)

Check:

$$C = 0.85f'_c ba = 0.85(4)(14)a = 47.6a$$

$$T = A_s f_y = 2.60(60) = 156 \text{ kips}$$

$$a = 156/47.6 = 3.28 \text{ in.}$$

$$M_n = T(d - a/2) = 156(17.5 - 1.64)\tfrac{1}{12} = 206 \text{ ft-kips}$$

$$\phi M_n = 0.90(206) = 186 \text{ ft-kips} > [M_u = 184 \text{ ft-kips}] \qquad \textbf{OK}$$

At section *B-B*, obtain by proportion

$$\text{required } R_n = \frac{M_u}{\phi bd^2} = 572\left(\frac{127}{184}\right) = 395 \text{ psi}$$

$$A_s \approx 2.59\left(\frac{395}{572}\right) = 1.79 \text{ sq in.}$$

Use 4-#6 bars ($A_s = 1.76$ sq in.)

Check:

$$C = 47.6a$$

$$T = 1.76(60) = 106 \text{ kips}$$

$$a = 106/47.6 = 2.22 \text{ in.}$$

$$\phi M_n = 0.90(106)(17.5 - 1.11)\tfrac{1}{12} = 130 \text{ ft-kips} > 127 \text{ ft-kips} \qquad \textbf{OK}$$

The reader is reminded that for a given R_n, the required ρ expression is a quadratic function; however, for practical use it may be approximated to be linear (see Fig. 3.8.1). A check of strength provided is then made for verification.

For simplicity in bar arrangement, no bending of bars is proposed. The arrangement of main reinforcement finally selected is shown in Fig. 6.22.2.

Note that in the design of the flexural reinforcement for the maximum negative moment, the center of support value of 127 ft-kips was used instead of the lesser value at the face of support, as could have been used in accordance with the discussion of the negative moment on *continuous* beams (Section 6.12). On this statically determinate structure, the authors prefer the more conservative approach of using the center of support value. For beams on wider supports, corrections to the face of support would be appropriate.

(b) Determine whether or not the crack control criterion of ACI-10.6.4 is satisfied at maximum positive moment region. Using ACI Formula (10-4) (see also Section 4.12),

$$z = f_s \sqrt[3]{d_c A}$$

where A is equal to the effective tension area of concrete surrounding the main tension reinforcing bars divided by the number of bars. Using 1.5 in. of clear cover,

$$d_c = 1.5 \text{ (cover)} + 0.375 \text{ (stirrup)} + 0.52 \text{ (avg radius)} = 2.40 \text{ in.}$$

$$\text{number of bars} = \frac{A_s}{\text{area of largest bar}} = \frac{2.60}{1.0} = 2.6$$

$$A = \frac{2(2.40)(14)}{2.6} = 25.8 \text{ sq in./bar}$$

$$f_s = 0.60 f_y = 0.60(60) = 36 \text{ ksi}$$

$$z = 36 \sqrt[3]{2.40(25.8)} = 36(3.96) = 143 \text{ kips/in.}$$

which is less than the limit of 145 allowed by ACI-10.6.4 for exterior exposure. If 2-in. cover is used (ACI-7.7.1), computed z is 162 kips/in., which would be unacceptable for exterior exposure. The actual service load stress is usually less than $0.60 f_y$ and the lesser value could be computed if needed to satisfy the crack control limitation (although little can be gained in this case because of the use of a load factor equal to 1.4 for all the loads).

(c) Development of reinforcement and bar cutoff. According to Eq. (6.7.1), the basic development lengths (ACI-12.2) are

$$L_{db} = 0.04 \frac{A_b f_y}{\sqrt{f_c'}} = 0.04 A_b \left(\frac{60{,}000}{\sqrt{4000}} \right) = 38 A_b$$

Thus,

$$L_{db} \, (\#6) = 38(0.44) = 16.7 \text{ in.}$$

$$L_{db} \, (\#7) = 38(0.60) = 22.8 \text{ in.}$$

$$L_{db} \, (\#9) = 38(1.00) = 38.0 \text{ in.}$$

For the 4-#6 in the negative moment zone, assuming no cutoffs are made, the clear spacing is

$$\frac{\text{clear}}{\text{spacing}} = \frac{14 - 2(1.5) - 2(0.375) - 4(0.75)}{3} = 2.42 \text{ in.}$$

which exceeds $3d_b$ of the #6 bar qualifying each for Category A development length, giving $\alpha_{abc} = 1.0$. This assumes minimum cover and at least minimum stirrups. The stirrup requirement will be satisfied on the 18 ft span side of the support; however, unless stirrups are extended to the end of the cantilever, Category A, item 1 (ACI-12.2.3.1a) would not be satisfied on the cantilever side. The use of #3 stirrups anywhere would make the top cover $1\frac{7}{8}$ in., which exceeds $2d_b$. Thus, in the region on the cantilever where no stirrups are used, Category A, item 2 (ACI-12.2.3.1d) is satisfied. Thus, $\alpha_{abc} = 1.0$. Categories D and E do not apply; $\alpha_{de} = 1.0$. Compare $\alpha_{abc} \alpha_{de} L_{db}$ with Min L_d,

$$\text{Min } L_d = 0.03 d_b \frac{f_y}{\sqrt{f_c'}} = \frac{0.03 d_b (40{,}000)}{\sqrt{3000}} = 21.9 d_b$$

which gives

$$\text{Min } L_d \, (\#6) = 21.9(0.75) = 16.4 \text{ in.}$$

$$\text{Min } L_d \, (\#7) = 21.9(0.875) = 19.2 \text{ in.}$$

$$\text{Min } L_d \, (\#9) = 21.9(1.128) = 24.7 \text{ in.}$$

For the #6 bars, $\alpha_{abc} \alpha_{de} L_{db} = 1.0(1.0)16.7 = 16.7$ in., which exceeds Min L_d of 16.4 in.; thus use "first-level L_d" as 16.7 in. These are top bars (ACI-12.2.4.1), making $\alpha_{top} = 1.3$. No other second-level multipliers are applicable, so

$$L_d \, (\#6) = \alpha_{top} \, (\text{first-level } L_d) = 1.3(16.7) = 21.7 \text{ in. } (1.8 \text{ ft})$$

For the #7 bar, one must first determine the bar arrangement before development length can be determined. Under the 1989 ACI Code, the development length is a function of the bar arrangement, cover, and possible confinement by stirrups. Therefore, the plan for bar cutoff must be established before the development length L_d is computed. In this case, examine the feasibility of cutting 1-#7 in the positive moment zone. The remaining 2-#9 would provide the following capacity,

$$C = 47.6a; \qquad T = 2(1.0)60 = 120 \text{ kips}$$

$$a = 2.52 \text{ in.}$$

$$\phi M_n = 0.90(120)(17.5 - 1.26)\tfrac{1}{12} = 146 \text{ ft-kips}$$

Based on moment capacity alone, the #7 bar could be cut at points A and B of Fig. 6.22.2; however, ACI-12.10.5 must also be satisfied to cut bars in the tension zone. This is examined in combination with designing stirrups [see item (f) below].

Regarding the #7 bar cut in the positive moment zone, if one treats that bar as existing alone in a 14 in. wide beam, clearly Category A, item 1 (ACI-12.2.3.1a), is satisfied. However, when the continuing #9 bars at the left end of the 18 ft span have their development length coming close or possibly overlapping the #7 bar development length, it may be prudent to consider the clear spacing of the bars in the region near point A where 2-#9 and 1-#7 exist. The clear spacing will be

$$\frac{\text{clear}}{\text{spacing}} = \frac{14 - 2(1.5) - 2(0.375) - 2(1.128) - 0.75}{2} = 3.62 \text{ in.}$$

This exceeds $3d_b$ for the #7 bar; thus, even taking the conservative approach, $\alpha_{abc} = 1.0$ for the #7. Thus, for the #7 bar $\alpha_{abc}\alpha_{de}L_{db} = 1.0(1.0)22.8 = 22.8$ in., which exceeds Min L_d of 19.2 in.; thus use "first-level L_d" as 22.8 in. This is a bottom bar and no second-level multipliers are applicable; thus,

$$L_d\,(\#7) = 22.8 \text{ in. (1.9 ft)}$$

Regarding the #9 bars extending into the supports, one could calculate the clear spacing of the 2-#9, or checking minimum width from Table 3.9.4 to satisfy $3d_b$ gives 9.8 in., which is less than the 14 in. actual width. Thus, $\alpha_{abc} = 1.0$. Thus, for the #9 bars, $\alpha_{abc}\alpha_{de}L_{db} = 1.0(1.0)38.0 = 38.0$ in., which exceeds Min L_d of 24.7 in.; thus use "first-level L_d" as 38.0 in. These are bottom bars and no second-level multipliers are applicable; thus,

$$L_d\,(\#9) = 38.0 \text{ in. (3.2 ft)}$$

In trying to cut the 2-#6 bars from the negative moment region near section B-B, the bars are extended into the span *to the left of the support* farther than 12 bar diameters or the effective depth d in order to have full development of the provided steel at the face of support. Since this potential cut at point C is not in a tension zone, no further investigation of this cutoff location is required. On determining that the remaining 2-#6 bars can be cut off only a short distance farther into the span at point D, the decision is made to cut all 4-#6 bars at point D. If for some reason, such as the desire to have compression steel for deflection control, two of the #6 bars were extended across the 18-ft span instead of being cut at D, then the other 2-#6 bars would be cut at point C. On the *right* side of the support in the negative moment region, any potential cut location would be so near the end of the cantilever that it was decided to run all 4-#6 bars to the end (say 1 in. clear).

(d) Check ACI-12.11.3 for development of reinforcement for positive moment reinforcement at the left simple support and at point of inflection near the right support. At the inflection point it is required that

$$\frac{M_n}{V_u} + L_a \geq L_d$$

Assuming that 1-#7 will be terminated (a conservative assumption for this computation), only the 2-#9 bars contribute to M_n. The actual L_a from the inflection point to the end of the bars exceeds d (17.5 in.); thus, use $L_a = 17.5$ in.

$$\frac{146(12)}{0.90(47.0)} + 17.5 = 58.9 \text{ in.} > [L_d = 38 \text{ in.}] \qquad \textbf{OK}$$

For the simple support, ACI-12.11.3 requires

$$\frac{1.30M_n}{V_u} + L_a \geq L_d$$

In this case, $L_a = 7.5$ in. since the bars extend this amount beyond the center of the support. Assuming the #7 bar does not extend into the support,

$$\frac{1.30(146)12}{0.90(47.0)} + 7.5 = 61.3 \text{ in.} > L_d = 38 \text{ in.} \qquad \textbf{OK}$$

(e) Shear reinforcement. At d from the face of right support on the 18-ft span,

$$V_u = 61.2 - 6.01\left(\frac{17.5 + 9}{12}\right) = 47.9 \text{ kips}$$

$$\phi V_c = \phi(2\sqrt{f_c'})b_w d$$

$$= 0.85\,(2\sqrt{4000})(14)(17.5)\tfrac{1}{1000} = 26.3 \text{ kips}$$

$$\text{max required } \phi V_s = V_u - \phi V_c = 47.9 - 26.3 = 21.6 \text{ kips}$$

$$\text{max permitted } \phi V_s\left(\text{for } s = \frac{d}{2}\right) = 2\phi V_c = 52.6 \text{ kips}$$

Note that maximum ϕV_s (ACI-11.5.4.3) is based on the nominal stress $v_s = 4\sqrt{f_c'}$, which gives maximum $\phi V_s = 2\phi V_c$ when the simplified V_c expression is used. For minimum percentage of stirrups (ACI-11.5.5.3),

$$\text{min } \phi V_s = \phi 50 b_w d = 0.85(50)(14)(17.5)\tfrac{1}{1000} = 10.4 \text{ kips}$$

For strength, using #3 U stirrups,

$$\phi V_s = \frac{\phi A_v f_y d}{s} = \frac{0.85(0.22)(60)(17.5)}{s} = \frac{196}{s}$$

s	ϕV_s
9.1 in.	21.6 kips max

Since the strength requirement permits stirrups at 9.1 in. and is less restrictive than the $d/2$ limit, the maximum spacing at $d/2 = 8.75$ in. controls wherever stirrups are required.

Use #3 U stirrups at 8 in. spacing, as dimensioned on the shear diagram in Fig. 6.22.2. (Ordinarily the stirrups should be dimensioned on the design sketch; that is, the side view of the beam in Fig. 6.22.2.)

(f) Check ACI-12.10.5 for cutting #7 bar at points A and B in the tension zone. Since ϕM_n for the continuing #9 bars *is not* twice the factored moment M_u at the cutoff point, ACI-12.10.5.3 cannot be satisfied. The actual cut locations 2.5 ft to point A and 5.5 ft to point B are used in this check. Check ACI-12.10.5.1, for #3 U stirrups at 8 in. as provided,

$$\phi V_s = \frac{196}{s} = \frac{196}{8} = 24.5 \text{ kips}$$

$$V_u \text{ (at } A) = 47.1 - 2.5(6.01) = 32.1 \text{ kips}$$

$$V_u \text{ (at } B) = 61.2 - 5.5(6.01) = 28.1 \text{ kips}$$

$$\text{percent stressed} = \frac{V_u}{\phi V_n} = \frac{V_u}{\phi(V_c + V_s)}$$

$$= \frac{32.1}{26.3 + 24.5} = 63\% < 67\% \qquad \textbf{OK}$$

Cut 1-#7 bar at points A and B (length = 10 ft).

(g) Moment capacity diagram. Figure 6.22.2 shows the comparison between the factored moment M_u diagram and the moment capacity ϕM_n diagram provided by the selected reinforcing bars. In computing the moment capacity, any effect of steel in the compression side of the beam has been neglected because it would have negligible effect since it was not required for strength. At the left end of the beam, the 2-#9 have been extended as far as feasible into the support in order to have the bars develop their full capacity at point A. The available embedment to the left of point A was almost exactly the 3.2 ft required.

(h) Deflection. Even though the reinforcement ratio ρ being used was expected to control the deflection, nevertheless the deflection must be investigated if excessive deflection may cause damage to partitions or other construction.

(i) Design sketch. The final arrangement of longitudinal steel and stirrups is shown in Fig. 6.22.2. The stirrup locations are dimensioned on the V_u diagram. Omitted from the elevation view of the beam are the nominal sized (say #3 or #4) longitudinal bars arbitrarily added for the stirrups to wrap around wherever the stirrups would otherwise have no support to hold them in vertical position. These pairs of bars would be located in this beam at both top and bottom faces of the beam where no longitudinal bars are shown in the figure. Cross-section views showing bars have been omitted here but are always part of the design sketch. ■

SELECTED REFERENCES

6.1.   ACI Committee 408. "Bond Stress—The State of the Art," *ACI Journal, Proceedings,* **63**, November 1966, 1161–1190. Disc. 1569–1570.

6.2.   LeRoy A. Lutz and Peter Gergely. "Mechanics of Bond and Slip of Deformed Bars in Concrete," *ACI Journal, Proceedings,* **64**, November 1967, 711–721. Disc., **65**, 412–414.

6.3.   C. O. Orangun, J. O. Jirsa, and J. E. Breen. "A Reevaluation of Test Data on Development Length and Splices," *ACI Journal, Proceedings,* **74**, March 1977, 114–122. Disc., **74**, September 1977, 470–475.

6.4.   Anders Losberg and Per-Ake Olsson. "Bond Failure of Deformed Reinforcing Bars Based on Longitudinal Splitting Effect of the Bars," *ACI Journal, Proceedings,* **76**, January 1979, 5–18.

6.5.   James O. Jirsa, LeRoy A. Lutz, and Peter Gergely. "Rationale for Suggested Development, Splice, and Standard Hook Provisions for Deformed Bars in Tension," *Concrete International,* **1**, July 1979, 47–61.

6.6.   D. Z. Yankelevsky. "Bond Action Between Concrete and a Deformed Bar—A New Model," *ACI Journal, Proceedings,* **82**, March–April 1985, 154–161.

6.7.   Emory L. Kemp. "Bond in Reinforced Concrete: Behavior and Design Criteria," *ACI Journal, Proceedings,* **83**, January–February 1986, 50–57.

6.8.   Herbert J. Gilkey, Stephen J. Chamberlain, and Robert W. Beal. "Bond Between Concrete and Steel," *Iowa Engineering Experiment Station Bulletin,* No. 147, Iowa State College, 1940.

6.9.   Arthur P. Clark. "Bond of Concrete Reinforcing Bars," *ACI Journal, Proceedings,* **46**, November 1949, 161–184.

6.10.   Phil M. Ferguson and J. Neils Thompson. "Development Length for Large High Strength Reinforcing Bars," *ACI Journal, Proceedings,* **62**, January 1965, 71–93. Disc. 1153–1156.

6.11.   Phil M. Ferguson, John E. Breen, and J. Neils Thompson. "Pullout Tests on High Strength Reinforcing Bars," *ACI Journal, Proceedings,* **62**, August 1965, 933–950.

6.12.   Ervin S. Perry and J. Neils Thompson. "Bond Stress Distribution on Reinforcing Steel in Beams and Pullout Specimens," *ACI Journal, Proceedings,* **63**, August 1966, 865–875.

6.13.   E. L. Kemp, F. S. Brezny, and J. A. Unterspan. "Effect of Rust and Scale on the Bond Characteristics of Deformed Reinforcing Bars," *ACI Journal, Proceedings,* **65**, September 1968, 743–756. Disc., **66**, 224–226.

6.14.   Saeed M. Mirza and Jules Houde. "Study of Bond Stress–Slip Relationships in Reinforced Concrete," *ACI Journal, Proceedings,* **76**, January 1979, 19–46.

6.15.   S. Soretz and H. Holzenbein. "Influence of Rib Dimensions of Reinforcing Bars on Bond and Bendability," *ACI Journal, Proceedings,* **76**, January 1979, 111–125.

6.16.   Raymond E. Untrauer and George E. Warren. "Stress Development of Tension Steel in Beams, " *ACI Journal, Proceedings*, **74**, August 1977, 368–372.

6.17.   E. L. Kemp and W. J. Wilhelm. "Investigation of the Parameters Influencing Bond Cracking," *ACI Journal, Proceedings*, **76**, January 1979, 47–72.

6.18.   Shiro Morita and Tetsuzo Kaku. "Splitting Bond Failures of Large Deformed Reinforcing Bars," *ACI Journal, Proceedings*, **76**, January 1979, 93–110.

6.19.   R. Jimenez, R. N. White, and P. Gergely. "Bond and Dowel Capacities of Reinforced Concrete," *ACI Journal, Proceedings*, **76**, January 1979, 73–92.

6.20.   ACI Committee 408. "Suggested Development, Splice, and Standard Hook Provisions for Deformed Bars in Tension," *Concrete International*, **1**, July 1979, 44–46.

6.21.   Phil M. Ferguson and Farid N. Matloob. "Effect of Bar Cutoff on Bond and Shear Strength of Reinforced Concrete Beams," *ACI Journal, Proceedings*, **56**, July 1959, 5–23.

6.22.   Anthony M. Kao and Raymond E. Untrauer. "Shear Strength of Reinforced Concrete Beams with Bars Terminated in Tension Zones," *ACI Journal, Proceedings*, **72**, December 1975, 720–722.

6.23.   Mete A. Sozen and Jack P. Moehle. *A Study of Experimental Data on Development and Lap-Splice Lengths for Deformed Reinforcing Bars in Concrete*. Report to The Portland Cement Association and The Concrete Reinforcing Steel Institute, August 1990.

6.24.   Paul R. Jeanty, Denis Mitchell, and M. Saeed Mirza. "Investigation of 'Top Bar' Effects in Beams," *ACI Structural Journal*, **85**, May–June 1988, 251–257.

6.25.   LeRoy A. Lutz. "Crack Control Factor for Bundled Bars and for Bars of Different Sizes," *ACI Journal, Proceedings*, **71**, January 1974, 9–10.

6.26.   N. W. Hanson and Hans Reiffenstuhl. "Concrete Beams and Columns with Bundled Reinforcement," *Journal of the Structural Division*, ASCE, **84**, ST6 (October 1958), 1–23.

6.27.   Frank D. Steiner. "Suggested Applications for Bundled Bars," *ACI Journal, Proceedings*, **64**, April 1967, 213–214.

6.28.   John A. Hribar and Raymond C. Vasko. "End Anchorage of High Strength Steel Reinforcing Bars," *ACI Journal, Proceedings*, **66**, November 1969, 875–883. Disc, **67**, 423–424.

6.29.   John Minor and James O. Jirsa. "Behavior of Bent Bar Anchorages," *ACI Journal, Proceedings*, **72**, April 1975, 141–149.

6.30.   Jose L. G. Marques and James O. Jirsa. "A Study of Hooked Bar Anchorages in Beam–Column Joints," *ACI Journal, Proceedings*, **72**, May 1975, 198–209.

6.31.   Robert L. Pine, Michael D. Watkins, and James O. Jirsa. "Strength of Hooked Bar Anchorages in Beam–Column Joints," CESRL Report No. 77-3, Department of Civil Engineering, University of Texas, Austin, Texas, November 1977.

6.32.   Raymond E. Untrauer and Robert L. Henry. "Influence of Normal Pressure on Bond Strength," *ACI Journal, Proceedings*, **62**, May 1965, 577–586.

6.33. Joint PCI/WRI Ad Hoc Committee on Welded Wire Fabric for Shear Reinforcement, PCI Technical Activities Committee (Leslie D. Martin, Chairman). "Welded Wire Fabric for Shear Reinforcement," *PCI Journal*, **25**, July-August 1980, 32–36.

6.34. Phil M. Ferguson and John E. Breen. "Lapped Splices for High Strength Reinforcing Bars," *ACI Journal, Proceedings*, **62**, September 1965, 1063–1078.

6.35. John P. Lloyd and Clyde E. Kessler. "Splices and Anchorages in One-Way Slabs Reinforced with Deformed Wire Fabric," *ACI Journal, Proceedings*, **67**, August 1970, 636–642.

6.36. M. A. Thompson, J. O. Jirsa, J. E. Breen, and D. F. Meinheit, "Behavior of Multiple Lap Splices in Wide Sections," *ACI Journal, Proceedings*, **76**, February 1979, 227–248.

6.37. Telvin Rezansoff, Jim A. Zacaruk, and Rob Topping. "Tensile Lap Splices in Reinforced Concrete Beams Under Inelastic Cyclic Loading." *ACI Structural Journal*, **85**, January–February 1988, 46–52.

PROBLEMS

All problems* are to be done in accordance with the strength method of the ACI Code, and all loads given are service loads, unless otherwise indicated. Unless otherwise indicated, assume all beams have at least minimum cover satisfying ACI-7.7.1, as well as minimum stirrups according to ACI-11.5.4 and ACI-11.5.5.3, but not sufficient to satisfy ACI Formula (12-1). All moment capacity ϕM_n diagrams used in these problems must be drawn to scale directly below a side view of the beam drawn to the same longitudinal scale *on the same page*. Moment capacity ϕM_n diagrams must have critical numerical values stated on the diagram, and horizontal distances to the critical points must be dimensioned with numerical values.

6.1 Draw the free-body diagram for the 3-in. slice shown crosshatched on the beam given in the accompanying figure. Using working stress assumptions, compute and show on the diagram values for the internal forces (tension, compression, and shear) on each side of the slice.

Problem 6.1

— — — —

* Many problems may be solved as problems stated in Inch-Pound units, or as problems in SI units using quantities in parenthesis at end of the statement. The SI conversions are approximate to avoid implying higher precision for the given information in SI units than that given for the Inch-Pound units.

(a) Compute the average flexural bond stress on the bars over the 3-in. slice.

(b) Compute the average flexural bond stress on the bars over the distance from section A to the left end of the beam.

(c) What is the average anchorage bond stress resisting the tensile force at section A?

(d) Explain what happens if the flexural bond stress in part **(a)** is so high that slippage occurs over the 3-in. slice. What determines the adequacy of the beam? Assume loading is within the usual service range.

6.2   For the simply supported beam shown, draw the moment capacity (ϕM_n) diagram. Assume that the cutoff location 3 ft from the support satisfies the requirements of ACI-12.10.5, and that ACI-12.11.3 is satisfied. What is the maximum uniformly distributed service load that the beam may be permitted to carry (assume 50% live load and 50% dead load)? Use $f'_c = 3500$ psi and $f_y = 40,000$ psi.

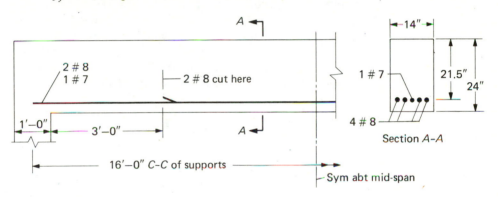

Problem 6.2

6.3   For the cantilever beam of the accompanying figure having $f'_c = 3000$ psi and $f_y = 40,000$ psi:

(a) Draw the moment capacity ϕM_n diagram. Be sure to include the 2-ft embedment into the support, and assume that the full cross-section capacity in that region is based on that at the support.

Problem 6.3

(b) Investigate the adequacy for development of reinforcement if the beam is subjected to uniform dead and live loads of 1.8 kips/ft (including weight of beam) and 2.1 kips/ft, respectively. Neglect any concern about cutting bars in the tension zone (ACI-12.10.5).

6.4   For the beam of Problem 6.3, but with 3-#7 bars as the reinforcement at section *A-A*, 1-#7 is cut at 4 ft–0 in. from support location, $f'_c = 4000$ psi, and $f_y = 60,000$ psi; investigate the adequacy for development of reinforcement if the beam is subjected to a uniform live load of 1.9 kips/ft and uniform dead load of 1.6 kips/ft (including beam weight). Compare factored moment M_u with the moment capacity ϕM_n by superimposing both diagrams. Consider all factors involved, including ACI-12.10.5 for cutting bars in the tension zone. Assume #3 U stirrups at 8-in. spacing are used in the vicinity of the cutoff point.

6.5   For the beam of the accompanying figure, investigate the adequacy for development of reinforcement if the beam must carry uniform dead and live loads of 30 kN/m (including weight of beam) and 34 kN/m, respectively. Compare factored moment M_u with the moment capacity ϕM_n by superimposing both diagrams. Consider all factors involved, including ACI-12.10.5 for cutting bars in the tension zone. Assume that #10M U stirrups at 200-mm spacings are used in the vicinity of the cutoff point. Use $f'_c = 30$ MPa and $f_y = 400$ MPa.

Problem 6.5

6.6   For the cantilever beam of the accompanying figure, determine the safe uniformly distributed load *w* (dead load, DL, plus live load, LL) that the beam may be permitted to carry, if the dead load to live load ratio is 0.8. Consider all factors involved, including ACI-12.10.5 for cutting bars in the tension zone. Use $f'_c = 4000$ psi and $f_y = 60,000$ psi. Show the comparison of factored moment M_u with the moment capacity ϕM_n diagram. (For SI problem use bars; 3-#25M; #10M stirrups; $f'_c = 30$ MPa; $f_y = 300$ MPa.)

Problem 6.6

6.7   If the beam of the accompanying figure is to carry uniformly distributed loads of 4.0 kips/ft live load and 1.9 kips/ft dead load (including beam weight), determine the adequacy of the bar cutoffs and the development of reinforcement. Stirrups are #3 at 6 in. spacing where bars A are cut, and #3 at 8 in. spacing where bars B are cut. As part of the solution, compare the factored moment M_u with the moment capacity ϕM_n by superimposing both diagrams. Use $f'_c = 5000$ psi and $f_y = 60,000$ psi.

Problem 6.7

6.8   Neglecting any compression reinforcement effect, plot the moment capacity ϕM_n diagram (positive moment over the 20-ft span and negative moment for the steel over the support) for the beam of the accompanying figure. Use $f'_c = 3000$ psi and $f_y = 40,000$ psi.

6.9   For the beam of Problem 6.8, check the development of reinforcement at the simply supported end and at the point of inflection closest to the right support on the 20-ft span. The loads are 1.4 kips/ft dead load (including beam weight) and 2.8 kips/ft live load.

6.10   For the beam of Problem 6.8 and the loading of Problem 6.9, determine the acceptability of the cut locations at points A and B near the right support.

(a) Assume #3 U stirrups are spaced at 10 in. in the vicinity of the cut locations.

(b) Assume #3 U stirrups are spaced at 12 in. in the vicinity of the cut locations.

6.11   Investigate the adequacy of the beam shown in the accompanying figure for (a) bar cutoffs and (b) stirrups. If not adequate, specify what changes are required. The uniform dead and live loading is 9.3 kN/m and 42 kN/m, respectively. Use $f'_c = 30$ MPa and $f_y = 400$ MPa.

Problem 6.11

6.12   A 14-in. wide by 24-in. deep beam (effective depth = 21.5 in.) is used as the section for a 20-ft simply supported span having an 8-ft cantilever at one end. The positive moment reinforcement is 4-#9 bars and the negative moment reinforcement is 4-#8 bars. The loading to be carried is a live load of 1.9 kips/ft and a dead load of 1.6 kips/ft (including beam weight). Determine the lengths of bars (3-in. increments) if two of the four bars in both the positive and the negative moment regions are to be terminated as soon as practicable. The remaining bars are to be extended as required by the ACI Code. Width of supports is 15 in., and #3 U stirrups spaced at 10 in. are used in any potential cutoff region. Verify your design by showing for one set of axes the factored moment M_u envelope and the provided moment capacity ϕM_n diagram. Use $f'_c = 3500$ psi and $f_y = 40,000$ psi.

6.13   Repeat Problem 6.12, except use a 12-in. beam width and the reinforcement is 2-#8 and 1-#9 for maximum negative moment and 2-#9 and 2-#8 for maximum positive moment. The factored moment and shear envelopes for uniform loading are as shown in the accompanying figure. Determine the lengths of bars if one #9 is to be cut off in the negative moment zone and two #8 are to be cut in the positive moment zone. Assume #3 U stirrups are spaced at 10 in. in any potential cutoff region. Use a moment capacity ϕM_n diagram to verify your answer.

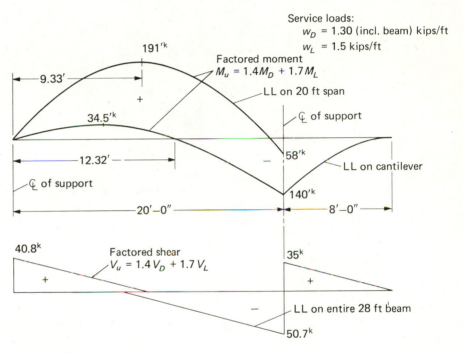

Problem 6.13

6.14 A 12-in. wide by 24-in. (overall size) beam is used as the section for a 20-ft simply supported span having an 8-ft cantilever at one end. Support widths are 12 in. The positive moment reinforcement is 2-#8 and 2-#7 and the negative moment reinforcement is 4-#7. The factored moment and shear envelopes for uniform loading are as shown in the accompanying figure. Assume that 2-#7 bars are to be bent up near the right support in the 20-ft span and are to be cut as soon as feasible on the cantilever. Locate the bend-up and bend-down points as well as the cut point on the cantilever. Assume #3 stirrups are spaced at 10 in. in the entire negative moment region. Verify your design by showing the moment capacity ϕM_n diagram. Use $f'_c = 3500$ psi and $f_y = 60,000$ psi.

Problem 6.14

6.15 In the accompanying figure, a cantilever slab (for example, for a retaining wall) varies in thickness from 24 in. at its supported end to 12 in. at its free end, with $2\frac{1}{2}$ in. clear cover over the reinforcement. Assuming that #7 bars at 6-in. spacing are effective at the supported end, at what distance from the supported end may every other #7 bar be cut off? No shear reinforcement is used. Verify your result by showing the provided moment capacity ϕM_n diagram superimposed on the factored moment M_u diagram. Use $f'_c = 3000$ psi and $f_y = 40,000$ psi. Assume the loading is entirely earth pressure and neglect the beam weight.

Problem 6.15

Problems Utilizing Concepts of Chapters 1 through 6

For the design of shear reinforcement, do *not* design to satisfy ACI Formula (12-1) merely for the purpose of reducing development length. Use at least one bar cutoff location, even if such cutoff seems impractical.

6.16   Design, including design sketch, a rectangular cantilever beam 14 ft long to carry a live load of 2.5 kips/ft. The beam size may not exceed 15 in. wide and 24 in. deep. Use $f'_c = 4000$ psi and $f_y = 60{,}000$ psi. (For SI problem, use length = 4.3 m; live load = 36 kN/m; $f'_c = 30$ MPa; $f_y = 400$ MPa.)

6.17   Design, including design sketch, a cantilever beam 14 ft long to carry a live load of 3.0 kips/ft. Without actually satisfying a deflection limit, design a rectangular beam of such size that no serious problem with excessive deflection is to be expected. Use $f'_c = 4000$ psi and $f_y = 60{,}000$ psi. (For SI problem, use live load = 43 kN/m; length = 4.3 m; $f'_c = 30$ MPa; $f_y = 400$ MPa.)

6.18   The floor system shown in the accompanying figure is given. The floor consists of 6-in. precast slab sections of 10 ft span. The live load is 50 psf, and the maximum depth available is 30 in. from the top of the floor slab. Completely design as a simply supported beam for the beam indicated on the figure. Use whole inch increments for beam depth and width. Use $f'_c = 4000$ psi and $f_y = 60{,}000$ psi.

Typical section through
30′–0″ span beams

Problem 6.18

6.19   Design, including design sketch, a simply supported rectangular beam to carry a dead load of 1.6 kips/ft (not including beam weight) and a live load of 3.4 kips/ft on a span of 22 ft. Use a reinforcement ratio roughly 0.3 of the value at the balanced strain condition. Assume support widths are 18 in. Use $f'_c = 4000$ psi and $f_y = 60,000$ psi. (For SI problem, use dead load = 23.3 kN/m; live load = 50 kN/m; span = 6.7 m; $f'_c = 30$ MPa; $f_y = 400$ MPa; support width = 460 mm.)

6.20   Repeat Problem 6.19, except use dead load 1.7 kips/ft instead of 1.6, and live load 3.8 kips/ft instead of 3.4, and a span of 16 ft. (For SI, dead load = 25 kN/m; live load = 55 kN/m; span = 5 m.)

6.21   Design, including design sketch, a simply supported rectangular beam on a span of 18 ft, and having a cantilever at the left end of span 6 ft. The dead load (not including beam weight) is 2.2 kips/ft and the live load is 3.5 kips/ft. Assume the support widths are 20 in. Use a reinforcement ratio for positive moment roughly 0.3 of the value at the balanced strain condition. Use $f'_c = 3000$ psi and $f_y = 60,000$ psi. (For SI, use dead load = 32 kN/m; live load = 52 kN/m; support widths = 500 mm; span = 5.5 m; cantilever = 1.8 m; $f'_c = 20$ MPa; $f_y = 400$ MPa.)

6.22   Repeat Problem 6.21, except use a 16-ft main span and an 8-ft cantilever. (For SI, use span = 5 m; cantilever = 2.4 m.)

CHAPTER 7

Continuity in Building Frames of Reinforced Concrete

7.1 COMMON BUILDING FRAMES

Reinforced concrete building construction commonly has floor slabs, beams, girders, and columns continuously placed to form a monolithic system. Consider the plan of a typical slab–beam–girder floor construction shown in Fig. 7.1.1. Section A-A through the slab shows that the slab is supported on 10 beams. Intermediate beams such as $B1$, $B2$, and $B3$ are supported on the girders, whereas beams on the column lines such as $B4$, $B5$, and $B6$ are supported directly by the columns. Girders such as $G1$, $G2$, and $G3$ also go directly into the columns.

Beams such as $B4$, $B5$, and $B6$ are not only continuous beams, but are also built integrally with the upper and lower columns. For correct analysis, then, the complete frame in this plane, which may consist of, say, 10 or 15 stories or more, should be analyzed as a rigid frame. In the analysis for gravity load on regular building frames, according to ACI-8.9.1 the beams and the adjacent columns may be isolated and treated as a subassembly (Fig. 7.1.2), with the far ends of the columns assumed as fixed. This assumption should not be used for wind load, however. For wind analysis, except for very tall structures, a simplified approximate method may be used [7.1]. Although titled "Equivalent Frame Analysis for Two-Way Floor Systems in Unbraced Frames," Chapter 17 gives the general treatment of building frame analysis for both vertical (gravity) and lateral (such as wind) loads by the matrix displacement method.

Girders $G1$, $G2$, and $G3$ may be treated in the same manner as the beams. Relative stiffnesses for columns, beams, and girders must first be assumed or established by preliminary design and later reviewed, as would be done in the analysis and design of any statically indeterminate rigid frame.

The continuous slab is supported at the beams, and the beams $B1$, $B2$, and $B3$ are supported at the girders. The supporting beams or girders possess torsional rigidity which may be approximated by using equivalent supporting columns with

Section A-A

Beams monolithic with girders

Beams monolithic with columns

Girders monolithic with columns

Figure 7.1.1 Slab–beam–girder floor.

Figure 7.1.2 Beams and adjacent columns.

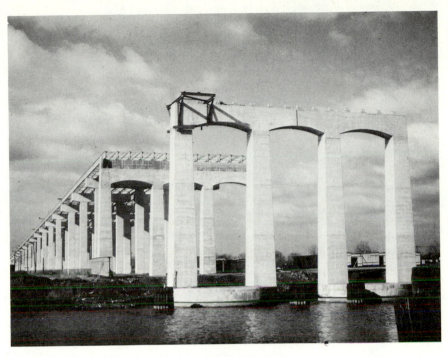

Rigid frame bridge piers, 1-75 near Detroit. (Photo by C. G. Salmon.)

fixed far ends. The stiffness factors to be used for such equivalent columns are discussed in Chapter 10 (Section 10.3).

7.2 POSITIONS OF LIVE LOAD FOR MOMENT ENVELOPE

The live load positions that cause the largest bending moments in slabs, beams, and girders are discussed in this section. The reader should review the subject of influence lines [7.2] in order to be able to establish the loading conditions appropriate for obtaining the bending moment and shear envelopes. Bending moments to be used in the design of columns are treated separately later.

Consider the continuous beam *ABCDEFGH* with its adjacent columns, as shown in Fig. 7.2.1(a). The influence lines for bending moment at any point in the central portion of span *CD* and at a section an infinitesimal distance to the left of support *D* are shown in Figs. 7.2.1(b) and (c), respectively. From these influence lines the following cases for uniform live load are indicated:

1. For maximum positive moment within a span, load that span and all other alternate spans.
2. For maximum negative moment within a span, load the two spans adjacent to that span and all other alternate spans (all the spans not loaded in 1).
3. For maximum negative moment at a support, load the two spans adjacent to that support and all other alternate spans.

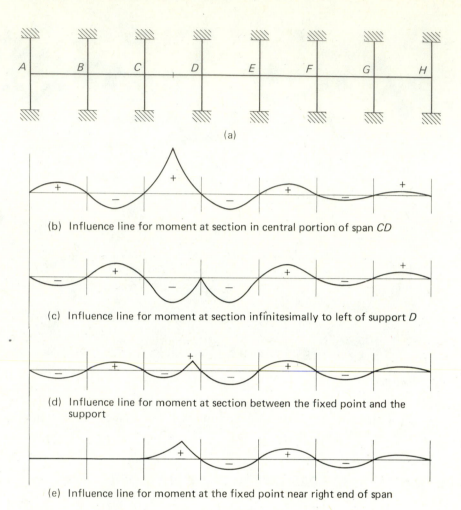

(a)

(b) Influence line for moment at section in central portion of span *CD*

(c) Influence line for moment at section infinitesimally to left of support *D*

(d) Influence line for moment at section between the fixed point and the support

(e) Influence line for moment at the fixed point near right end of span

Figure 7.2.1 Influence lines for continuous spans.

4. For maximum positive moment at a support, load the two spans beyond each of the two spans adjacent to that support and all other alternate spans (all the spans not loaded in 3).

It may be noted that loading cases 1 and 2 in the preceding list are complementary; that is, their combination results in *all* spans being loaded. Loading cases 3 and 4 are also complementary. It may be noted further that loading cases 1 and 3 are primary; that is, they result in moments of the same sign as dead load. Loading cases 2 and 4, on the other hand, are secondary; they result in moments opposite in sign to those due to dead load. When the secondary live load moment is numerically larger than the dead load moment, there is moment reversal.

The loading cases mentioned thus far involve loading entire spans, that is, no partial loading of a span. These full span loading cases correctly give maximum and minimum bending moment in the midspan region (roughly the middle 50 to 60%) of a span as well as *at* the support. The correct maximum (or minimum) values are not obtained, however, for approximately 20 to 25% of the span nearest the supports.

A qualitative examination of Figs. 7.2.1(b) and (c) shows that if an influence line were drawn for a series of specific points along span *CD* between the midspan and the support, the peak ordinate in the positive portion of such influence line would get smaller and smaller. Also, the *slope* of the influence line at point *C* goes upward to the right for the midspan influence line but goes downward to the right for the support *D* influence line.

As successive influence lines are drawn for the points along the span from midspan to the support, there must be some location for which the slope of the influence line at *C* is horizontal. Such point is called a *fixed point*, giving an influence line illustrated by Fig. 7.2.1(e). Any loading on spans to the left of the span under study will cause no moment at this fixed point. The fixed point is the closest location to the support for which full span loadings give the correct maximum or minimum bending moments. The influence line for a section between a fixed point and the support is as shown by Fig. 7.2.1(d), indicating partial loading for the span in question to obtain maximum or minimum bending moment. See Wang and Salmon [7.2] for an extended treatment of fixed points.

From a practical point of view, partial span loading for maximum or minimum bending moments is rarely actually done in the design of building frames because the effect of doing so on the design is too small to justify the effort. For the design of long spans, such as highway bridges, the full consideration of the proper loading (loading partial span as indicated by influence lines) for locations between the fixed point and the support would usually be made.

7.3 METHOD OF ANALYSIS

When using the strength method of design, the analysis of the continuous concrete structure is made using *factored loads*; that is, the service loads multiplied by appropriate overload factors in accordance with ACI-9.2. Thus the structural analysis is to be made assuming an elastic system even though the factored load may cause inelastic effects. After obtaining the moments and shears assuming an elastic structure under the factored load condition, each section is proportioned to provide adequate strength. Although the procedure may seem somewhat inconsistent, it has been found to provide safe and adequate designs. The other possible approach would be to utilize the true ultimate or collapse condition, using the so-called limit design (discussed in Section 10.12) for continuous beams or frames, and yield-line theory (Chapter 18) for slabs. These two methods are not incorporated in the ACI Code, although the Code does contain provisions for adjustments to the results of elastic analysis to reflect present knowledge of collapse behavior.

There are numerous methods of statically indeterminate structural analysis that may be used. Matrix methods will usually be preferred [7.2, 7.3] for practical office design. The reader is referred to Chapter 17 for the treatment of the global stiffness matrix, the load matrix, the displacement matrix, and the end-moment matrix. When using this method, either the subassembly or the entire frame may be analyzed. The subassembly analysis is useful in preliminary design, while the entire frame may be preferred for the final review of the design.

For computation using a spreadsheet program, such as Lotus 1-2-3, Borland Quattro, or Microsoft Excel, the moment distribution method is a practical approach for analyzing relatively simple structures, such as the rigid frame involving several continuous spans with far ends of upper and lower columns fixed. The moment distribution is described in most textbooks [7.2] on structural analysis, so it is not developed here. The application to a six-span continuous beam with upper and lower columns is illustrated in the following example.

■ **EXAMPLE 7.3.1**   Determine the maximum and minimum moments at the middle and the ends of each span in a rigid frame with six equal spans as shown in Fig. 7.3.1. Assume that the uniform live load w_L is twice the uniform dead load w_D, and that the stiffness factor K_c (representing $4EI/L$) of the column is twice the stiffness factor K_b of the beam span. Express all moments in terms of wL^2 in which $w = w_D + w_L$ and L is the span length.

Figure 7.3.1 A six-span continuous frame showing relative stiffnesses.

Solution: All moments will be expressed numerically in terms of $wL^2(10^{-4})$. The FEM (fixed-end moment) due to w_D and $w_L = \pm\frac{1}{12}wL^2 = \pm833wL^2(10^{-4})$, whereas the FEM due to w_D only $= \pm\frac{1}{12}w_DL^2 = \pm\frac{1}{36}wL^2 = \pm278wL^2(10^{-4})$.

Seven loading conditions as shown in Fig. 7.3.2 need to be investigated. Three cycles of moment distribution should provide sufficiently accurate results for design use. The moment distribution and the relative slope check are shown completely only for loading condition No. 1 in Table 7.3.1. For the other six loading conditions, only the FEM and the final end moments are shown in Tables 7.3.2 to 7.3.7.

For prismatic members, the final moment at end A of a member AB expressed in slope deflection terminology is

$$M_{ab} = \text{FEM}_{ab} + \left(\frac{4EI}{L}\right)\theta_a + \left(\frac{2EI}{L}\right)\theta_b$$

where FEM_{ab} is the original fixed-end moment at end A, and θ_a and θ_b are the slopes of the elastic curve at ends A and B, respectively. The quantity $4EI/L$ is the stiffness K at one end of a prismatic member with its far end fixed. The change from fixed-end moment to final moment, ΔM_{ab}, is

$$\Delta M_{ab} = M_{ab} - \text{FEM}_{ab} = \left(\frac{4EI}{L}\right)\theta_a + \left(\frac{2EI}{L}\right)\theta_b$$

Figure 7.3.2 Loading conditions.

Also

$$\Delta M_{ba} = M_{ba} - \text{FEM}_{ba} = \left(\frac{4EI}{L}\right)\theta_b + \left(\frac{2EI}{L}\right)\theta_a$$

Solving for θ_a and θ_b gives

$$\theta_a = \frac{\Delta M_{ab} - \frac{1}{2}\Delta M_{ba}}{3EI/L} \qquad \text{and} \qquad \theta_b = \frac{\Delta M_{ba} - \frac{1}{2}\Delta M_{ab}}{3EI/L}$$

Thus

$$\theta_a \text{ (relative)} = \left(\frac{\Delta M_{ab} - \frac{1}{2}\Delta M_{ba}}{K}\right) \tag{7.3.1}$$

$$\theta_b \text{ (relative)} = \left(\frac{\Delta M_{ba} - \frac{1}{2}\Delta M_{ab}}{K}\right) \tag{7.3.2}$$

Equations (7.3.1) and (7.3.2) are used to compute the relative slopes in the last four lines of Table 7.3.1.

The controlling values of final moments at the left and right ends of each span, M_L and M_R, for the various critical conditions are taken from Tables 7.3.1 through

Table 7.3.1 Loading Condition No. 1 (Fig. 7.3.2)

JOINT	A	B		C		D		E		F		G
MEMBER	AB	BA	BC	CB	CD	DC	DE	ED	EF	FE	FG	GF
Distribution factor	$\frac{1}{5}$	$\frac{1}{6}$	$\frac{1}{6}$	$\frac{1}{6}$	$\frac{1}{6}$	$\frac{1}{6}$	$\frac{1}{6}$	$\frac{1}{6}$	$\frac{1}{6}$	$\frac{1}{6}$	$\frac{1}{6}$	$\frac{1}{5}$
FEM	−833	+833	−278	+278	−833	+833	−278	+278	−833	+833	−278	+278
Balance	+167	−92	−92	+92	+92	−92	−92	+92	+92	−92	−92	−56
Carryover	−46	+83	+46	−46	−46	+46	+46	−46	−46	+46	−28	−46
Balance	+9	−22	−22	+15	+15	−15	−15	+15	+15	−3	−3	+9
Carryover	−11	+4	+8	−11	−8	+8	+8	−8	−1	+8	+4	−1
Balance	+2	−2	−2	+3	+3	−3	−3	+1	+1	−2	−2	0
Final M	−712	+804	−340	+331	−777	+777	−334	+332	−772	+790	−399	+184
Change	+121	−29	−62	+53	+56	−56	−56	+54	+61	−43	−121	−94
$-\frac{1}{2}$ change	+14	−60	−26	+31	+28	−28	−27	+28	+22	−30	+47	+60
sum	+135	−89	−88	+84	+84	−84	−83	+82	+83	−73	−74	−34
$\theta_{\text{rel}} = \text{sum}/K$	+135	−89	−88	+84	+84	−84	−83	+82	+83	−73	−74	−34

Table 7.3.2 Loading Condition No. 2 (Fig. 7.3.2)

MEMBER	AB	BA	BC	CB	CD	DC	DE	ED	EF	FE	FG	GF
FEM	−833	+833	−833	+833	−278	+278	−833	+833	−278	+278	−833	+833
Final M	−668	+911	−888	+729	−325	+337	−778	+777	−331	+340	−804	+712

Table 7.3.3 Loading Condition No. 3 (Fig. 7.3.2)

MEMBER	AB	BA	BC	CB	CD	DC	DE	ED	EF	FE	FG	GF
FEM	−278	+278	−278	+278	−833	+833	−278	+278	−833	+833	−278	+278
Final M	−227	+293	−241	+374	−785	+774	−333	+332	−772	+790	−399	+184

Table 7.3.4   Loading Condition No. 4 (Fig. 7.3.2)

MEMBER	AB	BA	BC	CB	CD	DC	DE	ED	EF	FE	FG	GF
FEM	−278	+278	−833	+833	−833	+833	−278	+278	−833	+833	−278	+278
Final M	−187	+390	−745	+878	−882	+732	−325	+336	−772	+790	−399	+184

Table 7.3.5   Loading Condition No. 5 (Fig. 7.3.2)

MEMBER	AB	BA	BC	CB	CD	DC	DE	ED	EF	FE	FG	GF
FEM	−833	+833	−278	+278	−278	+278	−833	+833	−278	+278	−833	+833
Final M	−709	+813	−385	+225	−228	+379	−786	+774	−330	+340	−804	+712

Table 7.3.6   Loading Condition No. 6 (Fig. 7.3.2)

MEMBER	AB	BA	BC	CB	CD	DC	DE	ED	EF	FE	FG	GF
FEM	−833	+833	−278	+278	−833	+833	−833	+833	−278	+278	−833	+833
Final M	−712	+805	−343	+323	−731	+883	−883	+731	−323	+343	−805	+712

Table 7.3.7   Loading Condition No. 7 (Fig. 7.3.2)

MEMBER	AB	BA	BC	CB	CD	DC	DE	ED	EF	FE	FG	GF
FEM	−278	+278	−833	+833	−278	+278	−278	+278	−833	+833	−278	+278
Final M	−184	+398	−787	+780	−378	+228	−228	+378	−780	+787	−398	+184

7.3.7 and entered in Table 7.3.8. Note that the designer's sign convention for bending moment, in which a positive moment causes compression on the top side of the beam, is used in Table 7.3.8.

The moment at the midspan M_s may be determined by superposition of the effect of end moments with that of the simple beam moment due to transverse loading,

$$M_s = M_0 - \tfrac{1}{2}(M_L + M_R)$$

in which M_0 is the moment at the midspan for a simple beam. When the end moments are not equal, the maximum moment in the span does not occur at midspan, but its value is close to that at midspan.

Table 7.3.8   Summary of Results of Moment Distribution Analysis

LINE NUMBER		SPAN$_e$	M_L	M_R	VALUES OF M_L AND M_R FROM TABLE NO.	$M_s = M_0 - \frac{1}{2}(M_L + M_R)$
1	For maximum	AB	-712	-804	7.3.1	$+492$
2	positive	BC	-790	-772	7.3.1	$+469 \; (M_0 = +1250)$
3	moment at midspan	CD	-777	-777	7.3.1	$+473$
4	For minimum	AB	-184	-399	7.3.1	$+125$
5	positive or	BC	-340	-331	7.3.1	$+81 \; (M_0 = +417)$
6	maximum negative moment at midspan	CD	-332	-334	7.3.1	$+84$
7	For maximum	A of AB	-712	(-804)	7.3.1	
8	negative	B of AB	(-668)	-911	7.3.2	
9	moment at	B of BC	-888	(-729)	7.3.2	
10		C of BC	(-745)	-878	7.3.4	
11		C of CD	-882	(-732)	7.3.4	
12		D of CD	(-731)	-883	7.3.6	
13	For minimum	A of AB	-184		7.3.1	
14	negative or	B of AB		-293	7.3.3	
15	maximum	B of BC	-241		7.3.3	
16	positive	C of BC		-225	7.3.5	
17	moment at	C of CD	-228		7.3.5	
18		D of CD		-228	7.3.7	

Note: Values of moments in $wL^2(10^{-4})$. Numbers in parenthesis are to be used in shear envelope computations (see Section 7.6).

For the purpose of illustration, the moment diagram for the maximum and minimum positive moments at midspan, using results of the first loading condition in Fig. 7.3.2, is shown in Fig. 7.3.3. First, the simple beam moment diagrams for the total load and for dead load only are drawn to scale in Fig. 7.3.3(a). Next, the end moments for each span are taken from Table 7.3.8 and plotted in Fig. 7.3.3(b). Note that, because of symmetry, only the moment diagrams for the first three spans are shown. The final moment diagrams in Figure 7.3.3(b) are drawn by rotating the base lines for zero end moments in Fig. 7.3.3(a) to those connecting the end moments in Fig. 7.3.3(b).

In this example, the dead load has been applied on all the spans in each of the seven loading conditions of Fig. 7.3.2. This has been done because an important purpose of the example is to justify the use of approximate moment coefficients as discussed in the next section. An alternative approach would be to use eight loading

(a) Simple beam moment diagrams

(b) Moment diagrams for maximum and minimum positive midspan moments

Figure 7.3.3 Moment diagrams for Example 7.3.1.

conditions; one condition for dead load only plus the same seven live load conditions but without the dead load applied simultaneously. Another alternative for the live load cases would be to load one span at a time and carry out the moment distribution. Because of symmetry, only three distributions would be required; live load in spans 1, 2, and 3 separately. Moments and shears due to load on spans 4, 5, or 6 can be deduced from load on spans 3, 2, or 1, respectively. The procedure of loading one span at a time with live load will require less moment distribution operations but more effort in combining the appropriate cases to obtain the maximum and minimum moments at the various locations. Separation of dead and live load reactions to columns is frequently desirable, because if a loading combination that includes wind or earthquake must be considered, different live load factors U must be used according to ACI-9.2 ∎

7.4 ACI MOMENT COEFFICIENTS

ACI-8.3 specifies that: (1) the theory of elastic analysis is to be used in analyzing frames or continuous construction; (2) except for prestressed concrete, approximate methods of frame analysis are permitted for buildings of usual types of construction, spans, and story heights; and (3) except for prestressed concrete, design for the moments and shears as listed in ACI-8.3.3 is satisfactory in the case of two or more approximately equal spans (the larger of two adjacent spans not exceeding the shorter by more than 20%) with loads uniformly distributed, where the unit live load does not exceed three times the unit dead load. One approximate method would be the method of isolating one floor at a time (with its upper and lower columns) as permitted by ACI-8.9.1 and discussed in Section 7.3. The ACI Code leaves open to the designer the precise definition of the term "approximate methods." The reader is referred to the subassembly model defined in Fig. 7.1.2.

It seems desirable to examine the moment coefficients in ACI-8.3.3. Observation of Table 7.3.8 in the preceding section shows that for a six-span frame in which the ratio of ΣK_{col} to K_{bm} is 4 and the ratio of w_L to w_D is 2, critical values of moments may vary within the following limits:

Exterior span:
  Exterior end $-0.0184wL^2$ and $-0.0712wL^2$
  Midspan $+0.0125wL^2$ and $+0.0492wL^2$
  Interior end $-0.0293wL^2$ and $-0.0911wL^2$

First interior span:
  Exterior end $-0.0241wL^2$ and $-0.0888wL^2$
  Midspan $+0.0081wL^2$ and $+0.0469wL^2$
  Interior end $-0.0225wL^2$ and $-0.0878wL^2$

Second interior span:
  Exterior end $-0.0228wL^2$ and $-0.0882wL^2$
  Midspan $+0.0084wL^2$ and $+0.0473wL^2$
  Interior end $-0.0228wL^2$ and $-0.0883wL^2$

Similar values may be worked out for other values of $(\Sigma K_{col})/K_{bm}$ and of w_L/w_D. It may be observed that the maximum positive moments in the first and second interior spans are about equal, that the maximum positive moment in the exterior span is higher than that in the interior spans, that the maximum negative moment at the interior end of the exterior span has the largest numerical value, and that the maximum negative moments at both ends of all interior spans are about equal. The moment coefficients in ACI-8.3.3 are in agreement with these observations.

As a matter of further justification of the ACI moment coefficients, a comparison of these values with the largest possible theoretical values [7.4] is shown in Table 7.4.1. Certainly the largest possible theoretical values will be for the case of $w_L/w_D = 3$, which is the limit set forth in ACI-8.3.3. In this instance, secondary live load moments with signs opposite to that of dead load occur infrequently; or, if they do occur, their values are small. Thus, as long as the ratio of live load to dead load is well within 3 and span lengths do not differ considerably, there will be no moment reversal so that the ACI moment coefficients are reasonably close and, in general, on the safe side. It may be noted, however, that the ACI moment coefficients are in terms of wL_n^2, in which L_n is the clear span for positive moment and the average of the two adjacent clear spans for negative moment—negative moments being those at the face of supports and not at the centerline of support. On the other hand, the theoretical coeffi-

Table 7.4.1   Comparison of ACI Moment Coefficients with Theoretical Values [7.4]

LOCATION OF SECTION	ACI	THEORETICAL COEFFICIENTS			
		VALUE	NUMBER OF SPANS	$(\Sigma K_{col})/K_{bm}$	w_L/w_D
Positive moment					
End spans					
If discontinuous end is unrestrained	$+\frac{1}{11}$	$+0.094$	3	0	3
If discontinuous end is integral with the support	$+\frac{1}{14}$	$+0.073$	3	0.5	3
Interior spans	$+\frac{1}{16}$	$+0.063$	4 or more	0.5	3
Negative moment at exterior face of first interior support					
Two spans	$-\frac{1}{9}$	-0.111	2	0.5	3
More than two spans	$-\frac{1}{10}$	-0.107	4 or more	0.5	3
Negative moment at other faces of interior supports	$-\frac{1}{11}$	-0.092	4 or more	2	3
Negative moment at face of all supports for (a) slabs with spans not exceeding 10 ft and (b) beams and girders where $(\Sigma K_{col})/K_{bm}$ exceeds 8 at each end of the span	$-\frac{1}{12}$	-0.083	any number	∞	any ratio
Negative moment at interior faces of exterior supports for members built integrally with their supports					
Where the support is a spandrel beam or girder	$-\frac{1}{24}$	-0.036	4 or more	0.5	3
		-0.050	4 or more	1	3
Where the support is a column	$-\frac{1}{16}$	-0.064	4 or more	2	3

cients are in terms of wL^2, in which L is the distance between centerlines of supports, and coefficients for negative moments refer to those at the centerlines of support. Although span lengths between centerlines of supports are always used in elastic analysis, ACI-8.7.3 states that for beams built integrally with supports, moments at faces of supports may be used for design.

Logically, any approximate analysis that provides adequate strength and no excessive deflection at service load should be acceptable. Furlong and Rezende [7.5] have proposed an alternative to the use of ACI-8.3. The alternate is based on limit analysis as discussed in Section 10.12, with requirements that every component be ductile and strong enough, and that no reinforcement yields at service load.

7.5 ACI MOMENT DIAGRAMS

In designing any span in a multispan continuous rigid frame subjected to live load where moment coefficients are used, two primary sets of shear and moment diagrams are inherently being assumed. In the general case, one will result from the loading position that causes maximum positive moment within the span, and the other will result from assuming that the maximum negative moments occur simultaneously at both ends. Actually, the loading position that causes maximum negative moment at

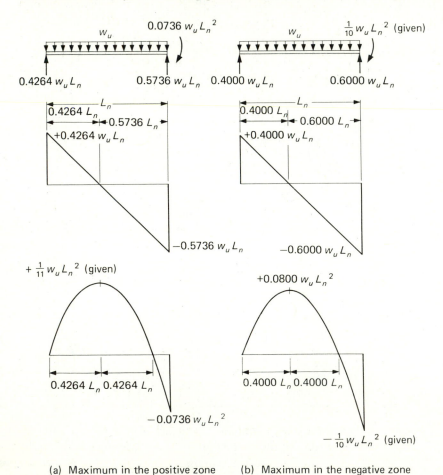

(a) Maximum in the positive zone   (b) Maximum in the negative zone

Figure 7.5.1 Exterior span with discontinuous end unrestrained.

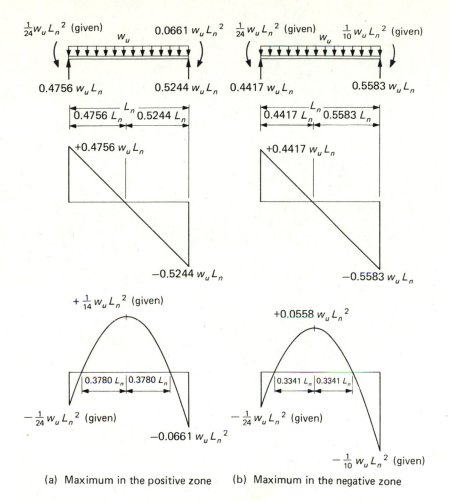

$\frac{1}{24} w_u L_n{}^2$ (given) $0.0661 \, w_u L_n{}^2$ $\frac{1}{24} w_u L_n{}^2$ (given) $\frac{1}{10} w_u L_n{}^2$ (given)

$0.4756 \, w_u L_n$ $0.5244 \, w_u L_n$ $0.4417 \, w_u L_n$ $0.5583 \, w_u L_n$

$0.4756 \, L_n$ $0.5244 \, L_n$ $0.4417 \, L_n$ $0.5583 \, L_n$

$+0.4756 \, w_u L_n$ $+0.4417 \, w_u L_n$

$-0.5244 \, w_u L_n$ $-0.5583 \, w_u L_n$

$+\frac{1}{14} w_u L_n{}^2$ (given) $+0.0558 \, w_u L_n{}^2$

$0.3780 \, L_n$ $0.3780 \, L_n$ $0.3341 \, L_n$ $0.3341 \, L_n$

$-\frac{1}{24} w_u L_n{}^2$ (given) $-\frac{1}{24} w_u L_n{}^2$ (given)

$-0.0661 \, w_u L_n{}^2$

$-\frac{1}{10} w_u L_n{}^2$ (given)

(a) Maximum in the positive zone   (b) Maximum in the negative zone

Figure 7.5.2 Exterior span with exterior support built integrally with spandrel beam or girder.

one end is different from that which causes maximum negative moment at the other end. However, by assuming that both maximum negative end moments occur simultaneously, a critical curve having greater magnitude than either of the two actual curves is obtained.

The ACI moment coefficients (ACI-8.3.3) shown in Table 7.4.1 are the common values from the two primary conditions as described in the preceding paragraph. No secondary moment coefficients are suggested by the Code, the reason being that as long as the design live to dead load ratio is limited to 3, no moment reversal will occur; that is, there can be only positive moment in the midspan region and only negative moment in the support region.

In Figs. 7.5.1 to 7.5.4, inclusive, are shown the two primary sets of shear and moment diagrams, for the various conditions, to be used in the design of continuous spans in accordance with the ACI moment coefficients. Note that these diagrams are

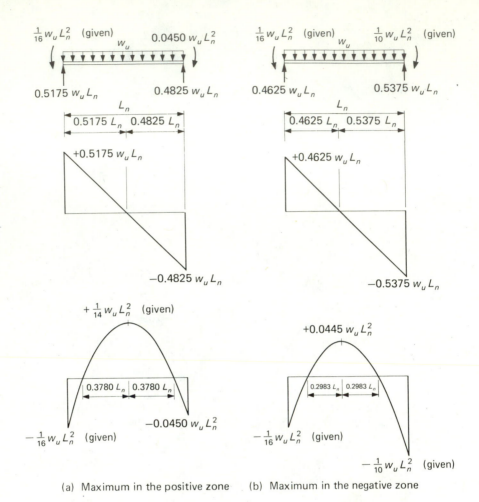

(a) Maximum in the positive zone   (b) Maximum in the negative zone

Figure 7.5.3 Exterior span with exterior support built integrally with column.

applicable to the actual *clear* span, which is also used to compute the positive moment. For negative moment, L_n is the average of adjacent clear spans (ACI-8.0).

The reader should utilize the fundamentals of shear and moment diagrams to verify the numerical ordinates on these diagrams. For instance, in case of maximum positive moment in Fig. 7.5.2(a), the distance x from the left support to the point of zero shear may be determined from the relationship that the change of moment between any two sections is equal to the area of the shear diagram between these two sections. Thus

$$\frac{w_u x^2}{2} = \left(\frac{1}{24} + \frac{1}{14}\right) w_n L_n^2$$

from which

$$x = 0.4756 L_n$$

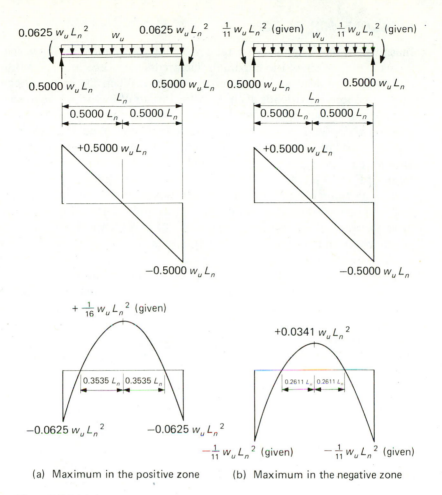

(a) Maximum in the positive zone   (b) Maximum in the negative zone

Figure 7.5.4 Interior span.

Also, the distance x between the section of maximum positive moment to the point of zero moment is

$$\frac{w_u x^2}{2} = \frac{1}{14} w_n L_n^2$$

from which

$$x = 0.3780 L_n$$

Any time a designer uses moment coefficients for determining the factored moments, as permitted by the approximate method of ACI-8.3.3, the moment diagrams that correspond to such coefficients should be used when establishing bar bend or cutoff locations. The use of a given moment coefficient implies a statically compatible moment diagram.

7.6 SHEAR ENVELOPE FOR DESIGN

Inasmuch as the design of shear reinforcement is dependent on the variation of shear forces along the span, the maximum factored shear force at any section due to the combination of dead and live loads is required. When the live load consists of important concentrated loads such as in the design of highway bridge spans, accurate calculations of such maximum shears at all sections must be performed. The proper position of live load for maximum shear at a section can be determined by examining the influence line for shear at that section along the span. For instance, the influence lines for end shear at C of span CD and for the shear at midspan of CD in the rigid frame of Fig. 7.6.1(a) are shown in Figs. 7.6.1(b) and (c). From Fig. 7.6.1(b), it is noted that the position of uniform live load for maximum end shear at C is identical with that for maximum negative moment at C [see Fig. 7.2.1(c)]. Likewise, the position of uniform live load for maximum end shear at D is identical with that for maximum negative moment at D. From Fig. 7.6.1(c), it is apparent that partial span loading of uniform live load is indicated to give maximum shear at any point within the span. The partial span loading to obtain the correct maximum shear at locations within the span has already been used in the shear strength design examples of Section 5.13 for those statically determinate situations.

In buildings of usual types of construction, spans, and story heights, wherein the idealized rigid frame, such as shown in Fig. 7.6.1(a), is taken into consideration, the use of partial span loading of uniform live load is commonly ignored, although, theoretically it is necessary for the computation of maximum shear at any section

(a)

(b) Influence line for maximum end shear at C of span CD

(c) Influence line for maximum shear at midspan of span CD

Figure 7.6.1 Influence lines for shear in rigid frames.

within the span. When partial span loading is not considered necessary, the maximum shears at the ends can be used alone to establish an approximate shear envelope. Note that the loading condition for maximum shear at one end is different from that for maximum shear at the other end. These two critical shear diagrams for each span, when a continuity analysis is performed, may be easily obtained by using the values of the negative end moments (including those in parenthesis) such as those contained in lines 7 through 12 of Table 7.3.8.

When the ACI moment coefficients are used, it is generally assumed that the shear diagrams accompanying the critical moment diagrams as shown in Figs. 7.5.1 to 7.5.4 may be used in the design.

SELECTED REFERENCES

7.1. Chu-Kia Wang. *Intermediate Structural Analysis*. New York: McGraw-Hill, 1983 (Chapter 14).

7.2. Chu-Kia Wang and Charles G. Salmon. *Introductory Structural Analysis*. Englewood Cliffs, New Jersey: Prentice-Hall, Inc., 1984 (Chapters 6 and 8).

7.3. Chu-Kia Wang. *Structural Analysis on Microcomputers*. New York: Macmillan, 1986.

7.4. A. J. Boase and J. T. Howell. "Design Coefficients for Building Frames," *ACI Journal, Proceedings*, **36**, September 1939, 21–36.

7.5. Richard W. Furlong and Carlos Rezende. "Alternate to ACI Analysis Coefficients," *Journal of the Structural Division*, ASCE, **105**, ST 11 (November 1979), 2203–2220.

PROBLEMS

7.1 Compute and draw to scale the bending moment and shear diagrams for the loading conditions 1 through 7 of Fig. 7.3.2; that is, verify the results given in Tables 7.3.2 through 7.3.7, by using any structural analysis method assigned by the instructor.

7.2 Compute and draw to scale the envelope of bending moments (diagram showing range over which bending moment may vary) due to factored loads for the beams of the frame of the accompanying figure, using a continuity analysis method such as moment distribution. The uniform factored dead load is 1 kip/ft and the uniform factored live load is 2 kips/ft.

7.3 For the beams of the frame in the accompanying figure, compute and draw the bending moment envelope using the coefficients of ACI-8.3.3. (If Problem 7.2 has also been solved, compare the moments by giving the percentage difference in the maximum values obtained by coefficients as compared with the more exact values of a continuity analysis.)

Problems 7.2 and 7.3

7.4   Consider an equal-span, uniform-section continuous beam over many supports. Compute and show diagrams for dead load coefficients of wL^2 for moments at critical locations in the exterior and first interior spans. Could the coefficients for the first interior span be applied appropriately to the other interior spans? Recommend dead load coefficients for equal spans.

7.5   Repeat Problem 7.4 for the case where alternate spans are 20% longer than the others ($1.2L$), taking the exterior span as a short one (L).

7.6   For an equal-span, uniform-section continuous beam over many supports, compute and show diagrams for live load coefficients in terms of wL^2 for maximum negative moments at: **(a)** the first interior support; **(b)** the second interior support; and **(c)** the typical interior support. Recommend live load coefficients.

7.7   For an equal-span, uniform-section continuous beam over many supports, compute and show diagrams for live load coefficients in terms of wL^2 for the maximum positive moment in: **(a)** the exterior span; **(b)** the first interior span; and **(c)** the typical interior span. Recommend live load coefficients.

Design of One-Way Slabs

8.1 DEFINITION

One of the most common types of floor construction is the slab–beam–girder system, as briefly described in Section 7.1. The slab panel, bounded on its two long sides by the beams and on its two short sides by the girders, is usually at least twice as long as it is wide. In such a condition, the dead and live load acting on the slab area may be considered as being entirely supported in the short direction by the beams, hence the term "one-way slab." Two-way floor systems with or without beams on column lines are treated in Chapter 16 and ribbed-joist floor construction is described in Sections 10.10 and 10.11.

The determination of an optimum floor framing plan—that is, the spacing of columns, beams, and girders—depends on both the functional and the structural requirements. In most cases preliminary calculations are necessary for several different layouts, and after comparison the most suitable and economical plan is chosen.

8.2 DESIGN METHODS

Since the loading on a one-way slab is nearly all transferred in the short direction, such a slab continuous over several supports may be treated as a beam. Because sufficiently accurate results are obtained, ACI-8.3.3 permits the use of moment and shear coefficients in the case of two or more approximately equal spans (the larger of two adjacent spans not exceeding the shorter by more than 20%) with loads uniformly distributed, where the unit live load does not exceed three times the unit dead load. These coefficients are in terms of clear span L_n and the values given are for critical locations, that is, faces of support for shears and negative moments and midspan regions for positive moments.

River City Apartments, Chicago, completed 1988; shows the curved shape and sculptured effect obtainable through the use of concrete. (Photo courtesy of Steven M. Baldridge.)

When the conditions of ACI-8.3.3 are not satisfied, an elastic analysis is required. Approximate methods are permitted in "buildings of usual types of construction, spans, and story heights." These approximate methods usually involve the use of tables and charts made for w_L to w_D ratio greater than 3. The Portland Cement Association publication, *Continuity in Concrete Building Frames* [8.1], provides moment coefficients that are nearly as exact as those which would be obtained by actually carrying out an elastic analysis.

Generally, span lengths between centers of supports are used in elastic analysis, notably in the computation for fixed-end moments and stiffness factors. Similarly, the results of an elastic analysis show only the negative bending moments at centers of supports. It will be shown that the negative bending moment at the face of support, the value of which should be used in the design of the member itself, may be obtained approximately by reducing from the maximum value at the center of support by a quantity equal to $Vb/3$, where V is the shear force at either the center or at the face of support, and b is the width of support.

Consider an ideal situation where a member AB is fixed at points A and B and subjected to a uniform load, as shown in Fig. 8.2.1. The elastic curve is shown as $AA'B'B$ in Fig. 8.2.1(a), where there are rotation and deflection at A' and B'. The

Figure 8.2.1 Single span fixed at points A and B.

bending moment at A' (Fig. 8.2.1) is

$$M = -\frac{wL^2}{12} + \frac{wL}{2}\left(\frac{b}{2}\right) - \frac{w(b/2)^2}{2} = -\left(\frac{wL^2}{12} - \frac{wLb}{4} + \frac{wb^2}{8}\right) \qquad (8.2.1)$$

Now, if the same member is considered to be fixed at A' and B' as shown in Fig. 8.2.2, the bending moment at A' (Fig. 8.2.2) is

$$M = -\frac{w(L-b)^2}{12} = -\left(\frac{wL^2}{12} - \frac{wLb}{6} + \frac{wb^2}{12}\right) \qquad (8.2.2)$$

The quantity in Eq. (8.2.2) is numerically larger than in Eq. (8.2.1) by an amount $wLb/12$, provided the terms involving b^2 are neglected. However, the condition in Fig. 8.2.2 is believed to be more correct because at a relatively stiff support, such as a column or girder, there can be very little change in rotation or deflection between the points A and A'. Thus the moment at the face of support is less than that at the center of support by a quantity equal to $Vb/3$ [Eq. (8.2.2)] rather than $Vb/2$ [Eq. (8.2.1)] if V is taken as $wL/2$ and, again, if the term involving b^2 is neglected (see also Reference 8.1).

On the other hand, the moment at the center of span in Fig. 8.2.2 is less than that in Fig. 8.2.1. Inasmuch as the elastic analysis results are associated with the concept of Fig. 8.2.1, midspan moments so obtained are on the safe side, and, in general, no attempt is necessary to correct them to the concept of Fig. 8.2.2.

As the span L becomes large compared to the support width b, the common practice of using the moment envelope value at the face of support (scaled, for example) is reasonable. That practice is equivalent to subtracting $Vb/2$ from the center of support value.

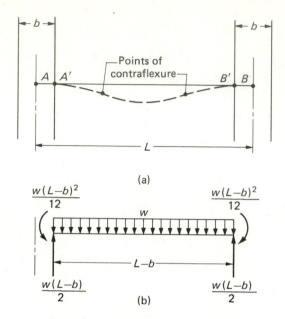

Figure 8.2.2 Single span fixed at points A' and B'.

8.3 THICKNESS OF SLAB

In designing a one-way slab, a typical imaginary strip 12 in. wide is usually considered. The continuous slab may then be designed as a continuous beam having a known width of 12 in.; the slab thickness is the only unknown.

The thickness of the slab depends on the deflection, bending, and shear requirements. Deflection requirements are imposed to prevent excessive deformations that might adversely affect the serviceability of the structure. According to ACI-Table 9.5a, one-way slabs must have at least a minimum slab thickness (for Grade 60 steel) of $L/20$, $L/24$, $L/28$, or $L/10$ depending on whether L is the length of a simply supported, a one end continuous, a both ends continuous, or a cantilever span. If the slab supports or is attached to construction likely to be damaged by large deflections, deflections must be computed and shown to satisfy the limits of ACI-Table 9.5b. An extended treatment of deflections will be found in Chapter 14.

The flexural strength M_n required to resist the factored moment M_u is provided in accordance with the principles of Chapter 3. The desired coefficient of resistance R_n can be computed for the desired reinforcement ratio ρ. Then the required $M_n = M_u/\phi = R_n bd^2$. In equal continuous spans the negative moment at the exterior face of the first interior support is the largest; therefore this negative moment should be used to establish the slab thickness.

The shear requirement does not usually control, but it must be checked. Because of practical space limitations, shear reinforcement is not used in a slab; thus the governing factored shear V_u, which in equal continuous spans occurs at the exterior face (the distance d therefrom) of the first interior support, must be kept below ϕV_c of ACI-11.2 for lightweight concrete and ACI-11.3 for nonprestressed normal-weight concrete.

According to ACI-7.7.1c, the concrete protective covering for reinforcement ($\#11$ and smaller) in slabs shall be not less than $\frac{3}{4}$ in. at surfaces not exposed directly to the ground or the weather. Also, when the top of a monolithic slab is the wearing surface and when unusual wear is expected as in buildings of the warehouse or industrial class, it has been customary to use an additional depth of $\frac{1}{2}$ in. of concrete protective covering over that required by the design of the member. The ACI Code does not require extra slab thickness for wearing surface, but ACI-8.12 *permits*, on the discretion of the designer, a monolithic floor finish to be considered as part of the structural member. For nonstructural purposes, any concrete floor finish may be considered as part of the required cover.

■ **EXAMPLE 8.3.1** Establish the thickness of the floor slab shown in the second-floor framing plan of Fig. 8.3.1 (see Fig. 8.4.2 for a section through slabs 2S1-2S2-2S3) for a service live load of 100 psf. Use $f'_c = 3000$ psi, $f_y = 40,000$ psi, and the strength method of the ACI Code. Assume an exterior staircase so that no openings are to be made in the slabs.

Figure 8.3.1 Second-floor framing plan. (For Section *A-A*, see Fig. 8.4.2.)

Solution: Since live load does not appear to exceed three times the dead load, the design will be done in accordance with ACI moment coefficients and their corresponding shear and moment diagrams (see Section 7.5).

(a) Minimum thickness. For spans with one end and both ends continuous, the respective minimum thicknesses h from ACI-Table 9.5a are $L/24$ and $L/28$. Since the table values are for $f_y = 60,000$ psi, they must be corrected for $f_y = 40,000$ psi by multiplying by 0.8.

$$\min h = \frac{L}{24}(0.8) = \frac{L}{30} = \frac{13(12)}{30} = 5.2 \text{ in.} \quad \text{(for 2S1)}$$

$$\min h = \frac{L}{28}(0.8) = \frac{L}{35} = \frac{13(12)}{35} = 4.5 \text{ in.} \quad \text{(for 2S2 and 2S3)}$$

Assume a $4\frac{1}{2}$-in. slab. The weight of the slab is $(4.5/12)(0.15) = 0.056$ kip/sq-ft. If deflection is of concern in the end span, that span could be made 5 in. thick or compression reinforcement could be used (see Example 14.12.2 for deflection calculation for a similar end span).

(b) Bending moment requirement. Assume width of supporting beams to be 13 in. Note that 13 in., an uncommon dimension for width, is used rather than 12 in. to permit easier following of the numerical calculations.

$$w_D = 0.056(1.4) = 0.079 \text{ kip/ft/ft of width}$$

$$w_L = 0.100(1.7) = 0.170 \text{ kip/ft/ft of width}$$

$$\text{clear span} = 13 - \tfrac{13}{12} = 11.92 \text{ ft}$$

$$M_u = \tfrac{1}{10}(0.079 + 0.170)(11.92)^2 = 3.54 \text{ ft-kips/ft of width}$$

Choose a reinforcement ratio ρ equal to about $0.375\rho_b$, or one-half the maximum permitted by the ACI Code, in order to have reasonable deflection control. From Table 3.6.1,

$$0.375\rho_b = 0.5(0.0278) = 0.0139$$

Then, using Eq. (3.8.4) or Fig. 3.8.1, find corresponding R_n,

$$m = \frac{f_y}{0.85f'_c} = \frac{40,000}{0.85(3000)} = 15.7$$

$$R_n = \rho f_y(1 - \tfrac{1}{2}\rho m)$$

$$= 0.0139(40,000)[1 - 0.5(0.0139)(15.7)] = 495 \text{ psi}$$

$$\text{required } d = \sqrt{\frac{M_u}{\phi R_n b}} = \sqrt{\frac{3.54(12,000)}{0.90(495)12}} = 2.82 \text{ in.}$$

Assume #5 bars ($d_b = 0.625$) and $\frac{3}{4}$-in. cover,

$$\text{required } h = 2.82 + 0.31 + 0.75 = 3.88 \text{ in.}$$

Use $h = 4\frac{1}{2}$ in.

$$\text{provided } d = 4.50 - 0.31 - 0.75 = 3.44 \text{ in.}$$

(c) Shear requirement.

$$\max V_u = 1.15 \frac{w_u L_n}{2} = 1.15 \frac{0.249(11.92)}{2} = 1.71 \text{ kips/ft of width}$$

The shear strength ϕV_c for a member without shear reinforcement is

$$\phi V_c = \phi[2\sqrt{f'_c}\, bd]$$
$$= 0.85(2\sqrt{3000})(12)(3.44)\tfrac{1}{1000} = 3.84 \text{ kips/ft} > 1.71 \text{ kips/ft}$$

The slab is acceptable for a member without stirrups. Note that to be strictly correct the shear force at a distance d from the face of support should have been used, in which case the factored shear would have been even less than 1.71 kips/ft. ■

8.4 CHOICE OF REINFORCEMENT

The choice of reinforcement depends primarily on the steel area and secondarily on development length requirements. The steel areas required at the principal sections, namely those at the middle and at the ends of each span, are first computed. Then a *tentative* choice of reinforcement may be made. One possible arrangement of reinforcement in floor slabs, though not commonly used, is as shown in Fig. 8.4.1, in which the straight and bent (trussed) bars are placed alternately. The straight bars across the bottom are ordinarily one size smaller than the bent bars. Frequently, in thin slabs of under 5 in. of thickness, straight bars are used in both the top and bottom in all spans. Because of economics, most designers use straight bars in all cases.

Development length requirements should next be examined. The positive moment bars that are bent up into the negative moment region have ample embedment. The bars that continue along the bottom of the slab and extend into the support at least 6 in. (ACI-12.11.1) must be checked for adequate development at points of inflection (ACI-12.11.3). In addition, the bars extending straight across the top beyond the bend-down points must be checked for satisfactory embedment (ACI-12.12.3).

Some other limitations affecting the choice of reinforcement in slabs are (1) that in structural slabs of uniform thickness (ACI-10.5.3) the minimum amount of reinforcement in the direction of the span shall not be less than that required for shrinkage and temperature reinforcement (ACI-7.12), and (2) that the principal reinforcement shall be centered not farther apart than three times the slab thickness nor more than 18 in. (ACI-7.6.5).

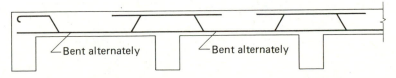

Figure 8.4.1 One possible arrangement of slab reinforcement. Note: Most designers use straight bars only because of economics.

■ **EXAMPLE 8.4.1**   Choose the arrangement of reinforcement in the $4\frac{1}{2}$-in. floor slab 2S1, 2S2, and 2S3, as designed in Example 8.3.1.

Solution: **(a)** Area requirements. The ACI moment coefficient, the bending moment, the steel area required, and the tentative choice of reinforcement at the critical section are shown in lines 1 to 6 of Table 8.4.1. The arrangement of reinforcement is shown in Fig. 8.4.2. Choice of this steel should begin at the typical interior support. In this case, #4 bars at 6 in. are chosen. At the middle of the first and typical interior spans, the required area would be furnished by #3 straight (st) bars at 12 in. in addition to the #4 bent (bt) bars at 12 in. which were already there. As a matter of practical judgment, however, some designers would not use bars smaller than #4 for main reinforcement. Since the decision has now been made on the 6-in. and 12-in. spacings, the bent-bar size in the exterior span follows automatically. These spacings satisfy the limitations of ACI-7.6.5.

(b) Development length requirements at inflection points. In order to confirm the choice of longitudinal reinforcement made on the basis of the required areas only, it is necessary to review the development length requirements. The requirements of ACI-12.11.3 must be checked at the exterior supported end (point 1 of Fig. 8.4.2) and at inflection points 2 to 6. In addition, embedment equal to the required development length must be provided in both directions from the maximum moment points at the faces of supports (points 7 through 12). Finally, embedment of the straight portion of the bars at the top of the slab beyond extreme points of inflection (points 13 through 17) must satisfy ACI-12.12.3.

From the ACI shear and moment diagrams in Figs. 7.5.2 and 7.5.4, the shears at inflection points on the typical 1-ft width of slab are

$$V_1 = V_2 = 0.3780 w_u L_n = 0.3780(0.249)(11.92) = 1.13 \text{ kips}$$

$$V_3 = V_4 = V_5 = V_6 = 0.3535 w_u L_n = 0.3535(0.249)(11.92) = 1.05 \text{ kips}$$

For the bars (#3) that extend past the inflection points into the supports, the basic development length L_{db} is (from Table 6.7.1 for $f'_c = 3000$ psi and $f_y = 40,000$ psi)

$$L_{db} (\#3) = 3.2 \text{ in.}$$

Since the clear spacing when #3 are placed on 12 in. centers will be greater than $3d_b$, the development length is in Category A with α_{abc} equal to 1.0. and because the

Figure 8.4.2 Longitudinal reinforcement in Example 8.4.1. Note: Most designers use straight bars only because of economics.

clear spacing exceeds $5d_b$, α_{de} is 0.80. Thus,

$$\alpha_{abc}\alpha_{de}L_{db} = 1.0(0.80)3.2 = 2.6 \text{ in.}$$

The minimum value of $\alpha_{abc}\alpha_{de}L_{db}$ is obtained from Table 6.7.2 as 8.2 in. The multipliers α_{top}, $\alpha_{\ell w}$, and α_{epx} are not applicable here; finally, the minimum L_d of 12 in. controls.

The requirement of ACI-12.11.3 at inflection points (such as point 2) is

$$\frac{M_n}{V_u} + L_a \geq L_d$$

For #3 @ 12 in.,

$$C = 0.85(3)12a = 30.6a$$

$$T = 0.11(40) = 4.4 \text{ kips}$$

$$a = \frac{4.4}{30.6} = 0.14 \text{ in.}$$

$$M_n = 4.4(3.44 - 0.07)\tfrac{1}{12} = 1.24 \text{ ft-kips/ft}$$

$$V_u = 1.13 \text{ kips (computed above)}$$

$$L_a = 12d_b \text{ max} = 4.5 \text{ in.}$$

$$\frac{1.24(12)}{1.13} + 4.5 = 13.1 + 4.5 = 17.6 \text{ in.} > L_d = 12 \text{ in.} \qquad \textbf{OK}$$

This calculation applies identically to other inflection points (3 through 6).

At the exterior end, which might be considered a simple support, the requirement of ACI-12.11.3

$$1.30 \frac{M_n}{V_u} + L_a \geq L_d$$

is satisfied by inspection.

(c) Cutoff points for negative moment reinforcement. The distance from the face of support to the cutoff location (beyond point 13, for instance) must be (1) greater than L_d, and (2) adequate to satisfy ACI-12.12.3.

For the #4 bars at 12 in. on centers, from Table 6.7.1,

$$L_{db} (\#4) = 5.8 \text{ in.}$$

Again, $\alpha_{abc} = 1.0$ and $\alpha_{de} = 0.80$; also, from Table 6.7.2, the minimum $\alpha_{abc}\alpha_{de}L_{db}$ is 11.0 in. Thus, again, the minimum L_d of 12 in. controls. Note that even though the #4 bars are in the top of the slab they are *not* top bars because there is less than 12 in. of concrete cast beneath them.

Table 8.4.1 Reinforcement for One-Way Slab of Example 8.4.1, Using ACI Moment Coefficients

LINE NUMBER	2S1			2S2			2S3		
	SUPPORT	MIDDLE	SUPPORT	SUPPORT	MIDDLE	SUPPORT	SUPPORT	MIDDLE	SUPPORT
1. ACI moment coefficient	$-\dfrac{1}{24}$	$+\dfrac{1}{14}$	$-\dfrac{1}{10}$	$-\dfrac{1}{11}$	$+\dfrac{1}{16}$	$-\dfrac{1}{11}$	$-\dfrac{1}{11}$	$+\dfrac{1}{16}$	$-\dfrac{1}{11}$
2. M_u (ft-kips) = line (1) $\times$ 0.249 (11.92)^2 = line (1) $\times$ 35.4	-1.48	$+2.53$	-3.54	-3.22	$+2.21$	-3.22	-3.22	$+2.21$	-3.22
3. Required $R_n = \dfrac{\text{line (2)} \times 12{,}000}{0.9(12)(3.44)^2}$ = line (2) $\times$ 94.0 (psi)	139	238	333	303	208	303	303	208	303
4. Required $\rho \approx \dfrac{\text{line (3)} \times 0.0139}{495}$ $= \dfrac{\text{line (3)}}{35{,}600}$	0.0039 (0.005 min.)	0.0067	0.0094	0.0085	0.0059	0.0085	0.0085	0.0059	0.0085
5. Required A_s = line (4) $\times$ 12(3.44) = line (4) $\times$ 41.3 (sq in./ft)	0.21	0.28	0.39	0.35	0.25	0.35	0.35	0.25	0.35
6. Provided A_s (sq in./ft)	#4 @ 12 st (0.20)	#4 @ 12 bt #3 @ 12 st (0.31)	#4 @ 12 bt (0.31)	#4 @ 12 bt #4 @ 12 bt (0.39)	#4 @ 12 bt #3 @ 12 st (0.31)	#4 @ 12 bt #3 @ 12 st (0.31)	#4 @ 12 bt #4 @ 12 bt (0.39)	#4 @ 12 bt #3 @ 12 st (0.31)	#4 @ 12 bt #3 @ 12 st (0.31)

To satisfy ACI-12.12.3, the moment diagram of Fig. 7.5.2(b) corresponding to the ACI coefficients may be used to locate the point of inflection. The required distance to the cut location is

$$(0.5583 - 0.3341)(11.92) + \frac{11.92}{16} = 3.42 \text{ ft}$$

which compares favorably with the commonly used value of 0.3 of the clear span, as suggested by the *ACI Detailing Manual—1988* [2.25] (see pp. 25–27).

(d) Bend-up or bend-down location. Since the bend slope is usually about 45°, the bend can be defined by locating either the bend-up or the bend-down location. The moment diagram of Fig. 7.5.2(a) can be used to determine the bend-up location near the right end of slab 2S1.

The percent of moment capacity remaining after the bend-up is approximately proportional to the percent of total positive moment steel area remaining in the bottom of the slab.

Moment corresponding to #3 @ 12 in.

$$= \frac{0.11}{0.31}\left(\frac{w_u L_n^2}{14}\right) = 0.0253 w_u L_n^2$$

The theoretical location, measured from the face of support, where the moment is $0.0253 w_u L_n^2$ is

$$0.5244 L_n - L_n \sqrt{(\tfrac{1}{14} - 0.0253)2} = 0.22 L_n$$

This compares reasonably close to the suggested value in the *ACI Detailing Manual—1988* [2.25] of 0.25 of the clear span from the face of support to the bend-up location. The theoretical bend-down location may be similarly determined by referring to the right end portion of the moment diagram shown in Fig. 7.5.2(b). ∎

8.5 CONTINUITY ANALYSIS

When the necessary conditions for using the ACI-8.3.3 moment and shear coefficients are not satisfied, an elastic analysis is required. As discussed in Section 8.2, several approaches to the elastic analysis requirement may be used. The adjustment of negative moments permitted (ACI-8.4) under certain conditions should not be applied with any of the moment coefficient methods used to design one-way slabs. The subject of redistribution of bending moment to conform partially with true ultimate load behavior is treated in Chapter 10.

In the interest of obtaining a comparison of answers, an elastic analysis will be made to determine the required steel area at all critical sections of the slab used in Examples 8.3.1 and 8.4.1. It is noted that the choice of reinforcement based on either the ACI coefficients or an elastic analysis will be practically identical, as should be expected for the conditions of this design.

■ **EXAMPLE 8.5.1** Using the results of the slightly approximate elastic method as given in *Continuity in Concrete Building Frames* [8.1], determine the required steel areas at all critical sections of the continuous slab of Examples 8.3.1 and 8.4.1.

Solution: The 12-span continuous slab is supported on 13 beams. The restraining action of the beams on the slab may be accounted for by approximating the torsional stiffness of the supporting beam as a column having half the flexural stiffness as the slab, as shown in Fig. 8.5.1. Using this equivalent stiffness assumption and a ratio of w_L to w_D equal to 2, the theoretical moment coefficients could be determined by a formal elastic analysis as demonstrated in Chapter 7. The results of such an analysis are available [8.1], as presented in Table 8.5.1. Note that these coefficients are in terms of wL^2, in which L is the distance between centers of supporting beams. Although the moments within the span may be thus computed, the moments at the face of support may be determined by deducting $Vb/3$ (see Fig. 8.5.2), as discussed in Section 8.2, from the negative moment at the center of support. The required computations are shown in Table 8.5.1.

The results generally show higher moments than those obtained by the ACI coefficients; the choice of reinforcement may be revised by changing the 6- and 12-in. spacings used in Example 8.4.1 to $5\frac{1}{2}$ and 11 in., respectively.

Figure 8.5.1 Equivalent stiffness assumption recommended [8.1] for one-way continuous slab on beam supports.

Figure 8.5.2 Shear diagram for Example 8.5.1.

Table 8.5.1 Reinforcement Areas Required Using Continuity Analysis

ITEM	2S1			2S2			2S3		
	SUPPORT	MIDDLE	SUPPORT	SUPPORT	MIDDLE	SUPPORT	SUPPORT	MIDDLE	SUPPORT
Theoretical coefficient from Table 8, *Continuity in Concrete Building Frames* [8.1]	−0.035	+0.070	−0.106	−0.103	+0.057 −0.007	−0.095	−0.096	+0.061 −0.004	−0.098
M_u (ft-kips) = coefficient × 0.249(13)^2 = 42.1	−1.47	+2.95	−4.47	−4.34	+2.40 −0.30	−3.58	−4.04	+2.57	−4.21
M_u (ft-kips) at face of support (see Fig. 8.5.2)	−1.07		−3.90	−3.83		−3.12	−3.54		−3.70
Required $R_n = M_u$ × 94.0 (see Table 8.4.1)	101	277	367	360	226	294	333	242	348
Required A_s (sq in./ft)	0.12	0.32	0.43	0.42	0.26	0.34	0.39	0.28	0.41

The moment reversal of 0.30 ft-kips may cause a tensile stress at the extreme top of the midspan in 2S2 of

$$\text{tensile stress} = \frac{0.30(12,000)}{(12)(4.5)^2/6} = 89 \text{ psi}$$

which is well below $5\phi\sqrt{f_c'} = 5(0.65)\sqrt{3000} = 178$ psi for plain concrete.*   ■

8.6 SHRINKAGE AND TEMPERATURE REINFORCEMENT

Reinforcement for shrinkage and temperature stresses normal to the principal reinforcement is required in structural floor and roof slabs, where the principal reinforcement extends in one direction only (ACI-7.12.1). Further, such reinforcement shall provide (ACI-7.12.2) for the following minimum ratios of reinforcement area to gross concrete area, but in no case shall such reinforcing bars be placed farther apart than five times the slab thickness nor more than 18 in. (ACI-7.12.2.2):

1. Slabs where Grades 40 or 50 deformed bars are used   0.0020
2. Slabs where Grade 60 deformed bars or welded wire fabric (smooth or deformed) are used   0.0018
3. Slabs where reinforcement with yield strength exceeding 60,000 psi measured at a yield strain of 0.35% is used. $\dfrac{0.0018(60,000)}{f_y}$

The above required reinforcement should develop the yield strength f_y in tension in accordance with ACI-Chapter 12 (ACI-7.12.2.3).

■ **EXAMPLE 8.6.1**   Design the shrinkage and temperature reinforcement in the floor slab of Examples 8.3.1 and 8.4.1.

Solution: Area of shrinkage and temperature reinforcement is

$$A_s = \rho bh = (0.0020)(12)(4.5) = 0.11 \text{ sq in.}$$

Use #3 bars at 12-in. spacing.   ■

8.7 BAR DETAILS

Consistent with the shear and moment diagrams given in Chapter 7 for the ACI Code coefficients, acceptable standard bar bend distances, extensions, and anchorage lengths have been developed. For instance, typical bar details in end and interior

---

* See *Building Code Requirements for Structural Plain Concrete* (ACI 318.1–89). Detroit: American Concrete Institute, 1989 (see Section 6.2.1).

spans of one-way slabs as presented in the *ACI Detailing Manual—1988* [2.25] are reproduced in Fig. 8.7.1.

It must be noted that although typical bar details are of importance in office practice, they must be used with discretion and care so that they conform with the prevailing shear and moment diagrams.

SINGLE SPAN, SIMPLY SUPPORTED

END SPAN, SIMPLY SUPPORTED

INTERIOR SPAN, CONTINUOUS

*MIN. 6", UNLESS OTHERWISE SPECIFIED BY THE ENGINEER

Figure 8.7.1 Typical details for one-way solid slabs.

SELECTED REFERENCES

8.1. *Continuity in Concrete Building Frames* (4th ed.). Chicago: Portland Cement Association, 1959.

PROBLEMS

All problems are to be worked in accordance with the ACI Code and all stated loads are service loads, unless otherwise indicated.

8.1 Design for a warehouse a continuous one-way slab supported on beams 12 ft on centers as shown in the accompanying figure. Assume that beam stems are 12 in. wide. The dead load is 25 psf in addition to the slab weight, and the live load is 200 psf. Use $f'_c = 3000$ psi, $f_y = 40,000$ psi, and the strength method. Use ACI coefficients if permissible, and use only straight reinforcing bars. (For SI problem, use 3.7-m spans, 300-mm beam widths; superimposed dead load = 1.2 kN/m; live load = 9.6 kN/m; $f'_c = 25$ MPa; $f_y = 300$ MPa.)

Problems 8.1 through 8.4

8.2 Repeat Problem 8.1 using a live load of 250 psf with $f'_c = 4000$ psi and $f_y = 60,000$ psi. (For SI: live load = 12.0 kN/m; $f'_c = 30$ MPa; $f_y = 400$ MPa.)

8.3 Repeat Problem 8.1 using a live load of 300 psf with $f'_c = 4000$ psi and $f_y = 60,000$ psi.

8.4 Repeat Problem 8.1 using a live load of 350 psf with $f'_c = 4000$ psi and $f_y = 60,000$ psi.

8.5 Design a one-way slab for the conditions shown in the accompanying figure. Assume that beam stems are 12 in. wide. The live load is 175 psf and the slab will not be the final wearing surface. Use $f'_c = 4000$ psi, $f_y = 60,000$ psi, and the strength method. Use alternate bent and straight bar reinforcement if it seems practical.

Problem 8.5

CHAPTER 9

T-Sections in Bending

9.1 GENERAL

The T-section treatment in this chapter provides the link between the basic concepts of flexural and shear strength in Chapters 3 through 6 and the monolithic statically indeterminate slab–beam–girder system in Chapter 10.

Chapter 7 contains a brief review of the necessary concepts of statically indeterminate analysis and positioning of live load for obtaining moment and shear envelopes, as well as the ACI moment coefficients for common situations. The design of the one-way slab in the slab–beam–girder system is treated in Chapter 8. Finally, the basic principles for the design of the beams (such as 2B1-2B2-2B1 in Fig. 8.3.1) is treated in this chapter. These beams are built monolithically with the slab—hence the name *T-beams*.

The monolithic multiple T-section in a slab–beam–girder system shown in Fig. 8.4.2 (section *A-A* of Fig. 8.3.1) has several stems and includes as its flange the entire one-way slab that spans transversely between the stems. For the purpose of design, the multiple T-section is divided into individual T-sections that have a portion of the slab projecting from each side of the stem as flanges. For negative bending moment the flange is on the tension side of the neutral axis, thus the T-section is in effect a rectangular section. For positive bending moment the flange does provide considerably more compression area than in the negative bending moment portion of the span. Just how much of the slab projecting from the stem may be considered as part of the individual T-section, as well as the methods of design and analysis for the flexural strength, appear in subsequent sections.

City Hall, Boston, Mass. (Photo by C. G. Salmon.)

9.2 COMPARISON OF RECTANGULAR AND T-SECTIONS

A comparison of the two sections shown in Fig. 9.2.1 indicates that the flexural strength of a rectangular section is identical with that of the T-section as long as they possess the same compression area above the neutral axis (NA) and the same steel

(a)   (b)

Figure 9.2.1 Two equivalent sections in bending.

N.A.

Figure 9.2.2 A T-section in bending.

area at the same effective depth. Thus, as far as bending is concerned, any T-section with a rectangular compression area, such as shown in Fig. 9.2.1(b) may be regarded as a rectangular section. When the neutral axis of a T-section is located below the flange, as shown in Fig. 9.2.2, computation of its flexural strength requires different treatment than for a rectangular section. Beams with T-sections may be individually built as such; however, they usually occur in the positive moment regions of floor beams and girders that are built monolithically with a slab.

9.3 EFFECTIVE FLANGE WIDTH

Very wide beams do not conform in behavior to the assumption of the elementary theory of bending. In ordinary theory, bending stresses are assumed not to vary across the beam width. Simple theory, therefore, would dictate a constant stress at, say, the extreme fiber over the entire flange width of a T-section, no matter how great the overhang from the stem. In reality, the stress based on the plate theory decreases the more distant a point is from the stem of the beam. Thus, for a flange of infinite width the compressive stress in the flange varies as shown in Fig. 9.3.1.

λ = equivalent width for uniform stress and same compressive force as actual stress distribution

Actual extreme fiber compressive stress f_c for infinitely wide flange

Figure 9.3.1 Actual and equivalent stress distribution over flange width.

Theoretical investigations for an infinitely long continuous beam on equidistant supports, with an infinitely large flange width and a small thickness compared to beam depth, have determined an effective flange width b_E over which the compressive stress may be considered constant. The total compressive force carried by the equivalent system is the same as that carried by the actual system. For such assumptions the equivalent width of overhang λ depends only on the type of loading and the span length of the beam. The theory, first developed by T. von Kármán, along with certain results is summarized by Timoshenko and Goodier [9.1] and Girkmann [9.2].

Whereas the aforementioned theory gives the equivalent width b_E for an infinite flange width as a function of the span length L, in practical situations other variables are important as well. These variables are the spacing of the beams, the width of the stem of the beam, and the relative thickness of the slab with respect to the total beam depth t/h. To illustrate the effect of loading, several cases (from Girkmann [9.2]) showing the variation of effective flange projection for a flange of zero thickness between beams ($t/h = 0$) are given in Fig. 9.3.2.

In the case of a floor slab built monolithically over floor beams, there is still transverse bending in the slab between beams, which also tends to reduce the effectiveness of the slab in carrying compression at points remote from the beam stem.

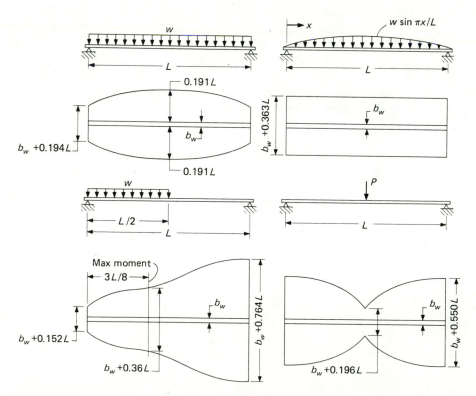

Figure 9.3.2 Equivalent flange widths for infinite actual width beams with a rib cross-sectional area of $0.1tL$ (adapted from Girkmann [9.2]).

Figure 9.3.3 Effective section in compression of a T-section in a floor system.

Thus there is a valid reason for using a conservatively low effective flange width. The effective section of a T-beam in a floor system is shown crosshatched in Fig. 9.3.3.

The ACI Code (ACI-8.10.2) prescribes a limit on the effective flange width b_E of *interior* T-sections to the *smallest* of the following:

1.
$$b_E = \frac{L}{4} \qquad (9.3.1a)$$

2.
$$b_E = b_w + 16t \qquad (9.3.1b)$$

3 $b_E =$ center-to-center spacing of beams $\qquad (9.3.1c)$

where L is the span length of the beam, and b_w and t are as shown in Fig. 9.3.3.

For *exterior* T-sections, ACI-8.10.3 prescribes for effective width b_E the *smallest* of the following:

1.
$$b_E = b_w + \frac{L}{12} \qquad (9.3.2a)$$

2.
$$b_E = b_w + 6t \qquad (9.3.2b)$$

3.
$$b_E = b_w + \frac{1}{2}(\text{clear distance to next beam}) \qquad (9.3.2c)$$

For *isolated* T-sections, ACI-8.10.4 gives

$$b_E \le 4b_w \qquad (9.3.3a)$$

and also requires that

$$t \ge \tfrac{1}{2}b_w \qquad (9.3.3b)$$

Since the true effective width is very much dependent on the type of loading and on the relationships (Fig. 9.3.3) t/h, L/b_w, and L/b_0, the ACI criteria are very much simplified. They seem properly adequate for certain loading cases and are unduly conservative for others. The European Concrete Committee [9.3, 9.4] has presented detailed procedures for determining the values of effective flange width.

9.4 INVESTIGATION OF T-SECTIONS IN BENDING—STRENGTH METHOD

In computing the strength of a T-section in bending, the neutral axis location determines whether the compression zone is T-shaped or rectangular. Since the option (ACI-10.2.6 and 10.2.7) of using Whitney's rectangular stress distribution (Section 3.3) is also available for non-rectangular sections, use of the rectangular stress distribution would mean that as long as its depth a does not exceed the flange thickness t, the beam strength would be the same as for a rectangular section with $b = b_E$.

The effective width b_E of the flange is an essential factor in the neutral axis location. Available information indicates that as the strain ϵ_c at the extreme compression fiber increases toward its ultimate value ϵ_{cu}, the width of the effective compression flange increases [9.4]. Therefore when applying the strength method using factored service loads, it is safe to use the smaller effective widths applicable under service load conditions. The ACI Code therefore, uses the same effective width for the strength method as it has for many years used for the working stress method.

The computation for the nominal moment strength M_n of a T-section occurs in two categories: (1) the depth a of the rectangular stress distribution is equal to or less than t; and (2) the depth a is greater than t.

Case 1: $a \leq t$ [Fig. 9.4.1(a)]. For this case to occur, the tension steel area A_s must not exceed

$$A_s \leq \frac{0.85 f'_c b_E t}{f_y} \quad \text{for} \quad x \leq \frac{t}{\beta_1} \qquad \textbf{(9.4.1a)}$$

(a) $a \leqslant t$

(b) $a > t$

Figure 9.4.1 Strength of T-sections in bending.

obtained from equating C to T, where

$$C = 0.85f'_c b_E a \tag{9.4.1b}$$

$$T = A_s f_y \tag{9.4.1c}$$

The analysis is exactly as presented for rectangular sections in Chapter 3, using b_E for b.

Case 2: $a > t$ [Fig. 9.4.1(b)].   For this case, where the neutral axis distance x is larger than t/β_1, the area on which the uniform stress $0.85f'_c$ acts is T-shaped. Thus it is desirable to separate the total compressive force into forces C_1 and C_2 with C_1 resulting from the stress on area A_1 and C_2 on area A_2. The moment arm for C_2 is equal to $d - t/2$ but that of C_1 is less than $d - t/2$. Thus

$$M_n = C_1 \left(d - \frac{a}{2} \right) + C_2 \left(d - \frac{t}{2} \right) \tag{9.4.2a}$$

in which

$$C_1 = 0.85f'_c b_w a \tag{9.4.2b}$$

$$C_2 = 0.85f'_c (b_E - b_w)t \tag{9.4.2c}$$

and

$$a = \frac{T - C_2}{0.85f'_c b_w} \tag{9.4.2d}$$

The tensile force T and the tension steel area A_s may also be separated into T_1 and T_2 and A_{s1} and A_{s2}, respectively.

Because the compressive strength of concrete is considered useful to a much greater extent in strength design than was the case for the working stress method, Case 2 seldom occurs in reinforced concrete building floor beams. This is due to the large equivalent width of the compression area that is available.

■ **EXAMPLE 9.4.1**   Determine the nominal moment strength M_n within the span of a floor beam (Fig. 9.4.2) whose projection below a $4\frac{1}{2}$-in. slab is 13×24 in. (effective depth is 25 in. for two layers of steel). Tension reinforcement is 8-#8 bars. The span

Figure 9.4.2 T-section for Example 9.4.1.

length of the beam is 26 ft and the beams are centered 13 ft apart. Use $f'_c = 3000$ psi and $f_y = 50,000$ psi.

Solution: **(a)** Determine the effective flange width. Following ACI-8.10.2, the effective flange width b_E is the smallest of $(26)(12)/4 = 78$ in.; $13 + 16(4.5) = 85$ in.; or $(13)(12) = 156$ in. Thus $b_E = 78$ in.

(b) Find A_s so that $a = 4\frac{1}{2}$ in.

$$C = 0.85f'_c b_E a = 0.85(3)(78)(4.5) = 895 \text{ kips}$$

$$A_s = \frac{T \text{ or } C}{f_y} = \frac{895}{50} = 17.9 \text{ sq in.}$$

Thus the depth a of the rectangular stress distribution is less than t for $A_s = 6.28$ sq in.

(c) Treat as a rectangular section.

$$T = f_y A_s = 50(6.28) = 314 \text{ kips}$$

$$a = \frac{T \text{ or } C}{0.85f'_c b_E} = \frac{314}{0.85(3)(78)} = 1.58 \text{ in.}$$

$$\text{moment arm} = d - \frac{a}{2} = 25 - 0.79 = 24.21 \text{ in.}$$

$$M_n = 314(24.21)\tfrac{1}{12} = 633 \text{ ft-kips} \quad \blacksquare$$

■ **EXAMPLE 9.4.2** Determine the nominal moment strength M_n of the isolated T-section shown in Fig. 9.4.3 when $A_s = 12.48$ sq in. (8-#11). Use $f'_c = 3000$ psi and $f_y = 50,000$ psi.

Figure 9.4.3 T-section for Examples 9.4.2 and 9.5.1.

Solution: **(a)** Check requirements for isolated T-sections. According to ACI-8.10.4, b_E cannot exceed $4b_w = 56$ in., and t must be at least $b_w/2 = 7$ in. Thus, flange thickness is satisfactory and $b_E = 30$ in. (the actual width b is less than $4b_w$).

(b) Find A_s so that $a = t$.

$$C = 0.85f'_c b_E a = 0.85(3)(30)(7) = 535 \text{ kips}$$

$$A_s = \frac{T \text{ or } C}{f_y} = \frac{535}{50} = 10.70 \text{ sq in.}$$

Thus, actual a exceeds t when $A_s = 12.48$ sq in.

(c) Treat by the two-couple method (Fig. 9.4.3).

$$T = A_s f_y = 12.48(50) = 624 \text{ kips}$$

$$C = C_1 + C_2 = 0.85f'_c A_1 + 0.85f'_c A_2$$

$$624 = 2.55(14a) + 2.55(16)(7)$$

$$a = 9.48 \text{ in.}$$

$$C_1 = 2.55(14)(9.48) = 338 \text{ kips}$$

$$C_2 = 2.55(16)(7) = 286 \text{ kips}$$

$$M_n = C_1[36 - 0.5(9.48)]\tfrac{1}{12} + C_2(36 - 3.50)\tfrac{1}{12}$$
$$= 882 + 774 = 1656 \text{ ft-kips} \quad \blacksquare$$

9.5 MAXIMUM TENSION REINFORCEMENT PERMITTED IN T-SECTIONS

As discussed in Section 3.13, the literal wording of ACI-10.3.3 will not be identical to a maximum $x = 0.75x_b$ for non-rectangular sections. The authors are convinced that the precise code wording was used for practical purposes and that different ductility requirements for different shapes of cross section were not intended. For practical purposes, the maximum tension reinforcement permitted for T-sections would rarely if ever be used, so that the argument as to what limit is specified or what is intended is purely academic.

For T-sections, placing the limit based on maximum $x = 0.75x_b$ is not as conservative as the literal use of maximum $\rho = 0.75\rho_b$. Thus, unless a higher maximum is truly appropriate for the T-section, the maximum tension reinforcement should be limited to $\rho = 0.75\rho_b$.

■ **EXAMPLE 9.5.1** For the T-section of Fig. 9.4.3, determine the maximum reinforcement ratio ρ and area A_s permitted by the strength method of the ACI Code. Use $f'_c = 3000$ psi and $f_y = 50,000$ psi.

Solution: **(a)** Determine the neutral axis distance x_b and the depth a_b of the rectangular stress distribution in the balanced strain condition. Referring to the strain diagram of Fig. 3.5.1,

$$x_b = \left(\frac{0.003}{0.003 + f_y/E_s}\right)d = \left(\frac{0.003}{0.003 + 0.00172}\right)d$$
$$= 0.635d = 0.635(36) = 22.9 \text{ in.}$$
$$a_b = \beta_1 x_b = 0.85(22.9) = 19.4 \text{ in.}$$

(b) Determine the amount of steel A_{sb} in the balanced strain condition. Since $a_b > t$, the two-couple approach may be used, using C_1 and C_2 acting on areas A_1 and A_2, respectively, similar to what is shown in Fig. 9.4.3, except that $x = x_b$ and $a = a_b$.

$$C_1 = C_{1b} = 0.85f'_c b_w a_b = 0.85(3)(14)(19.4) = 693 \text{ kips}$$

$$A_{s1b} = \frac{693}{50} = 13.9 \text{ sq in.}$$

$$C_2 = 0.85f'_c(b_E - b_w)t = 0.85(3)(16)(7) = 286 \text{ kips}$$

$$A_{s2} = \frac{286}{50} = 5.72 \text{ sq in.}$$

Note that $C_2 = 286$ kips even if $x < x_b$, as long as $a \geq t$.
    The balanced amount of steel is

$$A_{sb} = A_{s1b} + A_{s2} = 13.9 + 5.72 = 19.6 \text{ sq in.}$$

(c) Determine the maximum reinforcement permitted by ACI-10.3.3.

$$\max A_s = 0.75(19.6) = 14.7 \text{ sq in.}$$

$$\max \rho = \frac{A_s}{b_E d} = \frac{14.7}{30(36)} = 0.0136$$

(d) Compare the ductility requirement for a T-section with that for a singly reinforced beam as discussed in Section 3.5. Locate the neutral axis when the beam contains the maximum steel permitted by the ACI Code. Assume $a > t$,

$$C_1 = 0.85f'_c b_w a = 0.85(3)(14)a = 35.7a$$
$$C_2 = 0.85f'_c(b_E - b_w)t = 0.85(3)(16)7 = 286 \text{ kips}$$
$$T = A_s f_y = 14.7(50) = 735 \text{ kips}$$
$$C = T$$

$$a = \frac{735 - 286}{35.7} = 12.6 \text{ in.} > t, \quad \text{as assumed}$$

$$x = \frac{a}{\beta_1} = \frac{12.6}{0.85} = 14.8 \text{ in.}$$

$$\frac{x}{x_b} = \frac{14.8}{22.9} = 0.65$$

The neutral axis for this T-section actually is restricted to 0.65 of x_b, in effect requiring more ductility for the T-section than for singly or doubly reinforced rectangular beams.

Again, the reader is reminded that the beam of Fig. 9.4.2 is far more typical of T-sections than the beam of Fig. 9.4.3. Even for the beam of Fig. 9.4.3 it would be unlikely that more than the 8-#11 shown would be used. The maximum $A_s = 14.7$ sq in. would require three layers of steel. For the section of Fig. 9.4.2, the maximum $A_s = 0.75A_{sb}$ is 18.35 sq in., a very unlikely amount to be contemplated for use in that section. ■

9.6 DESIGN OF T-SECTIONS IN BENDING—STRENGTH METHOD

The design of T-shaped isolated beams involves the dimensions of the flange and web and the area of tension steel, or a total of five unknowns—one more if compression steel is used. Thus there are many possible solutions to the problem. Because of the large compression area in the flange, these T-sections, when used, are usually deep and wide enough so that the tension steel may be placed within the width of the web in not more than two or three layers.

The more common T-sections are those in the region of positive bending moment in continuous monolithic slab–beam–girder systems. The size of the available flange is therefore definitely known after the slab has been designed, and only the web size needs to be designed. The selection of web size for such beams is treated in Chapter 10.

Once the overall dimensions have been established, the design problem is to determine the amount of positive moment reinforcement. Thus it is important first to ascertain the location of the neutral axis associated with the positive moment. This may be done by first computing the moment strength at which the depth a of the rectangular compressive stress block is equal to the flange thickness t.

■ **EXAMPLE 9.6.1**   Determine the amount of tension steel required in the T-section of Fig. 9.4.3 to take a dead load moment of 370 ft-kips and a live load moment of 520 ft-kips, using $f'_c = 3000$ psi, $f_y = 50,000$ psi, and the strength method of the ACI Code.

Solution: (a) Determine the factored moment M_u.

$$M_u = 1.4(370) + 1.7(520) = 1402 \text{ ft-kips}$$

$$\text{required } M_n = \frac{M_u}{\phi} = \frac{1402}{0.90} = 1560 \text{ ft-kips}$$

(b) Determine if the depth a of the rectangular stress distribution will be greater than $t = 7$ in. For $a = t$,

$$C = 0.85f'_c b_E t = 0.85(3)(30)(7) = 536 \text{ kips}$$

$$M_n = C\left(d - \frac{t}{2}\right) = 536(36 - 3.50)\tfrac{1}{12} = 1450 \text{ ft-kips}$$

Since the required M_n exceeds 1450 ft-kips the actual a must exceed t.

(c) Use the two-couple method (Fig. 9.4.3) to obtain A_s.

$$M_n = 0.85f'_c A_1 \left(d - \frac{a}{2} \right) + 0.85f'_c A_2 \left(d - \frac{t}{2} \right)$$

$$1560(12) = 2.55(14a)\left(36 - \frac{a}{2} \right) + 2.55(16)(7)(36 - 3.50)$$

$$a^2 - 72a = -527$$

$$a = 8.3 \text{ in.}$$

$$C_1 = 0.85f'_c b_w a = 0.85(3)(14)(8.3) = 296 \text{ kips}$$

$$A_{s1} = \frac{T_1}{f_y} = \frac{296}{50} = 5.92 \text{ sq in.}$$

$$C_2 = 0.85f'_c (b_E - b_w)t = 0.85(3)(16)(7) = 286 \text{ kips}$$

$$A_{s2} = \frac{T_2}{f_y} = \frac{286}{50} = 5.72 \text{ sq in.}$$

$$A_s = 5.92 + 5.72 = 11.64 \text{ sq in.}$$

Since the 11.64 sq in. required is less than the maximum permitted amount ($0.75A_{sb} = 14.7$ sq in. from Example 9.5.1), the design would be acceptable. The only practical selection of steel is the 8-#11 originally used in the analysis for Example 9.4.2. ∎

■ **EXAMPLE 9.6.2** Design the steel reinforcement for the section of Fig. 9.4.2 to carry a factored moment M_u of 735 ft-kips. Use $f'_c = 3000$ psi and $f_y = 50,000$ psi.

Solution: **(a)** Determine whether or not the depth a of the rectangular stress distribution will be greater than t. For $a = t$,

$$C = 0.85f'_c b_E t = 0.85(3)(78)(4.5) = 895 \text{ kips}$$

$$M_n = C\left(d - \frac{t}{2} \right) = 895\left(25 - \frac{4.5}{2} \right)\frac{1}{12} = 1597 \text{ ft-kips}$$

$$\text{required } M_n = \frac{M_u}{\phi} = \frac{735}{0.90} = 817 \text{ ft-kips}$$

Since the required M_n is less than the amount necessary to cause $a = t$, the section can be designed as a rectangular section.

(b) Design as a rectangular section. The computation of the coefficient of resistance $R_n = M_u/(\phi b_E d^2)$ and the solving of the quadratic equation (or its equivalent as in item 4 of Section 3.8) for ρ may be made exactly as described in Chapter 3. However, because of the wide flange ($b_E = 78$ in.) and the likelihood of a relatively small value for a, a trial procedure may be preferred. First use $0.9d$ as a conservative (i.e., low) trial value for the moment arm $(d - a/2)$; then

$$\text{required } A_s = \frac{\text{required } M_n}{(d - a/2)f_y} = \frac{817(12)}{0.9(25)50} = 8.71 \text{ sq in.}$$

Try 4-#9 and 4-#10, $A_s = 9.08$ sq in.

Check:

$$C = 0.85f'_c b_E a = 0.85(3)(78)a = 198.9a$$

$$T = A_s f_y = 9.08(50) = 454 \text{ kips}$$

$$a = \frac{454}{198.9} = 2.28 \text{ in.}$$

$$\text{arm} = 25 - 0.5(2.28) = 23.86 \text{ in.} \quad (\text{i.e., } 0.95d)$$

A second trial gives

$$\text{required } A_s = \frac{817(12)}{23.86(50)} = 8.22 \text{ sq in.}$$

Revise to 4-#8 and 4-#10, $A_s = 8.24$ sq in. (minimum width = 12.9 in. from Table 3.9.2).

Check:

$$C = 198.9a \quad (\text{as before})$$

$$T = 8.24(50) = 412 \text{ kips}$$

$$a = \frac{412}{198.9} = 2.07 \text{ in.}$$

$$\text{arm} = 25 - 0.5(2.07) = 23.96 \text{ in.} \approx 23.86 \text{ in. used}$$

No further saving can be made.

$$M_n = 412(23.96)\tfrac{1}{12} = 823 \text{ ft-kips}$$

$$[\phi M_n = 0.90(823) = 740 \text{ ft-kips}] > [M_u = 735 \text{ ft-kips}] \qquad \textbf{OK}$$

Use 4-#8 and 4-#10. ■

9.7 INVESTIGATION OF T-SECTIONS IN BENDING—WORKING STRESS METHOD

T-sections in bending must be investigated by the working stress method when service load deflection calculations are needed. It is first necessary to determine whether or not the neutral or effective centroidal axis is below the flange. This may be done by comparing the first moment of the flange area about the base of the flange with that of the equivalent concrete area of steel. In the event that the neutral axis is within the flange, the T-section is to be treated as a rectangular section.

When it is found that the neutral axis is definitely below the flange, the exact transformed section should include the small compression area between the base of flange and the neutral axis. It will be shown in the following example, however, that this area may be ignored in ordinary cases without appreciably affecting the results.

■ **EXAMPLE 9.7.1** Determine the transformed cracked section moment of inertia I_{cr} under service load conditions for the floor beam in Example 9.4.1. Use $f'_c = 3000$ psi, $f_y = 50,000$ psi, and the ACI Code. Solve the problem by first considering and then ignoring the compression area in the web.

Solution: **(a)** Determine whether the neutral axis is above or below the bottom of the slab.

The effective flange width is 78 in. as shown in Fig. 9.4.2. The transformed area of steel is $(9)(6.28) = 56.52$ sq in. Comparing the first moment of the flange area about its base with that of the transformed area of steel (Fig. 9.7.1),

$$[(78)(4.5)(2.25) = 790] < [(56.52)(20.5) = 1159]$$

Therefore the neutral axis is below the flange.

(b) Compute I_{cr} considering the compression area in the web. Taking moments about the unknown actual neutral axis (Fig. 9.7.1),

$$\frac{13x^2}{2} + (65)(4.5)(x - 2.25) = 56.52(25 - x)$$

$$x^2 + 53.70x = 318.63$$

$$x = 5.39 \text{ in.}$$

The transformed cracked section moment of inertia may then be computed about the neutral axis,

$$I_{cr} = \frac{78(4.5)^3}{12} + 78(4.5)(3.14)^2 + \frac{13(0.89)^3}{3} + 56.52(19.61)^2 = 25,800 \text{ in.}^4$$

The serviceability check of deflection will utilize I_{cr} as is illustrated in Chapter 14 on deflections.

(c) Compute I_{cr} ignoring the compression area in the web. Taking moments about the unknown actual neutral axis (Fig. 9.7.1),

$$78(4.5)(x - 2.25) = 56.52(25 - x)$$

$$408x = 2203$$

$$x = 5.40 \text{ in.}$$

Figure 9.7.1 T-section for Example 9.7.1, including web contribution.

It is apparent that the neutral axis location remains essentially the same as when the compression area in the web was included. Computing the moment of inertia I_{cr}, also neglecting the compression area in the web, gives

$$I_{cr} = \frac{78(4.5)^3}{12} + 78(4.5)(3.15)^2 + 56.52(19.60)^2 = 25,800 \text{ in.}^4$$

For deflection computation purposes, $I_{cr} = 25,800$ in.^{4} is all that can be justified (three significant figures at most).

The service load moment capacity M_w according to the working stress method may be obtained by using either (a) the internal-force method or (b) the transformed section method with the flexure formula, the general principles of which have been treated in Chapter 4.

The applicable stress distribution is determined when the allowable stress is reached at one of the extreme fibers. Usually in T-sections the steel controls (i.e., meaning the concrete is underutilized) because of the abundance of concrete compression area. ∎

SELECTED REFERENCES

9.1.   S. Timoshenko and J. N. Goodier. *Theory of Elasticity* (2d ed.). New York: McGraw-Hill, 1951 (pp. 171–177).

9.2.   Karl Girkmann. *Flachentragwerke* (3d ed.). Vienna: Springer-Verlag, 1954 (pp. 116–123).

9.3.   Franco Leve. "Work of the European Concrete Committee," *ACI Journal, Proceedings,* **57,** March 1961, 1049–1054.

9.4.   Gottfried Brendel. "Strength of the Compression Slab of T-Beams Subject to Simple Bending," *ACI Journal, Proceedings,* **61,** January 1964, 57–76.

PROBLEMS

All problems* are to be done according to the ACI Code and all loads given are service loads unless otherwise indicated.

9.1   (a) Determine the nominal moment strength M_n for the beam cross section shown. (b) Determine the maximum tension reinforcement A_s permitted for this beam by ACI-10.3.3. What is the maximum x permitted, given in terms

— — — —

* Problems may be solved as problems stated in Inch-Pound units, or as problems in SI units using the quantities in parenthesis at the end of the statement. The SI values are approximate conversions to avoid implying higher precision for given information in metric units than for Inch-Pound units.

of a proportion of x_b? The beam span is 30 ft. Use $f'_c = 4000$ psi and $f_y = 60,000$ psi. (Span = 9.1 m; $f'_c = 30$ MPa; $f_y = 400$ MPa; slab $t = 125$ mm; beam $h = 900$ mm; $b_w = 380$ mm.)

Inch–Pound
$A_s = 9.08$ sq in.
(4 – # 9, 4 – # 10, two layers)
Span = 30 ft

SI
4 – # 30M
4 – # 35M

Problem 9.1

9.2 (a) Design the reinforcement for the beam shown, according to the strength method, if the dead load moment is 65 ft-kips and the live load moment is 100 ft-kips. (b) Determine the maximum tension reinforcement A_s permitted for this beam by ACI-10.3.3. What is the maximum x permitted, given in terms of a proportion of x_b? Use $f'_c = 3000$ psi and $f_y = 40,000$ psi. ($M_D = 88$ kN·m; $M_L = 135$ kN·m; $f'_c = 20$ MPa; $f_y = 300$ MPa.)

Problem 9.2

9.3 Make a partial design of a simply supported T-beam floor system to meet the conditions shown in the accompanying figure. Use $f'_c = 3000$ psi, $f_y = 40,000$ psi, and the strength method. ($f'_c = 20$ MPa; $f_y = 300$ MPa.)

$5\frac{1}{2}''$ slab (140 mm)   LL = 250 psf (12.0 kN/m²)

Beam span = 24 ft (7.3 m)

10′–0″ (typical) (3.0 m)

Problem 9.3

(a) Select the size of stem. Assume that a stem area consistent with a nominal shear stress $V_u/(\phi b_w d)$ equal to approximately $6\sqrt{f'_c}$ psi ($\sqrt{f'_c}/2$ with f'_c in MPa) will result in an economical section.

(b) Determine the longitudinal reinforcement, and determine bar lengths and cutoff or bends, if any.

(c) Determine the minimum spacing and size of U stirrups required, and comment on the suggested method of stem size selection given in part (a).

9.4 For the beam of Problem 9.1, use the transformed section method to determine the transformed cracked section moment of inertia I_{cr} (a) neglecting any compression in the web below the slab, and (b) including any compression in the web below the slab.

9.5 For the beam of Problem 9.1, use the transformed section method to compute the allowable service load moment M_w, using the principles of Chapter 4. Compare the result for M_w by using the internal force method, (a) neglecting any compression in the web below the slab, and (b) considering any compression in the web below the slab.

CHAPTER 10

Continuous Slab–Beam–Girder and Concrete Joist Floor Systems

10.1 INTRODUCTION

The design of rectangular and T-sections has been treated in Chapters 3, 4, and 9; shear strength along with stirrup design in Chapter 5; development of reinforcement in Chapter 6; and continuity analysis in Chapter 7. The slab–beam–girder type of floor construction has been described generally and the design of one-way slabs illustrated in Chapter 8. In this chapter, complete designs of a typical floor beam and girder in a monolithic slab–beam–girder system are shown. Primarily, what has been developed in the preceding chapters is applied. Thus the reader may consider the material of this chapter as an integrated review of the subjects in the aforementioned chapters.

Also included in this chapter is the design of one-way concrete joist floors. On continuous spans, the concrete joists should be regarded as having rectangular sections near the supports and T-sections in the positive moment region within the span.

It should be noted that in recent years a marked increase has occurred in the use of precast slabs, either conventionally reinforced or prestressed, placed on a monolithic beam and girder framing system. In such systems the slab rests on, but does not act with, the beams and girders, so T-sections are not involved. The procedures discussed in this chapter are also generally applicable to this simpler system of rectangular beams.

Reinforced concrete building design may be a complicated venture, involving irregular floor plans and intricate structural framing. The simple floor framing plan used in the examples does not necessarily reflect a typical practical situation, but it does serve to illustrate the basic essentials of design.

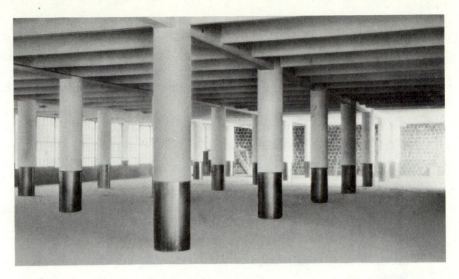

Monolithic slab–beam–girder system; Liggett and Myers Tobacco Company, Richmond, VA. (Courtesy of Portland Cement Association.)

10.2 SIZE OF BEAM WEB

The size of the beam web for continuous T-shaped sections is usually controlled by the flexural and shear strength requirements at the exterior face of the first interior support. For typical conditions of equal spans and uniform dead and live loads, a negative moment of $\frac{1}{10}wL_n^2$ and a shear of $0.60wL_n$ (see Fig. 7.5.1) or $0.575wL_n$ (see ACI-8.3) may be used in estimating the size of the beam. In the following discussion are given the detailed considerations involved in selecting the beam cross section based on the critical bending moment and shear.

Negative Moment Requirement. The section of the beam resisting the negative bending moment at the face of the support is a rectangular section, even in T-section construction, because compression is at the lower part of the section. The tension reinforcement is provided (a) usually by straight bars extending across the top of the beam as required for negative moment; (b) sometimes by bent bars from both adjacent spans; or (c) sometimes by a combination of bent bars and additional straight bars across the top. The development of positive moment reinforcement (ACI-12.11.1) requires some straight bars to continue along the bottom of the beam into the support; thus some portion of them could be utilized as compression reinforcement. Although compression reinforcement will rarely be required for strength, it is frequently desired for added ductility and for deflection control.

The guideline reinforcement ratio ρ to be used for deflection control of singly reinforced rectangular beams has been suggested (in Chapter 3) to be about $0.375\rho_b$ (one-half of the maximum permissible value). For the negative moment requirement on continuous T-sections, a higher value for ρ should be acceptable, because the gross moment of inertia of a typically proportioned T-section is roughly twice that of its rectangular portion. Thus it may be reasonable to design the negative moment region of a T-shaped section using twice the percentage of reinforcement that would be used for reasonable deflection control on a completely rectangular beam, that is, the full $0.75\rho_b$ permitted by the ACI Code (or more if compression steel is utilized).

Positive Moment Requirement.   In the positive moment region the flange of the T-section is in compression. Since the effective flange width is large, it is rare that the depth a of the Whitney rectangular stress distribution will extend below the bottom of the flange. On the tension side, the amount of steel required will be inversely proportional to the depth of the section.

For sizing the beam web, the only consideration utilizing the positive moment is to establish the width required to maintain adequate clearances for a given number of bars. This should be examined because a somewhat deeper or shallower beam might still permit the required steel to fit into one or two layers, as the case may be.

Shear Strength Requirement.   The designer may wish to establish the beam web size to achieve a certain maximum nominal shear stress. This may be desirable for economical stirrup size and spacings. The ranges of five categories of design for shear reinforcement are given in Section 5.11. The upper limit of category 4 ($V_s = 4\sqrt{f'_c}\,b_w d$) serves as a practical guideline for ordinary design, typically permitting stirrup spacing from 3 in. to a maximum of $d/2$.

When V_c is taken as $2\sqrt{f'_c}\,b_w d$ according to the simplified procedure, the total V_n at the upper limit in category 4 is

$$V_n = V_c + V_s$$
$$= 2\sqrt{f'_c}\,b_w d + 4\sqrt{f'_c}\,b_w d = 6\sqrt{f'_c}\,b_w d \tag{10.2.1}$$

or, for a given factored shear V_u, the $b_w d$ required to satisfy this guideline shear strength is

$$\text{required } b_w d = \frac{\text{required } V_n}{6\sqrt{f'_c}} = \frac{V_u}{\phi(6\sqrt{f'_c})} \tag{10.2.2}$$

Except for unusual conditions involving short spans, heavy concentrated loads, heavily doubly reinforced sections, or combinations of these, the maximum nominal shear stress $V_u/(\phi b_w d)$ should be about $6\sqrt{f'_c}$ for reasonable stirrup size and spacing.

■ **EXAMPLE 10.2.1**   Using the ACI strength method of design, establish the preliminary size for the floor beams 2B1-2B2-2B1 supported by girders as shown in the floor framing plan of Fig. 8.3.1. Use information previously described in Chapter 8.

Solution: (a) Negative moment requirement. For $f'_c = 3000$ psi and $f_y = 40,000$ psi, one may determine the maximum reinforcement ratio ($0.75\rho_b$) allowed by the ACI Code using basic principles, or from Table 3.6.1,

$$\text{max } \rho = 0.75\rho_b = 0.0278$$

Estimating the weight of the stem (portion of the web below the slab) at 0.3 kip/ft (2 sq ft of area), and applying the overload factors,

$$w_D = 1.4[0.056(13) + 0.3] = 1.44 \text{ kips/ft}$$

$$w_L = 1.7(0.100)(13) = 2.21 \text{ kips/ft}$$

Using basic principles, or Eq. (3.8.4) as follows, the R_n corresponding to $\rho = 0.0278$ is

$$n = \frac{f_y}{0.85f'_c} = 15.7$$

$$R_n = \rho f_y (1 - \tfrac{1}{2}\rho m) = 870 \text{ psi}$$

Assume width of supporting girders to be 18 in., in which case the clear span L_n is

$$L_n = 26 - 1.5 = 24.5 \text{ ft}$$

Using the moment coefficient from ACI-8.3,

$$\text{max } M_u = \tfrac{1}{10}(1.44 + 2.21)(24.5)^2 = 219 \text{ ft-kips}$$

$$\text{required } bd^2 = \frac{M_u}{\phi R_n} = \frac{219(12,000)}{0.90(870)} = 3360 \text{ in}^3.$$

If $b = 13$ in.,

$$\text{required } d = \sqrt{\frac{3360}{13}} = 16.1 \text{ in.}$$

The minimum effective size permitted is 13×16.1, for which the steel area required would be

$$A_s = 0.0278(13)(16.1) = 5.82 \text{ sq in.}$$

In the negative moment region the flange is available so that the steel does not have to fit within the web width.

Before making a decision, the shear requirement and the steel requirement for positive moment should be examined.

(b) Shear requirement. Using the shear coefficient from ACI-8.3,

$$\text{max } V_u = 1.15\left(\frac{wL_n}{2}\right) = 1.15\left(\frac{3.65}{2}\right)(24.5) = 51.4 \text{ kips}$$

If it is desired that the nominal shear stress $v_n = V_u/(\phi b_w d)$ does not exceed $6\sqrt{f'_c} = 329$ psi,

$$\text{required } b_w d = \frac{51,400}{0.85(329)} = 184 \text{ sq in.}$$

For $b = 13$ in., then $d = 14.1$ in. (Actually the maximum shear may be taken at a distance d from the face of the support.)

(c) Positive moment requirement. The effective flange width b_E is the smallest of the following:

1. $b_w + 16t = b_w + 16(4.5) = b_w + 72$

2. $\dfrac{L}{4} = 26\left(\dfrac{12}{4}\right) = 78$ in.   **(Controls)**

3. center–to–center spacing $= 13(12) = 156$ in.

Try $d = 20$ in.; estimate moment arm at 19 in.

$$M_u = \frac{1}{14}\,wL_n^2 = \frac{3.65}{14}\,(24.5)^2 = 156 \text{ ft-kips}$$

$$\text{required } A_s = \frac{M_u}{\phi f_y \,(\text{arm})} \approx \frac{156(12)}{0.90(40)(\approx 19)} = 2.74 \text{ sq in.}$$

This amount of steel should easily fit into a 13-in. wide beam in one layer.

(d) Minimum depth. According to ACI-Table 9.5a the depth cannot be less than

$$\min h = \frac{L}{18.5}\,(0.8) = \frac{26(12)(0.8)}{18.5} = 13.5 \text{ in.}$$

The negative moment region gives the most severe requirement. The designer must also keep in mind that if excessive deflection may cause damage, the deflection must be computed and satisfy ACI-Table 9.5b. The above check is only a minimum requirement where deflection is *not* likely to cause damage to nonstructural construction.

(e) Choice of size. Frequently the size is chosen larger than the requirements would dictate. Sometimes this occurs because of the designer's desire that a large number of beams have the same external dimensions for economy of forming or perhaps because of duct-work or pipes that are to pass through the beams, necessitating larger sizes.

   Use $b_w = 13$ in., $h = 22.5$ in. (which gives $d \approx 20$ in.). It is common to make the stem portion below the flange a whole inch increment, such as 18 in. for this case. The arbitrary size selected is larger than necessary; any size at least equal to that indicated by steps **(a)** through **(d)** would serve to illustrate the design procedure. ∎

10.3 CONTINUOUS FRAME ANALYSIS FOR BEAMS

The shear and moment diagrams to be used in the design of the floor beams in the preceding example could have followed the ACI moment coefficients. However, since these coefficients are more suitable for use in frames involving, say, more than four continuous spans, and as a matter of illustration, an elastic analysis will be used to obtain the bending moment and shear envelopes for beams 2B1-2B2-2B1.

(a) Actual continuous beam supported by girders

(b) Frame model to approximate floor beams
supported on girders

Figure 10.3.1 Actual continuous beam and analytical model.

Some designers would assume the girders to provide only vertical support to those floor beams supported on girders; thus a pure continuous beam analysis is made. The girders, however, have torsional stiffness that acts to restrain the rotation of the beam over these girder supports. Prior to 1971, neglect of torsional stiffness was permitted by the ACI Code where such stiffness did not exceed 20% of the flexural stiffness at the joint. Since 1971 the ACI Code requires that torsion be considered whenever the nominal torsional stress exceeds $1.5\sqrt{f'_c}$ psi ($0.12\sqrt{f'_c}$ for f'_c in MPa). In the case with relatively large and stiff girders, the stiffness of the beam may reasonably be taken as twice the torsional stiffness provided by the girder, or if one thinks in terms of an equivalent beam and column frame system, it is as shown in Fig. 10.3.1(b). More details with regard to torsion are to be found in Chapter 19.

For the beams that frame into columns, however, the combination of the bending stiffness of the columns plus the torsional stiffness of the girders means that the relative end restraint is greater than in the aforementioned case. To design such beams properly, the sizes of the members must be estimated and the $(\Sigma K_{col}/\Sigma K_{bm})$ ratio computed therefrom.

With regard to a T-section, there is considerable difference of opinion as to how its stiffness in a continuous beam should be computed. A T-section certainly provides more stiffness in the positive moment region where the flange is in compression than it does as a rectangular section in the negative moment region. The ACI Code prescribes no specific method but only that "... any set of reasonable assumptions is permitted for computing relative flexural and torsional stiffnesses of columns, walls, floors, and roof systems" (ACI-8.6.1). It has been common practice to use the gross moment of inertia, neglecting reinforcement, in computing the flexural stiffness of such elements. For T-sections in the positive moment region, the gross section usually includes the effective flange width.

One of the accepted methods [10.1] has been to use a T-section moment of inertia equal to $2(\frac{1}{12}b_w h^3)$, which is equivalent to using an effective flange width of about 6 times the web width. Since the true stiffness is that of a span with variable moment of inertia along its length with the T-section stiffness over, say, the middle one-half of the span and the rectangular section stiffness over the end quarters, the equivalent system is approximately obtained by using the T-section with a flange width only twice the web width over the entire span. Examination of analyses using various stiffness ratios will show that a fairly wide variation in stiffness ratio may be accommodated with relatively small changes in bending moments.

In the following example, only the analysis of the floor beams supported on girders is shown, using the analytical frame model in Fig. 10.3.1(b). An overestimate of the beam stiffness will give a conservative design for the beam; but such assumption should be reevaluated when designing the columns, if any.

■ **EXAMPLE 10.3.1** Using an elastic statically indeterminate analysis, obtain the shear and moment diagrams to be used in compiling the moment and shear envelopes for the design of 2B1 and 2B2 in Example 10.2.1.

Solution:

$$\text{stem weight} = \frac{13(18)}{144}(0.15) = 0.244 \text{ kip/ft}$$

$$w_D = 1.4[0.056(13) + 0.244] = 1.36 \text{ kips/ft}$$

$$w_L = 1.7(0.100)(13) = 2.21 \text{ kips/ft}$$

$$\text{FEM due to } w_D = \tfrac{1}{12}(1.36)(26)^2 = 76.6 \text{ ft-kips}$$

$$\text{FEM due to } w_L = \tfrac{1}{12}(2.21)(26)^2 = 124.5 \text{ ft-kips}$$

Loading conditions 1 through 5 as shown in Fig. 10.3.2 are established by visualizing the influence lines for positive moment in the midspan region of each span and for negative moment at each support.

The structural analysis may be performed by any method of statically indeterminate analysis, most often the matrix displacement method.* In fact, using three cycles of moment distribution by longhand is quite adequate for the present problem. Since the specific structural analysis technique is outside the scope of this text, only the end moments resulting from the analysis for each of the five loading cases are given in Table 10.3.1.

The primary shear and moment diagrams for maximum positive and negative moments for beams 2B1 and 2B2 are given individually in Figs. 10.3.3 and 10.3.4. The secondary shear and moment diagrams are shown in Figs. 10.3.5 and 10.3.6. The primary and secondary moment diagrams are drawn separately for clarity in the example. Ordinarily they are not drawn separately but instead superimposed to become the moment envelope as shown in Fig. 10.6.1. ■

---

* See, for example, Chu-Kia Wang and Charles G. Salmon. *Introductory Structural Analysis.* Englewood Cliffs, N.J.: Prentice-Hall, 1984; or Chu-Kia Wang. *Structural Analysis on Microcomputers.* New York: Macmillan, 1986.

Figure 10.3.2 Loading conditions.

Table 10.3.1 Final End Moments[a] (ft-kips) for the Five Loading Conditions of Fig. 10.3.2

JOINT	A	B		C		D	
MEMBER	AB	BA	BC	CB	CD	DC	
DF[b]	0.667	0.400	0.400	0.400	0.400	0.667	
Case 1, M =	−27.2	+90.0	−83.7	+83.7	−90.0	+27.2	Dead load only
Case 2, M =	−54.1	+89.4	−46.8	+46.8	−89.4	+54.1	Live load spans 1 and 3
Case 3, M =	+10.0	+56.8	−89.3	+89.3	−56.8	−10.0	Live load span 2
Case 4, M =	−42.5	+159.3	−156.1	+69.4	−43.5	−8.3	Live load spans 1 and 2
Case 5, M =	−1.7	−13.3	+19.9	+66.7	−102.6	+52.5	Live load span 3

[a] Clockwise moments acting on member ends are taken positive.

[b] DF = distribution factor.

(a) Maximum in positive region
(loading conditions 1 + 2)

(b) Maximum in negative region at *B*
(loading conditions 1 + 4)

Figure 10.3.3 Primary shear and moment diagrams for 2*B*1.

(a) Maximum in positive region
(loading conditions 1 + 3)

(b) Maximum in negative region at *B*
(loading conditions 1 + 4)

Figure 10.3.4 Primary shear and moment diagrams for 2*B*2.

(a) Minimum in positive region
(loading conditions 1 + 3)

(b) Minimum in negative region at B
(loading conditions 1 + 5)

Figure 10.3.5 Secondary shear and moment diagrams for 2*B*1.

(a) Minimum in positive region
(loading conditions 1 + 2)

(b) Minimum in negative region at B
(loading conditions 1 + 5)

Figure 10.3.6 Secondary shear and moment diagrams for 2*B*2.

10.4 CHOICE OF LONGITUDINAL REINFORCEMENT IN BEAMS

It will not be overemphasis to repeat here that the choice of longitudinal reinforcement depends on both the steel area and development length (or anchorage) requirements. The design of the main reinforcement in floor beams 2B1-2B2-2B1 is shown in the following example.

■ **EXAMPLE 10.4.1** Choose the arrangement of main reinforcement in the floor beams 2B1-2B2-2B1 of Example 10.2.1.

Solution: (a) Flexural requirements. The critical moment is corrected to the face of support by subtracting $\Delta M = Vb/3$ from the moment at the center of support, as discussed in Section 8.2. The shear V for this calculation is taken as that at the face of support. For sections A-A, C-C, and D-D (Fig. 10.4.1),

Section A-A (rectangular section):

$$M_u \text{ at face} = 81.3 - \frac{40.0(1.5)}{3} = 81.3 - 20.0 = 61.3 \text{ ft-kips}$$

$$\text{required } R_n = \frac{M_u}{\phi bd^2} = \frac{61.3(12,000)}{0.90(13)(20)^2} = 157 \text{ psi}$$

Using Eq. (3.8.5), or Fig. 3.8.1,

$$\text{required } \rho = \frac{1}{m}\left(1 - \sqrt{1 - \frac{2mR_n}{f_y}}\right)$$

$$m = \frac{f_y}{0.85f'_c} = \frac{40}{0.85(3)} = 15.7$$

$$\text{required } \rho = \frac{1}{15.7}\left(1 - \sqrt{1 - \frac{2(15.7)(157)}{40,000}}\right) = 0.004$$

$$\min \rho = \frac{200}{f_y} = \frac{200}{40,000} = 0.005 < \left[\frac{4}{3}(\text{required } \rho) = 0.0053\right]$$

Use $\rho = 0.005$.

$$\text{required } A_s = 0.005(13)(20) = 1.30 \text{ sq in.}$$

Section C-C (rectangular section):

$$M_u \text{ at face} = 249.3 - \frac{50.7(1.5)}{3} = 249.3 - 25.3 = 224 \text{ ft-kips}$$

$$\text{required } R_n = \frac{224(12,000)}{0.90(13)(20)^2} = 574 \text{ psi}$$

Figure 10.4.1 Longitudinal reinforcement areas required (sq in.) and bars selected for beams 2B1 and 2B2.

From Fig. 3.8.1, or Eq. (3.8.5),

$$\text{required } \rho = 0.0165 > \text{min } \rho \qquad \textbf{OK}$$

$$\text{required } A_s = 0.0165(13)(20) = 4.30 \text{ sq in.}$$

Section *D-D* (rectangular section):

$$M_u \text{ at face} = 239.8 - \frac{47.1(1.5)}{3} = 239.8 - 23.5 = 216 \text{ ft-kips}$$

$$\text{required } R_n = \frac{216(12,000)}{0.90(13)(20)^2} = 554 \text{ psi}$$

$$\text{required } A_s \approx 4.30\left(\frac{554}{574}\right) = 4.15 \text{ sq in.}$$

For sections *B-B* and *E-E* (Fig. 10.4.1),

$$\text{effective flange width } b_E = \frac{26(12)}{4} \quad \text{or} \quad 13 + 16(4.5) \quad \text{or} \quad 13(12)$$

$$b_E = 78 \text{ in.}$$

Section *B-B* (T-section):

$$\text{estimate moment arm} = 0.9d = 18 \text{ in.}$$

$$\text{required } A_s = \frac{M_u}{\phi f_y(\text{arm})} = \frac{173.3(12)}{0.90(40)(\approx 18)} = 3.21 \text{ sq in.}$$

Check:

$$C = 0.85f'_c b_E a = 0.85(3)(78)a = 199a$$

$$T = 3.23(40) = 129 \text{ kips}$$

$$a = \frac{129}{199} = 0.65 \text{ in.}$$

$$\text{arm} = 20 - \frac{0.65}{2} = 19.7 \text{ in.}$$

$$\text{revised required } A_s = \frac{173.3(12)}{0.90(40)19.7} = 2.93 \text{ sq in.}$$

Section *E-E* (T-section):

$$\text{required } A_s = \frac{128.7(12)}{0.90(40)(\approx 19.7)} = 2.18 \text{ sq in.}$$

$$\left(\rho_w = \frac{A_s}{b_w d} = \frac{2.18}{13(20)} = 0.0084 \right) > \left(\min \rho_w = \frac{200}{f_y} = 0.005 \right) \qquad \text{OK}$$

Check:

$$C = 199a$$

$$T = 2.20(40) = 88 \text{ kips}$$

$$a = \frac{88}{199} = 0.44 \text{ in.}$$

$$\text{arm} = 20 - \frac{0.44}{2} = 19.8 \text{ in.} \approx 19.7 \text{ in.} \qquad \text{OK}$$

On the basis of the areas required at sections *A-A*, *B-B*, *C-C*, *D-D*, and *E-E*, the arrangement of main reinforcement as shown in Fig. 10.4.1 is tentatively chosen. Note that the 2-#7 straight bars in 2B1 and the 2-#6 straight bars in 2B2 furnish the minimum one-fourth of the positive moment reinforcement that must be extended into the support in each span.

In order to confirm the choice of main reinforcement, the development length requirements (ACI-12.11.3) must be checked, and, if f_y were greater than 60,000 psi, the crack control provisions of ACI-10.6. Also the bars in the top of the beam at the face of the exterior support (location 4 in Fig. 10.4.1) should be fully developed.

(b) Basic development length L_{db} and minimum development length Min L_d. These values can be obtained without regard to the actual cover and bar spacing situation in a structure. However, the final development length L_d is a situation-specific value. From Tables 6.7.1 and 6.7.2,

$$L_{db} \text{ (for } \#6) = 12.9 \text{ in.} \qquad \text{Min } L_d \text{ (for } \#6) = 16.4 \text{ in.}$$

$$L_{db} \text{ (for } \#7) = 17.5 \text{ in.} \qquad \text{Min } L_d \text{ (for } \#7) = 19.2 \text{ in.}$$

$$L_{db} \text{ (for } \#9) = 29.2 \text{ in.} \qquad \text{Min } L_d \text{ (for } \#9) = 24.7 \text{ in.}$$

$$L_{db} \text{ (for } \#10) = 37.1 \text{ in.} \qquad \text{Min } L_d \text{ (for } \#10) = 27.8 \text{ in.}$$

(c) Development length for the #7 bars at locations 1 and 2 (Fig. 10.4.1). Here the 2-#7 extend into the support after 2-#9 are bent into the top face of the beam. For development length purposes, the 2-#7 are treated as alone in a 13 in. wide stem enclosed within at least minimum #3 U stirrups. Table 3.9.4 shows that the minimum beam width to provide $3d_b$ spacing is 8.8 in. Thus, the #7 are in Category A (satisfying ACI-12.2.3.1a) and $\alpha_{abc} = 1.0$. Though the #7 bar spacing exceeds $5d_b$, the side cover of 1.5 in. plus the $\frac{3}{8}$ in. stirrup is less than $2.5d_b$ required to qualify under ACI-12.2.3.4; therefore $\alpha_{de} = 1.0$. Then,

$$\left[\alpha_{abc}\alpha_{de}L_{db} = 1.0(1.0)17.5 = 17.5 \text{ in.}\right] < \left[\text{Min } L_d = 19.2 \text{ in.}\right]$$

Thus, the "first-level L_d" is 19.2 in. Finally, α_{top}, $\alpha_{\ell w}$, and α_{epx} are not applicable; therefore,

$$L_d \text{ (for #7)} = 19.2 \text{ in. (1.60 ft)}$$

which exceeds the 12 in. minimum.

At location 1 (Fig. 10.4.1): For 2-#7, extending 6 in. beyond face of support (actual length from inflection point $= 2.09 - 0.25 = 1.84$ ft),

$$C = 0.85f'_c ba = 0.85(3)(78)a = 199a$$

$$T = 2(0.60)40 = 48 \text{ kips}$$

$$a = \frac{48}{199} = 0.24 \text{ in.}$$

$$M_n = 48[20 - 0.5(0.24)]\tfrac{1}{12} = 79.5 \text{ ft-kips}$$

$$V_u = 42.6 - 3.57(2.09) = 35.1 \text{ kips}$$

$$L_a = d = 1.67 \text{ ft} < \text{actual length } 1.84 \text{ ft}$$

$$\frac{M_n}{V_u} + L_a = \frac{79.5}{35.1} + 1.67 = 3.93 \text{ ft} > L_d (\#7) \qquad \textbf{OK}$$

At location 2: For 2-#7, since M_n and V_u are identical to their values at location 1, this check is made by inspection.

(d) Development length for the #6 bars at location 3 (Fig. 10.4.1). The 2-#6 are treated as if alone in a 13 in. wide stem enclosed within at least minimum #3 U stirrups. From part **(c)** above, it is apparent that Category A also applies here; $\alpha_{abc} = 1.0$. In addition, since these are smaller bars and the clear spacing between them is at least $5d_b$, perhaps the side cover is at least $2.5d_b$, satisfying ACI-12.2.3.4. In this case, 1.5 in. cover plus $\frac{3}{8}$ in. stirrup gives 1.875 in.; exactly $2.5d_b$. Thus, $\alpha_{de} = 0.80$, making

$$\left[\alpha_{abc}\alpha_{de}L_{db} = 1.0(0.80)12.9 = 10.3 \text{ in.}\right] < \left[\text{Min } L_d = 16.4 \text{ in.}\right]$$

Thus, the "first-level L_d" is 16.4 in. Finally, α_{top}, $\alpha_{\ell w}$, and α_{epx} are not applicable; therefore,

$$L_d \text{ (for #6)} = 16.4 \text{ in. (1.36 ft)}$$

which exceeds the 12 in. minimum.

At location 3: For 2-#6,

$$C = 199a; \qquad T = 2(0.44)40 = 35.2 \text{ kips}; \qquad a = 0.18 \text{ in.}$$

$$M_n = 35.2[20 - 0.5(0.18)]\tfrac{1}{12} = 58.5 \text{ ft-kips}$$

$$V_u = 46.4 - 3.57(4.51) = 30.3 \text{ kips}$$

$$\frac{M_n}{V_u} + L_a = \frac{58.5}{30.3} + 1.67 = 3.60 \text{ ft} > L_d\,(\#6) \qquad\qquad \textbf{OK}$$

(e) Development length for the #10 bars extending into span 1 (Fig. 10.4.1). The 2-#10 are treated as if alone in a 13 in. wide stem enclosed within at least minimum #3 U stirrups. Table 3.9.4 shows that the minimum beam width to provide $3d_b$ spacing is 10.3 in. Thus, the #10 are in Category A and $\alpha_{abc} = 1.0$. Categories D and E do not apply; thus, $\alpha_{de} = 1.0$, making

$$[\alpha_{abc}\alpha_{de}L_{db} = 1.0(1.0)37.1 = 37.1 \text{ in.}] > [\text{Min } L_d = 27.8 \text{ in.}]$$

Thus, the "first-level L_d" is 37.1 in. Since the #10 are top bars, $\alpha_{top} = 1.3$. The other second-level multipliers $\alpha_{\ell w}$ and α_{epx} are not applicable; therefore,

$$L_d \text{ (for \#10)} = 1.3(37.1) = 48.2 \text{ in. (4.02 ft)}$$

which exceeds the 12 in. minimum.

(f) Development length for the 2-#9 at point 4 (Fig. 10.4.1) extending into the exterior support. The 2-#9 are enclosed within at least minimum #3 U stirrups in a 13 in. wide stem. Table 3.9.4 shows that the minimum beam width to provide $3d_b$ spacing is 9.8 in. Thus, the #9 bars are in Category A and $\alpha_{abc} = 1.0$. Categories D and E do not apply; thus, $\alpha_{de} = 1.0$, making

$$[\alpha_{abc}\alpha_{de}L_{db} = 1.0(1.0)29.2 = 29.2 \text{ in.}] > [\text{Min } L_d = 24.7 \text{ in.}]$$

Thus, the "first-level L_d" is 29.2 in. Since the #9 are top bars, $\alpha_{top} = 1.3$. The other second-level multipliers $\alpha_{\ell w}$ and α_{epx} are not applicable; therefore,

$$L_d \text{ (for \#9)} = 1.3(29.2) = 38.0 \text{ in. (3.16 ft)}$$

which exceeds the 12 in. minimum.

In order to develop properly 2-#9 at location 4, either a straight embedment of 38 in. is required or a standard hook must be used.

For a #9 standard hook, the basic development length L_{hb}, according to Eq. (6.11.1) (ACI-12.5), is

$$L_{hb} = \frac{1200 d_b}{\sqrt{f_c'}} = \frac{1200(1.128)}{\sqrt{3000}} = 24.7 \text{ in.}$$

which is to be modified by $f_y/60,000$. Assume concrete cover of at least $2\tfrac{1}{2}$ in. normal to the plane of the hook, *and* that on the bar extension beyond the 90° hook of at least 2 in. Thus, using modifications 1 and 2 from Table 6.11.1, the development length L_{dh} becomes

$$L_{dh} = 24.7\left(\frac{40,000}{60,000}\right)0.7 = 11.5 \text{ in.}$$

The available L_{dh} (see Fig. 6.11.2) is

$$\text{available } L_{dh} = 18(\text{support width}) - 2(\text{cover on tail}) = 16 \text{ in.}$$

Thus, the 2-#9 bars are assumed to be fully effective at point 4 of Fig. 10.4.1 when 90° hooks are used. Since a girder frames in at each side of beam 2B1 there will automatically be at least $2\frac{1}{2}$-in. side cover on the hooks.

(g) Recheck the development for the combination 2-#7 and 2-#9 that now extend along the bottom of the beam toward the exterior support once the decision is made to use straight hooked #9 at point 4 (Fig. 10.4.1). If all four bars extend into the support, check the minimum width to satisfy Category A. From Table 3.9.4, the minimum width for 2-#7 (i.e., the corner bars) is 8.8 in. Adding for the additional two bars two times $3d_b = 6.8$ in., plus the diameters for the two additional bars $= 2.256$ in. gives a 17.8 in. minimum width. This exceeds the actual width of 13 in.; Category A would not apply. Similarly, using Table 3.9.3, one can check whether the clear spacing exceeds $2d_b$. From that table,

$$\text{minimum width} = 7.9 + 2(2)(1.128) + 2(1.128) = 14.7 \text{ in.}$$

which exceeds the 13 in. provided. Thus, the development length is in Category C, making $\alpha_{abc} = 2.0$. Categories D and E do not apply; thus, $\alpha_{de} = 1.0$, making

$$[\alpha_{abc}\alpha_{de}L_{db} = 2.0(1.0)29.2 = 58.4 \text{ in.}] > [\text{Min } L_d = 24.7 \text{ in.}]$$

Thus, the "first-level L_d" is 58.4 in. Finally, α_{top}, $\alpha_{\ell w}$, and α_{epx} are not applicable; therefore,

$$L_d \text{ (for \#9)} = 58.4 \text{ in. (4.87 ft)}$$

which exceeds the 12 in. minimum. This would be the development length for the #9 when it is not bent into the top face.

   If on the other hand the 2-#9 are cut as soon as permitted, and the 2-#7 are run into the support, the development lengths may be different. Assuming the 2-#7 are fully developed before the 2-#9 are cut off, then the determination of the clear spacing is based on only two bars instead of the four bars as computed above. As shown in **(b)** and **(f)**, Category A will apply. Recognizing that these #9 are not top bars, the development lengths from parts **(b)** and **(f)** are

$$L_d \text{ (for \#7)} = 19.2 \text{ in. (1.60 ft)}$$

$$L_d \text{ (for \#9)} = 29.2 \text{ in. (2.43 ft)}$$

Even though it may be impractical to make such a cut, Fig. 10.6.1 shows the moment capacity diagram for cutting the 2-#9 at 1'-3" from the face of support.

(h) Crack control requirements. The most critical location to check is where the largest bars are put into the smallest area, that is, at section B-B (Fig. 10.4.1). Using Eq. (4.12.2) or ACI-10.6, and referring to Fig. 10.4.2,

$$z = f_s \sqrt[3]{d_c A} \leq 175 \text{ kips/in.} \qquad \text{(interior exposure)}$$

$$d_c = 1.5 \text{ (cover)} + 0.375 \text{ (stirrup)} + 0.4375 \text{ (bar radius)} = 2.31 \text{ in.}$$

Figure 10.4.2 Cross section for crack control investigation.

$$A = \frac{2d_c b_w}{\Sigma A_s / (A_b \text{ for } \#9)} = \frac{2(2.31)13}{(3.20/1.0)} = 14.4 \text{ sq in./bar}$$

$$z = 0.6(40)\sqrt[3]{2.31(14.4)} = 139 \text{ kips/in.} < 175 \qquad \textbf{OK}$$

Actually, because in this example f_y does not exceed 40,000 psi, the crack control investigation is not required. However, the calculation is presented to illustrate the procedure. ■

10.5 SHEAR REINFORCEMENT IN BEAMS

Although the portions of the bent bars in the body of the beam would provide some shear strength in their vicinity, it will generally be a more clean-cut procedure to depend on the vertical stirrups alone to provide the entire shear strength requirement and welcome the bent bars as giving additional assistance. The design of shear reinforcement in 2B1-2B2-2B1 is shown in the following example.

■ **EXAMPLE 10.5.1**   Design the shear reinforcement in the floor beams 2B1-2B2-2B1 of Example 10.2.1.

Solution: The factored shear V_u envelope for which shear reinforcement will be provided is taken from Figs. 10.3.3 and 10.3.4 and shown in Fig. 10.5.1. The maximum factored shear should be taken at a distance equal to the effective depth from the face of support. Let V_1, V_2, and V_3 be the maximum factored shear in regions 1, 2, and 3 as shown in Fig. 10.5.1; then using $w_u = 3.57$ kips/ft,

$$\phi V_c = \phi(2\sqrt{f'_c})\, b_w d = 0.85(2\sqrt{3000})(13)(20)\tfrac{1}{1000} = 24.2 \text{ kips}$$

$$V_1 = 3.57(9.52) = 34.0 \text{ kips}; \qquad x_1 = 114.2\left(\frac{34.0 - 24.2}{34.0}\right) = 32.9 \text{ in.}$$

$$V_2 = 3.57(12.51) = 44.7 \text{ kips}; \qquad x_2 = 150.1\left(\frac{44.7 - 24.2}{44.7}\right) = 68.8 \text{ in.}$$

$$V_3 = 3.57(11.51) = 41.1 \text{ kips}; \qquad x_3 = 138.1\left(\frac{41.1 - 24.2}{41.1}\right) = 56.8 \text{ in.}$$

Figure 10.5.1 Factored shear V_u envelope used for 2B1 and 2B2.

The shaded shear areas in Fig. 10.5.1 plus an additional distance of 40.7 in., equal to the distance to the location where $V_u = \phi V_c/2$, represent the portions of the beam where shear reinforcement is required. These portions are also designated as regions 1, 2, and 3 in Fig. 10.5.2.

Assume #3 vertical U stirrups. From Fig. 10.5.2 the maximum required ϕV_s anywhere on the beam is 20.5 kips. Since the largest required ϕV_s does not exceed

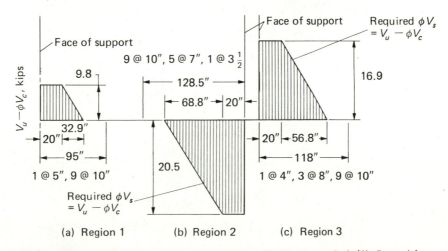

Figure 10.5.2 Portions of factored shear V_u in excess of ϕV_c (that is, required ϕV_s diagram) for beams 2B1 and 2B2.

that based on a nominal stress v_s of $4\sqrt{f'_c}$, the maximum permissible stirrup spacing may not exceed $d/2$.

$$\text{limit } \phi V_s = \phi(4\sqrt{f'_c}\, b_w d) = 2(\phi V_c) = 2(24.2) = 48.4 \text{ kips}$$

$$\text{max required } \phi V_s = 20.5 \text{ kips} < \text{limit } \phi V_s$$

Thus, the upper limit on stirrup spacing is $d/2 = 10$ in.
    For a minimum percentage of shear reinforcement,

$$\text{min } \phi V_s = \phi 50 b_w d = 0.85(50)(13)(20)\tfrac{1}{1000} = 11.1 \text{ kips}$$

The strength requirement to provide for the shear represented by the shaded areas of Fig. 10.5.2 is

$$V_s = \frac{A_v f_y d}{s}$$

$$s = \frac{\phi A_v f_y d}{\phi V_s} = \frac{0.85(0.22)(40)20}{\phi V_s} = \frac{149.6 \text{ kips in.}}{\phi V_s}$$

For region 1, a spacing of $d/2$ governs because the maximum required $\phi V_s = 9.8$ kips is less than the 11.1 kips required for minimum stirrups, and $\phi V_s = 11.1$ kips permits spacing of 13.5 in. which exceeds $d/2$.

<u>Use 1 @ 5 in. and 9 @ 10 in. (95 in. from face of support).</u>
    For region 2 the strength requirement is shown in Table 10.5.1.

Table 10.5.1 Distance x from Face of Support to Location Where ϕV_s Equals Required ϕV_s

s	ϕV_s	MAX REQUIRED ϕV_s $- \phi V_s$	x (DISTANCE FROM FACE OF SUPPORT) (in.)
7.3	20.5	$20.5 - 20.5 = 0$	
8	18.7	$20.5 - 18.7 = 1.8$	$20 + \dfrac{68.8}{20.5}(1.8) = 20 + 6 = 26$
9	16.6	$20.5 - 16.6 = 3.9$	$20 + \dfrac{68.8}{20.5}(3.9) = 20 + 13 = 33$
10	15.0	$20.5 - 15.0 = 5.5$	$20 + \dfrac{68.8}{20.5}(5.5) = 20 + 18 = 38$

<u>Use 1 @ $3\tfrac{1}{2}$ in.; 5 @ 7 in.; and 9 @ 10 in.</u> (128.5 in. from face of support).
    For region 3, to satisfy the strength requirement at the critical section,

$$\text{max } s = \frac{149.6}{16.8} = 8.9 \text{ in.}$$

A spacing of 10 in. can be used at $20 + (16.9 - 15.0)(56.8)/16.9 = 26$ in. from the face of support.

<u>Use 1 @ 4 in.; 3 @ 8 in.; and 9 @ 10 in.</u> (118 in. from face of support).

As a practical matter, many designers would place stirrups by scaling from the factored shear envelope (in terms of force or unit stress), rather than accurately compute distances as has been illustrated here. Further, the simplified procedure of ACI-11.3.1.1 for the value of V_c has been used in this example. The more detailed procedure involving $\rho V d/M$ does not seem appropriate unless an accurate shear envelope has been determined for the entire span. Such an envelope would include use of partial span loading for the maximum shear in the central portions of the beam spans, but it is not considered necessary for most ordinary beams in building frames. ■

10.6 DETAILS OF BARS IN BEAMS

For typical conditions of equal spans and uniform load where moment and shear coefficients of ACI-8.3 are used, standard bar details such as provided in the *ACI Detailing Manual—1988* [2.25] may be used. When the moment and shear envelopes are available, bar bend or cutoff locations should be determined therefrom, as illustrated for this example.

■ **EXAMPLE 10.6.1**   Determine the bar dimensions of the main reinforcing bars in 2B1 and 2B2 of Example 10.2.1 (see Fig. 10.6.1). $f_c' = 3000$ psi and $f_y = 40{,}000$ psi.

Solution: (a) Maximum moment capacities. In determining bar cutoff or bend locations it is necessary first to locate the theoretical points where the bars are no longer required. Lines representing full strength ϕM_n of the various bar combinations are computed. The effect of any compression steel will be small and is neglected.

At section *A-A*, 2-#9

$$C = 0.85(3)(13)a = 33.1a; \qquad T = 2(1.0)40 = 80.0 \text{ kips}$$

$$\phi M_n = 0.90(80.0)(20 - 1.20)\tfrac{1}{12} = 113 \text{ ft-kips}$$

Note that it is ϕM_n (not M_n) that must be compared with the factored moment M_u on Fig. 10.6.1.

At section *B-B*, 2-#7

$$C = 0.85(3)(78)a = 199a; \qquad T = 2(0.60)40 = 48 \text{ kips}$$

$$\phi M_n = 0.90(48)(20 - 0.12)\tfrac{1}{12} = 71.5 \text{ ft-kips}$$

At section *B-B*, 2-#7 and 2-#9

$$C = 199a; \qquad T = 3.20(40) = 128 \text{ kips}$$

$$\phi M_n = 0.90(128)(20 - 0.32)\tfrac{1}{12} = 189 \text{ ft-kips}$$

At sections *C-C* and *D-D*, 2-#10

$$C = 33.1a; \qquad T = 2.54(40) = 101.6 \text{ kips}$$

$$\phi M_n = 0.90(101.6)(20 - 1.53)\tfrac{1}{12} = 141 \text{ ft-kips}$$

Figure 10.6.1 Moment envelope, moment capacity diagram, and bar arrangement for beams 2B1 and 2B2 (all moments M_u and ϕM_n are given in ft-kips).

At sections C-C and D-D, 2-#9 and 2-#10

$$C = 33.1a; \qquad\qquad T = 4.54(40) = 181.6 \text{ kips}$$

$$\phi M_n = 0.90(181.6)(20 - 2.74)\tfrac{1}{12} = 235 \text{ ft-kips}$$

At section E-E, 2-#6

$$C = 199a; \qquad\qquad T = 2(0.44)40 = 35.2 \text{ kips}$$

$$\phi M_n = 0.90(35.2)(20 - 0.09)\tfrac{1}{12} = 52.5 \text{ ft-kips}$$

At section E-E, 2-#6 and 2-#9

$$C = 199a; \qquad\qquad T = 2.88(40) = 115.2 \text{ kips}$$

$$\phi M_n = 0.90(115.2)(20 - 0.29)\tfrac{1}{12} = 170 \text{ ft-kips}$$

(b) Extension of positive moment reinforcement into supports. In order to satisfy ACI-12.11.1, the #7 straight bars in 2B1 must extend at least 6 in. into the supports. Since there is no computed tensile requirement for these bars at the first interior support, the 6 in. minimum embedment is sufficient unless the bars are (1) to be used as compression reinforcement, or (2) required for "structural integrity" under ACI-7.13.2. Beams at the perimeter of a structure must have "... one-fourth of the positive moment reinforcement required at midspan made continuous around the perimeter and tied with closed stirrups." For non-perimeter beams, such requirement applies only when closed stirrups are not provided.

In this design, since the beam is *not* a perimeter beam, closed stirrups *are* used near the supports, and the bars are *not* utilized as compression reinforcement, an extension greater than the 6 in. minimum is not required. However, since the extra cost of providing more ductility and tying the structure together is minimal, the authors recommend that, in general, the positive moment bars required to extend into the support should develop f_y at the face of support, often requiring a Class A splice when the bars are not continuous through the support.

At the exterior support, the 6-in. embedment may not be adequate if the support is a column and the flexural member is part of the primary lateral load resisting system. ACI-12.11.2 requires development of the full yield stress in tension at the face of support for such cases. In this example, the supporting member is a girder and is not a part of the primary lateral load resisting system so that 6-in. embedment is sufficient. Similarly, the 2-#6 in beam 2B2 are extended into the supports 6 in. The total length of the #6 and #7 straight bars [24 ft 6 in. + 2(6 in.)] = 25 ft 6 in. (Bar lengths are usually specified in 3-in. increments.)

(c) Theoretical bend or cutoff locations. In detailing the bend-up and bend-down points for the #9 bars, and for establishing the cutoff for the #9 bars near section A-A, the distances x_1 through x_5 as marked in Fig. 10.6.1 [also shown as x_1, x_2 in Fig. 10.3.3(a), x_3 in Fig. 10.3.3(b), x_4 in Fig. 10.3.4(a), and x_5 in Fig. 10.3.4(b)] are in practice determined more frequently by graphical means, although they are computed in this example. Thus referring to Fig. 10.3.3(a),

$$\frac{3.57(11.19 - x_1)^2}{2} = \frac{3.57(13.31 - x_2)^2}{2} = 173.3 - 71.5 = 101.8 \text{ ft-kips}$$

which gives

$$x_1 = 3.64 \text{ ft}; \qquad x_2 = 5.76 \text{ ft}$$

In accordance with ACI-12.10.4, a distance of 12 bar diameters or the effective depth of the member, whichever is larger, must be subtracted from these distances. Thus the bend or cutoff locations could be

$$\text{bend-up or cutoff based on } x_1 = 3.64 - 1.67 = 1.97 \text{ ft}$$

$$\text{bend-up or cutoff based on } x_2 = 5.76 - 1.67 = 4.09 \text{ ft}$$

From Fig. 10.3.3(b),

$$\frac{3.57(14.18 - x_3)^2}{2} = 148.8 + 141 = 289.8 \text{ ft-kips}$$

$$x_3 = 1.44 \text{ ft}$$

$$\text{bend-down or cutoff based on } x_3 = 1.44 + 1.67 = 3.11 \text{ ft}$$

From Fig. 10.3.4(a),

$$\frac{3.57(12.25 - x_4)^2}{2} = 128.7 - 52.5 = 76.2 \text{ ft-kips}$$

$$x_4 = 5.72 \text{ ft}$$

$$\text{bend-up or cutoff based on } x_4 = 5.72 - 1.67 = 4.05 \text{ ft}$$

From Fig. 10.3.4(b),

$$\frac{3.57(13.18 - x_5)^2}{2} = 106.8 + 141 = 247.8 \text{ ft-kips}$$

$$x_5 = 1.40 \text{ ft}$$

$$\text{bend-down or cutoff based on } x_5 = 1.40 + 1.67 = 3.07 \text{ ft}$$

The distances x_1 through x_5 with the effective depth $d = 1.67$ ft either added or subtracted locate correctly the theoretical *cutoff* points. For bending bars, the objective is to have the resulting moment capacity diagram maintain an offset from the factored moment diagram equal to, in this case, 1.67 ft (effective depth). Thus the actual bend-up or bend-down points will be offset from the theoretical cutoff points an amount equal to one-half the horizontal projection of the sloping portion of the bar. If the bend is at 45°, the offset would be $(d - d')/2$. Typically, the bend is at an approximately 45° angle but the locations of the bend-up and bend-down points are made 3-in. increment dimensions from the face of support.

$$d - d' = 20 - 2.5 = 17.5 \text{ in. } (1.46 \text{ ft})$$

The bend-down position in beam 2B1 could be no closer than $x_3 + d - 1.46/2$, or 2.38 ft, from the face of support. The selected distance of 2.75 ft exceeds this and is acceptable. The bend-up position in beam $2B1$ must be no farther from the face of support than $x_2 - d + 1.46/2$, or 4.82 ft. The selected distance of 4.5 ft is closer than 4.82 ft and is therefore acceptable. The situation in beam $2B2$ is similar.

Use the bends shown in Fig. 10.6.1. The potential cutoff relating to x_1 requires further checking, as shown in part **(e)**, to satisfy ACI-12.10.5 for cutting bars in a tension zone.

(d) Inflection point extensions. ACI-12.12.3 requires an extension beyond the extreme point of inflection a distance equal to the largest of one-sixteenth of the clear span, the effective depth of the member, or 12 bar diameters. In this case, the effective depth of the member, 1.67 ft, controls. The #9 and #10 bars in member $2B1$ are terminated after the 1.67-ft extension. The cut bars develop their full capacity after an embedment equal to the development length L_d required for top bars, as computed in Example 10.4.1, parts **(e)** and **(f)**.

(e) Check cutoff for 2-#9 bars in the tension zone in the bottom of beam $2B1$. ACI-12.10.5 must be satisfied for the cutoff at $x_1 - d$ to be acceptable. Since the continuing bars (2-#7) provide more than double the area required for flexure at the potential cutoff point, it is necessary only to check whether or not the factored shear V_u exceeds three-fourths of the shear strength ϕV_n (ACI-12.10.5.3).

$$V_u = w_u (11.94 - 0.75 - x_1 + d)$$

$$V_u = 3.57(11.94 - 0.75 - 3.64 + 1.67) = 32.9 \text{ kips}$$

It is required that

$$V_u \leq [0.75\phi V_n = 0.75\phi(V_c + V_s)]$$

The strength provided, including stirrups, is

$$\phi V_c + \phi V_s = 24.2 + \frac{149.6}{10} = 39.2 \text{ kips}$$

$$\text{percent strength utilized} = \frac{32.9}{39.2} = 84\% > 75\% \text{ permitted} \qquad \textbf{NG}$$

If the cutoff is to be made at the indicated location, the stirrup spacing must be reduced so that V_s increases. Alternatively, the 2-#9 bars may be extended to the inflection point where they will no longer be in the tension zone. Thus the #9 bars are cut off at 1 ft–3 in. from the face of support. In this case, there is little material saved by cutting so close to the support, but it serves to illustrate the procedure.

The summary of bar arrangement, lengths, dimensions, and moment capacity provided is shown in Fig. 10.6.1. The cross sections for the final choice at each designated section of Fig. 10.6.1 are shown in Fig. 10.6.2.

(f) Stirrup development. For stirrup supports, 2-#3 bars are provided in the central part of beam $2B1$. The dimensions of the stirrups are shown in Fig. 10.6.3. ACI-12.13.2.1 provides that #5 and smaller stirrups are adequately developed when standard stirrup hooks, as defined in ACI-7.1.3, are used around longitudinal bars located near the tension face of a beam. ■

Figure 10.6.2 Typical sections for 2B1 and 2B2 (refer to Fig. 10.6.1).

Figure 10.6.3 Stirrup details in beams 2B1 and 2B2.

10.7 SIZE OF GIRDER WEB

For the purpose of determining the *size* of the girder web, the maximum negative moment at the exterior face of the first interior support may be used, taken as 0.8 of the maximum positive moment in a simple span with span length equal to the clear span of the girder and with loadings identical to those on the girder. The maximum shear at the same location may be taken as 1.20 times the reaction to the simple span described above. Note that the coefficients of ACI-8.3.3 cannot be used in lieu of an elastic analysis, since concentrated loads are involved.

The concentrated loads on the girder may be taken as one-half of the total dead and live loads on the clear span of the beams on both sides of the girder. The dead and live uniform load on the girder will be the weight of the concrete and the live load on the floor, respectively, both within the width of the girder. This loading transfer is a sufficiently good approximation, although it is probable that the slab weight and floor load on a narrow strip parallel and close to either edge of the girder would act on the girder as uniform load instead of being carried by the one-way slab to the adjacent beams and then to the girder as concentrated loads.

■ **EXAMPLE 10.7.1**   Design the floor girders 2G1-2G2-2G2-2G1 as shown in the floor framing plan of Fig. 8.2.1. Use information previously described in Chapter 8 and in the present chapter. Work through the choice of the size of the girder web in this example according to the strength method of the ACI Code ($f'_c = 3000$ psi; $f_y = 40,000$ psi).

Solution: The concentrated reactions from the beams using the 24.5 ft clear span are

$$\text{dead load} = 1.4(24.50)[(0.056)(13) + 0.244] = 33.4 \text{ kips}$$

$$\text{live load} = 1.7(24.50)(0.100)(13) = 54.2 \text{ kips}$$

$$\text{total load} = 33.4 + 54.2 = 87.6 \text{ kips}$$

Assuming 18×36 in. web, the uniform loads on the girder are

$$\text{uniform dead load} = 1.4\left[\frac{18(40.5)(0.150)}{144}\right] = 1.06 \text{ kips/ft}$$

$$\text{uniform live load} = 1.7(0.100)(1.5) = 0.255 \text{ kip/ft}$$

$$\text{uniform total load} = 1.06 + 0.255 = 1.315 \text{ kips/ft}$$

Maximum positive moment on a simple span (see Fig. 10.7.1) is

$$M_u = 87.6(12.25) + \tfrac{1}{8}(1.315)(37.5)^2 = 1306 \text{ ft-kips}$$

The estimated maximum negative moment in the girder at the exterior face of the first interior support is $0.8(1306) = 1045$ ft-kips, say 1050 ft-kips. From Fig. 10.7.1, estimate

$$d_{\text{neg}} = 40.50 - 4.75 = 35.75 \text{ in.}$$

$$d_{\text{pos}} = 40.50 - 3.50 = 37 \text{ in.}$$

The negative moment requirement is

$$\text{required } R_n = \frac{M_u}{\phi b d^2} = \frac{1050(12,000)}{0.90(18)(35.75)^2} = 609 \text{ psi}$$

which is less than the maximum $R_n = 870$ psi for $\rho = 0.75\rho_b$, and is an acceptable value. The girder system should be relatively stiff and will be so if this smaller R_n is used.

$$\text{required } A_s \text{ for } (-M) \approx 0.75\rho_b\left(\frac{609}{870}\right)bd$$

$$= 0.0278\left(\frac{609}{870}\right)(18)(35.75) = 12.5 \text{ sq in.}$$

For the positive moment requirement, using the coefficients of ACI-8.3.3 to obtain the approximate proportion between positive and negative moments,

$$\text{estimated } (+M) \approx \frac{10}{14}(-M) = \frac{10}{14}(1050) = 750 \text{ ft-kips}$$

$$\text{required } A_s \approx \frac{M_u}{\phi f_y \text{ (arm)}} = \frac{750(12)}{0.90(40)(\approx 33)} = 7.58 \text{ sq in.}$$

Figure 10.7.1 Loading information for floor girders 2G1 and 2G2.

This may well fit into one layer (6-#10 requires a beam width of 18 in. according to Table 3.9.2).

For the shear requirement,

$$\max V_u = 1.20[87.6 + 1.315(18.75)] = 135 \text{ kips}$$

$$v_n = \frac{135,000}{0.85(18)(35.75)} = 247 \text{ psi} = 4.5\sqrt{f'_c}$$

The stem could be made smaller; with this large depth it is unlikely that deflection would be excessive. The 18×36 stem appears to be somewhat large though it could certainly be used. In this case because of the large depth, reduce the stem size to 18×30. Estimated effective depths become

$$d_{\text{neg}} = 29.75 \text{ in.}$$

$$d_{\text{pos}} = 31 \text{ in.}$$

$$\text{revised girder weight} = 1.4 \left[\frac{18(34.5)}{144} (0.15) \right] = 0.91 \text{ kip/ft}$$

$$w_u = w_D + w_L = 0.91 + 0.255 = 1.165 \text{ kips/ft}$$

$$\text{revised } M_u \text{ (simple beam)} = 1075 + \tfrac{1}{8}(1.165)(37.5)^2 = 1280 \text{ ft-kips}$$

$$\text{required } R_n = \frac{0.8(1280)(12,000)}{0.90(18)(29.75)^2} = 860 \text{ psi} < 870 \text{ psi max} \qquad \textbf{OK}$$

$$v_n \text{ at support} = \frac{131,500}{0.85(18)(29.75)} = 289 \text{ psi} = 5.3\sqrt{f_c'}$$

Use the 18×30 stem section. ■

10.8 CONTINUOUS FRAME ANALYSIS FOR GIRDERS

Although floor girders subjected to large concentrated loads are structural members of common occurrence, the ACI Code makes no mention of moment coefficients for these cases. In the following example, elastic statically indeterminate structural analysis is used for 2G1-2G2-2G2-2G1.

■ **EXAMPLE 10.8.1**   By the use of an elastic statically indeterminate structural analysis, determine the shear and moment diagrams to be used in the design of 2G1 and 2G2 in Example 10.7.1 ($f_c' = 3000$ psi; $f_y = 40,000$ psi).

Solution: As discussed in Section 10.3, there are various ideas regarding what constitutes the correct stiffness for continuous T-sections. In accordance with the thought

Figure 10.8.1 T-section for stiffness computation.

that the true stiffness is that of a span with a variable cross section, the effective flange width for an equivalent uniform moment of inertia section may be assumed to be twice the web width (Fig. 10.8.1). The principal effect of changing the stiffness of the girder (K_{gr}) occurs at the exterior support where the relative stiffness of the columns (K_{col}) compared to that of the girder is greatest. The centroid of the gross area of the T-section is at

$$\bar{y} = \frac{18(30)15 - 36(4.5)(2.25)}{18(30) + 36(4.5)}$$

$$= \frac{7736}{702} = 11.0 \text{ in.}$$

The gross moment of inertia I_g is

$$I_g = \frac{36(4.5)^3}{3} + \frac{18(30)^3}{3} - 702(11.0)^2$$

$$= 78,100 \text{ in.}^4$$

$$K_{gr} = \frac{78,100}{39} = 2000 \text{ in.}^4/\text{ft}$$

If the size of the upper and lower columns were 18 × 18 in. and the column height were 15 ft,

$$K_{col} = \frac{18(18)^3/12}{15} = 583 \text{ in.}^4/\text{ft}$$

$$\frac{K_{gr}}{K_{col}} = \frac{2000}{583} = 3.43$$

In the rigid frame of Fig. 10.8.2 are shown the distribution factors

$$\frac{3.43}{3.43 + 1 + 1} = 0.632 \quad \text{and} \quad \frac{3.43}{3.43 + 3.43 + 1 + 1} = 0.387$$

The fixed-end moments are

$$\text{FEM due to dead load} = \frac{33.4(13)(26)}{39} + \frac{0.91(39)^2}{12}$$

$$= 405 \text{ ft-kips}$$

$$\text{FEM due to live load} = \frac{54.2(13)(26)}{39} + \frac{0.255(39)^2}{12}$$

$$= 502 \text{ ft-kips}$$

The results of moment distribution for the various loading conditions are given in Table 10.8.1. Note that identical results may be obtained by applying the matrix displacement method, or any other statically indeterminate structural analysis method.

Figure 10.8.2 Elastic analysis model of rigid frame system for 2G1-2G2-2G2-2G1.

Table 10.8.1 Summary of Moment Distribution Results Using Four Cycles

JOINT	A	B		C		D		E
MEMBER	AB	BA	BC	CB	CD	DC	DE	ED
DF	0.632	0.387	0.387	0.387	0.387	0.387	0.387	0.632
DEAD LOAD ONLY								
FEM	−405	+405	−405	+405	−405	+405	−405	+405
Total	−159	+487	−459	+380	−380	+459	−487	+159
LIVE LOAD ONLY SPANS 1 AND 3								
FEM	−502	+502	0	0	−502	+502	0	0
Total	−244	+376	−182	+146	−326	+384	−229	−48
LIVE LOAD ONLY SPANS 1, 2, AND 4								
FEM	−502	+502	−502	+502	0	0	−502	+502
Total	−188	+648	−641	+242	−96	+197	−385	+242
LIVE LOAD ONLY SPAN 3								
FEM	0	0	0	0	−502	+502	0	0
Total	−8	−44	+73	+230	−375	+369	−220	−46
LIVE LOAD ONLY SPANS 2 AND 3								
FEM	0	0	−502	+502	−502	+502	0	0
Total	+38	+176	−296	+605	−605	+296	−176	−38
LIVE LOAD ONLY SPANS 1 AND 4								
FEM	−502	+502	0	0	0	0	−502	+502
Total	−234	+429	−270	−133	+133	+270	−429	+234

Figure 10.8.3 Moment envelope and longitudinal bar arrangement for girder 2G1 (span 1).

374

The controlling moment envelopes for spans 2G1 and 2G2 are given in Figs. 10.8.3 and 10.8.4. In practice this composite diagram should be made instead of the individual moment and shear diagrams for each loading case.

For investigating development length requirements at points of inflection, for checking cutoff acceptability in the tension zone, and for designing stirrups, an approximate shear envelope (only full span loadings) is used, as given in Fig. 10.8.5. ∎

Figure 10.8.4 Moment envelope and longitudinal bar arrangement for girder 2G2 (span 2).

Figure 10.8.5
Shear envelopes for girders 2G1 and 2G2 (full-span loadings only).

10.9 CHOICE OF LONGITUDINAL REINFORCEMENT IN GIRDERS

Again, the choice of longitudinal reinforcement depends on both the steel area and the development length requirements as dictated by the moment and shear envelopes such as shown in Figs. 10.8.3 through 10.8.5. Attention is called to ACI-8.10.5, applicable when the main reinforcement in the slab is parallel to the girder. For this situation transverse steel must be provided in the top of the slab to carry the load on the portion of the slab acting effectively as the flange of the girder. The overhanging flange is assumed to act as a cantilever. The spacing of the transverse bars may not exceed five times the thickness of the flange or, in any case, 18 in.

■ **EXAMPLE 10.9.1** Choose the arrangement of the main reinforcement in the floor girders 2G1-2G2-2G2-2G1 of Example 10.7.1.

Solution: (a) Sections (rectangular sections) A-A, C-C, D-D, and F-F (Figs. 10.8.3, 10.8.4, and 10.9.1). Section A-A:

$$M_u \text{ at face} = 403 - \frac{97.5(1.5)}{3} = 403 - 49 = 354 \text{ ft-kips}$$

$$\text{required } R_n = \frac{M_u}{\phi b d^2} = \frac{354(12,000)}{0.90(18)(30.7)^2} = 278 \text{ psi}$$

Figure 10.9.1 Typical sections for 2G1 and 2G2 (refer to Figs. 10.8.3 and 10.8.4).

Either from Eq. (3.8.5), or from Fig. 3.8.1,

$$\text{required } \rho = 0.0075$$

$$\text{required } A_s = 0.0075(18)30.7 = 4.14 \text{ sq in.}$$

Sections *C-C* and *D-D*: The larger moment is at section *C-C*,

$$M_u \text{ at face} = 1135 - \frac{129.5(1.5)}{3} = 1070 \text{ ft-kips}$$

$$\text{required } R_n = \frac{1070(12,000)}{0.90(18)(29.3)^2} = 923 \text{ psi}$$

This value exceeds the maximum value of 870 psi allowed without using compression steel. In this case, only a small amount of compression steel is required; thus the positive moment steel extended into the support at the bottom of the beam will be developed by proper embedment in order to utilize it as compression steel. Since the use of compression steel influences the tension steel requirement hardly at all, the tension requirement is computed just as if no compression steel were to be used. Estimating the required percentage from Fig. 3.8.1,

$$\text{required } A_s \approx 0.030(18)(29.3) = 15.8 \text{ sq in.}$$

Section *F-F*:

$$M_u \text{ at face} = 985 - \frac{115.2(1.5)}{3} = 985 - 58 = 927 \text{ ft-kips}$$

$$\text{required } R_n = \frac{927(12,000)}{0.90(18)(29.6)^2} = 785 \text{ psi}$$

Using straightline approximation,

$$\text{required } A_s \approx 0.0278\left(\frac{785}{870}\right)(18)29.6 = 13.4 \text{ sq in.}$$

(b) Sections (T-sections) *B-B* and *E-E* (Figs. 10.8.3, 10.8.4, and 10.9.1): Available flange width b_E,

$$\frac{39(12)}{4} = 117; \quad \text{or} \quad \underline{18 + 16(4.5) = 90}; \quad \text{or} \quad (26)(12) = 312$$

It is likely that the depth *a* of the rectangular stress distribution will be less than the flange thickness *t*. Estimate $a = 2$ in. (about one-half flange thickness).
      Section *B-B*: estimated $d = 31.0$ in.

$$\text{required } A_s = \frac{M_u}{\phi f_y \,(\text{arm})} = \frac{781(12)}{0.90(40)(31.0 - 1)} = 8.7 \text{ sq in.}$$

Check:

$$C = 0.85(3)(90)a = 229.5a$$

$$T = 8.7(40) = 348 \text{ kips}; \quad a = 1.52 \text{ in.}$$

$$\text{revised required } A_s = \frac{781(12)}{0.90(40)(31.0 - 0.76)} = 8.6 \text{ sq in.}$$

Section *E-E*:

$$\text{required } A_s = \frac{585(12)}{0.90(40)(31.9 - 1)} = 6.3 \text{ sq in.}$$

Check:

$$C = 229.5a; \quad T = 6.3(40) = 252 \text{ kips}$$

$$a = 1.10 \text{ in.}$$

$$\text{revised required } A_s = \frac{585(12)}{0.90(40)(31.9 - 0.55)} = 6.2 \text{ sq in.}$$

(c) Confirmation of the tentative arrangement of main reinforcement as summarized in Fig. 10.9.2 awaits the check of development of reinforcement at the positive moment inflection points (ACI-12.11.3), crack control when f_y exceeds 40,000 psi (does not apply for this problem), and deflection if excessive deflection will cause cracking of attached nonstructural elements.
      Note in Fig. 10.9.2 that the requirements indicated in the compression zone at *B* and *C* are merely the minimums of one-fourth of the positive moment steel required

Figure 10.9.2 Longitudinal reinforcement areas (sq in.) required and bars selected for girders 2G1 and 2G2.

to satisfy ACI-12.11.1. It is likely that at B, which is the only place compression steel is required for strength, the minimum satisfying ACI-12.11.1 is sufficient.

The details of determining bar lengths are not shown; only the general arrangement is shown in Figs. 10.8.3 and 10.8.4.

(d) Check maximum percentage reinforcement in negative moment region at B (sections C-C and D-D), including effect of compression reinforcement. At balanced condition,

$$x_b = \left(\frac{0.003}{0.003 + f_y/E_s}\right)d = \left(\frac{0.003}{0.003 + 0.00138}\right)d$$

$$= 0.685d = 0.685(29.3) = 20.1 \text{ in.}$$

$$\max x = 0.75x_b = 0.75(20.1) = 15.1 \text{ in.}$$

$$a = 0.85x = 0.85(15.1) = 12.8 \text{ in.}$$

At max A_s condition,

$$\epsilon'_s = \left(\frac{15.1 - 2.6}{15.1}\right)(0.003) = 0.0025 > \epsilon_y$$

Thus compression steel yields at max A_s condition.

$$\max C_c = 0.85f'_c ba = 0.85(3)(18)(12.8) = 586 \text{ kips}$$

Assume 2-#10 that extend through the support region at C-C and D-D are the only bars developed for compression reinforcement.

$$C_s = A'_s(f_y - 0.85f'_c) = 2.54(40 - 2.55) = 95 \text{ kips}$$

$$\max T = \max C_c + C_s = 586 + 95 = 681 \text{ kips}$$

$$\max A_s = \frac{\max T}{f_y} = \frac{681}{40} = 17.0 \text{ sq in.}$$

Since the actual steel used (15.60) is less than max A_s, the design is acceptable.

(e) Check the development length requirement at the positive moment inflection points. From Fig. 10.8.5 (shear envelope), for span 1,

$$V_u \text{ (near left end)} = 97.5 - 3.6(1.165) = 93 \text{ kips}$$

$$V_u \text{ (near right end)} = 129.5 - 6.5(1.165) = 122 \text{ kips}$$

For span 2,

$$V_u \text{ (near left end)} = 121.6 - 7.0(1.165) = 113 \text{ kips}$$

$$V_u \text{ (near right end)} = 115.2 - 6.0(1.165) = 108 \text{ kips}$$

At the span 1 inflection points, 4-#10 bars continue along the bottom of the beam. The critical location is at the right inflection point where maximum V_u occurs.

$$V_u = 122 \text{ kips}$$

For 4-#10,

$$C = 0.85(3)(90)a = 229.5a$$

$$T = 5.08(40) = 203 \text{ kips}$$

$$a = 0.89 \text{ in.}$$

$$M_n = 203[31.9 - 0.5(0.89)]\tfrac{1}{12} = 532 \text{ ft-kips}$$

The development length L_d for the 4-#10 in the bottom of span 2G1 near the left support depends on the clear spacing of those bars. The basic development length L_{db} and the minimum L_d are 37.1 and 27.8 in., respectively, from Example 10.4.1, part **(b)**. From Table 3.9.4, the minimum width required to satisfy $3d_b$ is 20.5 in., which exceeds the 18 in. actual stem width. From Table 3.9.3, in order to satisfy $2d_b$ the minimum width is 16.7 in. Since the cover requirement of ACI-7.7.1 and at least minimum stirrups will be used, Category B applies; $\alpha_{abc} = 1.4$ in accordance with ACI-12.2.3.3. Categories D and E do not apply; thus, $\alpha_{de} = 1.0$, making

$$[\alpha_{abc}\alpha_{de}L_{db} = 1.4(1.0)37.1 = 51.9 \text{ in.}] > [\text{Min } L_d = 27.8 \text{ in.}]$$

Thus, the "first-level L_d" is 51.9 in. Finally, α_{top}, $\alpha_{\ell w}$, and α_{epx} are not applicable; therefore,

$$L_d \text{ (for #10)} = 51.9 \text{ in. (4.33 ft)}$$

which exceeds the 12 in. minimum.

$$\left[\frac{M_n}{V_u} + L_a = \frac{532}{122} + \frac{31.9}{12} = 7.02 \text{ ft}\right] > 4.33 \text{ ft} \qquad \textbf{OK}$$

Note that L_a = effective depth d for this situation.

In span 2, only 2-#10 continue past the inflection point and into the support. In this case, 2-#10 easily satisfy the $3d_b$ clear spacing in an 18 in. width, which exceeds the 10.3 in. minimum given by Table 3.9.4. Thus, Category A applies; $\alpha_{abc} = 1.0$. Thus,

$$L_d \text{ (for #10)} = L_{db} = 37.1 \text{ in. (3.09 ft)}$$

For 2-#10,

$$C = 229.5a; \qquad T = 2(1.27)(40) = 102 \text{ kips}$$

$$a = 0.44 \text{ in.}$$

$$M_n = 102[31.9 - 0.5(0.44)]\tfrac{1}{12} = 269 \text{ ft-kips}$$

$$\left[\frac{M_n}{V_u} + L_a = \frac{269}{113} + \frac{31.9}{12} = 5.04 \text{ ft}\right] > 3.09 \text{ ft} \qquad \textbf{OK}$$

Figure 10.9.3 Transverse steel across the top of girder.

(f) The transverse steel required in the top of the flange of the girder (ACI-8.10.5) may be computed as follows (Fig. 10.9.3):

$$M_u = \tfrac{1}{2}[0.100(1.7) + 0.056(1.4)](3)^2 = 1.12 \text{ ft-kips/ft}$$

$$\text{required } R_n = \frac{1.12(12{,}000)}{0.90(12)(2.94)^2} = 144 \text{ psi}$$

Since, from observing Fig. 3.8.1 this is less than the value for min ρ, the minimum value controls.

$$A_s = \frac{200}{f_y} bd = \frac{200}{40{,}000}(12)(2.94) = 0.18 \text{ sq in.}$$

Use #5 @ at 18 in. spacing. Generally the amount required for temperature and shrinkage in the slab according to ACI-7.12 is adequate for this purpose.

(g) The bar arrangement is shown in Figs. 10.8.3 and 10.8.4 and cross sections are shown in Fig. 10.9.1. The shear reinforcement and dimensions of longitudinal reinforcement are not shown, because little new will be involved in their presentation. ∎

10.10 ONE-WAY JOIST FLOOR CONSTRUCTION

One-way concrete joist floor construction (Fig. 10.10.1), sometimes called "ribbed-slab construction," consists of regularly spaced ribs monolithically built with a top floor slab and arranged to span in one direction. Such a system may also be designed as a two-way system (waffle slab) according to the procedures for two-way floor systems treated in Chapters 16 and 17. The dimensions of the one-way joist system are usually such that only temperature and shrinkage reinforcement is required in the slab. The slab is usually in the range of 2 to 4 in. (50 to 100 mm) thick but may occasionally be as much as 6 in. (150 mm). The ribs (joists) of at least 4 in. (100 mm)

Figure 10.10.1 Concrete joist floor construction.

width are usually tapered [Fig. 10.10.1(b)] and are spaced so that the clear spacing between adjacent ribs does not exceed 30 in. (760 mm). During construction removable and reusable form fillers are used in spaces between the joists. Such fillers may be standard-sized steel "pans" in 20 or 30 in. widths and 6, 8, 10, 12, 14, 16, and 20 in. depths. Sometimes form fillers are made from hardboard, fiberboard, glass-reinforced plastic, or corrugated cardboard. Occasionally, permanent fillers are used consisting of lightweight or normal-weight concrete blocks or clay tile blocks, as shown in Fig. 10.10.1(c).

In order to limit deflection on floor joist construction the minimum depth requirements of ACI-Table 9.5a for ribbed one-way slabs must be applied. Whenever excessive deflection may cause cracking or other adverse effects, deflections must be computed even if ACI-Table 9.5a has been satisfied.

Concrete joist floor construction is referred to in ACI-8.11. Some of the requirements are as follows:

1. The joists shall not be farther apart than 30 in. face to face. The ribs shall be not less than 4 in. wide and of a depth not more than $3\frac{1}{2}$ times the width.
2. The vertical shells of permanent fillers in contact with the joists may be included in strength calculations involving shear or negative bending moment provided the filler material has a unit compressive strength at least equal to that of the concrete in the joists. In this case the minimum slab thickness is $1\frac{1}{2}$ in. or $\frac{1}{12}$ of the clear distance between joists, whichever is smaller.
3. When removable forms or fillers having less compressive strength than required under (2) are used, the thickness of the concrete slab shall not be less than $\frac{1}{12}$ of the clear distance between joists, nor less than 2 in.

For more information on concrete joist construction see Chapter 10 of *Manual of Standard Practice* [10.1], Chapter 4 of Rice, Hoffman, Gustafson, and Gouwens [10.2], and Chapter 8 of *CRSI Handbook* [2.23].

10.11 DESIGN OF CONCRETE JOIST FLOORS

The design of concrete joist floors involves (1) the slab, (2) the joists, and (3) the girders.

Generally, shrinkage reinforcement is placed at right angles to the joists, and the concrete slab is treated as if it were of plain concrete. The short clear span between joists may be considered as being fixed at both ends.

The joist itself may be designed as a floor beam having a rectangular section in the region of negative bending and a T-section in the region of positive bending. The critical design moment curves for each span may either follow the ACI moment coefficients or be determined by a continuity analysis. Largely because of the interaction of slab with the closely spaced joists, ACI-8.11.8 permits the shear strength V_c provided by the concrete to be 10% higher than for regular beams.

The girder is designed as a floor girder, but the load from the joists may be considered as being uniformly distributed along the span.

In the case of joist floors over removable steel pans, tapered end forms are available that increase the effective joist width 2 in. on each side for 20-in. wide forms and $2\frac{1}{2}$ in. on each side for 30-in. wide forms in a distance of 3 ft from the end. This increased width may be necessary to take the large shear or negative bending moment near the end of the span.

■ **EXAMPLE 10.11.1** Design the typical interior span of a concrete joist floor, using 30-in. wide forms, for a center-to-center span of 26 ft and clear span of 24 ft–6 in. The live load is 80 psf and the additional superimposed dead load is 20 psf. Use $f'_c = 3000$ psi, $f_y = 60,000$ psi, and the strength method of the ACI Code. Use the moment and shear coefficients of ACI-8.3.

Solution: **(a)** Slab design. Assume a 3-in. slab for weight estimation purposes. The factored load is

$$w_u = 1.4\left[\frac{3.0(0.150)}{12}\right] + 1.4(0.020) + 1.7(0.080) = 0.22 \text{ ksf}$$

Assuming that the slab is fixed at its junction with the joist,

$$M_u \approx \frac{1}{12}(0.22)\left(\frac{30}{12}\right)^2 = 0.115 \text{ ft-kips/ft}$$

The design strength ϕM_n of plain concrete, computed using $f_r = 5\sqrt{f'_c}$ as the modulus of rupture* is

$$\phi M_n = \phi f_r\left(\frac{1}{6}bh^2\right) = 0.65(5\sqrt{f'_c})\left(\frac{1}{6}bh^2\right)$$

---

* Note that for a strength calculation the modulus of rupture is taken conservatively lower than the value $7.5\sqrt{f'_c}$ used in deflection calculations (ACI–9.5). The value $5\sqrt{f'_c}$ is given by Section 6.2.1, *Building Code Requirements for Structural Plain Concrete* (ACI 318.1–89) *and Commentary* (ACI 318.1R–89). Detroit: American Concrete Institute, 1989.

Setting $\phi M_n = M_u$ and $b = 12$ in. for a typical 1-ft strip, gives for the required slab thickness,

$$h = \sqrt{\frac{0.115(12,000)}{\frac{1}{6}(0.65)(5\sqrt{3000})12}} = 2.0 \text{ in.}$$

The 3-in. slab will be thick enough; perhaps $2\frac{1}{2}$ in. would be sufficient. *Use* $h = 3$ in. Shrinkage and temperature reinforcement (ACI-7.12) must be provided, equal to

$$A_s = 0.0018(12)(3.0) = 0.065 \text{ sq in./ft}$$

Use welded wire fabric, 4×12—W2.5 $\times$ W1.4 ($A_s = 0.075$ sq in.).

The selection is made from Table 10.11.1; A_s in direction perpendicular to joists is 0.075 sq in./ft, and A_s in direction parallel to joists is 0.014 sq in./ft. The "4×12" means 4-in. spacing of the wires running perpendicular to the joist and 12-in. spacing of the wires running parallel to the joist.

(b) Joist design. The overall depth of the joist floor must satisfy the minimum requirement of ACI-Table 9.5a unless deflections are computed; thus,

$$\text{min } h = \frac{L}{21} = \frac{26(12)}{21} = 14.9 \text{ in.}$$

If excessive deflection may cause damage to nonstructural elements, deflection must be computed in all cases. Assuming that deflection must later be computed it will be prudent to make joists deeper than indicated by Table 9.5a. Assume use of joists 12 in. deep below the bottom of the slab with a width of 5 in. at the bottom (i.e., average 6 in.). The weight of the joist will then be

$$w_D = [(3.0 + 12)6 + 30(3.0)]\frac{0.150}{144} = 0.19 \text{ kip/ft}$$

Table 10.11.1 Common Welded Wire Fabric for Temperature and Shrinkage Reinforcement

SPACING (in.) OF LONGITUDINAL[a] AND TRANSVERSE WIRES	DESIGNATION	
	WIRE SIZE DESIGNATION LONGITUDINAL $\times$ TRANSVERSE	A_s, LONGITUDINAL DIRECTION (sq in.)
4×12—	W1.4 $\times$ W1.4	0.042
4×12—	W2 $\times$ W1.4	0.060
4×12—	W2.5 $\times$ W1.4	0.075
4×12—	W3 $\times$ W2	0.090
4×12—	W3.5 $\times$ W2	0.105

[a] Longitudinal refers to the slab span, which is perpendicular to the joist span.

The factored load is

$$w_u = 1.4(0.19) + 1.4(0.020)\left(\frac{35}{12}\right) + 1.7(0.080)\left(\frac{35}{12}\right) = 0.744 \text{ kip/ft}$$

The maximum factored negative moment is

$$M_u = \tfrac{1}{11}(0.744)(24.5)^2 = 40.6 \text{ ft-kips}$$

Using the minimum cover of $\tfrac{3}{4}$ in. (ACI-7.7.1c) and assuming #5 bars for main steel,

$$d = 15 - 0.75 - 0.31 = 13.94 \text{ in.}$$

The required R_n then becomes

$$\text{required } R_n = \frac{M_u}{\phi b d^2} = \frac{40.6(12,000)}{0.90(5)(13.94)^2} = 557 \text{ psi}$$

Using Eq. (3.8.5) or Fig. 3.8.1,

$$\text{required } \rho = 0.011$$

This exceeds the minimum $200/f_y = 0.0033$ and does not exceed the maximum of $0.75\rho_b = 0.0160$; therefore it is acceptable. Thus,

$$\text{required } A_s \text{ (over support)} = 0.011(5)(13.94) = 0.77 \text{ sq in.}$$

With a reinforcement ratio ρ so low, deflection should not be a problem even if excessive deflection may cause cracking of nonstructural elements.

The shear at a distance d from the face of support is

$$V_u = 0.744(12.25 - 1.16) = 8.25 \text{ kips}$$

The shear strength of a joist without shear reinforcement (ACI-8.11.8) is

$$\phi V_c = \phi(1.10)(2\sqrt{f'_c})b_w d$$
$$= 0.85(1.10)(2\sqrt{3000})(5)(13.94)\tfrac{1}{1000} = 7.14 \text{ kips} < 8.25 \text{ kips} \qquad \textbf{NG}$$

"Widening the ends of ribs" is permitted by ACI-8.11.8, as shown in Fig. 10.11.1. When such a taper is used to satisfy the shear strength requirement, the 30 in. wide

Figure 10.11.1 Plan showing ends of tapered pan joists.

form tapers to 25 in. in the last 3 ft near the support. At d from the face, the beam width then becomes

$$b_w = 5 + \left(\frac{3.0 - 1.16}{3.0}\right)5 = 8.1 \text{ in.}$$

$$\phi V_c = 7.14\left(\frac{8.1}{5.0}\right) = 11.6 \text{ kips} > [V_u = 8.25 \text{ kips}] \qquad \textbf{OK}$$

Note that, though it is *not* an ACI requirement, the *minimum* width (instead of the average) of web should probably be used when the web is in compression.

The maximum factored positive moment near midspan is

$$M_u = \tfrac{1}{16}(0.744)(24.5)^2 = 27.9 \text{ ft-kips}$$

For the T-section, assume that the depth of the rectangular stress distribution falls within the flange; estimate $a \approx 1$ in. Then, by trial,

$$\text{required } A_s \approx \frac{M_u}{\phi f_y \text{ (arm)}} = \frac{27.9(12)}{0.90(60)(13.94 - 0.5)} = 0.46 \text{ sq in.}$$

Determine moment arm more accurately using b_E as 35 in., the smallest of $16t + b_w = 16(3) + 6 = 54$ in.; $L/4$; or c-c spacing $= 35$ in. (ACI-8.10.2),

$$C = 0.85f'_c ba = 0.85(3)(35)a = 89.3a$$

$$T = A_s f_y = 0.46(60) = 27.6 \text{ kips}$$

$$a = 27.6/89.3 = 0.31 \text{ in.}$$

$$\text{revised required } A_s = \frac{27.9(12)}{0.90(60)(13.94 - 0.16)} = 0.45 \text{ sq in.}$$

Since a is less than $t/2$, the effective section of the compression zone is rectangular. Use 1-#5 bottom bar, 1-#4 truss bar, and 1-#5 top bar (see Fig. 10.11.2).

$$\text{provided } A_s \text{ over support} = 0.71 \text{ sq in. } (2\text{-}\#4 + 1\text{-}\#5)$$

$$\text{provided } A_s \text{ at midspan} = 0.51 \text{ sq in. } (1\text{-}\#4 + 1\text{-}\#5)$$

Consistent with a loading for the maximum positive moment in an interior span, the inflection point is at $0.354L$ from centerline of span [see Fig. 7.5.4(a)]. Thus the development length requirement of ACI-12.11.3 must be checked. Only the #5 straight bar extends beyond the inflection point and into the support at least 6 in.

$$C = 0.85f'_c b_E a = 0.85(3)(35)a = 89.3a$$

$$T = 0.31(60) = 18.6 \text{ kips}$$

$$a = 0.21 \text{ in.}$$

$$M_n = 18.6[13.94 - 0.5(0.21)]\tfrac{1}{12} = 21.4 \text{ ft-kips}$$

$$V_u = 0.741(0.354)(24.5) = 6.4 \text{ kips}$$

$$L_a = \text{larger of } 12d_b \text{ or effective depth} = d = 13.9 \text{ in.}$$

Section A–A

Figure 10.11.2 Concrete joist floor using removable pans, Example 10.11.1.

The equivalent embedment length provided is

$$\frac{M_n}{V_u} + L_a = \frac{21.4(12)}{6.4} + 13.9 = 40.1 + 13.9 = 54.0 \text{ in.}$$

The development length L_d for the #5 that extends beyond the bend-up location of the #4 depends entirely on the amount of cover, because there are no other bars in the development length and no stirrups are used. The bottom cover is the minimum $\frac{3}{4}$ in., which exceeds d_b but is less than $2d_b$ for the #5 bar. Thus, Category B applies under ACI-12.2.3.3, making $\alpha_{abc} = 1.4$. Note that cover less than d_b would indicate Category C according to ACI-12.2.3.2, and more than $2d_b$ would satisfy Category A according to ACI-12.2.3.1d. Though not shown here, a check of the side cover would show it to be greater than $2d_b$; thus, the bottom cover controls. The basic development length L_{db} is 13.6 in., and Min L_d is 20.5 in. from Tables 6.7.1 and 6.7.2, respectively. Categories D and E do not apply; thus, $\alpha_{de} = 1.0$, making

$$[\alpha_{abc}\alpha_{de}L_{db} = 1.4(1.0)13.6 = 19.0 \text{ in.}] < [\text{Min } L_d = 20.5 \text{ in.}]$$

Thus, the "first-level L_d" is 20.5 in. Finally, α_{top}, $\alpha_{\ell w}$, and α_{epx} are not applicable; therefore,

$$L_d \text{ (for } \#5) = 20.5 \text{ in. } (1.71 \text{ ft})$$

which exceeds the 12 in. minimum.

Thus, the check of ACI-12.11.3 is completed by showing that the equivalent embedment provided of 54.0 in. exceeds L_d of 20.5 in. and thus is satisfactory.

Frequently a transverse distribution rib (Fig. 10.11.2) is used having a 4 in. minimum width and containing at least one #4 bar both top and bottom. Such a rib would be located at the third points of the span for spans greater than 30 ft. For spans between 20 and 30 ft one rib should be used near midspan, and for spans less than 20 ft the rib may be omitted [2.23]. ■

10.12 REDISTRIBUTION OF MOMENTS—INTRODUCTION TO LIMIT ANALYSIS

Methods of proportioning beams for flexure, shear, and bar development requirements according to the strength method have been discussed in Chapters 3, 5, and 6, and further illustrated in Sections 10.1 through 10.11. When the inelastic behavior of concrete at a particular location has been accounted for in the design of that cross section based on its strength, it may seem somewhat illogical to have used an elastic analysis to determine the design moments and shears. However, because the evaluation of the true ultimate strength, or limit strength, of an entire structure requires difficult and elaborate analysis, the present safe conservative procedure is used.

The concept of redistribution of moments first introduced into the ACI Code in 1963 was the result of considerable research [10.3–10.9] into the limit behavior of the entire structure (primarily continuous beams, at present) beyond the elastic range to the point where the collapse load is reached. Concrete design, therefore, has moved toward using limit analysis, or, as it is called in steel structures, "plastic" analysis.

The use of a plastic theory requires that the material involved actually behave plastically. Figure 10.12.1(a) shows the ideal relationship of moment M to curvature

Figure 10.12.1 Moment–curvature characteristics.

Figure 10.12.2 Limit condition for simply supported beam.

θ, where there is a perfectly elastic portion and an ideal plastic portion. Reinforced concrete exhibits an M/θ curve as shown in Fig. 10.12.1(b) which, although it differs markedly from the ideal, may be approximated by such an ideal system. It has been found that the lower the net reinforcement ratio $(\rho - \rho')$, the closer is the actual moment—rotation behavior to the ideal.

Consider the simply supported beam of Fig. 10.12.2. The limit load on such a system is reached when the rotation angle under the load reaches the value θ_u which corresponds to the extreme fiber of concrete in compression reaching the crushing strain ϵ_{cu} (taken by ACI as 0.003). In limit analysis, the moment strength achieved when ϵ_{cu} is reached at the extreme concrete fiber is referred to as M_n (nominal moment strength), which corresponds to M_p, the plastic moment in the ideal system. For such simply supported beams the achieving of M_p or M_n at one location along the span represents the limit of the system, and the beam becomes a mechanism, deforming further without inducing further resistance. It will be shown that in a continuous structure (statically indeterminate system), the material at a section where M_p is first reached must undergo additional strain before that structure achieves its limit condition. This additional strain, with regard to bending of a section, is referred to as "rotation capacity." Reinforced concrete has a reasonable amount of rotation capacity when the net reinforcement percentage ρ or $\rho - \rho'$ is 50% or less of the balanced amount ρ_b for a singly reinforced beam.

Next, consider the statically indeterminate fixed-end beam of Fig. 10.12.3(a), showing its moment diagram in the elastic range. As the load is increased, the moments at the fixed ends achieve M_p, while at all other points moments are still in the elastic range. Thus, at the point of reaching M_p at the ends,

$$\frac{w_y L^2}{12} = M_p \qquad \text{and} \qquad w_y = \frac{12 M_p}{L^2}$$

where w_y is the load carried at an end curvature θ_y (see Fig. 10.12.1) when M_p is just reached. In carrying this load, the beam is as stable as a simple beam and deflects less because of the end moments M_p. Figure 10.12.3(b) shows the limit condition of three plastic hinges (one more than the number of degrees of statical indeterminacy) and the associated moment diagram. The limit load may be computed using statics,

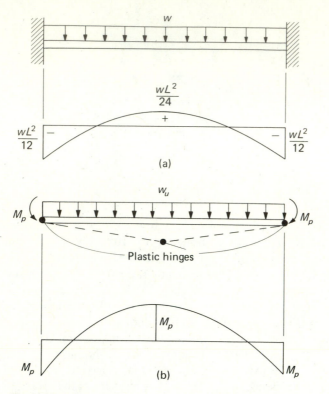

Figure 10.12.3 Limit condition for a fixed-end beam.

such as the midspan equilibrium requirement

$$+M_p = -M_p + \text{simple beam moment}$$

$$2M_p = \frac{w_u L^2}{8}$$

$$w_u = \frac{16M_p}{L^2}$$

Thus the capacity may be increased 33% after the plastic moment has been reached at the fixed ends, provided that sufficient additional deformation (rotation) can be accommodated at the fixed ends to permit development of the plastic moment at midspan. In other words, rotation capacity permitting, the positive and negative moments under any particular loading condition tend to equalize (assuming, of course, the strength of the section at the two regions is the same). For a thorough treatment of limit analysis the reader is referred to the work of Baker [10.8].

The aforementioned limit behavior has been thoroughly verified for structural steel, and its use under the term "plastic design" has become widespread. Steel being a very ductile material, rotation capacity in beams is available to a high degree. On the other hand, concrete, being a relatively brittle material, has traditionally been

Figure 10.12.4 Strain diagram for reinforced concrete beams wherein steel reaches ϵ_y before concrete $\epsilon_c = \epsilon_{cu}$.

considered to have no appreciable plastic deformability. Reinforced concrete beams, when the net percentage of reinforcement $\rho - \rho'$ is low, will have their moment strength controlled by yielding of the steel while the concrete strain is still of low magnitude (see Fig. 10.12.4). Reserve rotation $\theta_u - \theta_y$ is then available for a redistribution of moments to occur before the ultimate concrete strain of about 0.003 is reached.

Research has shown that such a redistribution does occur in reinforced concrete beams when certain rotation capacity related conditions are met; therefore the ACI Code (ACI-8.4) allows the negative moments at the supports of continuous flexural members, calculated by elastic theory, to be increased or decreased by not more than

$$20\left(1 - \frac{\rho - \rho'}{\rho_b}\right)\%$$

where ρ_b is the balanced percentage of reinforcement for a beam containing tension reinforcement only, given by Eq. (3.5.4) or ACI Formula (8-1),

$$\rho_b = \frac{0.85\beta_1 f'_c}{f_y}\left(\frac{87,000}{87,000 + f_y}\right)$$

The limits of applicability of this provision may be summarized as follows:

1. Application is limited to continuous flexural members.
2. No more than 20% of the negative moments for any given loading arrangement may be adjusted.

3. Bending moments used in such an adjustment must be obtained by an elastic analysis; moments from use of coefficients or other approximate methods may not be adjusted.
4. The *net* reinforcement ratio $\rho - \rho'$ at the cross section where the moment is *reduced* must not exceed one-half the balanced percentage ρ_b, as defined by ACI Formula (8-1).
5. Adjustment, when permitted, is made for each given loading condition. The envelope of the adjusted diagrams from all loading conditions is then used to proportion the members.

This method of partial redistribution of moments is generally conservative. Future codes may extend the method or go fully to a limit analysis approach once the plastic hinge behavior is more extensively understood. Such factors as shear, development of reinforcement, and deflection may still control the design; thus, limit analysis for flexure may offer little, if any, economic advantage.

In approaching a design, it may be advantageous to apply ACI-8.4 directly. It is likely, however, that more frequent use will occur when the design is in progress and the designer realizes that the conditions of the redistribution provision are met and savings appear possible. The following example illustrates an application of this provision.

■ **EXAMPLE 10.12.1**   Show the effects of redistribution on the moments obtained by elastic analysis for the floor beams $2B1$ and $2B2$ in Example 10.3.1.

Solution: The beams along with the critical moment diagrams are shown in Fig. 10.12.5. The elastic end moments M_u are those computed under factored loads and tabulated in Table 10.3.1. The elastic moment diagrams are shown in Figs. 10.3.3 through 10.3.6.

(a) Investigate negative moment region at B to determine the maximum percent moment adjustment. Referring to Fig. 10.4.1, and noting that $0.75\rho_b = 0.0278$ as used for design in Example 10.2.1,

$$\text{actual } \rho = \frac{4.54}{13(20)} = 0.0174$$

$$0.5\rho_b = 0.5\left(\frac{0.0278}{0.75}\right) = 0.0185 > 0.0174 \qquad \textbf{OK}$$

Moment redistribution is permissible at B.

$$\text{percent adjustment permitted} = 20\left(1 - \frac{\rho}{\rho_b}\right)$$

$$= 20\left(1 - \frac{0.0174}{2(0.0185)}\right) = 10.6\%$$

Since the steel in the compression zone was not fully developed at the faces of support (see Fig. 10.6.1), it may not be counted as compression steel. It might well be eco-

Figure 10.12.5 Redistribution of elastically computed moments according to ACI-8.4.

nomical to develop the capacity of the steel in the compression zone at the support. This would increase the ductility in that region and allow a higher percentage of moment redistribution.

(b) Make adjustments to elastic moments. Examine first the loading for maximum positive moment in span *AB*, Fig. 10.12.5(a). Increasing the negative moments by

10.6% reduces the maximum positive moment for this loading from 173 to 160 ft-kips. The increased negative moment for case (a) is still less than the negative moment occurring under the other loadings. The 10.6% adjustment for case (b) is made by reducing the negative moment at B and increasing those at A and C, thus minimizing the effect on the positive moments. The increased positive moment in span 1 is still slightly less than the reduced controlling positive moment of 160 ft-kips from case (a). In case (c), the negative moments are increased 10.6%, thus reducing the positive moment in span 2 from 130 to 112 ft-kips.

The adjustment of the negative moments may be either an increase or decrease so long as the positive moments are also adjusted to satisfy static equilibrium. The envelope of adjusted moments would then be used to design the sections by the strength method. The net effect on the envelope is a reduction for both positive and negative moments. This is not actually a reduction in the safety factor below that used for a simply supported beam. It is a reduction in the excess strength that the system has, by virtue of its continuity, one span with another. Since the redistribution does occur, partial utilization of it seems reasonable.

A simple application of limit design concepts to practical design has been presented by Furlong [10.10]. His method satisfies the ductility and strength requirements of limit design without the usual complexities that have impeded the acceptance of limit design into the ACI Code. Further discussion of Furlong's proposal has been presented [10.11] as a proposed alternate to the use of ACI-8.3.3. ∎

SELECTED REFERENCES

10.1.   CRSI. *Manual of Standard Practice* (25th ed.). Chicago: Concrete Reinforcing Steel Institute, 1990.

10.2.   Paul F. Rice, Edward S. Hoffman, David P. Gustafson, and Albert G. Gouwens. *Structural Design Guide to the ACI Building Code* (3rd ed.). New York: Van Nostrand Reinhold, 1985.

10.3.   A. H. Mattock. "Limit Design for Structural Concrete," *Journal of the Research and Development Laboratories*. Portland Cement Association, **1**, May 1959, 14–24.

10.4.   W. G. Corley. "Rotational Capacity of Reinforced Concrete Beams," *Journal of the Structural Division*, ASCE, **92**, ST5 (October 1966), 121–146. (Also PCA Development Department Bulletin D108.)

10.5.   ACI–ASCE Committee 428. "Progress Report on Code Clauses for 'Limit Design,'" *ACI Journal, Proceedings*, **65**, September 1968, 713–720. Disc, **66**, 221–223.

10.6.   M. Z. Cohn. "Limit Design for Reinforced Concrete Structures: An Annotated Bibliography," *ACI Bibliography No. 8*. Detroit: American Concrete Institute, 1970.

10.7.   Harold W. Conner, Paul H. Kaar, and W. Gene Corley. "Moment Redistribution in Precast Concrete Frame," *Journal of the Structural Division*, ASCE, **96**, ST3 (March 1970), 637–661.

10.8.   A. L. L. Baker. *Limit State Design of Reinforced Concrete*. London: Cement and Concrete Association, 1971.

10.9.   M. Z. Cohn. "Inelasticity of Reinforced Concrete and Structural Standards," *Journal of the Structural Division*, ASCE, **105**, ST11 (November 1979), 2221–2241.

10.10.   Richard W. Furlong. "Design of Concrete Frames by Assigned Limit Moments," *ACI Journal, Proceedings*, **67**, April 1970, 341–353.

10.11.   Richard W. Furlong and Carlos Rezende. "Alternate to ACI Analysis Coefficients," *Journal of the Structural Division*, ASCE, **105**, ST11 (November 1979), 2203–2220.

PROBLEMS

All problems* are to be worked in accordance with the strength method of the ACI Code, and all stated loads are service loads, unless otherwise indicated. A design sketch on $8\frac{1}{2} \times 11$ paper showing all design decisions is required for each problem. For verifying bar lengths, draw a moment capacity ϕM_n diagram directly below a side view of the beam showing the bars, as in Fig. 10.6.1.

Continuous Beam Problems

10.1   Design a rectangular beam continuous over three spans as shown in the accompanying figure. The live load is 2.75 kips/ft, and the dead load is 1.0 kip/ft in addition to the beam weight. The floor is to be a prefabricated system. Assume the supports to be 15 in. wide. Use $f'_c = 4000$ psi and $f_y = 60,000$ psi, and do not apply ACI-8.4 for moment redistribution. (Live load = 40 kN/m; dead load = 15 kN/m plus beam; supports 380 mm wide; $f'_c = 30$ MPa; $f_y = 400$ MPa.)

Problem 10.1

(a) Determine the moment envelope using factored loads.

(b) Determine the bar bends or cutoff locations, or both, directly from the moment envelope.

(c) Use only full-span loadings for computing the shear envelope and use U stirrups of #3 size if possible.

— — — —

* Many problems may be solved as problems stated in Inch-Pound units or as problems in SI units using quantities in parenthesis at the end of the statement. The SI conversions are approximate to avoid implying higher precision for the given information in metric units than that for the Inch-Pound units.

10.2 Redesign the beam of Problem 10.1 as a monolithic T-section floor system. The beams are spaced 8 ft on centers and the slab is 6 in. thick. The 1.0 kip/ft dead load includes the slab but not the beam stem. (Beam spacing = 2.4 m; slab thickness = 150 mm.)

10.3 Design the beam *ABC* of the frame shown in the accompanying figure in which the relative stiffnesses EI/L are given. The beams are T-sections having a 6-in. slab. The dead load is 0.40 kip/ft (not including beam stem or slab) and the live load is 3.75 kips/ft. Assume the supports to be 15 in. wide. Use $f'_c = 4000$ psi and $f_y = 60,000$ psi. (150 mm slab; dead load = 6 kN/m; live load = 55 kN/m; supports 380 mm wide; $f'_c = 30$ MPa; $f_y = 400$ MPa.)

Problem 10.3

10.4 Design the transverse beam indicated for the floor plan given in the accompanying figure. Assume that a warehouse live load of 375 psf is to be used. Assume a 5-in. slab placed monolithically with beams and girders, a width of support at longitudinal girders of 18 in., and that only a nominal minimum of moment restraint is provided by the exterior wall support (i.e., assume hinge for elastic analysis). Use $f'_c = 3500$ psi and $f_y = 60,000$ psi. For a comparison of the effect of considering torsional stiffness of longitudinal girders, divide the class into three parts, each using one of the following assumptions: **(a)** zero torsional stiffness of the two longitudinal girders; **(b)** torsional stiffness equal to 25% of the bending stiffness of the 21-ft span beam; and **(c)** torsional stiffness equal to 50% of the bending stiffness of the 21-ft span beam. (Live load = 18 kN/m²; 130 mm slab thickness; support width = 460 mm; $f'_c = 25$ MPa; $f_y = 400$ MPa.)

Note: For metric problems use all lengths in meters = 0.305 times lengths in feet rounded to two significant figures.

Section at wall

Problem 10.4

10.5   Redesign the transverse beams of Problem 10.4, except use spans 22-20-22 instead of the original spans 21-18-21, and use $f'_c = 4000$ psi and $f_y = 60,000$ psi. All other details are the same as in Problem 10.4. ($f'_c = 30$ MPa; $f_y = 400$ MPa.)

10.6   Redesign the transverse beams of Problem 10.4, except use spans 24-21-24, instead of original spans 21-18-21, and use $f'_c = 4000$ psi and $f_y = 60,000$ psi. All other details are the same as in Problem 10.4. ($f'_c = 30$ MPa; $f_y = 400$ MPa.)

10.7 Using the moment and shear envelopes of Figs. 10.8.3 and 10.8.5, design the actual lengths of reinforcement from the faces of support to the cutoff points for girder 2G1. Verify by drawing the moment capacity ϕM_n diagram superimposed on the factored moment M_u envelope.

10.8 Same as Problem 10.7 except use Fig. 10.8.4 for 2G2 instead of Fig. 10.8.3 for 2G1.

10.9 Design the four-span longitudinal girder indicated for the floor system of Problem 10.4. In lieu of using the more accurate loadings from the results of Problem 10.4, use concentrated dead and live loads of 15 and 60 kips, respectively, coming to the girder from *each* side, plus the weight of the girder. Assume that the columns are 18 in. square and 15 ft high and that the beam receives equivalent restraint from monolithic attachment to the 15-in. reinforced concrete exterior wall. Assume also that the columns are fixed at the far ends. Use $f'_c = 3500$ psi and $f_y = 60,000$ psi. (Concentrated dead and live loads from each side, 67 and 268 kN, plus girder; columns = 460 mm square and 4.5 m high; wall = 380 mm thick; $f'_c = 25$ MPa; $f_y = 400$ MPa.)

10.10 Redesign the floor beams 2B1-2B2-2B1 used in Example 10.2.1 and the following examples, taking the beam stem as 14 in. wide by 14 in. deep (i.e., use $h = 18.5$ in.).

10.11 Redesign the floor beams 2B1-2B2-2B1 used in Example 10.2.1 and the following examples, taking $f'_c = 4000$ psi and $f_y = 60,000$ psi. Make the stem size exactly as indicated by the text design guidelines (that is, do not make it arbitrarily larger as was done in the chapter example).

10.12 Design the continuous 26-ft span floor beams along a column line in Fig. 8.3.1. Assume the floor slab is 5 in. thick. Assume the columns are 15 in. square and 10 ft long. Use $f'_c = 4000$ psi, $f_y = 60,000$ psi, and the 100 psf live load. Make a preliminary design without structural analysis; then, using the preliminary size make the structural analysis to obtain the factored moment and factored shear envelopes for the final design.

More Extensive Continuous Beam (Frame) Problems

10.13 For the Case assigned by the instructor, use approximate procedures described in Section 10.2 to obtain preliminary size of the beam. If relative stiffnesses are given, assume you obtained them after obtaining the preliminary size of the beam, in which case use the given relative stiffnesses in making your structural analysis. If the relative stiffnesses are not given, use *your* computed relative stiffnesses from *your* preliminary design.

When the beams (girders) are supported on transverse beams (girders) rather than columns, assume that the torsional stiffness of the transverse beams (girders) may be computed according to ACI Formula (13-6) modified as follows to be used for a frame instead of a two-way slab system, or use an expression for K_t given by the instructor:

$$K_t = \sum \frac{E_c C}{2L_2}$$

Do the structural analysis (any method) to obtain the bending moment and shear envelopes. Use the correct placement of live loads to obtain the shear envelope, rather than the approximate full span loadings as was done in the chapter example.

A design objective is to obtain the shallowest beam that will satisfy the following deflection requirement (see Chapter 14):

$$(\Delta_i)_L + \Delta_{cp+sh} \le \frac{L}{480}$$

Assume none of the live load is sustained.

BASIC PROBLEM	LIVE LOAD	SLAB	f'_c	f_y
(a) Problem 10.3	3.75 kips/ft	6 in.	4000 psi	60,000 psi
(b) Problem 10.5	375 psf	6 in.	4000 psi	60,000 psi
(c) Problem 10.5	350 psf	7 in.	4000 psi	60,000 psi
(modified using spans 26-22-26; girder span 24 ft (beam spacing 8 ft)				
(d) Problem 10.6	375 psf	5 in.	4000 psi	60,000 psi
(e) Problem 10.9	conc loads	5 in.	3500 psi	60,000 psi
(f) Problem 10.11	100 psf	4.5 in.	4000 psi	60,000 psi
(g) Problem 10.12	100 psf	4.5 in.	4000 psi	60,000 psi

Concrete Joist Problems

10.14 Design a concrete joist, using 30-in. wide removable pans, for a typical interior span of 28 ft center to center of supporting girders. Assume a support width of 18 in. Use a live load of 100 psf, $f'_c = 4000$ psi, and $f_y = 40,000$ psi. (800-mm wide pans; span = 8.5 m; support width = 460 mm; $f'_c = 30$ MPa; $f_y = 300$ MPa.)

10.15 Determine the service live load capacity for a single-span joist of 20 ft clear span, using a $2\frac{1}{2}$-in. slab, 20 in. wide and 8-in. deep forms, and 4-in. wide joists with 2-#5 bars. No taper is used. Use $f'_c = 3000$ psi and $f_y = 40,000$ psi.

10.16 Determine the service live load capacity for an interior span joist of 32-ft span (clear span), using a 2-in. slab, 20-in. wide and 14-in. deep form, and 5-in. wide joists with #7 bars for the bottom, truss, and top steel. The bottom steel is properly embedded in the support to develop its compression capacity at the face of support. The joist has a standard taper. Use $f'_c = 3000$ psi and $f_y = 40,000$ psi.

10.17   Design an end-span joist for a continuous system to carry a live load of 225 psf, using 20-in. wide removable pans, for a clear span of 18 ft. Allow an extra $\frac{1}{2}$ in. of thickness for dead load purposes only, since the concrete slab is to serve as the final wearing surface. Use $f_c' = 4500$ psi and $f_y = 60,000$ psi.

Moment Redistribution Problem

10.18   Redesign the beam of Problem 10.1 taking into account permissible moment redistribution. Compare with Problem 10.1.

CHAPTER 11

Monolithic Beam-to-Column Joints

11.1 MONOLITHIC JOINTS

Considerable emphasis has been made concerning design of flexural members for bending, shear, and development of reinforcement in Chapters 3 through 10. Some attention has been given to development of reinforcement at exterior supports, including the use of hooks (Sections 6.11 and 6.14). The design of compression members is treated in Chapter 13. Often in design not enough attention is given to the details of connections—how the forces in beams and the forces in columns interact and get transmitted through the joint. The ACI Code provides little guidance specifically directed to joint details, except in Chapter 21—Special Provisions for Seismic Design, Section 21.6.

The state-of-the-art regarding the design of beam-to-column joints has been summarized by ACI-ASCE Committee 352, Joints and Connections in Monolithic Reinforced Concrete Structures [11.1]. In that report there are detailed provisions for the design of two classes of beam-to-column joints:

Type 1 joints, primarily for members designed to satisfy ACI Code strength requirements, where no significant inelastic deformations are expected; and

Type 2 joints, usually for members subject to earthquake or blast loading, where there is need for sustained strength under deformation reversals into the inelastic range.

In general, Type 1 joints require only nominal ductility, typical of those joints in a continuous moment-resisting structure designed to resist gravity and ordinary wind loads. The primary source of what follows is from the 1985 ACI-ASCE Committee 352 report [11.1], which also contains references to that date. Some of the more readily available references [11.2–11.5] are included at the end of this chapter, as well as references on the subject since 1985 [11.6–11.25].

Construction of tapered rigid frame knee for University of Wisconsin Stadium.
(Photo by C. G. Salmon.)

Some general concepts are discussed and two examples are presented using Committee 352 recommendations. A complete theoretical treatment is outside the scope of this book.

11.2 FORCES ACTING ON A JOINT

Just like the members themselves, the joints need to be designed for all forces that may act on them: axial load, bending moment, torsion, shear, as well as effects of creep, shrinkage, temperature, or settlement of supports. Assuming that the *members* have themselves been properly designed, the critical factor in joint design is the transmission into and through the joint of the forces that are present at the ends of the members. Figure 11.2.1 shows an interior joint with beams framing into it from all sides of a column.

Referring to Fig. 11.2.2, the forces T_1 and C_1 represent negative bending in a beam framing to a joint from the right side; the forces C_2 and T_2 represent positive bending in a beam framing in from the left side; the forces V_u (col.) represent the shears in the column just outside the joint. The shear within the joint that potentially

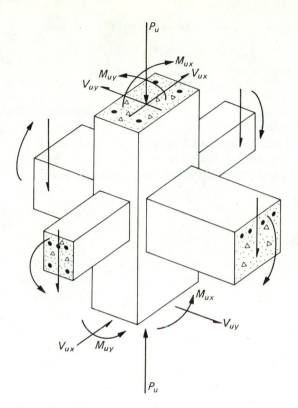

Figure 11.2.1 Forces on members at a joint.

Figure 11.2.2 Shear in a beam-to-column joint.

may cause the shear crack shown may be expressed

$$V_u = T_1 + T_2 - V_u \,(\text{col.})$$ **(11.2.1)**

or

$$V_u = f_y A_{st} + f_y A_{sb} - V_u \,(\text{col.})$$ **(11.2.2)**

ACI-ASCE Committee 352 [11.1, Section 3.2.1] recommends that the design forces used "should be determined from the nominal strengths of the members. . . . Strength reduction factors are not used." Thus, T_1 in Eq. (11.2.1) is $f_y A_{st}$ rather than the force computed from the bending moment determined by structural analysis at that location. In other words, the joint should be assumed to develop its full nominal strength.

11.3 COMPRESSION THROUGH A JOINT

In the design of a joint, it is difficult to have the column ties or spiral continue without interruption because of the crowded condition of the bars. When there is adequate lateral "confinement" by transverse members framing into a column at a *Type 1 joint,*

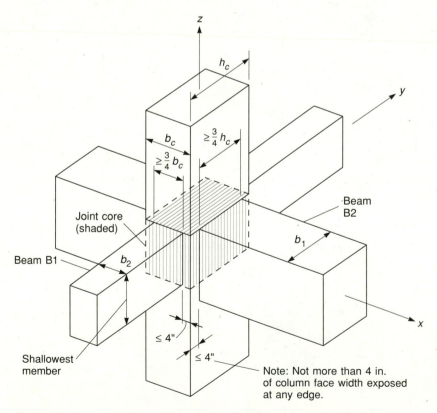

Figure 11.3.1 Confinement at a joint by members framing at all four column faces.

the transverse reinforcement (ties or spiral) may be omitted within the joint. ACI-ASCE Committee 352 [11.1, Section 4.2.1.2] has defined confinement as shown diagrammatically in Fig. 11.3.1. Within the depth of the *shallowest* member framing into a column is a region which may be called the *core*. The core is considered completely confined when beams frame in from all four sides and each beam has a width *b* at least three-fourths the column width and no more than 4 in. of column width is exposed on each side of the beam.

Where beams frame in from only two opposite sides, and their widths cover at least three-fourths of the column width at those sides, and not more than 4 in. of column width is exposed on each side of the beams, confinement occurs only at those faces. In this case, transverse reinforcement [Fig. 11.3.2] is needed through the joint only in the direction parallel to the two sides of the joint into which the beams frame. This provision has no effect when confinement is from a spiral, since to provide confinement the spiral must extend through the joint, thus providing *full* confinement (all directions) of the joint in any case. However, the restraint of ties to buckling of the column bars is directional. Conceivably, when using single hooked bars in addition to the closed ties going entirely around the column bars, fewer single hooked bars could be used in one direction than the other. Practically, in most cases the ties will have to extend through the joint unless there is confinement in the two orthogonal directions.

Column

Required transverse reinforcement (cross ties shown)

Required transverse reinforcement must be used through core at spacing along column axis not to exceed 12 in. for Type 1 joint.

≤4"

Figure 11.3.2 Confinement by members at two faces only.

When ties must extend through a joint, at least two layers of transverse reinforcement should be used between the top and bottom levels of beam reinforcement in the deepest member framing into the joint, and the center-to-center spacing should not exceed 12 in. [11.1, Section 4.2.1.3]. When the joint is part of the lateral load resisting system, the spacing should not exceed 6 in.

For a *Type 2 joint*, a minimum transverse reinforcement requirement applies whether or not adequate confinement is provided. In spirally reinforced columns, the volumetric ratio ρ_s must satisfy ACI-10.9.3, and must also satisfy the seismic requirement of ACI-21.4.4.1, as follows:

$$\rho_s \geq 0.12 \frac{f'_c}{f_{yh}} \tag{11.3.1}$$

When rectangular hoops and cross ties are used as transverse reinforcement as defined in ACI-21.4.4.1, the total cross-sectional area of such reinforcement required (see Fig. 11.3.3) is

$$A_{sh} \geq 0.3 s_h h_c \frac{f'_c}{f_{yh}} \left(\frac{A_g}{A_c} - 1 \right) \tag{11.3.2}$$

but not less than

$$A_{sh} = 0.09 s_h h_c \frac{f'_c}{f_{yh}} \tag{11.3.3}$$

Direction of shear being considered

$A_{sh} = 4(0.11) = 0.44$ sq in.
h_c = center-to-center of ties
d_h = center-to-center of ties
$A_{ch} = h_c d_h$

Figure 11.3.3 Rectangular hoop and cross-tie reinforcement in height s_h of column core.

where

A_{sh} = total area of all hoop and extra cross-tie legs crossing middepth of the section, within the spacing s_h in the direction considered

h_c = cross-sectional dimension of column core measured center-to-center of confining reinforcement, perpendicular to the transverse reinforcement area A_{sh} being designed

s_h = spacing of tie reinforcement measured along column bars (i.e., longitudinal axis of the column)

A_g = gross cross-sectional area of compression member

A_c = area of rectangular core measured out-to-out of transverse reinforcement

f_{yh} = yield stress of transverse reinforcement

11.4 SHEAR THROUGH A JOINT

As in any reinforced concrete design, design for shear strength must satisfy

$$\phi V_n \geq V_u \tag{11.4.1}$$

In the case of a joint, according to ACI-ASCE Committee Report [11.1, Section 4.3.1], the nominal shear strength V_n is

$$V_n = \gamma \sqrt{f'_c (\text{psi})} \, b_j h \tag{11.4.2}$$

where

b_j = effective *joint* width (see Fig. 11.4.1)

= column width b_c when design beam width b_b exceeds the column width

= $(b_b + b_c)/2$ when design beam width b_b is less than the column width

Plan views

Figure 11.4.1 Determination of effective joint width b_j. (From *Committee 352 Report* [11.1, Fig. 4.5].)

b_b = *design beam width* in the direction of loading. When there is a beam on only one face of the column in the direction of load, b_b is the width of that beam. When there are beams on the opposite faces of the column, b_b is the average of the widths of those beams.

h = thickness of the column in the direction of the load being considered

γ = constant that depends on the joint type and joint classification: for *Type 1* joints, $\gamma = 24$, 20, and 15 for interior, exterior, and corner joints, respectively; for *Type 2* joints, $\gamma = 20$, 15, and 12 for interior, exterior, and corner joints, respectively.

11.5 DESIGN EXAMPLES

Instead of completely listing all of the requirements indicated in the Committee 352 report, two examples will be presented to illustrate the recommendations as they apply to Type 1 joints used in ordinary building construction. The requirements are similar but more stringent for Type 2 joints, where ductility and dissipation of energy into the inelastic range are required.

■ **EXAMPLE 11.5.1**   Design the exterior beam–column joint shown in Fig. 11.5.1. The joint is to be a Type 1 joint where strength is the primary criterion. Use $f'_c = 4000$ psi and $f_y = 60,000$ psi.

Solution: **(a)** Examine the embedment situation for the 4-#10 bars in the main beam. From Table 3.9.4, the minimum width of beam is 20.5 in. to qualify this development length situation as Category A, and from Table 3.9.3 the minimum is 16.7 in. to qualify for Category B. Thus, Category C would apply; $\alpha_{abc} = 2.0$. From Tables 6.7.1 and 6.7.2, the basic development length L_{db} is 48.2 in. and the Min L_d is 36.1 in. Thus, for straight embedment, the "first-level L_d" is

$$\alpha_{abc}\alpha_{de}L_{db} = 2.0(1.0)48.2 = 96.4 \text{ in.}$$

Since the #10 are top bars, $\alpha_{top} = 1.3$. Thus, the straight bar development length L_d is

$$L_d = \alpha_{top} \times \text{(first-level } L_d) = 1.3(94.4) = 125 \text{ in.}$$

This is obviously impossible to provide in the 20 in. column. If the bars are to be developed, hooks must be used.

Referring to Fig. 6.11.1 for the definition of L_{dh}, the required development length L_{dh} equals a basic value L_{hb} multiplied by modification factors. For a #10 hooked bar, according to Eq. (6.11.1),

$$L_{hb} = \frac{1200d_b}{\sqrt{f'_c}} = \frac{1200(1.27)}{\sqrt{4000}} = 24.1 \text{ in.}$$

Figure 11.5.1 Design Example 11.5.1.

When minimum side cover of $2\frac{1}{2}$ in. and minimum cover of 2 in. on the tail of a 90°
hook are both available, the modification factor 0.7 may be used. For this joint (Fig.
11.5.1), since the 15-in. wide beam is much narrower than the 20-in. column, the side
cover will be satisfied, and the 2-in. cover on the tail (Fig. 11.5.2) also will be satisfied.

Figure 11.5.2 Dimensions for embedment into column.

Thus, the requirement is

$$L_{dh} = 24.1(0.7) = 16.9 \text{ in.}$$

If ties (vertical in this case) were to enclose the hooks in accordance with ACI-12.5.3.3, an additional modification factor of 0.8 could be used, reducing the required L_{dh} to $16.9(0.8) = 13.5$ in. Vertical ties would overcrowd this joint and should not be used. Allowing 2 in. for cover on the tail and assuming the critical section to be at the exterior face of the column steel (see Fig. 11.5.2) would leave available $L_{dh} = 20 - 4 = 16$ in., slightly less than the 16.9 in. required.

Try 5-#9 bars,

$$L_{hb} = \frac{1200(1.0)}{\sqrt{4000}} = 19.0 \text{ in.}$$

$$L_{dh} = L_{hb} (0.7) = 13.3 \text{ in.} < [\text{available } L_{dh} = 16 \text{ in.}] \qquad \textbf{OK}$$

Use 5-#9 instead of 4-#10

(b) Examine the shear on the column to be transmitted through the joint. Referring to Fig. 11.5.3, the moment on the columns may be assumed to be zero at midheight (or, preferably, the actual moment diagram would be used). In this case, with 12-ft and 10-ft column lengths, the factored column shear times 11 ft equals the factored moment at the end of the beam.

$$V_u \text{ (for column)} \left(\frac{h_1 + h_2}{2} \right) = M_u = 0.90 \, M_n \text{ (for beam)}$$

The strength of the beam (5-#9, $d = 21.6$ in.) is

$$C = 0.85(4)(15)a = 51.0a; \qquad T = 5(1.0)60 = 300 \text{ kips}$$

$$a = 5.88 \text{ in.}$$

$$M_n = 300(21.6 - 2.94)\tfrac{1}{12} = 466 \text{ ft-kips}$$

$$M_u = 0.90(466) = 420 \text{ ft-kips}$$

$$V_u \text{ (for column)} = \frac{420}{11} = 38.2 \text{ kips}$$

Figure 11.5.3 Column shear at joint.

Note that in cases where there is lateral loading on the building frames, V_u for column would have been obtained independently from structural analysis, and the V_u value above the joint may be different from the V_u value below the joint. The shear on the column through the joint is, in this case,

$$\text{joint } V_u = 0.90(300) - 38.2 = 232 \text{ kips}$$

The Committee 352 recommendation is

$$\text{joint } V_u = A_{st}f_y - V_u \text{ (for column)}$$
$$= (5.00)(60) - 38.2 = 262 \text{ kips}$$

The nominal strength of the joint is given by Eq. (11.4.2). The design beam width b_b *in the direction of loading* is 15 in. The effective joint width b_j when the design beam width b_b is less than the column width is

$$b_j = (b_b + b_c)/2 = (15 + 20)/2 = 17.5 \text{ in.}$$

The value γ is 20 for an exterior Type 1 joint. However, according to ACI-ASCE Committee 352 [11.1, 4.3.2], this joint is classified as a *corner* joint having γ of 15 rather than an exterior joint. Note that, in order for this joint to be classified as an exterior joint, the transverse spandrel beams must cover at least three-fourths of the width of the column and the total depth of such members must not be less than three-fourths of the total depth of the deepest member framing into the joint. Here the spandrel beams ($b = 10$ in.) do *not* cover three-fourths of the column width ($h = 20$ in.). Note particularly that the definition of an exterior joint does not relate to the definition of confinement. The confinement definition does not relate to the depth of the members framing into the joint; instead, the confinement region extends only to the upper and lower extremeties of the shallowest member.

The nominal shear strength is

$$V_n = \gamma\sqrt{f_c'(\text{psi})}\, b_j h$$
$$V_n = 15\sqrt{4000}(17.5)20\tfrac{1}{1000} = 332 \text{ kips}$$

[11.4.2]

Since this exceeds the joint V_u, the joint strength is adequate for shear.

(c) Horizontal column ties through the joint. As discussed in Section 11.3, a core is not considered confined unless beams frame in from all four sides and cover at least 0.75 of the column width, and not more than 4 in. of column width is exposed on each side of the beam. In this case, the joint is *not confined*. Thus the column ties designed to satisfy ACI-7.10 (or more closely spaced as discussed below) must extend through the joint. According to ACI-7.10.5.2, the spacing required for #3 ties is the smallest of

6 longitudinal bar diameters = 16(1.128) = 18 in.

48 tie diameters = 48(0.375) = 18 in.

least cross-sectional dimension of column = 20 in.

Figure 11.5.4 Joint for Example 11.5.1.

The above would not control for this joint, since an 18 in. spacing would indicate only one tie within the joint. Committee 352 [11.1, 4.2.1.3] recommends that *two layers* of transverse reinforcement be placed between the top and bottom levels of longitudinal reinforcement, and that the spacing not exceed 12 in.

When the joint is part of the primary lateral load resisting system, the spacing should not exceed 6 in. Assume this joint is part of the lateral load resisting system.

Use #3 ties @ 5 in. as detailed in Fig. 11.5.4. ■

■ **EXAMPLE 11.5.2**   For the interior joint shown schematically in Fig. 11.5.5, determine the shear reinforcement required if the joint is Type 1. Use f'_c = 4000 psi and f_y = 60,000 psi. Assume P_u of 250 kips exceeds $0.4P_b$ for the column.

Solution: (a) Development of reinforcement. All longitudinal steel is to be extended through the joint.

(b) Shear in direction of 16 × 24 beams. Due to lateral loading (from left to right) on the structure, the moments [Fig. 11.5.6(a)] M_{ut} and M_{ub} have the same rotational direction and give the highest shear through the joint.

Figure 11.5.5 Joint for Example 11.5.2.

Figure 11.5.6 Forces in direction of 16 × 24 beams, Example 11.5.2.

Again, it is assumed that the factored moment M_u acting equals the usable strength ϕM_n. The strength M_{nt} based on tension in the 4-#9 bars is

$$C = 0.85f'_c ba = 0.85(4)(16)a = 54.4a$$

$$T = 4(1.0)60 = 240 \text{ kips}$$

$$a = 4.41 \text{ in.}$$

$$M_{nt} = 240(21.5 - 2.20)\tfrac{1}{12} = 386 \text{ ft-kips}$$

$$M_{ut} = \phi M_{nt} = 0.90(386) = 347 \text{ ft-kips}$$

The strength M_{nb} based on tension in the 4-#8 bars is

$$C = 54.4a; \qquad T = 4(0.79)60 = 190 \text{ kips}$$

$$a = 3.48 \text{ in.}$$

$$M_{nb} = 190(21.5 - 1.74)\tfrac{1}{12} = 313 \text{ ft-kips}$$

$$M_{ub} = \phi M_{nb} = 0.90(313) = 282 \text{ ft-kips}$$

$$V_u \text{ (on col.)} = \frac{M_{ut} + M_{ub}}{11} = \frac{347 + 282}{11} = 57 \text{ kips}$$

The net shear through the joint [Fig. 11.5.6(b)] is

$$V_u = 190 + 240 - 57 = 373 \text{ kips}$$

The nominal strength of the joint is given by Eq. (11.4.2). The design beam width b_b *in the direction of loading* is the average width of the two beams; i.e., 16 in. since they are both 16 × 24. The effective joint width b_j when the design beam width b_b

is less than the column width is

$$b_j = (b_b + b_c)/2 = (16 + 20)/2 = 18 \text{ in.}$$

The value γ is 24 for an interior Type 1 joint. This joint is classified as an interior joint, according to ACI-ASCE Committee 352 [11.1, 4.3.2], not only because there are beams on all four sides of the column but also because the transverse beams *do* cover at least three-fourths of the width of the column and the total depth of such members is *not less than* three-fourths of the total depth of the deepest member framing into the joint. Here, the transverse beams ($b = 15$ in.) do cover three-fourths of the column width ($h = 20$ in.), and the 20 in. depth of such beams equals at least three-fourths of the deepest member (24 in.) framing into the joint.

The nominal shear strength is

$$V_n = \gamma \sqrt{f'_c(\text{psi})}\, b_j h$$
$$V_n = 24\sqrt{4000}(18)20\tfrac{1}{1000} = 546 \text{ kips}$$

[11.4.2]

Since this exceeds the joint V_u of 373 kips, the joint strength is adequate for shear.

(c) Shear in the direction of the 15×20 beam. If the lateral loading on the structure requires unbalanced moments in this direction, similar to the force system in Fig. 11.5.6, then the process used in step (b) would be repeated for this direction.

(d) Horizontal column ties through the joint. As discussed in Section 11.3, a core is considered confined when beams frame in from all four sides and cover at least 0.75 of the column width and no more than 4 in. of column width is exposed on each side of the beam. In this case, the joint *is* confined. Thus the column ties designed to satisfy ACI-7.10 need not extend through the joint. ∎

The factors relating to the design of joints intended to resist primarily static loads (Type 1) have been discussed and treated using two examples. For the design of Type 2 joints requiring greater ductility, the reader is referred to the Committee 352 report [11.1].

Beam–to–column connections resisting seismic loads have been given considerable attention by researchers, with the subject reviewed by Meinheit and Jirsa [11.5]. The behavior of beam–to–column connections in precast concrete has been studied by Pillai and Kirk [11.26]. Connections for composite construction have been studied by Sheikh, Deierlein, Yura, and Jirsa [11.27, 11.28].

SELECTED REFERENCES

11.1.   ACI–ASCE Committee 352. "Recommendations for Design of Beam–Column Joints in Monolithic Reinforced Concrete Structures," *ACI Journal, Proceedings*, **82**, May–June 1985, 266–283.

11.2.   John Minor and James O. Jirsa. "Behavior of Bent Bar Anchorages," *ACI Journal, Proceedings*, **72**, April 1975, 141–149.

11.3. James K. Wight and Mete A. Sozen. "Strength Decay of Reinforced Concrete Columns under Shear Reversals," *Journal of the Structural Division*, ASCE, **101**, ST5 (May 1975), 1053–1065.

11.4. Jose G. L. Marques and James O. Jirsa. "A Study of Hooked Bar Anchorages in Beam–Column Joints," *ACI Journal, Proceedings*, **72**, May 1975, 198–209.

11.5. Donald F. Meinheit and James O. Jirsa. "Shear Strength of R/C Beam–Column Connections," *Journal of the Structural Division*, ASCE, **107**, ST11 (November 1981), 2227–2244.

11.6. Erik Skettrup, Jørgen Strabo, Niels Houmark Anderson, and Troels Brøndum-Nielson. "Concrete Frame Corners," *ACI Journal, Proceedings*, **81**, November–December 1984, 587–593.

11.7. M. R. Ehsani and J. K. Wight. "Effect of Transverse Beams and Slab on Behavior of Reinforced Concrete Beam-to–Column Connections," *ACI Journal, Proceedings*, **82**, March–April 1985, 188–195.

11.8. Ahmad J. Durrani and James K. Wight. "Behavior of Interior Beam–to–Column Connections Under Earthquake-Type Loading," *ACI Journal, Proceedings*, **82**, May–June 1985, 343–349.

11.9. M. R. Ehsani and J. K. Wight. "Exterior Reinforced Concrete Beam–to–Column Connections Subjected to Earthquake-Type Loading," *ACI Journal, Proceedings*, **82**, July–August 1985, 492–499.

11.10. Prabhakara Bhatt and D. W. Kirk. "Tests on an Improved Beam Column Connection for Precast Concrete," *ACI Journal, Proceedings*, **82**, November–December 1985, 834–843.

11.11. T. Ueda, I. Lin, and N. M. Hawkins. "Beam Bar Anchorage in Exterior Column–Beam Connections," *ACI Journal, Proceedings*, **83**, May–June 1986, 412–422.

11.12. Charles W. Dolan, John F. Stanton, and Richard G. Anderson. "Moment Resistant Connections and Simple Connections," *PCI Journal*, **32**, March–April 1987, 62–74.

11.13. Ahmad J. Durrani and James K. Wight. "Earthquake Resistance of Reinforced Concrete Interior Connections Including a Floor Slab," *ACI Structural Journal*, **84**, September–October 1987, 400–406.

11.14. Daniel P. Abrams. "Scale Relations for Reinforced Concrete Beam–Column Joints," *ACI Structural Journal*, **84**, November–December 1987, 502–512.

11.15. S. E. El-Metwally and W. F. Chen. "Moment–Rotation Modeling of Reinforced Concrete Beam–Column Connections," *ACI Structural Journal*, **85**, July–August 1988, 384–394.

11.16. Parviz Soroushian and Ki-Bong Choi. "Local Bond of Deformed Bars with Different Diameters in Confined Concrete," *ACI Structural Journal*, **86**, March–April 1989, 217–222.

11.17. Olga Velez Ammerman and Catherine Wolfgram French. "R/C Beam–Column–Slab Subassemblages Subjected to Lateral Loads," *Journal of Structural Engineering*, ASCE, **115**, 6 (June 1989), 1289–1308.

11.18. Roberto T. Leon. "Interior Joints with Variable Anchorage Lengths," *Journal of Structural Engineering*, ASCE, **115**, 9 (September 1989), 2261–2275.

11.19. Thomas Paulay. "Equilibrium Criteria for Reinforced Concrete Beam–Column Joints," *ACI Structural Journal*, **86**, November–December 1989, 635–643.

11.20. Catherine Wolfgram French, Michael Hafner, and Viswanath Jayashankar. "Connections between Precast Elements—Failure within Connection Region," *Journal of Structural Engineering*, ASCE, **115**, 12 (December 1989), 3171–3192.

11.21. Roberto T. Leon. "Shear Strength and Hysteretic Behavior of Interior Beam–Column Joints," *ACI Structural Journal*, **87**, January–February 1990, 3–11.

11.22. S. J. Pantazopoulou and J. P. Moehle. "Truss Model for 3-D Behavior of R. C. Exterior Connections," *Journal of Structural Engineering*, ASCE, **116**, 2 (February 1990), 298–315.

11.23. M. R. Ehsani and J. K. Wight. "Confinement Steel Requirements for Connections in Ductile Frames," *Journal of Structural Engineering*, ASCE, **116**, 3 (March 1990), 751–767.

11.24. M. Seckin and H. C. Fu. "Beam–Column Connections in Precast Reinforced Concrete," *ACI Structural Journal*, **87**, May–June 1990, 252–261.

11.25. Philip K. C. Wong, M. J. N. Priestley, and R. Park. "Seismic Resistance of Frames with Vertically Distributed Longitudinal Reinforcement in Beams," *ACI Structural Journal*, **87**, July–August 1990, 488–498.

11.26. S. U. Pillai and D. W. Kirk. "Ductile Beam–Column Connection in Precast Concrete," *ACI Journal*, *Proceedings*, **78**, November–December 1981, 480–487.

11.27. Tauqir M. Sheikh, Gregory G. Deierlein, Joseph A. Yura, and James O. Jirsa. "Beam–Column Moment Connections for Composite Frames: Part 1," *Journal of Structural Engineering*, ASCE, **115**, 11 (November 1989), 2858–2877.

11.28. Gregory G. Deierlein, Tauqir M. Sheikh, Joseph A. Yura, and James O. Jirsa. "Beam–Column Moment Connections for Composite Frames: Part 2," *Journal of Structural Engineering*, ASCE, **115**, 11 (November 1989), 2877–2896.

PROBLEMS

All problems are to be worked using the ACI Code and the recommendations of ACI Committee 352 [11.1].

11.1 Check the anchorage of bars and determine the shear reinforcement for the exterior beam–column joint of the accompanying figure. Assume lateral loading provides the critical condition at the joint; assume columns bent in double curvature with an inflection point at midheight. Consider as a Type 1 joint and use $f'_c = 4000$ psi and $f_y = 60,000$ psi. Show detail of shear reinforcement in joint.

Problem 11.1

11.2   Repeat Problem 11.1, except consider the column to be 18 in. square, the spandrel 14×24 with 3-#9, and the beam 18×28 with 4-#10 bars.

11.3   Repeat Problem 11.1, except consider the joint to be an interior one with the 21×28 beam on both sides of the column. The spandrel then becomes an interior beam. Assume the 21×28 beam has 4-#9 in the bottom in addition to the bars in the top and that all bars are continuous through the column. The 18×30 beam has 3-#9 in the bottom in addition to the top bars, all bars being continuous through the column. The critical loading condition for the joint is with clockwise moments from the 21×28 beam acting on the joint at both sides of the column.

CHAPTER 12

Cantilever Retaining Walls

12.1 TYPES OF RETAINING STRUCTURES

Retaining structures hold back soil or other loose material and prevent its assuming the natural angle of repose at locations where an abrupt change in ground elevation occurs. The retained material exerts a push on a structure and thus tends to overturn or slide it or both. There may be several types of retaining structures (Fig. 12.1.1) as follows.

1. *Gravity Wall*. A gravity wall is usually of plain concrete and depends entirely on its weight for stability. It is used for walls up to about 10 ft high.

2. *Cantilever Retaining Wall*. The cantilever is the most common type of retaining structure and is used for walls in the range of 10 to 25 ft in height. The stem, heel, and toe of such a wall each acts as a cantilever beam.

3. *Counterfort Wall*. In the counterfort wall the stem and slab are tied together by counterforts, which are transverse walls spaced at intervals and act as tension ties to support the stem wall. Counterfort walls are often economical for heights over about 25 ft.

4. *Buttress Wall*. A buttress wall is similar to a counterfort wall except that the transverse support walls are located on the side of the stem opposite to the retained material and act as compression struts. Buttresses, as compression elements, are more efficient than the tension counterforts and are economical in the same height range. A counterfort is more widely used than a buttress because the counterfort is hidden beneath the retained material, whereas the buttress occupies what may otherwise be usable space in front of the wall.

5. *Bridge Abutment*. A wall-type bridge abutment acts similarly to a cantilever retaining wall except that the bridge deck provides an additional horizontal

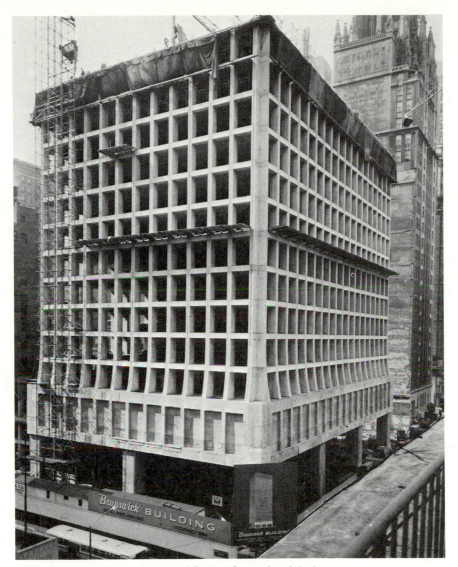

Brunswick Building, Chicago. (Courtesy of Portland Cement Association.)

restraint at the top of the stem. Thus this abutment is designed as a beam fixed at the bottom and simply supported or partially restrained at the top.

6. *Box Culvert.* The box culvert, with either single or multiple cells, acts as a closed rigid frame that must not only resist lateral earth pressure but also vertical load from either the soil that it supports or from both the soil and the highway vehicle loads.

(a) Gravity wall

(b) Cantilever retaining wall

(c) Counterfort wall

(d) Buttress wall

(e) Bridge abutment

(f) Box culvert

Figure 12.1.1 Types of retaining structures.

12.2 FORCES ON RETAINING WALLS

The magnitude and direction of the earth pressure that tends to overturn and slide a retaining wall may be determined by applying the principles of soil mechanics. Excellent texts are available, such as those of Terzaghi and Peck [12.1] and Huntington [12.2], for any extensive study of how to determine the soil pressure to be used in any given situation.

The pressure exerted by the retained material is proportional to the distance below the earth surface and to its unit weight. Analogous to the action of a fluid, the unit pressure p at a distance h below the earth surface may be expressed

$$p = Cwh \qquad (12.2.1)$$

where w is the unit weight of the retained material and C is a coefficient that depends on the physical properties of the material.

There are two categories of earth pressure: (1) the pressure exerted as the earth moves in the same direction as the retaining structure deflects, known as *active pressure*, and (2) the resistance developed as a structure moves against the earth, known as *passive pressure*. Passive pressure is several times larger than active pressure. Both active and passive pressures may be expressed in the form of Eq. (12.2.1), but using C_a and C_p as the active and passive pressure coefficient C, respectively.

The force P_a caused by active pressure on a wall of height h may be expressed

$$P_a = C_a w \frac{h^2}{2} \qquad (12.2.2)$$

and the force P_p developed in passive pressure may be expressed

$$P_p = C_p w \frac{h^2}{2} \qquad (12.2.3)$$

where $C_a w$ and $C_p w$ in Eqs. (12.2.2) and (12.2.3) may be considered equivalent fluid pressures. Typical values for C_a and C_p are 0.3 and 3.3, respectively, for granular material such as sand. Roughly, $C_p = 1/C_a$.

The proper evaluation of earth pressures on retaining structures is outside the scope of this book; the preceding brief comments are offered to provide some logic for the earth pressures that are used in the design example of this chapter.

The factors that may affect the pressure (active pressure) on a wall are as follows:

1. Type of backfill used.
2. Seasonal condition of the backfill material, such as wet, dry, or frozen.
3. Drainage of backfill material.
4. Possibility of backfill overload, such as trucks and equipment near the wall.
5. Degree of care exercised in backfilling.
6. Degree of rotational restraint between various components of the retaining structure.
7. Possibility of vibration in the vicinity of the wall (especially in the case of granular soil).
8. Type of material beneath the footing of the retaining structure.
9. Level of the water table.

Probably the most important single factor is that water must be prevented from accumulating in the backfill material. Walls are rarely designed to retain saturated material, which means that proper drainage must be provided.

When vehicles may travel near and exert their loads or when buildings are constructed near the top of a retaining wall, the lateral pressure against such a wall is increased. In the case of a fixed static load such as a building, the weight of the building can be converted into an additional height (surcharge) of backfill soil material. The effect of a highway or railroad passing over the retained material near the wall causes a dynamic reaction that cannot accurately be converted into a static effect. However, the AASHTO highway bridge specification (Section 3.20.3) [12.3] and the AREA railroad retaining wall specification [12.4] prescribe an equivalent static surcharge corresponding to a number of additional feet of backfill material.

12.3 STABILITY REQUIREMENTS

The first step in retaining wall design is to establish the proportions such that the stability of the structure (see Fig. 12.3.1) under active pressure is assured. Three requirements must be satisfied: (a) the overturning moment $P_{ah}(h'/3)$ must be more than balanced by the resisting moment $Wx_1 + P_{av}L$, so that an adequate factor of safety against overturning is provided, usually about 2.0; (b) sufficient frictional resistance F in combination with any reliable passive resistance P_p against the toe must provide an adequate factor of safety (usually 1.5) against sliding caused by P_{ah}; and (c) the base width L must be adequate to distribute the load R to the foundation soil without causing excessive settlement or rotation.

Typically, referring to the pressure distribution of Fig. 12.3.1, the overturning factor of safety (FS) would be computed

$$FS = \frac{\text{resisting moment}}{\text{overturning moment}} = \frac{Wx_1 + P_{av}L}{P_{ah}(h'/3)} \qquad (12.3.1)$$

Figure 12.3.1 Forces on retaining wall.

Table 12.3.1   Values of Coefficient of Friction Between Soil and Concrete, 1958 AREA Specification

SOIL	μ
Coarse-grained soils (without silt)	0.55
Coarse-grained soils (with silt)	0.45
Silt	0.35
Sound rock (with rough surface)	0.60

or, perhaps more frequently, neglecting the vertical component of P_a,

$$\text{FS} = \frac{Wx_1}{P_{ah}(h'/3)} \tag{12.3.2}$$

where W represents the weight of the concrete wall and footing and of the soil resting on the footing.

The factor of safety against sliding may be computed, using the notation of Fig. 12.3.1,

$$\text{FS} = \frac{\mu R + P_p}{P_{ah}} \tag{12.3.3}$$

where μ is the coefficient of friction between the soil and footing. Table 12.3.1 gives the coefficients of friction as recommended by the 1958 AREA specification [12.4] and these may serve as a guide of typical values in lieu of more accurate ones.

The inclusion of some passive resistance P_p on the toe of the footing may or may not be justified. Certainly, to actually develop passive pressure in the soil in front of the wall, the concrete must have been placed without using forms for the toe and without disturbing the soil against which the concrete is placed.

Referring to Fig. 12.3.2, the ordinary passive resistance against the toe is

$$P_{p1} = \tfrac{1}{2}C_p w h_1^2 \tag{12.3.4}$$

Figure 12.3.2 Passive resistance and effect of base key.

but frequently is neglected, and in nearly all cases at least the value of h_1 is reduced under the assumption that during construction, or after, the earth surface cannot be expected to remain undisturbed at its final designated elevation. If, when all reliable resistances have been included, the factor of safety remains inadequate, a base key (Fig. 12.3.2) may be used. Essentially, the base key, when placed in an unformed excavation against undisturbed material, may be expected to develop an additional passive force P_{p2} and shift the possible failure plane from line 1 to line 2.

The base key develops the additional resistance

$$P_{p2} = \tfrac{1}{2}C_p w(h_2^2 - h_1^2) \tag{12.3.5}$$

and also, an inert region, *bced* of Fig. 12.3.2, is created which moves the friction plane from *bd* to *ce*. Thus the frictional force developed along *ce* is based on the angle α, the angle of internal friction of the soil, rather than on the friction angle between soil and concrete. Normally, $\tan \alpha > \mu$ for granular material, so that, by making the base key sufficiently deep to create an inert-block, additional frictional resistance is developed.

Finally, the magnitude and distribution of the soil pressure requires investigation. Usual practice is to require the resultant vertical force R to be inside the middle third of the footing for sand and gravel subbases and within the middle half for rock subbase. In addition, the maximum pressure may not exceed the allowable value. Comments regarding allowable bearing capacity as well as some typical safe bearing values are to be found in Chapter 20.

Referring to Fig. 12.3.3(a), when the entire footing is under compression, the basic equation for combined bending and axial compression acting on a 1-ft strip along the wall is

$$p = \frac{R}{L} \pm \frac{Re(L/2)}{L^3/12} = \frac{R}{L}\left(1 \pm \frac{6e}{L}\right) \tag{12.3.6}$$

(a) Resultant within middle third

(b) Resultant outside middle third

Figure 12.3.3 Soil pressure distribution.

For the limiting condition of zero stress at the heel, $e = L/6$; thus Eq. (12.3.6) is valid for all positions of R within the middle third.

When the resultant R is outside the middle third [Fig. 12.3.3(b)], vertical force equilibrium requires

$$R = \tfrac{1}{2}p_t(3x_2)$$

$$p_t = \frac{2R}{3x_2} \qquad\qquad (12.3.7)$$

for $0 < 3x_2 < L$.

12.4 PRELIMINARY PROPORTIONING OF CANTILEVER WALLS

To begin the design it is necessary to apply certain "rules" along with an approximate use of statics and to estimate dimensions such as the length and thickness of the base footing and the relative position of the wall with respect to the footing.

Height of Wall. Since the bottom of the footing must be below the frost penetration depth, say 3 to 4 ft in the northern United States, the overall height equals the desired difference in elevation plus the frost penetration depth.

Position of Stem on the Base Footing. The following demonstration using approximate statics will show that the front face of the wall should coincide with the desired position of the resultant soil pressure beneath the base. Take the most typical situation of a vertical wall with level backfill, as shown in Fig. 12.4.1. Assume an average unit weight w for all material (concrete and earth) enclosed within $abcd$ and neglect entirely the concrete in the toe. Thus

$$R = W = wh\gamma L$$

Figure 12.4.1 Data for economical proportioning of wall.

Satisfying rotational equilibrium about the heel,

$$P_a \frac{h}{3} + W \frac{\gamma L}{2} = R\xi L$$

$$C_a \frac{wh^2}{2} \left(\frac{h}{3} \right) + wh\gamma L \frac{\gamma L}{2} = wh\gamma L(\xi L)$$

Solving for L/h gives

$$\frac{L}{h} = \sqrt{\frac{C_a}{3\gamma(2\xi - \gamma)}} \qquad\qquad \textbf{(12.4.1)}$$

The variable ξ must be selected by the designer, based on the type of soil and desired pressure distribution. For good granular soil and acceptance of a triangular pressure distribution with the resultant at the outer edge of the middle third, $\xi = \frac{2}{3}$. For clay where a uniform distribution might be desired, ξ would be $\frac{1}{2}$.

Minimizing the L/h in Eq. (12.4.1) such that the base width is a minimum,

$$\gamma = \xi \qquad\qquad \textbf{(12.4.2)}$$

Thus the front face of the wall should line up with the desired position of the soil pressure resultant.

Length of Base.   A preliminary length of base may be obtained by the application of static moments with respect to point b in Fig. 12.4.1, employing the same assumptions that underlie Eq. (12.4.1). By this process a more exact result may be obtained since one can easily estimate the stem thickness and footing thickness as fractions of L or of h.

Thickness of Footing.   The base thickness is usually 7 to 10% of the total height h, with a minimum of about 1 ft. It should be about equal to the base thickness of the stem.

Thickness of Stem.   The thickness at the top of the wall is arbitrary; however, cover requirements and construction constraints will dictate how thin it may be. Generally 10 or 12 in. minimum is preferred, though no minimum is prescribed by the ACI Code.

The base thickness of the stem is determined as required for bending moment and shear, though it may be estimated as 12 to 16% of the base width or 10 to 12% of the wall height. The stem thickness should not be too skimpy, since a thin wall deflects considerably at the top and savings in reinforcement due to larger effective depth will tend to offset the cost of any extra concrete used. It is recommended that a batter of $\frac{1}{4}$ in./ft of height be provided on the front face to offset deflection or forward tilting of the structure.

12.5 DESIGN EXAMPLE—CANTILEVER RETAINING WALL

Design a cantilever retaining wall to support a bank of earth 16 ft high above the final level of earth at the toe of the wall. The backfill is to be level, but a building is to be built on the fill. Assume that an 8-ft surcharge will approximate the lateral earth pressure effect. Data: weight of retained material = 120 pcf; equivalent fluid pressure = 32 pcf; angle of internal friction = 35°; coefficient of friction between masonry and soil = 0.40; for passive pressure use equivalent fluid pressure of 400 pcf; f'_c = 3000 psi; f_y = 40,000 psi; maximum soil pressure = 5 ksf (kips per square foot). Use the strength method of the ACI Code.

(a) Basic design data. For adequate deflection control on the cantilever wall, choose to use a reinforcement percentage ρ about one-half the maximum permitted by the ACI Code. Choose

$$\rho \approx 0.375\rho_b = 0.0139 \quad \text{(Table 3.6.1)}$$

This corresponds to R_n = 495 psi.

(b) Height of wall. Allowing 4 ft for frost penetration to the bottom of the footing in front of the wall, the total height becomes

$$h = 16 \text{ ft} + 4 \text{ ft} = 20 \text{ ft}$$

(c) Thickness of footing. The thickness may be estimated at this stage of design to be 7 to 10% of the overall height h. Assume a uniform footing thickness, t = 2 ft (about 10% of h).

(d) Base length. One could determine the base length by considering the equilibrium of factored loads (using the factor 1.7 for the horizontal earth pressure and using 0.9 for the earth and concrete weight); this practice will give a conservative result. However, it seems within the intent of the ACI Code to establish base dimensions for foundation structures using actual unfactored service loads (see ACI-15.2.2).

Using a 1-ft length of wall and letting the unit weight of material bounded by points a, b, c, and d in Fig. 12.5.1 equal 120 pcf,

$$P_1 = 0.256(20)(1) = 5.12 \text{ kips}$$

$$P_2 = \tfrac{1}{2}(0.640)(20)(1) = 6.40 \text{ kips}$$

$$W = 0.120(20 + 8)x = 3.36x \text{ kips}$$

Summation of moments about point b gives

$$W\left(\frac{x}{2}\right) = P_1(10.0) + P_2(6.67)$$

$$\frac{3.36x^2}{2} = 5.12(10.0) + 6.40(6.67)$$

$$1.68x^2 = 93.9$$

$$x = 7.47 \text{ ft}$$

Since for this granular material the resultant soil pressure is desired to be at the outer edge of the middle third of the footing,

$$\text{base length} = 1.5x = 1.5(7.47) = 11.2 \text{ ft}$$

Try 11 ft 3 in.

(e) Stem thickness. Prior to computing soil pressures and stability factors of safety, a more accurate knowledge of the concrete dimensions is necessary. The

Figure 12.5.1 Preliminary proportioning for cantilever wall in design example.

thickness of the base of the stem is selected with due regard for the bending moment and shear requirements.

Bending moment will normally provide the governing criterion, for which the general expression (y measured from the top of the wall) is

$$M_y = \frac{0.256y^2}{2} + \frac{0.032y^3}{6} = 0.128y^2 + 0.00533y^3 \qquad \textbf{(12.5.1)}$$

At the bottom of the assumed 18-ft high stem wall, using a 1.7 overload factor,

$$M_u = 1.7[0.128(18)^2 + 0.00533(18)^3] = 123 \text{ ft-kips}$$

For a desired $R_n = 495$ psi,

$$\text{required } d = \sqrt{\frac{M_u}{\phi R_n b}} = \sqrt{\frac{123(12,000)}{0.90(495)12}} = 16.6 \text{ in.}$$

Total thickness = 16.6 + 2 (cover) + 0.5 (estimated bar radius) = 19.1 in. Try 21 in. for the stem base thickness and select 12 in. as the practical minimum for the top of the wall. Note that the selected R_n of 495 psi is a *chosen* guideline value and need not be rigidly adhered to. In this case the 3-in. multiple of 21 in. is preferred to the computed value of 20 in., and the somewhat thicker stem will reduce deflection.

The critical section for shear strength is permitted to be taken at a distance d from the bottom of the stem (ACI-11.1.3). Conservatively, the shear at the base of the wall can be used. The general expression for shear is

$$V_y = 0.256y + \tfrac{1}{2}(0.032)y^2 = 0.256y + 0.016y^2 \qquad \textbf{(12.5.2)}$$

At the base of the stem ($y = 18$), using a 1.7 overload factor,

$$V_u = 1.7[0.256(18) + 0.016(18)^2] = 16.6 \text{ kips}$$

$$\phi V_c = \phi(2\sqrt{f_c'})bd$$

$$= 0.85(2\sqrt{3000})(12)(\approx 18)\tfrac{1}{1000} = 20.1 \text{ kips} > 16.6 \text{ kips} \qquad \textbf{OK}$$

Since $V_u < \phi V_c$, in accordance with ACI-11.5.5.1 shear reinforcement is not required for slabs. The cantilever is treated as a slab in applying any ACI Code limitations.

Where appearance on the front face is important, a batter of that face should be provided so as to counteract the effect of deflection. The usual batter is about $\tfrac{1}{4}$ in./ft of wall height. Thus the minimum batter here is

$$\tfrac{1}{4}(18) = 4\tfrac{1}{2} \text{ in.}$$

In this case the thickness increases by 9 in. from the top to the bottom of the wall, so batter the front face 5 in. and the rear face 4 in. When the wall is in place, it will deflect forward and become nearly vertical; therefore, the analysis from this point will consider the front face to be vertical. Economics may dictate having only one sloping face, or in some cases no sloping face.

(f) Factor of safety against overturning. Using the dimensions given in Fig. 12.5.2, locate the resultant of vertical forces (without overload factors) with respect to the heel, as in Table 12.5.1:

Figure 12.5.2 Dimensions for computing resultant soil pressure in design example.

Table 12.5.1

FORCE	ARM	MOMENT
$W_1 = 0.120(18 + 8)(6.5) = 20.3$	3.25	66
$W_2 = 0.030(\frac{1}{2})(18)(0.75) = 0.2$	6.25	1
$W_3 = 0.150(11.25)(2.0) = 3.4$	5.63	19
$W_4 = 0.150(1.0)(18) = 2.7$	7.00	19
Totals	26.6	105

$$\text{resultant, from heel} = \frac{105}{26.6} = 3.95 \text{ ft}$$

$$\text{resisting moment} = 26.6(11.25 - 3.95) = 194 \text{ ft-kips}$$

$$\text{overturning moment} = P_1(10) + P_2(6.67)$$
$$= 5.12(10) + 6.40(6.67)$$
$$= 93.9 \text{ ft-kips}$$

$$\text{FS against overturning} = \frac{194}{93.9} = 2.07 > 2.0 \qquad \textbf{OK}$$

Alternatively, if the stability check is to be made using factored loads, according to ACI-9.2.4,

$$U = 0.9D + 1.7H$$

where H is the horizontal earth pressure. In such a case, it would be required that

$$\text{resisting moment} \geq \text{overturning moment}$$

$$0.9(194) \geq 1.7(93.9)$$

$$175 \geq 160 \qquad \textbf{OK}$$

In effect the required factor of safety against overturning under service load conditions is $1.7/0.9 = 1.89$, which is not much less than the traditional value of 2.0.

(g) Location of resultant and footing soil pressures. Referring to Fig. 12.5.3, and using service loads because the maximum soil pressure limitation is given for that

Figure 12.5.3 Soil pressure and location of resultant under service load in design example.

condition,

$$R = 26.6 \text{ kips}$$

$$\bar{x} = \frac{105 + 93.9}{26.6} = 7.48 \text{ ft}$$

$$e = 7.48 - \frac{11.25}{2} = 1.85 \text{ ft}$$

$$\frac{6e}{\text{base length}} = \frac{6(1.85)}{11.25} = 0.99 < 1.0$$

The resultant lies just inside of the middle third.

The service load soil pressure diagram is essentially a triangle; thus

$$R = \tfrac{1}{2}(p_{max})(\text{base length})$$

$$26.6 = \tfrac{1}{2}(p_{max})(11.25)$$

$$p_{max} = 4.73 \text{ ksf} < 5 \text{ ksf} \qquad \textbf{OK}$$

(h) Factor of safety against sliding. Neglecting passive pressure against the toe of the footing and using service loads,

$$\text{force causing sliding} = P_1 + P_2 = 5.12 + 6.40 = 11.52 \text{ kips}$$

$$\text{frictional force} = \mu R = 0.40(26.6) = 10.64 \text{ kips}$$

$$\text{factor of safety} = \frac{10.64}{11.52} = 0.92 < 1.5 \qquad \textbf{NG}$$

Sliding resistance may also be checked using factored loads, $U = 0.9D + 1.7H$, in which case it is required that

$$\text{resisting force} \geq \text{sliding force}$$

$$0.9(10.64) \geq 1.7(11.52)$$

$$9.58 \text{ kips} < 19.6 \text{ kips} \qquad \textbf{NG}$$

Thus ACI-9.2.4 would require the same factor of safety against sliding as against overturning, that is, $1.7/0.9 = 1.89$. This exceeds the traditional factor to resist sliding of 1.5.

Since the resistance provided does not give an adequate safety factor, a key (Figs. 12.5.4 and 12.5.5) against sliding is required. Such a key is intended to develop passive pressure in the region in front of and below the bottom of the footing. The procedure for determining the required size of a key is one that is debated by designers. Generally it seems desirable to place the front face of the key about 5 in. in front of the back face of the stem. This will permit anchoring the stem reinforcement in the key.

With full realization of the tendency of practitioners to try to oversimplify this problem, various procedures may reasonably be used for granular cohesionless materials. The maximum effect of a key would be to develop the passive resistance

Figure 12.5.4 Passive resistance concept of frictional resistance.

over the depth BC of Fig. 12.5.4, with any resulting failure occurring along a curved path such as $C'C$. Fisher and Mains [12.5] have advocated the inert-block concept of frictional resistance. In this method, any key used must be extended deep enough below the footing to develop an inert block of soil, $BCDE$ of Fig. 12.5.5; it will then have a failure plane approximately as $C'CDGFH$, so that the passive resistance is developed over only the depth BC. In determining the passive resistance, the top 1 ft of overburden is usually neglected in the height h_1, of Fig. 12.5.4 or Fig. 12.5.5.

For this design, a factor of safety of 1.5 against sliding under service loads has been considered proper. When passive resistance against the toe is also included, a higher factor of safety, say 2.0, should be used.

Using the inert-block concept, the passive force P_p developed over the distance BC of Fig. 12.5.5 is

$$\text{equivalent fluid pressure} = 400 \text{ pcf} \quad \text{(given)}$$

$$P_p = \frac{0.400(h_1 + a)^2}{2} - \frac{0.400(h_1)^2}{2}$$

Figure 12.5.5 Inert-block concept of frictional resistance.

which for $h_1 = 3$ ft gives

$$P_p = 0.200(6a + a^2)$$

By inducing an inert-block, the frictional coefficient over the region CD becomes $\tan \alpha = \tan 35° = 0.7$, while the coefficient of 0.40 applies on DG and FH. Thus the frictional resistance

$$
\begin{aligned}
F &= \mu_1 R_1 + \mu_2 R_2 \\
&= 0.7(\tfrac{1}{2})(4.73 + 2.59)(5.08) + 0.4(\tfrac{1}{2})(2.59)(6.17) \\
&= 13.0 + 3.2 = 16.2 \text{ kips}
\end{aligned}
$$

Force equilibrium, incorporating a 1.5 factor of safety, gives

$$(P_1 + P_2)1.5 = P_p + F$$

$$11.52(1.5) = 0.200(6a + a^2) + 16.2$$

$$a^2 + 6a - 5.40 = 0$$

$$a = 0.8 \text{ ft}$$

Neglecting the frictional force in front of the key and considering the passive resistance developed below the toe as shown in Fig. 12.5.4, the depth of key required may be computed as

$$b = 5.08 \tan \alpha = 5.08(0.7) = 3.55 \text{ ft}$$

$$P_p = 0.400 \frac{(h_1 + a + b)^2}{2} - \frac{0.400(h_1)^2}{2}$$

$$= 0.200(a^2 + 13.1a + 33.9)$$

Force equilibrium, using a 1.5 safety factor, gives

$$(P_1 + P_2)1.5 = P_p + \mu_2 R_2$$

$$11.52(1.5) = 0.200(a^2 + 13.1a + 33.9) + 0.4(\tfrac{1}{2})(2.59)(6.17)$$

$$a^2 + 13.1a - 36.5 = 0$$

$$a = 2.4 \text{ ft}$$

This would indicate that the latter method is more conservative. The *CRSI Handbook* [2.23] arbitrarily uses a key whose depth equals two-thirds of the footing thickness. With this discussion and these computations to serve as a guide, make the key depth 18 in. The key may be made square; use 1 ft 6 in. × 1 ft 6 in. Reinforcement will rarely be required, but it is common practice to extend some of the stem steel down into the key.

(i) Design of heel cantilever. The loads considered are included in Fig. 12.5.6, where the effect of the base key is neglected.

For bending moment, the critical section is taken at the center of the stem steel. It seems that any possible plane of weakness will occur at the stem steel rather than at the face of support. Thus the downward uniform loading due to earth overburden,

Figure 12.5.6 Design of heel cantilever in design example.

surcharge, and footing concrete using an overload factor of 1.4 for the earth and footing concrete and 1.7 for the surcharge is, from Fig. 12.5.6,

$$w_u = 1.4(2.16 + 0.30) + 1.7(0.96) = 5.08 \text{ kips/ft}$$

$$M_u \text{ (downward)} = \tfrac{1}{2}(5.08)(5.96)^2 = 90.2 \text{ ft-kips}$$

The upward soil pressure would reduce this value. However, the upward soil pressure may not actually act in the linear fashion assumed; in fact it might not be there at all. Applying overload conditions under ACI-9.2.4, the most critical situation results when 0.9D (i.e., gravity dead load) and 1.7L (i.e., horizontal earth pressure) are considered. This would eliminate pressure under the heel entirely.

For shear, ACI-11.1.3 allows the critical section to be taken at the distance d from the face only when "... support reaction, in direction of applied shear, introduces compression into the end regions of member, ..." In this case there is tension induced in the concrete where the heel joins the stem and the inclined crack could extend into the region ahead of the back face of the stem. The factored shear at the face without including upward soil pressure is

$$V_u = 5.08(5.75) = 29.2 \text{ kips}$$

The design shear strength, unless shear reinforcement is used, is

$$\phi V_c = \phi(2\sqrt{f_c'})bd$$
$$= 0.85(2\sqrt{3000})(12)(21.5)\tfrac{1}{1000} = 24.0 \text{ kips} < 29.2 \text{ kips} \quad \textbf{NG}$$

Shear appears to control. The effective depth must be increased in the ratio 29.2/24.0 unless shear reinforcement is to be used.

$$\text{required } d = \frac{21.5(29.2)}{24.0} = 26.2 \text{ in.}$$

Use a heel thickness of 30 in. ($d \approx 27.5$ in.).

For flexural strength, using $M_u = 92.0$ ft-kips (corrected for the weight of the 30-in. footing),

$$\text{required } R_n = \frac{M_u}{\phi bd^2} = \frac{92.0(12,000)}{0.90(12)(27.5)^2} = 135 \text{ psi}$$

This is less than the R_n corresponding to the minimum percentage of reinforcement. Though ACI-10.5 exempts slabs of uniform thickness from the minimum requirement, the retaining wall is a major beamlike structure and the minimum is recommended to apply. Increasing the heel thickness from the preliminary 24 in. to 30 in. reduces the stem height and would permit reducing its thickness. The extra heel thickness will also improve stability against overturning.

From Fig. 3.8.1 or Eq. (3.8.5), the required reinforcement ratio is determined for $R_n = 135$ psi to be 0.00347, which is less than the minimum $200/f_y$ prescribed by ACI-10.5.1. Less than $200/f_y$ may be used provided the amount is one-third more than required for strength; that is, $1.33(0.00347) = 0.0046$.

$$\text{required } A_s = 0.0046(12)(27.5) = 1.52 \text{ sq in./ft}$$

Use #8 @ 6 ($A_s = 1.57$ sq in./ft).

Regarding the development length for the #8 bars, from Tables 6.7.1 and 6.7.2, the basic development length L_{db} and Min L_d are 23.1 and 21.9 in., respectively. Since the bars in the heel slab are 6 in. on centers, they do have at least $3d_b$ clear lateral spacing and also the 2 in. clear cover equals the minimum $2d_b$ of ACI-12.2.3.1(d); thus the bar development is in Category A with $\alpha_{abc} = 1.0$. The 6 in. center-to-center spacing give $5d_b$ clear lateral spacing, and assuming the side cover at the end of the wall is at least $2.5d_b$, Category D relating to wide lateral spacing is satisfied, giving $\alpha_{de} = 0.80$ in accordance with ACI-12.2.3.4. Thus,

$$[\alpha_{abc}\alpha_{de}L_{db} = 1.0(0.80)23.1 = 18.5 \text{ in.}] < [\text{Min } L_d = 21.9 \text{ in.}]$$

Thus, the "first-level L_d" is 21.9 in. Then, since these bars are cast with more than 12 in. of concrete beneath them, they are top bars according to ACI-12.2.4.1; thus, $\alpha_{top} = 1.3$. The other second-level adjustments, $\alpha_{\ell w}$, and α_{epx} are not applicable; therefore,

$$L_d \text{ (for #8)} = 1.3(21.9) = 28.5 \text{ in. (2.37 ft)}$$

This means the #8 bars in the heel must be embedded 28.5 in. to develop their full strength. This is measured as the distance beyond the main vertical reinforcement in the back face of the wall. Use an embedment of 3 ft from back face of wall.

(j) Design of toe cantilever. The toe of the footing is also treated as a cantilever beam, with the critical section for moment at the front face of the wall and the critical section for shear (inclined cracking) at a distance d (approximately 20.5 in.) from the front face of the wall (one-way action according to ACI-11.11.1.1 will govern). Shear will usually control the required toe thickness. The thickness need not be the same as the heel, though many engineers would make them the same. In this example the heel is unusually thick due to the heavy surcharge. Try a toe thickness somewhat less than the heel, say 2 ft.

Referring to Fig. 12.5.7 and neglecting the earth on the toe, the factored shear and bending moment are

$$V_u = 1.7\left(\frac{4.73 + 3.87}{2} - 0.300\right)(2.04) = 13.9 \text{ kips}$$

$$M_u = 1.7[\tfrac{1}{2}(4.73)(3.75)^2(\tfrac{2}{3}) + \tfrac{1}{2}(3.15)(3.75)^2(\tfrac{1}{3}) - \tfrac{1}{2}(0.300)(3.75)^2]$$

$$= 1.7\left[\frac{(3.75)^2}{6}(9.46 + 3.15 - 0.90)\right] = 46.7 \text{ ft-kips}$$

According to ACI-9.2.4, the 1.7 is to be used for the horizontal earth pressure and for live load, while 1.4 may be used for the weights of the concrete and the earth. Here it has been considered that the toe soil pressure is primarily the result of horizontal earth pressure; hence the conservative procedure of using the 1.7 factor.

$$\text{required } R_n = \frac{46.7(12,000)}{0.90(12)(20.5)^2} = 123 \text{ psi}$$

Figure 12.5.7 Design of toe cantilever in design example.

From Fig. 3.8.1, the required $\rho = 0.00318$ is less than $200/f_y$. Use $1.33(0.00318) = 0.0042$.

$$\text{required } A_s = 0.0042(12)(20.5) = 1.04 \text{ sq in./ft}$$

Use #7 @ 7 ($A_s = 1.03$ sq in./ft).
The shear strength is

$$\phi V_c = \phi(2\sqrt{f_c'})bd$$
$$= 0.85(2\sqrt{3000})(12)(20.5)\tfrac{1}{1000} = 22.9 \text{ kips} > 13.9 \text{ kips} \qquad \textbf{OK}$$

The #7 bars in the heel have basic development length L_{db} and Min L_d as 17.5 and 19.2 in., respectively, from Tables 6.7.1 and 6.7.2. The cover and clear lateral spacing satisfy Category A; $\alpha_{abc} = 1.0$. The 7 in. center-to-center spacing provides more than the required $5d_b$ clear required by Category D (ACI-12.2.3.4); thus, $\alpha_{de} = 0.80$. This gives

$$[\alpha_{abc}\alpha_{de}L_{db} = 1.0(0.80)17.5 = 14.0 \text{ in.}] < [\text{Min } L_d = 19.2 \text{ in.}]$$

Thus, the "first-level L_d" is 19.2 in. Then, since these bars are in the bottom of the toe, α_{top} is not applicable and second-level adjustments $\alpha_{\ell w}$, and α_{epx} are not applicable; therefore,

$$L_d \text{ (for #7)} = 19.2 \text{ in. (1.60 ft)}$$

Use embedment of 2 ft from front face of wall.

(k) Reinforcement for wall. The wall height (17.5 ft) is now 6 in. less than used in the preliminary design calculations. Retain the 21-in. thickness at the base of the stem. The factored moment diagram is shown in Fig. 12.5.8, using Eq. (12.5.1) with maximum $y = 17.5$ ft along with an overload factor of 1.7. The bending moment M_u diagram and the moment capacity ϕM_n diagram of the selected wall steel are shown in Fig. 12.5.8. The steel area required at the base of stem is

$$\text{required } R_n = \frac{115(12{,}000)}{0.90(12)(18.5)^2} = 373 \text{ psi}$$

$$\text{required } A_s = 0.0103(12)(18.5) = 2.29 \text{ sq in./ft}$$

Use #8 and #9 bars alternated @ $4\frac{1}{2}$ ($A_s = 2.38$ sq in./ft). The shear has previously been checked and found satisfactory.

The embedment of the #9 bars into the footing must equal or exceed the development length L_d. In order to determine the development length, the arrangement of bars must first be established. If all of the wall steel is run the entire height of the wall, dowels must be used since the bars cannot extend 18 ft or so from the footing before the wall is cast. For the #9 bars having the adjacent #8 bars centered 4.5 in. away,

$$\text{Clear spacing} = \frac{4.5 - 1.128/2 - 1.0/2}{1.128} \, d_b = 3.04 d_b$$

The required $3d_b$ of ACI-12.2.3.1(d) is satisfied. Note that the 2 in. clear cover (less than $2d_b$ for the #9 bar) at the back face of the wall is not an issue with regard to the

Figure 12.5.8 Determination of stem reinforcement for design example.

development length required into the footing. Thus, Category A applies; $\alpha_{abc} = 1.0$. Categories D and E are not applicable. From Tables 6.7.1 and 6.7.2, the basic development length L_{db} and Min L_d are 29.2 and 24.7 in., respectively. Thus,

$$[\alpha_{abc}\alpha_{de}L_{db} = 1.0(1.0)29.2 = 29.2 \text{ in.}] > [\text{Min } L_d = 24.7 \text{ in.}]$$

Thus, the "first-level L_d" is 29.2 in. Then, since these bars are vertically oriented, they are not top bars; thus, second-level adjustments α_{top}, $\alpha_{\ell w}$, and α_{epx} do not apply. Therefore,

$$L_d \text{ (for } \#9 \text{ into the footing)} = 29.2 \text{ in. } (2.43 \text{ ft)}$$

Use an embedment of 2 ft 6 in. into the footing for both the #8 and #9 bars. The base key is available for embedment of stem bars if it might be necessary.

In an endeavor to economize, the quantity of steel per foot of wall should be reduced in the upper parts of the wall so that the capacity provided approximately equals that required. Bar spacing and cutoff are done in accordance with the ACI Code, and the resulting design is analyzed by drawing the moment capacity ϕM_n diagram as described in Chapter 6.

In this design, proceeding from the stem base the #9 bars embedded in the footing should not be extended more than about 8 to 10 ft out of the footing, which is placed and cured first. When bars extend out too far, they are often bent or broken off during construction. Thus splice the #8 bars at the base of the stem.

For #8 and #9 alternated at $4\frac{1}{2}$ in., the moment capacity is

$$C = 0.85(3)(12)a = 30.6a$$

$$T = 2.38(40) = 95.2 \text{ kips}$$

$$a = 3.11 \text{ in.}$$

At top of wall,

$$\phi M_n = 0.90(95.2)[9.5 - 0.5(3.11)]\tfrac{1}{12} = 56.7 \text{ ft-kips}$$

At bottom of wall,

$$\phi M_n = 0.90(95.2)[18.5 - 0.5(3.11)]\tfrac{1}{12} = 121 \text{ ft-kips}$$

The two values of moment capacity computed above are used to locate the outer dashed line in Fig. 12.5.8.

Since it is proposed to lap the #8 bars at the base of the wall, ACI-12.15 must be applied to determine the lap distance required. When no more than one-half of the bar area is to be lap spliced within the lap length, the tension splice must meet the requirements for a Class A splice; in this case, less than one-half of the total bar area is to be spliced. Referring to Table 6.17.1 (which summarizes ACI-12.15.2), the lap required is

$$\text{required lap} = L_d$$

where L_d is the development length required for unspliced bars. The #8 bars to be spliced are located 9 in. on centers, as shown in Fig. 12.5.9. The #9 bars 4.5 in. away are not being developed within the splice region; thus, they do not influence the development length for the #8 bars. The clear spacing is

$$\text{Clear spacing} = \frac{9.0 - 2(1.0)}{1.00} \, d_b = 7.0 d_b$$

allowing an extra bar diameter for the lap splice. Thus, ACI-12.2.3.1(d) is satisfied (in this case, 2 in. clear cover equals the $2d_b$ required). Thus, $\alpha_{abc} = 1.0$. In addition, since the clear spacing exceeds $5d_b$, Category D applies; $\alpha_{de} = 0.80$. Then,

$$[\alpha_{abc}\alpha_{de}L_{db} = 1.0(0.80)23.1 = 18.5 \text{ in.}] < [\text{Min } L_d = 21.9 \text{ in.}]$$

Second-level adjustments α_{top}, $\alpha_{\ell w}$, and α_{epx} do not apply. Thus,

$$L_d \text{ (for #8)} = 21.9 \text{ in. (1.83 ft)}$$

and the required lap for a Class A splice also equals L_d.

Use a splice lap of 2 ft terminating at point A (Fig. 12.5.8).

Figure 12.5.9 Design sketch for cantilever retaining wall.

Next, using the moment diagram (Fig. 12.5.8), locate the point where the #9 bars may be terminated, leaving the remaining moment capacity based on #8 @ 9. This moment capacity represented by #8 @ 9 is shown in Fig. 12.5.8 by an inclined dashed line that is plotted by using the following moment capacities:

$$C = 30.6a, \qquad T = 42.0 \text{ kips}, \qquad a = 1.37 \text{ in.}$$

$$\phi M_n \text{ (at top)} = 0.90(42.0)[9.5 - 0.5(1.37)]\tfrac{1}{12} = 27.8 \text{ ft-kips}$$

$$\phi M_n \text{ (at bottom)} = 0.90(42.0)[18.5 - 0.5(1.37)]\tfrac{1}{12} = 56.1 \text{ ft-kips}$$

The termination point B is found by extending beyond the intersection of the remaining capacity line (#8 @ 9) with the factored moment diagram a distance of either the effective depth d or 12 bar diameters, whichever is greater. In this case the theoretical termination point and the practical location in 3-in. increments coincide.

The development length L_d for the #9 bars, measured from point B on Fig. 12.5.8, is based on the 9 in. center-to-center spacing and the 2 in. clear cover. The clear spacing of $(9.0 - 1.128)$ equals $6.98d_b$, which exceeds the $3d_b$ for Category A; however, the 2 in. cover is less than the required $2d_b$ for Category A of ACI-12.2.3.1(d). Category B applies; $\alpha_{abc} = 1.4$. Since the clear spacing exceeds $5d_b$, Category D applies; $\alpha_{de} = 0.80$. Using $L_{db} = 29.2$ in.,

$$[\alpha_{abc}\alpha_{de}L_{db} = 1.4(0.80)29.2 = 32.7 \text{ in.}] > [\text{Min } L_d = 24.7 \text{ in.}]$$

Thus, $L_d = 32.7$ in. for the #9 in the wall, since no second-level adjustments apply.

Whenever tension bars are terminated in the tension zone, stress concentrations occur; therefore a check of ACI-12.10.5 must also be made. Since stirrups are not used in retaining walls, either condition 1 or condition 3 of that section of the code must be satisfied. The conditions, one of which must be satisfied in this case, are (1) that the factored shear V_u at the cutoff point must not exceed two-thirds of the shear strength ϕV_n; and (2) that the continuing bars must provide at least twice the area required for bending moment at the cutoff point, *and* the factored shear V_u at the cutoff point must not exceed three-fourths of the shear strength ϕV_n.

From an inspection of Fig. 12.5.8, it is obvious that item (2) above is not satisfied. Check item (1). Using Eq. (12.5.2), the shear at $y = 10.25$ ft from the top is

$$V_u = 1.7[0.256(10.25) + 0.016(10.25)^2] = 7.32 \text{ kips}$$

$$\phi V_c = \phi(2\sqrt{f_c'})bd$$
$$= 0.85(2\sqrt{3000})(12)(15)\tfrac{1}{1000} = 16.8 \text{ kips}$$

$$\tfrac{2}{3}(16.8) > 7.32 \qquad\qquad \textbf{OK}$$

Thus terminate #9 bars at 7 ft 3 in. from the top of heel.

Additional economy may be achieved by cutting every other #8 bar, leaving #8 @ 18 to extend to the top of the wall. It should be noted that ACI-7.6.5 states that the principal reinforcement shall be spaced not farther apart than three times the wall or slab thickness nor more than 18 in. Figure 12.5.8 shows the actual cutoff point C for every other #8. In this case the extension d gives a point slightly below point C. A tension zone cutoff check (ACI-12.10.5) at point C ($y = 6.75$ ft from top) shows

$$V_u = 1.7[0.256(6.75) + 0.016(6.75)^2] = 4.18 \text{ kips}$$

$$\phi V_c = \phi(2\sqrt{f_c'})bd$$
$$= 0.85(2\sqrt{3000})(12)(13)\tfrac{1}{1000} = 14.5 \text{ kips}$$

$$\tfrac{2}{3}(14.5) > 4.18 \qquad\qquad \textbf{OK}$$

Terminate every other #8 at 10 ft 9 in. from the top of heel. Actual cut points are located to make bar lengths in the usual 3-in. increments. The complete bar arrangement for the stem is shown in Fig. 12.5.9.

The reader may note that the moment decreases rapidly from the base of the stem toward the top of the wall. At about 5 ft from the base, the factored moment has decreased about 50%. Somewhere in this vicinity the reinforcement ratio ρ that is required for strength equals the minimum of $200/f_y$. For the upper portion of the wall, ACI-10.5 would require the actual ρ to be either 0.005 (i.e., $200/f_y$) or four-thirds of the required ρ based on the factored moment. As an example, the moment capacity ϕM_n at point C is indeed approximately equal to four-thirds of the factored moment at C. Thus, in general, wherever actual ρ is less than $200/f_y$, the requirement of ACI-10.5 may be reviewed by seeing that the moment capacity diagram is offset at one-third or more from the factored moment curve.

It is further noted that the stem has been designed for bending moment and shear only. However, the weight of the stem causes compression in it, so that strictly speaking it might be treated as a compression member (under large moment) in accordance with the concepts of Chapter 13. In this design problem the compressive force P_u in the wall under factored loads would be

$$P_u = 1.4 \left(\frac{12 + 21}{12} \right) (0.5)(17.5)(0.15) = 5.1 \text{ kips}$$

The combination of this P_u with $M_u = 115$ ft-kips will give a design near the bottom of the "tension controls" region according to Chapter 13 (see Fig. 13.6.1). Although the action of the compressive force increases the nominal bending strength M_n, the ϕ factor will be slightly decreased from 0.90 toward 0.70. In general, treatment of such walls as compression members is rarely justified but, if done, would permit some slight reduction in reinforcement or thickness of the section.

(l) Temperature and shrinkage reinforcement. Horizontal bars along the length of the wall must be provided, in accordance with ACI-14.1.2 and 14.3.3. Accordingly, the total amount required is

$$A_s = 0.0025bh = 0.0025(12)(16.5) = 0.50 \text{ sq in./ft}$$

where the average wall thickness is used for h.

Since it is primarily the front face that is exposed to temperature changes, more of this reinforcement should be placed there. Thus it is suggested that about two-thirds be put in the front face and one-third in the rear face. Accordingly,

$$\tfrac{2}{3}A_s = \tfrac{2}{3}(0.50) = 0.33 \text{ sq in./ft}$$

$$\tfrac{1}{3}A_s = \tfrac{1}{3}(0.50) = 0.17 \text{ sq in./ft}$$

<u>*Use* on the front face #5 @ 10 ($A_s = 0.37$ sq in./ft). Use on the rear face #4 @ 12 ($A_s = 0.20$ sq in./ft).</u>

For vertical reinforcement on the front face, use any nominal amount (ACI-14.3.5 is probably a good guide) that is adequate for supporting the horizontal temperature and shrinkage steel in that face.

<u>*Use* #5 @ 1 ft 6 in. spacing.</u>

(m) Drainage and other details. Adequate drainage of backfill must be provided since the pressures used are for drained material. A common minimum provision is for weep holes (say 4-in. diameter tile) every 10 to 15 ft along the wall.

Construction of a retaining wall is accomplished in at least two stages; the footing is placed first and then the wall. A shear key is usually desirable for positive shear transfer between wall and footing (see Fig. 12.5.9). Such a key is made by embedding a beveled 2 × 4 or 2 × 6 timber in the top of the footing. This key may be designed using the shear-friction provisions of ACI-11.7 (discussed in Section 5.15). This is a situation in which it is not appropriate to consider shear as a measure of diagonal tension. The nominal shear strength of the shear key is based on a nominal stress of $0.2f'_c$ but not to exceed 800 psi, as given by ACI-11.7.5.

In this design no bent bars are used. Frequently, the bar arrangement in the toe can be conveniently bent up into the stem and meshed with additional stem reinforcement to give an economical system.

(n) Design sketch. The final details of this design are presented in a design sketch (Fig. 12.5.9) which must accompany any set of design computations.

SELECTED REFERENCES

12.1. Karl Terzaghi and Ralph B. Peck. *Soil Mechanics in Engineering Practice* (2nd ed.). New York: Wiley, 1968.

12.2. Whitney Clark Huntington. *Earth Pressures and Retaining Walls*. New York: Wiley, 1957.

12.3. AASHTO. *Standard Specifications for Highway Bridges* (14th ed.). Washington, D.C.: American Association of State Highway and Transportation Officials, 1989.

12.4. "Retaining Walls and Abutments," *American Railway Engineering Association Manual*, Vol. 1. Chicago: American Railway Engineering Association, 1958 (Chapter 8).

12.5. G. P. Fisher and R. M. Mains. "Sliding Stability of Retaining Walls," *Civil Engineering*, July 1952, 490.

PROBLEMS

Note: For all problems assume that frost penetration depth is 4 ft.

12.1 Given the cantilever retaining wall of the accompanying figure with the active pressure coefficient $C_a = 0.27$, the weight of retained soil = 100 pcf, and the coefficient of friction between concrete and earth = 0.40. Using service load conditions:

(a) Determine the factor of safety provided against overturning. Would you consider this to be adequate?

(b) Determine the factor of safety provided against sliding. Neglect any passive pressure resistance on the toe. Would you consider this factor of safety adequate?

(c) Determine the location of the resultant bearing pressure under footing. Is it within the middle third of the base? Would the position you determined probably be permissible if the foundation material is rock?

Problem 12.1

12.2 Determine the adequacy of the retaining wall of the accompanying figure with regard to stability (overturning, sliding, and soil pressure magnitude and distribution). Assume that the wall is on good granular soil with a maximum safe soil pressure of 5000 psf under service load conditions.

Problem 12.2

12.3 Determine the adequacy of the retaining wall of the accompanying figure with regard to stability if the backfill is level. Assume an allowable soil pressure of 4000 psf under service load conditions. Draw a conclusion regarding what would happen if the backfill became saturated to an equivalent fluid active pressure of 65 pcf. See the accompanying figure.

10″

Equivalent fluid *active* pressure = 35 pcf

Weight of retained material = 100 pcf

Equivalent fluid *passive* pressure = 285 pcf

15′–0″

17′–0″

2′–0″

1′–4″   4′–7″   1′–4″
2′–0″
3′–6″   5′–6″
11′–0″

Problem 12.3

12.4 Reconsider the retaining wall conditions of Section 12.5 if the surcharge is changed to 10 ft to approximate the effect of a railroad located parallel to and near the top of the wall. Sometimes under such conditions the lateral effect of a surcharge is included but the beneficial stabilizing effect of the vertical surcharge weight is omitted. Compare the wall proportions for both conditions using a minimum length of base and keeping the soil pressure resultant under service load conditions within the middle third of the base. Verify the adequacy of the dimensions, but do not design the reinforcement. Use $f'_c = 4000$ psi, $f_y = 60,000$ psi, and the ACI Code.

12.5 Assuming that the overall proportioning of the retaining wall of Problem 12.2 is adequate for earth stability, design the reinforcement for the wall cantilever. Besides dowels at the base, use three changes of reinforcement over the 20-ft wall height. Draw the resulting moment capacity ϕM_n diagram superimposed on the factored moment M_u diagram. Use $f'_c = 4000$ psi, $f_y = 60,000$ psi, and the strength method of the ACI Code.

12.6 Design the heel cantilever for the wall of Problem 12.2 if the backfill is level. Use $f'_c = 4000$ psi, $f_y = 60,000$ psi, and the strength method of the ACI Code. Revise the thickness from that shown, if it seems desirable.

12.7 Design the heel cantilever for the wall of Problem 12.3 if the backfill slopes 30° to the horizontal. Use $f'_c = 4000$ psi, $f_y = 60,000$ psi, and the strength method of the ACI Code. Revise the thickness if it seems desirable.

12.8 Design the toe cantilever for the wall of Problem 12.2 if the backfill is level. Use $f'_c = 4000$ psi, $f_y = 60,000$ psi, and the strength method of the ACI Code. Revise the thickness if it seems desirable.

12.9 Design the toe cantilever for the wall of Problem 12.3 if the backfill slopes at 30° to the horizontal. Use $f'_c = 4000$ psi, $f_y = 60,000$ psi, and the strength method of the ACI Code. Revise the thickness if it seems desirable.

12.10 Design a cantilever retaining wall to support a bank of earth 17 ft high above the final level of earth at the toe of the wall. The retained material is level and has a weight of 100 pcf with *active* and *passive* pressure coefficients of 0.28 and 3.6, respectively. The allowable soil pressure under service load is 5 ksf and the coefficient of friction between concrete and soil is 0.60. Use $f'_c = 3500$ psi, $f_y = 60,000$ psi, and the strength method of the ACI Code.

12.11 Design a cantilever retaining wall to support a bank of earth 22 ft high above the final level at the toe of the wall. The retained material is level and has a weight of 110 pcf and exerts an equivalent fluid active pressure of 30 pcf and an equivalent fluid passive pressure of 350 pcf. The coefficient of friction is 0.45 between concrete and soil. The maximum soil pressure under service load shall not exceed 3500 psf. Use $f'_c = 3000$ psi, $f_y = 60,000$ psi, and the strength method of the ACI Code.

12.12 Redesign the cantilever retaining wall of Problem 12.11 if the differential elevation is 24 ft instead of 22 ft.

CHAPTER 13

Members in Compression and Bending

13.1 INTRODUCTION

The compression member subjected to pure axial load rarely, if ever, exists. All columns are subjected to some bending moment, which may be caused by: (1) end restraint arising from the monolithic placement of floor beams and columns; (2) accidental eccentricity from imperfect alignment and variable materials; (3) unbalanced floor loads on both exterior and interior columns; (4) eccentric loads such as crane loads in industrial buildings, and (5) lateral loading such as from wind or earthquake. In determining bending moments in columns due to unbalanced floor loads, ACI-8.8 provides the following simplifying assumptions:

1. The far ends of columns that are monolithic with the structure may be considered fixed in continuity analysis for gravity loading.
2. Maximum bending in a column is due to factored loads on a single adjacent span of the floor under consideration. This bending is in addition to axial forces from factored loads on all floors.
3. The loading condition causing maximum ratio of bending moment to axial load shall also be considered.

Concrete construction is usually monolithic; thus reinforced concrete rigid frames and arches (Fig. 13.1.1) are common and advantageous. All sections in the two structures shown in Fig. 13.1.1 are subjected to combined bending and axial load. The vertical members in Fig. 13.1.1(a) and sections near the supports in Fig. 13.1.1(b) are subjected to a high ratio of axial force to bending moment, while the horizontal member in Fig. 13.1.1(a) and sections near the crown in Fig. 13.1.1(b) may be subjected to a low ratio of axial force to bending moment.

This chapter considers first the column having minimal bending moment, commonly called "axially" loaded, and later considers the effect of medium and large amounts of bending. Only the basic strength of short compression members is

Reinforced concrete tied columns under construction; Kurt F. Wendt Engineering Library, University of Wisconsin, Madison, Wis. (Photos by C. G. Salmon.)

Figure 13.1.1 Reinforced concrete rigid frame and arch.

considered, however. The reduction of the strength of compression members due to length effects is treated in Chapter 15.

13.2 TYPES OF COLUMNS

A column has been defined as a member used primarily to support axial compressive load with a ratio of height to least lateral dimension of 3 or greater (ACI-2.1). Shorter concrete compression members may be unreinforced and treated as pedestal footings. Reinforced concrete columns are principally of two types, classified according to the manner in which the longitudinal reinforcing bars are laterally supported. A *tied*

Figure 13.2.1 Types of columns.

column, usually of square, rectangular, or circular shape, is one in which the longitudinal reinforcing bars are held in position by separate lateral ties typically spaced about 12 to 24 in. (300 to 600 mm) apart, as shown in Fig. 13.2.1(a) and photos on page 448. A *spirally reinforced* column, usually of square or circular shape, is one in which the longitudinal reinforcing bars are arranged in a circle and wrapped by a continuous closely spaced spiral typically at a pitch of about 2 to 3 in. (50 to 75 mm), as shown in Fig. 13.2.1(b). A *composite* column is one in which a structural steel shape, pipe, or tubing is used, with or without additional longitudinal bars. One common composite column arrangement may contain a structural steel shape completely encased in concrete, which is further reinforced with both longitudinal and lateral reinforcement (spiral or ties) as shown in Fig. 13.2.1(c). In a second kind of composite column the steel may encase a concrete core, which may or may not contain longitudinal reinforcing bars, as shown in Fig. 13.2.1(d).

13.3 BEHAVIOR OF AXIALLY LOADED COLUMNS

Many tests were made in the early 1900s on reinforced concrete columns under axial loads [13.1–13.5], but loading was generally of short duration. As early as 1911, however, Withey [13.5] at the University of Wisconsin observed that as load increased beyond the service load range, a transfer of load from concrete to steel took place. In the early 1930s ACI Committee 105 reported [13.6] on the results of 564 column tests, primarily at Lehigh University and the University of Illinois, where attention was given to column size, quality of concrete, quality and amount of longitudinal and lateral reinforcement, rate of application of load, and shrinkage and creep under sustained loads. By 1940, design procedures for axially loaded columns became based on the ultimate strength results of the aforementioned extensive investigation. Joint ACI-ASCE Committee 441 has provided an excellent annotated bibliography of the early studies on reinforced concrete columns [13.7].

When concrete and steel act together in compression, the proportion of loading carried by each changes continuously during the time the load is acting. Initially, the stress in the steel is E_s/E_c times the stress in the concrete, according to the elastic theory. As the time-dependent effects of creep and shrinkage occur, the steel gradually takes over relatively more load than its elastic share. Creep and shrinkage deformations have been introduced in Section 1.10, discussed in regard to beams in Sections 4.8 and 4.9, and treated in detail with regard to deflections in Chapter 14.

Members that are subjected to axial compression, either alone or in combination with bending, frequently have a substantial portion of the total load sustained. Consequently the transfer of load to the steel from the concrete due to the time-dependent deformation is more pronounced in these members than in beams. However, even though actual stresses under service loads cannot be meaningfully computed, the strength can be determined. The experimenters verified that the nominal strength P_n for an axially loaded column may be properly expressed [13.6] by

$$P_n = k_c f'_c A_c + f_y A_{st} + k_s f_{sy} A_{sp} \tag{13.3.1}$$

where

P_n = nominal strength for a tied column (with third term omitted, i.e., no spiral)

or

P_n = yield strength for a spirally reinforced column

k_c = coefficient (0.85) to account for the difference between concrete in the column and that in a test cylinder

f'_c = standard 28-day cylinder strength

A_c = net area of concrete, based on gross area for tied column and core area for spirally reinforced column

A_{st} = area of longitudinal reinforcement

f_y = yield stress for longitudinal reinforcement

k_s = constant that varies from 1.5 to 2.5 with an average of 1.95

f_{sy} = yield stress of spiral steel

A_{sp} = volume of spiral steel per unit length of column

Equation (13.3.1) represents the yield load for the spirally reinforced column that exhibits a marked yielding followed by considerable deformation before complete failure, as shown in Figs. 13.3.1 and 13.3.2. On reaching the yield point, the shell spalls off and the spiral begins to act to confine the crushed concrete in the core. The spiral therefore adds little to the strength prior to reaching yield, but it provides ductility. From Eq. (13.3.1) it may be observed that spiral reinforcement is from 1.5 to 2.5 times as effective in adding strength as is the main longitudinal steel, but the spiral does not work at all until the deformation is excessive. Recent studies of spirally reinforced column ductility have been made by Priestley, Park, and Potangaroa [13.8], Ahmad and Shah [13.9], and for high-strength concrete by Martinez, Nilson, and Slate [13.10].

Figure 13.3.1 Typical load-deformation curves for tied and spirally reinforced columns.

Figure 13.3.2 Spirally reinforced column behavior. (Courtesy of Portland Cement Association.)

Figure 13.3.3 Tied column behavior. (Courtesy of Portland Cement Association.)

The tied column behaves exactly as a spirally reinforced column up to the yield point of the spirally reinforced column, at which point failure occurs suddenly in a manner similar to the failure in a compression cylinder test. At that instant, the longitudinal bars buckle between points that are restrained by lateral ties. The tied column behavior is illustrated in Figs. 13.3.1 and 13.3.3.

Studies by Sheikh and Uzumeri [13.11], Scott, Park, and Priestley [13.12], Park, Priestley, and Gill [13.13], Fafitis and Shah [13.14], Moehle and Cavanagh [13.15], Sheikh and Yeh [13.16, 13.20, 13.21], Ozcebe and Saatcioglu [13.17], Yong, Nour, and Nawy [13.18], and Abdel-Fattah and Ahmad [13.19], show how the ductility of tied columns varies with tie and longitudinal bar arrangements. Razvi and Saatcioglu [13.22] have studied the effects of confinement using welded wire fabric. Much of this work has been concerned with ductility under cyclic loading, such as for earthquake-resistant design.

As will be shown later, though both types of column have the same strength, a higher factor of safety should be provided for the tied column than for the spirally reinforced column because of the sudden failure and lack of toughness (energy absorption).

13.4 SAFETY PROVISIONS

The safety provisions for overload in the design of compression members as prescribed by the ACI Code are the same as for any other types of members; these are (ACI-9.2):

For gravity loads,

$$U = 1.4D + 1.7L \tag{13.4.1}$$

For loadings including wind,

$$U = 0.75(1.4D + 1.7L + 1.7W) \tag{13.4.2}$$

or

$$U = 0.9D + 1.3W \tag{13.4.3}$$

The most severe situation among Eqs. (13.4.1) through (13.4.3) will control the design.

Traditionally, under most building codes, members subject to stresses produced by wind or earthquake forces combined with other loads have been proportioned in the working stress method for unit stresses $33\frac{1}{3}\%$ greater than those specified, provided that the section thus required is not less than that required for the combination of dead and live load. In strength design, the infrequent occurrence of maximum wind load with maximum live load is taken into account by the smaller overload factors when wind effects are included. This reduced safety provision for the temporary effect of wind or earthquake, or both, appears in ACI-Appendix A.2 for the working stress method, and in ACI-9.2.2 and 9.2.3 for the strength method.

Additionally, the possibility of understrength is accommodated by the ϕ requirement of ACI-9.3. For spirally reinforced columns $\phi = 0.75$, and for tied columns $\phi = 0.70$, the difference being related partially to the reserve capacity to deform before failure which is exhibited by the spirally reinforced column.

The lower strength reduction factors ϕ for compression members arise from statistical variations in observed strength and are a rough indication that the strength variability to be expected in tied columns is slightly greater than in spiral columns and that the variability expected in columns is greater than in beams. The difference in the behavior of tied and spirally reinforced *axially* loaded columns is further accounted for by the use of a different maximum compression capacity as discussed in Section 13.11.

For combined compression and bending, the ϕ value is permitted to be variable and increase to 0.90 (for pure flexure) as the axial compression decreases to zero. This variation of ϕ is treated later in Section 13.17.

13.5 CONCENTRICALLY LOADED SHORT COLUMNS

In accordance with Eq. (13.3.1) without the term representing the contribution of the spiral, the maximum nominal strength $P_n = P_0$ for a concentrically loaded short column (Fig. 13.5.1) is

$$P_0 = 0.85f'_c(A_g - A_{st}) + f_y A_{st} \tag{13.5.1}$$

where A_g is the gross area bh and A_{st} is the total longitudinal reinforcement $(A_1 + A_2)$. This equation is also in agreement with the rectangular stress block assumptions of Section 3.4, where the entire cross section is subject to a failure compressive strain of 0.003. P_0 may also be expressed as

$$P_0 = A_g[0.85f'_c(1 - \rho_g) + f_y \rho_g] \tag{13.5.2}$$

where $\rho_g = A_{st}/A_g$.

When the terms including ρ_g are combined, Eq. (13.5.2) becomes

$$P_0 = A_g[0.85f'_c + \rho_g(f_y - 0.85f'_c)] \tag{13.5.3}$$

P_0 is located colinearly with the resultant of C_1, C_2, and C_3 at the plastic centroid

$C_2 = f_y A_1$ $C_3 = f_y A_2$

$C_1 = 0.85f_c'(A_g - A_{st})$

Figure 13.5.1 A concentrically loaded column.

13.6 STRENGTH INTERACTION DIAGRAM

When combined axial compression and bending moment act on a member having a low slenderness ratio (unbraced length L_u to radius of gyration r) where column buckling is not a possible mode of failure, the strength of the member is governed by the material strength (corresponding to yielding in a homogeneous elastic material) of the cross section. For this so-called short column the strength is achieved (according to ACI-10.2.3) when the extreme concrete compression fiber reaches the strain 0.003. Depending on the ratio of M_n to P_n (see Fig. 13.6.1), the strain diagram will exhibit two distinct categories. There may be (1) compression over most or all of the section such that the compressive strain in the concrete reaches 0.003 before the tension steel yields, known as the "compression controls" region, or (2) tension in a large portion of the section such that the strain in the tension steel is greater than the yield strain ($\epsilon_y = f_y/E_s$) when the compressive strain in the concrete reaches 0.003, known as the "tension controls" region. Separating the "compression controls" region and the "tension controls" region is the balanced strain condition wherein the concrete strain of 0.003 and the steel yield strain $\epsilon_y = f_y/E_s$ are reached simultaneously.

It is important to note that for the same cross section there is an infinite number of strength combinations at which P_n and M_n act together. These strength combinations lie on a curve as shown in Fig. 13.6.1, which is called the *strength interaction diagram*. The balanced strain condition in combined bending and axial load is represented by the point $P_n = P_b$ and $M_n = M_b$ on this diagram.

This maximum strength interaction relationship for short rectangular members has been verified by research [13.23–13.27]. Nonprismatic members have also been studied [13.28–13.29], as well as members subject to nonproportionally varying axial

Figure 13.6.1 Typical strength interaction diagram for axial compression and bending moment about one axis.

Figure 13.6.2 Eccentrically loaded column statically equivalent to member subject to axial load and bending moment.

load [13.30]. The major emphasis for this chapter is the analysis and design of sections whose nominal strength (P_n and M_n) lies at various points on the interaction diagram (Fig. 13.6.1).

In studying Fig. 13.6.1, the reader may note that radial lines from the origin ($P_n = 0$, $M_n = 0$) represent constant ratios of M_n to P_n; that is, they represent eccentricities e of the load P_n from the *plastic centroid*, which is defined in Fig. 13.5.1. Under usual conditions of symmetrical reinforcement, the plastic centroid coincides with the centroid of the gross section. The fact that e is equal to M/P may be observed from Fig. 13.6.2, where it is shown that a column subject to an eccentric load is statically equivalent to a member under the combined action of an axial load and a bending moment. Thus in Fig. 13.6.1 the vertical axis represents $e = 0$ and the horizontal axis represents $e = \infty$. This concept of replacing axial load and bending moment by a single eccentric load provides the basis for the practical approach to analysis and design computations in reinforced concrete.

13.7 LENGTH EFFECTS

When the height of an upright compression member does not exceed three times its least lateral dimension, it is considered a pedestal and should be designed as such.* For longer members, the effect of slenderness ratio (ratio of unbraced length L_u to

---

* According to *Building Code Requirements for Structural Plain Concrete* (ACI 318.1–89) *and Commentary* (ACI 318.1R–89). Detroit: American Concrete Institute, 1989 (Section 7.3).

radius of gyration r) must be considered. It is well known that buckling may control the strength of any compression member (see Fig. 15.2.2 for curve typical of reinforced concrete sections). The strength design method, together with better understanding of concrete compression member behavior, has produced slimmer members; thus the stability problem has become increasingly important. In addition to the stability problem, smaller members will deflect more under any primary bending moment and thus have a larger secondary moment, which is the product of the axial compression and deflection (as shown in Fig. 15.1.1).

The magnitude of the slenderness ratio determines whether the strength reduction is sufficiently important that it cannot be neglected. The reference condition for slenderness ratio is that of a column with hinged ends (i.e., no resistance to rotation at either end), as shown in Fig. 15.3.1(a). Equivalent pin-end lengths of columns with end restraints can be expressed by kL_u, where k is the effective length factor and L_u is the actual unsupported length.

A vital factor in the determination of the equivalent pin-end length is whether the structural system is *braced* so that relative movement of the ends of a compression member transverse to the axis of the member is prevented [see Figs. 15.3.1 and 15.3.3(a),(c)], or *unbraced* where such relative movement is possible and restraint is provided only by the rigidity of the joints and the stiffness of interacting beams and columns [see Figs. 15.3.2 and 15.3.3(b),(d)]. Without general derivation or proof, the following may be stated:

For *braced* systems, $k \leq 1.0$

For *unbraced* systems, $k \geq 1.0$

A qualitative explanation is available from the study of Figs. 15.3.1 to 15.3.3, but a theoretical development would require a study of structural stability (see Chapter 15, references 15.26 and 15.27).

The intent of the ACI Code is to permit the design of compression members as short columns, without strength reduction for slenderness effect, when the length effect consideration would result in a strength reduction not exceeding 5%. ACI-10.11.4 permits neglect of length effects when

$$\frac{kL_u}{r} < 34 - 12\frac{M_{1b}}{M_{2b}} \quad \text{(for } braced \text{ systems)} \tag{13.7.1}$$

where M_{1b} and M_{2b} are numerically the smaller and larger bending moments, respectively, at the ends of the member and the ratio M_{1b}/M_{2b} is positive for single curvature and negative for double curvature; and when

$$\frac{kL_u}{r} < 22 \quad \text{(for } unbraced \text{ systems)} \tag{13.7.2}$$

At the time these limits were established, a study of existing structures indicated [13.31] that over 90% of the columns in braced frames and over 40% of the columns in unbraced frames would fall within the limits of ACI-10.11.4 and thus strength reduction due to length effects could be ignored. Columns exceeding the ratios indicated by Eqs. (13.7.1) and (13.7.2) have become relatively more common, so that a higher percentage of designs today (1992) will likely require slenderness consideration.

The remainder of this chapter treats only the basic strength of short compression members. The detailed consideration of the strength reduction due to length effects in both unbraced and braced systems appears separately in Chapter 15.

13.8 LATERAL TIES

Lateral ties are used to hold the vertical bars in position, as shown by the photos on page 448, providing lateral support so that individual bars could have the tendency to buckle only *between* the tie supports. Ties do not contribute to the strength, as indicated by the column studies in the early 1930s [13.6]. Studies [13.11, 13.32] have indicated that present tie requirements are conservative for ordinary columns with Grade 40 reinforcement, but may not be conservative for columns with high-strength reinforcement, with large or bundled bars, or of unusual dimensions. An overall review of confinement in columns is provided by Sakai and Sheikh [13.36], and the subject of tie requirements for prestressed concrete columns is provided by Halvorsen and Carinci [13.37].

The effect of ties on the behavior of columns is complex. As a tied column is loaded to failure, the first occurrence is the spalling off of the exterior cover, which in turn causes transfer of load to the concrete core and the longitudinal steel. The loss of stiffness of the longitudinal steel, which begins to yield or to buckle outward, causes additional load on the concrete core. Once the core achieves its crushing strength, the column suddenly fails. The above sequence usually takes place rapidly, giving the so-called "sudden" failure. Ties placed at a sufficiently close spacing provide confinement and increase the strain at which concrete crushes to values well above the maximum of 0.003 used by the ACI Code [13.12, 13.38].

The following provisions have been prescribed for lateral ties in columns by the ACI Code (ACI-7.10.5):

1. All nonprestressed bars for tied columns shall be enclosed by lateral ties, at least #3 in size for longitudinal bars #10 or smaller, and at least #4 in size for #11, #14, #18, and bundled longitudinal bars.
2. The spacing of the ties shall not exceed 16 longitudinal bar diameters, 48 tie bar diameters, or the least dimension of the column.
3. The ties shall be so arranged that every corner and alternate longitudinal bar shall have lateral support provided by the corner of a tie having an included angle of not more than 135° and no bar shall be farther than 6 in. clear on either side from such a laterally supported bar.
4. Where the bars are located around the periphery of a circle, a complete circular tie may be used.

The ACI *Detailing Manual*—1988 [2.25] suggests tie arrangements for various numbers of bars, some common ones being shown in Fig. 13.8.1.

The provisions of ACI-7.10.5.3 permitting the included angle where a tie "corners" the longitudinal reinforcement (that is, a corner bar or every other bar) to be as

Figure 13.8.1 Common column tie arrangements (adapted from Reference 2.25).

large as 135° became part of the ACI Code as a result of the Bresler and Gilbert tests [13.33], a liberalization from earlier codes in which a 90° corner was required for every bar. However, since spliced bars and bundled bars have not been included in tests, ACI Commentary-R7.10.5 recommends [1.11] that "it would be prudent to provide at least a set of ties at each end of lap spliced bars, above and below end-bearing splices, and at minimum spacings immediately below sloping regions of offset bent bars."

It may also be noted that ACI-7.10.3 waives the lateral reinforcement requirements "where tests and structural analysis show adequate strength and feasibility of construction."

13.9 SPIRAL REINFORCEMENT AND LONGITUDINAL BAR PLACEMENT

The spiral provides the column with the ability to absorb considerable deformation prior to failure [13.8]. This toughness is the principal gain that is achieved by the use of spirally reinforced columns. The knowledge of spiral behavior is based on the column research of the early 1930s [13.6]. Although the spiral does actually contribute strength to the column (as early as 1903 Considère [13.1, 13.2] indicated that the spiral was 2.4 times as effective as longitudinal reinforcement in providing column capacity), the conservative policy of ACI specifications since about 1940 has been to provide spiral reinforcement sufficient to increase the capacity of the core by an amount equal to the capacity of the shell, thus maintaining the column yield capacity when the shell spalls off.

Using the third term of Eq. (13.3.1), with an average k_s of 2, the strength represented by the spiral reinforcement is

$$P_n = 2.0 f_{sy} A_{sp} \qquad (13.9.1)$$

An alternative approach to the acceptance of an experimental value for k_s has been presented by Huang [13.39], who considers the triaxial loading condition that exists when the spiral is acting in tension.

Let ρ_s be the ratio of the volume of spiral reinforcement to the total volume of the core (out-to-out of spirals), or $\rho_s = A_{sp}/A_c$. Equation (13.9.1) now becomes

$$P_n = 2.0 f_{sy} \rho_s A_c \qquad (13.9.2)$$

Equating Eq. (13.9.2) to the strength of the shell and taking the concrete strength of the shell as about 90% of that inside the core, or $0.75 f_c'$,

$$2.0 f_{sy} \rho_s A_c = 0.75 f_c'(A_g - A_c)$$

from which

$$\rho_s = 0.375 \left(\frac{A_g}{A_c} - 1 \right) \frac{f_c'}{f_{sy}} \qquad (13.9.3)$$

By providing an additional factor of safety of 1.20 to assure that the spiral strength exceeds the shell strength, Eq. (13.9.3) becomes

$$\rho_s = 0.45 \left(\frac{A_g}{A_c} - 1 \right) \frac{f_c'}{f_{sy}} \qquad (13.9.4)$$

which is specified in ACI-10.9.3. It is noted that the yield strength f_{sy} of the spiral may not exceed 60,000 psi. A review of the spiral steel requirement has been made by Gamble [13.40].

The design relationship may be obtained by referring to the definition of ρ_s following Eq. (13.9.1).

$$\rho_s = \frac{A_{sp}}{A_c} = \frac{\text{volume of spiral in one loop}}{\text{volume of core for a length } s}$$

$$= \frac{a_s \pi (D_c - d_b)}{(\pi D_c^2/4)s} \tag{13.9.5}$$

in which D_c is the diameter of the core, a_s is the area of the spiral, and d_b is the diameter of the spiral wire (Fig. 13.9.1).

According to ACI-7.10.4, the clear spacing between spirals must be at least 1 in. but shall not exceed 3 in. The spiral in cast-in-place construction shall have a diameter not less than $\frac{3}{8}$ in. Anchorage of spiral reinforcement shall be provided by $1\frac{1}{2}$ extra turns of spiral bar or wire at each end of the spiral unit. Splices, when necessary, shall be made in the spiral bar or wire by welding or by a lap of 48 spiral bar diameters minimum but not less than 12 in. The spiral reinforcement is to be protected by the usual $1\frac{1}{2}$ in. clear cover required by ACI-7.7.1 for nonprestressed beams and columns.

In order to hold the spiral in place and maintain the desired pitch, vertical spacer bars with small hooks at the desired pitch are usually used. However, the use of

$$A_c = \frac{\pi D_c^2}{4}$$

$$A_g = \frac{\pi h^2}{4}$$

a_s = area of spiral

Figure 13.9.1 Spirally reinforced column.

Figure 13.9.2 Bar arrangement in spirally reinforced columns.

spacers is no longer specified by the ACI Code, which now requires (ACI-7.10.4) only that spirals "shall be held firmly in place and true to line."

Since spirally reinforced columns are usually more heavily reinforced than tied columns, it sometimes becomes necessary at splice points to lap bars (see ACI-12.14.2.1, 12.15, and 12.16 on splices) inside the main circle of bars. This is done in order to maintain a desirable minimum center-to-center spacing of longitudinal bars at $2\frac{1}{2}$ times the bar diameter and a minimum clear distance (ACI-7.6.3) between individual longitudinal bars of $1\frac{1}{2}$ times the nominal bar diameter, but not less than $1\frac{1}{2}$ in. The preferred and alternate bar arrangements are shown in Fig. 13.9.2. Bars may also be butt spliced by welding or mechanical connections (ACI-12.14.3), thus permitting more utilization of available space. In a large column an inner core of bars wrapped with a spiral or by ties may also be used.

13.10 LIMITS ON PERCENTAGE OF REINFORCEMENT

The percentage of total longitudinal reinforcement A_{st} in terms of the gross cross-sectional area A_g must be between 1 and 8% (ACI-10.9.1); that is $\rho_g = A_{st}/A_g$ between 0.01 and 0.08. However, ACI-10.8.4 permits basing the percentage on a reduced concrete area A_g in cases where the gross concrete area is in excess of that needed for load considerations, but in no case may the reinforcement ratio ρ_g be less than 0.005 based on the gross area provided. The primary purpose of these provisions for minimum steel is to prevent the failure mode from becoming that of a plain concrete column, which might be more disastrous than the sudden failure of tied columns previously described. The upper limit on the amount of longitudinal reinforcement is a practical one in that if proper clearances are maintained between bars, little more than $\rho_g = 0.08$ can be put into the section. Thus the maximum ρ_g is, in a way, a double check on the minimum spacing restrictions of ACI-7.6.3.

There are no restrictions on bar size or on dimensions in order to allow for "wider utilization of reinforced concrete compression members in smaller size and lightly loaded structures, such as low rise residential and light office buildings" (ACI Commentary-R10.8). It is recommended (ACI Commentary-R10.8) that "the engineer should recognize the need for careful workmanship, as well as the increased significance of shrinkage stresses with small sections."

13.11 MAXIMUM STRENGTH IN AXIAL COMPRESSION—ACI CODE

Since a truly concentrically loaded column is rare, if not nonexistent, some minimal eccentricity should be provided for. This accidental eccentricity may occur due to end conditions, inaccuracy of manufacture, or variation in materials even when the load is theoretically concentric.

Prior to 1977, the ACI Code provided that no matter how small the computed e is from the actual loading, compression members had to be designed for an eccentricity not less than $0.05h$ for spirally reinforced or composite steel encased columns, or $0.10h$ for tied columns; but at least 1 in. in any case. The minimum eccentricity was measured with respect to either principal axis, with h defined as the overall dimension of the column.

This minimum eccentricity requirement meant that every compression member, even though carrying a small computed bending moment, must be designed to have a strength defined, for example, by point A of Fig. 13.11.1. In other words, not only could the axial strength not exceed that defined by point A but a bending moment M_B must be considered as acting simultaneously; that is, the horizontal portion of the strength curve through point A was not actually available in design.

The minimum eccentricity procedure was reasonable as long as the cross-sectional dimensions were small and the slenderness ratios were relatively large; a moment corresponding to 1 in. or so eccentricity on a 12-in. column was not unreasonable. However, an $e/h = 0.1$ minimum becomes unreasonable as accidental eccentricity if a column is a power plant machinery pedestal 6 ft square and 20 ft high. A 6-in.

Figure 13.11.1 Maximum axial strength—ACI Code prior to 1977.

accidental eccentricity is unlikely, although a maximum axial nominal strength less than P_0 may be reasonable.

Since 1977, the ACI Code prescribes (ACI-10.3.5) that for members where the effects of slenderness may be neglected, the maximum axial load nominal strength $P_{n(max)}$ may not exceed $0.80P_0$ for tied columns and $0.85P_0$ for spirally reinforced columns, with P_0 given by Eq. (13.5.1) or (13.5.2). This procedure makes available for design use the entire horizontal portion of the strength interaction diagram defined by $P_{n(max)}$. In other words, point C [Fig. 13.11.2(a)] becomes an acceptable point to use in design, which gives a maximum value of P_n comparable to that corresponding to a minimum $e/h = 0.1$ for tied columns or $e/h = 0.05$ for spirally reinforced columns; yet the design is not explicitly made for a bending moment of P_n times $0.1h$ or P_n times $0.05h$.

When the slenderness ratio is high enough to require consideration of the length effects, a minimum eccentricity of $(0.6 + 0.03h)$ is to be used in the evaluation of the magnified factored moment, but in no case may the resulting available nominal strength P_n be taken to exceed $P_{n(max)}$ (ACI-10.11.5.4).

In an unlikely situation as shown by Fig. 13.11.2(b), where $P_{n(max)}$ is vertically higher than point A, the value of P_n at point A on the interaction diagram must be used for design when the effects of slenderness must be considered; but where slenderness may be neglected, the horizontal portion through point C may be used in design for axial compression.

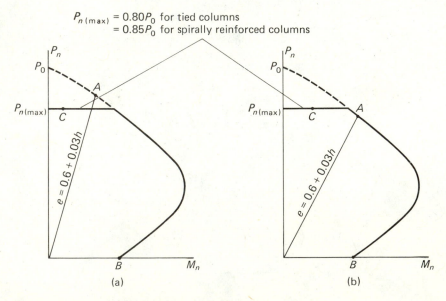

Figure 13.11.2 Maximum axial strength—1989 ACI Code.

13.12 BALANCED STRAIN CONDITION—RECTANGULAR SECTIONS

The balanced strain condition represents the dividing point between the compression controls and tension controls regions of the strength interaction diagram (Fig. 13.6.1). Defined in the same manner as in Chapter 3, it is the simultaneous occurrence of a strain of 0.003 in the extreme fiber of concrete and the strain $\epsilon_y = f_y/E_s$ on the tension steel.

It may be noted that in the case of bending moment *without axial load*, the balanced strain condition is not permitted by the ACI Code. However, in the case of combined bending and axial load, the balanced strain condition is only one point on an acceptable interaction diagram.

Referring to the rectangular section in Fig. 13.12.1, the balanced strain condition gives

$$\frac{x_b}{d} = \frac{0.003}{f_y/E_s + 0.003}$$

$$x_b = \frac{0.003}{f_y/[29(10^6)] + 0.003}\, d = \frac{87,000d}{f_y + 87,000} \tag{13.12.1}$$

Force equilibrium requires

$$P_b = C_c + C_s - T \tag{13.12.2}$$

where

$$C_c = 0.85f'_c ab = 0.85f'_c \beta_1 x_b b \tag{13.12.3}$$

$$T = A_s f_y \tag{13.12.4}$$

and if compression steel yields at balanced strain condition,

$$C_s = A'_s(f_y - 0.85f'_c) \tag{13.12.5}$$

Thus Eq. (13.12.2) becomes

$$P_b = 0.85f'_c \beta_1 x_b b + A'_s(f_y - 0.85f'_c) - A_s f_y \tag{13.12.6}$$

The eccentricity e_b is measured from the plastic centroid, which has been defined in Fig. 13.5.1. For symmetrical sections the plastic centroid is at the middepth of the section.

Rotational equilibrium of the forces in Fig. 13.12.1 is satisfied by taking moments about any point such as the plastic centroid,

$$P_b e_b = C_c\left(d - \frac{a}{2} - d''\right) + C_s(d - d' - d'') + Td'' \tag{13.12.7}$$

Equations (13.12.6) and (13.12.7) may be solved simultaneously to obtain P_b and e_b.

Figure 13.12.1 Balanced strain condition—rectangular section.

■ **EXAMPLE 13.12.1** Determine the eccentric compressive strength $P_n = P_b$ and the eccentricity e_b for a balanced strain condition on the section of Fig. 13.12.2. Use $f_c' = 3000$ psi, $f_y = 50,000$ psi, and the ACI Code.

Solution: **(a)** Locate the neutral axis for the balanced strain condition.

(a) Cross section

(b) Strain diagram

(c) Free body of
elemental length

Figure 13.12.2 Balanced strain conditon—Example 13.12.1.

$$x_b = \frac{0.003(21.6)}{0.00172 + 0.003} = 13.72 \text{ in.}$$

$$a_b = \beta_1 x_b = 0.85(13.72) = 11.67 \text{ in.}$$

The value β_1 is to be taken at 0.85 for $f'_c \leq 4000$ psi (ACI-10.2.7).

(b) Compute the forces C_c, C_s, and T.

$$C_c = 0.85(3.0)(11.67)(15) = 446 \text{ kips}$$

$$T = 50.0(2.37) = 118 \text{ kips}$$

$$\epsilon'_s = 0.003\left(\frac{13.72 - 2.4}{13.72}\right) = 0.00248 > \frac{f_y}{E_s}, \quad \text{compression steel yields}$$

$$C_s = 50.0(2.37) - 0.85(3.0)(2.37) = 118 - 6 = 112 \text{ kips}$$

(c) Compute P_b and e_b.

$$P_b = C_c + C_s - T = 446 + 112 - 118 = 440 \text{ kips}$$

For rotational equilibrium about the plastic centroid,

$$P_b e_b = [446(12 - 11.67/2) + 112(12 - 2.4) + 118(12 - 2.4)]\tfrac{1}{12}$$
$$= 229 + 90 + 95 = 414 \text{ ft-kips}$$

$$e_b = \frac{414(12)}{440} = 11.3 \text{ in.}$$

On the given section, if $P_n > 440$ kips (or $e < 11.3$ in.), the member is more a column than a beam and is referred to as a case where compression controls; if $P_n < 440$ kips (or $e > 11.3$ in.), the member is more a beam than a column and is referred to as a case where tension controls. ■

13.13 INVESTIGATION OF STRENGTH IN COMPRESSION CONTROLS REGION—RECTANGULAR SECTIONS

When the compression strength P_n exceeds the balanced strength P_b or when the eccentricity e is less than the balanced value e_b, the member acts more as a column than as a beam; this case is known as "compression controls." The tensile force T (see Fig. 13.13.1) will then be based on a strain less than ϵ_y (see Fig. 13.6.1) and may actually be a compressive force if the eccentricity is very small.

The nominal strength P_n for a given eccentricity $e < e_b$ may be obtained by considering the actual strain variation as the unknown and using the principles of statics. This is the most rational approach. As an alternative to the direct solution for the neutral axis as shown in the following examples, Reed [13.41] has presented an iterative procedure for analysis.

■ **EXAMPLE 13.13.1**   Determine the nominal compressive strength P_n for the section of Fig. 13.13.1 for an eccentricity $e = 8$ in. by direct application of statics. Use $f'_c = 3000$ psi, $f_y = 50,000$ psi, and the ACI Code.

Figure 13.13.1 Analysis in compression controls region—Example 13.13.1.

Solution: **(a)** Estimate whether the eccentricity will be in the compression controls or the tension controls region. Since the balanced strain condition was computed in Example 13.12.1 as

$$P_b = 440 \text{ kips}, \qquad e_b = 11.3 \text{ in.}$$

it is known that compression controls for $e < 11.3$ in. For $e = 8$ in., the position of the neutral axis x is not known.

(b) Determine the location of the neutral axis. Since the actual x for $e = 8$ in. should exceed the value of $x_b = 13.72$ in. (see Fig. 13.12.2) and the value of ϵ_s' exceeds ϵ_y at the balanced strain condition, it is certain that $\epsilon'_s > \epsilon_y$. Thus, referring to Fig. 13.12.2 for the forces,

$$C_s = A'_s(f_y - 0.85f'_c) = 2.37(50.0 - 2.55) = 112.5 \text{ kips}$$

$$C_c = 0.85f'_c b(0.85x) = 2.55(15)(0.85x) = 32.5x$$

$$T = A_s f_s = 2.37 \frac{(29{,}000)(0.003)(21.60 - x)}{x} = \frac{4450 - 206x}{x}$$

Taking moments about P_n,

$$0 = 112.5(4.0 - 2.40) - 32.5x\left(\frac{0.85x}{2} - 4\right) + \frac{4450 - 206x}{x}(21.6 - 4.0)$$

$$0 = x^3 - 9.41x^2 + 249.6x - 5673$$

$$x = 16.0 \text{ in.}$$

(c) Compute internal forces and strength P_n.

$$C_s = 112.5 \text{ kips}$$

$$C_c = 32.5(16) = 520 \text{ kips}$$

$$T = \frac{4450 - 206(16)}{16.0} = 72.2 \text{ kips}$$

$$P_n = 520 + 112.5 - 72.4 = 560 \text{ kips}$$

(d) Check by a moment equation about the plastic centroid.

$$560(8) \overset{?}{=} 112.5(9.6) + 72.2(9.6) + 520(12 - 6.80)$$

$$4480 \approx 4477 \quad \blacksquare \qquad\qquad\qquad\qquad\qquad\qquad\qquad \textbf{OK}$$

■ **EXAMPLE 13.13.2**   Repeat the calculation of the nominal strength P_n for the section of Fig. 13.13.1, except consider that 1-#8 is located in each of the 24-in. faces of the member, at the middepth 12 in. from the compression face of the member (see Fig. 13.13.2). The eccentricity of P_n is 8 in., $f'_c = 3000$ psi, and $f_y = 50{,}000$ psi.

Solution: **(a)** Estimate whether the eccentricity will be in the compression or tension controls region of the strength interaction diagram. Since the extra 2-#8 bars are at the center of the section, they are unlikely to have an important effect on e_b. Assume that the analysis will be in the compression controls region since e_b for Example 13.13.1 exceeds the 8-in. eccentricity. Note that even if a wrong assumption is made, the neutral axis location thus obtained will reveal $\epsilon'_s < \epsilon_y$ or $\epsilon_s > \epsilon_y$, in which case the revised expressions for C_s or T will have to be used in a new solution.

(b) Estimate for each layer of reinforcement whether it is in compression or tension, and whether the strain at the layer will be greater or less than $\epsilon_y = 0.00172$. The

Figure 13.13.2 Analysis of section containing intermediate layer of steel—Example 13.13.2.

assumptions are as follows:

1. 3-#8 at 2.4 in. from compression face; bars in compression and *assumed* to yield

$$C_{s1} = A'_{s1}(f_y - 0.85f'_c) = 2.37(50 - 2.55) = 112.5 \text{ kips}$$

2. 2-#8 at 12 in. from compression face; bars *assumed* in compression and *assumed* not to yield

$$\epsilon'_{s2} = \frac{0.003}{x}(x - 12.0)$$

$$C_{s2} = A'_{s2}\left[\frac{0.003}{x}(x - 12.0)29{,}000 - 0.85f'_c\right]$$

$$= 1.58\left[\frac{87}{x}(x - 12.0) - 2.55\right]$$

$$= 137.6 - \frac{1649.5}{x} - 4.03$$

3. 3-#8 at 21.6 in. from compression face; bars *assumed* in tension and *assumed* not to yield

$$T = A_s\left[\frac{0.003}{x}(21.6 - x)29{,}000\right]$$

$$= 2.37\left[\frac{87}{x}(21.60 - x)\right] = \frac{4453}{x} - 206.20$$

(c) Determine the compressive force C_c. From Example 13.13.1,

$$C_c = 32.5x$$

(d) Compute the neutral axis distance x. Taking moments about P_n,

$$0 = 112.5(4.0 - 2.40) - 32.5x\left(\frac{0.85x}{2} - 4\right)$$

$$- \frac{133.57x - 1649.5}{x}(12.0 - 4.0)$$

$$+ \frac{4453 - 206.2x}{x}(21.60 - 4.0)$$

$$0 = x^3 - 9.41x^2 + 327.1x - 6629$$

$$x = 15.63 \text{ in.}$$

(e) Verify assumptions (see Fig. 13.13.2).

1. 3-#8 at 2.4 in. from compression face,

$$\epsilon'_{s1} = \frac{0.003}{15.63}(15.63 - 2.40) = 0.00254 > \epsilon_y \quad \text{(as assumed)}$$

2. 2-#8 at 12 in. from compression face,

$$\epsilon'_{s2} = \frac{0.003}{15.63}(15.63 - 12.0) = 0.00070 < \epsilon_y \quad \text{(as assumed)}$$

$$f'_{s2} = \epsilon'_{s2}E_s = 0.00070(29{,}000) = 20.3 \text{ ksi}$$

These bars are in compression as assumed; thus, the correction for displaced concrete (i.e., the minus $0.85f'_c$ in the C_{s2} force) was appropriate.

3. 3-#8 at 21.60 in. from compression face,

$$\epsilon_s = \frac{0.003}{15.63}(21.60 - 15.63) = 0.00115 < \epsilon_y \quad \text{(as assumed)}$$

These bars are in tension as assumed.

(f) Compute internal forces and strength P_n.

$$C_{s1} = 112.5 \text{ kips}$$

$$C_{s2} = A'_{s2}(f'_{s2} - 0.85f'_c) = 1.58(20.3 - 2.55) = 28.0 \text{ kips}$$

$$T = A_s \epsilon_s E_s = 2.37(0.00115)29,000 = 79.0 \text{ kips}$$

$$C_c = 32.5x = 32.5(15.63) = 508.0 \text{ kips}$$

$$P_n = 112.5 + 28.0 - 79.0 + 508.0 = 569.5 \text{ kips}$$

(g) Check by a moment equation about the plastic centroid.

$$569.5(8) \overset{?}{=} 112.5(9.6) + 28.0(0) + 79.0(9.6) + 508.0(12 - 6.64)$$

$$4556 \approx 4561 \qquad\qquad\qquad\qquad \textbf{OK}$$

Note that the additional two bars placed at the plastic centroidal axis increase P_n from 560 kips (Example 13.13.1) to 569.5 kips (this example) for the same eccentricity of 8 in. ∎

Whitney Formula—Compression Controls Case. Approximate procedures may sometimes be desirable, particularly as an aid in selecting section sizes, as will be discussed in Sections 13.15 and 13.16.

One approximate procedure that may be applied to the case when the reinforcement is symmetrically placed in single layers parallel to the axis of bending is the one proposed by Whitney [3.5]. Taking moments of the forces in Fig. 13.13.3 about the tension steel gives

$$P_n\left(e + \frac{d - d'}{2}\right) = C_c\left(d - \frac{a}{2}\right) + C_s(d - d') \qquad \textbf{(13.13.1)}$$

In estimating the compressive force C_c in the concrete, Whitney used for the depth of the rectangular stress distribution an average value based on the balanced strain condition, $a = 0.54d$; thus,

$$C_c = 0.85f'_c ba = 0.85f'_c b(0.54d) = 0.459bdf'_c$$

and

$$C_c\left(d - \frac{a}{2}\right) = 0.459bdf'_c\left(d - \frac{0.54d}{2}\right) = \frac{1}{3}f'_c bd^2 \qquad \textbf{(13.13.2)}$$

Figure 13.13.3 Whitney formula—compression controls case.

When compression controls, compression steel usually yields when $\epsilon_{cu} = 0.003$ at the extreme fiber in compression. Neglecting displaced concrete,

$$C_s = A'_s f_y \tag{13.13.3}$$

Substituting Eqs. (13.13.2) and (13.13.3) in Eq. (13.13.1) gives

$$P_n = \frac{\frac{1}{3}f'_c b d^2}{e + \frac{1}{2}(d - d')} + \frac{A'_s f_y(d - d')}{e + \frac{1}{2}(d - d')}$$

from which

$$P_n = \frac{f'_c b h}{\dfrac{3he}{d^2} + \dfrac{3(d - d')h}{2d^2}} + \frac{A'_s f_y}{\dfrac{e}{d - d'} + \dfrac{1}{2}} \tag{13.13.4}$$

One of the boundary conditions of this relationship is that it must satisfy the condition

$$P_n = P_0 \qquad \text{at } e = 0 \tag{13.13.5}$$

in which

$$P_0 = 0.85 f'_c bh + 2 f_y A'_s \qquad \text{(13.13.6)}$$

if the correction for the displaced concrete is not made. Substituting the boundary condition as expressed by Eqs. (13.13.5) and (13.13.6) in Eq. (13.13.4) requires

$$\frac{3(d - d')h}{2d^2} = \frac{1}{0.85} = 1.18 \qquad \text{(13.13.7)}$$

Making this necessary substitution, Eq. (13.13.4) becomes

$$P_n = \frac{bhf'_c}{\dfrac{3he}{d^2} + 1.18} + \frac{A'_s f_y}{\dfrac{e}{d - d'} + 0.5} \qquad \text{(13.13.8)}$$

which is the Whitney formula for symmetrical steel with no correction for concrete displaced by compression steel.

   In using an approximate formula such as Eq. (13.13.8), it is desirable to be on the conservative side. This would be the case for small eccentricities, because the actual depth of the compressive stress block would then be larger than the assumed value of $0.54d$ at the balanced strain condition.

   A more useful expression for Eq. (13.13.8) in terms of dimensionless ratios may be obtained by letting $A_g = bh$, $\xi h = d$, $A_s = A'_s$ (for symmetrical reinforcement), $\rho_g = 2A'_s/A_g$, and $\gamma h = d - d'$; thus

$$P_n = A_g \left[\frac{f'_c}{\left(\dfrac{3}{\xi^2}\right)\left(\dfrac{e}{h}\right) + 1.18} + \frac{\rho_g f_y}{\left(\dfrac{2}{\gamma}\right)\left(\dfrac{e}{h}\right) + 1} \right] \qquad \text{(13.13.9)}$$

■ **EXAMPLE 13.13.3**   Determine the nominal compressive strength P_n for the section of Fig. 13.12.2 for an eccentricity of 8 in. by using the Whitney compression controls formula, Eq. (13.13.8). Use $f'_c = 3000$ psi, $f_y = 50,000$ psi, and the ACI Code.

Solution: Since this loading is in the compression controls region, Eq. (13.13.8) may be used to obtain an approximate strength.

$$
\begin{aligned}
P_n &= \frac{bhf'_c}{3he/d^2 + 1.18} + \frac{A'_s f_y}{e/(d - d') + 0.5} \\[2mm]
&= \frac{360(3.0)}{[3(24)(8)]/(21.6)^2 + 1.18} + \frac{2.37(50)}{8/(19.20) + 0.50} \\[2mm]
&= \frac{1080}{2.415} + \frac{118.5}{0.916} = 448 + 129 = 577 \text{ kips}
\end{aligned}
$$

   It may be noted that the Whitney formula value (577 kips) is not conservative in this case compared to the more exact statics solution (560 kips from Example 13.13.1). In general, Eq. (13.13.8) is conservative for small eccentricity loading and unconservative as the eccentricity approaches the balanced value, as schematically shown in Fig. 13.13.3. ■

13.14 INVESTIGATION OF STRENGTH IN TENSION CONTROLS REGION—RECTANGULAR SECTIONS

When the compressive strength P_n is less than the balanced value P_b, or the eccentricity e is greater than the balanced value e_b, the member acts more as a beam than as a column; this case is known as "tension controls." The strain in the steel most distant from the neutral axis (near the tension face) will then be greater than the strain $\epsilon_y = f_y/E_s$ (see Fig. 13.6.1).

The most rational approach is to consider the location of the actual neutral axis distance x as the unknown and use the principles of statics.

■ **EXAMPLE 13.14.1** Determine the nominal compressive strength P_n for the member shown in Fig. 13.14.1 for an eccentricity $e = 20$ in., using the method of statics for $f'_c = 3000$ psi, $f_y = 50{,}000$ psi, and the ACI Code.

Solution: From the result of Example 13.12.1 it is known that $e = 20$ in. exceeds the $e_b = 11.3$ in. for the balanced strain condition; therefore tension controls and the strain ϵ_s on the tension steel exceeds ϵ_y. It is assumed (initially) that the strain ϵ'_s on the compression steel is at least equal to the yield strain ϵ_y although the validity of the assumption must be verified before the solution is accepted. Referring to Fig. 13.14.1, the forces T, C_c, and C_s are

$$T = A_s f_y = 3(0.79)(50.0) = 118 \text{ kips}$$

$$C_c = 0.85 f'_c ab = 0.85(3.0)(0.85)(x)(15) = 32.5x$$

$$C_s = A'_s(f_y - 0.85f'_c) = 3(0.79)(50 - 2.55) = 112 \text{ kips}$$

Force equilibrium requires

$$P_n = C_c + C_s - T = 32.5x + 112 - 118 = 32.5x - 6.0$$

Taking moments arbitrarily about the tension steel, rotational equilibrium gives

$$P_n\left(e + \frac{d - d'}{2}\right) = C_c\left(d - \frac{a}{2}\right) + C_s(d - d')$$

$$(32.5x - 6.0)(20 + 9.60) = 32.5x(21.6 - 0.425x) + 112(19.20)$$

$$x^2 + 18.82x - 169.2 = 0$$

$$x = 6.64 \text{ in.}$$

Therefore

$$C_c = 32.5(6.64) = 216 \text{ kips}$$

$$P_n = 216 - 6.0 = 210 \text{ kips}$$

Figure 13.14.1 Analysis in tension controls region—Example 13.14.1.

Verifying the correctness of the strain condition on the compression steel,

$$\epsilon'_s = \epsilon_c \frac{x - d'}{x} = 0.003\left(\frac{6.64 - 2.40}{6.64}\right) = 0.00192$$

$$\epsilon_y = \frac{50}{29,000} = 0.00172 < 0.00192 \qquad\qquad \textbf{OK}$$

Therefore compression steel yields as assumed. When compression steel does yield, the solution using statics is the same as would be obtained from Eq. (13.14.7), which is derived below.

Figure 13.14.2 Strength interaction diagram for section of Fig. 13.14.1.

The complete strength interaction diagram for the section is given in Fig. 13.14.2. ∎

Approximate Formulas—Tension Controls Case for Rectangular Sections. Referring to Fig. 13.14.1 and assuming that the strain in the compression steel is larger than the yield strain, one finds that the forces T, C_s, and C_c are

$$T = A_s f_y$$

$$C_s = A_s'(f_y - 0.85f_c')$$

$$C_c = 0.85f_c'(\beta_1 x)b$$

From force equilibrium

$$P_n = 0.85f_c'\beta_1 xb + A_s'(f_y - 0.85f_c') - A_s f_y \qquad \textbf{(13.14.1)}$$

Let $m = f_y/(0.85f_c')$, $\rho = A_s/bd$, and $\rho' = A_s'/bd$; thus

$$P_n = 0.85f_c'[\beta_1 xb + \rho'(m-1)bd - \rho mbd]$$

$$= 0.85f_c'bd\left[\frac{\beta_1 x}{d} + \rho'(m-1) - \rho m\right] \qquad \textbf{(13.14.2)}$$

From moment equilibrium with respect to the tension steel,

$$P_n e' = 0.85 f_c'(\beta_1 x)b\left(d - \frac{\beta_1 x}{2}\right) + A_s'(f_y - 0.85 f_c')(d - d') \qquad \text{(13.14.3)}$$

Making the same substitutions as for Eq. (13.14.1) gives

$$P_n e' = 0.85 f_c' bd\left[\beta_1 x - \frac{(\beta_1 x)^2}{2d} + \rho'(m - 1)(d - d')\right] \qquad \text{(13.14.4)}$$

Substituting Eq. (13.14.2) in Eq. (13.14.4) gives

$$e'\left[\frac{\beta_1 x}{d} + \rho'(m - 1) - \rho m\right] = \left[\beta_1 x - \frac{(\beta_1 x)^2}{2d} + \rho'(m - 1)(d - d')\right]$$

or

$$x^2 + \left(\frac{2\beta_1 e'}{\beta_1^2} - \frac{2\beta_1 d}{\beta_1^2}\right)x + \frac{e'm(\rho' - \rho) - e'\rho' - \rho'(m - 1)(d - d')}{\beta_1^2}2d = 0$$

Solving the preceding quadratic equation for x gives

$$x = \frac{d - e'}{\beta_1} + \sqrt{\left(\frac{d - e'}{\beta_1}\right)^2 + \frac{2d[\rho'(m - 1)(d - d') + e'\rho' + e'm(\rho - \rho')]}{\beta_1^2}}$$

or

$$\frac{x}{d} = \frac{1 - e'/d}{\beta_1} + \sqrt{\left(\frac{1 - e'/d}{\beta_1}\right)^2 + \frac{2[\rho'(m - 1)(1 - d'/d) + \rho'(e'/d) + (e'/d)m(\rho - \rho')]}{\beta_1^2}}$$

$$\text{(13.14.5)}$$

Substituting Eq. (13.14.5) in Eq. (13.14.2) gives

$$P_n = 0.85 f_c' bd\left\{\rho'(m - 1) - \rho m + \left(1 - \frac{e'}{d}\right)\right.$$

$$\left. + \sqrt{\left(1 - \frac{e'}{d}\right)^2 + 2\left[\left(\frac{e'}{d}\right)(\rho m - \rho'm + \rho') + \rho'(m - 1)\left(1 - \frac{d'}{d}\right)\right]}\right\} \qquad \text{(13.14.6)}$$

For cases where the tension and compression faces are reinforced the same (i.e., $\rho' = \rho$), Eq. (13.14.6) reduces to

$$P_n = 0.85 f_c' bd\left\{-\rho + 1 - \frac{e'}{d}\right.$$

$$\left. + \sqrt{\left(1 - \frac{e'}{d}\right)^2 + 2\rho\left[(m - 1)\left(1 - \frac{d'}{d}\right) + \frac{e'}{d}\right]}\right\} \qquad \text{(13.14.7)}$$

When no compression reinforcement is present, Eq. (13.14.6) may be simplified by making $\rho' = 0$; thus

$$P_n = 0.85 f_c' bd\left[-\rho m + 1 - \frac{e'}{d} + \sqrt{\left(1 - \frac{e'}{d}\right)^2 + \frac{2e'\rho m}{d}}\right] \qquad \text{(13.14.8)}$$

13.15 DESIGN FOR STRENGTH—REGION I, MINIMUM ECCENTRICITY

The approach to design of compression members in bending in accordance with the strength design method of the ACI Code may be divided into three categories (see Fig. 13.15.1): (a) design in Region I for a member having small or negligible bending moment (i.e., maximum permitted axial strength $P_{n(\max)}$ governs); (b) design for Region II in which compression controls but the strength P_n is less than the $0.80P_0$ (tied) or $0.85P_0$ (spirally reinforced) maximum; and (c) design for Region III in which tension controls.

Design in Region I occurs under the following conditions:

1. For *braced* members (i.e., $k \leq 1$) having low slenderness ratio kL_u/r such that the *effects of slenderness may be neglected* according to ACI-10.11.4 (see also Section 13.7):

   (a) The member is subject to axial compression where the bending moment is considered negligible and is not computed.

   (b) The bending moment on the member is computed, but the corresponding eccentricity $e = M_u/P_u$ is less than $e_{\min}$ (Fig. 13.15.1) corresponding to the maximum axial capacity.

2. For *braced* members where the *effects of slenderness must be considered*, computation of bending moment is required, with the result magnified by the factor δ_b in accordance with ACI-10.11.5 (see Chapter 15).

   (a) When the eccentricity $e = M_u/P_u$ at any end of a member is computed and found to be less than the code-specified minimum $(0.6 + 0.03h)$ in.,

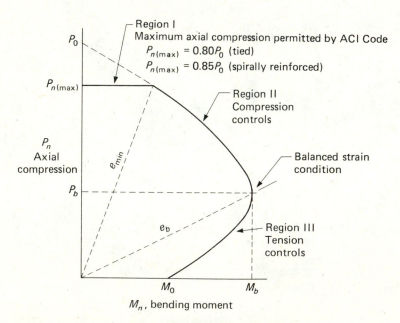

Figure 13.15.1 Design categories for strength of section under combined axial compression and bending moment.

the value $(0.6 + 0.03h)$ is to be magnified by the factor δ_b. When $\delta_b(0.6 + 0.03h)$ is less than e_{min} (Fig. 13.15.1), design is in Region I.

(b) When computed end eccentricity $e = M_u/P_u$ exceeds $(0.6 + 0.03h)$ in., the computed e is magnified to give $\delta_b e$. When $\delta_b e$ is less than e_{min} (Fig.13.15.1), design is in Region I.

An *unbraced* member (i.e., $k > 1.0$) will rarely if ever be a design in Region I; details of the design procedure for unbraced compression members are to be found in Chapter 15.

The reader may note particularly that the member designed in Region I has its axial compressive force P_u limited by the maximum given by ACI-10.3.5, but need not be explicitly designed to carry the moment based on e_{min} as was required prior to 1977. When slenderness effects can be neglected, there is no requirement for bending strength if the factored bending moment was not computed; but if the factored bending moment was computed, it must be provided for in design.

When design is in Region I and slenderness must be considered, the required bending strength for a section in a braced system is based on the magnified actual eccentricity $\delta_b e$ or the magnified minimum $\delta_b(0.6 + 0.03h)$, whichever is greater. Ordinarily any reasonable arrangement of steel will provide sufficient bending strength when design in Region I is based on just using the maximum axial strength prescribed by ACI-10.3.5.

■ **EXAMPLE 13.15.1**   Design an axially loaded spirally reinforced circular column for a gravity dead load of 320 kips and a live load of 360 kips using approximately $3\frac{1}{2}\%$ reinforcement. The column is of average height, and it will be assumed that there is no reduction in strength due to the effects of slenderness. Use $f'_c = 4000$ psi, $f_y = 60,000$ psi, and the ACI Code.

Solution: (a) Determine the required nominal strength P_n.

$$P_u = 1.4(320) + 1.7(360) = 448 + 612 = 1060 \text{ kips}$$

$$\text{required } P_n = \frac{P_u}{\phi} = \frac{1060}{0.75} = 1413 \text{ kips, say } 1410 \text{ kips}$$

(b) Determine the column size. ACI-10.3.5 gives the maximum nominal strength in axial compression as

$$P_{n(max)} = 0.85P_0$$

where P_0 is given by Eqs. (13.5.1) or (13.5.3). Thus

$$P_{n(max)} = 0.85A_g[0.85f'_c + \rho_g(f_y - 0.85f'_c)]$$

$$1410 = 0.85A_g[0.85(4) + 0.035(60 - 3.40)]$$

$$\text{required } A_g = \frac{1410}{4.57} = 308 \text{ sq in.,} \qquad h\text{(diameter)} = 19.8 \text{ in.}$$

Try $h = 20$ in. with $A_g = 314$ sq in.

(c) Determine reinforcement. Solve the $P_{n(max)}$ equation for ρ_g.

$$1410 = 0.85(314)[3.40 + \rho_g(60 - 3.40)]$$

$$\text{required } \rho_g = 0.0332$$

$$\text{required } A_{st} = \rho_g A_g = 0.332(314) = 10.45 \text{ sq in.}$$

Use 20 in. diameter column with 7-#11 bars ($A_s = 10.92$ sq in.).
 No further check is required since no moment has been computed and the maximum nominal axial strength given by ACI-10.3.5 governs in this case.

(d) Design the spiral reinforcement. Using Eq. (13.9.4) which is ACI Formula (10-5),

$$\rho_s = 0.45 \left(\frac{A_g}{A_c} - 1 \right) \frac{f'_c}{f_{sy}}$$

$$A_g = 314 \quad \text{and} \quad A_c = \frac{(20 - 3)^2 \pi}{4} = 227$$

$$\rho_s = 0.45 \left(\frac{314}{227} - 1 \right) \frac{4.0}{60.0} = 0.0115$$

Applying Eq. (13.9.5) gives

$$s_{max} = \frac{a_s \pi (D_c - d_b)}{\rho_s A_c} = \frac{a_s \pi (17 - d_b)}{0.0115(227)}$$

which gives the data in Table 13.15.1.

Table 13.15.1

BAR	a_s (sq. in.)	s_{max} (in.)	MAXIMUM CLEAR SPACING (in.)
#3	0.11	2.20	1.82
#4	0.20	3.97	3.47

Limitations (AC-7.10.4.3):

1. Clear spacing ≤ 3 in.
2. Clear spacing ≥ 1 in.

Use #3 spiral at 2 in. spacing. ∎

13.16 DESIGN FOR STRENGTH—REGION II, COMPRESSION CONTROLS ($e_{min} < e < e_b$)

■ **EXAMPLE 13.16.1** Design a square tied column with about 3% reinforcement for a dead load axial load of 214 kips and bending moment of 47 ft-kips, and a live load axial load of 132 kips and bending moment of 23 ft-kips. Use $f'_c = 3000$ psi, $f_y = 40,000$ psi, and the ACI Code.

Solution: (a) Compute the required nominal strength.

$$\text{required } P_n = \frac{P_u}{\phi} = \frac{1.4(214) + 1.7(132)}{0.70} = 750 \text{ kips}$$

$$\text{required } M_n = \frac{M_u}{\phi} = \frac{1.4(47) + 1.7(23)}{0.70} = 150 \text{ ft-kips}$$

(b) Compute the eccentricity for $P_u/\phi = 750$ kips.

$$e = \frac{M_u}{P_u} = \frac{\text{required } M_n}{\text{required } P_n} = \frac{150(12)}{750} = 2.4 \text{ in.}$$

Although it cannot be certain that $e = 2.4$ in. exceeds e_{min}, it is expected that e_{min} corresponding to $0.80P_0$ (Fig. 13.15.1) will be approximately $0.1h$. Thus for a section smaller than 24 in. the design will be in Region II.

(c) Determine approximate size if required $P_n = 750$ kips equal P_b.

$$x_b = \left(\frac{\epsilon_{cu}}{\epsilon_{cu} + \epsilon_y}\right)d = \left(\frac{0.003}{0.003 + 40/29,000}\right)d = 0.685d$$

$$P_b = C_c + C_s - T$$
$$= 0.85f'_c b\beta_1 x_b + A'_s f_y(\text{approx}) - A_s f_y$$

If the column is made symmetrical, then $A'_s = A_s$; therefore

$$P_b \approx 0.85f'_c b\beta_1 x_b = 0.85(3.0)b(0.85)(0.685d) = 1.483bd$$

For $P_b = $ required $P_n = 750$ kips,

$$\text{balanced } bd \approx \frac{750}{1.483} = 506 \text{ sq in.}$$

Assuming $d \approx 0.9h$,

$$A_g(\text{balanced}) = \frac{506}{0.9} = 562 \text{ sq in. (23.7 in. square)}$$

which means (referring to Fig. 13.16.1) that if an area larger than 562 sq in. is used, tension will probably control; and if an area less than 562 sq in. is provided, compression will probably control.

Since it is preferred to use a column smaller than the approximate balanced size, this design will probably be within Region II, in which compression controls. For the purpose of estimating dimensionless ratios, assume a section about 22 in. square.

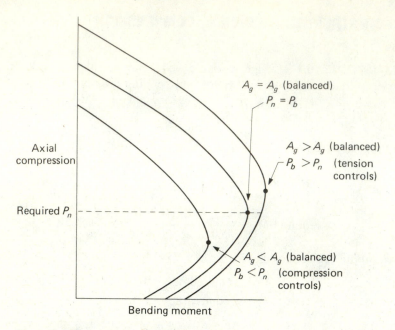

Figure 13.16.1 Variation of strength with gross concrete area, assuming constant percentage of steel reinforcement.

As an alternative to the design approach using the balanced strain condition in combination with the approximate Whitney formula, the reader may prefer the method of Young [13.42], wherein the moment is approximately converted into equivalent axial compression.

(d) Select size for about 3% reinforcement. Assuming compression controls for this small eccentricity and using Eq. (13.13.9),

$$P_n = A_g \left[\frac{f'_c}{(3/\xi^2)(e/h) + 1.18} + \frac{\rho_g f_y}{(2/\gamma)(e/h) + 1} \right]$$

$$\left(\frac{e}{h}\right)_{\text{approx}} = \frac{2.4}{22} = 0.109, \quad \text{estimated } \frac{e}{h} \approx 0.12$$

$$(d - d')_{\text{approx}} = 19.5 - 2.5 = 17.0, \quad \gamma \approx \frac{17}{22} \approx 0.77$$

$$\xi = \frac{d}{h} \approx \frac{19.5}{22} = 0.89$$

$$P_n = A_g \left[\frac{3.0}{(3/0.792)(0.12) + 1.18} + \frac{0.03(40)}{(2/0.77)(0.12) + 1} \right]$$

$$750 = A_g(1.845 + 0.915) = 2.76A_g$$

$$A_g = \frac{750}{2.76} = 272 \text{ sq in.}$$

Try a 17 in. square column.

(e) Estimate reinforcement.

$$A_{st} = 0.03(272) = 8.2 \text{ sq in.}$$

Try 10-#9 bars.

(f) Check design. An analysis may be made any one of the three ways: statics, approximate Eq. (13.13.8), or from nondimensional strength interaction diagrams, such as in the *ACI Design Handbook*, Volume 2 [2.22].

If an approximate formula is to be used as a check, one must be certain that the formula correctly applies. To do this, the balanced condition (P_b and e_b) must be determined for the selected section. In this example, following the procedure of Section 13.12, the balanced strain condition gives

$$P_b = 354.7 \text{ kips}, \qquad e_b = 11.03 \text{ in.}$$

Since $P_n > 354.7$ and $e < 11.03$, it is verified that compression controls. The approximate equation (13.13.8) gives for the nominal strength,

$$
\begin{aligned}
P_n &= \frac{bhf'_c}{3he/d^2 + 1.18} + \frac{A'_s f_y}{e/(d - d') + 0.5} \\[2mm]
&= \frac{289(3)}{3(17)(2.4)/(14.56)^2 + 1.18} + \frac{5(40)}{2.4/12.12 + 0.5} \\[2mm]
&= 493 + 286 = 779 \text{ kips} > 750 \text{ kips required} \qquad \textbf{OK}
\end{aligned}
$$

An analysis using basic statics on the section of Fig. 13.16.2, in the same manner as illustrated in Example 13.13.1, gives $x = 16.10$ in., and $P_n = 809$ kips for $e = 2.4$ in. Another analysis by statics for the section using 8-#9 bars shows $x = 15.90$ in. and

(a) Cross section

(b) Strength interaction diagram

Figure 13.16.2 Section for Example 13.16.1.

$P_n = 755$ kips; thus 8-#9 could have been used. Note that when x is larger than 14.56 in., both layers of steel are in compression, requiring the consideration of displaced concrete in both C_s forces.

Use 17 in. square column and 10-#9 bars, with 5 bars in each face (Fig. 13.16.2).

Note that a better arrangement of bars, especially when the eccentricity is small as in this case, is to distribute them on all four faces, preferably the same arrangement on all four faces to minimize construction errors.

(g) Select lateral ties. Applying the provisions of ACI-7.10.5.2, try #3 ties. Spacing limitations:

   1. Least lateral dimension = 17 in.
   2. 16 bar diameters = 16(1.128) = 18 in.
   3. 48 tie diameters = $48(\frac{3}{8})$ = 18 in.

Use #3 ties, 2 ties per set, at 17 in. spacing. The final cross section and the strength interaction diagram are shown in Fig. 13.16.2. ∎

■ **EXAMPLE 13.16.2**   Redesign the column of Example 13.16.1 ($f'_c = 3000$ psi; $f_y = 40{,}000$ psi) to be the smallest square tied column permitted by the ACI Code.

Solution: Use Eq. (13.13.9) to estimate size. Since $A_g < A_g$ (balanced), compression controls (see Fig. 13.16.1). In this equation, assume that $\xi^2 = (d/h)^2 \approx 0.65$ and $\gamma = (d - d')/h \approx 0.70$, which are reasonable values for $h < 16$ in. For the smallest column, the maximum ρ_g of 0.08 has been assumed.

$$\text{required } P_n = 750 \text{ kips,} \qquad e = 2.4 \text{ in.}$$

$$750 = A_g \left[\frac{3.0}{(3/0.65)(e/h) + 1.18} + \frac{0.08(40)}{(2/0.70)(e/h) + 1} \right]$$

$$750 = A_g \left[\frac{3.0}{4.6(e/h) + 1.18} + \frac{3.2}{2.9(e/h) + 1} \right]$$

Solving by trial:

$$\text{for } \frac{e}{h} = \frac{2.4}{14} = 0.17, \qquad A_g = \frac{750}{3.665} = 204, \qquad h = 14.3 \text{ in.}$$

Try $14\frac{1}{2}$ in. square column, $A_g = 210$ sq in. (Note: This is not a common size for columns and is used here only to illustrate the procedure.)

$$\text{required } A_{st} = 0.08(204) = 16.3 \text{ sq in.}$$

The only arrangement that can be used to get the steel into the two opposite faces is 4-#18. Try 4-#18, $A_s = 16.00$ sq in.

Figure 13.16.3 Trial section for Example 13.16.2.

Check (see Fig. 13.16.3):

$$\frac{e}{h} = \frac{2.4}{14.5} = 0.165$$

$$\xi^2 = \left(\frac{d}{h}\right)^2 = \left(\frac{11.50}{14.5}\right)^2 = 0.628$$

$$\gamma = \frac{d - d'}{h} = \frac{11.50 - 3.00}{14.5} = 0.586$$

$$P_n = A_g\left(\frac{f_c'}{(3/\xi^2)(e/h) + 1.18} + \frac{\rho_g A_g}{(2/\gamma)(e/h) + 1}\right)$$

$$= 210\left[\frac{3.0}{(3/0.628)(0.165) + 1.18} + \frac{0.0763(40)}{(2/0.586)(0.165) + 1}\right]$$

$$= 210(1.525 + 1.95) = 210(3.475) = 730 \text{ kips}$$

This is about 2.7% under the required capacity. If this is not acceptable, a more exact statics check could be made in an attempt to verify the adequacy. The practical solution is to increase the column size to 15 in. For ties, #4 size is minimum with #18 bars (ACI-7.10.5),

$$16 \text{ bar diameters} = 16(2.25) = 36 \text{ in.}$$

$$\text{least column dimension} = 15 \text{ in.}$$

$$48 \text{ tie diameters} = 48(\tfrac{1}{2}) = 24 \text{ in.}$$

Use a 15 × 15 in. square column, with 4-#18 bars, symmetrically placed. Provide #4 ties spaced at 15 in. ∎

Practical Design Approach.   In design practice, compression member design is rarely done in the detailed manner illustrated throughout this chapter. Instead, use is usually made of design aids giving the strength interaction diagram in nondimensional format. Various design aids are available, such as the *ACI Design Handbook, Vol. 2,*

Columns [2.22] (contains nondimensional interaction diagrams, including the variable ϕ for small P_n in Section 13.17), the very useful ACI predecessor document SP-7 by Everard and Cohen [13.43] (contains the nondimensional interaction diagrams with constant ϕ), the *CRSI Handbook* [2.23], and for SI, the *Canadian Metric Design Handbook* [13.44]. Typical interaction charts ($f_c' = 4000$ psi, $f_y = 60,000$ psi, and $\gamma = (d - d')/h = 0.6$ and 0.75) from Reference 2.22 are given as Fig. 13.16.4. For L-shaped columns, Marin [13.45] has provided design aids. Circular columns are treated in Section 13.18.

■ **EXAMPLE 13.16.3**   Design a square tied column containing about 2% reinforcement to carry a dead load axial compression of 710 kN and bending moment of 63 kN·m, and a live load axial compression of 510 kN and bending moment of 33 kN·m. Use $f_c' = 30$ MPa, $f_y = 400$ MPa, and the ACI Code.

Solution: **(a)** Apply the overload factors and compute the eccentricity.

$$P_u = 1.4(710) + 1.7(510) = 1861 \text{ kN}$$

$$M_u = 1.4(63) + 1.7(33) = 144 \text{ kN·m}$$

$$e = \frac{144(1000)}{1861} = 77.4 \text{ mm}$$

(b) Estimate e/h and use the strength interaction charts in Fig. 13.16.4 which include ϕ. Note that the chart values of $f_c' = 4000$ psi and $f_y = 60,000$ psi correspond closely to the given metric data, or the *Canadian Metric Design Handbook* [13.44] can be used. Try the chart for $\gamma = 0.75$. Try $e/h = 0.2$ with $\rho_g = 0.02$. From the chart, obtain

$$\frac{\phi P_n}{A_g} = 2.0 \text{ ksi}$$

$$\text{required } A_g = \frac{1861(1000)}{2.0(6.89)} = 135,000 \text{ mm}^2$$

where 6.89 is the conversion from ksi to MPa. Try a section 360 mm square ($A_g = 129,600 \text{ mm}^2$).

$$\frac{e}{h} = \frac{77.4}{360} = 0.215$$

$$\gamma = \frac{360 - \approx 2(65)}{360} = 0.64 < 0.75$$

The result so far obtained is slightly on the nonconservative side since it is based on a chart for which the bars at the faces are placed at a relatively farther distance apart than they actually are in this design.

$$\text{actual } \frac{P_u}{bh} = \frac{1861}{129.6(6.89)} = 2.08 \text{ ksi}$$

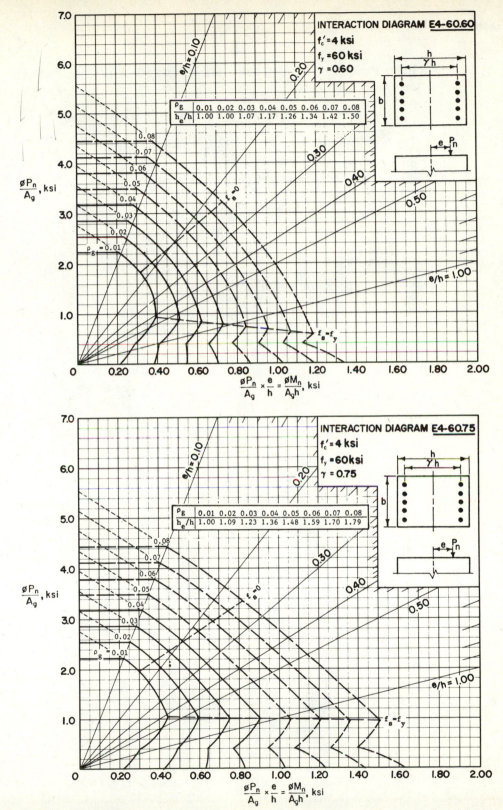

Figure 13.16.4 Strength interaction diagrams for uniaxial bending and compression, symmetrical reinforcement, $f'_c = 4$ ksi, $f_y = 60$ ksi, $\gamma = (d - d')/h = 0.60$ and 0.75 (from *Strength Design Handbook* [2.22]).

Enter Fig. 13.16.4 (interpolating for γ) with $e/h = 0.215$ and $\phi P_n/A_g = 2.08$; find $\rho_g \approx 0.024$.

$$\text{required } A_{st} = 0.024(129{,}600) = 3100 \text{ mm}^2$$

Try 6-#25M bars from Table 1.12.2, $A_{st} = 3000 \text{ mm}^2$.

(c) Check by statics. Based on the e/h ratio and the curves of Fig. 13.16.4, the solution is expected to be in the compression controls region. Referring to Fig. 13.16.5,

$$C_c = 0.85 f'_c ba = 0.85(30)(360)a \tfrac{1}{1000} = 9.18a$$

$$a = \beta_1 x = 0.85x \text{ (see footnote, Section 3.3)}$$

$$C_c = 7.80x$$

Assume compression steel yields,

$$C_s = 3(500)[400 - 0.85(30)]\tfrac{1}{1000} = 562 \text{ kN}$$

Figure 13.16.5 Section for Example 13.16.3.

Taking $E_s = 200,000$ MPa,

$$T = 3(500)\left(\frac{296 - x}{x}\right)(0.003)(200,000)\frac{1}{1000}$$

$$T = \frac{266,400 - 900x}{x}$$

Taking moments about P_n gives

$$T(296 - 102.6) + C_s(102.6 - 64) + C_c(102.6 - 0.425x) = 0$$

$$\frac{266,400 - 900x}{x}(193.4) + 562(38.6) + 7.80x(102.6 - 0.425x) = 0$$

$$x^3 - 241x^2 + 45,960x - 15,542,000 = 0$$

$$x = 277 \text{ mm}$$

Check strains:

$$\epsilon_y = \frac{f_y}{E_s} = \frac{400}{200,000} = 0.00200$$

At C_s,

$$\epsilon'_s = 0.003\left(\frac{277 - 64}{277}\right) = 0.00231 > \epsilon_y \quad \text{(compression steel yields)}$$

At T,

$$\epsilon_s = 0.003\left(\frac{296 - 277}{277}\right) = 0.00021 < \epsilon_y$$

This steel is in tension, but does not yield; the strength of the section is in the compression controls region.

Compute forces:

$$C_c = 7.80x = 7.80(277) \quad = 2161$$

$$C_s \quad\quad\quad\quad\quad\quad\quad = \quad 562$$

$$T = 3(500)(0.00021)(200) = -63$$

$$P_n = \overline{2660} \text{ kN}$$

$$\phi P_n = 0.70(2660) = 1862 \text{ kN} \approx P_u = 1861 \text{ kN required} \qquad \textbf{OK}$$

Check by taking moments about plastic centroid:

$$2660(77.4) = 2161[180 - 0.5(235)] + 562(116) + 63(116)$$

$$205,900 \approx 207,500 \qquad\qquad\qquad\qquad\qquad\qquad \textbf{OK}$$

Use 360 mm square column with 6-#25M bars, as shown in Fig. 13.16.5. ∎

13.17 DESIGN FOR STRENGTH—REGION III, TENSION CONTROLS ($e > e_b$)

The "tension controls" region contains the transition in strength from the condition of crushing failure of concrete as a column to that of ductile flexural failure as a beam. It follows that the safety provisions should logically provide for a gradual increase in the strength reduction factor ϕ from that of columns to that of beams.

The ACI Code requires the use of ϕ factors of 0.70 and 0.75 for tied and spirally reinforced columns, respectively, so long as the eccentric compressive load design strength ϕP_n is $0.10 f'_c A_g$ or larger. Then, as ϕP_n decreases from $0.10 f'_c A_g$ to zero, the ϕ factor may be linearly increased to 0.90.

In order to be safe, however, it is necessary that the value $\phi P_n = 0.10 f'_c A_g$ is within Region III; in other words, one must be certain that ϕP_n is less than ϕP_b (the value at the balanced strain condition). This will usually be the case when f_y does not exceed 60,000 psi, there is symmetrical reinforcement, and the distance $(d - d')$ (see Fig. 13.17.3) between tension and compression reinforcement is not less than $0.70h$. Under these conditions the strength reduction factors ϕ are (ACI-9.3.2) the following:

1. For *tied* compression members:

$$\phi = 0.90 - \frac{2.0 \phi P_n}{f'_c A_g} \geq 0.70 \tag{13.17.1}$$

which gives

$$\phi = \left[\frac{0.90}{1 + \dfrac{2.0 P_n}{f'_c A_g}} \right] \geq 0.70 \tag{13.17.2}$$

2. For *spirally reinforced* compression members:

$$\phi = 0.90 - \frac{1.5 \phi P_n}{f'_c A_g} \geq 0.75 \tag{13.17.3}$$

which gives

$$\phi = \left[\frac{0.90}{1 + \dfrac{1.5 P_n}{f'_c A_g}} \right] \geq 0.75 \tag{13.17.4}$$

The relationships expressed by Eqs. (13.17.1) and (13.17.3) are shown in Fig. 13.17.1(a); and those of Eqs. (13.17.2) and (13.17.4), in Fig. 13.17.1(b). The ϕ vs. P_u (or ϕP_n) curves in Fig. 13.17.1(a) are linear, but the ϕ vs. P_n curves in Fig. 13.17.1(b) are slightly nonlinear. The ACI Code rule was made with the intent to provide the *designer a practical* transition from the ϕ for a compression member to the ϕ for a beam; there is no theoretical or experimental reason for making ϕ vary linearly with ϕP_n but not with P_n. Equations (13.17.1) and (13.17.3) are useful in preliminary design, because with a trial value of A_g, a trial value of ϕ may be obtained by using the

Figure 13.17.1 Variation in ϕ for beam-columns having symmetrical reinforcement, $f_y \le 60{,}000$ psi, and $(d - d')/h \ge 0.70$.

factored force P_u for ϕP_n. Equations (13.17.2) and (13.17.4) are useful for investigating a compression member under bending already designed, because P_n is then available from which ϕ may be computed.

In the relatively few cases where ϕP_b is smaller than $0.10 f'_c A_g$, the increase in the ϕ factor should start from ϕP_b. Using ϕP_b for $0.10 f'_c A_g$ in Eqs. (13.17.1) and (13.17.4) gives

1. For *tied* compression members:

$$\phi = 0.90 - \frac{0.20P_n}{P_b} \geq 0.70 \qquad\qquad (13.17.5)$$

$$\phi = \left[\frac{0.90}{1 + \dfrac{0.20\phi P_n}{P_b}}\right] \geq 0.70 \qquad\qquad (13.17.6)$$

2. For *spirally reinforced* members:

$$\phi = 0.90 - \frac{0.15P_n}{P_b} \geq 0.75 \qquad\qquad (13.17.7)$$

$$\phi = \left[\frac{0.90}{1 + \dfrac{0.15\phi P_n}{P_b}}\right] \geq 0.75 \qquad\qquad (13.17.8)$$

For this situation, Eqs. (13.17.5) and (13.17.7) are useful for investigation of a compression member under bending already designed; and Eqs. (13.17.6) and (13.17.8) are useful in preliminary design because with a trial value of P_b, a trial value of ϕ may be obtained by using the factored force P_u for ϕP_n.

■ **EXAMPLE 13.17.1** Design a rectangular tied column, not over 14 in. wide with about 3% reinforcement, to carry a service dead load of $P = 43$ kips and $M = 96$ ft-kips and a service live load of $P = 32$ kips and $M = 85$ ft-kips. Use symmetrical reinforcement with $f'_c = 4500$ psi, $f_y = 50{,}000$ psi, and the ACI Code.

Solution: (a) Required nominal strengths and eccentricity. Assuming that $P_u \geq 0.10f'_c A_g$, the value of ϕ is 0.70.

$$\text{required } P_n = \frac{P_u}{\phi} = \frac{1.4(43) + 1.7(32)}{0.70} = \frac{114.6}{0.70} = 164 \text{ kips}$$

$$\text{required } M_n = \frac{M_u}{\phi} = \frac{1.4(96) + 1.7(85)}{0.70} = \frac{278.9}{0.70} = 398 \text{ ft-kips}$$

$$e = \frac{398(12)}{164} = 29.1 \text{ in.}$$

(b) Find the approximate size such that $P_n = P_b$ (see Fig. 13.16.1 and discussion relating to Example 13.16.1).

$$x_b = \frac{0.003d}{0.003 + 0.00172} = 0.635d$$

$$a = \beta_1 x_b = 0.825(0.635)d = 0.524d$$

$$P_b = 0.85(4.5)(0.524)bd + A'_s(f_y - 0.85f'_c) - A_s f_y$$

For $A'_s = A_s$, it is reasonable to assume that the two terms involving steel approximately cancel each other. Therefore

$$P_b \approx 2.00bd$$

$$\text{balanced } bd \approx \frac{164}{2.00} = 82.0 \text{ sq in.} \left(A_g \approx \frac{82.0}{0.8} = 103 \text{ sq in.} \right)$$

It is reasonably certain that an area larger than this must be used; therefore tension would control.

(c) Determine size required for reinforcement ratio ρ_g about 0.03. Using Eq. (13.14.7) to obtain preliminary size requirement,

$$P_n = 0.85f'_c bd$$

$$\times \left\{ -\rho + 1 - \frac{e'}{d} + \sqrt{ \left(1 - \frac{e'}{d}\right)^2 + 2\rho \left[(m-1)\left(1 - \frac{d'}{d}\right) + \frac{e'}{d} \right] } \right\} \qquad \textbf{[13.14.7]}$$

Estimate

$$\rho = 0.015 \text{ (one-half total percentage)}$$

$$\frac{e'}{d} = \frac{d - h/2 + e}{d} \approx 2.0$$

$$\frac{d'}{d} \approx 0.1$$

Also

$$m = \frac{f_y}{0.85f'_c} = \frac{50,000}{0.85(4500)} = 13.08$$

Thus

$$P_n = 3.82bd\{ -0.015 + 1 - 2 + \sqrt{(-1.0)^2 + 0.03[12.08(0.9) + 2.00]} \}$$
$$= 3.82bd(-1.015 + \sqrt{1.00 + 0.386})$$
$$= 3.82(0.163)bd = 0.623bd$$

$$\text{required } bd = \frac{164}{0.623} = 263 \text{ sq in.}$$

Try 14×18 column, $A_g = 252$ sq in. Rechecking several variables,

$$\frac{e'}{d} = \frac{d - h/2 + e}{d} = \frac{15.5 - 9 + 29.1}{15.5} = 2.30$$

$$\frac{d'}{d} = \frac{2.5}{15.5} = 0.16$$

$$P_n = 3.82bd(-1.315 + 1.445)$$

$$\text{required } bd = \frac{164}{0.497} = 330 \text{ sq in.}$$

Try 14 × 20 column, $A_g = 280$ sq in.

$$\frac{e'}{d} = \frac{17.5 - 10.0 + 29.1}{17.5} = 2.09$$

$$\frac{d'}{d} = \frac{2.5}{17.5} = 0.14$$

$$P_n = 3.82bd(-1.105 + 1.252)$$

$$\text{required } bd = \frac{164}{0.562} = 292 \text{ sq in.} \approx 280 \text{ sq in.}$$

The 14 × 20 section will be further investigated, and reinforcement will be determined by statics.

(d) Use approximate statics to determine reinforcement. Assuming that C_s and T are equal in Fig. 13.17.2,

$$P_n = C_c = 0.85f'_c ab$$

$$164 = 0.85(4.5)a(14)$$

$$a = 3.06 \text{ in.}$$

Equating the moments of the two couples,

$$P_n\left(19.1 + \frac{3.06}{2}\right) = A_s f_y(d - d')$$

$$A_s = \frac{164(20.63)}{50(15)} = 4.51 \text{ sq in.}$$

$$A_{st} = 9.02 \text{ sq in.}$$

$$\rho = \frac{9.02}{14(20)} = 0.0322$$

Try 6-#11 bars, $A_{st} = 9.36$ sq in.

Figure 13.17.2 Statics for Example 13.17.1.

(e) Check by statics (see Fig. 13.17.3). A statics check may be made (1) by determining the neutral axis location to satisfy $\Sigma F_y = 0$ for $P_u = \phi P_n = 114.6$ kips and then comparing the allowable eccentricity of this load with the required eccentricity, or (2) by determining the neutral axis location to satisfy the condition $e = 29.1$ in. and then comparing the design strength ϕP_n at this eccentricity with the factored eccentric load P_u. The first approach is taken here to illustrate an alternative procedure not previously used in this chapter.

For this section, using Eq. (13.17.1) assuming $\phi P_n = P_u$,

$$\phi = 0.90 - \frac{2.0(114.6)}{4.5(280)} = 0.90 - 0.182 = 0.718 > 0.70$$

Use $\phi = 0.718$.

$$\text{required } P_n = \frac{114.6}{0.718} = 160 \text{ kips}$$

Figure 13.17.3 Section for Example 13.17.1.

Assuming the neutral axis to be at $x = 4.50$ in. (somewhat larger than $x = a/\beta_1 = 3.06/0.825 = 3.70$ in.) from the extreme compression face,

$$\epsilon'_s \approx 0.003\left(\frac{4.50 - 2.5}{4.50}\right) \approx 0.0012 \qquad \text{say } f'_s = 35 \text{ ksi}$$

$$P_n = 160 \text{ kips}$$

$$T = 4.68(50) = 234 \text{ kips}$$

$$C_s = 4.68[35 - 0.85(4.5)] = 146 \text{ kips}$$

$$C_c = 160 + 234 - 146 = 248 \text{ kips}$$

$$C_c = 0.85f'_c ab$$

$$a = \frac{248}{0.85(4.5)14} = 4.63 \text{ in.}$$

$$\text{revised } x = \frac{a}{0.825} = 5.61 \text{ in.}$$

$$\epsilon'_s = 0.003\left(\frac{5.61 - 2.71}{5.61}\right) = 0.00155, \qquad f'_s = 45 \text{ ksi}$$

At this point it is known that the correct f'_s is between 35 and 45 ksi. The trial and error procedure is to compare a trial f'_s value with the corrected f'_s as follows:

$$\text{Revise } f'_s = 40 \text{ ksi,} \qquad C_s = 169.0 \text{ kips,} \qquad C_c = 225 \text{ kips}$$

$$a = 4.20 \text{ in.,} \qquad x = 5.09 \text{ in.,} \qquad \epsilon'_s = 0.00140$$

$$f'_s = 40.7 \text{ ksi}$$

$$\text{Revise } f'_s = 40.4 \text{ ksi,} \qquad C_s = 171 \text{ kips,} \qquad C_c = 223 \text{ kips}$$

$$a = 4.16 \text{ in.,} \qquad x = 5.05 \text{ in.,} \qquad \epsilon'_s = 0.00139$$

$$f'_s = 40.3 \text{ ksi} \qquad \qquad \qquad \qquad \textbf{Say OK}$$

Taking summation of moments about the center of the column,

$$160e = 223[10 - 0.5(4.16)] + (234 + 171)(7.29)$$

from which

$$e = 29.5 \text{ in.} > 29.1 \text{ in.} \qquad \qquad \textbf{OK}$$

Use 14 × 20 column, with 6-#11 bars, 3 bars in each face. ■

■ **EXAMPLE 13.17.2**   Design a rectangular tied column, not over 14 in. wide, to carry a service dead load of $P = 12$ kips and $M = 80$ ft-kips and a service live load of $P = 10$ kips and $M = 85$ ft-kips. Use $f'_c = 4500$ psi, $f_y = 50,000$ psi, and the ACI Code.

Solution: **(a)** Obtain eccentricity. Because the axial compression is of small magnitude relative to the bending moment, the correct ϕ factor quite likely will be between 0.70 and 0.90. The eccentricity may be obtained from the factored forces P_u and M_u; thus

$$P_u = 1.4(12) + 1.7(10) = 33.8 \text{ kips}$$

$$M_u = 1.4(80) + 1.7(85) = 256 \text{ ft-kips}$$

$$e = \frac{256(12)}{33.8} = 91.0 \text{ in.}$$

With this large eccentricity, the section is clearly near the bottom of the "tension controls" region on the strength interaction diagram.

(b) Approach the design as a compression member. Assuming $\phi \approx 0.80$,

$$\text{required } P_n = \frac{P_u}{\phi} = \frac{33.8}{0.8} = 42.2 \text{ kips}$$

Estimate size required using Eq. (13.14.7) for Region III (tension controls).

$$P_n = 0.85 f'_c bd \times \left\{ -\rho + 1 - \frac{e'}{d} + \sqrt{\left(1 - \frac{e'}{d}\right)^2 + 2\rho\left[(m-1)\left(1 - \frac{d'}{d}\right) + \frac{e'}{d}\right]} \right\}$$

Estimate a depth of say 20 in. and $\rho \approx 0.015$ ($\rho_g \approx 0.03$),

$$\frac{e'}{d} = \frac{d - h/2 + e}{d} \approx \frac{17.5 - 10 + 91}{17.5} = 5.6$$

$$\frac{d'}{d} \approx \frac{2.5}{17.5} = 0.143$$

$$m = 13.08 \quad \text{(see Example 13.17.1)}$$

$$P_n = 0.85(4.5)bd\{ -0.015 + 1 - 5.6$$
$$+ \sqrt{(1 - 5.6)^2 + 0.03[(13.08 - 1)(1 - 0.143) + 5.6]} \}$$
$$= 0.140bd$$

$$\text{required } bd = \frac{P_n}{0.140} = \frac{42.2}{0.140} = 301 \text{ sq in.}$$

which would indicate a gross section about 14×24.

(c) Alternative approach considering the member as a beam. Try about one-half the maximum percentage of reinforcement allowed by ACI Code. Using Eq. (3.8.4), or Fig. 3.8.1, find, for $\rho = 0.015$

$$R_n \approx 675 \text{ psi}$$

Estimate ϕ again at 0.80.

$$\text{required } M_n = \frac{M_u}{\phi} = \frac{256}{0.80} = 320 \text{ ft-kips}$$

$$\text{required } bd^2 = \frac{\text{required } M_n}{R_n} = \frac{320(12,000)}{675} = 5690 \text{ in}^3.$$

If $b = 14$ in., required $d = 20.5$ in. and $h = d + 2.5 = 20.5 + 2.5 = 23$, say 24 in. It appears from (b) and (c) that a section about 14×24 will work.

(d) Compute the correct value of ϕ to be used assuming that the 14×24 section is satisfactory.

$$0.10 f'_c A_g = 0.10(4.5)(14)(24) = 151 \text{ kips} > P_u$$

This exceeds ϕP_n, which here is assumed to equal P_u. Assuming one layer of steel in each face,

$$\gamma = \frac{d - d'}{h} \approx \frac{21.5 - 2.5}{24} = 0.79 > 0.70 \qquad \qquad \textbf{OK}$$

and

$$f_y < 60,000 \text{ psi}$$

Using Eq. (13.17.1) with $\phi P_n = P_u$,

$$\phi = 0.90 - 2.0\left(\frac{P_u}{f'_c A_g}\right)$$

$$= 0.90 - 2.0\left(\frac{33.8}{4.5(14)(24)}\right) = 0.90 - 0.04 = 0.86$$

A statics check may now be made to be certain the nominal strength of the section is at least

$$\text{required } P_n = \frac{P_u}{\phi} = \frac{33.8}{0.86} = 39.3 \text{ kips}$$

$$\text{required } M_n = \frac{M_u}{\phi} = \frac{256}{0.86} = 298 \text{ ft-kips}$$

Selection of bars and the check procedure offer no feature that has not already been treated elsewhere. ∎

13.18 CIRCULAR SECTIONS AS COMPRESSION MEMBERS WITH BENDING

The concepts presented for the calculation of the strength P_n (acting at an eccentricity e from the plastic centroid) for a rectangular section are equally applicable to a circular section. The rectangular stress distribution may be applied to the concrete area under compression according to ACI-10.2.6 and 10.2.7, though the use of statics requires knowing the area and centroid of area for circular segments. This information may be computed from formulas or by the use of coefficients for circular sections, as in Fig. 13.18.1. For a value on the strength interaction diagram in the tension controls zone where the portion of the circular section within the rectangular stress distribution may be small, the relationship between the centroid of a circular segment and its area given by Fig. 13.18.2 may be useful.

When the steel reinforcement is in a circular arrangement, each bar may be located and treated in the same manner as a layer of steel in a rectangular section.

$$A \text{ of segment} = \frac{h^2}{2} \int_0^\alpha \sin^2 \theta \, d\theta = h^2 \left[\frac{\alpha - \sin \alpha \cos \alpha}{4} \right]$$

$$Q_0 \text{ of segment} = \frac{h^3}{4} \int_0^\alpha \sin^2 \theta \cos \theta \, d\theta = h^3 \left[\frac{\sin^3 \alpha}{12} \right]$$

$$I_0 \text{ of segment} = \frac{h^4}{8} \int_0^\alpha \sin^2 \theta \cos^2 \theta \, d\theta = h^4 \left[\frac{4\alpha - \sin 4\alpha}{256} \right]$$

$\eta = 0.3 \quad 0.4 \quad 0.5 \quad 0.6 \quad 0.7$
$A = (\text{coefficient}) h^2$

$\eta = 0.3 \quad 0.4 \quad 0.5 \quad 0.6 \quad 0.7$
$Q_0 = (\text{coefficient}) h^3$

$\eta = 0.3 \quad 0.4 \quad 0.5 \quad 0.6 \quad 0.7$
$I_0 = (\text{coefficient}) h^4$

Figure 13.18.1 Properties of circular segments.

The analysis would require first an assumption regarding whether or not the bar (or bars) will yield when $\epsilon_{cu} = 0.003$, as for Example 13.13.2. When about eight or more bars are used the analysis may treat the bars as a steel tube.

Design aids for circular columns, both tied and spirally reinforced, are available from the Portland Cement Association [13.46], the American Concrete Institute [2.22], and the Concrete Reinforcing Steel Institute [2.23]. Mekonnen [13.47] has provided strength equations for circular beam-columns. The CRSI has presented the equations to use for the interaction diagram using a programmable calculator [13.48].

The figure shows:

$\bar{x}/h$

$(0.198, 0.324)$

Straight line approximation

$\dfrac{\bar{x}}{h} = 0.212 + \left(0.393 - \dfrac{A_{seg}}{h^2}\right)$

[Ref. 10]

$(0.293, 0.267)$

Actual curve

$(0.393, 0.212)$

$$\frac{\bar{x}}{h} = \frac{h^2 \sin^3 \alpha}{12 A_{seg}}$$

A_{seg}/h^2

Figure 13.18.2 Centroid of circular segment versus its area.

13.19 AXIAL TENSION AND BENDING MOMENT

In the relatively uncommon situation of axial tension in combination with bending moment, the strength interaction diagram can be considered to extend on the negative P_n side of the axis, as shown in Fig. 13.19.1. The strength analysis for tension values of P_n is similar to that used for compressive load in the "tension controls" region. When the eccentric load is tensile, the neutral axis distance x will be smaller than it is under pure bending (M_0).

When axial tension and bending moment exist, the usual tendency will be to proportion the section in a manner similar to the design as a beam. The subject of axial tension combined with bending moment has been discussed by Harris [13.49], Moreadith [13.50], and Villalta, Carreira, and Erler [13.51]. Since the usual situation will be a section having unsymmetrical reinforcement, the reader is alerted to the fact that the moment used on the interaction diagram must be clearly defined as to whether it is the moment of the tensile force about the centroid of the gross concrete section (this is probably the best choice), or about the "plastic centroid" in compression, or about the "plastic centroid" in tension. Furthermore, the direction of the moment in relation to the unsymmetrical reinforcement must be clearly defined. Villalta, Carreira, and Erler [13.51] have expanded the interaction diagram into all four quadrants of the P_n and M_n axes. In any event, the reader should review a section by using the strain diagram and verify that the external load eccentrically applied is balanced by the internal forces.

The following examples illustrate the calculation of points on the tension side of the interaction diagram (Fig. 13.19.1) for a section having symmetrical reinforcement.

Figure 13.19.1 Strength interaction diagram showing both axial compression and axial tension.

■ **EXAMPLE 13.19.1** Determine the maximum value for the tensile force P_n when no bending moment is acting on the section shown in Fig. 13.19.1.

Solution: Since the concrete will crack before the steel yields, only the steel participates in carrying axial tension. Thus

$$P_n = A_{st}f_y = 6(0.79)50 = 237 \text{ kips}$$

This is plotted on Fig. 13.19.1. ■

■ **EXAMPLE 13.19.2** Determine the axial tensile strength P_n on the section of Fig. 13.19.1 when the eccentricity is 20 in.

Solution: Referring to Fig. 13.19.2, the eccentric tensile force P_n must be acting on the tension side of the plastic centroid. In this case the neutral axis distance x is smaller than its value for pure bending. For bending alone,

$$C_c = 0.85f'_c ba = 0.85(3)(15)a = 38.3a$$

$$T = A_s f_y = 3(0.79)50 = 118.5 \text{ kips}$$

Assuming the compression steel yields,

$$C_s = 3(0.79)(50 - 2.55) = 112.5 \text{ kips}$$

$$C = T$$

$$a = \frac{118.5 - 112.5}{38.3} = 0.16; \qquad x = \frac{0.16}{0.85} < 2.4 \text{ in.}$$

Figure 13.19.2 Free-body diagram and assumed strain diagram for Example 13.19.2.

The above computation shows that compression steel does not yield under bending alone; thus it likely will not yield when P_n is tension at $e = 20$ in.

Thus for axial tension estimate C_s to be in compression but at less than yield stress. Neglecting any displaced concrete effect,

$$C_s = A_s' \epsilon_s' E_s = 3(0.79)\left(\frac{x - 2.4}{x}\right)(0.003)(29{,}000) = \frac{206x - 495}{x}$$

$$C_c = 0.85f_c'b(0.85x) = 0.85(3)(15)(0.85x) = 32.5x$$

$$T = A_s f_y = 3(0.79)50 = 118.5 \text{ kips}$$

Taking moments about P_n (refer to Fig. 13.19.2) gives

$$T(20 - 12 + 2.4) - C_c(20 + 12 - 0.425x) - C_s(20 + 9.6) = 0$$

$$x^3 - 75.3x^2 - 352x + 1060 = 0$$

$$x = 2.10 \text{ in.}$$

Check assumptions,

$$\epsilon_s' = \left(\frac{2.10 - 2.40}{2.10}\right)0.003 = -0.000429 < \epsilon_y$$

The steel represented by C_s actually is in tension and does not yield; there is no displaced concrete effect.

$$C_s = 3(0.79)(-0.000429)29,000 = - \; 29.4 \text{ kips}$$
$$C_c = 32.5(2.10) \qquad\qquad\quad = + \; 68.3 \text{ kips}$$
$$T = \qquad\qquad\qquad\qquad = -118.5 \text{ kips}$$
$$P_n = - \; 79.6 \text{ kips (tension)}$$

$$M_n = 79.6(20)\tfrac{1}{12} = 133 \text{ ft-kips}$$

This value is plotted as point A on Fig. 13.19.1. Check by taking moments about the plastic centroid:

$$133 = [(118.5 - 29.4)(9.6) + 68.3(12 - 0.89)]\tfrac{1}{12}$$

$$133 \approx 134 \qquad\qquad\qquad\qquad\qquad\qquad\qquad\qquad \textbf{OK}$$

The reader may note that when the axial tension is sufficiently high such that the neutral axis falls beyond the edge of the section, there is no longer a compression face of the member. However, the method illustrated in this example seems sufficient to show how points on the tension–bending moment interaction diagram may be obtained.

Note also that a ϕ factor of 0.90 is to be used for this entire region of axial tension. ∎

13.20 WORKING STRESS METHOD

In the working stress method, or "alternate method" of the ACI Code, the service load moment and axial force are used without applying the overload factors U or the strength reduction factors ϕ.

The allowable service load capacity (ACI-Appendix A.6.1) is taken as 40% of the nominal strength—that is, $0.40M_n$ and $0.40P_n$, subject to possible further reduction due to the effects of slenderness.

Note that limiting the service load capacity to 40% of the nominal strength is equivalent to using a U/ϕ value of 2.5. Even at the higher value of $U = 1.70$ for live load only, combined with the lower value of $\phi = 0.70$ for tied columns, the resulting U/ϕ value is $1.70/0.70 = 2.43$. This shows that design of columns by the ACI "alternate method" will always be more conservative than by the strength method.

13.21 BIAXIAL BENDING AND COMPRESSION

The investigation or design of a square or rectangular section subjected to an axial compression in combination with bending moments about both the x and y axes has received considerable attention [13.52–13.77].

One method of analysis is to use the basic principles of equilibrium with the same strength assumptions as were used earlier in this chapter for the case of axial compression and bending about one axis only. This method essentially involves a

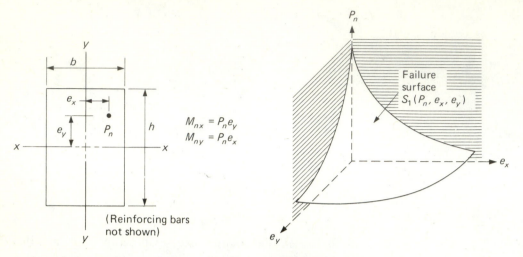

$$M_{nx} = P_n e_y$$
$$M_{ny} = P_n e_x$$

(Reinforcing bars not shown)

Figure 13.21.1 Notation.

Figure 13.21.2 Failure surface S_1 (P_n, e_x, e_y) (from Bresler [13.52]).

trial and error process for obtaining the position of an inclined neutral axis; hence any such method is sufficiently complex that no formula may be developed for practical use.

Failure Surfaces.   The concept of using failure surfaces has been presented by Bresler [13.52], Pannell [13.54], and more recently by Hsu [13.74]. The nominal strength of a section under biaxial bending and compression is a function of three variables, P_n, M_{nx}, and M_{ny}, which may also be expressed in terms of the axial force P_n acting at eccentricities $e_y = M_{nx}/P_n$ and $e_x = M_{ny}/P_n$ with respect to the x and y axes, respectively, as shown in Fig. 13.21.1.

Three types of failure surfaces may be defined. In the first type S_1, the variables used along the three orthogonal axes are P_n, e_x, and e_y, as shown in Fig. 13.21.2; in the second type S_2, the variables are $1/P_n$, e_x, and e_y, as shown in Fig. 13.21.3; and in the third type S_3, the variables are P_n, M_{nx}, and M_{ny}, as shown in Fig. 13.21.4. Bresler has developed a very useful analysis procedure [13.52] using the reciprocal surface S_2. The third type of failure surface S_3 is a three-dimensional extension of the interaction diagram for uniaxial bending and compression as used in the earlier part of this chapter. Bresler [13.52] and Parme, Nieves, and Gouwens [13.55] have suggested practical approaches to the use of the surface S_3. In the presentation that follows, two analysis methods are presented; the first using the reciprocal $1/P_n - e_x - e_y$ surface S_2 that gives a simple tool for investigating a design already made, and the second using the $P_n - M_{ny} - M_{nx}$ surface S_3 which is helpful in making a new design.

Bresler Reciprocal Load Method.   Bresler, in an attempt to develop a realistic procedure for investigation, suggested [13.52] approximating a point $(1/P_{n1}, e_{xA}, e_{yB})$ on the reciprocal failure surface S_2 by a point $(1/P_i, e_{xA}, e_{yB})$ on a *plane* S_2' passing through points A, B, and C (Fig. 13.21.5). Each point on the true surface is

Figure 13.21.3 Reciprocal failure surface S_2 $(1/P_n, e_x, e_y)$ (from Bresler [13.52]).

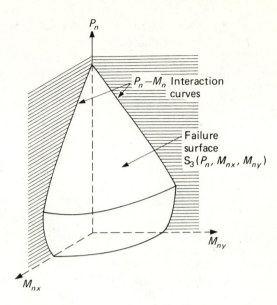

Figure 13.21.4 Failure surface $S_3(P_n, M_{ny}, M_{nx})$ (from Bresler [13.52]).

Figure 13.21.5 Graphical representation of the reciprocal load method (from Bresler [13.52]).

approximated by a different plane; that is, the entire failure surface is defined by an infinite number of planes.

The problem then is to determine the strength P_{n1} which exists with biaxial eccentricities e_{xA} and e_{yB} by assuming that P_{n1} equals the value P_i lying on the plane S_2' specifically established for it. The specific plane is defined by passing it through three points A, B, and C known to lie on the true failure surface S_2,

$$A\left(e_{xA}, \quad 0, \quad \frac{1}{P_y}\right)$$

$$B\left(0, \quad e_{yB}, \quad \frac{1}{P_x}\right)$$

$$C\left(0, \quad 0, \quad \frac{1}{P_0}\right)$$

where P_0 is the nominal strength under axial compression alone without any eccentricity; P_x is the nominal strength at the uniaxial eccentricity e_{yB} ($M_{nx} = P_x e_{yB}$); and P_y is the nominal strength at the uniaxial eccentricity e_{xA} ($M_{ny} = P_y e_{xA}$).

In other words, point A represents a point (P_y, M_{ny}) on the uniaxial P_n–M_n interaction diagram, such as Fig. 13.6.1, for bending about the y axis; point B represents a point (P_x, M_{nx}) on the uniaxial P_n–M_n interaction diagram for bending about the x axis; and point C is a point that is common to both of the uniaxial P_n–M_n interaction diagrams.

The equation of the plane S_2' may be defined in terms of the three points A, B, and C. By letting $x = e_x$, $y = e_y$, and $z = 1/P_n$, the general equation of a plane is

$$A_1 x + A_2 y + A_3 z + A_4 = 0 \tag{13.21.1}$$

Substitution of the coordinates of points A, B, and C (see Fig. 13.21.5) into Eq. (13.21.1) gives

$$A_1 e_{xA} + 0 + A_3 \frac{1}{P_y} + A_4 = 0 \tag{13.21.2a}$$

$$0 + A_2 e_{yB} + A_3 \frac{1}{P_x} + A_4 = 0 \tag{13.21.2b}$$

$$0 + 0 + A_3 \frac{1}{P_0} + A_4 = 0 \tag{13.21.2c}$$

Solving Eqs. (3.21.2abc) for A_1, A_2, and A_3 in terms of A_4,

$$A_1 = \frac{1}{e_{xA}}\left(\frac{P_0}{P_y} - 1\right) A_4 \tag{13.21.3a}$$

$$A_2 = \frac{1}{e_{yB}}\left(\frac{P_0}{P_x} - 1\right) A_4 \tag{13.21.3b}$$

$$A_3 = -P_0 A_4 \tag{13.21.3c}$$

Substitution of Eqs. (13.21.3abc) into Eq. (13.21.1) gives

$$A_4 \left[\frac{x}{e_{xA}} \left(\frac{P_0}{P_y} - 1 \right) + \frac{y}{e_{yB}} \left(\frac{P_0}{P_x} - 1 \right) - P_0 z + 1 \right] = 0 \qquad \textbf{(13.21.4)}$$

Dividing the above equation by P_0, the equation of the plane S_2' becomes

$$\frac{x}{e_{xA}} \left(\frac{1}{P_y} - \frac{1}{P_0} \right) + \frac{y}{e_{yB}} \left(\frac{1}{P_x} - \frac{1}{P_0} \right) - z + \frac{1}{P_0} = 0 \qquad \textbf{(13.21.5)}$$

At the point $(x = e_{xA}, y = e_{yB}, z = 1/P_i)$ on the plane that approximates the point $(x = e_{xA}, y = e_{yB}, z = 1/P_{n1})$ on the true failure surface, Eq. (13.21.5) becomes

$$\left(\frac{1}{P_y} - \frac{1}{P_0} \right) + \left(\frac{1}{P_x} - \frac{1}{P_0} \right) - \frac{1}{P_i} + \frac{1}{P_0} = 0$$

which reduces to the following expression for the value of P_i:

$$\frac{1}{P_i} = \frac{1}{P_x} + \frac{1}{P_y} - \frac{1}{P_0} \qquad \textbf{(13.21.6)}$$

Bresler [13.52] found computed values of P_i using Eq. (13.21.6) to be "in excellent agreement with test results, the maximum deviation being 9.4%, and the average deviation being 3.3%." By its 1989 specific reference for the first time in the ACI Commentary (R10.3.5 and R10.3.6), Eq. (13.21.6) is recognized as "a simple and somewhat conservative estimate of nominal strength." That Commentary further indicates Eq. (13.21.6) is "most suitable" when P_x and P_y are each less than P_b for the particular axis. Pannell [13.54] in his discussion of Reference 13.52 suggests that Eq. (13.21.6) is inappropriate when the ratio P_n/P_0 is in the range of 0.06 or less. In such low axial load cases, it is best to design the member for flexure only.

■ **EXAMPLE 13.21.1** Determine the adequacy of a 16-in. square tied column section containing 8-# 9 bars as shown in Fig. 13.21.6(a). The section is to carry factored loads, $P_u = 144$ kips, $M_{ux} = 120$ ft-kips, and $M_{uy} = 54$ ft-kips. The tied column has $f_c' = 3000$ psi and $f_y = 40,000$ psi.

Solution: The eccentricities are

$$e_y = \frac{M_{ux}}{P_u} = \frac{120(12)}{144} = 10.0 \text{ in.}$$

$$e_x = \frac{M_{uy}}{P_u} = \frac{54(12)}{144} = 4.5 \text{ in.}$$

In order to determine the values of P_x and P_y, the P_n–M_{nx} and P_n–M_{ny} interaction diagrams in uniaxial bending are needed for bending about the x and y axes, respectively. Because of the symmetry in this case, a single P_n–M_{nx} or M_{ny} diagram valid for both bending axes is all that is required, as shown in Fig. 13.21.6(b). The interaction information may be obtained by using equilibrium, by the use of approximate

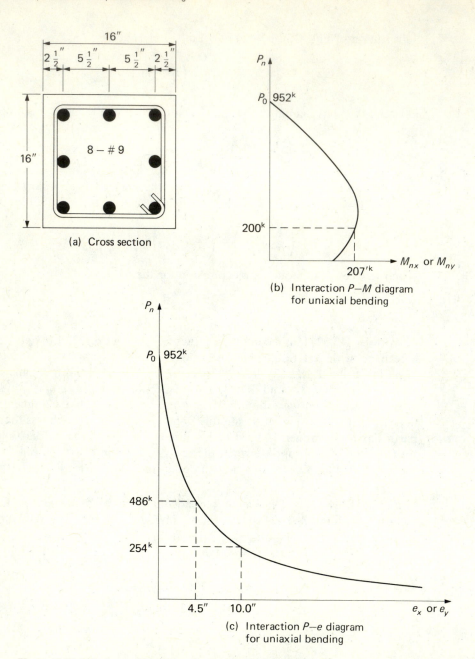

(a) Cross section

(b) Interaction P–M diagram for uniaxial bending

(c) Interaction P–e diagram for uniaxial bending

Figure 13.21.6 Section and interaction diagrams for Example 13.21.1.

formulas such as developed in Sections 13.13 and 13.14, or by means of nondimensionalized P_n–M_n diagrams such as provided in *ACI Design Handbook*, Vol. 2, *Columns* [2.22].

From the P_n–e_x or e_y interaction diagram, Fig. 13.21.6(c), determine P_x and P_y for uniaxial bending,

$$e_x = 4.5 \text{ in.,} \quad \text{find } P_y = 486 \text{ kips}$$

$$e_y = 10.0 \text{ in.,} \quad \text{find } P_x = 254 \text{ kips}$$

Then, using Eq. (13.21.6),

$$\frac{1}{P_n} \approx \frac{1}{P_i} = \frac{1}{P_x} + \frac{1}{P_y} - \frac{1}{P_0}$$

Multiplying by 1000 for convenience,

$$\frac{1000}{P_i} = \frac{1000}{254} + \frac{1000}{486} - \frac{1000}{952}$$

$$\frac{1000}{P_i} = 3.94 + 2.06 - 1.05 = 4.95$$

$$P_n \approx P_i = \frac{1000}{4.95} = 202 \text{ kips}$$

$$[\phi P_n = 0.70(202) = 141 \text{ kips}] \approx [P_u = 144 \text{ kips}] \qquad \textbf{OK}$$

One may note that the $1/P_i$ value on the S'_2 plane is higher than the $1/P_n$ value on the concave surface so that the approximate P_i value is lower than the P_n value; hence P_i is on the safe (low) side.

The more exact analysis of this section may be made by using statics on the assumption that there is a uniformly stressed compression zone, in the same way as for uniaxial bending and compression. In biaxial bending and compression, the uniform compressive stress $0.85f'_c$ is considered to act on a compression zone bounded by the edges of the cross section and a straight line at a distance $a = \beta_1 x$ from the fiber of maximum strain (corner of section) and parallel to the neutral axis. Such an analysis [13.76] gives the strength $P_n = 210$ kips with the neutral axis inclined at an angle of 28.7° clockwise with the x axis and the distance a equal to 8.90 in. from the extreme fiber in compression. This kind of detailed analysis is usually practical only with the aid of a computer. ■

Load Contour Method—Bresler Approach. The load contour method involves cutting the failure surface S_3 (Fig. 13.21.4) at a constant value of P_n to give a so-called "load contour" interaction relating M_{nx} and M_{ny}. In other words, the entire surface S_3 may be considered to include a family of curves (load contours) corresponding to constant values of P_n, which if drawn superimposed on one another in a single plane would be analogous to a contour map. A typical plane at constant P_n along with its load contour is shown in Fig. 13.21.7.

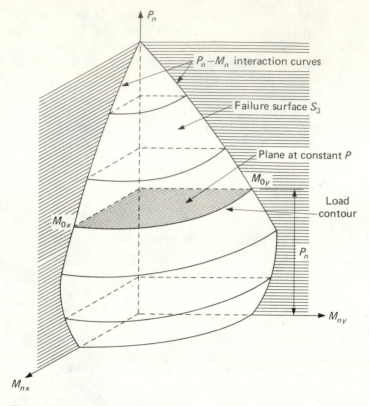

Figure 13.21.7 Load contours for constant P_n on failure surface S_3 (from Bresler [13.52]).

The general nondimensional equation for the load contour at constant P_n may be expressed [13.52] in the form

$$\left(\frac{M_{nx}}{M_{0x}}\right)^{\alpha_1} + \left(\frac{M_{ny}}{M_{0y}}\right)^{\alpha_2} = 1.0 \qquad \textbf{(13.21.7)}$$

where

$M_{nx} = P_n e_y; \ M_{ny} = P_n e_x$

$M_{0x} = M_{nx}$ capacity at axial load P_n when M_{ny} (or e_x) is zero

$M_{0y} = M_{ny}$ capacity at axial load P_n when M_{nx} (or e_y) is zero

and α_1 and α_2 are exponents that depend on the dimensions of the cross section, the reinforcement amount and location, concrete strength, steel yield stress, and amount of concrete cover.

Bresler [13.52] suggests that it is acceptable to take $\alpha_1 = \alpha_2 = \alpha$; then

$$\left(\frac{M_{nx}}{M_{0x}}\right)^{\alpha} + \left(\frac{M_{ny}}{M_{0y}}\right)^{\alpha} = 1 \qquad \textbf{(13.21.8)}$$

which is shown graphically in Fig. 13.21.8.

Figure 13.21.8 Interaction curves for Eq. (13.21.8) (from Bresler [13.52]).

In using Eq. (13.21.8) or Fig. 13.21.8, however, it is still necessary to have a value of α that is applicable to the particular column under investigation. Bresler [13.52] reported the calculated values of α to vary from 1.15 to 1.55.

For practical purposes, it seems satisfactory to take α as 1.5 for rectangular sections and between 1.5 and 2.0 for square sections.

Load Contour Method—Parme Approach.*   The approach described herein has been developed by Parme et al. [13.55] as an extension of the Bresler load contour method. The Bresler interaction equation (13.21.8) is assumed to be the basic strength criterion to define the typical load contour representing the intersection of the failure surface S_3 (Fig. 13.21.7) with a horizontal plane at a height P_n. Such a typical load contour is shown in Fig. 13.21.9. A change in the orientation of the M_{nx} and M_{ny} axes has been made in Fig. 13.21.9 to suit the two-dimensional representation.

In the Parme approach, a point B on the load contour is defined such that the biaxial moment strengths M_{nx} and M_{ny} at this point are in the same ratio as the

---

* This approach may also be referred to as the Portland Cement Association (PCA) method since Reference 13.55 is also available as PCA *Advanced Engineering Bulletin No. 18.*

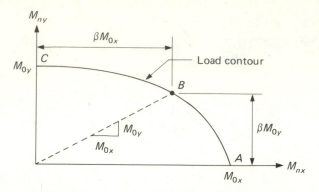

Figure 13.21.9 Load contour at plane of constant P_n cut through failure surface S_3 (Fig. 13.21.7).

uniaxial moment strengths M_{0x} and M_{0y}; thus at point B

$$\frac{M_{ny}}{M_{nx}} = \frac{M_{0y}}{M_{0x}} \qquad (13.21.9)$$

or

$$M_{nx} = \beta M_{0x}; \qquad M_{ny} = \beta M_{0y} \qquad (13.21.10)$$

When the load contour of Fig. 13.21.9 is adjusted to take the nondimensional form of Fig. 13.21.10, the point B will have the ratio β defined by Eq. (13.21.10) as its x and y coordinates. In the physical sense, the ratio β is that constant portion of

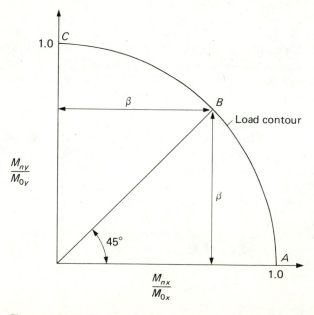

Figure 13.21.10 Nondimensional load contour at constant P_n.

the uniaxial moment strengths which may be permitted to act simultaneously on the column section. The actual value of β depends on the ratio of P_n to P_0, as well as the material and cross-sectional properties; however, the usual range is between 0.55 and 0.70 [13.55]. An average value of $\beta = 0.65$ is suggested for design. More accurate values of β have been computed using basic principles of equilibrium, and charts for β values have been presented in Ref. 13.55. These β-value charts appear as Fig. 13.21.11.

Once an empirical value of β has been ascertained for a given cross section and loading, the complete nondimensional load contour is defined if Eq. (13.21.8) is accepted as the correct relationship. The relationship between the α of Eq. (13.21.8) and β is obtained by using the coordinates of point B, which is known to lie on the contour. Thus substituting the coordinates of B into Eq. (13.21.8) gives

$$\left(\frac{\beta M_{0x}}{M_{0x}}\right)^{\alpha} + \left(\frac{\beta M_{0y}}{M_{0y}}\right)^{\alpha} = 1$$

$$\beta^{\alpha} = \tfrac{1}{2}$$

$$\alpha \log \beta = \log 0.5$$

$$\alpha = \frac{\log 0.5}{\log \beta} \qquad \textbf{(13.21.11)}$$

Thus Eq. (13.21.8) may be written

$$\left(\frac{M_{nx}}{M_{0x}}\right)^{\log 0.5/\log \beta} + \left(\frac{M_{ny}}{M_{0y}}\right)^{\log 0.5/\log \beta} = 1 \qquad \textbf{(13.21.12)}$$

Plots of Eq. (13.21.12) for different values of β are shown in Fig. 13.21.12.

Gouwens [13.63] has presented equations that may be used in place of the curves of Fig. 13.21.12.

For design purposes, the nondimensionalized load contour of Fig. 13.21.10 may be approximated by two straight lines AB and BC as shown in Fig. 13.21.13. When M_{ny}/M_{0y} exceeds M_{nx}/M_{0x}, the straightline approximation equation for BC is

$$\frac{M_{ny}}{M_{0y}} + \frac{M_{nx}}{M_{0x}}\left(\frac{1-\beta}{\beta}\right) = 1 \qquad \textbf{(13.21.13)}$$

When M_{ny}/M_{0y} is less than M_{nx}/M_{0x}, the straightline approximation equation for AB is

$$\frac{M_{nx}}{M_{0x}} + \frac{M_{ny}}{M_{0y}}\left(\frac{1-\beta}{\beta}\right) = 1 \qquad \textbf{(13.21.14)}$$

For design purposes, Eqs. (13.21.13) and (13.21.14) may be written

$$M_{ny} + M_{nx}\left(\frac{M_{0y}}{M_{0x}}\right)\left(\frac{1-\beta}{\beta}\right) = M_{0y}; \qquad \left[\text{for } \frac{M_{ny}}{M_{nx}} \geq \frac{M_{0y}}{M_{0x}}\right] \qquad \textbf{(13.21.15)}$$

$$M_{nx} + M_{ny}\left(\frac{M_{0x}}{M_{0y}}\right)\left(\frac{1-\beta}{\beta}\right) = M_{0x}; \qquad \left[\text{for } \frac{M_{ny}}{M_{nx}} \leq \frac{M_{0y}}{M_{0x}}\right] \qquad \textbf{(13.21.16)}$$

(a) 4 bar arrangement

(b) 8 bar arrangement

Figure 13.21.11 Biaxial bending design constants β (from Reference 13.55). For $0.6 \leq \gamma \leq 1.0$; $3000 \leq f_c' \leq 6000$; and $10 \leq h/b \leq 4.0$. (Figure is continued on right page.)

(c) 6, 8, and 10 bar arrangements

(d) 12 bar arrangement

Figure 13.21.11 (continued)

Figure 13.21.12 Biaxial bending interaction relationship (load contour) in terms of β values (from Reference 13.55).

$$\frac{M_{ny}}{M_{0y}} + \frac{M_{nx}}{M_{0x}}\left(\frac{1-\beta}{\beta}\right) = 1$$

$$\frac{M_{nx}}{M_{0x}} + \frac{M_{ny}}{M_{0y}}\left(\frac{1-\beta}{\beta}\right) = 1$$

Figure 13.21.13 Straightline approximation of load contour for design (from Reference 13.55).

Thus Eqs. (13.21.15) and (13.21.16) represent the alternate algebraic expressions to the exponential equation of Eq. (13.21.12) or Fig. 13.21.12.

When rectangular sections are used with reinforcement distributed uniformly along all faces, the ratio of M_{0y} to M_{0x} (i.e., M_{ny}/M_{nx} of Fig. 13.21.1) will be approximately equal to that of b to h; thus

$$\frac{M_{0y}}{M_{0x}} \approx \frac{b}{h}$$

which gives for Eqs. (13.21.15) and (13.21.16), respectively,

$$M_{ny} + M_{nx}\left(\frac{b}{h}\right)\left(\frac{1-\beta}{\beta}\right) \approx M_{0y} \quad \left[\text{for } \frac{M_{ny}}{M_{nx}} \geq \frac{b}{h}\right] \qquad (13.21.17)$$

$$M_{nx} + M_{ny}\left(\frac{h}{b}\right)\left(\frac{1-\beta}{\beta}\right) \approx M_{0x} \quad \left[\text{for } \frac{M_{ny}}{M_{nx}} \leq \frac{b}{h}\right] \qquad (13.21.18)$$

■ **EXAMPLE 13.21.2** Investigate the section of Example 13.21.1 (Fig. 13.21.6) using the Parme load contour method.

Solution: (a) Use Eq. (13.21.12) or Fig. 13.21.12. In order to make a check by this method, the correct β value must be determined.

$$\frac{\text{required } P_n}{P_0} = \frac{144/0.70}{952} = 0.216 \quad [P_0 \text{ from Fig. 13.21.6(b)}]$$

$$q = \rho_g \frac{f_y}{f_c'} = \frac{8(1.0)}{16(16)}\left(\frac{40}{3}\right) = 0.417$$

Find $\beta = 0.61$ from Fig. 13.21.11.

$$\frac{\text{required } M_{nx}}{M_{0x}} = \frac{120/0.70}{207} = 0.828$$

where M_{0x} is from Fig. 13.21.6(b).

Using $\beta = 0.61$ and $M_{nx}/M_{0x} = 0.828$, find from Fig. 13.21.12,

$$\frac{M_{ny}}{M_{0y}} = 0.37$$

$$M_{ny} = 0.37(207) = 77 \text{ ft-kips}$$

$$[\phi M_{ny} = 0.70(77) = 54 \text{ ft-kips}] \approx [M_{uy} = 54 \text{ ft-kips}] \qquad \textbf{OK}$$

(b) Use Eq. (13.21.14). The required biaxial moment strengths M_{nx} and M_{ny} give

$$\frac{\text{required } M_{nx}}{M_{0x}} = \frac{120/0.70}{207} = 0.828$$

$$\frac{\text{required } M_{ny}}{M_{0y}} = \frac{54/0.70}{207} = 0.373$$

Substituting $\beta = 0.61$ from part **(a)** in Eq. (13.21.14),

$$\frac{M_{nx}}{M_{0x}} + \frac{M_{ny}}{M_{0y}}\left(\frac{1-\beta}{\beta}\right) \leq 1$$

$$0.828 + 0.337\left(\frac{1-0.61}{0.61}\right) = 0.828 + 0.238 = 1.066$$

which would indicate an overstress. As is expected, the straightline relationship is conservative. ∎

Design Procedure.   The design procedure may be summarized as follows:

1. Estimate the value of β at 0.65, or use Fig. 13.21.11 to make an estimate.
2. If M_{ny} is larger than M_{nx}, compute the approximate equivalent uniaxial bending moment M_{0y} using Eq. (13.21.17); if M_{nx} is larger than M_{ny}, compute the approximate equivalent uniaxial bending moment M_{0x} using Eq. (13.21.18).
3. Design the section using the methods treated in Sections 13.15 through 13.17.
4. Check by using any one of the three approaches of this section: (a) the Bresler reciprocal load method, Eq. (13.21.6); (b) the Bresler load contour method, Eq. (13.21.8); (c) the Parme load contour method, Eq. (13.21.12), which is the same as Fig. 13.21.12; or Eqs. (13.21.13) and (13.21.14) which are straightline approximations to the load contour.

Several additional comments in regard to the design of compression members in biaxial bending may be made as follows:

1. Whenever possible, columns subjected to biaxial bending should be circular in cross section.
2. If rectangular or square columns are necessary, the reinforcement should be uniformly spread around the perimeter.
3. Circular or square columns can reasonably be designed to satisfy a given P_u by using the equation [13.52],

$$\left(\frac{M_{nx}}{M_{0x}}\right)^{1.75} + \left(\frac{M_{ny}}{M_{0y}}\right)^{1.75} = 1.0 \tag{13.21.19}$$

4. Rectangular columns can be approximately designed to satisfy a given P_u by using the equation [13.52],

$$\left(\frac{M_{nx}}{M_{0x}}\right)^{1.5} + \left(\frac{M_{ny}}{M_{0y}}\right)^{1.5} = 1.0 \tag{13.21.20}$$

∎ **EXAMPLE 13.21.3**   Select a rectangular cross section for a compression member subjected to biaxial bending to take the following service loads: dead load, $P = 80$ kips, $M_x = 40$ ft-kips, and $M_y = 20$ ft-kips; live load, $P = 32$ kips, $M_x = 48$ ft-kips, and $M_y = 16$ ft-kips. Use $f'_c = 4000$ psi and $f_y = 60,000$ psi.

Solution: **(a)** Determine factored axial force and moments.

$$P_u = 1.4(80) + 1.7(32) = 166 \text{ kips}$$

$$M_{ux} = 1.4(40) + 1.7(48) = 138 \text{ ft-kips}$$

for bending about the x axis; and

$$M_{uy} = 1.4(20) + 1.7(16) = 55 \text{ ft-kips}$$

for bending about the y axis.

(b) Obtain an approximate equivalent uniaxial factored moment. Estimating β at 0.65 to begin the design, and using the approximate strength relationship, Eq. (13.21.18),

$$\text{equivalent } M_{0x} \approx M_{nx} + M_{ny}\left(\frac{h}{b}\right)\left(\frac{1-\beta}{\beta}\right)$$ **[13.21.18]**

Using factored moments,

$$\text{equivalent } \phi M_{0x} = M_{ux} + M_{uy}\left(\frac{h}{b}\right)\left(\frac{1-\beta}{\beta}\right)$$

Assume h/b proportional to $M_{ux}/M_{uy} = 138/55 \approx 2.5$; the requirement is

$$\text{equivalent } \phi M_{0x} = 138 + 55(2.5)\left(\frac{1-0.65}{0.65}\right) = 212 \text{ ft-kips}$$

(c) Determine equivalent eccentricity for uniaxial bending and compression.

$$P_u = 166 \text{ kips}$$

$$M_u = \text{required equivalent } \phi M_{0x} = 212 \text{ ft-kips}$$

$$\text{equivalent } e_y = \frac{212(12)}{166} = 15.3 \text{ in.}$$

(d) Determine approximate requirements for a section that would have P_u/ϕ equal to the strength P_b at the balanced strain condition (see Section 13.16, Example 13.16.1),

$$x_b = \left(\frac{\epsilon_{cu}}{\epsilon_{cu} + \epsilon_y}\right)d = \left[\frac{0.003}{0.003 + (60/29{,}000)}\right]d = 0.592d$$

If symmetrical reinforcement is used, $\rho = \rho'$,

$$P_b \approx C_c = 0.85f_c'b\beta_1 x_b$$

$$= 0.85(4)b(0.85)(0.592d) = 1.71bd$$

For $P_b = P_u/\phi = 166/0.70 = 237 \text{ kips}$,

$$\text{balanced } bd \approx \frac{237}{1.71} = 139 \text{ sq in.}$$

and assuming $d \approx 0.9h$,

$$\text{balanced } A_g \approx \frac{139}{0.9} = 154 \text{ sq in.}$$

If $h/b = 2.5$,

$$\text{balanced } b = \sqrt{\frac{154}{2.5}} = 7.8$$

which gives a section about 8×20. Since an 8-in. section is narrower than generally desired, the section to be chosen will probably be larger than the balanced cross section. Thus, according to Fig. 13.16.1, the section will probably be in Region III where tension controls.

(e) Determine required size assuming tension controls. Use Eq. (13.14.7), and assume $\rho_g \approx 0.04$,

$$P_n = 0.85f'_c bd \left\{ -\rho + 1 - \frac{e'}{d} \right.$$
$$\left. + \sqrt{\left(1 - \frac{e'}{d}\right)^2 + 2\rho\left[(m-1)\left(1 - \frac{d'}{d}\right) + \frac{e'}{d}\right]} \right\}$$

Estimate

$$\rho = 0.02 \quad \text{(one-half total in each face)}$$

$$\frac{e'}{d} = \frac{d - h/2 + e}{d} \approx 1.3 \quad \text{(assuming } h \text{ as 18 to 22 in.)}$$

$$\frac{d'}{d} \approx 0.14$$

$$m = \frac{f_y}{0.85f'_c} = \frac{60}{0.85(4)} = 17.6$$

Thus

$$P_n = 3.4bd\{-0.32 + \sqrt{(0.3)^2 + 0.04[16.6(0.86) + 1.3]}\}$$
$$= 1.78bd$$

If $b = 12$ in. and $P_u/\phi = 237$ kips,

$$\text{required } d = \frac{237}{1.78(12)} = 11.1; \qquad h \approx 15 \text{ in.}$$

Revise equivalent M_{ox} to assume $h/b \approx 1.33$ (say 12×16),

$$\text{equivalent } \phi M_{ox} = 138 + 55(1.33)\left(\frac{1 - 0.65}{0.65}\right) = 177 \text{ ft-kips}$$

Using the required uniaxial strengths $M_{0x} = 177/0.70 = 253$ ft-kips and $P_n = 237$ kips,

$$e_y = \frac{253(12)}{237} = 12.8 \text{ in.}$$

Revise e'/d to 1.35 (assume $h \approx 16$ in.), and d'/d to 0.185,

$$P_n = 3.4bd\left\{-0.37 + \sqrt{(-0.35)^2 + 0.04[16.6(0.815) + 1.35]}\right\}$$
$$= 3.4bd(-0.37 + 0.85) = 1.63bd$$

If $b = 12$ in.,

$$\text{required } d = \frac{237}{1.63(12)} = 12.1 \text{ in.;} \qquad h = 16 \text{ in.}$$

A 12×16 section might work; but since the earlier preliminary indication was for a deeper section, try the deeper one. Try a section 12×18.

$$A_s \approx 0.04(12)(15.2) = 7.3 \text{ sq in.}$$

(f) Check the section of Fig. 13.21.14(a) using the Bresler reciprocal load method. The eccentricities are

$$e_y = \frac{M_{ux}}{P_u} = \frac{138(12)}{166} = 9.98 \text{ in.}$$

$$e_x = \frac{M_{uy}}{P_u} = \frac{55(12)}{166} = 3.98 \text{ in.}$$

Establish the strength P_n for the required values of e_y and e_x from uniaxial P_n–M_n interaction diagrams as in Fig. 13.21.14. The values of P_x and P_y for $e/h = 0.55$ and $e/b = 0.33$ may also be obtained directly using the *ACI Design Handbook*, Vol. 2 [2.22]. Thus

$$\text{for } e_y = 9.98 \text{ in.,} \qquad \text{find } P_x = 400 \text{ kips}$$

$$\text{for } e_x = 3.98 \text{ in.,} \qquad \text{find } P_y = 525 \text{ kips}$$

Using Eq. (13.21.6), with $P_0 = 1271$ kips (Fig. 13.21.14),

$$\frac{1}{P_i} = \frac{1}{P_x} + \frac{1}{P_y} - \frac{1}{P_0}$$

$$\frac{1000}{P_i} = \frac{1000}{400} + \frac{1000}{525} - \frac{1000}{1271}$$

$$= 2.50 + 1.90 - 0.79 = 3.61$$

$$\text{provided } P_n \approx P_i = \frac{1000}{3.61} = 277 \text{ kips}$$

$$[\phi P_n = 0.70(277) = 194 \text{ kips}] > [P_u = 166 \text{ kips}] \qquad \textbf{OK}$$

Use 12×18 section with 12-#8 bars, as in Fig. 13.21.14.

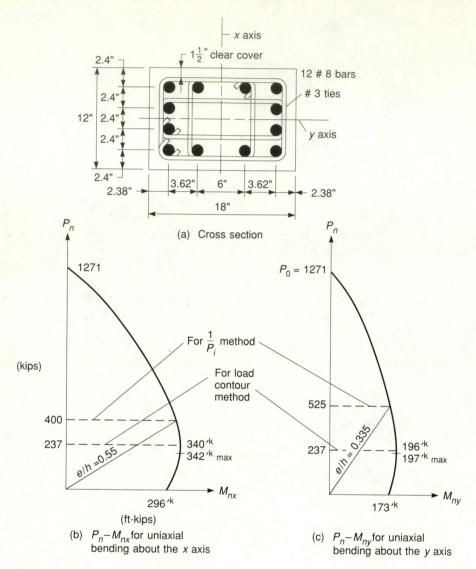

(a) Cross section

(b) P_n–M_{nx} for uniaxial bending about the x axis

(c) P_n–M_{ny} for uniaxial bending about the y axis

Figure 13.21.14 Data for Example 13.21.3.

(g) Check using the Parme load contour method. Using Fig. 13.21.11(d),

$$q = \rho_g \frac{f_y}{f_c'} = \frac{9.48}{12(18)} \left(\frac{60}{4}\right) = 0.66$$

$$\frac{\text{required } P_n}{P_0} = \frac{166/0.70}{1271} = \frac{237}{1271} = 0.19$$

Find $\beta = 0.56$ from Fig. 13.21.11(d).

Next, enter Fig. 13.21.12 with $M_{nx}/M_{0x} = (138/0.70)/340 = 0.58$ and $\beta = 0.56$. Find

$$\frac{\text{strength } M_{ny}}{M_{0y}} = 0.56$$

$$\text{strength } M_{ny} = 0.56(196) = 110 \text{ ft-kips}$$

$$[\phi M_{ny} = 0.70(110) = 77 \text{ ft-kips}] > [M_{uy} = 55 \text{ ft-kips}] \qquad \textbf{OK}$$

The basic statics analysis* for this section gives a strength P_n for $e_y = 9.98$ in. and $e_x = 3.98$ in. of 289 kips. From the published computer program of J. C. Smith [13.58] the statics analysis gives $P_n = 284$ kips, and from the program of C. K. Wang [13.76] $P_n = 278$ kips. The required strength P_n is 237 kips. ∎

SELECTED REFERENCES

Historical

13.1. A. Considère. "Compressive Resistance of Concrete Steel and Hooped Concrete, Part I," *Engineering Record*, December 20, 1902, 581–583; Part II, December 27, 1902, 605–606.

13.2. A. Considère. "Concrete-Steel and Hooped Concrete," *Reinforced Concrete*, 1903, p. 119.

13.3. A. N. Talbot. "Tests of Concrete and Reinforced Concrete Columns." *Bulletin*, No. 10, 1906, and No. 20, 1907, University of Illinois, Urbana.

13.4. M. O. Withey. "Tests of Plain and Reinforced Concrete Columns," *Engineering Record*, July 1909, 41.

13.5. M. O. Withey. "Tests on Reinforced Concrete Columns," *Bulletin*, No. 300, 1910, and No. 466, 1911, University of Wisconsin, Madison.

13.6. ACI Committee 105. "Reinforced Concrete Column Investigation," *ACI Journal, Proceedings*, **26**, April 1930, 601–612; **27**, February 1931, 675–676; **28**, November 1931, 157–158; **29**, September 1932, 53–56; February 1933, 275–284; **30**, September–October 1933, 78–90; November–December 1933, 153–156.

13.7. Joint ACI-ASCE Committee 441. *Reinforced Concrete Columns* (annotated bibliography), *ACI Bibliography* No. 5. Detroit: American Concrete Institute, 1965.

Confinement, Spirals, and Ties

13.8. M. J. N. Priestley, R. Park, and R. T. Potangaroa. "Ductility of Spirally-Confined Concrete Columns," *Journal of the Structural Division*, ASCE, **107**, ST1 (January 1981), 181–202.

----

* Unpublished program used by the junior author in the classroom; originally written by Jagdish Syal in 1967–68.

13.9.   S. H. Ahmad and S. P. Shah. "Stress–Strain Curves of Concrete Confined by Spiral Reinforcement," *ACI Journal, Proceedings,* **79**, November–December 1982, 484–490.

13.10.   Salvador Martinez, Arthur H. Nilson, and Floyd O. Slate. "Spirally Reinforced High-Strength Concrete Columns," *ACI Journal, Proceedings,* **81**, September–October 1984, 431–442.

13.11.   Shamim A. Sheikh and S. M. Uzumeri. "Strength and Ductility of Tied Concrete Columns," *Journal of the Structural Division,* ASCE, **106**, ST5 (May 1980), 1079–1102.

13.12.   B. D. Scott, R. Park, and M. J. N. Priestley. "Stress–Strain Behavior of Concrete Confined by Overlapping Hoops at Low and High Strain Rates," *ACI Journal, Proceedings,* **79**, January–February 1982, 13–27.

13.13.   Robert Park, M. J. Nigel Priestley, and Wayne D. Gill. "Ductility of Square-Confined Concrete Columns," *Journal of the Structural Division,* ASCE, **108**, ST4 (April 1982), 929–951.

13.14.   Apostolos Fafitis and Surendra P. Shah. "Predictions of Ultimate Behavior of Confined Columns Subjected to Large Deformations," *ACI Journal, Proceedings,* **82**, July–August 1985, 423–433.

13.15.   Jack P. Moehle and Terry Cavanagh. "Confinement Effectiveness of Crossties in RC," *Journal of Structural Engineering,* ASCE, **111**, 10 (October 1985), 2105–2120.

13.16.   Shamim A. Sheikh and C. C. Yeh. "Flexural Behavior of Confined Concrete Columns," *ACI Journal, Proceedings,* **83**, May–June 1986, 389–404.

13.17.   Guney Ozcebe and Murat Saatcioglu. "Confinement of Concrete Columns for Seismic Loading," *ACI Structural Journal,* **84**, July–August 1987, 308–315.

13.18.   Yook-Kong Yong, Malakah G. Nour, and Edward G. Nawy. "Behavior of Laterally Confined High-Strength Concrete under Axial Loads," *Journal of Structural Engineering,* ASCE, **114**, 2 (February 1988), 332–351.

13.19.   Hisham Adbel-Fattah and Shuaib H. Ahmad. "Behavior of Hoop-Confined High-Strength Concrete under Axial and Shear Loads," *ACI Structural Journal,* **86**, November–December 1989, 652–659.

13.20.   Shamim A. Sheikh and Ching-Chung Yeh. "Tied Concrete Columns under Axial Load and Flexure," *Journal of Structural Engineering,* ASCE, **116**, 10 (October 1990), 2780–2800.

13.21.   Shamim A. Sheikh and C. C. Yeh. "Concrete Strength in Tied Columns," *ACI Structural Journal,* **87**, July–August 1990, 379–385.

13.22.   Salim R. Razvi and Murat Saatcioglu. "Confinement of Reinforced Concrete Columns with Welded Wire Fabric," *ACI Structural Journal,* **86**, September–October 1989, 615–623.

Interaction Diagrams

13.23.   E. Hognestad. "A Study of Combined Bending and Axial Load in Reinforced Concrete Members," *Bulletin* No. 399, November 1951, Engineering Experiment Station, University of Illinois, Urbana (117 references).

13.24. ACI–ASCE Committee 327. "Report on Ultimate Strength Design," *Proceedings ASCE*, **81**, October 1955, Paper No. 809. See also *ACI Journal, Proceedings*, **52**, January 1956, 505–524.

13.25. A. H. Mattock, L. B. Kriz, and Eivind Hognestad, "Rectangular Concrete Stress Distribution in Ultimate Strength Design," *ACI Journal, Proceedings*, **57**, February 1961, 875–928.

13.26. E. O. Pfrang, C. P. Siess, and M. A. Sozen. "Load-Moment-Curvature Characteristics of Reinforced Concrete Cross Sections," *ACI Journal, Proceedings*, **61**, July 1964, 763–778. Disc., 1673–1683.

13.27. Parviz Soroushian and Kienuwa Obaseki. "Strain Rate-Dependent Interaction Diagrams for Reinforced Concrete Sections," *ACI Journal, Proceedings*, **83**, January–February 1986, 108–116.

13.28. Michael C. Head and J. Dario Aristizabal-Ochoa. "Analysis of Prismatic and Linearly Tapered Reinforced Concrete Columns," *Journal of Structural Engineering*, ASCE, **113**, 3 (March 1987), 575–589.

13.29. Abi O. Aghayere. "Design of Nonprismatic Concrete Columns by the Degree of Fixity Method," *ACI Structural Journal*, **87**, July–August 1990, 473–478.

13.30. M. Ala Saadeghvaziri and Douglas A. Foutch. "Behavior of RC Columns under Nonproportionally Varying Axial Load," *Journal of Structural Engineering*, ASCE, **116**, 7 (July 1990), 1835–1856.

13.31. James G. MacGregor, John E. Breen, and Edward O. Pfrang. "Design of Slender Concrete Columns," *ACI Journal, Proceedings*, **67**, January 1970, 6–28.

13.32. F. E. Richart, J. O. Draffin, T. A. Olson, and R. H. Heitman, "The Effect of Eccentric Loading, Protective Shells, Slenderness Ratios, and Other Variables in Reinforced Concrete Columns," *Bulletin* No. 368, 1947 (130 pp.), Engineering Experiment Station, University of Illinois, Urbana.

13.33. B. Bresler and P. H. Gilbert. "Tie Requirements for Reinforced Concrete Columns," *ACI Journal, Proceedings*, **58**, November 1961, 555–570; Disc., **58**, 897–907.

13.34. James F. Pfister. "Influence of Ties on the Behavior of Reinforced Concrete Columns," *ACI Journal. Proceedings*, **61**, May 1964, 521–537.

13.35. Edwin G. Burdette and Hubert K. Hilsdorf. "Behavior of Laterally Reinforced Concrete Columns," *Journal of the Structural Division*, ASCE, **97**, ST2 (February 1971), 587–602.

13.36. Koji Sakai and Shamim A. Sheikh. "What Do We Know about Confinement in Reinforced Concrete Columns? (A Critical Review of Previous Work and Code Provisions)," *ACI Structural Journal*, **86**, March–April 1989, 192–207.

13.37. Grant T. Halvorsen and Craig A. Carinci. "Tie Requirements for Prestressed Concrete Columns," *PCI Journal*, **22**, July–August 1987, 46–79.

13.38. J. S. Ford, D. C. Chang, and J. E. Breen. "Behavior of Concrete Columns Under Controlled Lateral Deformation," *ACI Journal, Proceedings*, **78**, January–February 1981, 3–20.

13.39. Ti Huang. "On the Formula for Spiral Reinforcement," *ACI Journal, Proceedings*, **61**, March 1964, 351–353. Disc., 1241–1248.

13.40. William L. Gamble. "Re-examination of Spiral Steel Requirements," *Concrete International*, **8**, November 1986, 33–34.

Design

13.41. Paul W. Reed. "Simplified Analysis for Thrust and Moment of Concrete Sections," *ACI Journal, Proceedings*, **77**, May–June 1980, 195–200.

13.42. Curtis J. Young. "Direct Selection of Concrete Dimensions in Columns," *Concrete International*, **3**, October 1981, 27–31.

13.43. N. J. Everard and Edward Cohen. *Ultimate Strength Design of Reinforced Concrete Columns* (SP-7). Detroit: American Concrete Institute, 1964. 182 pp.

13.44. *Metric Design Handbook for Reinforced Concrete Elements in Accordance with the Strength Design Methods of CSA Standard CAN3-A23.3-M77.* (Edited by Murat Saatcioglu). Ottawa, Ontario, Canada: Canadian Portland Cement Association, 1980.

13.45. Joaquin Marin. "Design Aids for L-Shaped Reinforced Concrete Columns," *ACI Journal, Proceedings*, **76**, November 1979, 1197–1216.

13.46. PCA. *Ultimate Load Tables for Circular Columns.* Chicago: Portland Cement Association, 1960.

13.47. Bekele Mekonnen. "Reinforced Concrete Column Design Equations," *ACI Journal, Proceedings*, **81**, May–June 1984, 242–250.

13.48. CRSI. *Interaction Diagrams for Uniaxial Bending Moment and Axial Force on Reinforced Concrete Circular Sections using a Programmable Calculator*, Structural Bulletin No. 12. Schaumburg, Illinois: Concrete Reinforcing Steel Institute, June 1985.

Axial Tension and Bending

13.49. E. C. Harris. "Design of Members Subject to Combined Bending and Tension," *ACI Journal, Proceedings*, **72**, September 1975, 491–495.

13.50. F. L. Moreadith. "Design of Reinforced Concrete for Combined Bending and Tension," *ACI Journal, Proceedings*, **75**, June 1978, 251–255. Disc., **75**, December 1978, 721–723.

13.51. Fernando Villalta, Domingo J. Carreira, and Bryan A. Erler. Discussion of "Design of Reinforced Concrete for Combined Bending and Tension," (*ACI Journal, Proceedings*, **75**, June 1978, 251–255), *ACI Journal, Proceedings*, **75**, December 1978, 721–723.

Biaxial Bending

13.52. Boris Bresler. "Design Criteria for Reinforced Columns under Axial Load and Biaxial Bending," *ACI Journal, Proceedings*, **57**, November 1960, 481–490. Disc., 1621–1638.

13.53. Richard W. Furlong. "Ultimate Strength of Square Columns under Biaxially Eccentric Loads," *ACI Journal, Proceedings,* **57**, March 1961, 1129–1140.

13.54. F. N. Pannell. "Failure Surfaces for Members in Compression and Biaxial Bending," *ACI Journal, Proceedings,* **60**, January 1963, 129–140.

13.55. Alfred L. Parme, Jose M. Nieves, and Albert Gouwens. "Capacity of Reinforced Rectangular Columns Subject to Biaxial Bending," *ACI Journal, Proceedings,* **63**, September 1966, 911–923.

13.56. Donald C. Weber. "Ultimate Strength Design Charts for Columns with Biaxial Bending." *ACI Journal, Proceedings,* **63**, November 1966, 1205–1320. Disc., 1583–1586.

13.57. Anis Farah and M. W. Huggins. "Analysis of Reinforced Concrete Columns Subjected to Longitudinal Load and Biaxial Bending," *ACI Journal, Proceedings,* **66**, July 1969, 569–575.

13.58. J. C. Smith. "Biaxially Loaded Concrete Interaction Curve," *Computers and Structures,* 3 (1973), 1461–1464.

13.59. K. N. Smith and W. H. Nelles. "Columns Subjected to Biaxial Bending—Preliminary Selection of Reinforcing," *ACI Journal, Proceedings,* **71**, August 1974, 411–413.

13.60. Peter D. Heimdahl and Albert C. Bianchini. "Ultimate Strength of Biaxially Eccentrically Loaded Concrete Columns Reinforced With High Strength Steel," *Reinforced Concrete Columns* (SP-50). Detroit: American Concrete Institute, 1975 (pp. 93–117).

13.61. S. I. Abdel-Sayed and N. J. Gardner. "Design of Symmetric Square Slender Reinforced Concrete Columns Under Biaxially Eccentric Loads," *Reinforced Concrete Columns* (SP-50). Detroit: American Concrete Institute, 1975 (pp. 149–164).

13.62. A. K. Basu and P. Suryanarayana. "Analysis of Restrained Reinforced Concrete Columns Under Biaxial Bending," *Reinforced Concrete Columns* (SP-50). Detroit: American Concrete Institute, 1975 (pp. 211–232).

13.63. Albert J. Gouwens. "Biaxial Bending Simplified," *Reinforced Concrete Columns* (SP-50). Detroit: American Concrete Institute, 1975 (pp. 233–261).

13.64. W. F. Chen and M. T. Shoraka. "Tangent Stiffness Method for Biaxial Bending of Reinforced Concrete Columns," *Publications,* International Association for Bridge and Structural Engineering, **35**–1, 1975, 23–44.

13.65. Richard W. Furlong. "Concrete Columns Under Biaxially Eccentric Thrust," *ACI Journal, Proceedings,* **76**, October 1979, 1093–1118.

13.66. M. Pinto de Magalhaes. "Biaxially Loaded Concrete Sections," *Journal of the Structural Division,* ASCE, **105**, ST12 (December 1979), 2639–2656.

13.67. L. N. Ramamurthy and T. A. Hafeez Khan. "L-Shaped Column Design for Biaxial Eccentricity," *Journal of Structural Engineering,* ASCE, **109**, 8 (August 1983), 1903–1917.

13.68. Michael A. Taylor. "Direct Biaxial Design of Columns," *Journal of Structural Engineering,* ASCE, **111**, 1 (January 1985), 158–173.

13.69. Cheng-Tzu Thomas Hsu. "Biaxially Loaded L-Shaped Reinforced Concrete Columns," *Journal of Structural Engineering*, ASCE, **111**, 12 (December 1985), 2576–2595.

13.70. Cheng-Tzu Thomas Hsu. "Reinforced Concrete Members Subject to Combined Biaxial Bending and Tension," *ACI Structural Journal*, **83**, January–February 1986, 137–144.

13.71. David A. Ross and J. Richard Yen. "Interactive Design of Reinforced Concrete Columns with Biaxial Bending," *ACI Structural Journal*, **83**, November–December 1986, 988–993.

13.72. Cheng-Tzu Thomas Hsu. "Channel-Shaped Reinforced Concrete Compression Members under Biaxial Bending," *ACI Structural Journal*, **84**, May–June 1987, 201–211.

13.73. Troels Brøndum-Nielsen. "Concrete Sections under Biaxial Bending," *Journal of Structural Engineering*, ASCE, **113**, 10 (October 1987), 2137–2144.

13.74. Cheng-Tzu Thomas Hsu. "Analysis and Design of Square and Rectangular Columns by Equation of Failure Surface," *ACI Structural Journal*, **85**, March–April 1988, 167–179.

13.75. Issam E. Harik and Hans Gesund. "Flowcharts for Biaxial Bending in R/C Tied Columns," *Journal of Structural Engineering*, ASCE, **114**, 6 (June 1988), 1230–1249; Disc. **116**, 2 (February 1990), 556–557.

13.76. Chu-Kia Wang. "Solving the Biaxial Bending Problem in Reinforced Concrete by a Three-Level Iteration Procedure," *Microcomputers in Civil Engineering*, **3** (1988), 311–320.

13.77. F. A. Zahn, R. Park, and M. J. N. Priestley. "Strength and Ductility of Square Reinforced Concrete Column Sections Subjected to Biaxial Bending," *ACI Structural Journal*, **86**, March–April 1989, 123–131.

PROBLEMS

All problems* are to be worked in accordance with the strength method of the ACI Code unless otherwise indicated, and all stated loads are service loads. Note that eccentricities e_x and e_y are measured along the x and y axes, respectively (see Fig. 13.21.1). For all design problems a design sketch (to scale) of the cross section is required, showing section dimensions, bars, location of bars, and tie or spiral size and spacing.

Problems on Rectangular Sections Subject to Uniaxial Bending

13.1 Calculate the nominal strength P_n for the section of the accompanying figure for an eccentricity e_y (measured along the y axis) $= 0.1h = 1.8$ in. Use $f'_c = 3000$ psi and $f_y = 40,000$ psi. Use basic principles of statics to obtain solution,

— — — —

* Many problems may be solved either as problems stated in Inch-Pound units, or as problems in SI units using quantities in parenthesis at the end of the statement. The SI conversions are approximate to avoid implying higher precision for the given information in metric units than that for the Inch-Pound units.

considering the effect of compression concrete displaced by steel; compare the result (P_n) with that obtained by the Whitney formula (Eq. 13.13.8) and compare with the maximun P_n given by ACI-10.3.5. ($e = 46$ mm; $f'_c = 20$ MPa; $f_y = 300$ MPa.)

Problems 13.1 and 13.2

13.2 Same as Problem 13.1 except $f'_c = 5000$ psi and $f_y = 60,000$ psi. ($f'_c = 35$ MPa; $f_y = 400$ MPa.)

13.3 Compute the nominal strength $P_n = P_b$ in the balanced strain condition for bending about the strong axis (measured as e_y) for the column of the accompanying figure. Compute also the eccentricity e_b. Use the basic statics method, including the effect of the compression concrete displaced by steel, $f'_c = 3000$ psi and $f_y = 40,000$ psi. ($f'_c = 20$ MPa; $f_y = 300$ MPa.)

13.4 Using the basic statics method and taking account of the compression concrete displaced by steel, calculate and plot the P_n vs M_n strength interaction diagram for the section of Problem 13.3. Take bending with respect to the strong axis ($h = 24$ in.). To obtain points for the diagram, compute P_n for the following cases in addition to e_b (see Problem 13.3):

(a) $e = 0$ (P_0)   (d) $e = 0.7h$
(b) $e = 0.1h$ (e) $e = h$
(c) $e = 0.3h$ (f) $e = \infty$ (M_0)

13.5 Same as Problem 13.3 except use $f'_c = 4000$ psi and $f_y = 60,000$ psi. ($f'_c = 30$ MPa; $f_y = 400$ MPa.)

13.6 Same as Problem 13.4 except use $f'_c = 4000$ psi and $f_y = 60,000$ psi. (For e_b, see Problem 13.5).

Problems 13.3, 13.4, 13.5, and 13.6

13.7 For the section of the accompanying figure, using basic statics compute and plot the strength interaction diagram of P_n vs M_n for bending about the x axis. Compute the balanced condition in addition to those points indicated for Problem 13.4. Use $f'_c = 3000$ psi and $f_y = 60,000$ psi. ($f'_c = 20$ MPa; $f_y = 400$ MPa.)

13.8 Repeat Problem 13.7 except consider bending about the y axis.

13.9 For the section of Problem 13.7, compute the nominal strength P_n for an eccentricity e_y of 5 in. with respect to the x axis. Use $f'_c = 5000$ psi and $f_y = 60,000$ psi. ($e = 125$ mm; $f'_c = 35$ MPa; $f_y = 400$ MPa.)

13.10 Repeat Problem 13.9, except use an eccentricity e_y of 22 in. with respect to the x axis. ($e_y = 560$ mm.)

Problems 13.7, 13.8, 13.9, and 13.10

13.11 For the section of the accompanying figure, compute the nominal strength P_n for the eccentricity given below (as assigned by the instructor) with respect to the x axis. Use $f'_c = 3000$ psi and $f_y = 40,000$ psi. ($f'_c = 20$ MPa; $f_y = 300$ MPa.)

(a) $e = 6$ in. (150 mm)   **(e)** $e = 30$ in. (750 mm)
(b) $e = 8$ in. (200 mm)   **(f)** $e = 40$ in. (1000 mm)
(c) $e = 12$ in. (300 mm)   **(g)** $e = 200$ in. (5 m)
(d) $e = 20$ in. (500 mm)   **(h)** $e = 2000$ in. (50 m)

Problem 13.11

13.12 For the section of the accompanying figure, compute the nominal strength P_n for the eccentricity given below (as assigned by the instructor) with respect to the x axis. Use $f'_c = 3000$ psi and $f_y = 40,000$ psi.

(a) $e = 2$ in.   **(e)** $e = 40$ in.
(b) $e = 3.6$ in.   **(f)** $e = 70$ in.
(c) $e = 10$ in.   **(g)** $e = 200$ in.
(d) $e = 15$ in.   **(h)** $e = 1000$ in.

Problem 13.12

13.13 For the section of the accompanying figure, using basic statics compute and plot the strength interaction diagram of P_n vs M_n for bending about the x axis. Compute the balanced condition in addition to enough other points to plot the curve. (*Hint:* By successively setting the neutral axis distance x at specific values the points can be computed without solving a cubic equation.) Use $f'_c = 4000$ psi and $f_y = 60,000$ psi. ($f'_c = 30$ MPa; $f_y = 400$ MPa.)

Problem 13.13

13.14 Using basic principles, compute the nominal strength P_n for an eccentricity e_y (as assigned by the instructor) on the column of the accompanying figure. Use $f'_c = 3500$ psi and $f_y = 40,000$ psi. ($f'_c = 25$ MPa; $f_y = 300$ MPa.)

(a) $e = 3$ in. (75 mm)   (f) $e = 30$ in. (750 mm)
(b) $e = 6$ in. (150 mm)   (g) $e = 60$ in. (1.5 m)
(c) $e = 9$ in. (220 mm)   (h) $e = 150$ in. (3.8 m)
(d) $e = 15$ in. (350 mm)   (i) $e = 300$ in. (7.5 m)
(e) $e = 20$ in. (500 mm)   (j) $e = 2000$ in. (50 m)

Problem 13.14

13.15 For the case assigned by the instructor, design the smallest (in whole inches) square tied column having not more than 4% reinforcement to carry an axial compression (i.e., $M = 0$ or neglected). Compute the strength P_n when $e = 0.10h$. *Comment!* Use $f'_c = 3500$ psi and $f_y = 60,000$ psi. ($f'_c = 25$ MPa; $f_y = 400$ MPa.)

	P_D (dead load)	P_L (live load)
(a)	100 kips (450 kN)	80 kips (350 kN)
(b)	200 kips (900 kN)	160 kips (700 kN)
(c)	350 kips (1500 kN)	150 kips (700 kN)
(d)	400 kips (1800 kN)	320 kips (1400 kN)
(e)	600 kips (2700 kN)	480 kips (2100 kN)

13.16 Repeat Problem 13.15, except use $f'_c = 4000$ psi and $f_y = 60,000$ psi. ($f'_c = 30$ MPa and $f_y = 400$ MPa.)

13.17 Repeat Problem 13.15, except use $f'_c = 3500$ psi and $f_y = 40,000$ psi. ($f'_c = 25$ MPa and $f_y = 300$ MPa.)

13.18 For the case assigned by the instructor, design a square tied column having symmetrical reinforcement of about $3\frac{1}{2}\%$ to carry service loads as given. Check final answer using basic statics. Use whole inches with $f'_c = 4000$ psi and $f_y = 50,000$ psi.

	P_D (dead load)	P_L (live load)	M_D (dead load)	M_L (live load)
(a)	100 kips	90 kips	40 ft-kips	30 ft-kips
(b)	200 kips	175 kips	75 ft-kips	65 ft-kips
(c)	300 kips	270 kips	120 ft-kips	100 ft-kips
(d)	300 kips	270 kips	250 ft-kips	200 ft-kips
(e)	300 kips	270 kips	500 ft-kips	400 ft-kips
(f)	600 kips	550 kips	125 ft-kips	100 ft-kips
(g)	600 kips	550 kips	250 ft-kips	200 ft-kips
(h)	600 kips	550 kips	500 ft-kips	400 ft-kips

13.19 Redesign the column of Problem 13.18 using a rectangular section not over 15 in. wide instead of a square section. If the depth h would be greater than 30 in., use $h/b = 2.0$ and exceed the 15 in. limit.

13.20 Design a square tied column with symmetrical reinforcement of about 4% to carry a dead load of $P = 225$ kips and $M = 220$ ft-kips and a live load of $P = 200$ kips and $M = 190$ ft-kips. Check final answer using basic statics. Use whole inches (20 mm for SI) for size with $f'_c = 4000$ psi and $f_y = 40{,}000$ psi. (DL: $P = 1000$ kN, $M = 300$ kN·m; LL: $P = 900$ kN, $M = 260$ kN·m; $f'_c = 30$ MPa; $f_y = 300$ MPa.)

13.21 Redesign the column of Problem 13.20 using a rectangular section not over 16 in. (400 mm) wide instead of a square section.

13.22 Design a square tied column with about 4% reinforcement to carry a dead load of $P = 250$ kips and $M = 90$ ft-kips, and a live load of $P = 185$ kips and $M = 70$ ft-kips. Use whole inches (20 mm for SI) for size with $f'_c = 4000$ psi and $f_y = 60{,}000$ psi. (DL: $P = 1100$ kN, $M = 120$ kN·m; LL: $P = 820$ kN, $M = 95$ kN·m; $f'_c = 30$ MPa; $f_y = 400$ MPa.)

13.23 Design a square tied column with symmetrical reinforcement of about 2% to carry a dead load of $P = 25$ kips and $M = 125$ ft-kips and a live load of $P = 10$ kips and $M = 50$ ft-kips. Use whole inches (20 mm for SI) for size with $f'_c = 4000$ psi and $f_y = 60{,}000$ psi. (DL: $P = 110$ kN, $M = 170$ kN·m; LL: $P = 45$ kN, $M = 68$ kN·m; $f'_c = 30$ MPa; $f_y = 400$ MPa.)

13.24 Redesign the column of Problem 13.23 as a rectangular column having a depth-to-width ratio of between 1.5 and 2.0 along with unsymmetrical reinforcement with respect to the bending axis.

13.25 Design a square tied column with symmetrical reinforcement of about $2\frac{1}{2}$% to carry a dead load of $P = 75$ kips and $M = 125$ ft-kips and a live load of $P = 30$ kips and $M = 50$ ft-kips. Use whole inches (20 mm for SI) for size with $f'_c = 4000$ psi and $f_y = 60{,}000$ psi. (DL: $P = 340$ kN, $M = 170$ kN·m; LL: $P = 140$ kN, $M = 70$ kN·m; $f'_c = 30$ MPa; $f_y = 400$ MPa.)

13.26 Redesign the column of Problem 13.25 using a rectangular member but still using symmetrical reinforcement.

Problems on Circular Sections

13.27 Determine the strength $P_n = P_b$ and the eccentricity e_b for a balanced strain condition on the section of the accompanying figure by using the method of statics and the rectangular stress distribution as for beams. Use $f'_c = 4000$ psi, $f_y = 50{,}000$ psi.

13.28 Determine the nominal compressive strength P_n for the section of Problem 13.27 for an eccentricity $e = 5$ in. by the direct application of statics. Use $f'_c = 4000$ psi, $f_y = 50{,}000$ psi.

Problems 13.27, 13.28, and 13.31

13.29 Compute the nominal strength P_n of the spirally reinforced column of the accompanying figure when the loading has an eccentricity $e = 0.05h = 0.9$ in. Use $f'_c = 3000$ psi and $f_y = 40,000$ psi. Use basic statics with the circular section, including the effect of compression concrete displaced by steel. Compare with maximum P_n obtained from ACI-10.3.5. ($e = 23$ mm; $f'_c = 20$ MPa; $f_y = 300$ MPa.)

13.30 Same as Problem 13.29, except use $f'_c = 5000$ psi and $f_y = 60,000$ psi. ($f'_c = 35$ MPa; $f_y = 400$ MPa.)

Problems 13.29 and 13.30

13.31 Determine the nominal compressive strength P_n for the section of Problem 13.27 for an eccentricity $e = 20$ in. by a direct application of statics. Use $f'_c = 4000$ psi, $f_y = 50,000$ psi, and the ACI Code.

13.32 For the column of the accompanying figure, determine the nominal strength P_n at an eccentricity of 21 in. Use $f'_c = 4000$ psi and $f_y = 50,000$ psi. Apply the basic principles of statics. ($e = 530$ mm; $f'_c = 30$ MPa; $f_y = 350$ MPa.)

Problem 13.32

13.33 Same as Problem 13.15, except use a spirally reinforced circular column.

13.34 Redesign the column of Problem 13.18 as a spirally reinforced circular column.

13.35 Redesign the column of Problem 13.20 as a spirally reinforced circular column.

13.36 Design a circular spirally reinforced column for the conditions given in Problem 13.22. Make statics check using the designed circular section.

13.37 Redesign the column of Problem 13.25 as a spirally reinforced circular column.

Problems on Biaxial Bending and Compression

13.38 Compute the nominal strength P_n for the column of Problem 13.3 under an eccentricity e_y of 8 in. about the strong axis and an eccentricity e_x of 5 in. about the weak axis. (*Hint:* Use the interaction curve of Problem 13.4 and calculate a similar interaction curve for the weak axis.) Compute strength using (a) Bresler reciprocal load method, and (b) Parme load contour method. Use $f'_c = 3000$ psi and $f_y = 40,000$ psi. ($e_y = 200$ mm; $e_x = 130$ mm; $f'_c = 20$ MPa; $f_y = 300$ MPa.)

13.39 Determine the adequacy of the rectangular tied column section of Problem 13.7 when subjected to an axial load of 90 kips, applied with an eccentricity e_y of 5 in. with respect to the x axis and with an eccentricity e_x of 3 in. with respect to the y axis. Use $f'_c = 3000$ psi and $f_y = 40,000$ psi (use information developed in Problems 13.7 and 13.8). Compare results using (a) Bresler reciprocal load method and (b) Parme load contour method.

13.40 Repeat Problem 13.39, except the axial load is 45 kips applied with eccentricities of 10 in. with respect to the x axis and 6 in. with respect to the y axis.

13.41 For the case assigned by the instructor, compute the nominal strength P_n for biaxial bending and compression for the section of Problem 13.7. Use both the Bresler reciprocal load and the Parme load contour methods. (If available, the strength interaction diagrams from Problems 13.7 and 13.8 may be used.) Use $f'_c = 3000$ psi and $f_y = 60,000$ psi.

(a) $e_y = 1.8$ in. $e_x = 1.4$ in.
(b) $e_y = 5.4$ in. $e_x = 1.4$ in.
(c) $e_y = 15$ in. $e_x = 2$ in.
(d) $e_y = 30$ in. $e_x = 3$ in.
(e) $e_y = 40$ in. $e_x = 6$ in.
(f) $e_y = 40$ in. $e_x = 20$ in.

13.42 Design a square tied column with about 4% reinforcement uniformly distributed around its sides. The loads are dead load: $P = 225$ kips, $M_x = 236$

ft-kips, $M_y = 108$ ft-kips, and live load: $P = 200$ kips, $M_x = 204$ ft-kips, $M_y = 64$ ft-kips. Select size in whole inches that are multiples of two, using $f'_c = 4000$ psi and $f_y = 40,000$ psi.

13.43 Repeat Problem 13.42, except use $f_y = 60,000$ psi.

13.44 Repeat Problem 13.42, except reduce moments M_x about the x axis to 215 ft-kips dead load and 162 ft-kips live load, and increase f_y to 60,000 psi.

CHAPTER 14

Deflections

14.1 DEFLECTIONS—GENERAL

Throughout the period from 1910 to 1956, while the working stress method was used almost exclusively, concrete with a compressive strength f'_c from 1500 to 3000 psi (approximately 10.5 to 21 MPa) and reinforcement with a yield stress from 33 to 40 ksi (approximately 230 to 280 MPa) were predominant. The use of these materials with conservative allowable stresses, along with the straightline working stress method, resulted in large stiff sections having small deflections. Ordinary reinforced concrete design involved little concern for deflections.

With the widespread use of the strength method and the realistic acceptance of the additional strength of concrete in compression according to the nonlinear relationship between stress and strain, sections could be made smaller. Such smaller sections deflect a greater amount than those designed under the working stress method.

The common use of 60,000 psi (420 MPa) yield strength steel and of concrete having strengths f'_c ranging from the ordinary value of 4000 psi (28 MPa) up to 10,000 psi (68 MPa) permits smaller sections than those resulting from the use of lower strength materials.

The permissible deflection is governed by the serviceability requirements for the structure, such as the amount of deformation that can be tolerated by the interacting components of the structure. Excessive deflection of the member may not in itself be detrimental, but the effect on structural components that are supported by the deflecting member frequently determines the acceptable amount. Both the short-time (instantaneous or immediate) and the long-time effects must be considered. The acceptable deflection depends on many factors, among which are the type of building (warehouse, school, factory, residence, etc.), the presence of plastered ceilings, the type and arrangement of partitions, the sensitivity of equipment to deflection, and the magnitude and duration of live load. Vibration and noise transmission are also

Exterior rigid frames; Water Tower Place, Chicago. (Courtesy of Portland Cement Association.)

serviceability concerns that depend on the stiffness of flexural members just as does deflection.

The maximum acceptable immediate deflections due to live load, for flat roofs or floors that do not support and are not attached to nonstructural elements such as plastered ceilings or frangible partitions likely to be damaged by large deflections, are prescribed by ACI-Table 9.5(b) to be (a) for flat roofs, $L/180$ and (b) for floors, $L/360$. Further, in recognition of the increase of deflection with time, ACI-Table 9.5(b) also prescribes limits for the sum of the *creep and shrinkage deflection due to all sustained loads plus any additional live load deflection*. The two stated limits for this deflection combination are (a) when nonstructural elements are likely to be damaged, $L/480$, and (b) when no damage to nonstructural elements is likely, $L/240$.

Limitations on deflection are somewhat arbitrary; historically, $L/360$ has been the accepted limit to prevent the cracking of plastered ceilings. Other limits should be considered as guidelines, with the designer having the responsibility for evaluating the possible adverse effects of excessive deflection in any given situation.

ACI Committee 435 has reported recommendations for allowable deflections for a great variety of situations [14.1].

The general concepts dealt with in this chapter are applicable to both one-way (beams and slabs) and two-way systems. Specific examples are applied only to one-way systems. The reader is referred to Section 16.20 for a brief discussion of two-way system deflections, along with specific references.

14.2 DEFLECTIONS FOR ELASTIC SECTIONS

Various methods are available in structural analysis for computing deflections on uniform and variable moment of inertia sections in statically determinate and indeterminate structures. In general, using any of the several methods, the maximum deflection in an elastic member may be expressed as

$$\Delta_{max} = \beta_a \frac{ML^2}{EI_c} \tag{14.2.1}$$

where

M = a reference value of bending moment such as the maximum positive value

L = span length

E = modulus of elasticity

I_c = moment of inertia of a standard section in the member

β_a = a coefficient that depends on the degree of fixity at supports, the variation in moment of inertia along the span, and the distribution of loading; such as $\frac{5}{48}$ for simply supported uniformly loaded beam; $\frac{1}{4}$ for uniformly loaded cantilever.

The deflection coefficient β_a for simple cases may be found in handbooks, such as *ACI Design Handbook*, Vol. 1 [2.21], or *Handbook of Concrete Engineering* [14.2].

The following example will demonstrate one elastic method, the conjugate beam method, while also deriving one of the most useful deflection expressions.

■ **EXAMPLE 14.2.1**   Using the conjugate beam method, derive the general expression for the elastic midspan deflection for a uniformly loaded span with unequal end moments, as shown in Fig. 14.2.1.

Solution: In the conjugate beam method, the deflection at a given point equals the bending moment at that point for a beam loaded with the M/EI diagram. Thus the system of Fig. 14.2.1 may be regarded as being composed of three separate conjugate beams, as shown in Fig. 14.2.2.

The midspan deflections, or the bending moments on the three conjugate beams, are

$$\text{uniform load, } \Delta_s = \frac{2}{3}\left(\frac{L}{2}\right)\left(\frac{M_0}{EI}\right)\left[\frac{L}{2} - \frac{3}{8}\left(\frac{L}{2}\right)\right] = \frac{5M_0L^2}{48EI}$$

$$\text{left end moment, } \Delta_a = \frac{1}{3}\left(\frac{-M_a}{EI}\right)\left(\frac{L}{2}\right)\left(\frac{L}{2}\right) - \frac{1}{2}\left(\frac{-M_a}{2EI}\right)\left(\frac{L}{2}\right)\left(\frac{1}{3}\right)\left(\frac{L}{2}\right)$$

$$= \frac{-M_aL^2}{16EI}$$

$$\text{right end moment, } \Delta_b = \frac{-M_bL^2}{16EI}$$

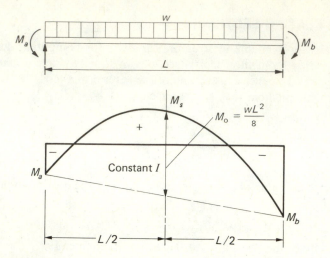

Figure 14.2.1 Typical bending moment diagram for a uniformly loaded span of a continuous beam.

Figure 14.2.2 Component conjugate beams.

The total midspan deflection Δ_m is

$$\Delta_m = \Delta_s + \Delta_a + \Delta_b$$

$$= \frac{5M_0 L^2}{48EI} - \frac{M_a L^2}{16EI} - \frac{M_b L^2}{16EI}$$

$$= \frac{L^2}{48EI}\left[5M_0 - 3(M_a + M_b)\right] \tag{14.2.2}$$

The net positive midspan moment M_s is

$$M_s = M_0 - \tfrac{1}{2}(M_a + M_b) \tag{14.2.3}$$

Then, upon substituting Eq. (14.2.3) in Eq. (14.2.2), one obtains

$$\Delta_m = \frac{L^2}{48EI}\left[5M_s + \frac{5}{2}(M_a + M_b) - 3(M_a + M_b)\right]$$

$$= \frac{5L^2}{48EI}\left[M_s - \frac{1}{10}(M_a + M_b)\right] \tag{14.2.4}$$

Equation (14.2.4) may be used with satisfactory results for practically all prismatic (i.e., constant EI_g) beams, even though the absolute maximum deflection will be obtained only when the loading is uniform and the end moments are equal. ■

Continuous Beams Having Variable Flexural Rigidity. When significant changes of gross cross section along the span length are involved, such changes should be included when performing the statically indeterminate analysis for end moments. When deflections are desired for such cases, the best approach is to account for the variable flexural rigidity EI explicitly, either mathematically in the component conjugate beams (see Fig. 14.2.2), or by the use of numerical integration. Section 14.4 gives further treatment of this subject when discussion is made on what should be the value of the moment of inertia to be used for reinforced concrete sections.

14.3 MODULUS OF ELASTICITY

As discussed in Section 1.9, variance in interpretation is encountered when referring to the modulus of elasticity of concrete. Ordinary beam theory presumes the modulus in tension to be the same as in compression for a homogeneous material. In a reinforced concrete section, creep affects the apparent modulus primarily on the compression side. Even the short-time loading modulus, measured as the secant modulus, is considerably more variant than the compressive strength f'_c. On the tension side there are cracks in the concrete in regions of high bending moment, while at low moment sections the concrete may not crack. The secant modulus in tension is essentially the same as in compression when the stress magnitude is low, but the modulus reduces markedly as the stress nears the cracking level. In other words, in both tension and compression the actual modulus of elasticity varies not only with the magnitude of stress from top to bottom at a section, but also along the span.

Further, creep and shrinkage over a period of time effectively reduce the modulus in compression and will generally magnify deflections by a factor of two or three. Referring to Fig. 1.10.1 one may note that the true elastic strain decreases with time, indicating that the modulus of elasticity increases. The modulus increases because it is dependent on f'_c which increases with age. In design practice, usually the apparent elastic strain is used as computed with a code-specified E_c, presumably corresponding to the 28-day age at loading. Beyond 28 days, however, the increase in modulus of elasticity is relatively slight.

14.4 MOMENT OF INERTIA

In addition to the modulus of elasticity, the flexural rigidity includes the cross-sectional property, moment of inertia. The moment of inertia, even for sections whose external dimensions are constant, varies considerably along a span. Consider a span with a T-shaped section in a continuous beam or frame, as shown in Fig. 14.4.1. In such a continuous structure, at or near the support the concrete slab is cracked, so that the effective section is section A—A; while in the positive moment zone the stem is cracked, so that the effective section is section B—B. Furthermore, near the points of inflection the entire section may be uncracked and fully effective, as shown in section C—C.

In an elastic continuity analysis as discussed in Chapters 7 and 10, only relative stiffness values are required, but in deflection computations, the absolute magnitudes

Figure 14.4.1 Effective moment of inertia for continuous T-shaped sections.

Figure 14.4.2 Typical variation of flexural rigidity with applied bending moment.

for E_c and I must first be determined or assumed. Moreover, the amount of deflection under each increment of load is not constant. The flexural rigidity $E_c I$ is greater at a low load level, since the fully uncracked section provides the greatest effective moment of inertia.

The subject of effective moment of inertia for computing deflections due to short-term loading has received significant study by Yu and Winter [14.3], and by Branson [14.4, 14.5]. It is known that the flexural rigidity $E_c I$ varies with the magnitude of bending moment in the general manner shown in Fig. 14.4.2. Of course, the moment of inertia I_{cr} of the transformed cracked section* increases roughly in proportion with an increase in the percentage of reinforcement. Sections with higher percentages of reinforcement exhibit less change in rigidity under increasing load than those with low percentages of reinforcement. Duan, Wang, and Chen [14.23] have recently reviewed and proposed a new expression for flexural rigidity.

For loads below the cracking load (see Fig. 14.4.3), deflections may be based on the gross concrete section, with generally a small difference arising from whether or not the transformed area of reinforcement is also included. However, as the load increases above the cracking load, the moment of inertia approaches that of the cracked transformed section, although it may be greater between cracks.

Generally, as shown by the typical load–deflection curve in Fig. 14.4.3, the use of gross section underestimates the deflection, and the use of transformed cracked section overestimates the deflection. However, the degree of accuracy is affected by the magnitude of the service load compared to the load which causes cracking.

ACI Effective Moment of Inertia I_e. In order to provide a smooth transition between the moment of inertia I_{cr} of the transformed cracked section and the moment of inertia I_g of the gross uncracked concrete section, since 1971 the ACI Code has used the

---

* See Sections 4.5, 4.6, and 4.10.

Figure 14.4.3 Typical load–deflection curve for reinforced concrete beams.

expression developed by Branson [14.4],

$$I_e = \left(\frac{M_{cr}}{M_{max}}\right)^3 I_g + \left[1 - \left(\frac{M_{cr}}{M_{max}}\right)^3\right] I_{cr} \leq I_g \qquad (14.4.1)$$

where

$M_{cr} = f_r I_g / y_t$ = cracking moment

M_{max} = maximum service load moment acting at the condition under which deflection is computed

I_g = moment of inertia of gross uncracked concrete section about the centroidal axis, neglecting reinforcement

I_{cr} = moment of inertia of transformed cracked section (see Sections 4.5, 4.6, and 4.10)

f_r = modulus of rupture of concrete (see Section 1.8), taken by ACI Code as $7.5\sqrt{f'_c}$ for normal-weight concrete; in general, may be taken [14.6–14.8] as $0.65\sqrt{w_c f'_c}$, where w_c is the unit weight of concrete

y_t = distance from neutral axis to extreme fiber of concrete in tension

Equation (14.4.1) is intended to be calculated at the location of maximum moment as a single value for the entire span in the case of simply supported beams, or as a single value between points of inflection in continuous beams. If one wishes to recognize the continuous variation of the moment of inertia along the span, Branson proposed using the fourth power instead of the third power in Eq. (14.4.1). In this case M_{cr} and M_{max} are the cracking moment and the applied moment, respectively, at each section along the span. The span may be broken into segments with each segment having a different moment of inertia and numerical integration used to compute the deflection.

Equation (14.4.1) was developed from a statistical study of 54 test specimens which had M_{max}/M_{cr} values from 2.2 to about 4 and I_g/I_{cr} values from 1.3 to 3.5.

The study included simple-span rectangular beams [14.9], T-beams [14.3], and two-span continuous rectangular beams [14.10]. Branson has provided an excellent summary [14.5, 14.7, 14.11] of background material relating to Eq. (14.4.1). Lutz [14.12] has provided an ingenious set of diagrams to allow graphical evaluation of Eq. (14.4.1).

The effective moment of inertia approach is also applicable to prestressed concrete (see Chapter 21); Branson and Trost [14.13] have presented unified procedures using the I_e method for partially cracked members, whether nonprestressed or prestressed.

Single Value of Effective Moment of Inertia for Practical Use.   As an *approximation*, a single value of effective moment of inertia is suggested for practical use when the variable I results from the variation in the extent of tension concrete cracking. Three methods have been suggested [14.14].

1. *Midspan value alone.* Recognized in ACI-9.5.2.4, this assumption is

$$I_e = I_m \tag{14.4.2}$$

where I_m is the effective moment of inertia at midspan for simply supported and continuous spans, and at the support section for cantilevers. This is the simplest method and when the inflection point is between about $0.2L$ and the support, the results are within $\pm 20\%$ of those obtained using variable I, as long as $0.33 \le \alpha \le 1.0$ [where $\alpha = (I_m$ at midspan$)/(I_e$ at end$)$]. When α lies between 0.50 and 1.0, the results are within $\pm 5\%$ of those obtained using variable I. Zuraski, Salmon, and Shaikh [14.15] have suggested that this method is satisfactory for ordinary design situations, and it is endorsed by ACI Committee 435 [14.8].

2. *Weighted average.* In this method the adjusted I is obtained by weighting the moments of inertia in accordance with the magnitudes of the end moments [14.14]. The following weighted average expression has been recommended by ACI Committee 435 [14.8] as giving a somewhat improved result over the use of the midspan value alone. For spans with *both* ends continuous,

$$\text{average } I_e = 0.70I_m + 0.15[I_{e1} + I_{e2}] \tag{14.4.3}$$

For spans with *one* end continuous,

$$\text{average } I_e = 0.85I_m + 0.15I_{e1} \tag{14.4.4}$$

3. *Simple average.* With this assumption the I to be used is the average I

$$\text{average } I_e = \frac{\frac{1}{2}(I_{e1} + I_{e2}) + I_m}{2} \tag{14.4.5}$$

where I_{e1} and I_{e2} are the effective moments of inertia at the two ends of the span. The use of both I_{e1} and I_{e2} is appropriate only when there are end moments at both ends. For spans having *one* end continuous, one should use

$$\text{average } I_e = 0.75I_m + 0.25I_{e1} \tag{14.4.6}$$

For uniform loading on continuous spans Eq. (14.4.3) is slightly more accurate (say on the order of 5%) than using only the midspan value, but for concentrated loads it is less accurate [14.15]. When an average value is used as *permitted* by

ACI-9.5.2.4 it should be done in accordance with Eq. (14.4.5), rather than taking the sum of I_m, I_{e1}, and I_{e2} and dividing by three [14.14].

For a single heavy concentrated load, averaging *reduces* accuracy [14.15]; Eq. (14.4.2) should be used in such cases.

14.5 INSTANTANEOUS DEFLECTIONS IN DESIGN

Throughout the history of reinforced concrete construction, computation of instantaneous (short-time) deflection has usually involved using either transformed cracked section or gross uncracked section. In either case, Eq. (14.2.1) is suitable after it is slightly rewritten using E_c and I_e

$$\Delta = \beta_a \left(\frac{ML^2}{E_c I_e} \right) \tag{14.5.1}$$

where

β_a = coefficient based on load and support conditions

I_e = effective moment of inertia

E_c = modulus of elasticity of concrete

For instantaneous (elastic) deflection and also generally for long-time deflection under sustained loads, the basic value of modulus of elasticity to be used for concrete is (ACI-8.5.1)

$$E_c = 33 w_c^{1.5} \sqrt{f_c'}$$

for concrete having a unit weight between 90 and 155 pcf. For normal-weight concrete,

$$E_c = 57,000 \sqrt{f_c'}$$

The generally accepted effective moment of inertia for use in Eq. (14.5.1) is Eq. (14.4.1), using a single value in accordance with Eqs. (14.4.2) through (14.4.5).

In order to compute deflection at different load levels, such as dead load or dead load plus live load, the effective moment of inertia I_e should be computed using Eq. (14.4.1) for that total load level in each case. The incremental deflection, such as for live load only, is then computed as the difference between the deflections due to dead plus live load and dead load only. It should be assumed that the live load cannot act in the absence of dead load.

Computation of the live load deflection as $\Delta_{D+L} - \Delta_D$ gives the live load deflection occurring during the *first* application of live load. Figure 14.5.1 shows the typical idealized load or moment versus deflection relationship [14.16]. For repeated loadings, the upper envelope of load–deflection curves is nearly the same as the single-loading curve for both reinforced and prestressed concrete members, even though increasing residual deflections occur due to creep and cracking effects [14.8, 14.17]. Thus it seems reasonable to compute short-time deflections using I_e as described

Figure 14.5.1 Typical idealized moment–deflection diagram for short-time loading (from References 14.7, 14.8, and 14.16).

above and the residual deflection separately as discussed in the sections on creep and shrinkage.

A comparison of the measured short-time deflections with the deflections computed by Eq. (14.4.1) [14.18] is shown in Fig. 14.5.2. A study by Committee 435 [14.19] indicates that by using the present ACI Code criteria for deflection for simply supported beams under *controlled laboratory conditions* "there is approximately a 90% chance that the deflections of a particular beam will be within the range of 20% less than to 30% more than the calculated value."

⊙ Simply supported rectangular beams
△ Simply supported T beams
⊡ 2 span continuous rectangular beams

Figure 14.5.2 Comparison of computed and measured short-time deflections (from Reference 14.18).

The following two examples demonstrate the computation of instantaneous deflection.

■ **EXAMPLE 14.5.1** Investigate the instantaneous (short-time loading) deflection for the simply supported beam of Fig. 14.5.3 over a span of 40 ft. Assume that the member has been designed by the strength method using $f'_c = 4000$ psi, $f_y = 60,000$ psi, and the ACI Code.

Solution: According to ACI-Table 9.5(a) (Table 14.10.2, p. 575) the minimum depth unless deflection is computed is $L/16 = 40(12)/16 = 30$ in. Thus a deflection computation is required regardless of whether or not excessive deflection is of concern.

(a) Dead load short-time deflection. The gross moment of inertia is

$$I_g = \tfrac{1}{12}(18)(24)^3 = 20,700 \text{ in.}^4$$

$$M_{max} = \frac{0.45(40)^2}{8} = 90 \text{ ft-kips}$$

(a) Midspan cross section

(b) Cracked section

(c) Service load bending moments

Figure 14.5.3 Beam for Example 14.5.1

For the transformed cracked section, as shown in Fig. 14.5.3(b), the neutral axis position is

$$\frac{18x^2}{2} = 57.3(20.7 - x)$$

$$x^2 + 6.37x = 131.8$$

$$x = 8.73$$

$$I_{cr} = \tfrac{1}{3}(18)(8.73)^3 + 57.3(20.7 - 8.73)^2 = 12{,}200 \text{ in.}^4$$

The effective moment of inertia I_e is dependent on the bending moment M_{cr} that causes cracking at the extreme tension fiber.

$$f_r = 7.5\sqrt{f'_c} = 7.5\sqrt{4000} = 474 \text{ psi} \quad \text{(ACI-9.5.2.3)}$$

$$M_{cr} = \frac{f_r I_g}{y_t} = \frac{0.474(20{,}700)}{12}\left(\frac{1}{12}\right) = 68 \text{ ft-kips}$$

Note that y_t is the distance $h/2$ for the 24 in. deep beam when the gross section is used and the reinforcement is neglected.

$$\frac{M_{cr}}{M_{max}} = \frac{68}{90} \text{ (dead load only)} = 0.756; \qquad \left(\frac{M_{cr}}{M_{max}}\right)^3 = 0.431$$

From ACI Formula (9-7), Eq. (14.4.1), the effective moment of inertia is

$$I_e = \left(\frac{M_{cr}}{M_{max}}\right)^3 I_g + \left[1 - \left(\frac{M_{cr}}{M_{max}}\right)^3\right] I_{cr}$$

$$= 0.431(20{,}700) + 0.569(12{,}200) = 15{,}860 \text{ in.}^4$$

$$E_c = 33w_c^{1.5}\sqrt{f'_c} = 33(145)^{1.5}\sqrt{4000} = 3.64 \times 10^6 \text{ psi}$$

$$(\Delta_i)_D = \frac{5WL^3}{384EI} = \frac{5(0.45)(40)(40)^3(1728)}{384(3.64)(10^3)(15{,}860)} = 0.45 \text{ in.}$$

This deflection may not be harmful because it may be accommodated by using a negative deflection (camber) in construction. Even if camber is not used, such dead load instantaneous deflection will not affect plastered ceilings or other items that are put into place after the immediate dead load deflection has taken place. Concern is primarily with the instantaneous deflection from live load and the long-term creep and shrinkage deflection from sustained loads.

(b) Dead load plus live load short-time deflection. The maximum service load moment at this load level is

$$M_{max} = \frac{25(40)}{4} + 90 = 340 \text{ ft-kips (for dead load plus live load)}$$

$$\frac{M_{cr}}{M_{max}} = \frac{68}{340}\left(\begin{array}{c}\text{dead load}\\ + \text{ live load}\end{array}\right) = 0.200; \qquad \left(\frac{M_{cr}}{M_{max}}\right)^3 = 0.008$$

$$I_e = (0.008)(20{,}700) + (0.992)(12{,}200) = 12{,}270 \text{ in.}^4$$

At this higher load level, I_e is only slightly larger than I_{cr}. Using $I_e = I_{cr} = 12,200$ in.^{4},

$$(\Delta_i)_{\text{beam weight}} = \frac{5(0.45)(40)(40)^3 1728}{384(3640)(12,200)} = 0.58 \text{ in.}$$

$$(\Delta_i)_{\text{conc. load}} = \frac{25(40)^3 1728}{48(3640)(12,200)} = 1.30 \text{ in.}$$

$$(\Delta_i)_{D+L} = 0.58 + 1.30 = 1.88 \text{ in.}$$

(c) Live load short-time deflection. Consistent logic dictates that live load deflection must be obtained indirectly as

$$(\Delta_i)_L = (\Delta_i)_{D+L} - (\Delta_i)_D$$
$$= 1.88 - 0.45 = 1.43 \text{ in.}$$

It is assumed that live load cannot act in the absence of dead load. Thus if the effective moment of inertia when dead load alone is acting is considerably different from that when dead load plus live load is acting, the live load deflection is properly obtained only by subtracting Δ_D from Δ_{D+L}. Even if the service live load has been preceded by a construction load of equal magnitude, References 14.7, 14.8, and 14.16 as shown in Fig. 14.5.1 indicate that the procedure is proper for repeated loadings.

From a practical viewpoint, $I_e = I_{cr}$ may be used whenever $(M_{cr}/M_{\max})^3$ is less than about 0.1. Assuming this to be a floor beam not supporting partitions, the acceptable immediate live load deflection from ACI-Table 9.5(b) is

$$\text{allowable } (\Delta_i)_L = \frac{L}{360} = \frac{40(12)}{360} = 1.33 \text{ in.} < 1.43 \text{ in.} \qquad \textbf{NG}$$

If plastered ceilings or frangible partitions are to be supported, the deflection due to long-time creep and shrinkage must be added to that due to live loads; the acceptable limit for such deflection is $L/480$ (ACI-Table 9.5b). It might have been expected that the deflection of this beam would be excessive since the reinforcement ratio ρ is 0.0192, which is $0.67\rho_b$. This exceeds the guideline value of about $0.5\rho_{\max}$, or $0.375\rho_b$, suggested in Chapter 3 as the maximum reinforcement ratio for deflection control. ∎

■ **EXAMPLE 14.5.2** Investigate the instantaneous deflection on the continuous beam of Fig. 14.5.4. Use $f'_c = 3000$ psi, $f_y = 40,000$ psi, and the ACI Code.

Solution: This continuous girder supports two smaller beams that frame to it and some uniform loading that comes directly to it. ACI-9.5.2.4 indicates that I_e may be computed as an average value for the positive and negative moment regions. For prismatic members, the effective moment of inertia I_e at the positive moment region may be used instead of the average value. In fact, for loading that is largely concentrated load, as in the situation for this example, ACI Committee 435 recommends [14.8] *against* using an average value. The results will be compared using the various assumptions for the single value of I_e, as discussed in Section 14.4.

$$\Delta = \frac{23}{648}\frac{PL^3}{E_cI} = P\frac{23(39)^3(1,728)}{648E_cI} = 3.64 \times 10^6 \frac{P}{E_cI}$$

$$\Delta = \frac{5}{384}\frac{wL^4}{E_cI} = \frac{5wL(39)^3(1,728)}{384E_cI} = 1.333 \times 10^6 \frac{wL}{E_cI}$$

$$\Delta \text{ (at midspan)} = \frac{M_aL^2}{16E_cI} = \frac{M_a(39)^2(1,728)}{16E_cI} = 0.164 \times 10^6 \frac{M_a}{E_cI}$$

$$\Delta \text{ (at midspan)} = \frac{M_bL^2}{16E_cI} = 0.164 \times 10^6 \frac{M_b}{E_cI}$$

Figure 14.5.4 Data for Example 14.5.2.

Figure 14.5.5 Gross section of beam for Example 14.5.2.

(a) Section at the left support. For the gross section neglecting reinforcement (see Fig. 14.5.5),

$$x = \frac{36(18)18 + 90(4.5)38.25}{36(18) + 90(4.5)} = \frac{27{,}160}{1053} = 25.79 \text{ in.}$$

$$I_g = \tfrac{1}{3}(18)[(25.79)^3 + (14.71)^3] + \tfrac{1}{12}(72)(4.5)^3 + 72(4.5)(12.46)^2$$
$$= 173{,}000 \text{ in.}^4$$

Had reinforcement been included, $x = 25.29$ in. and $I_g = 200{,}000$ in.^{4} It is generally accepted that I_g for use in ACI Formula (9-7) should not include steel reinforcement.

For the transformed cracked section [see Fig. 14.5.6(a)],

$$18x\left(\frac{x}{2}\right) + 40.6(x - 2.6) = 40.9(36.7 - x)$$

$$x^2 + 9.05x = 178.2$$

$$x = 9.56 \text{ in.}$$

$$I_{cr} = \tfrac{1}{3}(18)(9.56)^3 + 40.6(9.56 - 2.6)^2 + 40.9(36.7 - 9.56)^2$$
$$= 37{,}200 \text{ in.}^4$$

It is noted that the modulus of elasticity ratio n is used to transform the steel in the compression zone, which agrees with the concept of computing an elastic (short time) deflection.

Compute the cracking moment for the beam with tension on the flange.

$$f_r = 7.5\sqrt{f_c'} = 7.5\sqrt{3000} = 411 \text{ psi}$$

$$M_{cr} = \frac{f_r I_g}{y_t} = \frac{0.411(173{,}000)}{14.71}\left(\frac{1}{12}\right) = 402 \text{ ft-kips}$$

$$\frac{M_{cr}}{M_{max}} = \frac{402}{89.6}\left(\frac{\text{dead}}{\text{load}}\right) > 1; \quad \text{use } I_e = I_g = 173{,}000 \text{ in.}^4$$

$$\frac{M_{cr}}{M_{max}} = \frac{402}{200}\left(\frac{\text{dead load}}{+ \text{ live load}}\right) > 1; \quad \text{use } I_e = I_g = 173{,}000 \text{ in.}^4$$

(b) Midspan section [Fig. 14.5.6(b)]. Determine whether or not the neutral axis occurs in the flange by taking moments about the bottom of the flange,

$$90(4.5)(2.25) < 86.6(32.3)$$

Figure 14.5.6 Transformed cracked sections for Example 14.5.2.

Locate the neutral axis, including the effect of compression in the stem.

$$90(4.5)(x - 2.25) + 18(x - 4.5)^2(\tfrac{1}{2}) = 86.6(36.8 - x)$$

$$x^2 + 45.6x = 435$$

$$x = 8.09 \text{ in.}$$

$$I_{cr} = \tfrac{1}{12}(72)(4.5)^3 + 72(4.5)(5.84)^2 + \tfrac{1}{3}(18)(8.09)^3 + 86.6(36.8 - 8.09)^2$$
$$= 86,200 \text{ in.}^4$$

The cracking moment for the beam with tension in the stem is

$$M_{cr} = \frac{f_r I_g}{y_t} = \frac{0.411(173,000)}{25.79}\left(\frac{1}{12}\right) = 229.5 \text{ ft-kips}$$

$$\frac{M_{cr}}{M_{max}} = \frac{229.5}{229}\left(\begin{array}{c}\text{dead}\\\text{load}\end{array}\right) > 1; \qquad \text{use } I_e = I_g = 173,000 \text{ in.}^4$$

$$\frac{M_{cr}}{M_{max}} = \frac{229.5}{510}\left(\begin{array}{c}\text{dead load}\\+ \text{ live load}\end{array}\right) = 0.45; \qquad \left(\frac{M_{cr}}{M_{max}}\right)^3 = 0.091$$

$$I_e = \left(\frac{M_{cr}}{M_{max}}\right)^3 I_g + \left[1 - \left(\frac{M_{cr}}{M_{max}}\right)^3\right] I_{cr}$$
$$= 0.091(173,000) + 0.909(86,200) = 94,000 \text{ in.}^4$$

(c) Section at the right support [Fig. 14.5.6(c)]. Use transformed cracked section and locate the neutral axis,

$$18x\left(\frac{x}{2}\right) + 40.6(x - 2.6) = 122.6(35.7 - x)$$

$$x^2 + 18.13x = 498$$

$$x = 15.02 \text{ in.}$$

$$I_{cr} = \tfrac{1}{3}(18)(15.02)^3 + 122.6(35.7 - 15.02)^2 + 40.6(15.02 - 2.60)^2$$
$$= 79{,}000 \text{ in.}^4$$

$$M_{cr} = 402 \text{ ft-kips (same as left support)}$$

$$\frac{M_{cr}}{M_{max}} = \frac{402}{377}\left(\frac{\text{dead}}{\text{load}}\right) > 1; \quad \text{use } I_e = I_g = 173{,}000 \text{ in.}^4$$

$$\frac{M_{cr}}{M_{max}} = \frac{402}{591}\left(\frac{\text{dead load}}{+ \text{ live load}}\right) = 0.68; \quad \left(\frac{M_{cr}}{M_{max}}\right)^3 = 0.314$$

$$I_e = 0.314(173{,}000) + 0.686(79{,}000) = 109{,}000 \text{ in.}^4$$

(d) Summary of values for I_e. The values of effective moment of inertia are

	FOR DL	FOR DL + LL
Left end	$I_e = 173{,}000 \text{ in.}^4$	$I_e = 173{,}000 \text{ in.}^4$
Midspan	$I_e = 173{,}000 \text{ in.}^4$	$I_e = 94{,}000 \text{ in.}^4$
Right end	$I_e = 173{,}000 \text{ in.}^4$	$I_e = 109{,}000 \text{ in.}^4$

Having obtained the above values, usual practice is to use a single adjusted value of I_e as discussed in Section 14.4.

(e) Compute a single adjusted value for I_e using the various procedures of Section 14.4.

1. Midspan value:

$$I_e = 173{,}000 \text{ in.}^4 \text{ (for DL only)}$$

$$I_e = 94{,}000 \text{ in.}^4 \text{ (for DL + LL)}$$

2. Weighted average:

$$I_e = 0.70 I_m + 0.15(I_{e1} + I_{e2})$$

$$I_e = 173{,}000 \text{ in.}^4 \text{ (for DL only)}$$

$$I_e = 0.70(94{,}000) + 0.15(173{,}000 + 109{,}000)$$
$$= 108{,}100 \text{ in.}^4 \text{ (for DL + LL)}$$

3. Simple average:

$$I_e = 173{,}000 \text{ in.}^4 \text{ (for DL only)}$$

$$I_e = \frac{1}{2}\left(\frac{173{,}000 + 109{,}000}{2}\right) + 94{,}000$$

$$= 118{,}000 \text{ in.}^4 \text{ (for DL + LL)}$$

The task of determining the immediate deflection is actually one of analyzing a continuous beam with variable moment of inertia. Because the most "exact" computations at best give deflections within probably $\pm 20\%$, procedures more complex than those illustrated here are not justified.

(f) Immediate dead load deflection. I_e for dead load only is 173,000 in.^{4} regardless of whether midspan, weighted average, or simple average is used.

$$M_a = 89.6 \text{ ft-kips} \qquad M_b = 377 \text{ ft-kips}$$

$$w = 0.76 \text{ kip/ft} \qquad P = 24.48 \text{ kips}$$

It is to be noted that the span is taken as that measured between the centerlines of supports, and the end moments are those computed for the same locations. Equally acceptable results are obtained by using the clear span and the face-of-support moments. Referring to Fig. 14.5.4, the total midspan deflection is

$$\Delta = \frac{10^6}{E_c I_e}(3.64P + 1.333wL - 0.164M_a - 0.164M_b)$$

$$E_c = 33w_c^{1.5}\sqrt{f_c'} = 57{,}000\sqrt{3000} = 3.15 \times 10^6 \text{ psi (ACI-8.5.1)}$$

$$(\Delta_i)_D = \frac{1}{3.15(173)}[3.64(24.48) + 1.333(0.76)(39) - 0.164(89.6 + 377.0)]$$

$$= \frac{89.2 + 39.5 - 76.4}{545} = \frac{52.3}{545} = 0.10 \text{ in.}$$

As previously mentioned, dead load deflection will usually cause no difficulty, as most of it may be compensated for by the construction. However, it is used as a basis for determining the long-time creep and shrinkage deflection which is discussed in the next section.

(g) Immediate live load deflection using the midspan value of I_e. The midspan value is 94,000 in.^{4} as computed in part (e).

$$(\Delta_i)_D = 0.10 \text{ in.} \quad [\text{see part (f)}]$$

$$M_a = 200 \text{ ft-kips} \qquad M_b = 591 \text{ ft-kips}$$

$$w = 0.91 \text{ kip/ft} \qquad P = 56.33 \text{ kips}$$

$$(\Delta_i)_{D+L} = \frac{1}{3.15(94)}[3.64(56.33) + 1.333(0.91)(39) - 0.164(200 + 591)]$$

$$= \frac{205.0 + 47.0 - 129.7}{296} = \frac{122.3}{296} = 0.41 \text{ in.}$$

$$(\Delta_i)_L = 0.41 - 0.10 = 0.31 \text{ in.}$$

(h) Immediate live load deflection using the weighted average value of I_e.

$$(\Delta_i)_D = 0.10 \text{ in.} \quad [\text{see part (f)}]$$

weighted average value of $I_e = 108,100 \text{ in.}^4$

$$(\Delta_i)_{D+L} = 0.41\left(\frac{94,000}{108,100}\right) = 0.36 \text{ in.}$$

$$(\Delta_i)_L = 0.36 - 0.10 = 0.26 \text{ in.}$$

(i) Immediate live load deflection using the simple average value of I_e.

$$(\Delta_i)_D = 0.10 \text{ in.} \quad [\text{see part (f)}]$$

simple average value of $I_e = 118,000 \text{ in.}^4$

$$(\Delta_i)_{D+L} = 0.41\left(\frac{94,000}{118,000}\right) = 0.33 \text{ in.}$$

$$(\Delta_i)_L = (\Delta_i)_{D+L} - (\Delta_i)_D = 0.33 - 0.10 = 0.23 \text{ in.}$$

(j) Conclusion. The immediate live load deflection is computed to be 0.31 in., 0.26 in., and 0.23 in., respectively, depending on whether midspan, weighted average, or simple average value of I_e is used for dead and live load deflection. The use of midspan I_e seems appropriate; for this example one can conclude that the dead load deflection is about 0.1 in. and the live load deflection is about 0.3 in., using *one* significant figure.

If this is a floor that does not support frangible partitions, the limiting permissible deflection is $L/360$, or

$$\frac{L}{360} = \frac{39(12)}{360} = 1.3 \text{ in.} > 0.3 \text{ in. (computed)} \qquad \textbf{OK}$$

The reinforcement ratios used ($0.14\rho_b$ for the positive moment region and $0.37\rho_b$ for the negative moment regions) were below the guideline value suggested in Chapter 3 for deflection control; thus excessive deflection was not expected. ■

Recommended Values for Maximum Reinforcement Ratio ρ for Deflection Control.
ACI Committee 435 [14.8] has recommended the following values of the maximum reinforcement ratio ρ to be used in the positive moment zone for deflection control:

1. For members of normal-weight concrete *not supporting* or *not attached* to nonstructural elements likely to be damaged by large deflections,

Rectangular beams	$0.35\rho_b$
T-beams or box beams	$0.40\rho_b$

2. For members of normal-weight concrete *supporting* or *attached* to nonstructural elements likely to be damaged by large deflections,

Rectangular beams	$0.25\rho_b$
T-beams or box beams	$0.30\rho_b$

3. For members of lightweight concrete, use $0.05\rho_b$ *less* than that indicated in 1 and 2.

14.6 CREEP EFFECT ON DEFLECTIONS UNDER SUSTAINED LOAD

The total long-term deflection includes the instantaneous elastic deflection plus the contributions from creep and shrinkage. Creep is inelastic deformation with time under sustained loads at unit stresses within the accepted elastic range (say, below $0.5f_c'$), as shown in Fig. 14.6.1. This inelastic deformation increases at a decreasing rate during the time of loading.

Factors that affect the magnitude of creep deformation [14.7, 14.20] are: (1) the constituents—such as the composition and fineness of the cement, the admixtures, and the size, grading, and mineral content of the aggregates; (2) proportions, such as water content and water–cement ratio; (3) curing temperature and humidity; (4) relative humidity during storage; (5) size of the concrete member, particularly the thickness and the volume to surface ratio; (6) age at loading; (7) duration of loading; and (8) magnitude of stress.

Since, as seen from Fig. 14.6.1, the result of creep is an increase in strain with constant stress, one of the ways of accounting for it is by the use of a modified modulus of elasticity E_{ct}. An alternate and generally preferred procedure is to apply a multiplier C_t to the elastic deflection Δ_i.

In order to help understand the qualitative effect of creep on beam deformation, consider the singly reinforced beam of Fig. 14.6.2. It is noted that the strain at the tension steel is essentially unchanged because the concrete contributes little in taking tension and ordinary deformed steel reinforcement exhibits little creep. Since the neutral axis moves down, two observations may be made: (1) The concrete stress reduces at the compression face (i.e., same compressive force acting and x_{cp} exceeds x_i); and (2) the increase in compressive strain is much greater than the increase in curvature ϕ.

Figure 14.6.1 Typical stress–strain curves for instantaneous and long-time loading.

(a) Instantaneous strain
   due to initial loading

(b) Strain after creep
   has occurred

Figure 14.6.2 Creep effect on beam curvature.

For deflection purposes it is frequently desirable to use a creep coefficient C_t defined as the ratio of creep strain to elastic strain,

$$C_t = \frac{\epsilon_{cp}}{\epsilon_i}$$ (14.6.1)

and the deflection resulting from creep is

$$\Delta_{cp} = C_t(\Delta_i)_D$$ (14.6.2)

where $(\Delta_i)_D$ is the instantaneous deflection due to all sustained loads.

ACI Committee 209 has recommended [14.6] the hyperbolic-type equation of Branson et al. [14.6, 14.7] for the creep coefficient, as follows:

$$C_t = \left(\frac{t^{0.60}}{10 + t^{0.60}}\right)C_u$$ (14.6.3)

where

C_t = ratio of creep strain to elastic strain at any time t after a basic curing period

t = time in days after loading

C_u = ultimate creep coefficient; recommended *average* value is 2.35 for 40% humidity.

The general relationship of C_t/C_u appears in Fig. 1.10.2.

Equation (14.6.3) applies to the standard condition of 40% ambient relative humidity, 4 in. (100 mm) or less slump, average thickness of member 6 in. (150 mm), and loading age of 7 days for moist-cured concrete or 1 to 3 days for steam-cured concrete. For other conditions, the standard condition value is to be multiplied by the following correction factors (*CF*):

1. Age at loading. For moist-cured concrete,

$$(CF)_a = 1.25t_a^{-0.118}$$ (14.6.4a)

Table 14.6.1 Creep Correction Factor, $(CF)_a$ for Age at Loading, Eq. (14.6.4a,b)

t_a, AGE IN DAYS AFTER INITIAL CURING PERIOD	CORRECTION FACTOR, $(CF)_a$	
	MOIST CURED FOR 7 DAYS INITIAL CURING PERIOD	STEAM CURED FOR 1–3 DAYS INITIAL CURING PERIOD
10	0.95	0.90
20	0.87	0.85
30	0.83	0.82
60	0.77	0.76
90	0.74	0.74

For steam-cured concrete,

$$(CF)_a = 1.13 t_a^{-0.095} \qquad \textbf{(14.6.4b)}$$

In the above two equations, t_a is the age at loading, in days after the initial period of curing. Several useful values of Eq. (14.6.4ab) appear in Table 14.6.1.

2. Humidity. For $H \geq 40\%$,

$$(CF)_h = 1.27 - 0.0067H \qquad \textbf{(14.6.5)}$$

where H is the ambient relative humidity in percent. Values for this correction factor appear in Table 14.6.2.

3. Average thickness of member. Where the average thickness of the member in inches exceeds 6 in. (150 mm), a correction factor (reduction factor) may be applied. However, for most design purposes such a correction may be neglected. For members whose average thickness greatly exceeds 12 in. (300 mm), Meyers and Branson [14.21] provide a chart that may be used to correct for the effect of average thickness.

Table 14.6.2 Creep Correction Factor, $(CF)_h$ for Humidity, Eq. (14.6.5)

AMBIENT RELATIVE HUMIDITY, H, PERCENT	CORRECTION FACTOR $(CF)_h$
40 or less	1.00
50	0.94
60	0.87
70	0.80
80	0.73
90	0.67
100	0.60

4. Other correction factors. Additional correction factors are available [14.7, 14.21] to account for slump greater than 4 in., cement content, percent of fine aggregate, and air content. However, these tend to be either small or offset one another and may generally be neglected.

Compression Steel Effect on Creep.   The presence of compression steel decreases the deformation due to creep (and shrinkage as discussed in the next section). Evaluation of the effect of compression steel has been reported by Washa and Fluck [14.9], Yu and Winter [14.3], and Hollington [14.22], and a multiplier factor has been given by Branson [14.24] and recommended by Committee 435 [14.8], as follows:

$$k_r = \frac{0.85}{1 + 50\rho'} \tag{14.6.6}$$

where ρ' is A_s'/bd, the compression steel reinforcement ratio. Thus, the creep deflection Δ_{cp} would become

$$\Delta_{cp} = k_r C_t (\Delta_i)_D \tag{14.6.7}$$

instead of Eq. (14.6.2), when compression steel is present.

14.7 SHRINKAGE EFFECT ON DEFLECTIONS UNDER SUSTAINED LOAD

Shrinkage of concrete in beams may have a similar effect on the deflection as creep. Shrinkage of an isolated plain concrete member would merely shorten it without causing curvature. When steel reinforcement is added, however, bond between concrete and steel restrains the shrinkage. Thus, a singly reinforced beam, having its shrinkage restrained at the reinforced face and unrestrained at the unreinforced face, will have considerable curvature. Generally it is difficult to separate the effects of creep and shrinkage. Shrinkage occurs more pronounced during the first few months than does creep. Typically, 90% of the shrinkage will have occurred at the end of one year, whereas not until the end of five years will 90% of the creep have occurred. A number of investigators have studied shrinkage effects separately from those of creep [14.6, 14.7, 14.20, 14.21, 14.25–14.27].

If the free shrinkage strain is known, shrinkage curvature ϕ_{sh} must be determined as a function of shrinkage strain. Such curvature will be dependent on the relative amounts of compression and tension steel just as creep is so affected. Finally, the shrinkage deflection will involve the geometry of the support system. Shrinkage deflection Δ_{sh} may be expressed [14.6] as

$$\Delta_{sh} = \alpha_1 \phi_{sh} L^2 \tag{14.7.1}$$

where α_1 is a factor relating to the geometry of the support system and may be taken as the following:

$$\alpha_1 = 0.50 \text{ cantilever beams}$$
$$= 0.125 \text{ simply supported beams}$$
$$= 0.086 \text{ beams continuous at one end only}$$
$$= 0.063 \text{ beams continuous at both ends}$$

and L is the span length of the beam.

Shrinkage Strain, ϵ_{sh}. ACI Committee 209 has recommended [14.6] that the following expressions by Branson et al. [14.7] may be used for shrinkage strain ϵ_{sh}:

For any time t after age 7 days for moist-cured concrete,

$$\epsilon_{sh} = \frac{t}{35 + t}(\epsilon_{sh})_u \tag{14.7.2a}$$

For any time t after age 1 to 3 days for steam-cured concrete,

$$\epsilon_{sh} = \frac{t}{55 + t}(\epsilon_{sh})_u \tag{14.7.2b}$$

where

ϵ_{sh} = shrinkage strain at any time t after initial curing

t = time in days after initial curing

$(\epsilon_{sh})_u$ = ultimate shrinkage strains; average value suggested is 800×10^{-6} in./in. for 40% humidity.

Equation (14.7.2a) is shown graphically in Fig. 1.10.4. For conditions other than 40% ambient relative humidity, the standard condition value of Eqs. (14.7.2ab) is to be multiplied by the following correction factor (CF):

$$(CF)_h = 1.40 - 0.010H, \qquad 40 \le H \le 80\% \tag{14.7.3}$$

$$(CF)_h = 3.00 - 0.030H, \qquad H \ge 80\% \tag{14.7.4}$$

where H is the relative humidity in percent. Values for $(CF)_h$ appear in Table 14.7.1.

Other correction factors may normally be neglected. Should such factors be desired, corrections for average thickness other than 6 in., slump greater than 4 in., cement content, percentage of fines, and air content are available [14.7, pp. 45–47].

Shrinkage Curvature, ϕ_{sh}. Several investigators [14.4, 14.6, 14.27] have developed expressions for curvature due to warping that arises from nonuniform shrinkage. Reinforcement of different amounts in the two faces of a beam is the principal cause of shrinkage warping.

Table 14.7.1 Shrinkage Correction Factor $(CF)_h$ for Humidity, Eqs. (14.7.3) and (14.7.4)

AMBIENT RELATIVE HUMIDITY, H, PERCENT	CORRECTION FACTOR $(CF)_h$
40 or less	1.00
50	0.90
60	0.80
70	0.70
80	0.60
90	0.30
100	0

Figure 14.7.1 Shrinkage strain related to beam curvature for a singly reinforced beam (after Miller [14.27]).

Miller [14.27] established the following relationship for the singly reinforced beam. Referring to Fig. 14.7.1, by straightline proportion,

$$\phi_{sh} = \frac{\epsilon_{sh} - \epsilon_s}{d} = \frac{\epsilon_{sh}}{d}\left(1 - \frac{\epsilon_s}{\epsilon_{sh}}\right) \tag{14.7.5}$$

where ϵ_s is the compressive strain induced in the steel from shrinkage; ϵ_{sh} is the free shrinkage strain at the unreinforced face. Miller empirically established values for ϵ_s/ϵ_{sh} as a function of the percentage of reinforcement ρ.

Branson [14.4] modified Miller's equation and empirically extended the results to give equations including the effects of compression steel.

$$\phi_{sh} = 0.7 \frac{\epsilon_{sh}}{h} (\rho - \rho')^{1/3}\left(\frac{\rho - \rho'}{\rho}\right)^{1/2} \quad \text{for } (\rho - \rho') \leq 3\% \tag{14.7.6}$$

$$\phi_{sh} = \frac{\epsilon_{sh}}{h} \quad \text{for } (\rho - \rho') > 3\% \tag{14.7.7}$$

Note that ρ and ρ' are in percent, $100\,(A_s \text{ or } A_s')/(bd)$. Equations (14.7.6) and (14.7.7) are recommended [14.7, 14.8] as the most appropriate relationships.

Geometry of Warping from Shrinkage. In order to establish the factor α_1 in Eq. (14.7.1) for the four typical cases, the well-known moment area theorems for beam deflections may be used. Since the quantity $M/(EI)$ is in fact the curvature due to bending moment, the ϕ_{sh} diagrams in Fig. 14.7.2 may be regarded as the equivalent $M/(EI)$ diagrams. For the cantilever beam [Fig. 14.7.2(a)],

$$\Delta_{sh} = BB' = \text{moment of } \phi_{sh} \text{ diagram between } A \text{ and } B \text{ about } B$$

$$= (\phi_{sh}L)\left(\frac{L}{2}\right) = 0.50\phi_{sh}L^2$$

Figure 14.7.2 Geometry of warping due to shrinkage.

For the simply supported beam [Fig. 14.7.2(b)],

$$\theta_A = \text{area of } \phi_{sh} \text{ diagram between } A \text{ and } C$$

$$= \phi_{sh}\left(\frac{L}{2}\right)$$

$$\Delta_{sh} = CC' = CC_1 - C_1C'$$

$$= \theta_A\left(\frac{L}{2}\right) - (\text{moment of } \phi_{sh} \text{ diagram between } A \text{ and } C \text{ about } C)$$

$$= \phi_{sh}\left(\frac{L}{2}\right)\left(\frac{L}{2}\right) - \phi_{sh}\left(\frac{L}{2}\right)\left(\frac{L}{4}\right) = 0.125\phi_{sh}L^2$$

For the beam fixed at one end only [Fig. 14.7.2(c)],

deflection of B from tangent at A

= moment of ϕ_{sh} diagram between A and B about B

$$= -\phi_{sh}(L - x_1)\left(x_1 + \frac{L - x_1}{2}\right) + \phi_{sh}\left(\frac{x_1^2}{2}\right) = 0$$

from which

$$x_1 = \frac{\sqrt{2}L}{2}$$

In order that the tangent at C' be horizontal, the distances AD and DC must be equal; thus

$$AD = DC = L - x_1 = \left(1 - \frac{\sqrt{2}}{2}\right)L$$

$$\begin{aligned}
\Delta_{sh} = CC' &= \text{moment of } \phi_{sh} \text{ diagram between } A \text{ and } C \text{ about } C \\
&= \text{moment of a couple} = \phi_{sh}(L - x_1)^2 \\
&= \phi_{sh}L^2(1 - \tfrac{1}{2}\sqrt{2})^2 = \phi_{sh}L^2(1 - \sqrt{2} + \tfrac{1}{2}) \\
&= 0.086\phi_{sh}L^2
\end{aligned}$$

For the beam fixed at both ends [Fig. 14.7.2(d)], in order that the slope be horizontal at midspan for symmetry,

$$x_1 = \frac{L}{4}$$

Then

$$\begin{aligned}
\Delta_{sh} = CC' &= \text{moment of } \phi_{sh} \text{ diagram between } A \text{ and } C \text{ about } C \\
&= \text{moment of a couple} = \phi_{sh}\left(\frac{L}{4}\right)^2 \\
&= 0.063\phi_{sh}L^2
\end{aligned}$$

Compression Steel Effect on Combined Shrinkage and Creep. Whenever shrinkage is to be included in *combination with creep* in a deflection computation, a multiplier similar to that of Eq. (14.6.6) may be applied to the short-time deflection [14.6]. The following multiplier k_r has been recommended by ACI Committee 435 [14.8] as most appropriate to account for the compression steel effect on the long-time sustained load deflection, Δ_{cp+sh},

$$k_r = \frac{1}{1 + 50\rho'} \tag{14.7.8}$$

where $\rho' = A'_s/bd$. Branson [14.24] and Shaikh [14.28] have discussed the use of this expression to account for the compression steel effect. Equation (14.7.8) has been used in the ACI Code since 1983 as a part of the combined multiplier $\lambda = k_r\xi$, which combines the compression steel effect k_r and time-dependent effect ξ.

14.8 CREEP AND SHRINKAGE DEFLECTION—ACI CODE METHOD

In the ACI Code method, the creep and shrinkage deflection due to sustained load is obtained by multiplying the short-time (instantaneous) deflection by a factor λ (ACI-9.5.2.5). Thus

$$\Delta_{cp+sh} = k_r \xi (\Delta_i)_D = \lambda (\Delta_i)_D \qquad \textbf{(14.8.1)}$$

where

$$\lambda = k_r \xi = \frac{\xi}{1 + 50\rho'} \qquad \textbf{(14.8.2)}$$

and $(\Delta_i)_D$ is the instantaneous deflection due to all sustained loads (usually dead load).

The value of ξ is permitted by ACI-9.5.2.5 to be taken as the following in accordance with duration of sustained load:

5 years or more	2.0
1 year	1.4
6 months	1.2
3 months	1.0

■ **EXAMPLE 14.8.1**   For the beam of Example 14.5.1 determine the creep and shrinkage deflection according to the ACI Code. Assume only the dead load is sustained.

Solution: First it is necessary to compute the short-time (instantaneous) deflection due to all sustained loads, in this case the dead load. From Example 14.5.1, part **(a)** of the solution,

$$(\Delta_i)_D = 0.45 \text{ in.}$$

Since no compression steel is used, Eq. (14.8.2) gives for 5 years or more load duration,

$$\lambda = \frac{\xi}{1 + 50\rho'} = 2.0$$

Then from Eq. (14.8.1),

$$\Delta_{cp+sh} = \lambda(\Delta_i)_D = 2.0(0.45) = 0.90 \text{ in.}$$

If part of the live load were considered as sustained, such as certain types of equipment whose placement or installation is not expected to change for a period of 5 years or more, then it would be necessary to compute Δ_i for the dead load plus the sustained live load. Under the ACI Code, an additional effective moment of inertia I_e would be computed using M_{cr}/M_{max} where M_{max} is due to dead load plus sustained live load.   ■

14.9 CREEP AND SHRINKAGE DEFLECTION—ALTERNATE PROCEDURES

Separate Creep and Shrinkage Multiplier Procedure. A procedure for computing the deflections due to creep and shrinkage separately was recommended by ACI Committee 435 [14.8], based on the work of Branson [14.4], as modfied by recent improvement in the prediction of creep and shrinkage [14.6]. Thus

$$\Delta_{cp+sh} = \Delta_{cp} + \Delta_{sh} \tag{14.9.1}$$

where, using Eq. (14.7.1),

$$\Delta_{sh} = \alpha_1 \phi_{sh} L^2 \tag{14.9.2}$$

and

$$\Delta_{cp} = k_r C_t (\Delta_i)_D \tag{14.9.3}$$

For evaluation of Eqs. (14.9.2) and (14.9.3),

> α_1 = constant (see Eq. 14.7.1)
>
> ϕ_{sh} = shrinkage curvature [Eqs. (14.7.6) and (14.7.7)]
>
> L = span length
>
> k_r = compression steel factor [Eq. (14.6.6)]
>
> C_t = creep coefficient, using Eq. (14.6.3) with correction factors of Eqs. (14.6.4) and (14.6.5), or "For average conditions, ultimate $C_t = 1.6$ may be used." [14.8].
>
> $(\Delta_i)_D$ = instantaneous deflection due to all sustained loads

Combined Creep and Shrinkage Multiplier Procedure. This method is similar to the ACI Code method, except the time-dependent factor ξ can be more accurately evaluated [14.6, 14.7, 14.24]. It may be stated as

$$\Delta_{cp+sh} = k_r \xi (\Delta_i)_D \tag{14.9.4}$$

where

> $k_r = 1/(1 + 50\rho')$ (same as ACI) $\tag{14.9.5}$
>
> ξ = time-dependent coefficient (creep plus shrinkage), which may be taken from Table 14.9.1

■ **EXAMPLE 14.9.1**   For the beam of Example 14.5.1 determine the ultimate (i.e., 5 years duration of load) creep and shrinkage deflection using methods more accurate than the basic ACI method. Assume only the dead load is sustained, the ambient relative humidity is 70%, and the age at loading is 20 days after the initial moist-curing period.

Solution: It is noted that ACI-9.5.2.5 permits computation of long-time deflection by a "more comprehensive analysis," which could include either of these alternate methods.

Table 14.9.1   Time-Dependent Coefficient ξ Including Both Creep and Shrinkage Effects, for Both Normal-Weight and Lightweight Concrete Members of Common Types, Sizes, and Composition [from Branson (14.7, p. 278)][a]

CONCRETE STRENGTH f'_c AT 28 DAYS	AVERAGE RELATIVE HUMIDITY, AGE WHEN LOADED								
	100%			70%			50%		
	$\leq 7d$	$14d$	$\geq 28d$	$\leq 7d$	$14d$	$\geq 28d$	$\leq 7d$	$14d$	$\geq 28d$
2500 to 4000 psi (17 to 28 MPa)	2.0	1.5	1.0	3.0	2.0	1.5	4.0	3.0	2.0
>4,000 psi (28 MPa)	1.5	1.0	0.7	2.5	1.8	1.2	3.5	2.5	1.5

[a] It is suggested that the following percentages of the values in the table be used for sustained loads that are maintained for the periods indicated:

> 25% for 1 month or less
> 50% for 3 months
> 75% for 1 year
> 100% for 5 years or more

The 50% values may normally be used for average relative humidities lower than 50%, which might be the case in heated buildings, for example.

(a) Separate creep and shrinkage multiplier procedure.

$$\Delta_{cp+sh} = \Delta_{cp} + \Delta_{sh}$$

Using Eq. (14.9.3),

$$\Delta_{cp} = k_r C_t (\Delta_i)_D$$

$$k_r = \frac{0.85}{1 + 50\rho'} = 0.85 \quad \text{for } \rho' = 0$$

and from Eq. (14.6.3), for $t = 5(365)$ days,

$$C_t = \left(\frac{t^{0.60}}{10 + t^{0.60}} \right) C_u = 0.90 C_u$$

which for $C_u = 2.35$ as recommended for average conditions gives the basic value of C_t as

$$C_t = 2.12$$

Adjusting for 70% humidity, $(CF)_h = 0.80$ from Eq. (14.6.5) or Table 14.6.2, and for 20-day age of loading after initial moist-curing period, $(CF)_a = 0.87$. Thus the adjusted C_t is

$$C_t = 2.12(0.80)(0.87) = 1.47$$

From Example 14.5.1 using the ACI Code method,

$$(\Delta_i)_D = 0.45 \text{ in.}$$

Then

$$\Delta_{cp} = k_r C_t (\Delta_i)_D = 0.85(1.47)0.45 = 0.56 \text{ in.}$$

For shrinkage, from Eq. (14.7.1),

$$\Delta_{sh} = \alpha_1 \phi_{sh} L^2$$

where

$$\alpha_1 = 0.125 \quad \text{for simply supported beams}$$

For this beam, since $\rho = 1.92\%$ and $\rho' = 0$, Eq. (14.7.6) gives

$$\phi_{sh} = 0.7 \left(\frac{\epsilon_{sh}}{h} \right) \sqrt[3]{\rho}$$

Using ϵ_{sh} from Eq. (14.7.2a),

$$\epsilon_{sh} = \frac{t}{35 + t} (\epsilon_{sh})_u$$

which for $t = 5(365)$ days is, for average conditions,

$$\epsilon_{sh} \approx (\epsilon_{sh})_u = 800 \times 10^{-6} \text{ in./in.}$$

Adjusting for 70% humidity, $(CF)_h = 0.70$, from Eq. (14.7.3) or Table 14.7.1, the adjusted ϵ_{sh} is

$$\epsilon_{sh} = (800 \times 10^{-6})0.70 = 560 \times 10^{-6} \text{ in./in.}$$

Then,

$$\phi_{sh} = 0.7 \left(\frac{560 \times 10^{-6}}{24} \right) \sqrt[3]{1.92} = 20.3 \times 10^{-6} \text{ in./in.}$$

$$\Delta_{sh} = \alpha_1 \phi_{sh} L^2 = 0.125(20.3 \times 10^{-6})(480)^2 = 0.58 \text{ in.}$$

$$\Delta_{cp+sh} = 0.56 + 0.58 = 1.14 \text{ in.}$$

(b) Combined creep and shrinkage multiplier procedure. Using Eq. (14.9.4),

$$\Delta_{cp+sh} = k_r \xi (\Delta_i)_D$$

$$k_r = \frac{1}{1 + 50\rho'} = 1.0$$

$$\xi = \text{value from Table 14.9.1} \approx 1.8$$

Note that age at loading is *after* initial curing period.

$$(\Delta_i)_D = 0.45 \text{ in.} \quad \text{(from Example 14.5.1)}$$

$$\Delta_{cp+sh} = 1.0(1.8)0.45 = 0.81 \text{ in.}$$

A comparison of computation methods for creep and shrinkage deflection may be obtained from the following summary.

METHOD	Δ_{sh+cp}
1. ACI, using $\lambda = \xi/(1 + 50\rho') = 2.0$	0.90 in. (Example 14.8.1)
2. Separate creep and shrinkage as per Eqs. (14.9.1) to (14.9.3), etc.	1.14 in. (Example 14.9.1a)
3. Combined creep and shrinkage using $k_r\xi = 1.8$	0.81 in. (Example 14.9.1b)

14.10 ACI MINIMUM DEPTH OF FLEXURAL MEMBERS

According to the ACI Code, the minimum depths specified in ACI-9.5.2.1 shall apply to all case of "one-way construction . . . unless computation of deflection indicates a lesser thickness may be used without adverse effects." The minimum depth (thickness) values apply to members where large deflection is *not* likely to damage partitions, ceilings, or other frangible attachments. *When large deflection may cause such damage, deflections must be computed whether or not the minimum thickness requirement is satisfied.*

The minimum depths prescribed by any such table are arbitrary and not necessarily conservative. The following logic may be used to explain the limitation on span–depth ratio as an attempt to control deflection.

The deflection at midspan of a simply supported beam is

$$\Delta = \frac{5wL^4}{384EI} \tag{14.10.1}$$

where w is the *service* uniform load. The maximum bending moment is

$$M = \frac{wL^2}{8} = \frac{fI}{c} \tag{14.10.2}$$

where f is the service load stress and c is the distance from the neutral axis to the extreme fiber where f is computed.

Substituting Eq. (14.10.2) in Eq. (14.10.1) gives

$$\Delta = \frac{5L^2f}{48Ec} \tag{14.10.3}$$

Assuming that cracked section is effective at service load conditions,

$$\frac{f}{c} = \frac{f_s}{n(d-x)} = \frac{f_c}{x} \tag{14.10.4}$$

where x is the distance between the neutral axis and the extreme compression fiber. Assume the beam is fully stressed at $f_s = 24,000$ psi for Grade 60 steel, and that $x \approx 0.4h$.

$$\frac{f}{c} \approx \frac{24,000}{(E_s/E_c)(0.6h)} = \frac{40,000E_c}{hE_s} = \frac{E_c}{725h} \qquad \textbf{(14.10.5)}$$

for $E_s = 29 \times 10^6$ psi. Substituting Eq. (14.10.5) in Eq. (14.10.3) and calling $E = E_c$,

$$\frac{\Delta}{L} = \frac{5}{48}\left(\frac{1}{725}\right)\frac{L}{h}$$

$$\min h = \frac{1}{6950}\left(\frac{L}{\Delta}\right)L \qquad \textbf{(14.10.6)}$$

Equation (14.10.6) represents an approximate relationship between depth, span, and span-to-deflection ratio for a fully stressed section under *short-time loading*. If the member is under a reduced stress, the depth may be decreased proportionally to give the same short-time deflection. To account for the sustained load creep and shrinkage deflection, the depth must be increased. Table 14.10.1 shows the evaluation of Eq. (14.10.6) to give the minimum depth required for various deflection limitations under fully and partially stressed conditions. The last two columns in Table 14.10.1 assume that total deflection including creep and shrinkage effects is twice the immediate deflection.

Any selection of limiting values for minimum depth from Table 14.10.1 can only be a crude attempt to control deflection. Table 14.10.2 (ACI-Table 9.5a) is the result of a compromise between the relative conservative recommendations of ACI Committee 435 and the values practicing engineers believe suitable on the basis of experience.

When large deflections may cause cracking of partitions and other frangible attachments, the total deflection (ACI-9.5.2.6) that occurs after installation of such elements is limited to $L/480$. This shows that the minimum depths of ACI-Table 9.5a

Table 14.10.1   Minimum Depth h for Various Equivalent Immediate Deflections and Percentage Stressed[a]

PERCENT STRESSED	$\Delta = L/300$	$\Delta = L/360$	$\Delta = L/480$	$2\Delta = L/300$	$2\Delta = L/360$
100	$L/23.2$	$L/19.3$	$L/14.5$	$L/11.5$	$L/9.7$
67	$L/35$	$L/29$	$L/21.5$	$L/17.5$	$L/14.5$
60	$L/39$	$L/32$	$L/24$	$L/19.5$	$L/16$
50	$L/46.5$	$L/38.5$	$L/29$	$L/23$	$L/19.5$

[a] Assume $f_s = 24,000$ psi at 100% stressed.

Table 14.10.2 Minimum Depth *h* for Beams and One-Way Slabs, for Members *Not* Supporting or Attached to Partitions or Other Construction Likely to be Damaged by Large Deflections (from ACI-Table 9.5a).

TYPE OF MEMBERS		SIMPLE SUPPORT	ONE END CONTINUOUS	BOTH ENDS CONTINUOUS	CANTILEVER
Beams	$f_y = 60$ ksi	$L/16$	$L/18.5$	$L/21$	$L/8$
	$f_y = 40$ ski	$L/20$	$L/23$	$L/26$	$L/10$
One-way slabs (solid)	$f_y = 60$ ksi	$L/20$	$L/24$	$L/28$	$L/10$
	$f_y = 40$ ksi	$L/25$	$L/30$	$L/35$	$L/12.5$

Note: For structural lightweight concrete having unit weights w_c from 90–120 pcf, multiply table values by $1.65 - 0.005 \, w_c$ but not less than 1.09. (60 ksi = 420 MPa; 40 ksi = 280 MPa, approximately.)

are likely to be too low; hence that table *does not apply for such cases,* and deflection computations must be made.

In general, minimum depth as a proportion of span is an inadequate criterion for controlling deflection; computation of deflection should be made whenever deflection is of concern.

14.11 SPAN–TO–DEPTH RATIO TO ACCOUNT FOR CRACKING AND SUSTAINED LOAD EFFECTS

In order to illustrate the effects of the many variables on the span–to–depth ratio, the following general development, similar to that of Branson [14.32], is presented.

Grossman [14.29] has also provided a procedure for determining the minimum thickness that would approximately satisfy any deflection limitation given by ACI-Table 9.5b. The Grossman procedure has approximated the effective moment of inertia, thus eliminating the need to compute the moment of inertia I_{cr} of the cracked section. ACI Committee 318 seriously considered adopting the Grossman approach for the 1983 ACI Code; the approach is still being considered for a future code revision. The development in this section may be less practical than the Grossman method as a means to obtain an answer, but it is intended to illustrate how the variables interrelate, and to show why it is impossible to have a simple table of minimum thicknesses. The Grossman work [14.29] along with Branson's discussion [14.30] and Grossman's closure [14.29] provide the reader an interesting and useful discourse on the subject of deflection control. Rangan [14.31] has also presented a minimum thickness approach similar to that of the following development.

The short-time deflection Δ_i may be expressed, according to Eq. (14.5.1), as

$$\Delta_i = \beta_a \left(\frac{M_{max} L^2}{E_c I_e} \right) \qquad (14.11.1)$$

where

M_{max} = maximum moment at the stage for which deflection is desired

I_e = Eq. (14.4.1), [which is ACI Code Formula (9-7)]

$$= (M_{cr}/M_{max})^3 I_g + [1 - (M_{cr}/M_{max})^3]I_{cr} \leq I_g$$

M_{cr} = maximum moment to cause a beam to crack at the extreme fiber in tension = $f_r I_g / y_t$

f_r = modulus of rupture = $0.65\sqrt{w_c f'_c}$ (ACI uses $7.5\sqrt{f'_c}$ for normal-weight concrete)

y_t = distance from neutral axis to extreme fiber in tension

Multiplying Eq. (14.11.1) by $f_r I_g / (y_t M_{cr})$, which is equal to unity, gives

$$\Delta_i = \beta_a \left(\frac{M_{max}L^2}{E_c I_e} \right) \frac{f_r I_g}{y_t M_{cr}} \tag{14.11.2}$$

Solving Eq. (14.11.2) for L/y_t gives

$$\frac{L}{y_t} = \frac{\Delta_i}{L} \left(\frac{E_c}{f_r \beta_a} \right) \left(\frac{M_{cr}}{M_{max}} \right) \frac{I_e}{I_g} \tag{14.11.3}$$

Since both E_c and f_r are proportional to $\sqrt{f'_c}$, let

$$\beta_w = \frac{E_c}{f_r} = \frac{33 w_c^{1.5} \sqrt{f'_c}}{0.65 w_c^{0.5} \sqrt{f'_c}} = 50.8 w_c \tag{14.11.4}$$

For normal-weight concrete

$$\beta_w = 50.8 w_c = 50.8(145) = 7370$$

Conversion from normal-weight concrete to lightweight concrete may be made by multiplying L/y_t by the ratio of the unit weight of lightweight concrete to 145 pcf (2330 kg/m³).

Letting $\gamma = I_e/I_g$ and $\beta_w = E_c/f_r$ in Eq. (14.11.3),

$$\frac{L}{y_t} = \frac{\Delta_i}{L} \left(\frac{\beta_w}{\beta_a} \right) \left(\frac{M_{cr}}{M_{max}} \right) \gamma \tag{14.11.5}$$

where

$$\gamma = \frac{I}{I_g} = \left(\frac{M_{cr}}{M_{max}} \right)^3 + \left[1 - \left(\frac{M_{cr}}{M_{max}} \right)^3 \right] \frac{I_{cr}}{I_g} \tag{14.11.6}$$

Using the methods of Chapter 4, the ratio of the moment of inertia of the cracked section to that of the gross section may be computed for various shapes of beams. As an example, for a singly reinforced rectangular beam assuming $d = 0.9h$.

$$\frac{I_{cr}}{I_g} = \frac{bx^3/3 + nA_s(d-x)^2}{bh^3/12}$$

$$= 8.75 \left[\frac{(x/d)^3}{3} + n\rho \left(1 - \frac{x}{d} \right)^2 \right] \tag{14.11.7}$$

where

$$x/d = \sqrt{(\rho n)^2 + 2\rho n} - \rho n \tag{14.11.8}$$

Charts are given by Lutz [14.12, 14.8 pp 266–267] to obtain I_g and I_{cr} for T-sections.

For dead load deflection, $M_{max} = M_D$, which makes $\gamma = \gamma_D$; Eq. (14.11.5) then becomes

$$\frac{(\Delta_i)_D}{L} = \frac{\beta_a}{\beta_w}\left(\frac{M_D}{M_{cr}}\right)\left(\frac{L}{y_t}\right)\left(\frac{1}{\gamma_D}\right) \tag{14.11.9}$$

For dead load plus live load,

$$\frac{(\Delta_i)_{D+L}}{L} = \frac{\beta_a}{\beta_w}\left(\frac{M_{D+L}}{M_{cr}}\right)\left(\frac{L}{y_t}\right)\left(\frac{1}{\gamma_{D+L}}\right) \tag{14.11.10}$$

As discussed in Section 14.5, since live load cannot act in the absence of dead load, the live load deflection must be obtained indirectly,

$$\frac{(\Delta_i)_L}{L} = \frac{(\Delta_i)_{D+L}}{L} - \frac{(\Delta_i)_D}{L} \tag{14.11.11}$$

which, using Eqs (14.11.9) and (14.11.10), and letting $C_L = M_L/M_D$, gives

$$\frac{(\Delta_i)_L}{L} = \frac{\beta_a}{\beta_w}\left(\frac{L}{y_t}\right)\left(\frac{M_{D+L}}{M_{cr}}\right)\left[\frac{1}{\gamma_{D+L}} - \frac{1}{\gamma_D(1 + C_L)}\right] \tag{14.11.12}$$

When excessive deflection may cause damage to partitions and other nonstructural construction, it is the sum of deflections due to live load plus creep and shrinkage that is of concern. The instantaneous (short-time) dead load deflection will have occurred when forms are removed and before any breakable attachments are put in place. Thus using Eq. (14.8.1), and the ACI time-dependent multiplier λ,

$$\Delta_{cp+sh} = \lambda(\Delta_i)_D$$

Finally, the deflection to be controlled to minimize possible damage is

$$\begin{aligned}
\frac{(\Delta_i)_L}{L} + \frac{\Delta_{cp+sh}}{L} &= \frac{\beta_a}{\beta_w}\left(\frac{L}{y_t}\right)\left(\frac{M_{D+L}}{M_{cr}}\right)\left[\frac{1}{\gamma_{D+L}} + \frac{(\lambda - 1)}{\gamma_D(1 + C_L)}\right] \\
&= \frac{\beta_a}{\beta_w}\left(\frac{L}{y_t}\right)\left(\frac{M_{D+L}}{M_{cr}}\right)\left[\frac{1 + C_L + (\lambda - 1)(\gamma_{D+L}/\gamma_D)}{1 + C_L}\right]\frac{1}{\gamma_{D+L}}
\end{aligned} \tag{14.11.13}$$

Solving for L/y_t gives

$$\left(\frac{L}{y_t}\right)_{\substack{\text{limit for} \\ L+cp+sh}} = \left(\frac{\Delta}{L}\right)\left(\frac{\beta_w}{\beta_a}\right)\left(\frac{M_{cr}}{M_{D+L}}\right)\left[\frac{C_L + 1}{C_L + 1 + (\lambda - 1)(\gamma_{D+L}/\gamma_D)}\right]\gamma_{D+L} \tag{14.11.14}$$

For instantaneous dead load plus live load, Eq. (14.11.10) gives

$$\left(\frac{L}{y_t}\right)_{\substack{\text{limit for} \\ D+L}} = \left(\frac{\Delta}{L}\right)\left(\frac{\beta_w}{\beta_a}\right)\left(\frac{M_{cr}}{M_{D+L}}\right)\gamma_{D+L} \tag{14.11.15}$$

As shown below, if the span-to-depth ratio limit is available for short-time deflection under dead load plus live load, such as from Table 14.10.1, the effect of creep and shrinkage can be obtained by the use of a multiplier. Comparing Eqs. (14.11.14) and

(14.11.15), in which Δ/L may be taken as a stated limit,

$$\left(\frac{L}{y_t}\right)_{L+cp+sh} = \left(\frac{L}{y_t}\right)_{D+L}\left[\frac{C_L + 1}{C_L + 1 + (\lambda - 1)(\gamma_{D+L}/\gamma_D)}\right] \qquad \textbf{(14.11.16)}$$

For the situation in which the I_e under dead load only is approximately the same as I_e under dead plus live load, $\gamma_{D+L} \approx \gamma_D$, in which case Eq. (14.11.16) becomes

$$\left(\frac{L}{y_t}\right)_{L+cp+sh} = \left(\frac{L}{y_t}\right)_{D+L}\left(\frac{C_L + 1}{C_L + \lambda}\right) \qquad \textbf{(14.11.17)}$$

Charts are available [14.32] for the L/h ratio for short-time effects of dead load plus live load, Eq. (14.11.10). The chart for singly reinforced rectangular beams ($y_t = 0.5h$) is given in Fig. 14.11.1.

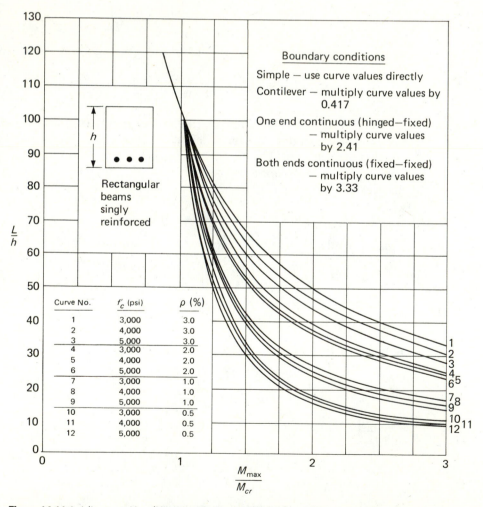

Figure 14.11.1 L/h versus M_{max}/M_{cr} curves for the conditions: $\Delta = L/360$, normal-weight concrete, and uniformly distributed short-time loading, and for different boundary conditions, steel percentages, and concrete strengths (from Branson [14.32]).

■ **EXAMPLE 14.11.1**   Determine the depth of beam required for the loading conditions of Example 14.5.1 (Figure 14.5.3) if the sum of the immediate live load plus creep and shrinkage deflection must not exceed $L/480$. Assume only the dead load is sustained and the sustained load factor $\lambda = 2$ as given by ACI-9.5.2.5.

Solution: **(a)** Use ACI-Table 9.5a.

$$\frac{L}{h} = 16$$

This considers average conditions and includes some effect of sustained load deflection.

$$\min h = \frac{480}{16} = 30 \text{ in.}$$

(b) Use more accurate procedure with Fig. 14.11.1.

$$\rho = 0.0192$$

$$M_{cr} = \frac{f_r I_g}{y_t} = \frac{7.5 \sqrt{4000}(20,700)}{12(12,000)} = 68 \text{ ft-kips}$$

$$\frac{M_{max}}{M_{cr}} = \frac{250 + 90}{68} = 5 > 3; \quad \text{use } 3$$

$$C_L = \frac{M_L}{M_D} = \frac{250}{90} = 2.78$$

From Fig. 14.11.1 for $M_{max}/M_{cr} = 3$ and $\rho = 0.0192$, find

$$\left(\frac{L}{h}\right)_{D+L} = 23, \quad \text{for } \frac{\Delta}{L} = \frac{1}{360}$$

$$\left(\frac{L}{h}\right)_{D+L} = 23\left(\frac{360}{480}\right) = 17.3, \quad \text{for } \frac{\Delta}{L} = \frac{1}{480}$$

$$\left(\frac{L}{h}\right)_{L+cp+sh} = 17.3\left(\frac{C_L + 1}{C_L + \lambda}\right) = 17.3\left(\frac{2.78 + 1}{2.78 + 2}\right) = 13.7$$

$$\min h = \frac{480}{13.7} = 35 \text{ in.}$$

which is somewhat more severe than ACI-Table 9.5a.

(c) Required depth and adequacy of beam in Example 14.5.1. For the given beam with $h = 24$ in.,

$$(\Delta_i)_L = 1.43 \text{ in.} \quad \text{(Example 14.5.1)}$$

$$\Delta_{cp+sh} = 0.90 \text{ in.} \quad \text{(Example 14.8.1)}$$

$$\text{allowable } \Delta = \frac{L}{480} = \frac{480}{480} = 1 \text{ in.}$$

$$\text{actual } \Delta = 1.43 + 0.90 = 2.23 \text{ in.}$$

If the beam were 28 in. deep, with the same loading it would be approximately carrying only (24/28 = 0.855) 85.5% of its capacity, that is, stressed to only 85.5%. In which case,

$$\text{min } h = 35(0.855) = 29.9 \text{ in.}$$

It would appear that 29 or 30 in. is required to satisfy the deflection limit. Unless the width is reduced, which seems undesirable here, increasing the depth will increase dead load moment, so the 30-in. minimum given by ACI-Table 9.5a seems in this case to be about right. ∎

14.12 ACI CODE DEFLECTION PROVISIONS—BEAM EXAMPLES

The ACI Code provisions (ACI-9.5) regarding deflection computations may be summarized as follows. In both the strength method and the working stress method (ACI "alternate" method), deflections *under service load conditions* must be computed

1. Whenever excessive deflection may adversely affect the strength or serviceability of the structure at service loads (ACI-9.5.1).
2. When minimum thickness used is *less* than that given by ACI-Table 9.5a for beams and one-way slabs (see also Table 14.10.2 in this chapter).

The use of minimum thicknesses from ACI-Table 9.5a eliminates the need for computation of deflection *only for those cases where partitions, ceilings, and other nonstructural elements are not being supported.* Beams or one-way slabs supporting such frangible elements are in category (1) above, and deflections must be computed whether or not the minimum thickness limits of Table 9.5a have been met.

■ **EXAMPLE 14.12.1** Investigate the deflection for the beam of Fig. 14.12.1 (same as obtained in Example 3.8.1) used on a simple span of 25 ft. The maximum bending moments under service load are 158 ft-kips dead load and 105 ft-kips live load, which

Figure 14.12.1 Section for Example 14.12.1.

correspond to $M_u = 400$ ft-kips as used for design by the strength method. Assume that all loading is uniformly distributed and that none of the live load is sustained. The beam supports partitions and other construction likely to be damaged by large deflections. Use $f'_c = 4000$ psi, $f_y = 40,000$ psi, and the ACI Code.

Solution: **(a)** Check minimum thickness (ACI-Table 9.5a), or from Table 14.10.2 for $f_y = 40$ ksi,

$$\text{min } h = \frac{L}{20} = \frac{25(12)}{20} = 15 \text{ in.} < 25 \text{ in. used}$$

This would make it seem that deflection is not likely to be excessive; however, deflection computations must be made because of the frangible items the beam supports.

(b) Examine reinforcement ratio ρ

$$\rho = \frac{A_s}{bd} = \frac{7.22}{14(21.5)} = 0.024$$

This exceeds $0.5\rho_{max} = 0.375\rho_b = 0.0185$ suggested in Chapter 3 as a guideline for deflection control; thus from this check one may predict that deflection may be a problem.

(c) Determine the moment of inertia I_g for gross uncracked section without steel and I_{cr} for the cracked transformed section.

For gross uncracked section.

$$I_g = \tfrac{1}{12}(14)(25)^3 = 18,200 \text{ in.}^4$$

For cracked section, locate the neutral axis under service loads,

$$\frac{14x^2}{2} = 7.22(8)(21.5 - x)$$

$$x = 9.82 \text{ in.}$$

$$I_{cr} = \tfrac{1}{3}(14)(9.82)^3 + 7.22(8)(11.68)^2 = 12,300 \text{ in.}^4$$

(d) Determine the effective moment of inertia I_e (ACI Formula 9-7). The cracking moment is

$$M_{cr} = \frac{f_r I_g}{y_t} = \frac{7.5\sqrt{4000}(18,200)}{12.5(12,000)} = 57.6 \text{ ft-kips}$$

For dead load deflection,

$$\frac{M_{cr}}{M_{max}} = \frac{57.6}{158} = 0.365; \qquad \left(\frac{M_{cr}}{M_{max}}\right)^3 = 0.05$$

$$I_e = \left(\frac{M_{cr}}{M_{max}}\right)^3 I_g + \left[1 - \left(\frac{M_{cr}}{M_{max}}\right)^3\right] I_{cr}$$

$$= 0.05(18,200) + 0.95(12,300) = 12,600 \text{ in.}^4$$

For dead load plus live load deflection,

$$\frac{M_{cr}}{M_{max}} = \frac{57.6}{263} = 0.22; \qquad \left(\frac{M_{cr}}{M_{max}}\right)^3 = 0.01$$

$$I_e \approx I_{cr} = 12,300 \text{ in.}^4$$

(e) Compute immediate deflections.

$$E_c = 57,000\sqrt{f_c'} = 57,000\sqrt{4000} = 3.6 \times 10^6 \text{ psi}$$

For dead load,

$$(\Delta_i)_D = \frac{5wL^4}{384EI} = \frac{5ML^2}{48EI} = \frac{5(158)(12)(300)^2}{48(3.6)(10^3)(12,600)} = 0.39 \text{ in.}$$

For dead load plus live load,

$$(\Delta_i)_{D+L} = \frac{5(263)(12)(300)^2}{48(3.6)(10^3)(12,300)} = 0.67 \text{ in.}$$

Then the immediate live load deflection is

$$(\Delta_i)_L = (\Delta_i)_{D+L} - (\Delta_i)_D = 0.67 - 0.39 = 0.28 \text{ in.}$$

(f) Compute creep and shrinkage deflection. From ACI-9.5.2.5 the multiplier is

$$\lambda = k_r \xi = \frac{2.0}{1 + 50\rho'} = 2.0$$

for sustained load at 5 years or more.

$$\Delta_{cp+sh} = \lambda(\Delta_i)_D = 2.0(0.39) = 0.78 \text{ in.}$$

(g) Check limitation of ACI-Table 9.5b. For roof or floor construction supporting or attached to nonstructural elements likely to be damaged by large deflection,

$$(\Delta_i)_L + \Delta_{cp+sh} \le \frac{L}{480}$$

This limit is for the deflection that is estimated to occur *after* the nonstructural elements are put in place. Whatever portion, if any, of the live load or creep and shrinkage deflection has occurred prior to the placement of nonstructural elements may be excluded from the $L/480$ limitation. Further, if adequate measures are taken to prevent damage to supported or attached elements, the $L/480$ limit may be exceeded (footnote, ACI-Table 9.5b).

For this example,

$$\left[0.28 + 0.78 = 1.06 \text{ in.} \right] > \left[\frac{L}{480} = \frac{300}{480} = 0.63 \text{ in.} \right]$$

which is not acceptable. This shows that the minimum thickness indicated by ACI-Table 9.5a was not even close. ∎

■ **EXAMPLE 14.12.2** Investigate the deflection for the one-way continuous slab shown in Fig. 14.12.2. The slab is 5 in. thick. Assume that 60% of the 100 psf live load is sustained. Use $f'_c = 3000$ psi and $f_y = 40,000$ psi.

Solution: Use may be made of Eq. (14.2.4) to compute maximum deflection:

$$\Delta_m = \frac{5L^2}{48EI}\left[M_s - \frac{1}{10}(M_a + M_b)\right]$$ **[14.2.4]**

Center-to-center of supports will be used as the span L. However, when coefficients are used to compute moments as discussed in Chapter 7 using clear span L_n, the L_n should be used in computing deflections [14.8]. In this mathematical model, a conservative assumption of zero moment at the exterior support is used. Though Eq. (14.2.4) provides *midspan* deflection, maximum deflection occurs between the location of maximum moment and midspan. It is reasonable and practical to use maximum moment for M_s in that equation.

$$E_c = 57,000\sqrt{f'_c} = 57,000\sqrt{3000} = 3.12 \times 10^6 \text{ psi}$$

(a) Determine I_g and I_{cr}, using a 1-ft width of section.

$$I_g = \tfrac{1}{12}(12)(5.0)^3 = 125 \text{ in.}^4$$

Figure 14.12.2 End-span details for continuous slab Example 14.12.2.

For the positive moment region, $d = 3.94$ in.,

$$\frac{12x^2}{2} = 9(0.37)(3.94 - x)$$

$$x = 1.23 \text{ in.}$$

$$I_{cr} = \tfrac{1}{3}(12)(1.23)^3 + 9(0.37)(2.44 - 1.23)^2 = 31.9 \text{ in.}^4$$

Note that only the positive moment region properties need to be used according to ACI-9.5.2.4.

(b) Determine effective moment of inertia I_e.

$$f_r = 7.5\sqrt{f'_c} = 7.5\sqrt{3000} = 411 \text{ psi}$$

$$M_{cr} = \frac{f_r I_g}{y_t} = \frac{0.411(125)}{2.50(12)} = 1.71 \text{ ft-kips}$$

$$I_e = \left(\frac{M_{cr}}{M_{max}}\right)^3 I_g + \left[1 - \left(\frac{M_{cr}}{M_{max}}\right)^3\right] I_{cr}$$

In the positive moment region, for dead load

$$\frac{M_{cr}}{M_D} = \frac{M_{cr}}{M_{max}} = \frac{1.71}{0.86} > 1; \qquad I_e = I_g = 125 \text{ in.}^4$$

and for dead load plus live load,

$$\frac{M_{cr}}{M_{D+L}} = \frac{M_{cr}}{M_{max}} = \frac{1.71}{2.17} = 0.788; \qquad \left(\frac{M_{cr}}{M_{max}}\right)^3 = 0.49$$

$$I_e = 0.49(125) + 0.51(31.9) = 77.5 \text{ in.}^4$$

(c) Consider the effect of sustained live load that contributes to the creep and shrinkage deflection. The immediate deflection due to all sustained loads is required as the base value on which to apply the time-dependent multiplier.

For the positive moment region, using sustained load moment of $0.86 + 0.6(1.31) = 1.65$ ft-kips/ft,

$$\frac{M_{cr}}{M_{max}} = \frac{1.71}{1.65} > 1.0 \qquad I_e = I_g = 125 \text{ in.}^4$$

(d) Immediate live load deflection.

$$(\Delta_i)_L = (\Delta_i)_{D+L} - (\Delta_i)_D$$

$$(\Delta_i)_{D+L} = \frac{5(12)^2 144}{48(3.12)(10^3)77.5}\left(2.17 - \frac{1.65}{10}\right)(12) = 0.21 \text{ in.}$$

$$(\Delta_i)_D = \frac{5(12)^2 144}{48(3.12)(10^3)125}\left(0.86 - \frac{0.59}{10}\right)(12) = 0.05 \text{ in.}$$

$$(\Delta_i)_L = 0.21 - 0.05 = 0.16 \text{ in.}$$

(e) Creep and shrinkage deflection. The immediate deflection due to sustained loads,

$$(\Delta_i)_{D+\text{sust } L} = \frac{5(12)^2 144}{48(3.12)(10^3)125}\left(1.65 - \frac{1.23}{10}\right)(12) = 0.10 \text{ in.}$$

where the sustained moment at the support is $0.59 + 0.6(1.06) = 1.23$ ft-kips/ft. Considering 5 years or more load duration, and with no compression steel, $\lambda = 2.0$

$$\Delta_{cp+sh} = \lambda(\Delta_i)_{D+\text{sust } L} = 2.0(0.10) = 0.20 \text{ in.}$$

(f) Check deflection if the limit of $L/480$ in ACI-Table 9.5b applies,

$$(\Delta_i)_L + \Delta_{cp+sh} = 0.16 + 0.20 = 0.36 \text{ in.}$$

$$\text{allowable } \Delta = \frac{L}{480} = \frac{12(12)}{480} = 0.30 \text{ in.}$$

which appears unacceptable. If none of the live load was considered sustained,

$$(\Delta_i)_L + \Delta_{cp+sh} = 0.16 + 2(0.05) = 0.26 \text{ in.} < 0.30 \text{ in.}$$

The result is acceptable. Actually, since the maximum bending moments of 0.86 and 1.31 ft-kips were used as midspan moments in Eq. (14.2.4), the computed deflection is slightly on the high side. Furthermore, considering the accuracy within which deflection may be predicted (say, $\pm 50\%$), the original calculation (0.36 in. compared to 0.30 in. limit) could be considered acceptable. ∎

The decision regarding whether or not part of the live load is sustained should be made by considering the actual loading and its duration. For instance, a floor system supporting library stacks as the live load might well be considered to have a significant part of the live load treated as being sustained. In most situations it is acceptable to consider that only the dead load is sustained and thereby affects the magnitude of creep and shrinkage deflection.

In spite of the fact that the deflection calculations have been illustrated throughout this chapter without shortcutting any steps in the formal procedure, deflection computations should be made keeping practicality in mind. The effective moment of inertia I_e theoretically should be computed at each total load level for which deflection is of concern; usually dead load, dead load plus live load, and dead load plus sustained live load. I_e should be computed at each end and at the midspan region in order to obtain an average as is encouraged by the ACI Code.

However, when the true accuracy obtainable from a deflection computation is recognized, the designer should use the I_e equation only when the result will be significantly different from using either I_{cr} or I_g. Furthermore, for most situations only the midspan I_e need be computed; an average does not measurably improve accuracy. The authors believe these practical suggestions satisfy the spirit of the ACI Code.

For additional practical deflection examples based on the ACI Code and ACI Committee 435 recommendations, including composite beams and two-way systems, see the chapter by Branson in *Handbook of Concrete Engineering* [14.2] and the *PCA Notes* [14.33].

SELECTED REFERENCES

14.1. ACI Committee 435, Subcommittee 1. "Allowable Deflections," *ACI Journal, Proceedings*, **65**, June 1968, 433–444. Disc., 1037–1038.

14.2. Mark Fintel (Ed.). *Handbook of Concrete Engineering*. New York: Van Nostrand, 1974 (801 pp.) (Chap. 2, "Deflections," p. 49).

14.3. Wei-Wen Yu and George Winter. "Instantaneous and Long-Time Deflections of Reinforced Concrete Beams Under Working Loads," *ACI Journal, Proceedings*, **57**, July 1960, 29–50. Disc., 1165–1171.

14.4. Dan E. Branson. "Instantaneous and Time-Dependent Deflections of Simple and Continuous Reinforced Concrete Beams," Part 1, Report No. 7. Alabama Highway Research Report, Bureau of Public Roads, August 1963 (1965) (pp. 1–78).

14.5. Dan E. Branson. Discussion of "Variability of Deflections of Simply Supported Reinforced Concrete Beams," by ACI Committee 435, *ACI Journal, Proceedings*, **69**, July 1972, 449–451.

14.6. ACI Committee 209. "Prediction of Creep, Shrinkage, and Temperature Effects in Concrete Structures," *Designing for Effects of Creep, Shrinkage, and Temperature in Concrete Structures* (SP-27). Detroit: American Concrete Institute, 1971 (pp. 51–93).

14.7. Dan E. Branson. *Deformation of Concrete Structures*. New York: McGraw-Hill, 1977.

14.8. ACI Committee 435. "Proposed Revisions By Committee 435 to ACI Building Code and Commentary Provisions on Deflections," *ACI Journal, Proceedings*, **75**, June 1978, 229–238.

14.9. G. W. Washa and P. G. Fluck. "Effect of Compressive Reinforcement on the Plastic Flow of Reinforced Concrete Beams," *ACI Journal, Proceedings*, **49**, October 1952, 89–108.

14.10. G. W. Washa and P. G. Fluck. "Plastic Flow (Creep) of Reinforced Concrete Continuous Beams," *ACI Journal, Proceedings*, **52**, January 1956, 549–561.

14.11. Dan E. Branson. Discussion of "Proposed Revision of ACI 318-63 Building Code Requirements for Reinforced Concrete," *ACI Journal, Proceedings*, **67**, September 1970, 692–693.

14.12. LeRoy A. Lutz. "Graphical Evaluation of the Effective Moment of Inertia for Deflection," *ACI Journal, Proceedings*, **70**, March 1973, 207–213. Disc., **70**, September 1973, 662–663.

14.13. Dan E. Branson and Heinrich Trost. "Unified Procedures for Predicting the Deflection and Centroidal Axis Location of Partially Cracked Nonprestressed and Prestressed Concrete Members," *ACI Journal, Proceedings*, **79**, March–April 1982, 119–130.

14.14. ACI Committee 435, Subcommittee 7. "Deflections of Continuous Beams," *ACI Journal, Proceedings*, **70**, December 1973, 781–787.

14.15. P. D. Zuraski, C. G. Salmon, and A. Fattah Shaikh. "Calculation of Instantaneous Deflections for Continuous Reinforced Concrete Beams," *Deflections of Concrete Structures* (SP-43). Detroit: American Concrete Institute, 1974 (pp. 315–331).

14.16. N. H. Burns and C. P. Siess. "Repeated and Reverse Loading in Reinforced Concrete," *Journal of the Structural Division*, ASCE, **92**, ST5 (October 1966), 65–78.

14.17. Koladi M. Kripanarayanan and Dan E. Branson. "Short-Time Deflections of Beams Under Single and Repeated Load Cycles," *ACI Journal, Proceedings*, **69**, February 1972, 110–117.

14.18. ACI Committee 435. "Deflections of Reinforced Concrete Flexural Members," *ACI Journal, Proceedings*, **63**, June 1966, 637–674.

14.19. ACI Committee 435. "Variability of Deflections of Simply Supported Reinforced Concrete Beams," *ACI Journal, Proceedings*, **69**, January 1972, 29–35.

14.20. ACI Committee 209. "Effects of Concrete Constituents, Environment, and Stress on the Creep and Shrinkage of Concrete," *Designing for Effects of Creep, Shrinkage, and Temperature in Concrete Structures* (SP-27). Detroit: American Concrete Institute, 1971 (pp. 1–42).

14.21. B. L. Meyers and D. E. Branson. "Design Aid for Predicting Creep and Shrinkage Properties of Concrete," *ACI Journal, Proceedings*, **69**, September 1972, 551–555.

14.22. M. R. Hollington. *A Series of Long-Term Tests to Investigate the Deflection of a Representative Precast Concrete Floor Component*, Technical Report TRA 442. London: Cement and Concrete Association, April 1970 (43 pp.).

14.23. Lian Duan, Fu-Ming Wang, and Wai-Fah Chen. "Flexural Rigidity of Reinforced Concrete Members," *ACI Structural Journal*, **86**, July–August 1989, 419–427. Disc. **87**, May–June 1990, 364–365.

14.24. Dan E. Branson. "Compression Steel Effect on Long-Time Deflection," *ACI Journal, Proceedings*, **68**, August 1971, 555–559.

14.25. T. C. Hansen and A. H. Mattock. "Influence of Size and Shape of Member on Shrinkage and Creep of Concrete," *ACI Journal, Proceedings*, **63**, February 1966, 267–289.

14.26. Hans Gesund. "Shrinkage and Creep Influence on Deflections and Moments of Reinforced Concrete Beams," *ACI Journal, Proceedings*, **59**, May 1962, 689–704.

14.27. Alfred L. Miller. "Warping of Reinforced Concrete Due to Shrinkage," *ACI Journal, Proceedings*, **54**, May 1958, 939–950.

14.28. A. F. Shaikh. Discussion of "Proposed Revision of ACI 318–63: Building Code Requirements for Reinforced Concrete," *ACI Journal, Proceedings*, **67**, September 1970, 722–723.

14.29. Jacob S. Grossman. "Simplified Computations for Effective Moment of Inertia I_e and Minimum Thickness to Avoid Deflection Computations," *ACI Journal, Proceedings*, **78**, November–December 1981, 423–439. Disc., **79**, September–October 1982, 413–419.

14.30. Dan E. Branson. Discussion of "Simplified Computations for Effective Moment of Inertia I_e and Minimum Thickness to Avoid Deflection Computations," (*ACI Journal, Proceedings*, **78**, November–December 1981, 423–439), *ACI Journal, Proceedings*, **79**, September–October 1982, 413–414.

14.31. B. Vijaya Rangan. "Control of Beam Deflections by Allowable Span–Depth Ratios," *ACI Journal, Proceedings*, **79**, September–October 1982, 372–377.

14.32. Dan E. Branson. "Design Procedures for Computing Deflections," *ACI Journal, Proceedings*, **65**, September 1968, 730–742.

14.33. PCA. *Notes on ACI 318-89 Building Code Requirements for Reinforced Concrete*, 5th ed., ed. by S. K. Ghosh and Basile G. Rabbat. Skokie, Illinois: Portland Cement Association, 1990, Chapter 8.

PROBLEMS

All problems* are to be done in accordance with the ACI Code unless otherwise indicated.

14.1 Compute the immediate deflections due to dead load and live load on the beam of the accompanying figure. The span of the uniformly loaded, simply supported beam is 30 ft, and the maximum service load moments are 80 ft-kips for dead load and 95 ft-kips for live load. Use $f'_c = 3500$ psi ($n = 8.5$) and $f_y = 60,000$ psi. (Span = 9.2 m; $M_D = 110$ kN·m; $M_L = 130$ kN·m; $f'_c = 25$ MPa; $f_y = 400$ MPa.)

Inch – Pound units

SI units

Problems 14.1, 14.7, and 14.14

14.2 Compute the immediate deflections due to dead load and live load on the beam of the accompanying figure. The uniformly loaded, simply supported beam on a span of 29 ft must resist maximum service load moments of 300 ft-kips dead load and 500 ft-kips live load. Use $f'_c = 4000$ psi ($n = 8$) and $f_y = 40,000$ psi. (Span = 8.9 m; $M_D = 407$ kN·m; $M_L = 680$ kN·m; $f'_c = 30$ MPa; $f_y = 300$ MPa.)

- - - -

* Most problems may be solved as problems stated in Inch-Pound units, or as problems in SI units using quantities in parenthesis at the end of the statement. The SI conversions are approximate to avoid implying higher precision for given information in metric units than that for the Inch-Pound units.

Problems 14.2, 14.15, and 14.16

14.3 Compute the immediate deflections due to dead load and live load on the beam of the accompanying figure. The uniformly loaded, simply supported beam on a span of 30 ft must resist maximum service load moments of 20 ft-kips dead load and 14 ft-kips live load. Use $f'_c = 3000$ psi ($n = 9$) and $f_y = 40,000$ psi. (Span $= 9.1$ m; $M_D = 27$ kN·m; $M_L = 19$ kN·m; $f'_c = 20$ MPa; $f_y = 300$ MPa.)

Problems 14.3, 14.4, 14.17, and 14.18

14.4 Repeat Problem 14.3, except the service load moments are 29 ft-kips dead load and 19 ft-kips live load, and $f_y = 60,000$ psi. ($M_D = 40$ kN·m; $M_L = 26$ kN·m; $f_y = 400$ MPa.)

14.5 Investigate the acceptability of the beam of the accompanying figure for immediate live load deflection if the limitation is $L/360$. Use $f'_c = 3500$ psi ($n = 8.5$) and $f_y = 40,000$ psi.

Section A—A   Section B—B

Problems 14.5 and 14.19

14.6   Investigate the acceptability of the beam of the accompanying figure for immediate live load deflection if the limitation is $L/360$. Use $f'_c = 3000$ psi $(n = 9)$ and $f_y = 60,000$ psi.

Problems 14.6 and 14.20

14.7   Investigate the acceptability of the beam of Problem 14.1 if the limit for live load plus creep and shrinkage deflection is $L/360$. Consider that none of the live load is sustained. Use data computed in Problem 14.1 if that problem was previously assigned.

14.8   Investigate the acceptability of the deflection due to live load plus creep and shrinkage on the 24-ft main span for the beam of Problem 3.13, Case 2, if the cross section is 12 × 24 with 6-#10 (3 in each of two layers) in the positive moment zone and 3-#9 (one layer) in the negative moment zone. Assume #3 stirrups are used. Only the dead load is sustained. Use $f'_c = 3500$ psi and $f_y = 40,000$ psi. Assume none of the steel is compression steel.

14.9   Investigate the acceptability of the deflection for the cantilever portion of the beam of Problem 14.8.

14.10   Investigate the acceptability of the deflection due to live load plus creep and shrinkage on the 20 ft span of the beam of Problem 3.13, Case 5, if the cross section is 12 × 18 with 3-#9 (one layer) as positive moment reinforcement and 2-#9 (one layer) as negative moment reinforcement. Assume #3 stirrups are used. Assume also only the dead load is sustained. Use $f'_c = 3500$ psi and $f_y = 60,000$ psi. Assume none of the steel is compression steel.

14.11   Investigate the acceptability of the deflection for the cantilever of the beam of Problem 14.10.

14.12   Repeat Problem 14.10 if the beam cross section is 16 × 28 having 4-#9 for positive moment reinforcement and 5-#6 for negative moment reinforcement.

14.13   Investigate the acceptability of the deflection for the cantilever of the beam of Problem 14.12.

14.14   Repeat Problem 14.7, except use the separate creep and shrinkage multiplier procedure instead of ACI method. Assume humidity is 80%, age at loading is 28 days after initial curing period, and duration of sustained load is 5 years.

14.15   Investigate the acceptability of the beam of Problem 14.2 if the beam supports partitions and other nonstructural construction likely to be damaged by large deflection. Consider that 20% of the live load is sustained.

14.16   Repeat Problem 14.15, except use the separate creep and shrinkage multiplier procedure instead of the ACI method. The deflection limit is the same as for ACI Code, however. Assume humidity is 90%, age at loading is 60 days after initial curing period, and duration of sustained load is 5 years.

14.17   Investigate the acceptability of the beam Problem 14.3 if the beam supports partitions, etc., which limit the maximum deflection due to live load plus creep and shrinkage to $L/360$. Assume none of the live load is sustained, the relative humidity is 50%, age at loading is 20 days after initial curing, and sustained load will be in place for 1 year.

    (a) Use ACI method.

(b) Use separate creep and shrinkage multiplier procedure.

(c) Use combined creep and shrinkage multiplier procedure.

14.18 Investigate the acceptability of the beam of Problem 14.4 if the beam supports partitions, etc., which limit the maximum deflection due to live load plus creep and shrinkage to $L/480$. Assume none of the live load is sustained, the relative humidity is 90%, age at loading is 30 days after initial curing period, and duration of sustained load is 5 years or more.

(a) Use ACI method.

(b) Use separate creep and shrinkage multiplier procedure.

14.19 Investigate the acceptability of the beam of Problem 14.5 if the live load plus creep and shrinkage deflection is limited to $L/250$ and 10% of the live load is considered sustained. Assume relative humidity is 50%, age at loading is 10 days after initial curing period, and duration of sustained loading is 1 year.

(a) Use ACI Code method.

(b) Use separate creep and shrinkage multiplier procedure.

14.20 Investigate the acceptability of the beam of Problem 14.6 if the beam supports partitions and other nonstructural construction likely to be damaged by large deflections. Assume that 10% of the live load is sustained.

14.21 Investigate the acceptability of an interior 45 ft span (see the accompanying figure) of a continuous T-section with regard to deflection. The service load moments at midspan are 40 ft-kips dead load and 100 ft-kips live load; and at both supports are 50 ft-kips dead load and 114 ft-kips live load. Assume that the $L/480$ limit of ACI-Table 9.5b applies and that none of the live load should be considered as sustained. Use $f'_c = 3500$ psi $(n = 8.5)$ and $f_y = 60,000$ psi.

End sections
Regions of negative moment

Center section
Region of positive moment

Problem 14.21

14.22 Use the span-to-depth ratio approach described in Section 14.11 to check the deflection for Problem 14.8.

14.23 Use the span-to-depth ratio approach described in Section 14.11 to check the deflection for Problem 14.10.

CHAPTER 15

Length Effects
on Columns

15.1 GENERAL

In the basic treatment of compression members in Chapter 13, the assumption was made that the effects of buckling and lateral deflection on strength were small enough to be neglected. Short compression members—that is, those having a low slenderness ratio L/r (L = column height and r = radius of gyration = $\sqrt{I/A}$)—when overloaded experience a material failure (crushing of concrete) prior to reaching a buckling mode of failure. Further, the lateral deflections of short compression members subjected to bending moments are small, thus contributing little secondary bending moment $P\Delta$ as shown in Fig. 15.1.1. It is these buckling and deflection effects that reduce the strength of a compression member below the value computed according to the principles of Chapter 13.

The adoption of the strength method for design, along with the use of higher strength steel and concrete, has led to the increased use of slender members. A stocky member having an L/r less than 20 will achieve essentially the strength discussed in Chapter 13, whereas a member having L/r greater than about 70 will have a considerable reduction in strength, both due to the likelihood of buckling and because of secondary bending moment. To permit the greatest flexibility in structural design, specifications must provide for adequate determination of strength with any slenderness ratio. Thus the provisions of ACI-10.10 and 10.11 take into account the length effects on long compression members.

In computing r, ACI-10.11.3 endoreses a simple value for the rectangular column having b as the width and h as the depth

$$r = \sqrt{\frac{I}{A}} = \sqrt{\frac{\frac{1}{12}bh^3}{bh}} = 0.288h \approx 0.30h$$

Long columns, Denver, Colorado. (Photo by C. G. Salmon.)

Figure 15.1.1 Primary and secondary moment for beam-columns.

and for the circular column having h as the diameter

$$r = \sqrt{\frac{I}{A}} = \sqrt{\frac{\pi h^4 (4)}{64 \pi h^2}} = 0.25h$$

These values actually should be slightly larger due to the effect of the reinforcement.

In the ACI Code, the evaluation of the effect of slenderness may be approximated by using the moment magnifier approach, whereby the sum of the primary and secondary moments (Fig. 15.1.1) is treated as being equal to the product of the primary moment and a magnification factor δ. The general idea relating to this approach is derivable from the differential equation of the beam-column.

In the next several sections, the general concepts relating to the effect of slenderness on the strength of compression members are presented.

15.2 BUCKLING OF CONCENTRICALLY LOADED COLUMNS

Over 200 years ago Leonhard Euler derived the well-known Euler formula [15.1] for concentrically loaded columns stressed below the proportional limit. Engesser [15.2] in 1889 proposed the tangent modulus modification of the Euler formula. In 1910, von Kármán [15.3] performed a series of careful tests verifying Engesser's assumptions. The tangent modulus formula has now been accepted as representing the lower bound for buckling strength of concentrically loaded columns,

$$P_c = \frac{\pi^2 E_t I}{(kL_u)^2} \tag{15.2.1}$$

where

P_c = buckling load

E_t = tangent modulus of elasticity of concrete at the buckling load

I = moment of inertia of the effective section

kL_u = equivalent pin-end length (L_u = actual unbraced length)

Although von Kármán was the first to use a rational analytical method for the inelastic buckling of long slender columns, his work did not consider reinforced concrete.

The fact that concentrically loaded columns rarely, if ever, exist in reinforced concrete structures led investigators [15.4–15.22] during the past 25 years to focus attention on the interaction of long columns with beams in frame structures, resulting in the more rational provisions for length effects on compression members beginning with the 1971 ACI Code.

The basic design limitation of a maximum axial strength equal to 80 or 85% of the concentric capacity P_0 (see Fig. 13.11.2) means, of course, that from a practical viewpoint the concentrically loaded column is considered not to exist. However, to help the reader understand the effect of the slenderness ratio on the behavior of beam-columns over the entire range from $P_n = P_0$ with $M_n = 0$ to $P_n = 0$ with $M_n = M_0$ (see Fig. 13.6.1), the limiting case of the concentrically loaded column will be considered first.

In order to apply Eq. (15.2.1), a realistic expression for E_t of concrete must be used. Since buckling may occur at practically any value of concrete strain, it is necessary to know as accurately as possible the stresses at all strain levels.

The idealized stress–strain diagram for steel was shown in Chapter 3 (Fig. 3.2.2), where the modulus of elasticity is taken at 29,000,000 psi. One of the more realistic stress–strain diagrams for concrete in compression is that of Hognestad [13.23] shown in Fig. 15.2.1 in which the initial modulus of elasticity for concrete is taken as

$$E_c = 1,800,000 + 500f_c'' \quad \text{psi} \tag{15.2.2}$$

in which $f_c'' = 0.85f_c'$. Such a stress–strain relationship may be used along with the assumptions that (1) concrete resists no tensile stress, (2) linear strains exists across the section, and (3) the deflected shape is part of a sine wave. On the basis of these assumptions, evaluation of Eq. (15.2.1) gives the typical column strength curves for concentric loading, such as those of Fig. 15.2.2.

A study of Fig. 15.2.2 shows that in the curves for $f_y = 40,000$ psi there occurs a flat leveling off (such as portion BC of Fig. 15.2.2) of the curve, indicating yielding of the steel with a sudden drop in E_s from 29×10^6 psi to zero. In such cases where $\epsilon_y < \epsilon_0$, the concrete may still increase the capacity (such as portion AB in Fig. 15.2.2) with an increased strain up to ϵ_0. For $f_y = 50,000$ psi and $f_c' = 4000$ psi, the strains $\epsilon_y = 0.00173$ and $\epsilon_0 = 0.00194$ are nearly equal so that little increase above the plateau of yielding in the steel can occur. The effect of creep on long-time loading may be noted by the crosshatched region.

Figure 15.2.1 Hognestad's stress–strain diagram for flexure (Reference 13.23).

Figure 15.2.2 Strength curves for reinforced concrete ($f'_c = 4000$ psi) (28 MPa) concentrically loaded pin-end columns (adapted from Reference 15.23).

■ **EXAMPLE 15.2.1** Calculate the ordinate and abscissa of points A, B, and C of Fig. 15.2.2. Use $f'_c = 4000$ psi, $f_y = 40,000$ psi, and the steel area of $A_s = 0.02bh$ ($0.01bh$ in each face located at $0.45h$ from the center). Use the stress–strain diagrams of Figs. 15.2.1 and 3.2.2.

Solution: The basic quantities of the concrete stress–strain diagram are computed first:

$$f''_c = 0.85f'_c = 0.85(4) = 3.4 \text{ ksi}$$

$$E_c = 1800 + 500f''_c = 1800 + 500(3.4) = 3500 \text{ ksi}$$

$$\epsilon_0 = \frac{2f''_c}{E_c} = 2\left(\frac{3.4}{3500}\right) = 0.00194$$

$$\epsilon_y = \frac{f_y}{E_s} = \frac{40}{29{,}000} = 0.00138$$

(a) Point A, maximum strength of section according to the principles of Chapter 13 (upper limit, $\epsilon_c = \epsilon_0 > \epsilon_y$),

$$P_n = 0.85f'_c bh + A_s f_y$$

without correcting for the displaced concrete.

$$P_n = 0.85f'_c bh + 0.02bh(40)$$

$$= 0.85f'_c bh\left[1 + \frac{0.02(40)}{0.85(4)}\right]$$

$$\frac{P_n}{0.85f'_c bh} = 1.235 \quad \text{(ordinate of point } A\text{)}$$

(b) Point B, $\epsilon_c = \epsilon_y = 0.00138 < \epsilon_0$, $E_s = 0$.

$$f_c = f''_c\left[2\left(\frac{\epsilon}{\epsilon_0}\right) - \left(\frac{\epsilon}{\epsilon_0}\right)^2\right]$$

$$= 3.4\left[2\left(\frac{1.38}{1.94}\right) - \left(\frac{1.38}{1.94}\right)^2\right]$$

$$= 3.4[2(0.712) - 0.508] = 3.12 \text{ ksi}$$

$$P_n = f_c bh + A_s f_y = 3.12bh + 0.02bh(40) = 3.92bh$$

$$\frac{P_n}{0.85f'_c bh} = \frac{3.92}{3.4} = 1.153 \quad \text{(ordinate of point } B\text{)}$$

Applying Eq. (15.2.1) using for I the uncracked transformed section,

$$E_t = \frac{df_c}{d\epsilon} = f''_c\left(\frac{2}{\epsilon_0} - \frac{2\epsilon}{\epsilon_0^2}\right) = E_c - \frac{E_c \epsilon}{\epsilon_0}$$

$$= 3500\left(1 - \frac{1.38}{1.94}\right) = 1008 \text{ ksi}$$

$$I = \frac{bh^3}{12} + 0.02bh\left(\frac{E_s}{E_t}\right)(0.45h)^2$$

$$= bh^3\left[\frac{1}{12} + 0.02\left(\frac{0}{1008}\right)(0.45)^2\right] = 0.0833bh^3$$

$$(kL_u)^2 = \frac{\pi^2 E_t I}{P_c} \qquad\qquad [15.2.1]$$

$$(kL_u)^2 = \frac{\pi^2(1008)(0.0833bh^3)}{3.92bh} = 212h^2$$

$$\frac{kL_u}{h} = 14.55$$

For a rectangular section, $r = h/\sqrt{12}$,

$$\frac{kL_u}{r} = 14.55\sqrt{12} = 50.3 \quad \text{(abscissa of point } B\text{)}$$

(c) Point C, ϵ_c at an infinitesimal amount less than ϵ_y; $E = 29,000$ ksi.

$$I = bh^3\left[\frac{1}{12} + 0.02\left(\frac{29,000}{1008}\right)(0.45)^2\right]$$

$$= bh^3(0.0833 + 0.1167) = 0.2000bh^3$$

$$(kL_u)^2 = \frac{\pi^2(1008)(0.200bh^3)}{3.92bh} = 507h^2$$

$$\frac{kL_u}{h} = 22.5$$

$$\frac{kL_u}{r} = 22.5\sqrt{12} = 78 \quad \text{(abscissa of point } C\text{)} \quad \blacksquare$$

15.3 EQUIVALENT PIN-END LENGTHS

For conditions other than pin ends where the factor k in Eq. (15.2.1) is 1.0, the equivalent pin-end length (also called *effective* length) factor k must be determined for various rotational and translational end restraint conditions. Where translation at both ends is adequately prevented, the distance between points of inflection is shown in Fig. 15.3.1. For all such cases the equivalent pin-end length is less than the actual unbraced length (i.e., k is less than one).

If sidesway or joint translation is possible, as in the case of the unbraced frame, the equivalent pin-end length exceeds the actual unbraced length (i.e., k is greater than one), as shown in Fig. 15.3.2.

(a) End rotations unrestrained

(b) End rotations fully restrained

(c) One end restrained, other unrestrained

(d) Partially restrained at each end

Figure 15.3.1 Equivalent pin-end (i.e., effective) lengths; no joint translation.

As reinforced concrete columns are in general part of a larger frame, it is necessary to understand the concepts of a *braced frame* (where joint translation is prevented by rigid bracing, shear walls, or attachment to an adjoining structure) and the *unbraced frame* (where stability is dependent on the stiffness of the beams and

(a) End rotation fully restrained

(b) One end rotation fully restrained, other unrestrained

(c) One end rotation partially restrained, other end unrestrained

Figure 15.3.2 Equivalent pin-end (i.e., effective) lengths; joint translation possible.

(a) Braced frame, hinged base

(b) Unbraced frame, hinged base

(c) Braced frame, fixed base

(d) Unbraced frame, fixed base

Figure 15.3.3 Equivalent pin-end (i.e., effective) lengths for frames.

columns that constitute the frame). As shown in Figs. 15.3.3(a) and (c), the effective length kL_u for cases where joint translation is prevented may never exceed the actual length L_u. In an unbraced frame [Fig. 15.3.3(b) and (d)] instability results in a sidesway type of buckling with the effective length kL_u always exceeding the actual length L_u. Values for effective length factor k are discussed in Section 15.11.

15.4 MOMENT MAGNIFICATION—SIMPLIFIED TREATMENT FOR MEMBERS IN SINGLE CURVATURE WITHOUT END TRANSLATION (i.e., NO SIDESWAY)

As stated previously, nearly all compression members are simultaneously subjected to some bending moment that causes lateral deflections. Any deflected compression member is further subjected to a secondary bending moment $P\Delta$, as shown in Fig. 15.1.1. One may consider this as a magnification of the applied bending moment. An approximate determination of the magnifying effect may be made by considering the member as finally achieving a deflection Δ_{max}, which is composed of the deflection Δ_0 due to the primary applied bending moment and the additional deflection Δ_1 due to the secondary moment from axial compression (see Fig. 15.4.1). It may be assumed

Figure 15.4.1 Primary and secondary bending moment.

that the secondary bending moment takes the shape of a sine curve (very nearly exact for members with no end restraint and whose primary bending moment and deflection are both maximum at midspan). The midspan deflection Δ_1 equals the moment of the $M/(EI)$ diagram (for secondary bending moment) between the support and midspan taken about the support, according to the moment area principle. Thus

$$\Delta_1 = \frac{P}{EI}(\Delta_0 + \Delta_1)\left(\frac{L}{2}\right)\frac{2}{\pi}\left(\frac{L}{\pi}\right) = (\Delta_0 + \Delta_1)\frac{PL^2}{\pi^2 EI} \tag{15.4.1}$$

from which

$$\Delta_1 = \Delta_0\left[\frac{PL^2/(\pi^2 EI)}{1 - PL^2/(\pi^2 EI)}\right] = \Delta_0\left(\frac{\alpha}{1 - \alpha}\right) \tag{15.4.2}$$

where $\alpha = PL^2/(\pi^2 EI)$. Since $\Delta_{\max}$ is the sum of Δ_0 and Δ_1,

$$\Delta_{\max} = \Delta_0 + \Delta_1 = \Delta_0 + \Delta_0\left(\frac{\alpha}{1 - \alpha}\right) = \frac{\Delta_0}{1 - \alpha} \tag{15.4.3}$$

The maximum bending moment, including the effect of axial load, becomes

$$M_{\max} = M_m + P\Delta_{\max} \tag{15.4.4}$$

Substituting the expression for $\Delta_{\max}$ of Eq. (15.4.3) and making $P = \alpha\pi^2 EI/L^2$, Eq. (15.4.4) becomes

$$M_{\max} = M_m\left(\frac{C_m}{1 - \alpha}\right) = M_m\delta \tag{15.4.5}$$

where

$$\delta = \frac{C_m}{1 - \alpha} = \text{magnification factor} \qquad (15.4.6)$$

and

$$C_m = 1 + \left(\frac{\pi^2 EI\Delta_0}{M_m L^2} - 1\right)\alpha \qquad (15.4.7)$$

Thus, for common cases of single curvature deflection, the magnification factor to be applied to the primary bending moment is equal to $C_m/(1 - \alpha)$. For Cases 1 to 7 shown in Table 15.4.1, rigorous solution of the differential equation may be

Table 15.4.1 Suggested Values of C_m for Common Situations with No Joint Translation

	CASE	C_m (POSITIVE MOMENT)	C_m (NEGATIVE MOMENT)	PRIMARY BENDING MOMENT
1	M, P ... M, P	$1 + 0.2\alpha$	—	$+$ M_m
2	P w P	1.0	—	M_m $+$
3	P ⊢$L/2$⊣ Q P	$1 - 0.2\alpha$	—	M_m $+$
4	P w P	$1 - 0.3\alpha$	$1 - 0.4\alpha$	M_m $+$ $-$
5	P w P	$1 - 0.2\alpha$	$1 - 0.4\alpha$	$-$ M_m $+$ $-$
6	P ⊢$L/2$⊣ Q P	1.0	$1 - 0.3\alpha$	M_m $+$ $-$
7	P ⊢$L/2$⊣ Q P	$1 - 0.2\alpha$	$1 - 0.2\alpha$	$-$ M_m $+$ $-$
8	P q P M_A $M_A \neq M_B$ M_B	Eq. (15.4.7)	not available	M_m $+$

made including the term Py in the expression for $EI\ (d^2y/dz^2)$. The approximate values of C_m for positive moment shown in this table for Cases 1–3 are computed using Eq. (15.4.7) and they are in general agreement with theoretical results. Derivations for approximate values of C_m for negative moment are presented by Iwankiw [15.24]. Positive moments for the restrained end cases (4–7) are given to correlate with theoretical solutions [15.24]. Inherent in the theoretical solutions for these restrained cases is the assumption that the slope at a restrained support is *zero*, the practicality of which can be argued. One may note that for all cases this C_m value will be close to 1.0, because in actual concrete structures α rarely exceeds about 0.3. The approximate treatment of slenderness in ACI-10.11.5 conservatively requires that C_m shall be taken as 1.0 for all cases with transverse loading between supports.

15.5 MOMENT MAGNIFICATION—MEMBERS SUBJECT TO END MOMENTS ONLY; NO JOINT TRANSLATION

Consider the general case shown in Fig. 15.5.1 wherein the end moments M_1 and M_2 constitute the primary bending moment M_i which is a function of z. The sum of primary and secondary moments causes the member to have a deflection y which gives rise to the secondary moment Py. Stating the total moment M_z at the section z of Fig. 15.5.1 gives

$$M_z = M_i + Py = -EI\frac{d^2y}{dz^2} \qquad (15.5.1)$$

for sections with constant EI, and dividing by EI gives

$$\frac{d^2y}{dz^2} + \frac{P}{EI}y = -\frac{M_i}{EI} \qquad (15.5.2)$$

For design purposes, the general expression for moment M_z is of greater importance than the deflection y. Differentiating Eq. (15.5.2) twice gives

$$\frac{d^4y}{dz^4} + \frac{P}{EI}\frac{d^2y}{dz^2} = -\frac{1}{EI}\frac{d^2M_i}{dz^2} \qquad (15.5.3)$$

Figure 15.5.1 Beam-column having end moments without transverse loading.

From Eq. (15.5.1),

$$\frac{d^2y}{dz^2} = -\frac{M_z}{EI} \quad \text{and} \quad \frac{d^4y}{dz^4} = -\frac{1}{EI}\frac{d^2M_z}{dz^2}$$

Substitution into Eq. (15.5.3) gives

$$-\frac{1}{EI}\frac{d^2M_z}{dz^2} + \frac{P}{EI}\left(-\frac{M_z}{EI}\right) = -\frac{1}{EI}\frac{d^2M_i}{dz^2}$$

Simplifying and letting $\lambda^2 = P/EI$,

$$\frac{d^2M_z}{dz^2} + \lambda^2 M_z = \frac{d^2M_i}{dz^2} \tag{15.5.4}$$

which is the same form as the deflection differential equation, Eq. (15.5.2).

The homogeneous solution for Eq. (15.5.4) is

$$M_z = A \sin \lambda z + B \cos \lambda z \tag{15.5.5}$$

To this must be added the particular solution that will satisfy the right-hand side of the differential equation. In the special case of unequal end moments acting without transverse loading,

$$M_i = M_1 + \frac{M_2 - M_1}{L} z \tag{15.5.6}$$

Since

$$\frac{d^2M_i}{dz^2} = 0$$

Eq. (15.5.4) becomes a homogeneous equation, in which case Eq. (15.5.5) represents the entire solution.

In order to determine the maximum moment,

$$\frac{dM_z}{dz} = 0 = A\lambda \cos \lambda z - B\lambda \sin \lambda z \tag{15.5.7}$$

or

$$\tan \lambda z = \frac{A}{B} \tag{15.5.8}$$

At maximum M_z,

$$\sin \lambda z = \frac{A}{\sqrt{A^2 + B^2}} \qquad \cos \lambda z = \frac{B}{\sqrt{A^2 + B^2}} \tag{15.5.9}$$

Substitution of Eq. (15.5.9) in Eq. (15.5.5) gives

$$M_{max} = \frac{A^2}{\sqrt{A^2 + B^2}} + \frac{B^2}{\sqrt{A^2 + B^2}}$$

$$= \sqrt{A^2 + B^2} \tag{15.5.10}$$

Now the constants A and B are evaluated by applying the boundary conditions to Eq. (15.5.5). The conditions are

(1) at $z = 0$, $M_z = M_1$

$$\therefore B = M_1$$

(2) at $z = L$, $M_z = M_2$

$$M_z = A \sin \lambda L + M_1 \cos \lambda L$$

$$\therefore A = \frac{M_2 - M_1 \cos \lambda L}{\sin \lambda L}$$

so that

$$M_z = \left(\frac{M_2 - M_1 \cos \lambda L}{\sin \lambda L} \right) \sin \lambda z + M_1 \cos \lambda z \qquad \textbf{(15.5.11)}$$

and

$$M_{max} = \sqrt{\left(\frac{M_2 - M_1 \cos \lambda L}{\sin \lambda L} \right)^2 + M_1^2}$$

$$= M_2 \sqrt{\frac{1 - 2(M_1/M_2) \cos \lambda L + (M_1/M_2)^2}{\sin^2 \lambda L}} \qquad \textbf{(15.5.12)}$$

For the general case of a beam-column subject to end moments, the maximum moment may be either (1) the larger end moment M_2 at the braced (supported) location, [Fig. 15.5.2(a)] or (2) the magnified moment given by Eq. (15.5.12) that occurs at a variable location out along the span [Fig. 15.5.2(b)], depending on the ratio M_1/M_2 and the value of α, where α is the ratio of P to the critical load $\pi^2 EI/L^2$, making $\lambda L = \pi \sqrt{\alpha}$. In order to investigate the strength of a beam-column, one needs to know whether the maximum moment occurs at a location between the supports, and if so, the correct *distance*. To eliminate the need for such information, the concept of equivalent uniform moment [Fig. 15.5.2(c)] is used. Thus, for the case with unequal end moments, use of the equivalent moment assumes that M_{max} is at midspan.

To establish the equivalent moment, let $M_1 = M_2 = M_{equiv}$ in Eq. (15.5.12),

$$M_{max} = M_{equiv} \sqrt{\frac{2(1 - \cos \lambda L)}{\sin^2 \lambda L}} \qquad \textbf{(15.5.13)}$$

Equating Eq. (15.5.12) and Eq. (15.5.13) gives

$$M_{equiv} = M_2 \sqrt{\frac{(M_1/M_2)^2 - 2(M_1/M_2) \cos \lambda L + 1}{2(1 - \cos \lambda L)}} \qquad \textbf{(15.5.14)}$$

According to the procedure of Section 15.4, the approximate expression for maximum moment was shown to be

$$M_{max} = M_m \delta = M_m \left(\frac{C_m}{1 - \alpha} \right) \qquad \textbf{(15.5.15)}$$

(a) Max moment at end

(b) Max moment *not* at end

(c) Equivalent uniform moment with max magnified moment at midspan

Figure 15.5.2 Combined primary and secondary bending moment diagrams for beam-columns having end moments without transverse loading.

For the case of uniform moment ($M_1 = M_2 = M_{equiv}$),

$$\Delta_0 = \frac{M_{equiv}L^2}{8EI}$$

$$M_m = M_{equiv}$$

$$C_m = 1 + \left[\left(\frac{\pi^2 EI}{L^2}\right)\frac{M_{equiv}L^2}{8EIM_{equiv}} - 1\right]\alpha \approx 1$$

Thus

$$M_{max} = M_{equiv}\left(\frac{1}{1-\alpha}\right) \tag{15.5.16}$$

Substitution of Eq. (15.5.14) into Eq. (15.5.16) gives

$$M_{max} = M_2\left(\frac{C_m}{1-\alpha}\right) \tag{15.5.17}$$

in which

$$C_m = \sqrt{\frac{(M_1/M_2)^2 - 2(M_1/M_2)\cos\lambda L + 1}{2(1 - \cos\lambda L)}} \tag{15.5.18}$$

Comparing Eq. (15.5.17) with Eq. (15.5.16), $C_m M_2$ may be considered to be the equivalent uniform moment along the span.

Equation (15.5.18) assumes that the strength of the beam-column is limited by excessive deflection *in the plane of bending*. Also, it does not fully cover the double-curvature cases where M_1/M_2 lies between -0.5 and -1.0. The actual failure mode of members bent in double curvature with such bending moment ratios is generally one of "unwinding" from double to single curvature in a sudden type of buckling.

Massonnet [15.25] and the LRFD steel design specification of AISC [2.18] have suggested expressions for C_m to be used for design in place of Eq. (15.5.18); ACI-10.11.5.3 has adopted the expression long used by AISC. The comparison of Eq. (15.5.18) with the recommendations of Massonnet and the ACI expression is shown in Fig. 15.5.3. The reader should note that for a given value of α, the curve shown terminates when the moment M_2 at the end of the member exceeds the magnified moment. The straight line adopted by ACI-10.11.5.3 falls near the upper limit for

Figure 15.5.3 Comparison of theoretical C_m with design recommendations for members subject to end moments only, without joint translation.

C_m at any given bending moment ratio, and thus seems to be a realistic and simple approximation. In the ACI strength method α is computed by using for P the required nominal strength $P_n = P_u/\phi$. Thus ACI-10.11.5.3, Formula (10-12), for members braced against sidesway and without transverse loads between supports, is

$$C_m = 0.6 + 0.4\,\frac{M_{1b}}{M_{2b}} \geq 0.4 \qquad (15.5.19)$$

where the additional subscript b is used to denote that these moments are those acting on a "braced" compression member.

The 1986 AISC *LRFD Specification* [2.18] has removed the 0.4 lower limit on the basis that such a limit "was intended to apply to lateral-torsional buckling not second order in plane bending strength." [2.18, Commentary H1]. The practical situation is that when C_m is 0.4 (or less if it were permitted) the moment at the end of the member governs over the magnified moment out in the span. Thus, having the 0.4 lower limit has essentially no practical effect.

15.6 MEMBERS WITH SIDESWAY POSSIBLE—UNBRACED FRAMES

As shown in Figs. 15.3.2 and 15.3.3(b) and (d), an unbraced frame (i.e., a frame subject to sidesway) must rely on the interaction of its beams and columns to limit the horizontal displacement resulting from lateral loads, such as wind or earthquake. Under lateral loads, a "braced" frame will resist the lateral force by such components as diagonal bracing or shear walls, so that any lateral distortion will be small.

In order to explain the concept of the unbraced frame, refer to Fig. 15.6.1(a) where the moments M_{1s} and M_{2s} (the subscript "s" denotes "sway") are the moments resulting from a first-order analysis, and from equilibrium,

$$M_{1s} + M_{2s} = H_u h_s \qquad (15.6.1)$$

The model of Fig. 15.6.1(a) acted upon by H_u deflects relatively by the amount Δ_{1u}, which means the gravity load ΣP_u is acting with an eccentricity Δ_{1u}. This eccentricity in turn increases the lateral load moment $H_u h_s$ by the amount $\Sigma P_u \Delta_{1u}$. Since the total moment now acting is $H_u h_s + \Sigma P_u \Delta_{1u}$ the relative lateral deflection will finally be Δ_{2u} (larger than Δ_{1u}) when the structure reaches equilibrium in the final displaced position, as shown in Fig. 15.6.1(b). Note that the subscripts "1" and "2" in Δ_{1u} and Δ_{2u} represent correctly the meanings of "first-order" and "second-order" analysis. The total sum of moments at the column ends (primary plus secondary) then may be expressed as

$$\delta_s(M_{1s} + M_{2s}) = H_u h_s + \Sigma P_u \Delta_{2u} \qquad (15.6.2)$$

As shown by the right side of Eq. (15.6.2), the total moment can be thought of as the sum of the primary plus secondary part, or as on the left side it can be thought of as the primary moment magnified by the factor δ_s. Detailed treatment of braced and unbraced elastic frames is available elsewhere [15.26, 15.27].

In a braced system, the secondary bending moments $P\Delta$ from any small amount of sidesway (the P–Δ effect) may ordinarily be neglected. However, for unbraced

(a)   First-order analysis: For equilibrium, $M_{1s} + M_{2s} = H_u h_s$

(b)   Second-order analysis: For equilibrium, $\delta_s (M_{1s} + M_{2s}) = H_u h_s + \Sigma P_u \Delta_{2u}$

Figure 15.6.1 Summation of forces acting on all columns in one story of a multistory building frame.

frames, the relatively larger sidesway deflection Δ gives secondary moments that must be provided for in design. Thus, an unbraced frame requires an analysis to accomplish the following tasks:

1. Provide strength under factored loads to resist gravity load, neglecting any sidesway effect except in rare cases of unbalanced loading or unsymmetrical structural configuration as to cause "appreciable" sidesway. ACI Commentary 10.11.5.1 indicates that "appreciable" would be where the calculated sway Δ (from a first-order frame analysis) divided by the clear height L_u exceeds 1/1500. Randomly occurring construction inaccuracies cause columns to be

"out-of-plumb"; that is, the centroid of the top of the column is not directly over the centroid of the bottom of the column. MacGregor [15.28] indicates that Δ/h of 1/450 may be expected due to out-of-plumbness, and further that out-of-plumbness typically increases first-order moments by about $6\frac{1}{2}\%$; whereas a second-order analysis performed on a structure where out-of-plumbness was neglected increases first-order moments by only about $4\frac{1}{2}\%$. In other words, out-of-plumbness typically has more effect than the P–Δ effect from a second-order analysis.

2. Provide strength under factored loads to resist lateral load (i.e., factored wind or earthquake load). The moments under lateral load include the primary moments from first-order analysis plus secondary moments due to the P–Δ effect.

3. Provide stiffness such that the relative horizontal deflection between adjacent floors, and for the entire frame, is within specified limits (usually, say, equal to the clear height L_u divided by 400 or 500).

MacGregor and Hage [15.29] have provided an excellent summary of the various methods and have suggested practical approaches to accomplish the above tasks.

In general, the primary moments M_{1s} and M_{2s} are obtained from a first-order elastic analysis of the unbraced frame. ACI-10.11.5.1 uses M_{2s} to be the "larger factored end moment on a compression member due to loads which result in appreciable sidesway ...."

As shown by Eq. (15.6.2), there are two ways by which magnified column end moments $\delta_s M_{1s}$ and $\delta_s M_{2s}$ may be obtained:

1. Moment magnifier method. As previously described for braced frames, the magnified moment $\delta_s M_{2s}$ may be treated as the primary moment M_{2s} multiplied by a magnification factor δ_s in accordance with ACI-10.11.5.1. This method is described in Sections 15.11 and 15.12.

2. Second-order analysis. Alternatively, the magnified moments $\delta_s M_{1s}$ and $\delta_s M_{2s}$ may be computed directly from a second-order analysis of the structure in accordance with ACI-10.10.1. This procedure involves utilization of the final displaced position [the Δ_{2u} position of Fig. 15.6.1(b)] of the structure in the frame analysis. A rigorous solution may involve either several applications of the first-order analysis wherein the lateral load $H_u h_s$ is incremented by $\Sigma P_u \Delta_{1u}$ each time, or use a bona fide second-order analysis computer program [15.30] wherein reduction in the relative sidesway resistance for the appropriate elements is used in the global stiffness matrix.

No matter which way the "P–Δ effect" is accounted for, a lateral load analysis must be performed to obtain at least the primary moments.

15.7 INTERACTION DIAGRAMS—EFFECT OF SLENDERNESS

In order to understand the ACI Code procedure and its approximations as discussed in later sections of this chapter, the general approach is described for determining a point on the P_n–M_n interaction diagram for kL/r not equal to zero. In Chapter 13

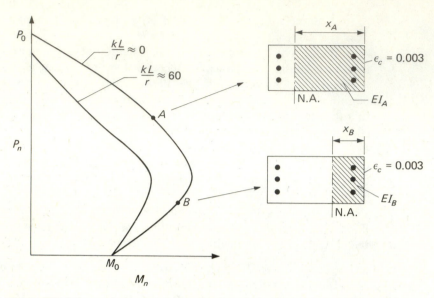

Figure 15.7.1 Beam-column strength interaction diagram, including effective cross section for moment of inertia.

the basic strength of a section with zero kL/r was treated, giving an interaction diagram such as Fig. 13.6.1, and designated by $kL/r = 0$ in Fig. 15.7.1.

Points A and B represent combinations of P_n and M_n with the neutral axes (NA) located at x_A and x_B, respectively, from the extreme compression fiber whose crushing strain is taken as the ACI Code prescribed value 0.003. In the following development it will be shown that when kL/r is not zero, the total nominal strength M_n to resist primary and secondary moment may be achieved when the strain at the extreme compression fiber is less than 0.003. The curve labeled $kL/r \approx 60$ in Fig. 15.7.1 represents a typical strength interaction curve that includes slenderness effects.

Pfrang and Siess [15.6], Pfrang [15.12], and MacGregor, Breen, and Pfrang [15.31] have provided excellent discussions of the slenderness effects on interaction diagrams. A long column may fail in one of two ways: (1) it may fail by reaching a combined $P_n–M_n$ that exceeds the cross-section strength computed by the methods of Chapter 13; (2) it may fail by instability when an infinitesimal increase in axial load results in additional deflection such that equilibrium cannot be achieved.

Referring to Fig. 15.7.2(a), the member of large slenderness, say $kL/r = 100$, will generally follow the loading path up to point D where the material strength is reached. Point D is on the short column ($kL/r = 0$) interaction diagram but is at a smaller axial load P_{long} than it would be [P_{short} in Fig. 15.7.2(a)] if kL/r were actually zero. If the column fails by instability, it would follow the path (dashed) up to point E; that is, it would be unable to reach the material strength interaction diagram (for $kL/r = 0$).

Generally, columns in braced frames are capable of achieving a "material failure," while the "instability failure" is not common but may occur in unbraced frames.

Figure 15.7.2 Slender column interaction diagrams (adapted from Reference 15.31).

In the construction of interaction diagrams, as shown in Fig. 15.7.2b, a material failure occurring at D on the $kL/r = 0$ curve due to an axial load P_1 plus a magnified moment $M_m\delta$ (equal to $M_m + P_1\Delta_{max}$) may be plotted *for the particular loading arrangement* at point C on the primary moment radial line. Whatever the primary moment loading arrangement—such as equal end eccentricities of axial load, unequal end eccentricities of axial load, or lateral transverse loading—the deflection of the member will differ so that the secondary bending moment $P\Delta$ will differ. This means that different types of primary moments will cause a member of $kL/r = 100$ to follow a curved path to intersect the $kL/r = 0$ material strength interaction diagram at different locations such as point D. The radial line through point C is a function of only the primary moment, which is the same for all slenderness ratios.

The development of a correct interaction diagram for members of large slenderness, such as Fig. 15.7.2(b), would require an elaborate analysis for each structure taking into account such factors as the following: (1) a realistic moment–curvature relationship; (2) the time-dependent and cracking effects on deflections; and (3) the influence of axial load on the flexural stiffness of members. Mockry and Darwin [15.32] have made such an analysis and prepared a practical set of design charts. Parme [15.33] has treated the subject in a similar manner and the Portland Cement Association has published design aids [15.34].

15.8 ACI CODE—GENERAL ANALYSIS METHOD

The ACI Code gives the highest priority to a general frame analysis taking into account the "influence of axial loads and variable moment of inertia on member stiffness and fixed-end moments, effect of deflections on moments and forces, and the effects of duration of loads" (ACI-10.10.1). Whether a frame is braced or unbraced, the general analysis including the second-order effects gives "a better approximation to the real moments" than does the use of the magnifier approach (ACI Commentary-R10.10.1).

The following requirements are suggested by the ACI Commentary-R.10.10.1 as minimum for an adequate rational analysis.

1. Realistic moment–curvature or moment–end rotation relationships should be used to provide accurate values of deflections and secondary moments (P–Δ moments). Since column design is based on strength and stability under factored loads, the stiffnesses used (even though in an elastic analysis) should be consistent with the strength limit state. ACI Commentary accepts as satisfactory the following:

$$\text{column } EI = E_c I_g (0.2 + 1.2 \rho_g E_s / E_c) \tag{15.8.1}$$

$$\text{beam } EI = 0.5 E_c I_g \tag{15.8.2}$$

These are as suggested by MacGregor, Oelhafen, and Hage [15.35].

2. The effect of foundation rotations on the lateral deformations should be considered.

3. The influence of axial load on the flexural stiffness must be considered for very slender columns when L_u/r exceeds 45. For instance, the flexural stiffness of a prismatic member without axial load is $4EI/L$. When axial compression is present, the 4 becomes reduced. The general idea relating to the effect of axial compression on flexural stiffness of elastic members can be found in Salmon and Johnson [15.26].

4. The effects of creep should be considered, particularly in frames subject to sustained lateral loads, such as a building resisting a horizontal reaction from an arch or an unbalanced horizontal earth force. Creep must also be considered in frames where unbalanced dead load gives rise to differential shortening of the two sides of a building resulting in lateral deflection.

5. The maximum moment in the compression member should be determined considering effects of the lateral deflections of the frame and the deflection of the compression member itself. This means that a second-order analysis must be made to include the P–Δ effects. The methods of MacGregor and Hage [15.29], Wood, Beaulieu, and Adams [15.36, 15.37], and Scholz [15.38, 15.39] all are directed at making the second-order analysis, using various degrees of precision.

15.9 ACI CODE—MOMENT MAGNIFIER METHOD FOR BRACED FRAMES

It has been shown in Sections 15.4 and 15.5 that the maximum moment in an elastic beam-column is given by Eq. (15.4.4)

$$M_{\max} = M_m + P\Delta_{\max} = M_m + \frac{P\Delta_0}{1 - \alpha} \tag{15.4.4}$$

where $\alpha = P/P_c$ and $P_c = \pi^2 EI/L^2$. Furthermore, Eqs. (15.4.5) and (15.5.17), applicable to braced frames, indicate that the maximum moment may also be expressed as the maximum primary moment M_m times a magnification factor δ_b (where the subscript "b" denotes "braced").

$$M_{\max} = \left(\frac{C_m}{1 - \alpha}\right) M_m \qquad (15.9.1)$$

$$= \delta_b M_m \qquad (15.9.2)$$

A number of studies have shown [15.18, 15.31, 15.40] that this approach is acceptable for reinforced concrete compression members using the strength design method. Design of the members is based on the factored axial load P_u combined with the corresponding factored moment M_m from an elastic frame analysis magnified by the factor δ_b. M_m is the maximum moment acting on the member and it may occur at either end, or if there is transverse loading, in the midspan region.

The required nominal strength of the designed section must be

$$P_n = \frac{P_u}{\phi} \quad \text{and} \quad M_n = \frac{M_m}{\phi}\left(\frac{C_m}{1 - P_n/P_c}\right) \qquad (15.9.3)$$

For practical purposes the strength reduction factor ϕ in the denominator may be treated as shown, or as a multiplier on the other side of the equation. However, since P_n appears in the magnifier, the expression for δ_b contains ϕ in any case,

$$\delta_b = \frac{C_m}{1 - \dfrac{P_u}{\phi P_c}} \geq 1.0 \qquad (15.9.4)$$

which is ACI Formula (10-7).

Factor C_m. The quantity C_m has two basic meanings: (1) for braced frames with transverse loading and single-curvature deflection, it is truly a part of the moment magnifier; and (2) for braced frames with end moments alone acting, the factor C_m is really not part of the magnifier—rather $C_m M_m$ gives an equivalent uniform moment which is then magnified by multiplying by $1/[1 - P_u/(\phi P_c)]$. For the first meaning of C_m, given by Eq. (15.4.6), ACI-10.11.5.3 states "C_m shall be taken as 1.0." This is a conservative approach since the correct C_m will usually be between 0.9 and 1.0. Thus the use of C_m under ACI-10.11.5 may be summarized as follows for *braced* frames:

   1. Transverse loading,

$$C_m = 1.0 \qquad (15.9.5)$$

   2. End moments only, use Eq. (15.5.19)

$$C_m = 0.6 + 0.4 \frac{M_{1b}}{M_{2b}} \geq 0.4 \qquad (15.9.6)$$

which is ACI Formula (10-12). Note that M_{2b} is larger than M_{1b} and the ratio M_{1b}/M_{2b} is positive when the member is bent in single curvature.

Stiffness Parameter _EI_. The other quantity required for evaluating the moment magnifier δ_b is

$$P_c = \frac{\pi^2 EI}{(kL_u)^2} \tag{15.9.7}$$

where L used in Sections 15.4 and 15.5 becomes kL_u, which is the unsupported length L_u of the member in a reinforced concrete frame modified for the pin-end condition by the effective length factor k.

The principal difficulty with the magnifier method is that it requires a value for EI, which varies due to cracking, time-dependent effects, and nonlinearity of the concrete stress–strain curve. MacGregor, Breen, and Pfrang [15.31] proposed to use the larger of two simple expressions _when more precise values are not available_. These appear in ACI-10.11.5.2 as Formulas (10-10) and (10-11), respectively,

$$EI = \frac{0.2E_c I_g + E_s I_{se}}{1 + \beta_d} \tag{15.9.8}$$

or

$$EI = \frac{0.4E_c I_g}{1 + \beta_d} \tag{15.9.9}$$

where

E_c = concrete modulus of elasticity = $57{,}000\sqrt{f_c'}$ for normal-weight concrete (ACI-8.5)

I_g = gross moment of inertia of concrete section, neglecting reinforcement

I_{se} = moment of inertia of reinforcement

β_d = proportion of the factored _axial load_ that is considered sustained so as to contribute to time-dependent deformations (usually factored axial dead load to total factored axial load)

The larger of Eqs. (15.9.8) and (15.9.9) is appropriate for use and is still an underestimate of the correct EI. Note that the β_d definition was revised in 1989 to reflect that creep is more related to axial load than to bending moment in beam-columns. This clarifies that $\beta_d = 0$ should be used in a lateral load analysis for wind or earthquake alone acting on a structure. However, in the rare situation when _sustained_ lateral load is acting on a story of an unbraced frame, instead of zero β_d would be the ratio of the maximum factored sustained lateral load to the maximum total factored lateral load in that story.

The relative accuracy of these EI expressions is shown in Fig. 15.9.1 from Ref. 15.31. The theoretical values were for the situation of no sustained load ($\beta_d = 0$) [15.31]. In the study of Eqs. (15.9.8) and (15.9.9), EI values for about 100 cases were estimated using theoretical load–moment–curvature diagrams, for columns of various dimensions, strengths, and steel percentages. Effective EI values were also computed for the University of Texas frame tests [15.5, 15.8, 15.9], and for a series of frames simulated by the computer.

Figure 15.9.1 Comparison of equations for EI with EI values from moment-curvature diagrams, for short-duration loading ($\beta_d = 0$) (adapted from Reference 15.31).

Alternate expressions for EI have been proposed by MacGregor, Oelhafen, and Hage [15.35] and Medland and Taylor [15.41]. As a result of these studies it seemed appropriate to divide only the concrete term in Eq. (15.9.8) by the factor $(1 + \beta_d)$. However, when this is done, the number of cases where the ratio of theoretical EI to formula computed EI is less than 1.0 becomes three times more than otherwise, particularly when the reinforcement ratios are low. Thus, ACI Committee 318 concluded that the $(1 + \beta_d)$ factor should be retained as the divisor for both the steel and concrete terms. Recent studies of flexural stiffness expressions for EI include those of Diaz and Roesset [15.42] and Mirza [15.43] for rectangular sections, and those of Sigmon and Ahmad [15.44] and Ehsani and Alameddine [15.45] for circular sections.

The sustained load effect on slender columns has also been studied by Goyal and Jackson [15.46], Drysdale and Huggins [15.47], and Rangan [15.48].

15.10 ACI CODE—MOMENT MAGNIFIER METHOD FOR UNBRACED FRAMES

In recognition that the general analysis prescribed in ACI-10.10 is practical only with the aid of a sophisticated second-order analysis computer program, ACI-10.10.2 permits the approximate moment magnifier method in accordance with ACI-10.11. Such a method requires a first-order "conventional elastic frame analysis" using factored loads. Design of a compression member in a frame involves treating the member as an isolated one acted upon by factored axial force P_u and factored end moments M_{u1} and M_{u2}, where the subscript 2 indicates the larger magnitude end moment. It is noted again that values of P_u, M_{u1}, and M_{u2} are results of a first-order elastic analysis using factored loads. In strength design, the factored axial force P_u should be considered to act in combination with a magnified factored moment M_c defined by ACI Formula (10-6),

$$M_c = \delta_b M_{2b} + \delta_s M_{2s} \qquad (15.10.1)$$

As has been stated previously, the subscript "b" refers to "braced," that is, moments resulting from "no appreciable" sidesway; and the subscript "s" refers to "sway," that is, moments resulting from a sidesway analysis. As defined in ACI-10.0,

δ_b = moment magnification factor for frames braced against sidesway to reflect effects of member curvature between ends of a compression member

δ_s = moment magnification factor for frames not braced against sidesway to reflect lateral drift resulting from lateral and gravity loads

M_{2b} = value of larger factored end moment on compression member due to loads which result in no appreciable sidesway, calculated by conventional elastic frame analysis

M_{2s} = value of larger factored end moment on compression member due to loads which result in appreciable sidesway, calculated by conventional elastic frame analysis

This approach of separating the braced and unbraced portions of the frame action is based on the recommendations of Ford, Chang, and Breen [15.49]. The first-order analysis of an unbraced structure subject to gravity plus wind, for instance, would involve making two analyses—one for gravity load alone, and a second for lateral load alone. For the gravity load alone acting on a typical relatively symmetrical frame, the sidesway effect may be considered negligible. ACI Commentary-R10.11.5.1 states, "Unless the lateral deflection due to vertical load is appreciable, the minor effect of this sway component can be neglected and the corresponding moments can be considered as nonsway moments . . ." The word "appreciable" is indicated to mean approximately lateral sway to story height of 1/1500.

The reader is reminded that in the *braced* system, the magnified moment *may* occur at the end of the member or at an unknown location away from the ends. The gravity moment M_g must therefore be magnified by the braced frame magnifier δ_b. To conservatively allow for the separate magnification effects in the case of combined gravity and lateral loads, ACI Formula (10-6), Eq. (15.10.1), may for practical purposes be written as

$$M_c = \delta_b M_g + \delta_s M_{2s} \qquad (15.10.2)$$

according to ACI Commentary-R10.11.5.1, Eq. (4).

The magnifier δ_b for the braced frame has been described as Eq. (15.9.4); the magnifier δ_s for the unbraced frame is

$$\delta_s = \frac{1}{1 - \dfrac{\Sigma P_u}{\phi \Sigma P_c}} \geq 1.0 \qquad (15.10.3)$$

in which

$$P_c = \frac{\pi^2 EI}{(kL_u)^2} \qquad (15.10.4)$$

and ΣP_u and ΣP_c are the summations for all columns resisting sway in a story.

An approximate procedure in which a quantity Q_u is used in place of $(\Sigma P_u/\phi)/\Sigma P_c$ is suggested by MacGregor and Hage [15.29] and described in Section 15.12.

15.11 ALIGNMENT CHARTS FOR EFFECTIVE LENGTH FACTOR *k*

The most commonly used procedure for obtaining the equivalent pin-end (effective) length is to use the alignment charts from the Structural Stability Research Council Guide [15.27], originally developed by O. J. Julian and L. S. Lawrence, and presented in detail by T. C. Kavanagh [15.50]. The charts are shown in Fig. 15.11.1.

The effective length factor k is a function of the end restraint factors ψ_A and ψ_B, for the top and bottom joints at the ends of the column respectively, defined as

$$\psi = \frac{\Sigma EI/L \text{ for column members in the plane of bending}}{\Sigma EI/L \text{ for beam members in the plane of bending}} \qquad \textbf{(15.11.1)}$$

which for a hinged end gives $\psi = \infty$ and for a fixed end, $\psi = 0$. Since a frictionless hinge cannot exist in practical construction, ψ is to be taken equal to 10 for an end assumed as hinged in the analysis.

One nomogram (or alignment chart), Fig. 15.11.1(a), is for braced frames where sidesway (joint translation) is prevented, and the other, Fig. 15.11.1(b), is for the unbraced frame where sidesway is possible, being restrained only by the stiffness of interacting beams and columns.

An effective length procedure has been adopted by ACI-10.11 in the approximate evaluation of slenderness effects. The alignment charts are implicitly endorsed for determining the k factor by their inclusion in the ACI Commentary (ACI-R10.11.2).

The assumptions inherent in the development of the alignment chart for the braced frame [Fig. 15.11.1(a)] are as follows [15.50]:

1. All columns reach their respective critical loads simultaneously.
2. The structure is assumed to consist of symmetrical rectangular frames.
3. At any joint, the restraining moment provided by the beams is distributed among the columns in proportion to their stiffnesses.
4. The beams are elastically restrained at their ends by the columns, and at the onset of buckling the rotations of the beam at its ends are equal and opposite (i.e., the beams are deflected in single curvature). See Fig. 15.3.3(a).
5. The beams carry no axial loads.

For the unbraced frame alignment chart [Fig. 15.11.1(b)], assumptions (1) through (3) and (5) are unchanged; however, the beams are assumed to be deflected in double curvature, where the rotations of the ends are equal in magnitude and direction. See Fig. 15.3.3(b).

By means of the alignment charts one may determine the k factor for a column of constant cross section in a multistory, multibay frame. With steel frames where the material is homogeneous and isotropic, the modulus of elasticity E is constant for all members, and the moment of inertia I is computed for the gross cross section. In

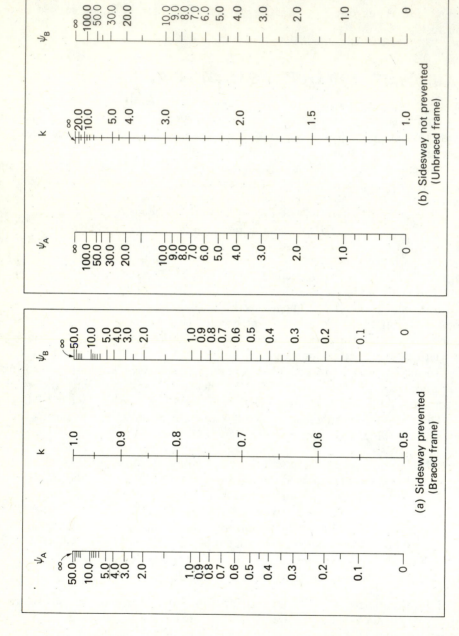

Figure 15.11.1 Alignment charts for effective length factor for columns in continuous frames (from Reference 15.50), where

$$\psi = \frac{\Sigma EI/L, \text{ columns}}{\Sigma EI/L, \text{ beams}}$$

reinforced concrete, E varies with concrete strength and magnitude of loading, while I also varies depending on the degree of cracking and the reinforcement percentage. ACI-10.11.2.2 requires that the effective length factor k for the *unbraced* frame "shall be determined with due consideration of effects of cracking and reinforcement on relative stiffness, and shall be greater than 1.0." For the *braced* frame, ACI-10.11.2.1 merely says the effective length factor "shall be taken as 1.0, unless analysis shows that a lower value is justified." It is believed that an appropriate use of the alignment chart would constitute an "analysis" as required by the ACI Code.

Thus, for the purpose of evaluating the end restraint factor ψ, cracked section moment of inertia should be used for the beams whereas gross moment of inertia is probably satisfactory for the columns [15.51] (or ACI Formula 10-10 with $\beta_d = 0$, as discussed in Section 15.9). Recognition of different behavior in beams and columns is particularly necessary when the reinforcement ratio ρ is significantly different, such as with $\rho_g = 0.08$ in the columns and $\rho = 0.005$ in the beams. The ACI Commentary-R10.11.2 indicates that when kL_u/r is less than 60, the use of $0.5I_g$ as an approximation of the cracked transformed section moment of inertia for the beam "... will usually result in reasonable member sizes for columns ..."

As an alternative to actually using the nomograms of Figure 15.11.1, some approximate formulas for the effective length factor k have been proposed and are endorsed by the ACI Commentary.

For members in *braced* frames, the 1972 British Code of Standard Practice [15.52] suggests that an upper bound for k is obtained by using the *smaller* of the following two equations:

$$k = 0.7 + 0.05(\psi_A + \psi_B) \le 1.0 \qquad (15.11.2)$$

$$k = 0.85 + 0.05\psi_{\min} \le 1.0 \qquad (15.11.3)$$

where ψ_A and ψ_B are the ψ values at the two ends of the member, and $\psi_{\min}$ is the smaller of the two values.

For members in *unbraced* frames, Furlong [15.53] proposed for members restrained at both ends:

When $\psi_{\text{avg}} < 2$ (i.e., high end restraint),

$$k = \frac{20 - \psi_{\text{avg}}}{20} \sqrt{1 + \psi_{\text{avg}}} \qquad (15.11.4)$$

When $\psi_{\text{avg}} \ge 2$ (i.e., moderate to low end restraint),

$$k = 0.9\sqrt{1 + \psi_{\text{avg}}} \qquad (15.11.5)$$

where ψ_{avg} is the average of the ψ values at the two ends of the member. Equations (15.11.4) and (15.11.5) give k values that are within 2% of those obtained by the nomograms.

For members in *unbraced* frames, when hinged at one end, the British Code of Standard Practice [15.52] proposes

$$k = 2.0 + 0.3\psi \qquad (15.11.6)$$

where ψ is the value at the restrained end.

15.12 MacGREGOR–HAGE MOMENT MAGNIFIER METHOD

Even in the moment magnifier method of ACI-10.11, a first-order analysis is required. The details of such an analysis are not self-explanatory, particularly for the unbraced frame. MacGregor and Hage [15.29] have presented a procedure for the unbraced frame that provides not only the moments needed for the magnifier method but also examines indirectly the *stability of the frame* in terms of the magnitude of a parameter Q_u.

The expression $(\Sigma P_u/\phi)/\Sigma P_c$ in Eq. (15.10.3) may be substituted by the quantity Q_u, where

$$Q_u = \frac{\Sigma P_u \Delta_{1u}}{H_u h_s}$$

(15.12.1)

and therefore

$$\delta_s = \frac{1}{1 - Q_u}$$

(15.12.2)

The derivation of Eqs. (15.12.1) and (15.12.2) may be made as follows. Referring to Figs. 15.6.1(a) and (b) and Eqs. (15.6.1) and (15.6.2), and using a proportionality factor η, let

$$\Delta_{1u} = \eta H_u$$

(15.12.3)

which is valid for linear first-order analysis. The equivalent magnified lateral load in Fig. 15.6.1(b) may be taken as

$$\text{equivalent lateral load} = H_u + \frac{\Sigma P_u \Delta_{2u}}{h_s}$$

(15.12.4)

Thus,

$$\Delta_{2u} = \eta \,(\text{equivalent lateral load})$$

$$= \eta\left(H_u + \frac{\Sigma P_u \Delta_{2u}}{h_s}\right)$$

$$= \Delta_{1u} + \frac{\Delta_{1u} \Sigma P_u \Delta_{2u}}{H_u h_s}$$

$$= \Delta_{1u} + Q_u \Delta_{2u}$$

from which

$$\Delta_{2u} = \frac{\Delta_{1u}}{1 - Q_u}$$

(15.12.5)

Comparing Eq. (15.6.2) with Eq. (15.6.1),

$$\delta_s = \frac{H_u h_s + \Sigma P_u \Delta_{2u}}{H_u h_s}$$

(15.12.6)

Substituting Eq. (15.12.5) into Eq. (15.12.6) gives

$$\delta_s = 1 + \frac{\Sigma P_u}{H_u h_s}\frac{\Delta_{1u}}{1 - Q_u} = 1 + \frac{Q_u}{1 - Q_u} = \frac{1}{1 - Q_u}$$

(15.12.7)

Using Eqs. (15.12.1) and (15.12.2) as the basic formulas, MacGregor and Hage have suggested the following steps in the application of the ACI moment magnifier method:

1. Perform a first-order analysis for *service* lateral loads to obtain unfactored sway moments. Apply factors from ACI-9.2 appropriate for lateral load to obtain factored lateral load moments M_s. In this step no gravity loads are acting. Compute Δ_{1s}/h_s due to *service* load for each story and for the overall frame, and compare with the desired limit if any is desired.
2. Perform a factored gravity load analysis to obtain moments M_g.
3. Compute Q_u for each story:

$$Q_u = \frac{\Sigma P_u \Delta_{1u}}{H_u h_s} \qquad [15.12.1]$$

in which ΣP_u is the sum of the factored axial loads for all columns in the story obtained from Step 2 above, H_u is the factored lateral force at the story, and Δ_{1u} is the relative factored load deflection in the story. Note that Δ_{1u} and H_u are *both* for the same load level, either service load or factored load. For Q_u exceeding 0.04 but less than 0.2,

$$\delta_s = \frac{1}{1 - Q_u} \qquad [15.12.2]$$

4. Combine the effects of gravity and lateral loads for the design of the member.

$$P_u = \text{factored gravity load on member}$$

$$M_c = \delta_b M_g + \delta_s M_{2s}$$

where

$$\delta_b = \text{Eq. (15.9.4)}$$
$$M_g = \text{moments from Step 2}$$
$$\delta_s = \text{Eq. (15.12.2)}$$
$$M_{2s} = \text{larger of the end moments on the member, computed in Step 1}$$
$$\text{and increased by the appropriate factors from ACI-9.2}$$

15.13 MacGREGOR-HAGE AMPLIFIED LATERAL LOAD METHOD

In this method, the lateral loads are amplified (magnified) before the first-order structural analysis is performed, and the resulting moments are used directly in the design. The following steps may be followed:

1. Establish whether the frame is to be treated as *braced* or *unbraced*. According to 1983 ACI Commentary-10.11.2, par. 4, a compression member in a story may be considered braced when the following is satisfied within the story:

$$\frac{\Sigma K \text{ of walls or lateral bracing}}{\Sigma K \text{ of columns in parallel frames}} \geq 6 \qquad (15.13.1)$$

where $K = $ stiffness of element.

2. Establish the maximum Δ_{2s}/h_s (say 1/400 or 1/500) that can be tolerated in any story under *service* load. The deflection Δ_{2s} is the total including second-order deflection. [MacGregor and Hage [15.29] used Δ_s/h. It was not clear whether Δ_s meant Δ_{1s} or Δ_{2s}; however, any limitation Δ_s/h established by the designer would likely be one related to the *total* (including second-order) deflection.] There are no prescribed ACI Code limits on Δ_s/h_s.

3. Estimate the drift index Δ_{2u}/h_s at factored loads. Δ_{2u} would be the total primary plus secondary deflection (Δ_{2u} of Fig. 15.6.1). MacGregor and Hage [15.29] have indicated that if $Q_u \leq 0.2$ the service load deflections (Δ_{1s} or Δ_{2s}) will be about 40% of the factored load deflections (Δ_{1u} or Δ_{2u}). Further, Δ/h_s (either Δ_{1s} or Δ_{2s}) for an entire multistory frame will be 85% or less of the maximum sway (first- or second-order, respectively) in one story. Thus, the Δ_{2u}/h_s at factored load may be estimated for the overall frame to be

$$\text{estimated } \frac{\Delta_{2u}}{h_s} \approx \left[\left(\frac{\Delta_{2s}}{h_s} \right)_{\substack{\text{service} \\ \text{load limit}}} \right] \left(\frac{0.85}{0.40} \right)$$

$$= 2.125 \left[\left(\frac{\Delta_{2s}}{h_s} \right)_{\substack{\text{service} \\ \text{load limit}}} \right] \qquad \textbf{(15.13.2)}$$

4. Compute the estimated Q_u for the overall frame, and for each story,

$$Q_u = \frac{\Sigma P_u}{H_u} \left(\frac{\Delta_{2u}}{h_s} \right) \qquad \textbf{(15.13.3)}$$

where

Δ_{2u}/h_s = the value from Eq. (15.13.2) for the overall frame and 1/0.85 of that value for each story

H_u = factored lateral force within the story

h_s = story height

ΣP_u = sum of the factored gravity loads acting on the columns of the story

The value of Q_u must not exceed 0.2.

5. Compute the approximate magnified lateral loads. The magnified lateral forces H_{2u} to use in the lateral load analysis are, following the concept of Eq. (15.12.7),

$$H_{2u} = H_u \left(\frac{1}{1 - Q_u} \right) \qquad \textbf{(15.13.4)}$$

6. Apply the magnified factored lateral forces H_{2u} (omitting all gravity load) and perform a first-order analysis. For moment of inertia I, the use of $0.4I_g$ for beams and $0.8I_g$ for columns is suggested [15.29]. Obtain the moments $\delta_s M_s$ directly from the analysis, as well as the total deflection Δ_{2u} at each story level.

7. Check that Δ_{2u}/h_s computed for the entire frame does not exceed the limit given by Eq. (15.13.2), and that Δ_{2u}/h_s for any story does not exceed 1/0.85 times the value given by Eq. (15.13.2). If any of the limits is exceeded the structure must be stiffened.
8. Perform an elastic analysis under factored gravity load. Obtain the moments M_g.
9. Obtain the factored moments for the design of the member.

$$M_c = \delta_b M_g + \delta_s M_{2s} \qquad\qquad \text{[15.10.2]}$$

where

$$\delta_b = \frac{C_m}{1 - \dfrac{P_u}{\phi P_c}} \geq 1.0$$

$\delta_s M_{2s}$ = values from analysis in Step 6

M_g = factored gravity load moments from Step 8

P_u = factored gravity axial compression in member

15.14 FURLONG'S RATIONAL METHOD FOR UNBRACED FRAMES

Furlong [15.54] has proposed guidelines for a rational analysis that is intended to satisfy the intent of ACI-10.10.1 within the complexity implied by the five requirements of the ACI Commentary referred to previously in Section 15.8 for the general analysis. The recommended procedure applies to *unbraced* frames. For such frames, the primary slenderness effects are the products of the column load P_u times the sway deflection Δ.

Furlong indicates that lateral displacements may be obtained from either a second-order analysis or an iterative first-order analysis that uses a "dummy horizontal load" to give the second-order result, provided the appropriate "model" for the structure is used. The effects of the actual nonlinear moment–curvature behavior may be accounted for by the use in the model of prismatic member *EI* values that give moment–rotation characteristics *no stiffer* than real members after concrete cracks and steel yields prior to nominal strength being reached. This will be accomplished by following ACI Commentary recommendations for *EI* values, as given by Eqs. (15.8.1) and (15.8.2).

The Furlong recommended [15.54] procedure for *unbraced* frames is as follows:

1. Estimate member sizes and compute *EI* values for each component of the frame, using $0.5E_c I_g$ for beams and $0.3E_c I_g$ for columns [Eq. (15.8.1) gives about $0.3E_c I_g$ for $\rho_g = 0.01$].
2. Perform factored gravity load analyses for the following cases:
   (a) $1.4D + 1.7L$ in every span of every floor
   (b) $1.4D + 1.7L$ in alternate spans on every floor, beginning with the end span
   (c) $1.4D + 1.7L$ in alternate spans on every floor, beginning with the second span from end

When live load reductions are appropriate for multistory frames or for large floor areas (see for example, ANSI [2.10, Section 4.7]), the reduced live load is to be used in loading cases (a), (b), and (c).

3. Compute factored axial loads acting on all columns at each floor level using $1.4D + 1.7L$, where L is the *total* live load tributory to each column.

4. Perform lateral load analysis, including second-order analysis, using factored wind ($1.7W$) for horizontal forces and nodal point factored gravity loads obtained in Step 3. There are iterative methods (such as that of Wood, Beaulieu, and Adams [15.36]) for determining the cumulative amount of additional moments so that a first-order analysis procedure could be used.

5. Select beam reinforcement for the more severe of the following:
   (a) Factored gravity load moments from Step 2
   (b) 0.75 (factored gravity load moments from Step 2 + factored lateral load moments from Step 4)
   This latter combination corresponds to ACI Formula (9-2).

6. Select column reinforcement for the more severe of the following:
   (a) Factored axial forces from Step 3 + factored gravity load moments from the most severe condition of Step 2
   (b) 0.75 (Step 3 factored axial forces + Step 4 lateral load factored moments + Step 2 gravity load factored moments)

7. Examine individual columns for possible slenderness effect strength reduction as a *braced* member. For every interior column having a ratio L_u/h greater than 8, and every exterior column having a ratio L_u/h greater than 12, compute and use moment magnification factor δ_b for the factored gravity load moments of Step 2.

8. The overall stiffness of the unbraced frame may be evaluated by *estimating* that 25% of the factored load horizontal deflection (lateral drift) computed in Step 4 will be the traditional service load drift computed using gross properties. If 25% of the factored load drift exceeds the desired limit of 1/400 to 1/500, the members should probably be stiffened. There are no ACI Code serviceability limits on drift.

15.15 MINIMUM ECCENTRICITY IN DESIGN

Braced Frames. When computations indicate only a small moment to be acting on a member such that the eccentricity M_u/P_u is less than $(0.6 + 0.03h)$ in., the primary moment M_m in Eq. (15.9.2) is to be computed as $P_u(0.6 + 0.03h)^*$ (ACI-10.11.5.4). Although ACI-10.11.5.4 refers only to *end* eccentricities, the authors believe the intent is to use this code-stated minimum primary moment as the basis for applying the magnifier whether the small primary moment actually acts at the end of a member or at some point between the ends.

----

* For SI, ACI 318M–89 gives

$$e_{min} = (15 + 0.03h) \text{ mm}$$

Further, when computed *end* moments are small (or zero) for a *braced* frame member having only end moments acting, ACI-10.11.5.4 specifies determining the ratio M_{1b}/M_{2b} for the C_m equation by either of the following:

1. When actual computed end moments give eccentricities less than a minimum of $(0.6 + 0.03h)$ in., such actual end moments may be used for M_{1b}/M_{2b} in Eq. (15.9.6).

2. If computations indicate essentially no moment at both ends—that is, the member is considered to be axially loaded—single curvature is to be assumed with $C_m = 1$.

Unbraced Frames. When computations indicate that both end moments M_{1s} and M_{2s} are less than those based on an end eccentricity $e_{\min}$ of $(0.6 + 0.03h)$ in., M_{2s} in ACI Formula (10-6), Eqs. (15.10.1) or (15.10.2) must be based on $e_{\min} = (0.6 + 0.03h)$ in.*

When the minimum eccentricity requirements control, the provisions are intended to be applied to bending about only one axis at a time, *not* as a case of biaxial bending.

15.16 BIAXIAL BENDING AND AXIAL COMPRESSION

For compression members subject to bending about both principal axes, the moment about *each* axis is to be magnified by the factor δ which is computed from the restraint conditions for that axis (ACI-10.11.7). In general, the effective length factor k, the stiffness factor EI, and C_m may differ for each bending axis. The member may also be considered as part of a braced system in one direction and unbraced in the other. Additional data on biaxial bending of slender members may be found in References 15.47 and 15.55, as well as References 13.52 to 13.78.

15.17 ACI CODE—SLENDERNESS RATIO LIMITATIONS

Although the slenderness ratio is never zero in actual structures, there are certain limits for kL_u/r below which the reduction in strength may reasonably be neglected. The ACI Code provisions are based on the assumption that a strength loss of up to 5% can be tolerated without the designer having to consider the slenderness effect; thus a significant number of ordinary columns can be designed considering only the provisions of Chapter 13.

ACI-ASCE Committee 441 surveyed typical reinforced concrete buildings to determine the normal range of variables found in columns of such buildings [15.31]. A great variety of buildings were studied, including towers (braced frames) as high as 33 stories and an unbraced frame 20 stories high. The total number of columns exceeded 20,000. The following results were reported [15.31]. For braced frames, 98% of the columns had L/h less than 12.5 ($L/r \approx 42$) and e/h less than 0.64. For unbraced frames, 98% of the columns had L/h less than 18 ($L/r \approx 60$) and e/h less than 0.84. Furthermore, it was found that the practical upper limit on the slenderness ratio kL/r

---

* For SI, ACI 318M–89 gives

$$e_{\min} = (15 + 0.03h) \text{ min}$$

is about 70 in building columns. In general, these limits for the variables provide a guide to the range of variables that are used in any approximate method.

Taking the idea that attainment of at least 95% of the material strength of a short column is acceptable, the effects of slenderness may be neglected when:

For *braced* frame members (ACI-10.11.4.1),

$$\frac{kL_u}{r} < 34 - 12\,\frac{M_{1b}}{M_{2b}} \tag{15.17.1}$$

where M_{1b} is the smaller and M_{2b} the larger of end moments on the member.

For single curvature cases the ratio M_{1b}/M_{2b} is positive, while for double curvature the ratio is negative. When the member is subject to large transverse loading (other than end moments alone), the ratio M_{1b}/M_{2b} should probably be taken as $+1$.

For *unbraced* frame members (ACI-10.11.4.2),

$$\frac{kL_u}{r} < 22 \tag{15.17.2}$$

For all compression members with kL_u/r exceeding 100, a more elaborate analysis as discussed in Sections 15.8 and 15.12 to 15.14 is required (ACI-10.11.4.3).

Comparison of the slenderness limits of Eqs. (15.17.1) and (15.17.2) with actual columns in existing buildings indicates [15.31] that over 90% of the columns in braced frames and over 40% of the columns in unbraced frames will fall within the limits of those equations and allow neglect of the slenderness effect.

15.18 RESTRAINING EFFECT OF BEAMS

The restraining effect of beams has a major effect on column behavior. The problem is discussed in detail by Pagay, Ferguson, and Breen [15.15]; Okamura, Pagay, Breen, and Ferguson [15.16]; and Breen, MacGregor, and Pfrang [15.31].

Braced Frames.   Consider the portion of a braced frame shown in Fig. 15.18.1. The primary moment M_c on the column, computed from a nominal first-order analysis using gross moment of inertia I_g, depends on the relative stiffnesses of beam and column. The moment M_{ext} applied to the cantilever at A is resisted by the beam and the column; thus

$$M_{ext} = M_b + M_c \tag{15.18.1}$$

Since the column deflects, there will be an additional moment $P\Delta$ on it, such that

$$M_{max} = M_c + P\Delta \tag{15.18.2}$$

Solving for M_c in Eq. (15.18.1) and substituting into Eq. (15.18.2) gives

$$M_{max} = M_{ext} - M_b + P\Delta \tag{15.18.3}$$

As the column deflects laterally due to $M_c + P\Delta$, joint A rotates forcing more and more of the applied moment to be resisted by the beam. This effect is further increased by the reduction in column stiffness due to axial load. However, beam deflection due to creep and shrinkage gives the reverse effect by putting moment back on the column.

Figure 15.18.1 Braced frame—restraining effect of beam.

When beam is relatively stiff (high reinforcement ratio ρ) compared to the column (i.e., slender column), the primary moment on the column is overestimated by the nominal analysis and the moment on the beam is underestimated. This is not a problem because the slender column will be conservatively designed, and underestimating the beam end moment will increase the positive moment and also the chance that beam deflection will control.

When the beam is relatively flexible (low reinforcement ratio ρ) compared to the column (i.e., short stiff column), the primary moment on the column is underestimated by the nominal analysis and the beam end moment is overestimated. In this case the column could be significantly underdesigned. For the beam design, it makes little difference because the beam having low ρ is ductile and redistribution of moments (see Section 10.12) can occur. In other words, on the ductile beam it is not too essential whether the moment strength is somewhat larger at the support or at midspan as long as the total load is capable of being carried.

A multiplier to increase the design moment on short columns has been suggested by Okamura, Pagay, Breen, and Ferguson [15.16]. For the *single-curvature* case in *braced* frames, the nominal e/h should be multiplied by

$$\text{multiplier} = 1.38 + 5.5(\rho_g - 5.5\rho) \tag{15.18.4}$$

where

ρ_g = ratio of column steel to gross area bh

ρ = ratio of beam tension steel to effective area bd

The relationship of Eq. (15.18.4) was developed for an average end restraint factor ψ of Eq. (15.11.1) for the column equal to 1.0, which corresponds to an effective length factor k about 0.8. It would seem Eq. (15.18.4) could be used for any short column in a braced frame.

Unbraced Frames. In unbraced frames, the beam may be inadequately designed for moment at its junction with the column, when a lateral load such as wind is applied. The beam moment must be equal to the magnified moment on the column. When the moment magnifier method or the general analysis is used, no problem arises as long as one recognizes that the moment, $M_m + P_u\Delta$, or $\delta_b M_{2b} + \delta_s M_{2s}$, acting at the end of the column must be carried by the beams in a manner such that equilibrium of the joint is maintained.

Moment of Inertia for Restraining Beam. Even though gross section has been used for a nominal elastic frame analysis, the adjustment to the nominal e/h as given by Eq. (15.18.4) may be eliminated if cracked transformed section is used for the moment of inertia in computing the end restraint factor ψ of Eq. (15.11.1) [15.15]. In general, more accurate results are obtained by using Eq. (15.9.8) or (15.9.9) with $\beta_d = 0$ for the moment of inertia of the column members and transformed cracked section for the moment of inertia of the beam members in determining the effective length factor k.

15.19 EXAMPLES

In the preceding sections of this chapter, basic concepts underlying the ACI Code provisions relating to length effects on columns have been discussed. Details are given for the moment magnifier method as well as suggestions for a second-order analysis of the unbraced frame. In order to facilitate easy reference to the actual quantitative procedures or formulas, most of which appear in the ACI Code or Commentary, Table 15.19.1 is presented on pages 632 and 633, summarizing the information needed for solving practical problems. For illustration, the following six examples are presented.

■ **EXAMPLE 15.19.1** Determine the adequacy of the interior top floor column (column A) of the *braced* frame of Fig. 15.19.1. The column is 10×10 with 4-#8 and 4-#9 bars ($f_y = 50$ ksi and $f'_c = 3$ ksi) and is to carry a service axial compression of 108 kips live load and 36 kips dead load. The bending moments that may act in combination with the axial load have been computed and found to be negligible. If the member is not adequate, revise the design so that it satisfies the moment magnifier method of the ACI Code.

Solution: **(a)** Determine slenderness ratio. Unless a rational evaluation of end restraint is made, ACI-10.11.2.1 requires taking the effective length factor k for a braced frame equal to 1.0. The radius of gyration may be taken as $0.3h$ according to ACI-10.11.3. The clear height L_u is

$$L_u = 12 - \frac{20}{12} = 10.33 \text{ ft}$$

Then

$$\frac{kL_u}{r} = \frac{1.0(10.33)(12)}{0.3(10)} = 41.3$$

Figure 15.19.1 Rigid frame for Examples 15.19.1 and 15.19.2.

(b) Slenderness ratio limits. Since the end moments are negligible, the minimum eccentricity provisions of ACI-10.11.5.4 govern the design. Accordingly, the deformation should be considered as single curvature with $M_{1b}/M_{2b} = 1.0$. The slenderness limit is

$$\left(\frac{kL_u}{r}\right)_{limit} = 34 - 12\frac{M_{1b}}{M_{2b}} = 22 < 41.3$$

Thus slenderness effects must be considered.

(c) Braced frame moment magnifier δ_b.

$$\delta_b = \frac{C_m}{1 - \dfrac{P_u}{\phi P_c}}$$

where

$C_m = 1.0$ for single-curvature member in braced frame

$P_u = 1.4(36) + 1.7(108) = 234$ kips

$\phi = 0.7$ for this tied column

$P_c = \dfrac{\pi^2 EI}{(kL_u)^2}$

For the stiffness parameter EI using Eqs. (15.9.8) or (15.9.9),

$$E_c = 57,000\sqrt{f'_c} = 3120 \text{ ksi}, \qquad I_g = \frac{10(10)^3}{12} = 833 \text{ in.}^4$$

$$E_s = 29,000 \text{ ksi}, \qquad I_s = 2(2.79)(2.59)^2 = 37.4 \text{ in.}^4$$

$$0.2E_c I_g + E_s I_s = 0.2(3120)(833) + 29,000(37.4)$$
$$= 520,000 + 1,090,000 = 1,610,000 \text{ kip in.}^2$$

$$0.4E_c I_g = 1,040,000 \text{ kip in.}^2$$

Table 15.19.1   Length Effects on Columns—Summary of Useful Formulas

1. Definitions

$$\psi = \frac{\Sigma EI/L \text{ of columns}}{\Sigma EI/L \text{ of beams}},$$

$\psi = 0$ (theoretical fixed end)
$\psi = 1$ (practical fixed end)
$\psi = 10$ (column end supported on footing)
$\psi = \infty$ (theoretical hinged end)   **[15.11.1]**

$$P_c = \frac{\pi^2 EI}{(kL_u)^2}$$ **[15.9.7]**

2. Slenderness Ratio Limitation (No Length Effect)

Braced frames:
$$\frac{kL_u}{r} < 34 - 12\frac{M_{1b}}{M_{2b}}$$ **[15.17.1]**

$|M_{1b}| < |M_{2b}|$; M_{1b}/M_{2b} positive for single curvature

Unbraced frames:
$$\frac{kL_u}{r} < 22$$ **[15.17.2]**

Approximate values of r: $0.30h$ for rectangular columns   **[Sec. 15.1]**
$0.25h$ for circular columns   **[Sec. 15.1]**

3. Stiffness Factor EI in Moment Magnifier Method

Larger of
$\begin{cases} EI = \dfrac{0.2E_cI_g + E_sI_{se}}{1 + \beta_d} & \text{ACI Formula (10-10)} \quad \textbf{[15.9.8]} \\[3mm] EI = \dfrac{0.4E_cI_g}{1 + \beta_d} & \text{ACI Formula (10-11)} \quad \textbf{[15.9.9]} \end{cases}$

$$E_c = 57{,}000\sqrt{f'_c} \qquad \beta_d = \frac{\text{factored sustained axial load}}{\text{factored total axial load}}$$

For sustained lateral load on unbraced frame,

$$\beta_d = \frac{\text{factored sustained lateral load}}{\text{factored total lateral load}}$$

$$EI_{cr} \text{ of beam} \approx \frac{EI_g}{2}\left(\text{for } \frac{kL_u}{r} < 60; \text{ ACI Commentary-R10.11.2}\right)$$ **[Sec. 15.11]**

4. Effective Length Factor k in Moment Magnifier Method
Braced frames:

(a) Nomogram (alignment chart)   **[Fig. 15.11.1(a)]**

(b) 1972 British Code $k = 0.7 + 0.05(\psi_A + \psi_B) \leq 1.0$ **[15.11.2]**
$\left.\vphantom{\begin{matrix}a\\b\end{matrix}}\right\}$ (use smaller)
$k = 0.85 + 0.05\psi_{min} \leq 1.0$ **[15.11.3]**

(c) ACI Code $k = 1.0$ (on safe side)   **[Sec. 15.11]**

Unbraced frames:

(a) Nomogram (alignment chart)   **[Fig. 15.11.1(b)]**

(b) Furlong $k = \dfrac{20 - \psi_{avg}}{20}\sqrt{1 + \psi_{avg}};$ $\psi_{avg} < 2$ **[15.11.4]**

$k = 0.9\sqrt{1 + \psi_{avg}};$ $\psi_{avg} > 2$ **[15.11.5]**

(c) 1972 British Code $k = 2.0 + 0.3\psi;$ one hinged end   **[15.11.6]**

(continued)

Table 15.19.1 (continued)

---

5. C_m *Factor in Moment Magnifier Method for Braced Frames*

Transverse loading, $\quad C_m = 1.0$ [15.9.5]

End moments only, $\quad C_m = 0.6 + 0.4 \dfrac{M_{1b}}{M_{2b}} \geq 0.4$ [15.9.6]

$\qquad |M_{1b}| < |M_{2b}|; \qquad M_{1b}/M_{2b}$ positive for single curvature

---

6. *Moment Magnifier Method*

Braced frames: $\quad M_c = \delta_b M_{2b} \qquad (M_{2b} = $ larger end moment$)$

$\qquad\qquad\qquad M_c = \delta_b M_m \qquad (M_m = $ primary bending moment [15.9.2]
$\qquad\qquad\qquad\qquad\qquad\qquad\qquad$ due to transverse loading$)$

$$\delta_b = \frac{C_m}{1 - \dfrac{P_u}{\phi P_c}} \geq 1.0 \qquad [15.9.4]$$

Unbraced frames: $\quad M_c = \delta_b M_{2b} + \delta_s M_{2s}$ [15.10.1]
$\qquad\qquad\qquad (M_{2b} = $ larger end moment in braced case$)$
$\qquad\qquad\qquad (M_{2s} = $ larger end moment in sway case$)$

$\qquad\qquad\qquad M_c = \delta_b M_g + \delta_s M_{2s}$ [15.10.2]
$\qquad\qquad\qquad (M_g = $ gravity load end moment$)$
$\qquad\qquad\qquad (M_{2s} = $ larger end moment in sway case$)$

$$\delta_s = \frac{1}{1 - \dfrac{\Sigma P_u}{\phi \Sigma P_c}} \geq 1.0 \qquad [15.10.3]$$

---

7. *Second-Order Analysis Method*

Unbraced frames:

$$Q_u = \text{stability index} \approx \frac{\Sigma P_u}{\phi \Sigma P_c}$$

(a) $\quad Q_u = \dfrac{\Sigma P_u \Delta_{1u}}{H_u h_s} \leq 0.04; \qquad \left(\dfrac{\Delta_{1u}}{h_s} = \dfrac{1}{400} \text{ or } \dfrac{1}{500} \right)$ [15.12.1]

Second-order analysis may be neglected.

(b) $\quad 0.04 < Q_u \leq 0.2$

Second-order analysis must be performed.

$$\delta_s = \frac{1}{1 - Q_u} \qquad [15.12.2]$$

(c) Stiffness factor EI
From ACI Commentary:

$\qquad\qquad$ column $EI = E_c I_g (0.2 + 1.2 \rho_g E_s / E_c)$ [15.8.1]

$\qquad\qquad$ beam $EI = 0.5 E_c I_g$ [15.8.2]

---

8. *Restraining Effect of Beams*

Braced frames: $\qquad$ multiplier for $\dfrac{e}{h} = 1.38 + 5.5(\rho_g - 5.5\rho)$ [15.18.4]

$\left(\begin{array}{c} \text{single curvature} \\ \text{short columns} \end{array} \right)$ $\quad \rho_g(\text{column}) = A_{st}/(bh); \; \rho \,(\text{beam}) = A_s/(bd)$

Unbraced frames: $\qquad$ magnified moment at beam end
$\qquad\qquad\qquad\qquad\quad = $ magnified moment at column end

---

The EI values are to be divided by $(1 + \beta_d)$ to account for time-dependent deflection due to creep and shrinkage. Use the larger of the two values of EI, and divide by $(1 + \beta_d)$, where β_d is the proportion of factored axial load that is sustained. In this case,

$$\beta_d = \frac{36(1.4)}{108(1.7) + 36(1.4)} = 0.215$$

$$EI = \frac{1{,}610{,}000}{1 + 0.215} = 1{,}330{,}000 \text{ kips in.}^2$$

$$P_c = \frac{\pi^2(1{,}330{,}000)}{[1.0(10.33)(12)]^2} = 854 \text{ kips}$$

$$\frac{P_u}{\phi P_c} = \frac{234}{0.7(854)} = 0.391$$

$$\delta_b = \frac{1.0}{1 - 0.391} = 1.64$$

In this case, the minimum eccentricity $(0.6 + 0.03h)$ should be magnified so that

$$\text{required } e = 1.64(e_{min}) = 1.64(0.6 + 0.03h) = 0.98 + 0.049h \text{ in.}$$

Note that even the magnified eccentricity might not exceed the eccentricity corresponding to the maximum axial compressive strength of $0.80P_0$ ($P_{n(max)}$ of Fig. 13.15.1). When the magnified e is less than e_{min} of Fig. 13.15.1, there is still no reduction in strength due to the slenderness effect.

(d) Rational analysis for effective length factor k. For the beam the cracked section moment of inertia is recommended. An approximation is $I_{cr} = I_g/2 = 4000 \text{ in.}^4$

$$(EI)_{bm} = E_c I_{cr} = 3120(4000) = 12{,}500{,}000 \text{ kip in.}^2$$

$$(EI)_{col} = 1{,}610{,}000 \text{ kip in.}^2$$

End restraint factors,

$$\psi_A(\text{top}) = \frac{\Sigma EI/L \text{ for cols}}{\Sigma EI/L \text{ for beams}} = \frac{1610/12}{2(12{,}500)/24} = 0.13$$

$$\psi_B(\text{bottom}) = \frac{(1610 + 3230)/12}{2(12{,}500)/24} = 0.39$$

Since the 12×12 column below has not been designed, its EI value is taken as $0.6E_c I_g$, which is approximately the general expression obtained for the 10×10 column by the ACI formula. From Fig. 15.11.1(a), $k = 0.62$. The more correct effective slenderness ratio is

$$\frac{kL_u}{r} = \frac{0.62(10.33)12}{0.3(10)} = 25.6$$

The magnification factor is also affected,

$$P_c = \frac{\pi^2(1,330,000)}{[0.62(10.33)(12)]^2} = 2220 \text{ kips}$$

$$\frac{P_u}{\phi P_c} = \frac{234}{0.7(2220)} = 0.151$$

$$\delta_b = \frac{1.0}{1 - 0.151} = 1.18$$

$$\text{required } e = 1.18(e_{\min}) = 0.71 + 0.035h \text{ in.}$$

In this case the beams are very stiff compared to the columns. Using cracked section for the beams and the ACI EI formula for the columns gave little difference in the result than would be obtained by using gross section.

(e) Check capacity. The strength of the section may be checked by the methods of Chapter 13.

$$\text{required } P_n = \frac{P_u}{\phi} = \frac{234}{0.7} = 334 \text{ kips}$$

$$\text{required } e = 0.71 + 0.035(10) = 1.06 \text{ in. } (0.106h)$$
$$[\text{according to (d) above}]$$

The actual nominal strength P_n at $e = 1.06$ in. is 456 kips (a convenient source is the *ACI Design Handbook* [2.22]). Even when $e = 0.98 + 0.049h = 1.47$ in. as obtained in (c), the nominal strength P_n is 382 kips. So this section is adequate as a braced frame column.

Note that the strength P_n may not be taken in design greater than $0.80P_0$ according to ACI-10.3.5,

$$\begin{aligned}
P_{n(\max)} &= 0.80[0.85f'_c(A_g - A_{st}) + f_y A_{st}] \\
&= 0.80[0.85(3)(100 - 7.16) + 50(7.16)] \\
&= 476 \text{ kips} > 431 \text{ kips} \qquad\qquad \textbf{OK}
\end{aligned}$$

Thus, $P_n = 456$ kips at $e = 1.06$ in. is the correct strength, a situation corresponding to point A of Fig. 13.11.2(b). ∎

■ **EXAMPLE 15.19.2** Repeat Example 15.19.1, except consider the frame as unbraced.

Solution: In general, members in unbraced frames will have end moments on the members. One might assume that this case was the result of gravity load analysis that happened to give negligible column moments. In some cases where beams are unusually stiff, the behavior of the unbraced frame is little different from that of the braced frame.

(a) Effective pin-end length. From part (d) of Example 15.19.1 the end restraint factors are

$$\psi_A(\text{top}) = 0.13 \qquad \psi_B(\text{bottom}) = 0.39$$

From Fig. 15.11.1(b), $k = 1.07$.

(b) Compute magnification factor δ_s for sidesway. From part (c) of Example 15.19.1,

$$EI = 1{,}330{,}000 \text{ kip in.}^2 \quad \text{(for column)}$$

which includes the effect of 21.5% sustained factored load.

$$P_c = \frac{\pi^2 EI}{(kL_u)^2} = \frac{\pi^2(1{,}330{,}000)}{[1.07(10.33)(12)]^2} = 746 \text{ kips}$$

Assuming P_u/P_c is the same for all columns in the story, $\Sigma P_u/\Sigma P_c = P_u/P_c$,

$$\frac{P_u}{\phi P_c} = \frac{234}{0.7(746)} = 0.448$$

Using Eq. (15.10.3) according to ACI-10.11.5.3,

$$\delta_s = \frac{1.0}{1 - \Sigma P_u/(\phi \Sigma P_c)} = \frac{1.0}{1 - 0.448} = 1.81$$

(c) Compute magnified factored moment M_c. ACI-10.11.5.1, or Eq. (15.10.1), prescribes

$$M_c = \delta_b M_{2b} + \delta_s M_{2s} \qquad \qquad \textbf{[15.10.1]}$$

Assuming this is the gravity dead and live load case where the moment $M_{2b} = M_g$ is negligible, M_{2s} may also be considered negligible, and design is nominally based on minimum eccentricity e_{min}. The magnified factored moment M_c is

$$M_c = \delta_s M_{2s} = \delta_s P_u e_{min}$$
$$= 1.81 P_u (0.6 + 0.03h)$$

which for $h = 10$ in. gives

$$M_c = 1.63 P_u$$

$$\text{required } e \text{ (for short column)} = \frac{M_c}{P_u} = 1.63 \text{ in.}$$

The strength ϕP_n for the member with $e = 1.63$ in. is 255 kips, which exceeds the requirement of $P_u = 234$ kips, and is acceptable. ∎

■ **EXAMPLE 15.19.3** Determine the adequacy of the square tied column (17 in. square, with 10-#9 bars, $f'_c = 3000$ psi, $f_y = 40{,}000$ psi) which is an exterior first-floor column in the frame of Fig. 15.19.2. Assume that this frame is braced sufficiently to prevent relative translation of its joints. Also assume 40% of the factored axial load is sustained.

Figure 15.19.2 Rigid frame for Example 15.19.3.

Solution: **(a)** Effective length. In accordance with the more conservative procedure of ACI-10.11.2.1 that allows the option of not doing an analysis to determine a k value less than 1.0 for a braced frame, use

$$kL_u = L_u = 12 - 2 = 10 \text{ ft}$$

(b) Slenderness ratio limits. The actual column slenderness ratio is

$$\frac{kL_u}{r} = \frac{120}{0.3(17)} = 23.6$$

Slenderness effects may be neglected when

$$\frac{kL_u}{r} < 34 - 12\frac{M_{1b}}{M_{2b}}$$

In this case, $M_{1b}/M_{2b} = 0$,

$$\left(\frac{kL_u}{r}\right)_{\text{limit}} = 34 > 23.6$$

Slenderness effects may be neglected. In the following, the method will be illustrated even though it would not be required by the ACI Code.

(c) Braced frame moment magnifier δ_b.

$$E_c = 57\sqrt{f_c'} = 3120 \text{ ksi}$$

$$I_g = \tfrac{1}{12}(17)(17)^3 = 6950 \text{ in.}^4$$

$$I_{se} = 2(5)(6)^2 = 360 \text{ in.}^4$$

Using ACI Formulas (10-10) and (10-11), Eqs. (15.9.8) and (15.9.9),

$$EI = 0.2E_c I_g + E_s I_{se}$$
$$= 0.2(3120)(6950) + 29,000(360)$$
$$= 4,340,000 + 10,440,000 = 14,800,000 \text{ kip in.}^2$$

or

$$EI = 0.4E_c I_g = 8,680,000 \text{ kip in.}^2$$

Using the larger value of EI and applying the factor $(1 + \beta_d)$ to account for sustained load,

$$\frac{EI}{1 + \beta_d} = \frac{14,800,000}{1.40} = 10,500,000 \text{ kip in.}^2$$

$$P_c = \frac{\pi^2 EI}{(kL_u)^2} = \frac{\pi^2(10,500,000)}{[1.0(10.0)(12)]^2} = 7200 \text{ kips}$$

$$P_u = 525 \text{ kips} \quad [\text{from Fig. 15.19.2(c)}]$$

$$\frac{P_u}{\phi P_c} = \frac{525}{0.70(7200)} = 0.104$$

$$C_m = 0.6 + 0.4\frac{M_{1b}}{M_{2b}} = 0.6$$

$$\delta_b = \frac{C_m}{1 - P_u/\phi P_c} = \frac{0.6}{1 - 0.104} = 0.67 < 1.0$$

In this case the magnified moment out in the span is less than that at the braced point. The factored loads to be carried are $P_u = 525$ kips and $M_u = 105$ ft-kips at the top of the column. The strength ϕP_n is found to be 566 kips at $e = 2.4$ in. ■

■ **EXAMPLE 15.19.4** Determine the adequacy of the square tied column (20 in. square, with 12-#10 bars, $f_c' = 4000$ psi, $f_y = 60,000$ psi) which is an exterior first-floor column in the unbraced frame of Fig. 15.19.3. Assume 20% of the factored axial load is sustained. The ΣP_u on the four columns in the lowest story is 3150 kips.

Solution: (a) Factored load combinations. The design of beam-columns usually involves consideration of at least two loading combinations, according to ACI-9.2,

(a) Frame dimensions

(b) Cross section

(c) Member A

Figure 15.19.3 Rigid frame for Example 15.19.4.

as follows:

$$U = 1.4D + 1.7L \qquad \text{ACI Formula (9-1)}$$

$$U = 0.75(1.4D + 1.7L + 1.7W) \quad \text{ACI Formula (9-2)}$$

Assume that for this example, the forces given in Fig. 15.19.3 are from the factored load combination including wind, ACI Formula (9-2); and that wind causes insignificant axial compression (assumed zero here) on the member being studied. The unbraced frame loading from ACI Formula (9-2) is then divided into a *nonsway part* due to $(1.05D + 1.275L)$ having factored first-order loads of $P_u = 525$ kips and $M_{2b} = 70$ ft-kips, and a *sway part* from wind loading alone due to $1.275W$ giving factored first-order loads of $P_u = 0$ kips along with $M_{2s} = 290$ ft-kips. Only the check of strength required by ACI Formula (9-2) is illustrated in this example.

(b) Effective length and slenderness ratio. The end restraint factors ψ must be determined. Assuming the cracked section moment of inertia for the beam to be half of the gross moment of inertia,

$$I_{cr} \approx \frac{I_g}{2} = \frac{14(24)^3/12}{2} = 8070 \text{ in.}^4$$

When the beam reinforcement is known, the actual I_{cr} for the beam should be used here. For the column, either I_g or the larger EI from ACI Formulas (10-10) or (10-11) should be used.

For the 20-in. square column,

$$I_g = \tfrac{1}{12}(20)(20)^3 = 13{,}333 \text{ in.}^4$$

$$I_s = 1.27(2)[4(7.5)^2 + 2(2.5)^2] = 603 \text{ in.}^4$$

$$E_c = 57\sqrt{f'_c} = 57\sqrt{4000} = 3605 \text{ ksi}$$

$$
\begin{aligned}
EI &= 0.2E_cI_g + E_sI_s \\
&= 0.2(3605)(13{,}333) + 29{,}000(603) \\
&= 9{,}610{,}000 + 17{,}500{,}000 \\
&= 27{,}110{,}000 \text{ kip in.}^2 \text{ (larger than } 0.4E_cI_g)
\end{aligned}
$$

The EI for obtaining k is taken *without* the sustained load effect.

For the unbraced frame, the magnifier δ_s involves the *sum* of the Euler loads P_c for all columns participating in the sidesway resistance at the story level. Thus, the effective length factor k is needed for each of these columns. In this example,

$$
\begin{aligned}
\psi_A(\text{top of exterior column}) &= \frac{\Sigma EI/L \text{ for columns}}{\Sigma EI/L \text{ for beams}} \\
&= \frac{2(27{,}110{,}000)/12}{3605(8070)/30} = 4.66
\end{aligned}
$$

$$\psi_A(\text{top of interior column}) = 2.33$$

For this calculation use of center–to–center span distances is recommended as being consistent with the nominal frame analysis using those distances. Theoretically, at the fixed end ψ equals zero; however, the Structural Stability Research Council [15.27] recommends that for practical purposes ψ should not be taken smaller than 1.0. Thus,

$$\psi_B(\text{bottom}) = 1.0$$

Using Fig. 15.11.1(b), find

$$k \text{ of exterior column} = 1.68, \qquad k \text{ of interior column} = 1.50$$

The effective slenderness ratio for the exterior column being investigated is

$$\frac{kL_u}{r} = \frac{1.68(10.0)(12)}{0.3(20)} = 33.6$$

which exceeds the limit of 22 given by ACI-10.11.4.2 for unbraced frames. Slenderness effects must be considered.

(c) Magnification factors. For the unbraced frame, the ACI Code conservatively requires adding the magnified moments *that would exist if the frame were braced* to the frame sidesway moments magnified by the P–Δ effect. Thus, ACI Formula (10-6) states

$$M_c = \delta_b M_{2b} + \delta_s M_{2s} \qquad \text{[15.10.1]}$$

For this example, the end moment M_{2b} under factored gravity load is given as 70 ft-kips and the end moment M_{2s} caused by factored lateral load is given as 290 ft-kips.

1. For the *braced frame magnifier* δ_b:

$$\delta_b = \frac{C_m}{1 - \dfrac{P_u}{\phi P_c}}$$

$$C_m = 0.6 + 0.4\frac{M_{1b}}{M_{2b}} = 0.6 - 0.4\left(\frac{35}{70}\right) = 0.4$$

Note that 0.4 is the minimum permissible value for C_m.

Next, the Euler load P_c must be computed.

$$P_c = \frac{\pi^2 EI}{(kL_u)^2}$$

and for the 20-in. square column,

$$EI = \frac{27,110,000}{1 + \beta_d}$$

Since β_d is the proportion of factored axial load that is sustained, and is given here as 0.2,

$$EI = \frac{27,110,000}{1 + 0.2} = 22,600,000 \text{ kip in.}^2$$

Note that for the braced frame multiplier, the k is for a braced frame; for $\psi_A = 4.66$ and $\psi_B = 1.0$, Fig. 15.11.1(a) gives $k = 0.85$. Then,

$$P_c = \frac{\pi^2(22,600,000)}{[0.85(10)(12)]^2} = 21,400 \text{ kips}$$

$$\delta_b = \frac{C_m}{1 - \dfrac{P_u}{\phi P_c}} = \frac{0.4}{1 - \dfrac{525}{0.70(21,400)}} = 0.41 < 1.0$$

When δ_b is less than 1.0 it means that the maximum moment occurs at the end of the member instead of out in the span; thus, use $\delta_b = 1.0$.

2. For the *unbraced frame magnifier* δ_s:

$$\delta_s = \frac{1}{1 - \dfrac{\Sigma P_u}{\phi \Sigma P_c}}$$

In the sway analysis, $\beta_d = 0$ since there is no sustained axial load when the wind alone is acting (and no sustained lateral load either). Thus, $EI = 27,110,000$ kip in.^{2} as computed in part (b).

For the 20-in. square exterior column,

$$P_c = \frac{\pi^2(27,110,000)}{[(1.68)(10)(12)]^2} = 6580 \text{ kips}$$

For the 20-in. square interior column,

$$P_c = \frac{\pi^2(27,110,000)}{[(1.50)(10)(12)]^2} = 8260 \text{ kips}$$

The P_c is needed for each of the first-story columns; in this case,

$$\Sigma P_c = 2(6580 + 8260) = 29,680 \text{ kips}$$

The total factored load ΣP_u coming to the four columns is given as 3150 kips. Thus,

$$\frac{\Sigma P_u}{\Sigma P_c} = \frac{3150}{29,680} = 0.108$$

$$\frac{\Sigma P_u}{\phi \Sigma P_c} = \frac{0.108}{0.70} = 0.152$$

This value is the value of Q_u (see Section 15.13) and was recommended by MacGregor and Hage to not be permitted to exceed 0.2, ensuring overall frame stability. The magnifier δ_s for the $P-\Delta$ effect is

$$\delta_s = \frac{1}{1 - \dfrac{\Sigma P_u}{\phi \Sigma P_c}} = \frac{1}{1 - 0.152} = 1.18$$

(d) Compute the required short column strength P_n and M_n for the member. Using ACI Formula (10-6), Eq. (15.10.1),

$$M_c = \delta_b M_{2b} + \delta_s M_{2s}$$
$$= 1.0(70) + 1.18(290) = 412 \text{ ft-kips}$$

The strength required from the 20-in. exterior square column having 12-#10 is

$$\text{required } P_n = \frac{P_u}{\phi} = \frac{525}{0.70} = 750 \text{ kips}$$

$$\text{required } M_n = \frac{M_c}{\phi} = \frac{412}{0.70} = 589 \text{ ft-kips}$$

$$\text{required } e = \frac{M_n}{P_n} = \frac{589(12)}{750} = 9.42 \text{ in.}$$

Using the principles of Chapter 13, the nominal strength $P_n = 790$ kips at $e = 9.42$ in. ∎

■ **EXAMPLE 15.19.5** Determine the adequacy of the 14 × 20 in. compression member designed without regard to length effects in Section 13.17, Example 13.17.1 (6-#11 bars, $f'_c = 4500$ psi, and $f_y = 50,000$ psi). The member serves as an exterior column in a braced frame, with loading and factored moment diagram as shown in Fig. 15.19.4, and having a clear height of 22 ft 6 in.

Solution: **(a)** For possible instability in the plane of the frame, design as a short column neglecting slenderness, as demonstrated in Example 13.17.1, has resulted in a member that is in the tension controls zone and has ϕ slightly greater than 0.70. For this braced frame member, $kL_u = L_u = 22.5$ ft.

$$\frac{kL_u}{r_x} = \frac{1.0(22.5)12}{0.3(20)} = 45$$

which exceeds the maximum value of 22 (used conservatively for a braced frame having this irregular moment diagram) for which slenderness effects may be neglected according to ACI-10.11.4. When the bending moment diagram has the largest moment at a location other than at an end, M_{1b}/M_{2b} should be taken conservatively as 1.0. Note that the moment diagram is similar to the case of transverse loading and should be similarly treated.

Figure 15.19.4 Member and loading for Example 15.19.5.

$$I_g = \tfrac{1}{12}(14)(20)^3 = 9330 \text{ in.}^4$$

$$I_{se} = 2(3)(1.56)(7.29)^2 = 497 \text{ in.}^4$$

$$E_c = 57\sqrt{4500} = 3820 \text{ ksi}$$

Assuming no sustained load, $\beta_d = 0$,

$$EI = 0.2E_cI_g + E_sI_{se}$$
$$= 0.2(3820)(9330) + 29{,}000(497) = 21{,}500{,}000 \text{ kip in.}^2$$

or

$$EI = 0.4E_cI_g = 0.4(3820)(9330) = 14{,}300{,}000 \text{ kip in.}^2$$

$$C_m = 1.0$$

$$P_c = \frac{\pi^2 EI}{(kL_u)^2} = \frac{\pi^2(21{,}500{,}000)}{[1.0(22.5)(12)]^2} = 2910 \text{ kips}$$

$$\delta_b = \frac{C_m}{1 - (P_u/\phi)/P_c} = \frac{1.0}{1 - (115/0.70)/2910} = 1.06$$

The required capacity of the member *in the plane of the frame* (i.e., the strong orientation of the member), using the $\phi = 0.718$ computed in Example 13.17.1, is

$$\text{required } P_n = \frac{P_u}{\phi} = \frac{115}{0.718} = 160 \text{ kips}$$

$$\text{required } M_n = \frac{\delta_b M_u}{\phi} = \frac{1.06(279)}{0.718} = 412 \text{ ft-kips}$$

A statics analysis of this section using an eccentricity of

$$e = \frac{412(12)}{160} = 30.9 \text{ in.}$$

gives $P_n = 151$ kips, which is probably not close enough to 160 kips to be acceptable.
(b) Buckling transverse to the plane of the frame. The slenderness ratio is

$$\frac{kL_u}{r_y} = \frac{22.5(12)}{0.3(14)} = 64.4$$

which exceeds the limiting value of 22 for which the effect of slenderness may be neglected. Again, as in part (a), M_{1b}/M_{2b} should be conservatively taken as 1.0. Since slenderness effects must be considered, the minimum $e = 0.6 + 0.03h$ must be magnified by the factor δ_b.

$$I_g = \tfrac{1}{12}(20)(14)^3 = 4570 \text{ in.}^4$$

$$I_{se} = 2(2)(1.56)(4.29)^2 = 115 \text{ in.}^4$$

$$EI = 0.2E_cI_g + E_sI_{se}$$
$$= 0.2(3820)(4570) + 29{,}000(115) = 6{,}830{,}000 \text{ kip in.}^2$$

or

$$EI = 0.4E_cI_g = 0.4(3820)(4570) = 6,980,000 \text{ kip in.}^2$$

$$C_m = 1.0$$

$$P_c = \frac{\pi^2 EI}{(kL_u)^2} = \frac{\pi^2(6,980,000)}{[1.0(22.5)(12)]^2} = 945 \text{ kips}$$

For this axis, a small eccentricity makes $\phi = 0.70$.

$$\delta_b = \frac{C_m}{1 - (P_u/\phi)/P_c} = \frac{1.0}{1 - (115/0.70)/945} = 1.21$$

Thus, in the weaker direction the member must have the strength $P_n = P_u/0.70 = 164$ kips at an eccentricity

$$e = 1.21(0.6 + 0.03h) = 0.72 + 0.036h$$
$$= 0.72 + 0.036(14) = 1.22 \text{ in.}$$

A statics analysis indicates that the nominal strength P_n is 1205 kips at this eccentricity with respect to the weaker axis.

The strength P_n may not be taken greater than $0.80P_0$,

$$P_{n(\text{max})} = 0.80[0.85(4.5)(280 - 9.36) + 50(9.36)] = 1200 \text{ kips}$$

which is approximately the same as the capacity at the minimum eccentricity of 1.22 in. ■

■ **EXAMPLE 15.19.6** Design column A for the unbraced frame of Fig. 15.19.5 for the dead load plus live load plus wind load case of ACI-9.2.2. The service axial compression is 106 kips dead load and 106 kips live load, and service bending moment is 32 ft-kips dead load, 32 ft-kips live load, and 150 ft-kips wind load. Consider only the dead load to be sustained. The factored axial load P_u on the interior columns is 1.8 times that on the exterior columns. Select a square member to contain approximately $2\frac{1}{2}\%$ reinforcement. Use $f'_c = 5000$ psi, $f_y = 60,000$ psi, and the strength method of the ACI Code.

Solution: (a) Factored loads. There are two major loading combinations from ACI-9.2 that must be considered in the design of a compression member.
   1. Gravity load alone according to ACI-9.2.1,

$$P_u = 1.4(106) + 1.7(106) = 329 \text{ kips}$$
$$M_u = 1.4(32) + 1.7(32) = 99 \text{ ft-kips}$$

   2. Gravity load plus wind according to ACI-9.2.2,

$$P_u = 0.75[1.4(106) + 1.7(106)] = 246 \text{ kips}$$
$$M_u = 0.75[1.4(32) + 1.7(32) + 1.7(150)] = 266 \text{ ft-kips}$$

Figure 15.19.5 Date for Example 15.19.6.

The gravity plus wind combination commonly controls the design. Proceed with the design using this combination. The gravity alone combination can be checked before making the final choice.

(b) Estimate size at balanced condition. Using the procedure described in Sections 13.16 and 13.17, assume that the axial force contributions of the steel in the two faces are approximately equal, $T \approx C_s$; then

$$C_c \approx P_b = \text{required } P_n$$

$$0.85 f'_c \beta_1 x_b b = \frac{246}{\phi}$$

$$x_b = \left(\frac{0.003}{0.003 + 60/29{,}000}\right) d = 0.592 d$$

$$0.85(5)(0.8)(0.592d)b = \frac{246}{0.7} = 351$$

$$\text{balanced } bd = \frac{351}{2.01} = 175 \text{ sq in.}$$

Assume $d \approx 0.85h$

$$\text{balanced } bh = \frac{175}{0.85} = 206 \text{ sq in.,} \quad \text{say 15 in. square}$$

(c) Estimate the required eccentricity e. The moment magnifier requirement is

$$M_c = \delta_b M_{2b} + \delta_s M_{2s}$$

The braced magnifier δ_b is assumed to be 1.0. Since the member is relatively long, a sway magnifier δ_s of 1.2 to 1.5 may be expected; try $\delta_s = 1.3$. The factored primary moments are next separated into the braced frame nonsway moment M_{2b} and the sway moment M_{2s}. The assumption here is that the gravity load portion of this gravity plus wind combination acts essentially in a symmetrical manner on a generally symmetrical frame such that the sway under gravity alone is not "appreciable" (ACI-R10.11.5.1). Thus,

$$M_{2b} = 0.75[1.4(32) + 1.7(32)] = 75 \text{ ft-kips}$$

$$M_{2s} = 0.75(1.7)(150) = 191 \text{ ft-kips}$$

$$\text{estimated } M_c = 1.0(75) + 1.3(191) = 323 \text{ ft-kips}$$

$$\text{required } e \text{ (estimated)} = \frac{323(12)}{246} = 15.8 \text{ in.}$$

For an eccentricity this large, it is likely the design will be in the tension controls region and the section will be larger than the 15-in. section for the balanced condition. Try an 18-in. section to obtain a first approximation for the magnifier δ_s. The magnifier δ_b is likely to be 1.0.

(d) Estimate effective length factor k for the unbraced frame to be used in computing the sidesway magnifier δ_s.

$$I_g \text{ for column} = \tfrac{1}{12}(18)(18)^3 = 8750 \text{ in.}^4$$

$$I_g \text{ for beam} = \tfrac{1}{12}(16)(28)^3 = 29{,}200 \text{ in.}^4$$

Since cracked section should be used for the beam moment of inertia, assume I_{cr} for the beam to be about 15,000 in.^{4}

$$\psi \text{ (top of exterior column)} = \frac{\Sigma EI/L, \text{ columns}}{\Sigma EI/L, \text{ beams}}$$

$$= \frac{8750/14 + 8750/21}{15{,}000/30} = \frac{1042}{500} \approx 2.0$$

$$\psi \text{ (top of interior column)} = \frac{1042}{1000} \approx 1.0$$

Using the alignment chart of Fig. 15.11.1(b), with ψ (bottom) = 1.0 (given), find k of exterior column = 1.45 and k of interior column = 1.30.

(e) Obtain revised sway magnifier δ_s based on preliminary 18 in. square section. Since reinforcement is not yet selected, the following approximation [somewhat between

ACI Formulas (10-10) and (10-11)] is used,

$$EI \approx \frac{0.5E_c I_g}{1 + \beta_d}$$

$$= \frac{0.5(57\sqrt{5000})8750}{1.0} = 17,630,000 \text{ kip in.}^2$$

where $\beta_d = 0$ since no gravity load acts in the wind-only analysis and wind is not sustained load.

$$P_c = \frac{\pi^2 EI}{(kL_u)^2} = \frac{\pi^2(17,630,000)}{[1.45(18.67)(12)]^2} = 1650 \text{ kips} \quad \text{(exterior column)}$$

$$P_c = \frac{\pi^2(17,630,000)}{[1.30(18.67)(12)]^2} = 2050 \text{ kips} \quad \text{(interior column)}$$

where $L_u = 21.0 - 2.33 = 18.67$ ft.

$$\delta_s = \frac{1.0}{1 - \dfrac{\Sigma P_u}{\phi \Sigma P_c}}$$

$$\Sigma P_c = 2(1650) + 3(2050) = 9450 \text{ kips}$$

$$\Sigma P_u = 246[2 + 3(1.8)] = 1820 \text{ kips}$$

$$\delta_s = \frac{1}{1 - \dfrac{1820}{0.70(9450)}} = 1.38$$

(f) Revised required eccentricity.

$$\text{estimated } M_c = 1.0(75) + 1.38(191) = 339 \text{ ft-kips}$$

$$\text{required } e = \frac{339(12)}{246} = 16.5 \text{ in.}$$

(g) Determine column size considering $e = 16.5$ in. and $\rho_g = 0.025$, and assuming tension controls. Use Eq. (13.14.7),

$$P_n = 0.85f'_c bd \left\{ -\rho + 1 - \frac{e'}{d} + \sqrt{\left(1 - \frac{e'}{d}\right)^2 + 2\rho\left[(m-1)\left(1 - \frac{d'}{d}\right) + \frac{e'}{d}\right]} \right\}$$

Estimate (see Fig. 13.14.1 for e') using $h = 18$ in.,

$$\frac{e'}{d} = \frac{16.5 + 6.5}{15.5} = 1.48, \qquad \frac{d'}{d} = \frac{2.5}{15.5} = 0.16$$

$$m = \frac{f_y}{0.85f'_c} = \frac{60}{0.85(5)} = 14.1, \qquad \rho = 0.5\rho_g = 0.0125$$

$$P_n = 0.85(5)bd\{-0.0125 + 1 - 1.48$$
$$+ \sqrt{(1 - 1.48)^2 + 2(0.0125)[13.1(1 - 0.16) + 1.48]}\}$$
$$= 4.25(0.243)bd = 1.031bd$$

Assuming $d \approx 0.85h$

$$\text{required } bh = \frac{246/0.70}{1.031(0.85)} = 401 \text{ sq in.}$$

An 18-in. square section is not acceptable. A second iteration of the formula with $h = 20$ in. gives required $bh = 336$ sq in. Try 20 in. square with 10-#9 bars ($\rho_g = 0.025$).

(h) Recheck effective length factor k. Use ACI Formula (10-10) without sustained load factor β_d for EI.

$$EI \text{ for column} = 0.2E_cI_g + E_sI_{se}$$
$$I_g = \tfrac{1}{12}(20)(20)^3 = 13,300 \text{ in.}^4$$
$$E_c = 57\sqrt{5000} = 4030 \text{ ksi}$$
$$I_{se} = 2(4)(7.5)^2 = 450 \text{ in.}^4$$
$$EI = 0.2(4030)(13,300) + 29,000(450)$$
$$= 10,700,000 + 13,100,000 = 23,800,000 \text{ kips in.}^2 > 0.4E_cI_g$$

For the beam, cracked transformed section should be used. In lieu of using the general method of Chapter 4, Eqs. (14.11.7) and (14.11.8) may be used with $n = 7$ for $f'_c = 5000$ psi.

$$\frac{x}{d} = \sqrt{(\rho n)^2 + 2\rho n} - \rho n, \qquad \rho n = 0.015(7) = 0.105$$

$$\frac{x}{d} = \sqrt{(0.105)^2 + 0.21} - 0.105 = 0.365$$

$$\frac{I_{cr}}{I_g} = 8.75\left[\frac{(x/d)^3}{3} + \rho n\left(1 - \frac{x}{d}\right)^2\right]$$

$$= 8.75\left[\frac{(0.365)^3}{3} + 0.105(0.635)^2\right] = 0.512$$

$$EI \text{ for beam} = 0.512E_cI_g$$
$$= 0.512(4030)(29,200) = 60,300,000 \text{ kip in.}^2$$

$$\psi \text{ (top of exterior column)} = \frac{23.8/14 + 23.8/21}{60.3/30} = \frac{2.833}{2.01} = 1.4$$

$$\psi \text{ (top of interior column)} = \frac{2.833}{4.02} = 0.7$$

Using Fig. 15.11.1(b), find $k = 1.37$ for exterior column and $k = 1.27$ for interior column.

(i) Recompute sway magnifier δ_s.

$$\frac{EI}{1 + \beta_d} = \frac{23{,}800{,}000}{1.0} = 23{,}800{,}000 \text{ kip in.}^2$$

$$P_c = \frac{\pi^2 EI}{(kL_u)^2} = \frac{\pi^2(23{,}800{,}000)}{[1.37(18.67)(12)]^2} = 2490 \text{ kips} \quad \text{(exterior column)}$$

$$P_c = \frac{\pi^2(23{,}800{,}000)}{[1.27(18.67)(12)]^2} = 2900 \text{ kips} \quad \text{(interior column)}$$

$$\Sigma P_c = 2(2490) + 3(2900) = 13{,}680 \text{ kips}$$

$$\Sigma P_u = 1820 \text{ kips} \quad [\text{from (e) above}]$$

$$\frac{\Sigma P_u}{\phi \Sigma P_c} = \frac{1820}{0.70(13{,}680)} = 0.190$$

$$\delta_s = \frac{1.0}{1 - \dfrac{\Sigma P_u}{\phi \Sigma P_c}} = \frac{1.0}{1 - 0.190} = 1.23$$

(j) Recheck capacity by approximate formula, Eq. (13.14.7). Using the 20-in. square section with 10-#9 (4 in each face) as shown in Fig. 15.19.6, the more accurate distance from face of concrete to center of bars is

$$d' = d_s = 1.5 \text{ (cover)} + 0.375 \text{ (#3 tie)} + 0.564 \text{ (bar radius)} = 2.44 \text{ in.}$$

instead of 2.5 in. used in the preliminary computations.

$$\text{required } e = \frac{\delta_b M_{2b} + \delta_s M_{2s}}{P_u}$$

$$= \frac{[1.0(75) + 1.23(191)]12}{246} = 15.1 \text{ in.}$$

$$\frac{e'}{d} = \frac{15.1 + 7.56}{17.56} = 1.29, \qquad \frac{d'}{d} = \frac{2.44}{17.56} = 0.139$$

$$\rho = 0.01 \quad \text{(using only the bars in one face)}$$

$$m = 14.1, \qquad m' = m - 1 = 13.1$$

$$P_n = 4.25bd\{-0.01 + 1 - 1.29$$

$$+ \sqrt{(1 - 1.29)^2 + 2(0.01)[13.1(1 - 0.139) + 1.29]}\,\}$$

$$= 4.25(0.279)bd = 1.187bd$$

$$P_n = 1.187(20)(17.56) = 417 \text{ kips}$$

$$\phi P_n = 0.70(417) = 292 \text{ kips} > [P_u = 246 \text{ kips}] \qquad \textbf{OK}$$

A statics analysis of the selected section indicates a strength $P_n = 475$ kips at $e = 15.1$ in.

Figure 15.19.6 Section selected for Example 15.19.6.

(k) Check magnification factor δ_b. For a fixed base member having no transverse load,

$$C_m = 0.6 + 0.4\left(\frac{M_{1b}}{M_{2b}}\right)$$

$$= 0.6 + 0.4(\approx -\tfrac{1}{2}) = 0.4 \quad \text{(the minimum value)}$$

For the braced frame $k = 1.0$ (conservatively). This time $\beta_d \neq 1.0$ since the axial dead load is sustained,

$$\beta_d = \frac{\text{Factored axial dead load}}{\text{Factored total axial load}} = \frac{0.75[1.4(106)]}{0.75[1.4(106) + 1.7(106)]} = 0.45$$

$$EI = \frac{23,800,000 \text{ [from part (h)]}}{1 + \beta_d} = \frac{23,800,000}{1.45} = 16,400,000 \text{ kip in.}^2$$

$$P_c = \frac{\pi^2 EI}{(kL_u)^2} = \frac{\pi^2(16,400,000)}{[1.0(18.67)(12)]^2} = 3220 \text{ kips}$$

$$\delta_b = \frac{C_m}{1 - \dfrac{P_u}{\phi P_c}} = \frac{0.4}{1 - \dfrac{246}{0.70(3220)}} = 0.45 < 1.0$$

Since computed $\delta_b < 1.0$, this means that support conditions control; thus, $\delta_b = 1.0$ which confirms the assumption made in part (c).

(l) Check strength and stability for loading case 1. Check the strength, first assuming this to be a nonsway case. Thus,

$$\text{required } e = \frac{M_u}{P_u} = \frac{99(12)}{329} = 3.6 \text{ in.}$$

$$\delta_b = \frac{C_m}{1 - \dfrac{329}{0.70(3220)}} = 0.47 < 1.0; \quad \text{use } 1.0.$$

The nominal strength P_n when $e = 3.6$ in. is 1470 kips, giving $\phi P_n = 0.70(1470) = 1029$ kips, well above the 329 kips required.

Regarding stability, there is no ACI Code provision specifically for this. Under gravity load alone there is presumably no "appreciable" sway (ACI-R10.11.5.1) in this particular design. Thus, ordinarily the sway magnifier δ_s under this loading case would not be computed. However, one should note that very long slender columns could be unstable under the maximum factored gravity load case. The authors recommend that in the gravity load alone $(1.4D + 1.7L)$ case, in lieu of doing a stability analysis, an indirect indication of stability can be obtained by noting the size of the sway magnifier. A reasonably stiff structure will have a sway magnifier δ_s not greater than say 2 to 2.5. For the gravity load case (loading case 1),

$$\Sigma P_u = 329[2 \text{ exterior columns} + 1.8(3 \text{ interior columns})] = 2435 \text{ kips}$$

$$\beta_d = 0.45 \quad \text{[from part (k)]}$$

$$EI = 16{,}400{,}000 \text{ kip in}^2. \quad \text{[from part (k)]}$$

$$k = 1.37 \text{ (exterior column)}; \qquad k = 1.27 \text{ (interior column)} \quad \text{[from part (i)]}$$

$$P_c = \frac{\pi^2 EI}{(kL_u)^2} = \frac{\pi^2(16{,}400{,}000)}{[1.37(18.67)12]^2} = 1720 \text{ kips (exterior)}$$

$$P_c = \frac{\pi^2 EI}{(kL_u)^2} = \frac{\pi^2(16{,}400{,}000)}{[1.27(18.67)12]^2} = 2000 \text{ kips (interior)}$$

$$\Sigma P_c = 2(1720) + 3(2000) = 9440 \text{ kips}$$

$$\delta_s = \frac{1.0}{1 - \dfrac{\Sigma P_u}{\phi \Sigma P_c}} = \frac{1.0}{1 - \dfrac{2435}{0.70(9440)}} = 1.58$$

One may note that here δ_s is larger than it was in the gravity plus wind loading case (loading case 2). In this example, it does not exceed about 2; thus, stability under gravity load alone "nonsway" case is considered satisfactory.

Use 20 in. square section with 10-#9 bars as shown in Fig. 15.19.6.   ■

SELECTED REFERENCES

15.1.   L. Euler, *DeCurvis Elasticis, Additamentum I, Methodus Inveniendi Lineas Curvas Maximi Minimive Proprietate Gaudentes.* Lausanne and Geneva: 1744, (pp. 267–268); and "Sur la Force des Colonnes," *Mémoires de l'Académie de Berlin,* Vol. 13. Berlin: 1759 (pp. 252–282).

15.2.   F. Engesser. "Ueber die Knickfestigkeit gerader Stäbe," *Zeitschrift des Architekten- und Ingenieur-Vereins zu Hannover,* 35(1889), 455–462. Also "Die Knickfestigkeit gerader Stäbe," *Zentralblatt der Bauverwaltung,* Berlin, Dec. 5, 1891, p. 483.

15.3.   T. von Kármán. "Die Knickfestigkeit gerader Stäbe," *Physikalische Zeitschrift,* Vol. 9, 1908 (p. 136). Also, "Untersuchungen über Knickfestigkeit," *Mitteilungen über Forschungsarbeiten auf dem Gebiete des Ingenieurwesens,* No. 81. Berlin: 1910.

15.4.   Luis P. Sáenz and Ignacio Martín. "Test of Reinforced Concrete Columns with High Slenderness Ratios," *ACI Journal, Proceedings*, **60**, May 1963, 589–616.

15.5.   John E. Breen and Phil M. Ferguson. "The Restrained Long Concrete Column As a Part of A Rectangular Frame," *ACI Journal, Proceedings*, **61**, May 1964, 563–587.

15.6.   Edward O. Pfrang and Chester P. Siess. "Behavior of Restrained Reinforced Concrete Columns," *Journal of the Structural Division*, ASCE, **90**, ST5 (October 1964), 113–135; Disc., **91**, ST3 (June 1965), 280–287.

15.7.   J. G. MacGregor. Discussion of "Behavior of Restrained Reinforced Concrete Columns," by E. O. Pfrang and C. P. Siess, *Journal of the Structural Division*, ASCE, **91**, ST3 (June 1965) 280–287.

15.8.   Richard W. Furlong and Phil M. Ferguson. "Tests of Frames with Columns in Single Curvature," *Symposium on Reinforced Concrete Columns* (SP-13). Detroit: American Concrete Institute, 1966 (pp. 55–73).

15.9.   Phil M. Ferguson and John E. Breen. "Investigation of the Long Column in a Frame Subject to Lateral Loads," *Symposium on Reinforced Concrete Columns* (SP-13). Detroit: American Concrete Institute, 1966 (pp. 75–119).

15.10.   Ignacio Martín and Elmer Olivieri. "Test of Slender Reinforced Concrete Columns Bent in Double Curvature," *Symposium on Reinforced Concrete Columns* (SP-13). Detroit: American Concrete Institute, 1966 (pp. 121–138).

15.11.   J. G. MacGregor and S. L. Barter. "Long Eccentrically Loaded Concrete Columns Bent in Double Curvature," *Symposium on Reinforced Concrete Columns* (SP-13). Detroit: American Concrete Institute, 1966 (pp. 139–156).

15.12.   Edward O. Pfrang. "Behavior of Reinforced Concrete Columns with Sidesway," *Journal of the Structural Division*, ASCE, **92**, ST3 (June 1966), 225–252.

15.13.   R. Green and John E. Breen. "Eccentrically Loaded Concrete Columns Under Sustained Load," *ACI Journal, Proceedings*, **66**, November 1969, 866–874.

15.14.   John E. Breen and Phil M. Ferguson. "Long Cantilever Columns Subject to Lateral Forces," *ACI Journal, Proceedings*, **66**, November 1969, 884–893.

15.15.   Shriniwas N. Pagay, Phil M. Ferguson, and John E. Breen. "Importance of Beam Properties on Concrete Column Behavior," *ACI Journal, Proceedings*, **67**, October 1970, 808–815.

15.16.   Hajime Okamura, Shriniwas N. Pagay, John E. Breen, and Phil M. Ferguson. "Elastic Frame Analysis—Corrections Necessary for Design of Short Concrete Columns in Braced Frames," *ACI Journal, Proceedings*, **67**, November 1970, 894–897.

15.17.   Phil M. Ferguson, Hajime Okamura, and Shriniwas N. Pagay. "Computer Study of Long Columns in Frames," *ACI Journal, Proceedings*, **67**, December 1970, 955–958.

15.18.   G. A. Blomier and J. E. Breen. "Effect of Yielding of Restraints on Slender Concrete Columns With Sidesway Prevented," *Reinforced Concrete Columns* (SP-50). Detroit: American Concrete Institute, 1975 (pp. 41–65).

15.19.   F. N. Rad and R. W. Furlong. "Behavior of Unbraced Reinforced Concrete Frames," *ACI Journal, Proceedings*, **77**, July–August 1980, 269–278.

15.20. J. S. Ford, D. C. Chang, and J. E. Breen. "Behavior of Concrete Columns Under Controlled Lateral Deformation," *ACI Journal, Proceedings,* **78**, January–February 1981, 3–20.

15.21. J. S. Ford, D. C. Chang, and J. E. Breen. "Experimental and Analytical Modeling of Unbraced Multipanel Concrete Frames," *ACI Journal, Proceedings,* **78**, January–February 1981, 21–35.

15.22. J. S. Ford, D. C. Chang, and J. E. Breen. "Behavior of Unbraced Multipanel Concrete Frames," *ACI Journal, Proceedings,* **78**, March–April 1981, 99–115.

15.23. Bengt Broms and I. M. Viest. "Ultimate Strength Analysis of Long Hinged Reinforced Concrete Columns," *Journal of the Structural Division,* ASCE, **84**, ST1 (January 1958), (Paper No. 1510).

15.24. Nestor R. Iwankiw. "Note on Beam-Column Moment Amplification Factor," *Engineering Journal,* AISC, **21**, 1(1st Quarter 1984), 21–23.

15.25. Charles Massonnet. "Stability Considerations in the Design of Steel Columns," *Journal of the Structural Division,* ASCE, **85**, ST7 (September 1959), 75–111.

15.26. Charles G. Salmon and John E. Johnson. *Steel Structures–Design and Behavior; Emphasizing Load and Resistance Factor Design,* 3rd ed. New York: Harper & Row, Publishers, Inc., 1990.

15.27. Theodore V. Galambos, ed. *Guide to Stability Design Criteria for Metal Structures,* 4th ed. New York: John Wiley & Sons, 1988.

15.28. J. G. MacGregor. "Out-of-Plumb Columns in Concrete Structures," *Concrete International,* **1**, June 1979, 26–31.

15.29. James G. MacGregor and Sven E. Hage. "Stability Analysis and Design of Concrete Frames," *Journal of the Structural Division,* ASCE, **103**, ST10 (October 1977), 1953–1970.

15.30. Chu-Kia Wang and Carlito P. Talaboc. "Second-Order Analysis of Multistory Building Frames by Marching Technique," *Microcomputers in Structural Engineering,* **1**, 3 (December 1986), 209–232.

15.31. James G. MacGregor, John E. Breen, and Edward O. Pfrang. "Design of Slender Concrete Columns," *ACI Journal, Proceedings,* **67**, January 1970, 6–28.

15.32. Eldon F. Mockry and David Darwin. "Slender Column Interaction Diagrams," *Concrete International,* **4**, June 1982, 44–50.

15.33. Alfred L. Parme. "Capacity of Restrained Eccentrically Loaded Long Columns," *Symposium on Reinforced Concrete Columns* (SP-13). Detroit: American Concrete Institute, 1966 (pp. 325–367).

15.34. PCA. *Design Constants for Rectangular Long Columns.* Advanced Engineering Bulletin No. 12. Skokie, Illinois: Portland Cement Association, 1964 (36 pp.).

15.35. J. G. MacGregor, U. H. Oelhafen, and S. E. Hage. "A Reexamination of the *EI* Value for Slender Columns," *Reinforced Concrete Columns* (SP-50). Detroit: American Concrete Institute, 1975 (pp. 1–40).

15.36. Brian R. Wood, Denis Beaulieu, and Peter F. Adams. "Column Design by P Delta Method," *Journal of the Structural Division*, ASCE, **102**, ST2 (February 1976), 411–427.

15.37. Brian R. Wood, Denis Beaulieu, and Peter F. Adams. "Further Aspects of Design by P–Delta Method," *Journal of the Structural Division*, ASCE, **102**, ST3 (March 1976), 487–500.

15.38. H. Scholz. "A Novel P–Delta Method for Reinforced Concrete Frames Designed by LRFD Approaches," *ACI Journal, Proceedings*, **83**, July–August 1986, 633–641.

15.39. H. Scholz. "Alternative P–Delta Method for Prestressed Members Subjected to Axial Load and Bending," *ACI Structural Journal*, **87**, July–August 1990, 464–472.

15.40. James Colville. "Slenderness Effects in Reinforced Concrete Square Columns," *Reinforced Concrete Columns* (SP-50). Detroit: American Concrete Institute, 1975 (pp. 165–191).

15.41. Ian C. Medland and Donald A. Taylor. "Flexural Rigidity of Concrete Column Sections," *Journal of the Structural Division*, ASCE, **97**, ST2 (February 1971) 573–586.

15.42. M. A. Diaz and J. M. Roesset. "Evaluation of Approximate Slenderness Procedures for Nonlinear Analysis of Concrete Frames," *ACI Structural Journal*, **84**, March–April 1987, 139–148.

15.43. S. A. Mirza. "Flexural Stiffness of Rectangular Reinforced Concrete Columns," *ACI Structural Journal*, **87**, July–August 1990, 425–435.

15.44. Gregory R. Sigmon and Shuaib H. Ahmad. "Flexural Rigidity of Circular Reinforced Concrete Sections," *ACI Structural Journal*, **87**, September–October 1990, 548–556.

15.45. M. R. Ehsani and F. Alameddine. "Refined Stiffness of Slender Circular Reinforced Concrete Columns," *ACI Structural Journal*, **84**, September–October 1987, 419–427.

15.46. Brij B. Goyal and Neil Jackson. "Slender Concrete Columns under Sustained Load," *Journal of the Structural Division*, ASCE, **97**, ST11 (November 1971), 2729–2750.

15.47. Robert G. Drysdale and Mark W. Huggins. "Sustained Biaxial Load on Slender Concrete Columns," *Journal of the Structural Division*, ASCE, **97**, ST5 (May 1971), 1423–1443.

15.48. B. Vijaya Rangan. "Lateral Deflection of Slender Reinforced Concrete Columns under Sustained Load," *ACI Structural Journal*, **86**, November–December 1989, 660–663.

15.49. J. S. Ford, D. C. Chang, and J. E. Breen. "Design Indications from Tests of Unbraced Multipanel Concrete Frames," *Concrete International*, **3**, March 1981, 37–47.

15.50. Thomas C. Kavanagh. "Effective Length of Framed Columns," *Transactions ASCE*, **127**, Part II, 1962, 81–101.

15.51. John E. Breen, James G. MacGregor, and Edward O. Pfrang. "Determination of Effective Length Factors for Slender Concrete Columns," *ACI Journal, Proceedings*, **69**, November 1972, 669–672.

15.52. BSI. *Code of Practice for the Structural Use of Concrete*, Part I. *Design, Materials and Workmanship*. London: British Standards Institution, 1972.

15.53. Richard W. Furlong. "Column Slenderness and Charts for Design," *ACI Journal, Proceedings,* **68,** January 1971, 9–18.

15.54. Richard W. Furlong. "Rational Analysis of Multistory Concrete Structures," *Concrete International,* **3,** June 1981, 29–35.

15.55. S. I. Abdel-Sayed and N. J. Gardner. "Design of Symmetric Square Slender Reinforced Concrete Columns under Biaxially Eccentric Loads," *Reinforced Concrete Columns* (SP-50). Detroit: American Concrete Institute, 1975 (pp. 149–164).

PROBLEMS

All problems* are to be done in accordance with the strength method of the ACI Code unless otherwise indicated. All loads given are service loads unless otherwise indicated. Even though the slenderness effect may be neglected according to the ACI Code in a given problem, consider the length effect anyway. For design problems, use whole inches for member size.

15.1 Determine the adequacy of the section including length effect, for a 16-in. square tied column that has a clear height of 18 ft and serves as an interior member of a braced frame. The member is designed as axially loaded with the following service loads: live load, 130 kips; dead load, 200 kips. Use $f'_c = 4000$ psi and $f_y = 60,000$ psi. Without actually selecting bars, assume about $2\frac{1}{2}\%$ total reinforcement equally divided in the opposite faces of the member. Use ACI moment magnifier method. (400 mm square section; 5.5 m clear height; LL = 580 kN; DL = 890 kN; $f'_c = 30$ MPa; $f_y = 400$ MPa.)

15.2 Determine the adequacy, including length effects, for a 14 in. diameter spirally reinforced column (assume about $2\frac{1}{2}\%$ reinforcement) which is an interior second-floor column (column A) in the braced frame of the accompanying figure. The member is to carry an axial service load of 87 kips dead load and 133 kips live load, with only the dead load considered sustained. The 14×22 beams contain $\rho = 0.015$ for negative moment tension steel. Use $f'_c = 5000$ psi and $f_y = 60,000$ psi. Use the ACI moment magnifier method. (350 mm diameter; $P_n = 2200$ kN; 350 mm × 560 mm beams; $f'_c = 35$ MPa; $f_y = 300$ MPa.)

15.3 Assume the frame of Problem 15.2 is *unbraced* and consists of 4 bays symmetrical about the middle. The gravity axial loads are 44 kips dead load and 66 kips live load. The largest moments in column A are 15 ft-kips dead load, 12 ft-kips live load, and 58 ft-kips wind load. An exterior column carries one-half the

---

* Most problems may be solved as problems stated in Inch-Pound units, or as problems in SI units using quantities in parenthesis at the end of the statement. The SI conversions are approximate to avoid implying higher precision for the given information in metric units than that given for the Inch-Pound units.

factored axial load of an interior column. Investigate the adequacy of the interior column *A*. If not adequate, redesign it to make it satisfactory according to the moment magnifier method of the ACI Code.

15.4   Determine the adequacy of the exterior square column (column *B*) of the figure for Problem 15.2, if the member is carrying a gravity axial compression of 217 kips dead load and 145 kips live load and primary bending moments of 27 ft-kips dead load and 18 ft-kips live load. Assume that the primary bending moment varies from $+M$ at the top of the member linearly to $-M/2$ at the bottom, with joint translation adequately prevented (i.e., a braced frame). Use $f'_c = 5000$ psi and $f_y = 60,000$ psi. The 14×22 beams contain $\rho = 0.015$ for negative moment tension steel. Use ACI moment magnifier method. ($P_D =$ 965 kN; $P_L = 645$ kN; $M_D = 37$ kN·m; $M_L = 24$ kN·m; $f'_c = 35$ MPa; $f_y =$ 400 MPa; 350 mm $\times$ 560 mm beams.)

15.5   Repeat Problem 15.4, except consider the primary bending moment constant over the height of the column.

15.6   Assume the member (column *B*) in Problem 15.4 is part of an *unbraced* frame, and that maximum moments M_2 of 28 ft-kips dead load, 20 ft-kips live load, and 85 ft-kips wind load must be carried. The axial compressive loads are

Problems 15.2 to 15.8

30 kips dead load and 40 kips live load. The total factored load ΣP_u to all five columns in the story is 590 kips for ACI Formula (9-2). Investigate the adequacy of the 14-in. square exterior column (column B). If found to be inadequate, redesign the member assuming the loadings are not affected by a change in member stiffness.

15.7 Determine the adequacy of the 16 in. square first-floor column (column C) of the figure for Problem 15.2, which is carrying service axial load of 370 kips and a primary bending moment of 121 ft-kips (assume loads 70% dead load). The primary bending moment varies from a maximum at the top to zero at the bottom and joint translation is adequately prevented. Use $f'_c = 5000$ psi and $f_y = 60,000$ psi. Use ACI moment magnifier method. ($P = 1645$ kN; $M = 164$ kN·m; $f'_c = 35$ MPa; $f_y = 400$ MPa.)

15.8 Assume the member (column C) of Problem 15.7 is part of an *unbraced* frame. The axial loads are 45 kips dead load and 60 kips live load. Assume the 14 × 22 beams contain negative moment reinforcement $\rho = 0.015$. The bending moments are 30 ft-kips dead load, 22 ft-kips live load, and 110 ft-kips wind load. Investigate the adequacy according to the moment magnifier method of the ACI Code. If not adequate, redesign assuming the loadings are not affected by changes in member stiffnesses. The total factored load ΣP_u to all columns in the lower story is 885 kips for ACI Formula (9-2).

15.9 Determine the adequacy of an 18 in. square tied member in a *braced* frame where the member has end flexibilities $\psi = 2.0$ at its top and 10.0 at its bottom and has a clear height of 16 ft. Use rational analysis to establish k. The member is required to carry a factored axial load $P_u = 70$ kips, and the maximum factored primary bending moment $M_u = 168$ ft-kips. Assume primary bending moment is constant over the height of the member, and that 70% of the loads are dead load, and the remainder live load. Use $f'_c = 3000$ psi and $f_y = 40,000$ psi. The member reinforcement consists of 8-#9 bars, four in each face centered $2\frac{1}{2}$ in. (or 64 mm) from edge of member. Use ACI moment magnifier method. (460 mm square member; clear height = 5 m; $P_u = 312$ kN; $M_u = 228$ kN·m; $f'_c = 20$ MPa; $f_y = 300$ MPa; reinforcement, 4-#25M and 4-#35M bars.)

15.10 Repeat Problem 15.9 if bending moment arises from uniform lateral load, and apply rational analysis to determine k (less than 1.0) and for determination of C_m (assume the member fixed at one end and simply supported at the other). Use moment magnifier method.

15.11 **(a)** Redesign the column used in Example 15.19.6, except for service loads use an axial compression of 132 kips dead load and 132 kips live load, and bending moments M_2 of 26 ft-kips dead load, 26 ft-kips live load, and 123 ft-kips wind load. The length for column A is 20 ft instead of 21 ft. ($P_D = 580$ kN; $P_L = 580$ kN; $M_D = 35$ kN·m; $M_L = 35$ kN·m; $M_W = 167$ kN·m; length of column above = 4.3 m; length of column $A = 6.1$ m; beam = 400 mm × 700 mm on 9 m span; $f'_c = 35$ MPa; $f_y = 400$ MPa.)

(b) How much smaller could the member have been if the frame were adequately braced to prevent joint translation?

15.12 (a) Redesign the columns used in Example 15.19.6 except use 18 ft (5.5 m) instead of 21 ft for the length of column *A*.

(b) How much smaller could the member have been if the frame were adequately braced to prevent joint translation?

15.13 through 15.19 An unbraced multistory frame has been analyzed using a conventional first-order elastic analysis. The service load results of a gravity load analysis to get dead load and live load values, and the separately computed results of a lateral load analysis to obtain the wind load moments (M_w) are given in the accompanying table. The rotational moments given are in the same rotational direction at each end of the member. For each design assume

Table for Problems 15.13 Through 15.19

PROBLEM NO.	L_u (ft)	P_D (kips)	P_L (kips)	M_D (ft-kips)	M_L (ft-kips)	M_W (ft-kips)	LOCATION
15.13	14	48	13	Negl.	32	7.5	Top of column
				Negl.	46	2.0	Bottom of column
$\Sigma P_D = 250$ kips; $\Sigma P_L = 75$ kips							
15.14	14	423	234	Negl.	15	99	Top of column
				Negl.	30	80	Bottom of column
$\Sigma P_D = 2120$ kips; $\Sigma P_L = 1100$ kips							
15.15	14	352	202	39	26	62	Top of column
				39	26	32	Bottom of column
$\Sigma P_D = 1050$ kips; $\Sigma P_L = 600$ kips							
15.16	14	352	202	Negl.	13	83	Top of column
				Negl.	13	62	Bottom of column
$\Sigma P_D = 1050$ kips; $\Sigma P_L = 600$ kips							
15.17	14	176	133	18	13	59	Top of column
				18	13	37	Bottom of column
$\Sigma P_D = 880$ kips; $\Sigma P_L = 650$ kips							
15.18	14	176	154	20	13	26	Top of column
				20	13	10	Bottom of column
$\Sigma P_D = 880$ kips; $\Sigma P_L = 650$ kips							
15.19	22	500	320	15	13	41	Top of column
				30	26	316	Bottom of column
$\Sigma P_D = 2800$ kips; $\Sigma P_L = 1760$ kips							

the column being designed is one of five identical ones resisting sidesway in that story; assume the one you are designing is an interior column. For the beams, assume the ratio of gross moment of inertia I to span L is 30 in.^{3} for each beam. Assume columns in the storys above and below the story in question are the same as the column being designed.

For the loading condition given in the accompanying table, design a square tied column (use whole inches for size) having symmetrical reinforcement with an approximate gross reinforcement ratio ρ_g of 0.010. Design formulas from Chapter 13 may be used to size the section, but the final check of strength is to be done using statics with the rectangular stress distribution. Use $f'_c = 4000$ psi and $f_y = 60,000$ psi.

CHAPTER 16

Design of Two-Way Floor Systems

16.1 GENERAL DESCRIPTION

In reinforced concrete buildings, a basic and common type of floor is the slab—beam—girder construction, which has been treated in Chapters 8, 9, and 10. As shown in Fig. 16.1.1(a), the shaded slab area is bounded by the two adjacent beams on the sides and portions of the two girders at the ends. When the length of this area is two or more times its width, almost all of the floor load goes to the beams, and very little, except some near the edge of the girders, goes directly to the girders. Thus the slab may be designed as a one-way slab as treated in Chapter 8, with the main reinforcement parallel to the girder and the shrinkage and temperature reinforcement parallel to the beams. The deflected surface of a one-way slab is primarily one of single curvature.

When the ratio of the long span L to the short span S as shown in Fig. 16.1.1(b) is less than about 2, the deflected surface of the shaded area becomes one of double curvature. The floor load is carried in both directions to the four supporting beams around the panel; hence the panel is a *two-way slab*. Obviously, when S is equal to L, the four beams around a typical interior panel should be identical; for other cases the long beams take more load than the short beams.

Both the *flat slab* and *flat plate* floors shown in Figs. 16.1.2(a) and (b) are characterized by the absence of beams along the interior column lines, but edge beams may or may not be used at the exterior edges of the floor. Flat slab floors differ from flat plate floors in that flat slab floors provide adequate shear strength by having either or both of the following: (a) drop panels (i.e., increased thickness of slab) in the region of the columns; or (b) column capitals (i.e., tapered enlargement of the upper ends of columns). In flat plate floors a uniform slab thickness is used and the shear strength is obtained by the embedment of multiple-U stirrups or structural steel devices known as *shearhead reinforcement* [see Fig. 16.16.1(b)] within the slab

Flat slab (waffle slab) with capitals in the Fisher Cleveland Plant. (Courtesy of Portland Cement Association.)

of uniform thickness. Relatively speaking, flat slabs are more suitable for larger panel size or heavier loading than flat plates.

Historically, flat slabs predate both two-way slabs on beams and flat plates. Flat slab floors were originally patented by O. W. Norcross [16.2] in the United States on April 29, 1902. Several systems of placing reinforcement have been developed and patented since then—the four-way system, the two-way system, the three-way system, and the circumferential system. C. A. P. Turner [16.2] was one of the early advocates of a flat slab system known as the "mushroom" system. About 1908, the flat slab began being recognized as an acceptable floor system, but for many years designers were confronted with difficulties of patent infringements.

Figure 16.1.1 One-way versus two-way slabs.

(a) Flat slab floor   (b) Flat plate floor

Figure 16.1.2 Flat slab and flat plate floor construction.

Actually the terms *two-way slab* [Fig. 16.1.1(b)], *flat slab* [Fig. 16.1.2(a)], and *flat plate* [Fig. 16.1.2(b)] are arbitrary, because there is in fact two-way action in all three types and a flat (usually nearly square) ceiling area usually exists within the panel in all three types. Following tradition, the implication is that there are beams between columns in two-way slabs; but no such beams, except perhaps edge beams along the exterior sides of the entire floor area, are used in flat slabs or flat plates. From the viewpoint of structural analysis, however, the distinction as to whether or not there are beams between columns is not pertinent, because if beams of any relative size could be designed to interact with the slab, use of beams of zero size would be only the limit condition.

If methods of structural analysis and design are developed for two-way slabs with beams, many of these general provisions should apply equally well to flat slabs or flat plates. Until 1971 the design of two-way slabs supported on beams has, historically, been treated separately from the flat slabs or flat plates without beams. Various empirical procedures have been proposed and used [16.6–16.8]. Chapter 13 of the present ACI Code takes an integrated view and refers to two-way slab *systems* with or without beams. In addition to solid slabs, hollow slabs with interior voids to reduce dead weight, slabs (such as waffle slabs) with recesses made by permanent or removable fillers between joists in two directions, and slabs with paneled ceilings near the central portion of the panel are also included in this category (ACI-13.1.3).

Thus the term two-way *floor* systems (rather than the term two-way *slab* systems as in the ACI Code) is used in this book to include all three systems: the two-way slab with beams, the flat slab, and the flat plate.

16.2 GENERAL DESIGN CONCEPT OF ACI CODE

The basic approach to the design of two-way floor systems involves imagining that vertical cuts are made through the entire building along lines midway between the columns. The cutting creates a series of frames whose width lies between the centerlines of the two adjacent panels as shown in Fig. 16.2.1. The resulting series of rigid frames, taken separately in the longitudinal and transverse directions of the building, may be treated for gravity loading floor by floor as would generally be acceptable for a rigid frame structure consisting of beams and columns, in accordance with ACI-8.9.1. A typical rigid frame would consist of (1) the columns above and below the floor, and (2) the floor system, with or without beams, bounded laterally between the centerlines of the two panels (one panel for an exterior line of columns) adjacent to the line of columns.

Thus the design of a two-way floor system (including two-way slab, flat slab, and flat plate) is reduced to that of a rigid frame; hence the name "equivalent frame method."

As in the case of design of actual rigid frames consisting of beams and columns, approximate methods of analysis may be suitable for many usual floor systems, spans, and story heights. As treated in Chapter 7, the analysis for actual frames could be (a) approximate using the moment and shear coefficients of ACI-8.3, or (b) more accurate using structural analysis after assuming the relative stiffnesses of the members. For gravity load only and for floor systems within the specified limitations, the moments and shears on these equivalent frames may be determined (a) approximately using moment and shear coefficients prescribed by the "direct design method" of ACI-Chapter 13, or (b) by structural analysis in a manner similar to that for actual frames using the special provisions of the "equivalent frame method" of ACI-Chapter 13. An elastic analysis (such as by the equivalent frame method) must be used for

Figure 16.2.1 Tributary floor area for an interior equivalent rigid frame of a two-way floor system.

lateral load even if the floor system meets the limitations of the direct design method for gravity load.

The equivalent rigid frame is the structure being dealt with whether the moments are determined by the "direct design method (DDM)" or by the "equivalent frame method (EFM)." These two ACI Code terms describe two ways of obtaining the longitudinal variation of bending moments and shears.

When the "equivalent frame method" is used for obtaining the longitudinal variation of moments and shears, the relative stiffness of the columns, as well as that of the floor system, can be assumed in the preliminary analysis and then reviewed, as is the case for the design of any statically indeterminate structure. Design moment envelopes may be obtained for dead load in combination with various patterns of live load, as described in Chapter 7 (Section 7.2). In lateral load analysis, moment magnification in columns due to sidesway of vertical loads must be taken into account as prescribed in ACI-10.11.5 and 10.12.

Once the longitudinal variation in factored moments and shears has been obtained, whether by ACI "DDM" or "EFM," the moment across the entire width of the floor system being considered is distributed laterally to the beam, if used, and to the slab. The lateral distribution procedure and the remainder of the design is essentially the same whether "DDM" or "EFM" has been used.

The accuracy of analysis methods utilizing the concept of dividing the structure into equivalent frames has been verified for *gravity load* analysis by tests [16.12–16.25] and analytical studies [16.26–16.35]. For *lateral load* analysis where there is less agreement on procedure, various studies have been made, including those of Pecknold [16.38], Allen and Darvall [16.39, 16.47], Vanderbilt [16.32, 16.40], Elias [16.41–16.43], Fraser [16.44], Vanderbilt and Corley [16.45], Lew and Narov [16.46], Pavlovic and Poulton [16.48], Moehle and Diebold [16.49], Hsu [16.50], and Cano and Klingner [16.51].

16.3 TOTAL FACTORED STATIC MOMENT

Consider two typical interior panels *ABCD* and *CDEF* in a two-way floor system, as shown in Fig. 16.3.1(a). Let L_1 and L_2 be the panel size in the longitudinal and transverse directions, respectively. Let lines 1–2 and 3–4 be centerlines of panels *ABCD* and *CDEF*, both parallel to the longitudinal direction. Isolate as a free body [see Fig. 16.3.1(b)] the floor slab and the included beam bounded by the lines 1–2 and 3–4 in the longitudinal direction and the transverse lines 1'–3' and 2'–4' at the faces of the columns in the transverse direction. The load acting on this free body [see Fig. 16.3.1(c)] is $w_u L_2$ per unit distance in the longitudinal direction. The total upward force acting on lines 1'–3' or 2'–4' is $w_u L_2 L_n/2$, where w_u is the factored load per unit area and L_n is the clear span in the longitudinal direction between faces of supports (as defined by ACI-13.6.2.5).

If M_{neg} and M_{pos} are the numerical values of the total negative and positive bending moments along lines 1'–3' and 5–6, then moment equilibrium of the free body of Fig. 16.3.1(d) requires

$$M_{\text{neg}} + M_{\text{pos}} = \frac{w_u L_2 L_n^2}{8} \tag{16.3.1}$$

Figure 16.3.1 Statics of a typical *interior* panel in a two-way floor system.

For a typical exterior panel, the negative moment at the interior support would be larger than that at the exterior support, as has been shown in Section 7.5. The maximum positive moment would occur at a section to the left of the midspan, as shown in Fig. 16.3.2(c). In practical design, it is customary to use M_{pos} at midspan for determining the required positive moment reinforcement. For this case,

$$\frac{M_{neg}(\text{left}) + M_{neg}(\text{right})}{2} + M_{pos} = \frac{w_u L_2 L_n^2}{8} \qquad \textbf{(16.3.2)}$$

A proof for Eq. (16.3.2) can be obtained by writing the moment equilibrium equation about the left end of the free body shown in Fig. 16.3.2(a),

$$M_{neg}(\text{left}) + M_{pos} = \frac{w_u L_2 L_n}{2}\left(\frac{L_n}{4}\right) - V_{\mathbb{C}}\left(\frac{L_n}{2}\right)$$

and, by writing the moment equilibrium equation about the right end of the free body shown in Fig. 16.3.2(b),

$$M_{neg}(\text{right}) + M_{pos} = \frac{w_u L_2 L_n}{2}\left(\frac{L_n}{4}\right) + V_{\mathbb{C}}\left(\frac{L_n}{2}\right)$$

Figure 16.3.2 Statics of a typical *exterior* panel in a two-way floor system.

Equation (16.3.2) is arrived at by adding the two preceding equations and dividing by 2 on each side. Note that Eq. (16.3.2) may also be obtained, as shown in Fig. 16.3.2(c), by the superposition of the simple span uniform loading parabolic positive moment diagram over the trapezoidal negative moment diagram due to end moments.

ACI-13.6.2 uses the symbol M_0 to mean $w_u L_2 L_n^2/8$ and calls M_0 the *total factored static moment*. It states, "Absolute sum of positive and average negative factored moments in each direction shall not be less than M_0"; or

$$\frac{M_{\text{neg}}(\text{left}) + M_{\text{neg}}(\text{right})}{2} + M_{\text{pos}} \geq M_0 = \frac{w_u L_2 L_n^2}{8} \tag{16.3.3}$$

in which

w_u = factored load per unit area

L_n = clear span in the direction moments are being determined, measured face to face of supports (ACI-13.6.2.5), but not less than $0.65L_1$

L_1 = span length in the direction moments are being determined, measured center to center of supports

L_2 = transverse span length, measured center to center of supports

Equations (16.3.1) and (16.3.2) are theoretically derived on the basis that $M_{\text{neg}}(\text{left})$, M_{pos}, and $M_{\text{neg}}(\text{right})$ occur simultaneously for the same live load pattern on the adjacent panels of the equivalent rigid frame defined in Fig. 16.2.1. If the live load is relatively heavy compared with dead load, then different live load patterns should be used to obtain the critical positive moment at midspan and the critical negative moments at the supports. In such a case, the "equal" sign in Eqs. (16.3.1) and (16.3.2) becomes the "greater" sign. This is the reason why ACI-13.6.2 states "absolute sum . . . shall not be *less* than M_0" as the design requirement. The designer should keep this in mind when steel reinforcement is selected for positive and negative bending moment in two-way floors when the direct design method is used for gravity load. To avoid the use of excessively small values of M_0 in the case of short spans and large columns or column capitals, the clear span L_n to be used in Eq. (16.3.3) is not to be less than $0.65L_1$ (ACI-13.6.2.5).

When the limitations for using the direct design method are met, it is customary to divide the value of M_0 into M_{neg} and M_{pos}, if the restraints at each end of the span are identical (Fig. 16.3.1); or into $[M_{neg}\,(\text{left}) + M_{neg}\,(\text{right})]/2$ and M_{pos} if the span end restraints are different (Fig. 16.3.2). Then the moments $M_{neg}\,(\text{left})$, M_{neg} (right), and M_{pos} must be distributed transversely along the lines $1'-3'$, $2'-4'$, and $5-6$, respectively. This last distribution is a function of the relative flexural stiffness between the slab and the included beam.

Total Factored Static Moment in Flat Slabs.   The ability of flat slab floor systems to carry load has been substantiated by numerous tests of actual structures [16.2]. However, the amount of reinforcement used, say, in a typical interior panel, was less than what it should be to satisfy an analysis by statics, as is demonstrated in this section. This led to some controversy [16.1], but after studies by Westergaard and Slater [16.3], a provision was adopted (about 1921) into the code that a reduction of moment coefficient from the statically required value of 0.125 to 0.09 may be made. This reduction was not regarded as a violation of statics but was used as a way of permitting an increase in the usable strength. The reduction, moreover, was applicable only to flat slabs that satisfied the limitations then specified in the code. Over the years these limitations had been liberalized, but at the same time the moment coefficient was raised to values closer to 0.125. The present ACI Code logically stipulates the use of the full statically required coefficient of 0.125.

The statical analysis of a typical interior panel was first made in 1914 by Nichols [16.1] and further developed later by Westergaard and others [16.3–16.5].

Consider the typical interior panel of a flat slab floor subjected to a factored load of w_u per unit area, as shown in Fig. 16.3.3(a). The total load on the panel area (rectangle minus four quadrantal areas) is supported by the vertical shears at the four quadrantal arcs. Let M_{neg} and M_{pos} be the total negative and positive moments about a horizontal axis in the L_2 direction along the edges of $ABCD$ and EF, respectively. Then

$$\text{load on area } ABCDEF = \text{sum of reactions at arcs } AB \text{ and } CD$$

$$= w_u\left(\frac{L_1 L_2}{2} - \frac{\pi c^2}{8}\right)$$

Figure 16.3.3 Statics of a typical interior panel in a flat slab floor system.

Considering the half-panel $ABCDEF$ as a free body, recognizing that there is no shear at the edges BC, DE, EF, and FA, and taking moments about axis 1-1,

$$M_{\text{neg}} + M_{\text{pos}} + w_u\left(\frac{L_1 L_2}{2} - \frac{\pi c^2}{8}\right)\left(\frac{c}{\pi}\right) - \frac{w_u L_1 L_2}{2}\left(\frac{L_1}{4}\right) + \frac{w_u \pi c^2}{8}\left(\frac{2c}{3\pi}\right) = 0$$

Letting $M_0 = M_{\text{neg}} + M_{\text{pos}}$,

$$M_0 = \frac{1}{8}w_u L_2 L_1^2\left(1 - \frac{4c}{\pi L_1} + \frac{c^3}{3L_2 L_1^2}\right) \approx \frac{1}{8}w_u L_2 L_1^2\left(1 - \frac{2c}{3L_1}\right)^2 \quad \text{(16.3.4)}$$

Actually, Eq. (16.3.4) may be more easily visualized by inspecting the equivalent interior span as shown in Fig. 16.3.3(b).

ACI-13.6.2.5 states that circular or regular polygon shaped supports shall be treated as square supports having the same area. For flat slabs, particularly with column capitals, the clear span L_n computed from using equivalent square supports should be compared with that indicated by Eq. (16.3.4), which is L_1 minus $2c/3$. In some cases the latter value is larger and should be used, consistent with the fact that ACI-13.6.2.2 does express its intent in an inequality.

Design Examples. In an effort to present, explain, and illustrate the design procedure for the three types of two-way floor systems, identified in this chapter as two-way slabs (with beams), flat slabs, and flat plates, it will be necessary to assume

$f_c' = 3$ ksi
$f_y = 40$ ksi
Service LL = 120 psf
Story height = 12 ft

Assume:
  Slab thickness = $6\frac{1}{2}''$
  Long beams 14 × 28
    overall
  Short beams 12 × 24
    overall
  Columns 15" × 15"

Figure 16.3.4 Floor plan in design example for two-way slab with beams.

that preliminary dimensions and sizes of the slab (and drop, if any), beams, and columns (and column capitals, if any) are available. In the usual design processes, not only the preliminary sizes may need to be revised as they are found unsuitable, but also designs based on two or three different relative beam sizes (when used) to slab thickness should be made and compared. Preliminary data for the three types of two-way floor systems to be illustrated are as follows.

Data for Two-Way Slab (with Beams) Design Example.   Figure 16.3.4 shows a two-way slab floor with a total area of 12,500 sq ft. It is divided into 25 panels with a panel size of 25 ft × 20 ft. Concrete strength is $f_c' = 3000$ psi and steel yield strength is $f_y = 40,000$ psi. Service live load is to be taken as 120 psf. Story height is 12 ft. The preliminary sizes are as follows: slab thickness is $6\frac{1}{2}$ in.; long beams are 14 × 28 in. overall; short beams are 12 × 24 in. overall; upper and lower columns are 15 × 15 in. The four kinds of panels (corner, long-sided edge, short-sided edge, and interior) are numbered 1, 2, 3, and 4 in Fig. 16.3.4.

■ **EXAMPLE 16.3.1**   For the two-way slab (with beams) design example, determine the total factored static moment in a loaded span in each of the four equivalent rigid frames whose widths are designated A, B, C, and D in Fig. 16.3.5.

Solution: The factored load w_u per unit floor area is

$$w_u = 1.4w_D + 1.7w_L = 1.4(6.5)(150/12) + 1.7(120)$$
$$= 114 + 204 = 318 \text{ psf}$$

Note that for two-way slabs *supported on beams* the definition of clear span L_n is different in ACI-9.0 for establishing minimum thickness for deflection control from that in ACI-13.6.2.5 for obtaining the statical moment on the slab-frame segment

Figure 16.3.5 Equivalent rigid frame notations in the two-way slab (with beams) design example.

where the supports are columns. ACI-9.0 logically states L_n is measured "face-to-face of beams" for the span relating to deflection of the slab panel, but ACI-13.6.2.5 states that L_n "shall extend from face-to-face of columns, capitals, brackets, or walls" as the support span for moments in the frame. Thus,

$$\text{for frame } A, \quad M_0 = \frac{1}{8}w_u L_2 L_n^2 = \frac{1}{8}(0.318)(20)(25 - 1.25)^2 = 448 \text{ ft-kips}$$

$$\text{for frame } B, \quad M_0 = 224 \text{ ft-kips}$$

$$\text{for frame } C, \quad M_0 = \frac{1}{8}w_u L_2 L_n^2 = \frac{1}{8}(0.318)(25)(20 - 1.25)^2 = 349 \text{ ft-kips}$$

$$\text{for frame } D, \quad M_0 = 175 \text{ ft-kips} \quad \blacksquare$$

Data for Flat Slab Design Example. Figure 16.3.6 shows a flat slab floor with a total area of 12,500 sq ft. It is divided into 25 panels with a panel size of 25 × 20 ft. Concrete strength is $f_c' = 3000$ psi and steel yield strength is $f_y = 40,000$ psi. Service live load is 120 psf. Story height is 10 ft. Exterior columns are 16 in. square and interior columns are 18 in. round. Edge beams are 14 × 24 in. overall. Thickness of slab is $7\frac{1}{2}$ in. outside of drop panel and $10\frac{1}{2}$ in. through the drop panel. Sizes of column capitals and drop panels are as shown in Fig. 16.3.6.

■ **EXAMPLE 16.3.2** Compute the total factored static moment in the long and short directions of an interior panel in the flat slab design example as shown in Fig. 16.3.6. Compare the results obtained by using Eqs. (16.3.3) and (16.3.4).

Solution: Neglecting the weight of the drop panel, the service dead load is $(150/12)(7.5) = 94$ psf; thus

$$w_u = 1.4w_D + 1.7w_L = 1.4(94) + 1.7(120) = 132 + 204 = 336 \text{ psf}$$

Figure 16.3.6 Flat slab design example.

Using Eq. (16.3.4),

$$M_0 = \frac{1}{8} w_u L_2 L_1^2 \left(1 - \frac{2c}{3L_1} \right)^2 = \frac{1}{8}(0.336)(20)(25)^2 \left[1 - \frac{2(5)}{3(25)} \right]^2 = 395 \text{ ft-kips}$$
(in long direction)

$$M_0 = \frac{1}{8} w_u L_2 L_1^2 \left(1 - \frac{2c}{3L_1} \right)^2 = \frac{1}{8}(0.336)(25)(20)^2 \left[1 - \frac{2(5)}{3(20)} \right]^2 = 292 \text{ ft-kips}$$
(in short direction)

The equivalent square area for the column capital (ACI-13.6.2.5) has its side equal to 4.43 ft; then, using Eq. (16.3.3), with L_n measured to the face of capital (i.e., equivalent square),

$$M_0 = \tfrac{1}{8} w_u L_2 L_n^2 = \tfrac{1}{8}(0.336)(20)(25 - 4.43)^2 = 356 \text{ ft-kips}$$
(in long direction)

$$M_0 = \tfrac{1}{8} w_u L_2 L_n^2 = \tfrac{1}{8}(0.336)(25)(20 - 4.43)^2 = 255 \text{ ft-kips}$$
(in short direction)

Insofar as flat slabs with column capitals are concerned, it appears that the larger values of 395 ft-kips and 292 ft-kips should be used because Eq. (16.3.4) is specially

Figure 16.3.7 Flat plate design example.

suitable; in particular, ACI-13.6.2.2 states that the total factored static moment shall *not be less than* that given by Eq. (16.3.3). ■

Data for Flat Plate Design Example. Figure 16.3.7 shows a flat plate floor with a total area of 4500 sq ft. It is divided into 25 panels with a panel size of 15 × 12 ft. Concrete strength is $f'_c = 4000$ psi and steel yield strength is $f_y = 50,000$ psi. Service live load is 60 psf. Story height is 9 ft. All columns are rectangular, 12 in. in the long direction and 10 in. in the short direction. Preliminary slab thickness is set at $5\frac{1}{2}$ in. No edge beams are used along the exterior edges of the floor.

■ **EXAMPLE 16.3.3** Compute the total factored static moment in the long and short directions of a typical panel in the flat plate design example as shown in Fig. 16.3.7.

Solution: The dead load for a $5\frac{1}{2}$ in. slab is

$$w_D = (5.5/12)(150) = 69 \text{ psf}$$

The factored load per unit area is

$$w_u = 1.4w_D + 1.7w_L = 1.4(69) + 1.7(60) = 96 + 102 = 198 \text{ psf}$$

Using Eq. (16.3.3), with clear span L_n measured face-to-face of columns,

$$M_0 = \tfrac{1}{8}(0.198)(12)(15 - 1)^2 = 58.2 \text{ ft-kips}$$
(in long direction)

$$M_0 = \tfrac{1}{8}(0.198)(15)(12 - 0.83)^2 = 46.3 \text{ ft-kips}$$
(in short direction) ■

16.4 RATIO OF FLEXURAL STIFFNESSES OF LONGITUDINAL BEAM TO SLAB

When beams are used along the column lines in a two-way floor system, an important parameter affecting the design is the relative size of the beam to the thickness of the slab. This parameter can best be measured by the ratio α of the flexural rigidity (called flexural stiffness by the ACI Code) $E_{cb}I_b$ of the beam to the flexural rigidity $E_{cs}I_s$ of the slab in the transverse cross section of the equivalent frame shown in Fig. 16.4.1. The separate moduli of elasticity E_{cb} and E_{cs}, referring to the beam and slab, provide for different strength concrete (and thus different E_c values) for the beam and slab. The moments of inertia I_b and I_s refer to the gross sections of the beam and slab within the cross section of Fig. 16.4.1(c). ACI-13.2.4 permits the slab on each side of the beam web to act as a part of the beam, this slab portion being limited to a distance equal to the projection of the beam above or below the slab, whichever is greater, but not greater than four times the slab thickness, as shown in Fig. 16.4.2. More accurately, the small portion of the slab already counted in the beam should not be used in I_s, but ACI permits the use of the total width of the equivalent frame

Figure 16.4.1 Plan, elevation, and cross section of equivalent frame in a two-way floor system.

Figure 16.4.2 Cross sections for moment of inertia of a flanged section.

in computing I_s. Thus,

$$\alpha = \frac{E_{cb}I_b}{E_{cs}I_s} \qquad (16.4.1)$$

The moment of inertia of a flanged beam section about its own centroidal axis (Fig. 16.4.2) may be shown to be

$$I_b = k\,\frac{b_w h^3}{12} \qquad (16.4.2a)$$

in which

$$k = \frac{1 + \left(\dfrac{b_E}{b_w} - 1\right)\left(\dfrac{t}{h}\right)\left[4 - 6\left(\dfrac{t}{h}\right) + 4\left(\dfrac{t}{h}\right)^2 + \left(\dfrac{b_E}{b_w} - 1\right)\left(\dfrac{t}{h}\right)^3\right]}{1 + \left(\dfrac{b_E}{b_w} - 1\right)\left(\dfrac{t}{h}\right)} \qquad (16.4.2b)$$

where

h = overall beam depth

t = overall slab thickness

b_E = effective width of flange

b_w = width of web

Equation (16.4.2b) expresses the nondimensional constant k in terms of (b_E/b_w) and (t/h). Typical values of k are tabulated in Table 16.4.1 and three curves are plotted

Table 16.4.1   Values of k in Terms of (b_E/b_w) and (t/h) in Eq. (16.4.2b)

b_E/b_w	0.1	0.2	0.3	0.4	0.5	0.6	0.7	0.8	0.9	1.0
2	1.222	1.328	1.366	1.372	1.375	1.396	1.454	1.565	1.743	2.000
3	1.407	1.564	1.605	1.608	1.625	1.694	1.844	2.098	2.477	3.000
4	1.564	1.744	1.777	1.781	1.825	1.956	2.212	2.621	3.209	4.000

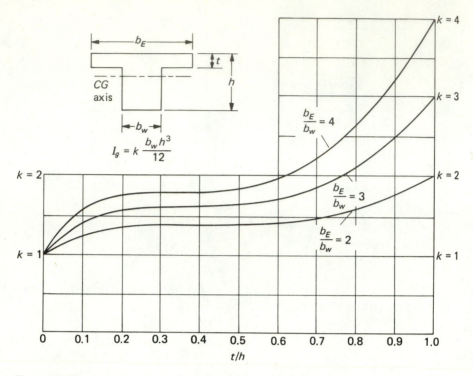

Figure 16.4.3 Values of k in terms of b_E/b_W and t/h.

in Fig. 16.4.3. The values of k are about 1.4, 1.6, and 1.8, respectively, for b_E/b_w values of 2, 3, and 4, when t/h values are between 0.2 and 0.5. Thus

$$k \approx 1.0 + 0.2\left(\frac{b_E}{b_w}\right) \quad \text{for } 2 < \frac{b_E}{b_w} < 4 \quad \text{and} \quad 0.2 < \frac{t}{h} < 0.5 \qquad \textbf{(16.4.2c)}$$

■ **EXAMPLE 16.4.1**   For the two-way slab (with beams) design example described in Section 16.3, compute the ratio α of the flexural stiffness of the longitudinal beam to that of the slab in the equivalent rigid frame, for all the beams around panels 1, 2, 3, and 4 in Fig. 16.4.4.

Solution: (a) B1–B2. Referring to Fig. 16.4.4, the effective width b_E for B1–B2 is the smaller of $14 + 2(21.5) = 57$ in. and $14 + 8(6.5) = 66$ in.; thus $b_E = 57$ in. Using Eq. (16.4.2b),

$$\frac{b_E}{b_w} = \frac{57}{14} = 4.07, \qquad \frac{t}{h} = \frac{6.5}{28} = 0.232$$

$$k = 1.774, \qquad I_b = 1.774\,\frac{14(28)^3}{12} = 45{,}400 \text{ in.}^4$$

Figure 16.4.4 Computation of α values in Example 16.4.1.

A slightly higher value of k would have been obtained using Eq. (16.4.2c). Using Eq. (16.4.1), where $E_{cb} = E_{cs}$,

$$I_s = \tfrac{1}{12}(240)(6.5)^3 = 5490 \text{ in.}^4, \qquad \alpha = \frac{E_{cb}I_b}{E_{cs}I_s} = \frac{45,400}{5490} = 8.27$$

(b) $B3$–$B4$. Referring to Fig. 16.4.4, the effective width b_E for $B3$–$B4$ is the smaller of $14 + 21.5 = 35.5$ in. and $14 + 4(6.5) = 40$ in.; thus $b_E = 35.5$ in. Using Eq. (16.4.2b),

$$\frac{b_E}{b_w} = \frac{35.5}{4} = 2.54, \qquad \frac{t}{h} = \frac{6.5}{28} = 0.232$$

$$k = 1.484, \qquad I_b = 1.484 \frac{14(28)^3}{12} = 38,000 \text{ in.}^4$$

Using Eq. (16.4.1),

$$I_s = \tfrac{1}{12}(120)(6.5)^3 = 2745 \text{ in.}^4, \qquad \alpha = \frac{E_{cb}I_b}{E_{cs}I_s} = \frac{38,000}{2745} = 13.83$$

(c) *B5–B6.* Referring to Fig. 16.4.4, the effective width b_E for *B5–B6* is the smaller of $12 + 2(17.5) = 47$ in. and $12 + 8(6.5) = 64$ in.; thus $b_E = 47$ in. Using Eq. (16.4.2b),

$$\frac{b_E}{b_w} = \frac{47}{12} = 3.92, \qquad \frac{t}{h} = \frac{6.5}{24} = 0.271$$

$$k = 1.762, \qquad I_b = 1.762 \frac{12(24)^3}{12} = 24,400. \text{ in.}^4$$

Using Eq. (16.4.1),

$$I_s = \tfrac{1}{12}(300)(6.5)^3 = 6870 \text{ in.}^4, \qquad \alpha = \frac{E_{cb}I_b}{E_{cs}I_s} = \frac{24,000}{6870} = 3.55$$

(d) *B7–B8.* Referring to Fig. 16.4.4, the effective width b_E for *B7–B8* is the smaller of $12 + 17.5 = 29.5$ in. and $12 + 4(6.5) = 38$ in.; thus $b_E = 29.5$ in. Using Eq. (16.4.2b),

$$\frac{b_E}{b_w} = \frac{29.5}{12} = 2.46, \qquad \frac{t}{h} = 0.271$$

$$k = 1.480, \qquad I_b = 1.480 \frac{12(24)^3}{12} = 20,500 \text{ in.}^4$$

Using Eq. (16.4.1),

$$I_s = \tfrac{1}{12}(150)(6.5)^3 = 3435 \text{ in.}^4, \qquad \alpha = \frac{E_{cb}I_b}{E_{cs}I_s} = \frac{20,500}{3435} = 5.96$$

The resulting α values for *B1* through *B8* around panels 1, 2, 3, and 4 are shown in Fig. 16.4.4. For this design, the α values vary between 3.55 and 13.83; thus the equivalent rigid frames have their substantial portion along or close to the column lines, even though their widths vary from 10 to 25 ft. ∎

16.5 MINIMUM SLAB THICKNESS FOR DEFLECTION CONTROL

Control of deflections in two-way floor systems is dealt with in ACI-9.5.3. When deflections are computed according to ACI-9.5.3.6, then ACI-Table 9.5(b) states the maximum permissible computed deflection. To compute deflections, the use of the effective moment of inertia I_e, Eq. (14.4.1), is endorsed unless computed deflections using other procedures are "in reasonable agreement with results of comprehensive tests." Various methods for obtaining deflections of two-way floor systems have been proposed [16.52–16.71]; however, no specific procedure is given by the ACI Code or Commentary. Computation of two-way floor system deflections is outside the scope of this book.

To aid the designer, ACI-9.5.3.2 provides a minimum thickness table for slabs without interior beams. Also, ACI-9.5.3.3 has minimum thickness equations applicable to all two-way floor systems with or without interior beams, wherein the ratio of long to short span does not exceed 2. Only in cases where no interior beams exist, although there may or may not be edge beams, does the ACI Code provide the *option* of obtaining the minimum thickness from a table [ACI-Table 9.5(c)] instead of from the equations. This table and the formulas give slab thicknesses that, from experience, are considered satisfactory.

Four parameters affect the equations of ACI-9.5.3.3; they are (1) L_n, the longer clear span of the slab panel, (2) L_n/S_n (ACI calls it β), the ratio of longer clear span to shorter clear span of the slab panel, (3) f_y, the yield strength of the steel reinforcement; and (4) α_m, average for the four α values for relative stiffness of a panel perimeter beam compared to the slab, as described in Section 16.4. In terms of these parameters, ACI-9.5.3.3 requires the following for two-way floor systems:

$$\min t = \frac{L_n(0.8 + 0.2f_y/40,000)}{36 + 5\alpha_m\, L_n/S_n - 0.6(1 + L_n/S_n)} \qquad \text{ACI Formula (9-11)} \qquad \textbf{(16.5.1)}$$

but not less than

$$\min t = \frac{L_n(0.8 + 0.2f_y/40,000)}{36 + 9L_n/S_n} \qquad \text{ACI Formula (9-12)} \qquad \textbf{(16.5.2)}$$

and need not be more than

$$\min t = \frac{L_n(0.8 + 0.2f_y/40,000)}{36} \qquad \text{ACI Formula (9-13)} \qquad \textbf{(16.5.3)}$$

The rationale for the ACI formulas may be examined by observing the plot of the formulas in Fig. 16.5.1. The quantity obtained is the long direction clear span to thickness ratio L_n/t, similar to the span to depth ratio limitations that are used for beams (see Sections 14.10 and 14.11).

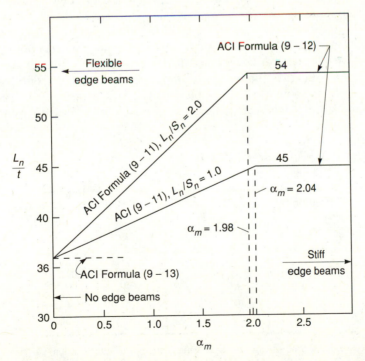

Figure 16.5.1 ACI minimum slab thickness formulas (for Grade 40 steel). For Grade 60, divide L_n/t by 1.1; for Grade 50, divide L_n/t by 1.05.

Figure 16.5.1 includes the full feasible range of parameters: (1) the panel proportions L_n/S_n ranging from square to two-to-one rectangular, and (2) the α_m ranging from zero with no edge beams to 2.5 or so for very stiff edge beams. The parameter f_y is included in the footnote that the allowable L_n/t must be divided by 1.1 for $f_y = 60,000$ psi, or by 1.05 for $f_y = 50,000$ psi, which means that the required minimum slab thickness is larger for a given L_n when f_y exceeds 40,000 psi. The parameter L_n is included in the diagram of Fig. 16.5.1 for maximum L_n/t that may be allowed.

The following logical results are indicated from a study of Fig. 16.5.1:

1. The average stiffness of the panel perimeter beams is the predominant factor; the no-beam condition requires the thickest slab ($L_n/t = 36$ max), from which the thickness required decreases until the panel perimeter beams provide an average stiffness corresponding to $\alpha_m = 1.92 + 0.12S_n/L_n$, which has a maximum value of 2.04 for combinations of variables within the applicable range.
2. The ratio L_n/S_n of the plan dimensions of the panel affects the thickness by requiring the thickest slab (lowest L_n/t) for the square panel. As the panel becomes more rectangular, more load is transferred to the supports in the short direction; thus, with the shorter span becoming more effective the thickness can be reduced (L_n/t increased).
3. The steel yield strength f_y influences the thickness, since the higher the yield stress, the higher the stress and strain at service load, the more the cracking, and the greater the deflection. The L_n/t values of Fig. 16.5.1 are for Grade 40 steel and are to be divided by 1.1 and 1.05 for Grade 60 steel and Grade 50 steel, respectively.

In flat slab and flat plate two-way systems, there may or may not be edge beams but there are definitely no interior beams in such systems. Thus, for the flat slab and flat plate, ACI provides the option of obtaining the minimum slab thickness from ACI-Table 9.5(c), shown here as Table 16.5.1. The parameters involved are (1) the yield strength f_y, (2) whether qualified drop panels as defined in ACI-13.4.7.1 and

Table 16.5.1   Minimum Thickness of Slab Without Interior Beams [ACI-Table 9.5(c)]

f_y^* (ksi)	WITHOUT DROP			WITH DROP		
	EXTERIOR PANEL		INTERIOR PANEL	EXTERIOR PANEL		INTERIOR PANEL
	$\alpha^{\ddagger} = 0$	$\alpha \geq 0.8$		$\alpha = 0$	$\alpha \geq 0.8$	
40	$\dfrac{L_n}{33}$	$\dfrac{L_n}{36}$	$\dfrac{L_n}{36}$	$\dfrac{L_n}{36}$	$\dfrac{L_n}{40}$	$\dfrac{L_n}{40}$
60	$\dfrac{L_n}{30}$	$\dfrac{L_n}{33}$	$\dfrac{L_n}{33}$	$\dfrac{L_n}{33}$	$\dfrac{L_n}{36}$	$\dfrac{L_n}{36}$

* For f_y between 40 and 60 ksi, min t is to be obtained by linear interpolation.

‡ For corner panels, α refers to beam having lower flexural stiffness.

13.4.7.2 are present, and (3) whether the panel is exterior or interior. In the 1983 ACI Code, a parameter β_s representing the proportion of the panel perimeter that is continuous with an adjacent panel was included in Eqs. (16.5.1) and (16.5.2). The parameter β_s was eliminated in the 1989 ACI Code by substitution of a compromise numerical value. It is noted again that Table 16.5.1 is an *optional* alternative to using Eqs. (16.5.1) to (16.5.3) for flat slabs and flat plates.

16.6 NOMINAL REQUIREMENTS FOR SLAB THICKNESS AND SIZE OF EDGE BEAMS, COLUMN CAPITAL, AND DROP PANEL

Whether the ACI "direct design method" or the "equivalent frame method" is used for determining the longitudinal distribution of moments, certain nominal requirements for slab thickness and size of edge beams, column capital, and drop panel must be fulfilled. These requirements are termed "nominal" because they are code prescribed. It should be realized, of course, that the code provisions are based on a combination of experience, judgment, tests, and theoretical analysis.

Slab Thickness. While ACI Formulas (9-11), (9-12), and (9-13), as shown in Eqs. (16.5.1) to (16.5.3), set limits for slab thickness in two-way floor systems, other nominal and practical requirements included in ACI-9.5.3.2 and 9.5.3.3 are:

1. For flat plates and flat slabs without drop panels,   5 in.
2. For two-way slabs on shallow beams having $\alpha_m < 2.0$,   5 in.
3. For slabs without beams but with properly sized drop panels.   4 in.
4. For two-way slabs on stiff beams having $\alpha_m \geq 2.0$, $3\frac{1}{2}$ in.

Edge Beams. When edge beams are not used, or if the edge beam is so small that α is less than 0.80, ACI-9.5.3.5 states that the requirement for slab thickness expressed by Eqs. (16.5.1) to (16.5.3) must be increased by 10% in the panel having the discontinuous edge.

Column Capital. Used in flat slab construction, the column capital (Fig. 16.6.1) is an enlargement of the top of the column as it meets the floor slab or drop panel. Since no beams are used, the purpose of the capital is to gain increased perimeter around the column to transmit shear from the floor loading and to provide increasing thickness as the perimeter decreases near the column. Assuming a maximum 45° line for distribution of the shear into the column, ACI-13.1.2 requires that the effective column capital for strength considerations be within the largest circular cone, right pyramid, or tapered wedge with a 90° vertex that can be included within the outlines of the actual supporting element (see Fig. 16.6.1). The diameter of the column capital is usually about 20 to 25% of the average span length between columns.

Drop Panel. Used in flat slab and flat plate construction, the drop panel (Fig. 16.1.2) is an area of increased slab thickness surrounding the column. When drop panels extend from the centerline of supports a minimum distance of one-sixth of the span length measured from center to center in each direction, and when the projection below the slab is at least one-fourth of the slab thickness outside of the drop, ACI-9.5.3.4 permits the minimum slab thickness required by Eqs. (16.5.1) to (16.5.3) to be

Figure 16.6.1 Effective dimension of column capital.

reduced by 10%. For determining reinforcement, ACI-13.4.7.3 requires that the thickness of the drop below the slab be assumed at a value no larger than one fourth of the distance between the edge of the drop panel and the edge of the column capital. Because of this requirement, there is little reason to use a drop panel of larger thickness.

■ **EXAMPLE 16.6.1** For the two-way slab (with beams) design example described in Section 16.3, determine the minimum thickness requirement for deflection control; and compare it with the preliminary thickness of $6\frac{1}{2}$ in.

Solution: The average ratios α_m for panels 1, 2, 3, and 4 may be computed from the α values shown in Fig. 16.4.4; thus

$$\alpha_m \text{ for panel 1} = \tfrac{1}{4}(5.96 + 8.27 + 3.55 + 13.83) = 7.90$$

$$\alpha_m \text{ for panel 2} = \tfrac{1}{4}(3.55 + 8.27 + 3.55 + 13.83) = 7.30$$

$$\alpha_m \text{ for panel 3} = \tfrac{1}{4}(5.96 + 8.27 + 3.55 + 8.27) = 6.51$$

$$\alpha_m \text{ for panel 4} = \tfrac{1}{4}(3.55 + 8.27 + 3.55 + 8.27) = 5.91$$

Since the α_m values for all four panels are well above 2, Fig. 16.5.1 shows that Eq. (16.5.2), which is ACI Formula (9-12), applies. The minimum thickness for all panels, using $L_n = 24$ ft, $S_n = 18.83$ ft, and $f_y = 40,000$ psi, becomes

$$\min t = \frac{L_n(0.8 + 0.2f_y/40,000)}{36 + 9L_n/S_n} = \frac{24(12)1.0}{36 + 9(24)/18.83} = 6.07 \text{ in.}$$

If a uniform slab thickness for the entire floor area is to be used, the minimum for deflection control is 6.07 in., which compares well with the $6\frac{1}{2}$-in. preliminary thickness. ■

■ **EXAMPLE 16.6.2**   Review the slab thickness and other nominal requirements for the dimensions in the flat slab design example described in Section 16.3.

Solution: **(a)** Stiffness of edge beams. Before using either Table 16.5.1 or Eqs. (16.5.1) to (16.5.3), the α values for the edge beams are needed. The moment of inertia of the edge beam section shown in Fig. 16.6.2(b) is 22,900 in.[4] Thus the α value for the long

(a) Plan view

(b) Edge beam section

Figure 16.6.2 Computation of α values in Example 16.6.2.

edge beam is

$$\alpha = \frac{I_b}{I_s} = \frac{22,900}{120(7.5)^3/12} = \frac{22,900}{4220} = 5.42$$

and for the short edge beam, it is

$$\alpha = \frac{I_b}{I_s} = \frac{22,900}{150(7.5)^3/12} = \frac{22,900}{5270} = 4.34$$

These α values are entered on Fig. 16.6.2(a). The average α (i.e., α_m) for panels 1, 2, and 3 are, respectively, 2.44, 1.36, and 1.08.

(b) Minimum slab thickness using Table 16.5.1 or ACI-Table 9.5(c). The long and short clear spans for deflection control are:

$$L_n = 25 - 4.43 = 20.57 \text{ ft}; \qquad S_n = 20 - 4.43 = 15.57 \text{ ft}$$

from which

$$\frac{L_n}{S_n} = \frac{20.57}{15.57} = 1.32$$

For $f_y = 40$ ksi, a flat slab with drop panel, and $\alpha =$ smaller of 4.34 and 5.42, Table 16.5.1 gives

$$\min t = \frac{L_n}{40} = \frac{20.57(12)}{40} = 6.17 \text{ in.}$$

for both exterior and interior panels.

(c) Minimum slab thickness using Eqs. (16.5.1) to (16.5.3); that is, ACI Formulas (9-11) to (9-13). For flat slabs without interior beams, it is more practical to use ACI-Table 9.5(c), as in part (b), although the Code does not preclude use of the formulas. Neither ACI-Table 9.5(c) nor ACI Formulas (9-11) to (9-13) should be considered more "accurate" than the other; both approaches are intended to provide a practical way of obtaining a satisfactory thickness. The results of applying the formulas are shown in Table 16.5.2, which shows that the minimum slab thickness is 5.15 in., 5.66 in., 5.9 in., and 6.86 in., for panels 1, 2, 3, and 4, respectively. The 6.86 in. is probably conservative because the limiting case of $\alpha_m = 0$ is meant more for flat plates with neither capitals nor drop panels. However, applying ACI-9.5.3.5, the

Table 16.5.2 Minimum Slab Thickness for Example 16.6.2, Using Eqs. (16.5.1) to (16.5.3)

PANEL	1	2	3	4
α_m	2.44	1.36	1.08	0
Min·t, inches				
Eq. (16.5.1), at least	4.87	5.66	5.91	7.13
Eq. (16.5.2), not less than	5.15	5.15	5.15	5.15
Eq. (16.5.3), need not exceed	6.86	6.86	6.86	6.86
CONTROLLING	5.15	5.66	5.91	6.86

6.86 in. may be reduced by 10% to obtain 6.17 in., the same result as obtained in part (a).

(d) Nominal requirement for slab thickness. Considering the results of parts (b) and (c), and the practical decision to use the same thickness for all panels, the controlling minimum thickness would be 6.17 in. The $7\frac{1}{2}$ in. slab thickness used is more than ample; $6\frac{1}{2}$ in. should probably have been used.

(e) Thickness of drop panel. Reinforcement within the drop panel must be computed on the basis of the $10\frac{1}{2}$ in. thickness actually used or $7\frac{1}{2}$ in. plus one-fourth of the projection of the drop beyond the column capital, whichever is smaller. In order that the full 3-in. projection of the drop below the $7\frac{1}{2}$ in. slab is usable in computing reinforcement, the 6 ft 8 in. side of the drop is revised to 7 ft so that one-fourth of the distance between the edges of the 5-ft column capital and the 7-ft drop is just equal to $(10.5 - 7.5) = 3$ in. ■

■ **EXAMPLE 16.6.3** Review the slab thickness and other nominal requirements for the dimensions in the flat plate design example described in Section 16.3.

Solution: **(a)** Minimum slab thickness from ACI-Table 9.5(c). For $f_y = 50$ ksi, for a flat plate which inherently has $\alpha = 0$, and $L_n = 15 - 1 = 14$ ft, from Table 16.5.1,

$$\min t = \text{linear interpolation between } f_y = 40 \text{ ksi and } f_y = 60 \text{ ksi}$$

$$= \frac{1}{2}\left(\frac{L_n}{33} + \frac{L_n}{30}\right) = \frac{1}{2}(168)\left(\frac{1}{33} + \frac{1}{30}\right) = 5.34 \text{ in.} \quad \text{(exterior panel)}$$

and

$$\min t = \frac{1}{2}\left(\frac{L_n}{36} + \frac{L_n}{33}\right) = \frac{1}{2}(168)\left(\frac{1}{36} + \frac{1}{33}\right) = 4.88 \text{ in.} \quad \text{(interior panel)}$$

(b) Minimum slab thickness using ACI Formulas (9-11) to (9-13). One may easily determine by referring to Fig. 16.5.1 that the controlling equation for the flat plate is Formula (9-13), Eq. (16.5.3). By definition the flat plate has no edge beams along panel boundaries; therefore, $\alpha = 0$. Thus,

$$\min t = \frac{L_n(0.8 + 0.2f_y/40,000)}{36} = \frac{14(1.05)12}{36} = 4.90 \text{ in.}$$

which applies to panel 4. For panels 1, 2, and 3 which have one or more discontinuous edges, the 4.90 in. must be increased by 10% to 5.39 in. according to ACI-9.5.3.5.

(c) The ACI Code permits the option of using the results in either parts (a) or (b) above. The $5\frac{1}{2}$ in. slab thickness as used for all panels satisfies these requirements and it exceeds the nominal minimum of 5 in. for slabs without drop panels and without beams. ■

16.7 LIMITATIONS OF DIRECT DESIGN METHOD

Over the years the use of two-way floor systems has been extended from one-story or low-rise to medium or high-rise buildings. For the common cases of one-story or low-rise buildings, lateral load (wind or earthquake) is of lesser concern; thus most of the ACI Code refers only to gravity load (dead and live uniform load). In particular, when the dimensions of the floor system are quite regular and when the live load is not excessively large compared to the dead load, the use of a set of prescribed coefficients to distribute longitudinally the total factored static moment M_0 seems reasonable. As shown in Figs. 16.3.1 and 16.3.2, for each clear span in the equivalent rigid frame, the equation

$$\frac{M_{neg}(\text{left}) + M_{neg}(\text{right})}{2} + M_{pos} \geq \left[M_0 = \frac{w_u L_2 L_n^2}{8} \right] \qquad [16.3.3]$$

is to be satisfied.

In order that the designer may use the direct design method, in which a set of prescribed coefficients give the negative end moments and the positive moment within the span of the equivalent rigid frame, ACI-13.6.1 imposes the following limitations:

1. There is a minimum of three continuous spans in each direction.
2. Panels must be rectangular with the ratio of longer to shorter span center-to-center of supports within a panel not greater than 2.0.
3. The successive span lengths center-to-center of supports in each direction do not differ by more than one-third of the longer span.
4. Columns are not offset more than 10% of the span in the direction of the offset.
5. The load is due to gravity only and is uniformly distributed over an entire panel, and the service live load does not exceed 3 times the service dead load.
6. The relative stiffness ratio of L_1^2/α_1 to L_2^2/α_2 must lie between 0.2 and 5.0, where α is the ratio of the flexural stiffness of the included beam to that of the slab.

Though the design of two-way floor systems is to a large extent empirical, the ACI limitations conform to the experimental results that are available [16.15–16.22] and to many years of experience with slabs in actual structures. The "direct design method" can also be used when it can be demonstrated that variations from any of the six limitations will still produce a slab system that satisfies the conditions of equilibrium and geometric compatibility and provides adequate strength and serviceability. Van Buren [16.28] has provided such an analysis for staggered columns in flat plates.

■ **EXAMPLE 16.7.1**   Show that for the two-way slab (with beams) design example described in Section 16.3 the six limitations of the direct design method are satisfied.

Solution: The first four limitations are satisfied by inspection. For the fifth limitation,

$$\text{service dead load } w_D = 6.5\left(\frac{150}{12}\right) = 81 \text{ psf}$$

$$\text{service live load } w_L = 120 \text{ psf}$$

$$\frac{w_L}{w_D} = \frac{120}{81} < 3 \qquad\qquad \textbf{OK}$$

For the sixth limitation, referring to Fig. 16.4.4 and taking L_1 and L_2 in the long and short directions, respectively,

Panel 1,

$$\frac{L_1^2}{\alpha_1} = \frac{625}{0.5(13.83 + 8.27)} = 56.6$$

$$\frac{L_2^2}{\alpha_2} = \frac{400}{0.5(5.96 + 3.55)} = 84.0$$

Panel 2,

$$\frac{L_1^2}{\alpha_1} = \frac{625}{0.5(13.83 + 8.27)} = 56.6$$

$$\frac{L_2^2}{\alpha_2} = \frac{400}{3.55} = 112.7$$

Panel 3,

$$\frac{L_1^2}{\alpha_1} = \frac{625}{8.27} = 75.6$$

$$\frac{L_2^2}{\alpha_2} = \frac{400}{0.5(5.96 + 3.55)} = 84.0$$

Panel 4,

$$\frac{L_1^2}{\alpha_1} = \frac{625}{8.27} = 75.6$$

$$\frac{L_2^2}{\alpha_2} = \frac{400}{3.35} = 112.7$$

All ratios of L_1^2/α_1 to L_2^2/α_2 lie between 0.2 and 5. ∎

16.8 DIRECT DESIGN METHOD—LONGITUDINAL DISTRIBUTION OF MOMENTS

In the "direct design method," moment curves in the direction of span length need not be computed by an elastic analysis of the equivalent rigid frame subjected to various pattern loadings; instead they are nominally defined for regular situations, with additional prescribed adjustments for pattern loading effects.

Figure 16.8.1 shows the longitudinal moment diagram for the typical interior span of the equivalent rigid frame in a two-way floor system, as prescribed by

Figure 16.8.1 Longitudinal moment diagram for interior span.

ACI-13.6.3.2. Later in Section 16.12, the positive moment $0.35M_0$ or the negative moment $0.65M_0$ is to be distributed transversely to the slab having width L_2 and to the included beam (if any) having clear span L_n. Note that

$$M_0 = \tfrac{1}{8}w_u L_2 L_n^2 \qquad\qquad \textbf{[16.3.3]}$$

Figure 16.8.2 Longitudinal moment diagram for exterior span.

For a span that is completely fixed at both ends, the negative moment at the fixed end is twice as large as the positive moment at midspan. For a typical interior span satisfying the limitations for the direct design method, the specified negative moment of $0.65M_0$ is a little less than twice the specified positive moment of $0.35M_0$, which is fairly reasonable because the restraining effect of the columns and adjacent panels is definitely less than that of a completely fixed-ended beam.

For the exterior span, ACI-13.6.3.3 provides the longitudinal moment diagram for each of the five categories as described in Fig. 16.8.2. Upon examination of these diagrams, one sees that the negative moment at the exterior support increases from 0 to $0.65M_0$, the positive moment within the span decreases from $0.63M_0$ to $0.35M_0$, and the negative moment at the interior support decreases from $0.75M_0$ to $0.65M_0$, all gradually as the restraint at the exterior support increases from the case of a masonry wall support to that of a reinforced concrete wall built monolithically with the slab. ACI Commentary-13.6.3.3 states that high positive moments are purposely assigned into the span since design for exterior negative moment will be governed by minimum reinforcement to control cracking.

Regarding the ACI Code suggested moment diagrams of Figs. 16.8.1 and 16.8.2, ACI-13.6.7 permits these moments to be modified by 10% provided the total factored static moment M_0 for the panel is statically accommodated.

16.9 DIRECT DESIGN METHOD—EFFECT OF PATTERN LOADINGS ON POSITIVE MOMENT

To understand the effect of pattern loadings on the longitudinal moment values in multiple panel two-way floor systems, it is convenient to review some aspects of the continuity analysis of the usual column–beam type of rigid frames discussed earlier in Chapter 7. Some of the findings, which might be visualized by those knowledgeable in influence lines and maximum moment envelopes due to dead and live load combinations, are as follows: (1) the higher the ratio of column stiffness to beam stiffness, the smaller the effect of pattern loadings, because the ends of the span are closer to the fixed condition and less effect is exerted on the span by loading patterns on adjacent spans; (2) the lower the ratio of dead load to live load, the larger the effect of pattern loadings, because dead load exists constantly on all spans and the pattern is related to live load only; and (3) maximum negative moments at supports are less affected by pattern loadings than maximum positive moments within the span.

Based on studies by Jirsa, Sozen, and Siess [16.36], it is recommended that in order to limit the increase in positive moment within the two-way floor system caused by pattern loadings to such a tolerance (33% according to ACI Commentary-13.6.10) that the moment values in Fig. 16.8.2 (ACI-13.6.3) may be used, the ratio of column stiffness to slab stiffness must have at least the values shown in Table 16.9.1 (from ACI-Table 13.6.10). If the ratio α_c of column stiffness to slab and beam stiffness is less than the minimum value $\alpha_{\min}$ shown in Table 16.9.1, then the positive moments in Figs. 16.8.1 and 16.8.2 must be multiplied by the coefficient δ_s given by the following

Table 16.9.1 Minimum Ratio α_{min} of Column Stiffness to Combined Beam and Slab Stiffness to Qualify for Positive Moment Multiplier $\delta_s = 1$ (α = Ratio of Beam to Slab Stiffness)[a]

$\beta_a = \dfrac{DL}{LL}$	ASPECT RATIO L_2/L_1	RATIO OF BEAM TO SLAB STIFFNESS, α				
		0	0.5	1.0	2.0	4.0
2.00	0.50–2.00	0	0	0	0	0
1.00	0.50	0.6	0	0	0	0
	0.80	0.7	0	0	0	0
	1.00	0.7	0.1	0	0	0
	1.25	0.8	0.4	0	0	0
	2.00	1.2	0.5	0.2	0	0
0.50	0.50	1.3	0.3	0	0	0
	0.80	1.5	0.5	0.2	0	0
	1.00	1.6	0.6	0.2	0	0
	1.25	1.9	1.0	0.5	0	0
	2.00	4.9	1.6	0.8	0.3	0
0.33	0.50	1.8	0.5	0.1	0	0
	0.80	2.0	0.9	0.3	0	0
	1.00	2.3	0.9	0.4	0	0
	1.25	2.8	1.5	0.8	0.2	0
	2.00	13.0	2.6	1.2	0.5	0.3

[a] From ACI–Table 13.6.10.

formula (ACI Formula 13-5),

$$\delta_s = 1 + \frac{2 - \beta_a}{4 + \beta_a}\left(1 - \frac{\alpha_c}{\alpha_{min}}\right) \tag{16.9.1}$$

in which

β_a = ratio of service dead load to service live load

α_c = ratio of column stiffness to slab and beam stiffness

   = $\Sigma K_c / (\Sigma K_s + \Sigma K_b)$

α_{min} = value shown in Table 16.9.1

For the half column strip parallel to an exterior edge, it is conservative, and therefore permissible to use the value of α_c computed for the adjacent interior column if it is equal in size to the exterior column. The graph shown in Fig. 16.9.1 is not so much to save the arithmetic involved in Eq. (16.9.1) as to show the variation of δ_s with respect to α_c/α_{min} and β_a.

$$\delta_s = 1 + \frac{2-\beta_a}{4+\beta_a}\left(1 - \frac{\alpha_c}{\alpha_{min}}\right)$$

$$\beta_a = \frac{DL}{LL}$$

Figure 16.9.1 Direct design method—effect of pattern loadings.

16.10 DIRECT DESIGN METHOD—PROCEDURE FOR COMPUTATION OF LONGITUDINAL MOMENTS

The background explanation for the distribution of the total static moment M_0 in the longitudinal direction and for the modification of positive moment due to the effect of pattern loadings has been presented in the two preceding sections. Utilizing this information, the procedure for the computation of longitudinal moments by the "direct design method" may be summarized as follows:

1. Check if the six limitations for the "direct design method" listed in Section 16.7 are satisfied.
2. Compute the total static moment $M_0 = w_u L_2 L_n^2/8$ as stated by Eq. (16.3.2). Note that L_n is not to be taken less than $0.65L_1$. For flat slabs, Eq. (16.3.4) should preferably be used for computing M_0.
3. Compute the slab stiffness

$$K_s = \frac{4EI_s}{L_1}, \qquad I_s = \sum L_2\left(\frac{t^3}{12}\right)$$

4. Compute the moment of inertia I_b of the longitudinal beam (if any) (ACI-13.2.4) and its flexural stiffness by

$$K_b = \frac{4E_{cb}I_b}{L_1}$$

5. Compute the column stiffness

$$\Sigma K_c = K_{c1} + K_{c2} = \frac{4EI_{c1}}{L_{c1}} + \frac{4EI_{c2}}{L_{c2}}$$

6. Compute the ratio α of the longitudinal beam (ACI-13.2.4) stiffness to that of the slab by $\alpha = (E_{cb}I_b)/(E_{cs}/I_s)$ and with the value of $\beta_a = $ (service DL/ service LL), obtain the value of α_{min} from Table 16.9.1 or ACI-Table 13.6.10.

7. Compute for the exterior and interior spans the ratio α_c of the flexural stiffnesses of the upper and lower columns to the combined flexural stiffnesses of the slab and beam (if any) on both sides of the columns by the following:

$$\alpha_c = \frac{K_{c1} + K_{c2}}{\Sigma K_s + \Sigma K_b}$$

8. If the value of α_c computed in Step 7 does not reach the α_{min} shown in Table 16.9.1, the positive moments M_{pos} must be increased for pattern loading effects by the multiplier δ_s, Eq. (16.9.1). If a_c/α_{min} exceeds one, $\delta_s = 1$ (i.e., no pattern loading effect).

9. Obtain the three critical ordinates on the longitudinal moment diagrams for the exterior and interior spans using Figs. 16.8.1 and 16.8.2, but multiplying the positive moments by the δ_s for that span if δ_s is larger than 1.

■ **EXAMPLE 16.10.1** For the two-way slab (with beams) design example described in Section 16.3, determine the longitudinal moments in frames A, B, C, and D, as shown in Figs. 16.3.5 and 16.10.1.

Figure 16.10.1 Longitudinal moments for two-way slab (with beams) design example.

Solution: **(a)** Check the six limitations for the direct design method. These limitations have been checked in Example 16.7.1.

(b) Total factored static moment M_0. The total factored static moments M_0 for the equivalent rigid frames A, B, C, and D have been computed in Example 16.3.1; they are

$$M_0 \text{ (frame } A) = 448 \text{ ft-kips}$$

$$M_0 \text{ (frame } B) = 224 \text{ ft-kips}$$

$$M_0 \text{ (frame } C) = 349 \text{ ft-kips}$$

$$M_0 \text{ (frame } D) = 175 \text{ ft-kips}$$

(c) Compute the slab stiffness, beam stiffness, and column stiffness. Since all columns are of the same size, 15 by 15 in. and 12 ft long,

$$K_{c1} = K_{c2} = \frac{4E(15)(15)^3/12}{144} = 117E$$

For frame A,

$$K_s = \frac{4E(240)(6.5)^3/12}{300} = 73.2E$$

$$K_b = \frac{4EI_b \text{ of B1--B2 (Example 16.4.1)}}{300} = \frac{4E(45,400)}{300} = 605E$$

$$\alpha = \frac{K_b}{K_s} = \frac{605E}{73.2E} = 8.27 \quad \text{(same as in Fig. 16.4.4)}$$

$$\alpha_c = \frac{234E}{605E + 73.2E} = 0.344 \quad \text{(at exterior column)}$$

$$\alpha_c = 0.172 \quad \text{(at interior column)}$$

For frame B,

$$K_s = 36.6E$$

$$K_b = \frac{4EI_b \text{ of B3--B4 (Example 16.4.1)}}{300} = \frac{4E(38,000)}{300} = 507E$$

$$\alpha = \frac{507E}{36.6E} \approx 13.83 \quad \text{(same as in Fig. 16.4.4)}$$

$$\alpha_c = \frac{234E}{507E + 36.6E} = 0.430 \quad \text{(at exterior column)}$$

$$\alpha_c = 0.215 \quad \text{(at interior column)}$$

For frame C,

$$K_s = \frac{4E(300)(6.5)^3/12}{240} = 114E$$

$$K_b = \frac{4EI_b \text{ of } B5-B6 \text{ (Example 16.4.1)}}{240} = \frac{4E(24,400)}{240} = 406E$$

$$\alpha = \frac{406E}{114E} \approx 3.55 \quad \text{(same as in Fig. 16.4.4)}$$

$$\alpha_c = \frac{234E}{406E + 114E} = 0.450 \quad \text{(at exterior column)}$$

$$\alpha_c = 0.225 \quad \text{(at interior column)}$$

For frame D,

$$K_s = 57E$$

$$K_b = \frac{4EI_b \text{ of } B7-B8 \text{ (Example 16.4.1)}}{240} = \frac{4E(20,500)}{240} = 341E$$

$$\alpha = \frac{341E}{57E} \approx 5.96 \quad \text{(same as in Fig. 16.4.4)}$$

$$\alpha_c = \frac{234E}{341E + 57E} = 0.587 \quad \text{(at exterior column)}$$

$$\alpha_c = 0.294 \quad \text{(at interior column)}$$

(d) Minimum ratio α_{min}. The ratio β_a of service dead to live load is

$$\beta_a = \frac{6.5(150/12)}{120} = \frac{81}{120} = 0.67$$

The minimum ratios α_{min} required for making positive moment multiplier $\delta_s = 1.0$ are tabulated below.

FRAME	A	B	C	D
$\alpha = \dfrac{E_{cb}I_b}{E_{cs}I_s}$ from part **(c)** above or Fig. 16.4.4	8.27	13.83	3.55	5.96
L_2/L_1		0.80	0.80	1.25	1.25
α_{min} from Table 16.9.1	0	0	0	0

Note that in Table 16.9.1 when α is larger than 4, α_{min} is zero (thus the positive moment multiplier δ_s is 1.0) except when $L_2/L_1 = 2.00$ and $\beta_a = 0.33$.

(e) Positive moment multiplier δ_s. Since there is sufficient stiffness in the columns to minimize the pattern loading effects for all frames, $\delta_s = 1.0$ in all cases.

(f) Longitudinal moments in the frames. The longitudinal moments in frames A, B, C, and D are computed using Case 2 of Fig. 16.8.2 for the exterior span and Fig. 16.8.1 for the interior span. The computations are as shown below, and the results are summarized in Fig. 16.10.1.

For frame A, $\qquad\qquad M_0 = 448$ ft-kips

M_{neg} at exterior support	$= 0.16(448) = 72$ ft-kips
M_{pos} in exterior span	$= 1.0(0.57)(448) = 255$ ft-kips
M_{neg} at first interior support	$= 0.70(448) = 313$ ft-kips
M_{neg} at typical interior support	$= 0.65(448) = 291$ ft-kips
M_{pos} in typical interior span	$= 1.0(0.35)(448) = 157$ ft-kips

For frame B, $\qquad\qquad M_0 = 224$ ft-kips

M_{neg} at exterior support	$= 0.16(224) = 36$ ft-kips
M_{pos} in exterior span	$= 1.0(0.57)(224) = 128$ ft-kips
M_{neg} at first interior support	$= 0.70(224) = 157$ ft-kips
M_{neg} at typical interior support	$= 0.65(224) = 146$ ft-kips
M_{pos} in typical interior span	$= 1.0(0.35)(224) = 78$ ft-kips

For frame C, $\qquad\qquad M_0 = 349$ ft-kips

M_{neg} at exterior support	$= 0.16(349) = 56$ ft-kips
M_{pos} in exterior span	$= 1.0(0.57)(349) = 199$ ft-kips
M_{neg} at first interior support	$= 0.70(349) = 244$ ft-kips
M_{neg} at typical interior support	$= 0.65(349) = 227$ ft-kips
M_{pos} in typical interior span	$= 1.0(0.35)(349) = 122$ ft-kips

For frame D, $\qquad\qquad M_0 = 175$ ft-kips

M_{neg} at exterior support	$= 0.16(175) = 28$ ft-kips
M_{pos} in exterior span	$= 1.0(0.57)(175) = 100$ ft-kips
M_{neg} at first interior support	$= 0.70(175) = 123$ ft-kips
M_{neg} at typical interior support	$= 0.65(175) = 114$ ft-kips
M_{pos} in typical interior span	$= 1.0(0.35)(175) = 61$ ft-kips

■ **EXAMPLE 16.10.2** For the flat slab design example described in Section 16.3, compute the longitudinal moments in frames A, B, C, and D as shown in Figs. 16.3.6 and 16.10.2.

Solution: **(a)** Check the five limitations (the sixth limitation does not apply here) for the direct design method. These five limitations are all satisfied.

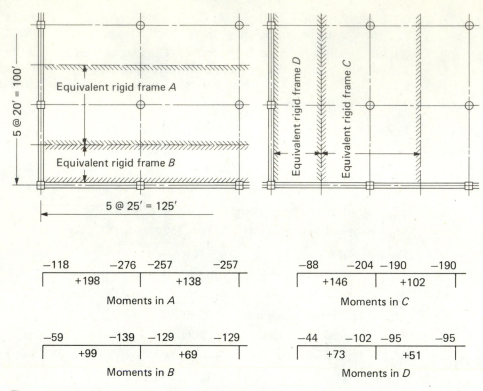

Figure 16.10.2 Longitudinal moments for flat slab design example.

(b) Total factored static moment M_0. Referring to the equivalent rigid frames A, B, C, and D in Fig. 16.10.2, the total factored static moment may be taken from the results of Example 16.3.2; thus

$$M_0 \text{ for } A = 395 \text{ ft-kips}$$

$$M_0 \text{ for } B = \tfrac{1}{2}(395) = 198 \text{ ft-kips}$$

$$M_0 \text{ for } C = 292 \text{ ft-kips}$$

$$M_0 \text{ for } D = \tfrac{1}{2}(292) = 146 \text{ ft-kips}$$

(c) Slab stiffness K_s.

$$K_s \text{ for } A = \frac{4EI_s}{L_1} = \frac{4E(240)(7.5)^3/12}{25(12)} = 112.5E$$

$$K_s \text{ for } B = 56.25E$$

$$K_s \text{ for } C = \frac{4EI_s}{L_1} = \frac{4E(300)(7.5)^3/12}{20(12)} = 176E$$

$$K_s \text{ for } D = 88E$$

(d) Column stiffness K_c.

$$\Sigma K_c \text{ (upper and lower interior columns)} = \frac{2(4EI_c)}{L} = \frac{2(4E)\pi(18)^4/64}{10(12)} = 343E$$

$$\Sigma K_c \text{ (upper and lower exterior columns)} = \frac{2(4EI_c)}{L} = \frac{2(4E)(16)^4/12}{10(12)} = 364E$$

(e) Beam stiffness K_b. From Example 16.6.2, the gross moment of inertia of the edge beam shown in Fig. 16.6.2 about the centroidal axis is 22,900 in.[4]

$$K_b \text{ for } A = 0$$

$$K_b \text{ for } B = \frac{4E(22,900)}{300} = 305E$$

$$K_b \text{ for } C = 0$$

$$K_b \text{ for } D = \frac{4E(22,900)}{240} = 381E$$

(f) Compute positive moment multipliers δ_s. The ratio β_a of service dead to live load is

$$\beta_a = \frac{7.5(150/12)}{120} = 0.78$$

The α_{min} values required for $\delta_s = 1.0$ are

FRAME	A	B	C	D
α (from Fig. 16.6.2)	0	5.42	0	4.34
L_2/L_1	0.80	0.80	1.25	1.25
α_{min} from Table 16.9.1	1.05	0	1.28	0

For frame A,

$$\alpha_c \text{ (exterior)} = \frac{364E}{0 + 112.5E} = 3.24 > \alpha_{min} = 1.05; \qquad \delta_s = 1.0$$

$$\alpha_c \text{ (interior)} = \frac{343E}{0 + 225E} = 1.52 > \alpha_{min} = 1.05; \qquad \delta_s = 1.0$$

For frame B,

$$\alpha_c \text{ (exterior)} = \frac{364E}{305E + 56.25E} = 1.01 > \alpha_{min} = 0; \qquad \delta_s = 1.0$$

$$\alpha_c \text{ (interior)} = \frac{364E}{610E + 112.5E} = 0.50 > \alpha_{min} = 0; \qquad \delta_s = 1.0$$

Table 16.10.1 Longitudinal Moments (ft-kips) for the Flat Slab Design Example

FRAME	A	B	C	D
M_0	395	198	292	146
M_{neg} at exterior support, $0.30M_0$	118	59	88	44
M_{pos} in exterior span, δ_s $(0.50M_0)$	198 ($\delta_s = 1.0$)	99 ($\delta_s = 1.0$)	146 ($\delta_s = 1.0$)	73 ($\delta_s = 1.0$)
M_{neg} at first interior support, $0.70M_0$	276	139	204	102
M_{neg} at typical interior support, $0.65M_0$	257	129	190	95
M_{pos} in typical interior span, δ_s $(0.35M_0)$	138 ($\delta_s = 1.0$)	69 ($\delta_s = 1.0$)	102 ($\delta_s = 1.06$)	51 ($\delta_s = 1.0$)

For frame C,

$$\alpha_c \text{ (exterior)} = \frac{364E}{0 + 176E} = 2.07 > \alpha_{min} = 1.28; \qquad \delta_s = 1.0$$

$$\alpha_c \text{ (interior)} = \frac{343E}{0 + 352E} = 0.97 < \alpha_{min} = 1.28; \qquad \delta_s > 1.0$$

$$\delta_s \text{ (interior)} = \frac{2 - 0.78}{4 + 0.78} \left(1 - \frac{0.97}{1.28}\right) = 1.06$$

For frame D,

$$\alpha_c \text{ (exterior)} = \frac{364E}{381E + 88E} = 0.78 > \alpha_{min} = 0; \qquad \delta_s = 1.0$$

$$\alpha_c \text{ (interior)} = \frac{364E}{762E + 176E} = 0.39 > \alpha_{min} = 0; \qquad \delta_s = 1.0$$

Only the positive moment in the interior span of frame C needs to be multiplied by 1.06.

(g) Longitudinal moments in the frames. The longitudinal moments in frames A, B, C, and D are computed using Case 4 of Fig. 16.8.2 for the exterior span and Fig. 16.8.1 for the interior span. The computations are shown in Table 16.10.1 and the results are summarized in Fig. 16.10.2. ∎

■ **EXAMPLE 16.10.3** For the flat plate design example described in Section 16.3, compute the longitudinal moments in frames A, B, C, and D as shown in Figs. 16.3.7 and 16.10.3.

Figure 16.10.3 Longitudinal moments for flat plate design example.

Solution: **(a)** Check the five limitations (the sixth limitation does not apply here) for the direct design method. These limitations are all satisfied.

(b) Total factored static moment M_0 from the results of Example 16.3.3.

$$M_0 \text{ for } A = 58.2 \text{ ft-kips}$$

$$M_0 \text{ for } B = \tfrac{1}{2}(58.2) = 29.1 \text{ ft-kips}$$

$$M_0 \text{ for } C = 46.3 \text{ ft-kips}$$

$$M_0 \text{ for } D = 23.1 \text{ ft-kips}$$

(c) Slab stiffness K_s.

$$K_s \text{ for } A = \frac{4EI_s}{L_1} = \frac{4E(144)(5.5)^3/12}{15(12)} = 44.4E$$

$$K_s \text{ for } B = 22.2E$$

$$K_s \text{ for } C = \frac{4EI_s}{L_1} = \frac{4E(180)(5.5)^3/12}{12(12)} = 69.4E$$

$$K_s \text{ for } D = 34.7E$$

(d) Column stiffness K_c.

$$\Sigma K_c \text{ for } A \text{ and } B = \frac{2(4EI_c)}{L} = \frac{2(4E)(10)(12)^3/12}{9(12)} = 106.7E$$

(upper and lower columns)

$$\Sigma K_c \text{ for } C \text{ and } D = \frac{2(4EI_c)}{L} = \frac{2(4E)(12)(10)^3/12}{9(12)} = 74.1E$$

(upper and lower columns)

(e) Compute positive moment multipliers δ_s. The ACI Code will be followed to obtain δ_s. Alternate equations for the multiplier δ_s have been proposed by Jofriet and McNeice [16.37]. The ratio β_a of service dead to live load is

$$\beta_a = \frac{5.5(150/12)}{60} = 1.15$$

The α_{min} values required for $\delta_s = 1.0$ are, at $\alpha = E_{cb}I_b/E_{cs}I_s = 0$,

FRAME	A	B	C	D
L_2/L_1	0.80	0.80	1.25	1.25
α_{min} from Table 16.9.1	0.60	0.60	0.68	0.68

For frame A,

$$\alpha_c \text{ (exterior)} = \frac{106.7E}{0 + 44.4E} = 2.40 > 0.60; \qquad \delta_s = 1.0$$

$$\alpha_c \text{ (interior)} = \frac{106.7E}{0 + 88.8E} = 1.20 > 0.60; \qquad \delta_s = 1.0$$

For frame B,

$$\alpha_c \text{ (exterior)} = \frac{106.7E}{0 + 22.2E} = 4.81 > 0.60; \qquad \delta_s = 1.0$$

$$\alpha_c \text{ (interior)} = \frac{106.7E}{0 + 44.4E} = 2.40 > 0.60; \qquad \delta_s = 1.0$$

For frame C,

$$\alpha_c \text{ (exterior)} = \frac{74.1E}{0 + 69.4E} = 1.07 > 0.68; \qquad \delta_s = 1.0$$

$$\alpha_c \text{ (interior)} = \frac{74.1E}{0 + 138.8E} = 0.53 < 0.68; \qquad \delta_s > 1.0$$

$$\delta_s \text{ (interior)} = 1 + \frac{2 - 1.15}{4 + 1.15}\left(1 - \frac{0.53}{0.68}\right) = 1.04$$

Table 16.10.2   Longitudinal Moments (ft-kips) for the Flat Plate Design Example

FRAME	A	B	C	D
M_0	58.2	29.1	46.3	23.1
M_{neg} at exterior support, $0.26M_0$	15.1	7.6	12.0	6.0
M_{pos} in exterior span, $\delta_s (0.52M_0)$	30.3 ($\delta_s = 1.0$)	15.1 ($\delta_s = 1.0$)	24.1 ($\delta_s = 1.0$)	12.0 ($\delta_s = 1.0$)
M_{neg} at first interior support, $0.70M_0$	40.7	20.4	32.4	16.2
M_{neg} at typical interior support, $0.65M_0$	37.8	18.9	30.1	15.0
M_{pos} in typical interior span, $\delta_s (0.35M_0)$	20.4 ($\delta_s = 1.0$)	10.2 ($\delta_s = 1.0$)	16.9 ($\delta_s = 1.04$)	8.1 ($\delta_s = 1.0$)

For frame *D*,

$$\alpha_c \text{ (exterior)} = \frac{74.1E}{0 + 34.7E} = 2.14 > 0.68; \qquad \delta_s = 1.0$$

$$\alpha_c \text{ (interior)} = \frac{74.1E}{0 + 69.4E} = 1.07 > 0.68; \qquad \delta_s = 1.0$$

Only the positive moment in the interior spans of frame *C* needs to be multiplied by 1.04.

(f) Longitudinal moments in the frames. The longitudinal moments in frames *A*, *B*, *C*, and *D* are computed using Case 3 of Fig. 16.8.2 for the exterior span and Fig. 16.8.1 for the interior span. The computations are as shown in Table 16.10.2 and the results are summarized in Fig. 16.10.3. ■

16.11 TORSION CONSTANT *C* OF THE TRANSVERSE BEAM

One important parameter useful in the analysis (that is, the application of the equivalent frame method described in Chapter 17) and design (that is, for the transverse distribution of the longitudinal moment) is the torsional constant *C* of the transverse beam spanning from column to column. Even if there is no such beam (as defined by projection above or below the slab) actually visible, for the present use one still should imagine that there is a beam made of a portion of the slab having a width equal to that of the column, bracket, or capital in the longitudinal direction (ACI-13.7.5.1a). When there is actually a transverse beam web above or below the slab, the cross section of the transverse beam should include the portion of slab within the width of column, bracket, or capital described above plus the projection of beam web above or below the slab (ACI-13.7.5.1b). As a third possibility, the transverse beam may include that portion of slab on each side of the beam web equal to its projection

Figure 16.11.1 Definition of cross sections for transverse beams in torsion. [Projection of slab beyond beam in Case (c) is allowed on each side for interior beam.]

above or below the slab, whichever is greater, but not greater than four times the slab thickness (ACI-13.7.5.1c). The largest of the three definitions as shown in Fig. 16.11.1 may be used.

The torsional constant C of the transverse beam equals (ACI Formula 13-7),

$$C = \sum \left(1 - 0.63\frac{x}{y}\right)\left(\frac{x^3 y}{3}\right) \qquad (16.11.1)$$

where

x = shorter dimension of a component rectangle

y = longer dimension of a component rectangle

and the component rectangles should be taken in such a way that the largest value of C is obtained (ACI Commentary-R13.7.5). Equation (16.11.1) is identical to Eq. (19.3.5), for which there is additional discussion in Chapter 19.

■ **EXAMPLE 16.11.1**   For the two-way slab (with beams) design example, compute the torsional constant C for the edge and interior beams in the short and long directions.

Solution: Each cross section shown in Fig. 16.11.2 may be divided into component rectangles in two different ways and the larger value of C is to be used. For the short direction,

$$C\left(\begin{array}{c}\text{edge} \\ \text{beam}\end{array}\right) = \left[1 - \frac{0.63(6.5)}{29.5}\right]\frac{29.5(6.5)^3}{3} + \left[1 - \frac{0.63(12)}{17.5}\right]\frac{17.5(12)^3}{3}$$
$$= 2325 + 5725 = 8050 \text{ in.}^4$$

or

$$C\left(\begin{array}{c}\text{edge} \\ \text{beam}\end{array}\right) = \left[1 - \frac{0.63(6.5)}{17.5}\right]\frac{17.5(6.5)^3}{3} + \left[1 - \frac{0.63(12)}{24}\right]\frac{24(12)^3}{3}$$
$$= 1230 + 9470 = 10,700 \text{ in.}^4 \qquad \textit{Use}$$

(a) Short direction

(b) Long direction

Figure 16.11.2 Effective cross sections of transverse beams resisting torsion, in two-way slab (with beams) design example.

$$C\left(\frac{\text{interior}}{\text{beam}}\right) = \left[1 - \frac{0.63(6.5)}{47}\right]\frac{47(6.5)^3}{3} + \left[1 - \frac{0.63(12)}{17.5}\right]\frac{17.5(12)^3}{3}$$
$$= 3925 + 5725 = 9650 \text{ in.}^4$$

or

$$C\left(\frac{\text{interior}}{\text{beam}}\right) = 2(1230) + 9470 = 11{,}930 \text{ in.}^4 \qquad Use$$

For the long direction,

$$C\left(\frac{\text{edge}}{\text{beam}}\right) = \left[1 - \frac{0.63(6.5)}{35.5}\right]\frac{35.5(6.5)^3}{3} + \left[1 - \frac{0.63(14)}{21.5}\right]\frac{21.5(14)^3}{3}$$
$$= 2900 + 11{,}600 = 14{,}500 \text{ in.}^4$$

or

$$C\left(\frac{\text{edge}}{\text{beam}}\right) = \left[1 - \frac{0.63(6.5)}{21.5}\right]\frac{21.5(6.5)^3}{3} + \left[1 - \frac{0.63(14)}{28}\right]\frac{28(14)^3}{3}$$
$$= 1600 + 17{,}500 = 19{,}100 \text{ in.}^4 \qquad Use$$

$$C\left(\frac{\text{interior}}{\text{beam}}\right) = \left[1 - \frac{0.63(6.5)}{57}\right]\frac{57(6.5)^3}{3} + \left[1 - \frac{0.63(14)}{28}\right]\frac{28(14)^3}{3}$$
$$= 4800 + 11{,}600 = 16{,}400 \text{ in.}^4$$

or

$$C\left(\frac{\text{interior}}{\text{beam}}\right) = 2(1600) + 17{,}500 = 20{,}700 \text{ in.}^4 \qquad Use \quad \blacksquare$$

$35''$ = distance from outer edge of exterior column to inner edge of square capital (i.e. $2'$–$3''$ + half the $16''$ column)

(a) Short or long edge beam

$8.5''$ = weighted slab thickness

$12 (4.43') = 53.2''$

(b) Short or long interior beam

Figure 16.11.3 Cross sections of torsional transverse beams in flat slab design example.

■ **EXAMPLE 16.11.2** For the flat slab design example, compute the torsional constant C for the edge beam and the interior beam in the short and long directions.

Solution: For the short or long edge beam [Fig. 16.11.3(a)], the torsional constant C is computed on the basis of the cross section shown in Fig. 16.11.3(a).

$$C = \left[1 - \frac{0.63(7.5)}{51.5}\right]\frac{(7.5)^3(51.5)}{3} + \left[1 - \frac{0.63(14)}{16.5}\right]\frac{(14)^3(16.5)}{3}$$
$$= 6575 + 7025 = 13{,}600 \text{ in.}^4$$

or

$$C = \left[1 - \frac{0.63(14)}{24}\right]\frac{(14)^3(24)}{3} + \left[1 - \frac{0.63(7.5)}{37.5}\right]\frac{(7.5)^3(37.5)}{3}$$
$$= 13{,}890 + 4610 = 18{,}500 \text{ in.}^4 \qquad Use$$

For the short or long interior beam [Fig. 16.11.3(b)], a weighted slab thickness of 8.5 in. is used, on the assumption that one-third of the span has a $10\frac{1}{2}$-in. thickness and the remainder has a $7\frac{1}{2}$-in. thickness. (Actually, the ratio is not exactly so because the drop width has been revised from 6 ft 8 in. to 7 ft.)

$$C = \left(1 - 0.63\frac{x}{y}\right)\frac{x^3 y}{3} = \left[1 - \frac{0.63(8.5)}{12(4.43)}\right]\left[\frac{(8.5)^3(12)(4.43)}{3}\right] = 9800 \text{ in.}^4 \quad ■$$

■ **EXAMPLE 16.11.3** For the flat plate design example, compute the torsional constant C for the short and long beams.

Figure 16.11.4 Cross sections of torsional transverse beams in flat plate design example.

Solution: Since no actual edge beams are used, the torsional member is, according to Fig. 16.11.4, equal to the slab thickness t by the column width c_1.

$$C \text{ for short beams} = \left[1 - \frac{0.63(5.5)}{12}\right] \frac{(5.5)^3(12)}{3} = 474 \text{ in.}^4$$

$$C \text{ for long beams} = \left[1 - \frac{0.63(5.5)}{10}\right] \frac{(5.5)^3(10)}{3} = 363 \text{ in.}^4 \quad \blacksquare$$

16.12 TRANSVERSE DISTRIBUTION OF LONGITUDINAL MOMENT

The longitudinal moment values, whether those of the "direct design method" shown in Figs. 16.8.1 and 16.8.2 or those obtained by structural analysis using the "equivalent frame method" (Chapter 17), are for the entire width (sum of the two half panel widths in the transverse direction, for an interior column line) of the equivalent rigid frame. Each of these moments is to be divided, on the basis of studies by Gamble, Sozen, and Siess [16.12], between the column strip and the two half middle strips as defined in Fig. 16.12.1. If the two adjacent transverse spans are each equal to L_2, the width of the column strip is then equal to one-half of L_2, or one-half of the longitudinal span L_1, whichever is smaller (ACI-13.2.1). This seems reasonable, since when the longitudinal span is shorter than the transverse span, a larger portion of the moment across the width of the equivalent frame might be expected to concentrate near the column centerline.

The transverse distribution of the longitudinal moment to column and middle strips is a function of three parameters, using L_1 and L_2 for the longitudinal and transverse spans, respectively: (1) the aspect ratio L_2/L_1; (2) the ratio $\alpha_1 = E_{cb}I_b/(E_{cs}I_s)$ of the longitudinal beam stiffness to slab stiffness; and (3) the ratio $\beta_t = E_{cb}C/(2E_{cs}I_s)$ of the torsional rigidity of edge beam section to the flexural rigidity of a width of slab equal to the span length of the edge beam. According to ACI-13.6.4, the column

Figure 16.12.1 Definition of column and middle strips.

strip is to take the percentage of the longitudinal moment as shown in Table 16.12.1. As may be seen from Table 16.12.1, only the first two parameters affect the transverse distribution of the negative moments at the first and typical interior supports as well as the positive moments in exterior and interior spans, but all three parameters are involved in the transverse distribution of the negative moment at the exterior support.

Regarding the distributing percentages shown in Table 16.12.1, the following observations may be made:

1. In general, the column strip takes more than 50% of the longitudinal moment.
2. The column strip takes a larger share of the negative longitudinal moment than the positive longitudinal moment.
3. When no longitudinal beams are present, the column strip takes the same share of the longitudinal moment, irrespective of the aspect ratio. The reader

Table 16.12.1 Percentage of Longitudinal Moment in Column Strip (ACI-13.6.4)

ASPECT RATIO L_2/L_1			0.5	1.0	2.0
Negative moment at exterior support	$\alpha_1 L_2/L_1 = 0$	$\beta_t = 0$	100	100	100
		$\beta_t \geq 2.5$	75	75	75
	$\alpha_1 L_2/L_1 \geq 1.0$	$\beta_t = 0$	100	100	100
		$\beta_t \geq 2.5$	90	75	45
Positive moment	$\alpha_1 L_2/L_1 = 0$		60	60	60
	$\alpha_1 L_2/L_1 \geq 1.0$		90	75	45
Negative moment at interior support	$\alpha_1 L_2/L_1 = 0$		75	75	75
	$\alpha_1 L_2/L_1 \geq 1.0$		90	75	45

may note, however, that the column strip width is a fraction of L_1 or L_2 ($0.25L_1$ or $0.25L_2$ on each side of column line), whichever is smaller.

4. In the presence of longitudinal beams, the larger the aspect ratio, the smaller the distribution to the column strip. This seems consistent because the same reduction in the portion of moment going into the slab is achieved by restricting the column strip width to a fraction of L_1 when L_2/L_1 is greater than one.
5. The column strip takes a smaller share of the exterior moment as the torsional rigidity of the edge beam section increases.

When the exterior support consists of a column or wall extending for a distance equal to or greater than three-fourths of the transverse width, the exterior negative moment is to be uniformly distributed over the transverse width (ACI-13.6.4.3).

The procedure for distributing the longitudinal moment across a transverse width to the column and middle strips may be summarized as follows:

1. Divide the total transverse width applicable to the longitudinal moment into a column strip width and two half middle strip widths, one adjacent to each side of the column strip. For an exterior column line, the column strip width is $\frac{1}{4}L_1$ or $\frac{1}{4}L_2$, whichever is smaller; for an interior column line, the column strip width is $\Sigma(\frac{1}{4}L_1$ or $\frac{1}{4}L_2$, whichever is smaller, of the panels on both sides).
2. Determine the ratio $\beta_t = E_{cb}C/(2E_{cs}I_s)$ of edge beam torsional rigidity to slab flexural rigidity. (*Note:* The 2 arises from approximating the shear modulus of elasticity in the numerator as $E_{cb}/2$.)
3. Determine the ratio $\alpha_1 = E_{cb}I_b/(E_{cs}I_s)$ of longitudinal beam flexural stiffness to slab flexural stiffness.
4. Divide the longitudinal moment at each critical section into two parts according to the percentage shown in Table 16.12.1: one part to the column strip width; and the remainder to the half middle strip for an exterior column line, or to the half middle strips on each side of an interior column line.
5. If there is an exterior wall instead of an exterior column line, the strip ordinarily called the exterior column strip will not deflect and therefore no moments act. In this case there can be no longitudinal distribution of moments; thus there is no computed moment to distribute laterally to the half middle strip adjacent to the wall. This half middle strip should be combined with the next adjacent half middle strip, which itself receives a lateral distribution in the frame of the first interior column line. The total middle strip in this situation is designed for twice the moment in the half middle strip from the first interior column line (ACI-13.6.6.3).

Distribution of Moment in Column Strip to Beam and Slab. When a longitudinal beam exists in the column strip along the column centerline, the column strip moment as determined by the percentages in Table 16.12.1 (ACI-13.6.4) should be divided between the beam and the slab. ACI-13.6.5 states that 85% of this moment be taken by the beam if $\alpha L_2/L_1$ is equal to or greater than 1.0, and for values of $\alpha L_2/L_1$ between 1.0 and 0, the proportion of moment to be resisted by the beam is to be obtained by linear interpolation between 85 and 0%. In addition, any beam must be designed to carry its own weight (projection above and below the slab), and any concentrated or linear loads applied directly on it (ACI-13.6.5.3).

■ **EXAMPLE 16.12.1**   For the two-way slab (with beams) design example described in Section 16.3, distribute the longitudinal moments computed for frames A, B, C, and D (see Fig. 16.10.1) into three parts—namely, for the longitudinal beam, for the column strip slab, and for the middle strip slab.

Solution: The values for the total longitudinal moments in frames A, B, C, and D at the five critical sections are taken from Example 16.10.1 and shown again in Table 16.12.4. The results of transverse distribution of these moments are also shown in this table.

(a) Negative moment at face of exterior support. For frame A, $L_2/L_1 = 0.80$; $\alpha_1 = 8.27$ (Fig. 16.4.4); $\alpha_1 L_2/L_1 = 6.61$; $C = 10,700$ in.^{4} (Example 16.11.1); $I_s = 240(6.5)^3/12 = 5490$ in.^{4}; and $\beta_t = C/(2I_s) = 10,700/[2(5490)] = 0.98$. Table 16.12.2 shows the linear interpolation for obtaining the column strip percentage from the prescribed limits of Table 16.12.1. The total moment of 72 ft-kips is divided into three parts, 92.6% to column strip (of which 85% goes to the beam and 15% to the slab since $\alpha_1 L_2/L_1 = 6.61 \geq 1.0$) and 7.4% to the middle strip slab. The results are shown in Table 16.12.4.

For frame B, $L_2/L_1 = 0.80$; $\alpha_1 = 13.83$ (Fig. 16.4.4); $\alpha_1 L_2/L_1 = 11.1$; $\beta_t = 0.98$, the same as for frame A; and column strip moment percentage $= 92.6\%$, the same as for frame A.

For frame C, $L_2/L_1 = 1.25$; $\alpha_1 = 3.55$ (Fig. 16.4.4); $\alpha_1 L_2/L_1 = 4.44$; $C = 19,100$ in.^{4} (Example 16.11.1); $I_s = 300(6.5)^3/12 = 6870$ in.^{4}; and $\beta_t = C/(2I_s) = 19,100/[2(6870)] = 1.39$. Table 16.12.3 shows the linear interpolation for obtaining the column strip percentage from the prescribed limits of Table 16.12.1. The total moment of 56 ft-kips is divided into three parts, 81.9% to column strip (of which

Table 16.12.2   Linear Interpolation for Column Strip Percentage of Exterior Negative Moment—Frame A

L_2/L_1		0.5	0.8	1.0
	$\beta_t = 0$	100%	100%	100%
$\alpha_1 L_2/L_1 = 6.61$	$\beta_t = 0.98$	96.1%	92.6%	90.2%
	$\beta_t \geq 2.50$	90%	81%	75%

Table 16.12.3   Linear Interpolation for Column Strip Percentage of Exterior Negative Moment—Frame C

L_2/L_1		1.0	1.25	2.0
	$\beta_t = 0$	100%	100%	100%
$\alpha_1 L_2/L_1 = 4.44$	$\beta_t = 1.39$	86.1%	81.9%	69.4%
	$\beta_t \geq 2.50$	75%	67.5%	45%

Table 16.12.4 (also see Fig. 16.10.1) Transverse Distribution of Longitudinal Moments in Two-Way Slab (with Beams) Design Example

FRAME *A*
TOTAL WIDTH = 20 ft, COLUMN STRIP WIDTH = 10 ft, MIDDLE STRIP WIDTH = 10 ft

	EXTERIOR SPAN			INTERIOR SPAN	
	EXTERIOR NEGATIVE	POSITIVE	INTERIOR NEGATIVE	NEGATIVE	POSITIVE
Total moment	−72	+255	−313	−291	+157
Moment in beam	−57	+176	−216	−200	+108
Moment in column strip slab	−10	+31	−38	−36	+19
Moment in middle strip slab	−5	+49	−60	−55	+30

FRAME *B*
TOTAL WIDTH = 10 ft, COLUMN STRIP WIDTH = 5 ft, HALF MIDDLE STRIP WIDTH = 5 ft

	EXTERIOR SPAN			INTERIOR SPAN	
	EXTERIOR NEGATIVE	POSITIVE	INTERIOR NEGATIVE	NEGATIVE	POSITIVE
Total moment	−36	+128	−157	−146	+78
Moment in beam	−28	+88	−108	−101	+54
Moment in column strip slab	−5	+16	−19	−17	+9
Moment in middle strip slab	−3	+25	−30	−28	+15

FRAME *C*
TOTAL WIDTH = 25 ft, COLUMN STRIP WIDTH = 10 ft, MIDDLE STRIP WIDTH = 15 ft

	EXTERIOR SPAN			INTERIOR SPAN	
	EXTERIOR NEGATIVE	POSITIVE	INTERIOR NEGATIVE	NEGATIVE	POSITIVE
Total moment	−56	+199	−244	−227	+122
Moment in beam	−39	+114	−140	−130	+70
Moment in column strip slab	−7	+20	−25	−23	+13
Moment in middle strip slab	−10	+65	−79	−74	+40

FRAME *D*
TOTAL WIDTH = 12.5 ft, COLUMN STRIP WIDTH = 5 ft, HALF MIDDLE STRIP WIDTH = 7.5 ft

	EXTERIOR SPAN			INTERIOR SPAN	
	EXTERIOR NEGATIVE	POSITIVE	INTERIOR NEGATIVE	NEGATIVE	POSITIVE
Total moment	−28	+100	−123	−114	+61
Moment in beam	−20	+57	−71	−65	+35
Moment in column strip slab	−3	+10	−12	−12	+6
Moment in middle strip slab	−5	+33	−40	−37	+20

85% goes to the beam and 15% to the slab since $\alpha_1 L_2/L_1 = 4.44 \geq 1.0$) and 18.1%
to the middle strip slab.

For frame D, $L_2/L_1 = 1.25$; $\alpha_1 = 5.96$ (Fig. 16.4.4), $\alpha_1 L_2/L_1 = 7.45$; $\beta_t = 1.39$, the
same as for frame C; and column strip moment percentage $= 81.9\%$, the same as
for frame C.

(b) Negative moments at exterior face of first interior support and at face of typical
interior support. For frame A, $L_2/L_1 = 0.80$ and $\alpha_1 L_2/L_1 = 6.61 > 1.0$. Using the
prescribed values in Table 16.12.1, the proportion of moment going to the column
strip is determined to be 81% by linear interpolation:

L_2/L_1	0.5	0.8	1.0
$\alpha_1 L_2/L_1 = 6.61$	90%	81%	75%

For frame B, $L_2/L_1 = 0.80$ and $\alpha_1 L_2/L_1 = 11.1$. The proportion of moment is
again 81% for the column strip, the same as for frame A.

For frame C, $L_2/L_1 = 1.25$ and $\alpha_1 L_2/L_1 = 4.44$. Using the prescribed values in
Table 16.12.1, the proportion of moment going to the column strip is determined to
be 67.5% by linear interpolation:

L_2/L_1	1.0	1.25	2.0
$\alpha_1 L_2/L_1 = 4.44$	75%	67.5%	45%

For frame D, $L_2/L_1 = 1.25$ and $\alpha_1 L_2/L_1 = 7.47$. The proportion of moment is
again 67.5% for the column strip, the same as for frame C.

(c) Positive moments in exterior and interior spans. Since the prescribed limits for
$\alpha_1 L_2/L_1 \geq 1.0$ are the same for positive moment and for negative moment at interior
support, the percentages of column strip moment for positive moments in exterior
and interior spans are identical to those for negative moments as determined in part
(b) of this example. ■

■ **EXAMPLE 16.12.2**   Divide the five critical moments in each of the equivalent rigid
frames A, B, C, and D in the flat slab design example, as shown in Fig. 16.10.2, into
two parts: one for the half column strip (for frames B and D) or the full column strip (for
frames A and C), and the other for the half middle strip (for frames B and D) or the
two half middle strips on both sides of the column line (for frames A and C).

Solution: The percentages of the longitudinal moments going into the column strip
width are shown in lines 10 to 12 of Table 16.12.5. Note that the column strip width
shown in line 2 is one-half of the shorter panel dimension for both frames A and C,

Table 16.12.5   Transverse Distribution of Longitudinal Moment for Flat Slab Design Example

LINE NUMBER	EQUIVALENT RIGID FRAME	A	B	C	D
1	Total transverse width (in.)	240	120	300	150
2	Column strip width (in.)	120	60	120	60
3	Half middle strip width (in.)	2@60	60	2@90	90
4	C (in.^{4}) from Example 16.11.2	18,500	18,500	18,500	18,500
5	I_s (in.^{4}) in β_t	8,440	8,440	10,600	10,600
6	$\beta_t = E_{cb}C/(2E_{cs}I_s)$	1.10	1.10	0.87	0.87
7	α_1 from Example 16.10.2(f)	0	5.42	0	4.34
8	L_2/L_1	0.80	0.80	1.25	1.25
9	$\alpha_1 L_2/L_1$	0	4.33	0	5.43
10	Exterior negative moment, percent to column strip	89.0%	91.6%	91.3%	88.7%
11	Positive moment, percent to column strip	60.0%	81.0%	60.0%	67.5%
12	Interior negative moment, percent to column strip	75.0%	81.0%	75.0%	67.5%

and one-fourth of this value for frames B and D. Note also that the sum of the values on lines 2 and 3 should be equal to that on line 1, for each respective frame.

The moment of inertia of the slab equal in width to the transverse span of the edge beam is

$$I_s \text{ in } \beta_t \text{ for } A \text{ and } B = \frac{240(7.5)^3}{12} = 8440 \text{ in.}^4$$

and

$$I_s \text{ in } \beta_t \text{ for } C \text{ and } D = \frac{300(7.5)^3}{12} = 10,600 \text{ in.}^4$$

These values are shown in line 5 of Table 16.12.5.

The percentages shown in lines 10 to 12 are obtained from Table 16.12.1, by interpolation (as illustrated in Tables 16.12.2 and 16.12.3) if necessary. Having these percentages, the separation of each of the longitudinal moment values shown in Fig. 16.10.2 into two parts is a simple matter and thus is not shown further. ■

■ **EXAMPLE 16.12.3**   Divide the five critical moments in each of the equivalent rigid frames A, B, C, and D in the flat plate design example, as shown in Fig. 16.10.3, into two parts: one for the half column strip (for frames B and D) or the full column strip (for frames A and C), and the other for the half middle strip (for frames B and D) or the two half middle strips on both sides of the column line (for frames A and C).

Solution: The percentages of the longitudinal moments going into the column strip width are shown in lines 10 to 12 of Table 16.12.6. Explanations are identical to those for the preceding example.

Table 16.12.6 Transverse Distribution of Longitudinal Moment for Flat Plate Design Example

LINE NUMBER	EQUIVALENT RIGID FRAME	A	B	C	D
1	Total transverse width (in.)	144	72	180	90
2	Column strip width (in.)	72	36	72	36
3	Half middle strip width (in.)	2@36	36	2@54	54
4	C(in.^{4}) from Example 16.11.3	474	474	363	363
5	I_s(in.^{4}) in β_t	2,000	2,000	2,500	2,500
6	$\beta_t = E_{cb}C/(2E_{cs}I_s)$	0.118	0.118	0.073	0.073
7	α_1	0	0	0	0
8	L_2/L_1	0.80	0.80	1.25	1.25
9	$\alpha_1 L_2/L_1$	0	0	0	0
10	Exterior negative moment, percent to column strip	98.8%	98.8%	99.3%	99.3%
11	Positive moment, percent to column strip	60%	60%	60%	60%
12	Interior negative moment, percent to column strip	75%	75%	75%	75%

16.13 DESIGN OF SLAB THICKNESS AND REINFORCEMENT

Slab Thickness. Ordinarily the minimum thickness specified in ACI-9.5.3 controls the thickness for design. Of course, reinforcement for bending moment must be provided, but the reinforcement ratio ρ required is usually well below $0.5\rho_{max}$; thus, it does not dictate slab thickness. For flat slabs, flexural strength must be provided both within the drop panel and outside its limits. In evaluating the strength within a drop panel, the drop width should be used as the transverse width of the compression area, because the drop is usually narrower than the width of the column strip. Also, the effective depth to be used should not be taken greater than what would be furnished by a drop thickness below the slab equal to one-fourth the distance from the edge of drop to the edge of column capital.

The shear requirement for two-way slabs (with beams) may be investigated by observing strips 1-1 and 2-2 in Fig. 16.13.1. Beams with $\alpha_1 L_2/L_1$ values larger than 1.0 are assumed to carry the loads acting on the tributary floor areas bounded by 45° lines drawn from the corners of the panel and the centerline of the panel parallel to the long side (ACI-13.6.8.1). If this is the case, the loads on the trapezoidal areas E and F of Fig. 16.13.1 go to the long beams, and those on the triangular areas G and H go to the short beams. The shear per unit width of slab along the beam is highest at the ends of slab strips 1-1 and 2-2, which, considering the increased shear at the exterior face of the first interior support, is approximately equal to

$$V_u = 1.15\left(\frac{w_u S}{2}\right) \tag{16.13.1}$$

If $\alpha_1 L_2/L_1$ is equal to zero, there is, of course, no load on the beams (because there are no beams). When the value of $\alpha_1 L_2/L_1$ is between 0 and 1.0, the percentage of

Figure 16.13.1 Load transfer from floor area to beams.

the floor load going to the beams should be obtained by linear interpolation. In such a case, the shear expressed by Eq. (16.13.1) would be reduced, but the shear around the column due to the portion of the floor load going directly to the columns by two-way action should be investigated as for flat plate floors.

The shear strength requirement for flat slab and flat plate systems (without beams) is treated separately in Sections 16.15, 16.16, and 16.18.

Reinforcement. When the nominal requirements for slab thickness as discussed in Section 16.6 are satisfied, no compression reinforcement will likely be required. The tension steel area required within the strip being considered can then be obtained by the following steps:

1. required $M_n = \dfrac{\text{factored moment } M_u \text{ in the strip}}{(\phi = 0.90)}$

2. $m = \dfrac{f_y}{0.85 f'_c}, \qquad R_n = \dfrac{M_n}{bd^2},$

$$\rho = \frac{1}{m}\left(1 - \sqrt{1 - \frac{2mR_n}{f_y}}\right), \qquad A_s = \rho bd$$

Instead of using the equation for ρ in Step 2, the curves in Fig. 3.8.1 may be used. Note also that the values of b and d to be used in Step 2 for negative moment in a column strip with drop are the drop width for b, and for d the smaller of the actual effective depth through the drop and that provided by a drop thickness below the slab at no more than one-fourth the distance between the edges of the column capital and the drop. For positive moment computation, the full column strip width should be used for b, and the effective depth in the slab for d. After obtaining the steel area A_s required within the strip, a number of bars may be chosen so that they provide either the area required for strength or the area required for shrinkage and temperature reinforcement, which is $0.002bt$ for Grades 40 and 50 steel, but somewhat less for higher grades (see ACI-7.12). The spacing of reinforcing bars must not exceed 2 times the slab thickness (ACI-13.4.2), except in slabs of cellular or ribbed construction where the requirement for shrinkage and temperature reinforcement governs (i.e., 5 times the slab thickness but not greater than 18 in.).

Reinforcement Details in Slabs Without Beams. ACI-13.4.8, in particular ACI-Fig. 13.4.8, provides detailed dimensions for minimum extensions required for each portion of the total number of bars in the column and middle strips. Since the forces acting in the bars were empirically determined, there is no practical means to evaluate the distances required to develop the reinforcement. Thus, past practice and engineering judgment were used in preparing ACI-Fig. 13.4.8. In 1989, that figure omitted details for bent bars because they are seldom used, although their use is still permitted by ACI-13.4.8.3 when the depth–span ratio allows bends at 45° or less.

For unbraced frames, reinforcement lengths must be determined by analysis but not be less than those prescribed in ACI-Fig. 13.4.8. Also, ACI-13.4.8.5 requires the use of "integrity steel", which consists of a minimum of two of the column strip bottom bars passing continuously (or spliced with Class A splices or anchored within support) through the column core in each direction. The purpose of this integrity steel is to provide some residual strength following a single punching shear failure.

Corner Reinforcement for Two-Way Slab (With Beams). It is well known from plate bending theory that a transversely loaded slab simply supported along four edges will tend to develop corner reactions as shown in Fig. 16.13.2, for which reinforcement must be provided. Thus in slabs supported on beams having a value of α greater than 1.0, special reinforcement (Fig. 16.13.3) shall be provided at exterior corners in both the bottom and top of the slab. This reinforcement (ACI-13.4.6) is to be provided for a distance in each direction from the corner equal to one-fifth the longer span. The reinforcement in both the top and bottom of the slab must be sufficient to resist

Figure 16.13.2 Edge reactions for simply supported two-way slab.

Figure 16.13.3 Corner reinforcement in two-way slab.

a moment equal to the maximum positive moment per foot of width in the slab, and it may be placed in a single band parallel to the diagonal in the top of the slab and perpendicular to the diagonal in the bottom of the slab, or in two bands parallel to the sides of the slab.

Crack Control.   In addition to deflection control, crack control is the other major serviceability requirement usually considered in the design of flexural members. ACI-10.6 gives criteria for beams and one-way slabs to ensure distribution of flexural reinforcement to minimize crack width under service loads. No ACI Code provisions are given for two-way floor (or roof) systems; however, ACI Committee 224, Cracking, has suggested a formula to predict the possible crack width in two-way acting slabs, flat slabs, and flat plates. The recommendations are based on the work of Nawy et al. [16.72–16.75]. When the predicted crack width is considered excessive (there are no ACI Code limits for slabs), the distribution (size and spacing) of flexural reinforcement may be adjusted [16.75] to decrease predicted crack width. Ordinarily crack width is not a problem on two-way acting slabs, but when steel with f_y equal to 60,000 psi or higher is used, crack control should be considered.

■ **EXAMPLE 16.13.1**   Investigate if the preliminary slab thickness of $6\frac{1}{2}$ in. in the two-way slab (with beams) design example described in Section 16.3 is sufficient for resisting flexure and shear.

Solution: For each of the equivalent frames A, B, C, and D, the largest bending moment in the slab occurs at the exterior face of the first interior support in the middle strip slab. From Table 16.12.4, this moment is observed to be 60/10, 30/5, 79/15, or 40/7.5 ft-kips per ft of width in frames A, B, C, and D, respectively. Taking the effective depth to the contact level between the reinforcing bars in the two directions, and assuming #5 bars,

$$\text{average } d = 6.50 - 0.75 - 0.63 = 5.12 \text{ in.}$$

The largest R_n required is

$$R_n = \frac{M_u}{\phi b d^2} = \frac{6000(12)}{0.90(12)(5.12)^2} = 254 \text{ psi}$$

From Fig. 3.8.1, the reinforcement ratio ρ for this value of R_n is 0.007, which is well below $0.375\rho_b = 0.0139$. Hence excessive deflection should not be expected; this is further verification of the minimum thickness formulas given in ACI-9.5.3.

The factored floor load w_u is

$$w_u = 1.4w_D + 1.7w_L = 318 \text{ psf}$$

Since all $\alpha L_2/L_1$ values are well over 1.0, take V from Eq. (16.13.1) as

$$V_u = \frac{1.15w_u S}{2} = \frac{1.15(0.318)(20)}{2} = 3.66 \text{ kips}$$

$$V_c = 2\sqrt{f'_c}\,b_w d = 2\sqrt{3000}(12)(5.12)\tfrac{1}{1000} = 6.73 \text{ kips}$$

$$\phi V_c = 0.85(6.73) = 5.72 \text{ kips} > [V_u = 3.66 \text{ kips}] \qquad \textbf{OK}$$

Note that the factored shear 3.66 kips is the maximum at strip 1-1 of Fig. 16.13.1; actually the average for all such strips will be lower. ∎

■ **EXAMPLE 16.13.2** Design the reinforcement in the exterior and interior spans of a typical column strip and a typical middle strip in the short direction of the flat slab design example. As described earlier in Section 16.3, $f'_c = 3000$ psi and $f_y = 40,000$ psi.

Solution: (a) Moments in column and middle strips. The typical column strip is the column strip of equivalent rigid frame C of Fig. 16.10.2; but the typical middle strip is the sum of two half middle strips, taken from each of the two adjacent equivalent rigid frames C. The factored moments in the typical column and middle strips are shown in Table 16.13.1.

(b) Slab thickness for flexure. For $f'_c = 3000$ psi and $f_y = 40,000$ psi, the maximum percentage for tension reinforcement only is $0.75\rho_b = 0.0278$ (Table 3.6.1). The actual percentages used (line 6 of Tables 16.13.2 and 16.13.3) are nowhere near this maximum.

Table 16.13.1 Factored Moments in a Typical Column Strip and Middle Strip, Example 16.13.2 (Flat Slab)

LINE NUMBER	MOMENTS AT CRITICAL SECTION (ft-kips)	EXTERIOR SPAN			INTERIOR SPAN		
		NEGATIVE MOMENT	POSITIVE MOMENT	NEGATIVE MOMENT	NEGATIVE MOMENT	POSITIVE MOMENT	NEGATIVE MOMENT
1	Total M in column and middle strips (Fig. 16.10.2) (rigid frame C)	-88	$+146$	-204	-190	$+102$	-190
2	Percentage to column strip (Table 16.12.5)	91.3%	60%	75%	75%	60%	75%
3	Moment in column strip	-80	$+88$	-153	-142	$+61$	-142
4	Moment in middle strip	-8	$+58$	-51	-48	$+41$	-48

Table 16.13.2   Design of Reinforcement in Column Strip, Example 16.13.2 (Flat Slab) ($f_y = 40,000$ psi, $f'_c = 3000$ psi)

LINE NUMBER	CRITICAL SECTION	EXTERIOR SPAN			INTERIOR SPAN		
		NEGATIVE MOMENT	POSITIVE MOMENT	NEGATIVE MOMENT	NEGATIVE MOMENT	POSITIVE MOMENT	NEGATIVE MOMENT
1	Moment, Table 16.13.1, line 3 (ft-kips)	−80	+88	−153	−142	+61	−142
2	Width b of drop or strip (in.)	100	120	100	100	120	100
3	Effective depth d (in.)	8.81	6.44	8.81	8.81	6.44	8.81
4	M_u/ϕ (ft-kips)	−89	+98	−170	−158	+68	−158
5	R_n(psi) $= M_u/(\phi bd^2)$	138	236	263	244	164	244
6	ρ, Eq. (3.8.5) or Fig. 3.8.1	0.35%	0.62%	0.70%	0.64%	0.42%	0.64%
7	$A_s = \rho bd$	3.08	4.79	6.17	5.63	3.24	5.63
8	$A_s = 0.002bt$*	2.40	1.80	2.40	2.40	1.80	2.40
9	N = larger of (7) or (8)/0.31	9.9	15.5	19.9	18.2	10.5	18.2
10	N = width of strip/(2t)	5	8	5	5	8	5
11	N required, larger of (9) or (10)	10	16	20	19	11	19

* $bt = 100(10.5) + 20(7.5) = 1200$ in.^{2} for negative moment region.

Table 16.13.3   Design of Reinforcement in Middle Strip, Example 16.13.12 (Flat Slab) ($f_y = 40,000$ psi, $f'_c = 3000$ psi)

LINE NUMBER	CRITICAL SECTION	EXTERIOR SPAN			INTERIOR SPAN		
		NEGATIVE MOMENT	POSITIVE MOMENT	NEGATIVE MOMENT	NEGATIVE MOMENT	POSITIVE MOMENT	NEGATIVE MOMENT
1	Moment, Table 16.13.1, line 4 (ft-kips)	−8	+58	−51	−48	+41	−48
2	Width of strip, b (in.)	180	180	180	180	180	180
3	Effective depth d (in.)	6.44	5.81	6.44	6.44	5.81	6.44
4	M_u/ϕ (ft-kips)	−9	+64	−57	−53	+46	−53
5	R_n(psi) $= M_u/(\phi bd^2)$	14	126	92	85	91	85
6	ρ, Eq. (3.8.5) or Fig. 3.8.1	0.04%	0.32%	0.23%	0.22%	0.23%	0.22%
7	$A_s = \rho bd$	0.46	3.30	2.67	2.53	2.41	2.53
8	$A_s = 0.002bt$	2.70	2.70	2.70	2.70	2.70	2.70
9	N = larger of (7) or (8)/0.31*	8.7	10.6	8.7	8.7	8.7	8.7
10	N = width of strip/(2t)	12	12	12	12	12	12
11	N required, larger of (9) or (10)	12	12	12	12	12	12

* A mixture of #5 and #4 bars could have been selected.

Figure 16.13.4 Effective depths provided at critical sections in flat slab design example.

Thus there is ample compressive strength in the slab. This phenomenon is usual because of the deflection control exerted by the minimum slab thickness requirements.

(c) Design of reinforcement. The design of reinforcement for the typical column strip is shown in Table 16.13.2; for the typical middle strip, it is shown in Table 16.13.3. Because the moments in the long direction are larger than those in the short direction, the larger effective depth is assigned to the long direction wherever the two layers of steel are in contact. This contact at crossing occurs in the top steel at the intersection of column strips and in the bottom steel at the intersection of middle strips. Assuming #5 bars and $\frac{3}{4}$ in. clear cover, the effective depths provided at various critical sections of the long and short directions are shown in Fig. 16.13.4. ∎

16.14 BEAM (IF USED) SIZE REQUIREMENT IN FLEXURE AND SHEAR

The size of the beams along the column centerlines in a two-way slab (with beams) should be sufficient to take the bending moments and shears at the critical sections.

For approximately equal spans, the largest bending moment should occur at the exterior face of the first interior column where the available section for strength computation is rectangular in nature because the effective slab projection is on the tension side. Then with the preliminary beam size the required reinforcement ratio ρ may be determined and compared with $0.75\rho_b$, the maximum value permitted. Deflection is unlikely to be a problem with T-sections, but must be investigated if excessive deflection may cause difficulty.

The maximum shear in the beam should also occur at the exterior face of the first interior column. The shear diagram for the exterior span may be obtained by placing the negative moments already computed for the beam at the face of the column at each end and loading the span with the percentage of floor load interpolated (ACI-13.6.8) between $\alpha_1 L_2/L_1 = 0$ and $\alpha_1 L_2/L_1 \geq 1.0$. As discussed in Section 10.2, the stem (web) $b_w d$ should for practicality be sized such that nominal shear stress $v_n = V_u/(\phi b_w d)$ does not exceed about $6\sqrt{f'_c}$ at the critical section d from the face of support.

■ **EXAMPLE 16.14.1**   Investigate if the preliminary overall sizes of 14 × 28 in. for the long beam and 12 × 24 in. for the short beam are suitable for the two-way slab (with beams) design example.

Solution: Since the values of α, or of $\alpha_1 L_2/L_1$, are considerably larger than 1.0 for all beam spans, there is to be no reduction of the floor load going into the beams from the tributary areas (ACI-13.6.8). As shown in Fig. 16.14.1, the most critical span is $B1$ for the long direction and $B5$ for the short direction. Actually, the load acting on the clear span of the beam should include the floor load (including the weight of the beam stem itself or any other load) directly over the beam stem width plus the floor load on the tributary areas bounded by the 45° lines from the corner of the panel. Also for practical purposes it is acceptable to consider the shear due to floor load at the face of column equal to one-half of the floor load on the tributary areas between column centerlines, as shown in Fig. 16.14.1.

(a)   Size of long beam $B1$. The negative moments at the face of supports, 57 and 221 ft-kips, are taken from Table 16.12.2, frame A.

$$\text{weight of beam stem} = \frac{14(21.5)}{144}(150) = 314 \text{ lb/ft}$$

$$\text{maximum negative moment} = \tfrac{1}{10}(1.4)(0.314)(23.75)^2 + 221$$
$$= 25 + 221 = 246 \text{ ft-kips}$$

$$b = 14 \text{ in.} \qquad d = 28 - 2.5 \text{ (assume one layer of steel)} = 25.5 \text{ in.}$$

$$R_n = \frac{M_u}{\phi b_w d^2} = \frac{246(12{,}000)}{0.90(14)(25.5)^2} = 360 \text{ psi}$$

Figure 16.14.1 Beams around the two-way slab panel.

From Fig. 3.8.1, $\rho = 0.010$, which is well below $0.75\rho_b^* = 0.027$. Perhaps the beam size should be reduced. From Fig. 16.14.1,

$$\text{total factored floor load on } B1 = 6.36(15) = 95.4 \text{ kips}$$

$$\max V_u = 1.15(1.4)(0.314)\frac{23.75}{2} + \frac{1}{2}(95.4) + \frac{221 - 57}{23.75}$$

$$= 6.0 + 47.7 + 6.9 = 60.6 \text{ kips}$$

$$v_n = \frac{V_u}{\phi b_w d} = \frac{60,600}{0.85(14)(25.5)} = 200 \text{ psi} = 3.7\sqrt{f_c'} \qquad\qquad \textbf{OK}$$

(b) Size of short beam $B5$. The negative moments at the face of supports, 39 and 141 ft-kips, are taken from Table 16.12.2, frame C.

$$\text{weight of beam stem} = \frac{12(17.5)}{144}(150) = 219 \text{ lb/ft}$$

$$\text{maximum negative moment} = \tfrac{1}{10}(1.4)(0.219)(18.75)^2 + 141$$

$$= 11 + 141 = 152 \text{ ft-kips}$$

$$b = 12 \text{ in.} \qquad d = 24 - 2.5 \text{ (assume one layer of steel)} = 21.5 \text{ in.}$$

$$R_n = \frac{M_u}{\phi b_w d^2} = \frac{152(12,000)}{0.90(12)(21.5)^2} = 365 \text{ psi}$$

From Fig. 3.8.1, $\rho = 0.0105$, which is well below $0.75\rho_b = 0.027$. From Fig. 16.14.1,

$$\text{total factored floor load on } B5 = 6.36(10) = 63.6 \text{ kips}$$

$$\max V_u = 1.15(1.4)(0.219)\frac{18.75}{2} + \frac{1}{2}(63.6) + \frac{141 - 39}{18.75}$$

$$= 3.3 + 31.8 + 5.4 = 40.5 \text{ kips}$$

$$v_n = \frac{V_u}{\phi b_u d} = \frac{40,500}{0.85(12)(21.5)} = 185 \text{ psi} = 3.4\sqrt{f_c'} \qquad\qquad \textbf{(OK)}$$

As mentioned above, the size of beams in both the long and short directions should probably be reduced; the nominal shear stress v_n is well below $6\sqrt{f_c'}$ at the face of support and is even lower at d therefrom. ∎

16.15 SHEAR STRENGTH IN TWO-WAY FLOOR SYSTEMS

The shear strength of a flat slab or flat plate floor around a typical interior column under dead and full live loads is analogous to that of a square or rectangular spread footing subjected to a concentrated column load, except each is an inverted situation of the other. The area enclosed between the parallel pairs of centerlines of the adjacent panels of the floor is like the area of the footing, because there is no shear force along the panel centerline of a typical interior panel in a floor system. Consequently the discussion here is essentially identical to what is included in Chapter 20 on footings.

The shear strength of two-way slab systems without shear reinforcement has been studied by many investigators [16.76–16.92]. An excellent summary is provided by ASCE-ACI Task Committee 426 under Chairman N. M. Hawkins [16.83].

Wide-Beam Action. The shear strength of the flat slab or flat plate should be first investigated for wide-beam action and then for two-way action (ACI-11.12). In the wide-beam action, the critical section is parallel to the panel centerline in the transverse direction and extends across the full distance between two adjacent longitudinal panel centerlines. As in beams, this critical section of width b_w times the effective depth d is located at a distance d from the face of the equivalent square column capital or from the face of the drop panel, if any. The nominal strength in usual cases where no shear reinforcement is used is

$$V_n = V_c = 2\sqrt{f_c'}\,b_w d \qquad \textbf{(16.15.1)}$$

according to the simplified method of ACI-11.3.1.1. Alternatively, V_c may be determined using the more detailed expression (ACI-11.3.2.1) involving $\rho V_u d/M_u$, Eq. (5.10.5).

Two-Way Action. Potential diagonal cracking may occur along a truncated cone or pyramid around columns, concentrated loads, or reactions. The critical section is located so that its periphery b_0 is at a distance $d/2$ (that is, one-half the effective depth) outside a column, concentrated load, or reaction. In the 1989 ACI Code, clarification was made (ACI-11.12.1.2) that investigations are required at potential critical sections located at $d/2$ outside changes in slab thickness, such as capitals and drop panels. Furthermore, since the Code states that b_0 is a "minimum but need not approach closer than $d/2$," some confusion arose as to whether the "minimum" at $d/2$ would require using a perimeter around a square or rectangular column that had curved corners. Since exactness is neither improved nor reduced by calculating the critical section b_0 by such elaborate procedures, the 1989 ACI Code specifically permits the critical section for square or rectangular loaded areas to "have four straight lines."

When shear reinforcement is not used, the nominal shear strength $V_n = V_c$, which is given by ACI-11.12.2.1 as the smallest of

$$V_c = \left(2 + \frac{4}{\beta_c}\right)\sqrt{f_c'}\,b_0 d, \qquad \text{ACI Formula (11-36)} \qquad \textbf{(16.15.2a)}$$

$$V_c = \left(\frac{\alpha_s}{b_0/d} + 2\right)\sqrt{f_c'}\,b_0 d, \qquad \text{ACI Formula (11-37)} \qquad \textbf{(16.15.2b)}$$

and

$$V_c = 4\sqrt{f_c'}\,b_0 d, \qquad \text{ACI Formula (11-38)} \qquad \textbf{(16.15.2c)}$$

where

b_0 = perimeter of critical section

β_c = ratio of long side to short side of the column

α_s = 40 for interior columns, 30 for edge columns, and 20 for corner columns

Equation (16.15.2a) recognizes that there should be a transition between, say, a square column ($\beta_c = 1$) where V_c might be based on $4\sqrt{f_c'}$ for two-way action, and a wall ($\beta_c = \infty$) where V_c should be based on the $2\sqrt{f_c'}$ used for one-way action as for beams. However, unless β_c is larger than 2.0, Eq. (16.5.2a) does not control.

Equation (16.15.2b) is new with the 1989 ACI Code. Though designers have generally been investigating critical sections around the perimeter at changes in slab thickness, such as at the edges of capitals and drop panels, it was not generally realized that the strength at such locations may be less than that based on $4\sqrt{f_c'}$. The two-way action strength may reduce, even for a square concentrated load area, both (1) as the distance to the critical section from the concentrated load increases, such as for drop panels, and (2) as the perimeter becomes large compared to the slab thickness, such as for example a 6 in. slab supported by a 10 ft square column ($b_0/d \approx 80$). The new equation accounts for this reduced strength.

In the application of Eqs. (16.15.2), b_0 is the perimeter of the critical section at a distance $d/2$ from the edge of column capital or drop panel. For Eq. (16.15.2b), α_s for an "interior column" applies when the perimeter is four-sided, for an "edge column" when the perimeter is three-sided, and for a "corner column" when the perimeter is two-sided. As shown by Fig. 16.15.1, Eq. (16.15.2b) will give a V_c smaller than $4\sqrt{f_c'}b_0d$ for large columns (or very thin slabs), such as a square interior column having side larger than $4d$, a square edge column having side larger than $4.33d$, and a square corner column having side larger than $4.5d$. Thus, the nominal shear strength V_c in a two-way system is generally set by Eq. (16.5.2c), that is $V_c = 4\sqrt{f_c'}b_0d$, unless either of Eq. (16.5.2a) or Eq. (16.5.2b) gives a lesser value.

Shear Reinforcement.   Even when shear reinforcement is used (ACI-11.12.3.2), the nominal strength is limited to a maximum of

$$V_n = V_c + V_s \leq 6\sqrt{f_c'}b_0d \qquad \textbf{(16.15.3)}$$

$$b_0 = 4(s + d)$$

If $s = 4d$, $\dfrac{40}{b_0/d} = 2$

(a) Interior column

$$b_0 = 3s + 2d$$

If $s = 4.33d$, $\dfrac{30}{b_0/d} = 2$

(b) Edge column

$$b_0 = 2s + d$$

If $s = 4.5d$, $\dfrac{20}{b_0/d} = 2$

(c) Corner column

Figure 16.15.1 Minimum size of square columns for $V_c = 4\sqrt{f_c'}b_0d$.

Further, in the design of any shear reinforcement, the portion of the strength V_c may not exceed $2\sqrt{f'_c}b_0d$ (ACI-11.12.3.1). If shearhead reinforcement such as described in Section 16.16 is used (ACI-11.12.4.8), the maximum V_n in Eq. (16.15.3) is $7\sqrt{f'_c}b_0d$.

The investigation for concentric shear (without moment) transfer from slab to column is shown in the following two examples, for the flat slab and flat plate design examples, respectively. When there must be transfer of both shear and moment from the slab to the column, ACI-11.12.6 applies, as will be discussed in Section 16.18.

■ **EXAMPLE 16.15.1** Investigate the shear strength in wide-beam and two-way actions in the flat slab design example for an interior column with no bending moment to be transferred. Note that $f'_c = 3000$ psi.

Solution: **(a)** Wide-beam action. Investigation for wide-beam action is made for sections 1-1 and 2-2 in the long direction, as shown in Fig. 16.15.2(a). The short direction has a wider critical section and shorter span; thus it does not control. For

(a) Wide-beam action

(b) Two-way action

Figure 16.15.2 Critical sections for shear in flat slab design example.

section 1-1, if the entire width of 20 ft is conservatively assumed to have an effective depth of 6.12 in.,

$$V_u = 0.336(20)(9.52) = 64 \text{ kips} \quad (\text{section 1-1})$$

$$V_n = V_c = 2\sqrt{f'_c}(240)(6.12)\tfrac{1}{1000} = 161 \text{ kips}$$

$$\phi V_n = 0.85(161) = 137 \text{ kips} > V_u \qquad\qquad\qquad \textbf{OK}$$

If, however, b_w is taken as 84 in. and d as 9.12 in. on the contention that the increased depth d is only over a width of 84 in.,

$$V_n = V_c = 2\sqrt{f'_c}(84)(9.12)\tfrac{1}{1000} = 84 \text{ kips}$$

This latter value is probably unrealistically low. For section 2-2, the shear resisting section has a constant d of 6.12 in.; thus

$$V_u = 0.336(20)(7.82) = 53 \text{ kips} \quad (\text{section 2-2})$$

$$\phi V_n = 137 \text{ kips} > V_u \qquad\qquad\qquad \textbf{OK}$$

It will be rare that wide-beam (one-way) action will govern.

(b) Two-way action. The critical sections for two-way action are the circular section 1-1 at $d/2 = 4.56$ in. from the edge of the column capital and the rectangular section 2-2 at $d/2 = 3.06$ in. from the edge of the drop, as shown in Fig. 16.15.1(b). Since there are no shearing forces at the centerlines of the four adjacent panels, the shear forces around the critical sections 1-1 and 2-2 in Fig. 16.15.1(b) are

$$V_u = 0.336\left[500 - \frac{\pi(5.76)^2}{4}\right] + 1.4(0.038)\left[7(8.33) - \frac{\pi(5.76)^2}{4}\right]$$

$$= 159.2 + 1.7 = 161 \text{ kips} \quad (\text{section 1-1})$$

In the second term, the 0.038 is the weight of the 3-in. drop in ksf.

$$V_u = 0.336[500 - 8.84(7.51)] = 146 \text{ kips} \quad (\text{section 2-2})$$

Compute the shear strength at section 1-1 around the perimeter of the capital (Fig. 16.15.2b),

$$b_0 = \pi(5.76)12 = 217.1 \text{ in.}; \qquad \frac{b_0}{d} = \frac{217.1}{9.12} = 23.8$$

Since $b_0/d > 20$, and $\beta_c = 1$, Eq. (16.5.2b) controls. Thus,

$$\phi V_n = \phi V_c = \phi\left(\frac{40}{23.8} + 2\right)\sqrt{f'_c}\,b_0 d = \phi(3.68\sqrt{f'_c}\,b_0 d)$$

$$= 0.85(3.68\sqrt{f'_c})(217.1)(9.12)\tfrac{1}{1000} = 339 \text{ kips}$$

At section 2-2, Fig. 16.15.2(b),

$$b_0 = [2(8.84) + 2(7.51)]12 = 392.4 \text{ in.}; \qquad \frac{b_0}{d} = \frac{392.4}{6.12} = 64.1$$

and since $b_0/d > 20$, Eq. (16.5.2b) controls. Thus,

$$\phi V_n = \phi V_c = \phi\left(\frac{40}{64.1} + 2\right)\sqrt{f_c'}\,b_0 d = \phi(2.62\sqrt{f_c'}\,b_0 d)$$

$$= 0.85(2.62\sqrt{f_c'})(392.4)(6.12)\tfrac{1}{1000} = 293 \text{ kips}$$

Though both sections 1-1 and 2-2 have ϕV_n significantly greater than V_u, the section around the drop panel is loaded to a slightly higher percentage of its strength (50% for section 2-2 vs 47% for section 1-1). Prior to the 1989 ACI Code, using $4\sqrt{f_c'}\,b_0 d$, the shear strength at the drop panel perimeter would rarely have been of concern. Shear reinforcement is not required at this interior location. ∎

■ **EXAMPLE 16.15.2** Investigate the shear strength in wide-beam and two-way actions in the flat plate design example for an interior column with no bending moment to be transferred. Note that $f_c' = 4000$ psi.

Solution: **(a)** Wide-beam action. Assuming $\frac{3}{4}$ in. clear cover and #4 bars, the average effective depth when bars in two directions are in contact is

$$\text{avg } d = 5.50 - 0.75 - 0.50 = 4.25 \text{ in.}$$

Referring to Fig. 16.15.3(a),

$$V_u = 0.198(12)6.65 = 15.8 \text{ kips}$$

$$\phi V_n = \phi V_c = 0.85(2\sqrt{4000})(12)(12)4.25\tfrac{1}{1000} = 65.8 \text{ kips} \qquad \textbf{OK}$$

(b) Two-way action. Referring to Fig. 16.15.3(b),

$$V_u = 0.198[15(12) - 1.35(1.19)] = 35.3 \text{ kips}$$

The perimeter of the critical section at $d/2$ around the column is

$$b_0 = 2(16.25) + 2(14.25) = 61.0 \text{ in.}; \qquad \frac{b_0}{d} = \frac{61.0}{4.25} = 14.3 < 20$$

(a) Wide-beam action   (b) Two-way action

Figure 16.15.3 Critical sections for shear in flat plate design example.

Since $b_0/d < 20$, and $\beta_c = 1$, Eq. (16.15.2c) controls. Thus,

$$V_c = 4\sqrt{f'_c}b_0d \qquad\qquad \textbf{[16.15.2c]}$$

$$\phi V_n = \phi V_c = 0.85(4\sqrt{4000})(61.0)4.25\tfrac{1}{1000} = 55.7 \text{ kips} \qquad \textbf{OK}$$

Shear reinforcement is not required at this interior location. ∎

16.16 SHEAR REINFORCEMENT IN FLAT PLATE FLOORS

In flat plate floors where neither column capitals nor drop panels are used, shear reinforcement is frequently necessary. In such cases, two-way action usually controls. The shear reinforcement may take the form of properly anchored bars or wires placed in vertical sections around the column [Fig. 16.16.1(a)], or consist of shearheads, which are steel I- or channel-shaped sections fabricated by welding into four (or three for an exterior column) identical arms at right angles and uninterrupted within the column section [Fig. 16.16.1(b)]. A summary of shear reinforcement for flat plates has been provided by Dilger and Ghali [16.99]. The strength of two-way slab systems with shear reinforcement has been summarized by Hawkins [16.94]. Corley and Hawkins [16.93, 16.108] have studied shearhead reinforcement. Other studies of shear reinforcement in flat plates have been made by Ghali, Dilger et al. [16.95–16.100], and Pillai, Kirk, and Scavuzzo [16.113].

When *bar or wire shear reinforcement* is used, the nominal strength is

$$V_n = V_c + V_s = 2\sqrt{f'_c}b_0d + \frac{A_v f_y d}{s} \qquad\qquad \textbf{(16.16.1)}$$

(a) Bar reinforcement   (b) Shearhead reinforcement

Figure 16.16.1 Bar and shearhead reinforcement in flat plate floors.

where b_0 is the periphery around the critical section for two-way shear action and A_v is the total stirrup bar area around b_0. Such bar or wire reinforcement is required wherever V_u exceeds ϕV_c based on V_c of Eqs. (16.15.2a to c). However, in the design of shear reinforcement, V_c for Eq. (16.16.1) equals $2\sqrt{f'_c}b_0d$, and the maximum strength V_s from the shear reinforcement may not exceed $4\sqrt{f'_c}b_0d$ according to ACI-11.12.3.2.

Shear strength may be provided by *shearheads* under ACI-11.12.4 whenever V_u/ϕ at the critical section is between that permitted by Eqs. (16.15.2a to c) and $7\sqrt{f'_c}b_0d$. These provisions, based on the tests of Corley and Hawkins [16.93], apply only where shear alone (i.e., no bending moment) is transferred at an interior column. When there is moment transfer to columns, ACI-11.12.6.3 applies, as is discussed in Section 16.18.

With regard to the size of the shearhead, it must furnish a ratio α_v of 0.15 or larger (ACI-11.12.4.5) between the stiffness for each shearhead arm (E_sI_x) and that for the surrounding composite cracked slab section of width $(c_2 + d)$, or

$$\text{min } \alpha_v = \frac{E_sI_x}{E_c \text{ (composite } I_s)} = 0.15 \qquad \textbf{(16.16.2)}$$

The steel shape used must not be deeper than 70 times its web thickness, and the compression flange must be located within $0.3d$ of the compression surface of the slab (ACI-11.12.4.2 and 11.12.4.4). In addition, the plastic moment capacity M_p of the shearhead arm must be at least (ACI-11.12.4.6).

$$\text{min } M_p = \frac{V_u}{2\eta\phi}\left[h_v + \alpha_v\left(L_v - \frac{c_1}{2}\right)\right] \qquad \textbf{(16.16.3)}$$

where

η = number (usually 4) of identical shearhead arms

V_u = factored shear around the periphery of column face

h_v = depth of shearhead

L_v = length of shearhead measured from column centerline

ϕ = 0.90, the strength reduction factor for flexure

Equation (16.16.3) is to ensure that the required shear strength of the slab is reached before the flexural strength of the shearhead is exceeded.

The length of the shearhead should be such that the nominal shear strength V_n will not exceed $4\sqrt{f'_c}b_0d$ computed at a peripheral section located at $\frac{3}{4}(L_v - c_1/2)$ along the shearhead but no closer elsewhere than $d/2$ from the column face (ACI-11.12.4.7 and 11.12.4.8). This length requirement is shown in Fig. 16.16.1(b).

When a shearhead is used, it may be considered to contribute a resisting moment (ACI-11.12.4.9)

$$M_v = \frac{\phi\alpha_vV_u}{2\eta}\left(L_v - \frac{c_1}{2}\right) \qquad \textbf{(16.16.4)}$$

to each column strip, but not more than 30% of the total moment resistance required in the column strip, nor the change in column strip moment over the length L_v, nor the required M_p given by Eq. (16.16.3).

■ **EXAMPLE 16.16.1**   Using the dimensions of the flat plate design example but changing the live load to 200 psf, investigate the shear strength for wide-beam and two-way actions around an interior column. If the required nominal shear strength V_n for two-way action is between that permitted by Eqs. (16.15.2) and $6\sqrt{f'_c}b_0d$, determine the A_v/s requirement for shear reinforcement at the peripheral critical section and show the nominal shear stress (which is factored shear V_u divided by $\phi b_0 d$) variation from the critical section to the panel centerline. Use $f'_c = 4000$ psi and $f_y = 50,000$ psi; assume #5 slab reinforcement.

Solution: **(a)** Wide-beam action.

$$w_u = 1.4w_D + 1.7w_L = 1.4(150)(5.5/12) + 1.7(200)$$
$$= 96 + 340 = 436 \text{ psf}$$

avg d in column strip $= 5.50 - 0.75 - 0.63 = 4.12$ in.

For a 12-in. wide strip along section 1-1 of Fig. 16.16.2,

$$v_n = \frac{V_u}{\phi b_w d} = \frac{436(6.66)}{0.85(12)(4.12)} = 69 \text{ psi} < (2\sqrt{f'_c} = 126 \text{ psi}) \qquad \textbf{OK}$$

(b) Two-way action. Referring to Section 2-2 of Fig. 16.16.2,

$$b_0 = 2(16.12) + 2(14.12) = 60.48 \text{ in.}; \qquad \frac{b_0}{d} = \frac{60.48}{4.12} = 14.7 < 20$$

With a rectangular perimeter b_0 having long to short side ratio less than 2 (meaning β_c less than 2), and b_0/d less than 20 for an interior column, Eq. (16.15.2c) controls; thus, the strength without shear reinforcement is $V_c = 4\sqrt{f'_c}b_0d$. Using nominal stress $v_n = V_u/\phi b_0 d$,

$$v_n = \frac{V_u}{\phi b_0 d} = \frac{436[180 - 1.34(1.18)]}{0.85(60.48)(4.12)} = 368 \text{ psi}$$

Figure 16.16.2 Critical sections for shear, Example 16.16.1.

Figure 16.16.3 Variation of two-way nominal shear stress ($V_u/\phi b_0 d$), Example 16.16.1.

Since the maximum nominal shear stress of 368 psi exceeds $4\sqrt{f'_c} = 253$ psi but not the maximum $6\sqrt{f'_c} = 380$ psi permitted when bar or wire shear reinforcement is used, shear reinforcement is required to take the excess stress v_n which exceeds $2\sqrt{f'_c} = 126$ psi. The shear reinforcement in this case may consist of properly anchored bars or wires and need not be a shearhead. The A_v/s requirement around the critical section of 60.48 in. periphery is, from applying Eq. (16.16.1) with required $V_n = V_u/\phi$,

$$\frac{A_v}{s} = \frac{V_u/\phi - (2\sqrt{f'_c})b_0 d}{f_y d} = \frac{(v_n - 2\sqrt{f'_c})b_0}{f_y}$$

$$= \frac{(368 - 126)(60.48)}{50,000} = 0.292 \text{ in.}$$

Assuming $s = d/2 \approx 2$ in. spacing,

$$A_v = 0.58 \text{ sq in.}$$

If two double #3 U stirrups are used at each of the four sides,

$$\text{provided } A_v = 4(4)(0.11) = 1.76 \text{ sq in.}$$

The variation of the nominal shear stress v_n from the maximum value of 368 psi to zero at the panel centerline over the equally spaced points 2 to 6 is shown in Fig. 16.16.3. The nominal shear stress ($V_u/\phi b_0 d$) drops to 126 psi in a rather steep manner so that the number and spacing of these U stirrups can be laid out by the aid of the excess ($v_n - 2\sqrt{f'_c}$) diagram. ■

■ **EXAMPLE 16.16.2** Design the shearhead reinforcement for the two-way shear action of Example 16.16.1.

Solution: (a) Two-way action. Since the maximum nominal shear stress $v_n = V_u/\phi b_0 d$ of 368 psi is between $4\sqrt{f'_c} = 253$ psi and the maximum of $7\sqrt{f'_c} = 442$ psi when a shearhead is used, shearhead reinforcement for the interior column (having no moment transfer to column) is permitted to be designed according to ACI-11.12.4.

(b) Length of shearhead. The length of shearhead should be such that the nominal shear stress is less than $4\sqrt{f'_c}$, computed around a periphery passing through points at $\frac{3}{4}(L_v - c_1/2)$ from but no closer than $d/2$ to the column faces. Assuming a square as the critical periphery since the shearhead is to have four identical arms (ACI-11.12.4.1), the required b_0 (ft) may be computed from Fig. 16.16.4; thus,

$$4\sqrt{4000} = \frac{436[180 - (b_0/4)^2]}{0.85(12b_0)(4.12)}$$

Neglecting the $(b_0/4)^2$ in the numerator,

$$b_0 = \frac{436(180)}{253(0.85)(4.12)(12)} = 7.4 \text{ ft (88.5 in.)}$$

The required distance L_v may be computed from the following geometric considerations. From right triangle oab,

$$\left[\frac{3}{4}\left(L_v - \frac{c}{2}\right) + \frac{c}{2}\right]\sqrt{2} = \frac{b_0}{4}$$

which gives, based on leg ob,

$$L_v = \left(\frac{88.5}{4\sqrt{2}} - 5\right)\frac{4}{3} + 5 = 19.2 \text{ in.}$$

Figure 16.16.4 Required length of shearhead, Example 16.16.2.

and, based on leg *oa*,

$$L_v = \left(\frac{88.5}{4\sqrt{2}} - 6\right)\frac{4}{3} + 6 = 18.8 \text{ in.}$$

For the periphery *abcd* not to approach closer than $d/2$ to the periphery of the column section,

$$oa = ob = 8.06 + 7.06 = 15.12 \text{ in.}$$

But,

$$oa = 6 + \tfrac{3}{4}(L_v - 6) \qquad \text{and} \qquad ob = 5 + \tfrac{3}{4}(L_v - 5)$$

which gives

$$L_v = (15.12 - 6)\tfrac{4}{3} + 6 = 18.2 \text{ in.}$$

$$L_v = (15.12 - 5)\tfrac{4}{3} + 5 = 18.5 \text{ in.}$$

<u>*Use* $L_v = 20$ in.</u>

(c) Size of shearhead. The shearhead stiffness must be at least 0.15 of that of the composite cracked slab section of width $(c_2 + d)$. It can be shown that 14-#5 bars and 10-#5 bars are required for negative slab reinforcement in the 72-in. wide column strips of the long and short directions, respectively. The composite cracked section across width *A-A* in Fig. 16.16.5 should be used because there is more steel in the slab in the long direction. The steel area A_s in section *A-A*, of width $c_2 + d$ is

$$A_s = \frac{10 + d}{72}(14)(0.31) = \frac{14.12}{72}(14)(0.31) = 0.85 \text{ sq in.}$$

Figure 16.16.5 Cracked slab section of width $(c_2 + d)$, Example 16.16.2.

Assume a $S3 \times 5.7$ section for the shearhead placed as shown in Fig. 16.16.5. The $S3 \times 5.7$ is the shallowest available rolled steel I- or channel-shaped section. With $\frac{3}{4}$-in. cover at the top face of slab and #5 bars for top reinforcement in the two orthogonal directions, average d will be $4\frac{1}{8}$ in., but the cover to the compression face (bottom) of the rolled shape will be only $\frac{1}{2}$ in. Even $\frac{3}{4}$-in. cover at the compression face would require that all bottom slab steel be cut short. If the $\frac{1}{2}$-in. cover over the rolled shape is not deemed adequate, either a thicker slab must be used or a shallower shearhead fabricated (welded) from three plates would have to be used.

The centroidal axis of the composite cracked section may be obtained by equating the static moments of the compression and tension transformed areas,

$$\frac{14.12x^2}{2} = 13.4(2.0 - x) + 6.8(4.12 - x)$$

$$x = 1.70 \text{ in.}$$

$$\text{composite } I_s = \frac{14.12(1.70)^3}{3} + n(I_x \text{ of steel section}) + 13.4(0.30)^2 + 6.8(2.42)^2$$

$$= 23.1 + 8(2.52) + 1.2 + 39.8 = 87.3 \text{ in.}^4$$

$$\text{provided } \alpha_v = \frac{E_s(2.52)}{E_c \text{ (composite } I_s)} = \frac{8(2.52)}{87.3} = 0.23 > 0.15 \qquad \textbf{OK}$$

The plastic section modulus of the $S3 \times 5.7$ is given by the *AISC Manual**[*]* as 1.95 in.[3] Using A36 steel, the provided M_p is

$$\text{provided } M_p = 36(1.95) = 70.2 \text{ in.-kips}$$

The required M_p is computed from using Eq. (16.16.3) as

$$\text{required } M_p = \frac{V_u}{8\phi}\left[h_v + \alpha_v\left(\text{required } L_v - \frac{c_1}{2}\right)\right]$$

$$= \frac{0.436(180)}{8(0.90)}[3 + 0.23(19.2 - 5)]$$

$$= 68.3 \text{ in.-kips} < 70.2 \text{ in.-kips} \qquad \textbf{OK}$$

(d) Shearhead contribution to resist negative moment in slab. The negative moments at the face of column in the 72-in. column strip width in the long and short directions are $(436/198)(0.75)$ times those for equivalent rigid frames A and C in Fig. 16.10.3, wherein $(436/198)$ is the ratio of factored loads (using 200 psf compared to using 60 psf live load) on the slab and 0.75 is the factor for transverse distribution shown in line 12 of Table 16.12.6. Thus

$$\text{column strip moment in long direction} = \frac{436}{198}(0.75)(37.8) = 62.4 \text{ ft-kips}$$

$$\text{column strip moment in short direction} = \frac{436}{198}(0.75)(30.1) = 49.7 \text{ ft-kips}$$

--- --- --- ---

[*] See *Manual of Steel Construction, Load and Resistance Factor Design*, 1986; or *Allowable Stress Design* (9th ed.) 1989. Chicago: American Institute of Steel Construction.

The resisting moment of the shearhead may be computed from Eq. (16.16.4),

$$M_c = \frac{\phi \alpha_v V_u}{8} \left(L_v - \frac{c_1}{2} \right)$$

$$= \frac{0.90(0.23)(0.436)(180)}{8(12)} \left[(20 - 6) \quad \text{or} \quad (20 - 5) \right]$$

$$= 2.37 \text{ or } 2.54 \text{ ft-kips}$$

Thus the contribution is rather small and the revision of slab reinforcement is unnecessary. ∎

16.17 DIRECT DESIGN METHOD—MOMENTS IN COLUMNS

The moments in columns due to unbalanced loads on adjacent panels are readily available when an elastic analysis is performed on the equivalent rigid frame for the various pattern loadings. In the "direct design method," wherein the six limitations listed in Section 16.7 are satisfied, the longitudinal moments in the slab are prescribed by the provisions of the Code (ACI-13.6.3). In a similar manner, the Code prescribes the unbalanced moment at an interior column as follows [ACI Formula (13-4)]:

$$M = 0.07 \left[\left(w_D + \frac{1}{2} w_L \right) L_2 L_n^2 - w_D' L_2' (L_n')^2 \right] \tag{16.17.1}$$

where

$$w_D = \text{factored dead load per unit area}$$
$$w_L = \text{factored live load per unit area}$$
$$w_D', L_2', L_n' = \text{quantities referring to shorter span}$$

The moment is yet to be distributed between the two ends of the upper and lower columns meeting at the joint.

The rationale for Eq. (16.17.1) may be observed from the stiffness ratios at a typical interior joint shown in Fig. 16.17.1(a), wherein the distribution factor for the sum

(a) Interior joint   (b) Exterior joint

Figure 16.17.1 Direct design method—moments in columns.

of the column end moments is taken as $\frac{7}{8}$ and the unbalanced moment in the column strip is taken to be $0.080/0.125$ times the difference in the total static moments due to dead plus half live load on the longer span and dead load only on the shorter span.

For the edge column, ACI-13.6.3.6 requires using the column strip nominal moment strength *provided* as the moment to be transferred. Alternatively, one might apply Eq. (16.17.1) taking the shorter span as zero, as shown in Fig. 16.17.1(b).

■ **EXAMPLE 16.17.1**   Obtain the moments in the interior and exterior columns in each direction for the flat plate design example.

Solution: **(a)** Exterior column, long direction.

$$w_D = 1.4(150)(5.5/12) = 1.4(69) = 96 \text{ psf}$$

$$w_L = 1.7(60) = 102 \text{ psf}$$

The factored moment M_u to be transferred to the columns is, according to Eq. (16.17.1),

$$M_u = 0.07(0.096 + 0.051)(12)(15 - 1)^2 = 24.2 \text{ ft-kips}$$

This moment is to be divided between upper and lower columns in proportion to their stiffnesses (in this case, equally).

The column strip nominal moment strength *provided* (ACI-13.6.3.6) is not available (i.e., reinforcement not selected) in this example. However, the required amount would be based on M_u equal to 98.8% (Table 16.12.6 for frame A) of the 15.1 ft-kips from Table 16.10.2 for frame A. This moment is much lower than the 24.2 ft-kips from ACI Formula (13-4). Conservatively, the larger of the slab requirement or ACI Formula (13-4) should be used.

(b) Interior column, long direction. The factored moment to be transferred to the columns is

$$M_u = 0.07[(0.096 + 0.051)(12)(15 - 1)^2 - 0.096(12)(15 - 1)^2]$$
$$= 0.07(0.051)(12)(14)^2 = 8.4 \text{ ft-kips}$$

This moment is to be divided between upper and lower columns.

(c) Exterior column, short direction. The factored moment to be transferred to the columns is, according to Eq. (16.17.1),

$$M_u = 0.07(0.096 + 0.051)(15)(12 - 0.83)^2 = 19.3 \text{ ft-kips}$$

This moment is to be divided between upper and lower columns. The 19.3 ft-kips computed above is more conservative than the column strip moment M_u (99.3% of 12.0 ft-kips) indicated for the exterior support for frame C (Tables 16.10.2 and 16.12.6).

(d) Interior column, short direction. The factored moment to be transferred to the columns is

$$M_u = 0.07(0.051)(15)(12 - 0.83)^2 = 6.7 \text{ ft-kips}$$

This moment is to be divided between the upper and lower columns.   ■

16.18 TRANSFER OF MOMENT AND SHEAR AT JUNCTION OF SLAB AND COLUMN

Inasmuch as the columns meet the slab at monolithic joints, there must be moment as well as shear transfer between the slab and the column ends. The moments may arise out of lateral loads due to wind or earthquake effects acting on the multistory frame, or they may be due to unbalanced gravity loads as considered in Section 16.17. In addition, the shear forces at the column ends and throughout the columns must be considered in the design of lateral reinforcement (ties or spiral) in the columns (ACI-11.11). The transfer of moment and shear at the slab–column interface is extremely important in the design of flat plates and has been the subject of numerous research studies [16.101–16.130]. Particularly, the current status is presented by ACI-ASCE Committee 352 in its *Recommendations for Design of Slab–Column Connections in Monolithic Reinforced Concrete Structures* [16.126], and the background explanation by Moehle, Kreger, and Leon [16.125].

Let M_u be the total factored moment that is to be transferred to both ends of the columns meeting at an exterior or an interior joint. Test results by Hanson and Hanson [16.102] have shown that about 60% of the moment is transferred by flexure and the remainder by unbalanced shear stresses around the critical periphery located at $d/2$ from the column faces. The ACI Code requires the total factored moment M_u to be divided into M_{ub} "transferred by flexure" (ACI-13.3.3) and M_{uv} "transferred by shear" (ACI-11.12.6) such that

$$M_{ub} = \frac{M_u}{1 + \frac{2}{3}\sqrt{\frac{b_1}{b_2}}} \tag{16.18.1}$$

where

b_1 = critical section dimension in the longitudinal direction

$\quad = c_1 + d/2$ for exterior columns (Fig. 16.18.1a)

$\quad = c_1 + d$ for interior columns (Fig. 16.18.1b)

b_2 = critical section dimension in the transverse direction

$\quad = c_2 + d$ (Fig. 16.18.1)

and

$$M_{uv} = M_u - M_{ub} \tag{16.18.2}$$

The moment M_{ub} is considered to be transferred within an effective slab width equal to $(c_2 + 3t)$ at the column (ACI-13.3.3), where t is the slab or drop panel thickness. The moment strength for M_{ub} is achieved by using additional reinforcement and closer spacing within the width $(c_2 + 3t)$.

If $b_2 = b_1$, Eq. (16.18.1) becomes

$$M_{ub} = 0.60 M_u$$

If $b_2 = 1.5 b_1$, Eq. (16.18.1) becomes

$$M_{ub} = 0.648 M_u$$

It appears reasonable that when b_2 in the transverse direction is larger than b_1 in the longitudinal direction, the moment transferred by flexure is greater because the effective slab width $(c_2 + 3t)$ resisting the moment is larger.

Figure 16.18.1 Shear transfer of moment to columns.

The moment M_{uv} transferred by shear acts in addition to the associated shear force V_u at the centroid of the shear area around the critical periphery located at $d/2$ from the column faces, as shown in Fig. 16.18.1. Referring to that figure, the factored shear *stress* is

$$v_{u1} = \frac{V_u}{A_c} - \frac{M_{uv}x_1}{J_c} \qquad (16.18.3)$$

$$v_{u2} = \frac{V_u}{A_c} + \frac{M_{uv}x_2}{J_c} \qquad (16.18.4)$$

By using a section property J_c analogous to the polar moment of inertia of the shear areas along the critical periphery taken about the z-z axis, it is assumed that there are both horizontal and vertical shear stresses on the shear areas having dimensions b_1 by d in Fig. 16.18.1. The z-z axis is perpendicular to the longitudinal axis of the equivalent frame; that is, in the transverse direction, and located at the centroid of the shear area.

For an exterior column, x_1 and x_2 are obtained by locating the centroid of the channel-shaped vertical shear area represented by the dashed line $(b_1 + b_2 + b_1)$

shown in Fig. 16.18.1(a), and

$$A_c = (2b_1 + b_2)d \qquad\qquad (16.18.5)$$

$$x_2 = \frac{b_1^2 d}{A_c} \qquad\qquad (16.18.6)$$

$$J_c = d\left[\frac{b_1^3}{3} - (2b_1 + b_2)x_2^2\right] + \frac{b_1 d^3}{6} \qquad\qquad (16.18.7)$$

For an interior column, referring to Fig. 16.18.1(b),

$$A_c = 2(b_1 + b_2)d \qquad\qquad (16.18.8)$$

$$J_c = d\left[\frac{b_1^3}{6} + \frac{b_2 b_1^2}{2}\right] + \frac{b_1 d^3}{6} \qquad\qquad (16.18.9)$$

Equations (16.18.5) to (16.18.9) are derivable by letting the shear stress at any location resulting from M_{uv} alone be proportional to the distance from the centroidal axis z-z to the two shear areas b_1 by d, and either (a) to the one shear area b_2 by d for an exterior column as shown in Fig. 16.18.2, or (b) to the two shear areas b_2 by d for the interior column.

According to ACI-11.12.6.2, the larger factored shear stress v_{u2} shown in Fig. 16.18.1 must not exceed the stress $\phi v_n = \phi V_c/b_0 d$ obtained from ACI Formulas (11-36) to (11-38), that is, Eqs. (16.15.2a, b, and c), otherwise shear reinforcement as described in Section 16.16 is required.

(a) Shear areas to resist M_{uv}

(b) Vertical resisting shear stresses

(c) Horizontal resisting shear stresses

Figure 16.18.2 Resisting shear stresses due to M_{uv} acting on an exterior column.

■ **EXAMPLE 16.18.1** For the flat plate design example, investigate the transfer of the unbalanced gravity load moments in the long direction, as already computed in Example 16.17.1, to the exterior and interior columns, respectively. Compare the moment transferred by flexure in the critical slab width with the total moment in the column strip computed in Examples 16.10.3 and 16.12.3. Compute the shear stresses around the critical periphery due to the moment transferred by shear.

Solution: **(a)** Exterior column (long direction), transfer by flexure. From Example 16.17.1, the moment to be transferred is

$$M_u = 24.2 \text{ ft-kips}$$

From Eq. (16.18.1), using the average effective depth for #4 slab reinforcement of 4.25 in.,

$$M_{ub} = \frac{M_u}{1 + \frac{2}{3}\sqrt{\dfrac{c_1 + d/2}{c_2 + d}}} = \frac{24.2}{1 + \frac{2}{3}\sqrt{\dfrac{12 + 2.125}{10 + 4.25}}} = 0.602(24.2)$$

$$= 14.6 \text{ ft-kips}$$

As shown by Fig. 16.18.3, this moment is in a critical slab width of 26.5 in. From Table 16.12.6 and Fig. 16.10.3, the total amount in the 72-in. wide column strip is

$$M \text{ in column strip} = 0.988(15.1) = 15 \text{ ft-kips}$$

t = slab or drop panel thickness

Figure 16.18.3 Transfer of moments at exterior column, Example 16.18.1.

If the slab reinforcement is placed at equal spacing in the column strip, additional reinforcement is needed in the 26.5 in. width for a moment of

$$M_{ub} - 15\left(\frac{26.5}{72}\right) = 14.5 - 5.5 = 9 \text{ ft-kips}$$

(b) Exterior column (long direction), transfer by eccentricity of shear. The factored shear force V_u is taken as w_u times the floor area, 12 ft × 7.5 ft, tributary to the exterior column.

$$V_u = 0.198(12)7.5 = 17.8 \text{ kips}$$

$$M_{uv} = M_u - M_{ub} = 24.2 - 14.6 = 9.6 \text{ ft-kips}$$

From Fig. 16.18.3,

$$x_2 = \frac{2(14.12)7.06}{28 + 14.25} = 4.69 \text{ in.}$$

$$A_c = 4.25(28.24 + 14.25) = 180 \text{ sq in.}$$

$$J_c = 4.25\left[\frac{2(14.12)^3}{3} - 42.49(4.69)^2\right] + \frac{14.12(4.25)^3}{6}$$

$$= 4004 + 181 = 4185 \text{ in.}^4$$

$$v_{u1} = \frac{17,800}{180} - \frac{9600(12)9.43}{4185} = 99 - 260 = -161 \text{ psi}$$

$$v_{u2} = \frac{17,800}{180} + \frac{9600(12)4.69}{4185} = 99 + 129 = +228 \text{ psi}$$

The capacity $\phi v_n = \phi v_c = \phi(4\sqrt{f_c'}) = 0.85(253) = 215$ psi when no shear reinforcement is provided. Additional shear reinforcement must be provided (or the slab made thicker) as was done in Example 16.16.1.

The horizontal shear stress v_{uh} at the upper or lower edge of the two shear areas b_1 by d is

$$v_{uh} = \frac{9600(12)(4.25/2)}{4185} = 58 \text{ psi}$$

The v_{uh} of 58 psi, v_{u1} of -161 psi, and v_{u2} of 228 psi may be drawn on a sketch like that of Fig. 16.18.2, and by basic statics computation of the resultant upward force should equal V_u of 17.8 kips and the resultant moment about the z-z axis should equal M_{uv} of 9.6 ft-kips.

(c) Interior column (long direction), transfer by flexure. From Example 16.17.1, the moment to be transferred is

$$M_u = 8.4 \text{ ft-kips}$$

$$M_b = \frac{M_u}{1 + \dfrac{2}{3}\sqrt{\dfrac{12 + 4.25}{10 + 4.25}}} = 0.584 M_u$$

$$M_b = 0.584 M_u = 0.584(8.4) = 4.8 \text{ ft-kips}$$

From Table 16.12.6 and Fig. 16.10.3, the total moment in the 72-in. wide column strip is

$$M \text{ in column strip} = 0.75(37.8) = 28.4 \text{ ft-kips}$$

Since the column strip moment in the 26.5-in. width of $26.5(28.4)/72 = 10.5$ ft-kips is larger than 4.8 ft-kips, no additional reinforcement is needed.

(d) Interior column (long direction), transfer by eccentricity of shear. The factored shear force V_u is taken as w_u times the tributary floor area of 12×15 ft,

$$V_u = 0.198(12)15 = 35.6 \text{ kips}$$

$$M_{uv} = M_u - M_{ub} = 8.4 - 4.9 = 3.5 \text{ ft-kips}$$

From Fig. 16.18.4,

$$A_c = 4.25(32.50 + 28.50) = 259 \text{ sq in.}$$

$$J_c = 4.25\left[\frac{(16.25)^3}{6} + \frac{14.25(16.25)^2}{2}\right] + \frac{16.25(4.25)^3}{6}$$

$$= 11{,}040 + 210 = 11{,}250 \text{ in.}^4$$

$$v_{u1} = \frac{35{,}600}{259} - \frac{3500(12)8.12}{11{,}250} = 137 - 30 = -107 \text{ psi}$$

$$v_{u2} = \frac{35{,}600}{259} + \frac{3500(12)8.12}{11{,}250} = 137 + 30 = +167 \text{ psi}$$

Figure 16.18.4 Transfer of moments in the long direction at interior column, Example 16.18.1.

The capacity $\phi v_n = \phi v_c = \phi(4\sqrt{f_c'}) = 0.85(253) = 215$ psi when no shear reinforcement is provided.

Again, by basic statics the sum of the factored load vertical shear stresses on the two areas b_1 by d plus that on the two areas b_2 by d should add to give V_u of 35.6 kips. Likewise, the moment of these shear stress resultants along with that of the horizontal shear stresses on the two faces b_1 by d should add up to 3.5 ft-kips. The horizontal shear stress at the upper or lower edge of the faces b_1 by d is

$$v_{uh} = \frac{3500(12)(4.25/2)}{11,250} = 8 \text{ psi} \quad \blacksquare$$

Moment Transfer from Flat Plate to Column When Shearheads Are Used. Tests [16.93, 16.108] have indicated that shear stresses computed for factored loads at the critical section distance $d/2$ from the column face are appropriate for transfer of $M_{uv} = M_u - M_{ub}$ as described above, even when shearheads are used. However, the critical section for V_u is at a periphery passing through points at $\frac{3}{4}(L_v - c_1/2)$ from but no closer than $d/2$ to the column faces. When there are both V_u and M_u to be transferred, ACI-11.12.6.3 requires that the sum of the shear stresses computed for M_{uv} and V_u at their respective locations not exceed $\phi(4\sqrt{f_c'})$. The reason for this apparent inconsistency (ACI Commentary-R11.12.6.3) is that these two critical sections are in close proximity at the column corners where the failures initiate (under factored loads, of course).

■ **EXAMPLE 16.18.2** Recompute the periphery length b_0 required, located at $\frac{3}{4}(L_v - c/2)$ but no closer than $d/2$ from the column face for the shearhead in Example 16.16.2 when there is unbalanced moment at the interior column equal to that of Eq. (16.17.1), ACI Formula (13-4).

Solution: (a) Determine whether or not additional reinforcement is necessary for moment transfer. Referring to Example 16.16.1, part (a) of solution,

$$w_u = 436 \text{ psf}$$

Referring to Example 16.16.2, part (d) of solution,

$$\text{column strip moment in long direction} = 62.4 \text{ ft-kips}$$

The moment to be transferred to the column is, using Eq. (16.17.1),

$$M_u = 0.07[(0.096 + 0.170)(12)(15 - 1)^2 - 0.096(12)(15 - 1)^2]$$
$$= 0.07(0.170)(12)(14)^2 = 28.0 \text{ ft-kips}$$

Referring to Example 16.18.1, part (c) of solution,

$$M_{ub} = 0.584 M_u = 0.584(28.0) = 16.4 \text{ ft-kips}$$

Column strip moment in 26.5-in. width of Fig. 16.18.4 is

$$62.4\left(\frac{26.5}{72}\right) = 23.0 \text{ ft-kips}$$

Additional reinforcement is needed to take $(28.0 - 23.0) = 5.0$ ft-kips within the 26.5-in. width unless $28.0/62.4 = 44.9\%$ of the total column strip reinforcement is concentrated in the 26.5-in. width.

(b) Compute factored load shear stress at critical section of Fig. 16.18.4 due to M_{uv} only.

$$M_{uv} = 28.0 - 16.4 = 11.6 \text{ ft-kips}$$

Using critical section properties in Example 16.18.1, part (d) of solution,

$$v_u = \frac{11,600(12)8.12}{11,250} = 100 \text{ psi}$$

(c) Compute the required periphery b_0. Referring to Example 16.16.2, part (b) of solution,

$$\phi(4\sqrt{4000}) - [100 \text{ from part (b)}] = \frac{436[180 - (b_0/4)^2]}{12b_0(4.12)}$$

Neglecting the $(b_0/4)^2$ in the numerator,

$$b_0 = \frac{436(180)}{115(12)(4.12)} = 13.8 \text{ ft}$$

Placing $b_0 = 13.8$ ft in the numerator and solving for b_0 again

$$b_0 = \frac{436[180 - (3.45)^2]}{115(12)(4.12)} = 12.9 \text{ ft}$$

(d) Dicussion. This example is for illustration of the procedure only. Actually when the service live load is increased from 60 psf in the original flat plate design example to 200 psf, the slab thickness of $5\frac{1}{2}$ in. would have to be increased if the spans are not reduced. The requirement of ACI-11.12.6.3 is expected to be more controlling for the shearhead in the exterior column. ∎

16.19 OPENINGS AND CORNER CONNECTIONS IN FLAT SLABS

When openings and corner connections are present in flat slab floors, designers must make sure that adequate provisions are made for them. The ASCE-ACI Joint Task Committee [16.83] has summarized available information. Tests by Roll et al. [16.79] have provided additional data for treating openings, while Zaghlool et al. [16.105, 16.106] have provided data for corner connections.

ACI-13.5.1 first prescribes in general that openings of any size may be provided if it can be shown by analysis that all strength and serviceability conditions, including the limits on the deflections, are satisfied. However, in common situations (ACI-13.5.2) a special analysis need not be made for slab systems not having beams when (1) openings are within the middle half of the span in each direction, provided the total amount of reinforcement required for the panel without the opening is maintained; (2) openings in the area common to two column strips do not interrupt more than one-eighth of the column strip width in either span, and the equivalent of reinforce-

Figure 16.19.1 Effect of openings and free edges on critical periphery of two-way shear action (from ACI Commentary-R11.12.5).

ment interrupted is added on all sides of the openings; (3) openings in the area common to one column strip and one middle strip do not interrupt more than one-fourth of the reinforcement in either strip, and the equivalent of reinforcement interrupted is added on all sides of the openings.

In regard to nominal shear strength in two-way action, the critical section for slabs without shearhead is not to include that part of the periphery which is enclosed by radial projections of the openings to the center of the column (ACI-11.12.5). For slabs with shearhead, the critical periphery is to be reduced only by one-half of what is cut away by the radial lines from the center of the column to the edges of the opening. Some critical sections with cutaways by openings are shown in Fig. 16.19.1.

16.20 EQUIVALENT FRAME METHOD FOR GRAVITY AND LATERAL LOAD ANALYSIS

For gravity load analysis the "equivalent frame method" prescribed by the ACI Code differs from the "direct design method" *only* in the way by which the longitudinal moments along the spans of the equivalent rigid frame (as defined in Section 16.2) are obtained. In either method of analysis, the transverse distribution of the longitudinal moments may be carried out as described in Section 16.12, except in the case of the two-way slab *supported on beams*, where the beams must be sufficiently stiff (limitation No. 6 of the direct design method discussed in Section 16.7) (ACI-13.7.7.5) to serve as boundary supports for the slab.

When lateral (wind) load needs to be considered, an elastic analysis must be made of the structure under lateral load and the results combined with those due to gravity load. Consistent with the tradition under ACI Code of using an equivalent frame for gravity load analysis, a logical extension is to use an equivalent frame approach to lateral load analysis. ACI-13.3.1.2 does not prescribe an equivalent frame method for lateral load analysis, but rather only refers to "unbraced frames" and requires taking into account "effects of cracking and reinforcement on stiffness of frame members." It is suggested in ACI Commentary-R13.3.1.2 that the effective moment of inertia I_e given by ACI Formula (9-7), or just the moment of inertia of the cracked slab-beam section may be used.

The maximum positive moments (and reversals) within the span and negative moments at the supports should be obtained for various combinations of gravity load patterns with lateral load as indicated by ACI-9.2. When the equivalent frame method *is* used for gravity load analysis of two-way floor systems meeting the limitations of the direct design method, the resulting factored moments may be reduced in a proportion such that the absolute sum of positive and average negative moments used in design need not exceed $w_u L_2 L_n^2 / 8$ (ACI-13.7.7.4).

The elastic analysis for the equivalent rigid frame is treated separately in Chapter 17. In this section, considerations are given to the determination of the flexural stiffnesses of the columns and of the slab-beam within the width of the equivalent rigid frame, the torsional stiffness of the transverse beam, and the fixed-end moments due to gravity load. These values are the required input data for the analysis procedure presented in Chapter 17.

Torsional Stiffness of the Transverse Beam.

The structure enclosed between the two parallel centerlines of two adjacent panels in a multistory two-way floor system is a three-dimensional structure. The equivalent rigid frame described in Section 16.2 approximates the three-dimensional structure by a series of two-dimensional ones. But the columns stand on or provide support for only a small portion of the width of the equivalent rigid frame. Hence, either the column stiffness has to be spread thinly over the entire width (denoted by L_2) of the equivalent rigid frame, or the slab-beam has to be shrunk to the narrow transverse width (denoted by c_2) of the columns. Corley, Sozen, and Siess [16.26] first developed the idea of attaching a torsional member to the column (a cutaway from the three-dimensional structure) in the transverse direction and in essence shifting the flexural stiffness of the slab-beam to the end of the torsional member away from the columns (see Fig. 17.3.1). Thus the effectiveness of the column to restrain the ends of the slab-beam is reduced; hence the name of a less effective "equivalent column." As shown by Fig. 16.20.1, under gravity loading, the restraint at the column is more like a fixed end to the slab-beam but the restraint away from the column tends to approach that of a simple support.

Corley and Jirsa [16.27] developed a formula for the torsional stiffness K_t of the attached torsional member so that results of the equivalent frame analysis are close

Figure 16.20.1 Attached torsional member for the columns.

to those of tests, as follows [ACI Formula (13-6)]:

$$K_t = \sum \frac{9E_{cs}C}{L_2\left(1 - \dfrac{c_2}{L_2}\right)^3}\left(\frac{I_{sb}}{I_s}\right) \tag{16.20.1}$$

in which

C = torsional constant of the transverse beam (see Section 16.11)

E_{cs} = modulus of elasticity of slab concrete

I_s = moment of inertia of slab over width of equivalent frame

I_{sb} = moment of inertia of entire T-section (if so) within the width of the equivalent rigid frame

L_2 = span of member subject to torsion

The summation sign is for the transverse spans (denoted by L_2) on each side of the column.

This parameter K_t is the most influential parameter to relate results of the theoretical elastic analysis to those of tests for gravity (or lateral) load. As test results become more available, especially for lateral load, the coefficient "9" in Eq. (16.20.1) might be adjusted in the future.

Treatment of Flexural Element Having Variable Moment of Inertia. The flexural stiffness of a flexural element ij having variable moment of inertia can be expressed by two near-end stiffnesses S_{ii} and S_{jj} and a cross stiffness S_{ij}. In applying the moment distribution method, the carry-over factor from i to j is S_{ji}/S_{ii} and the carry-over factor from j to i is S_{ij}/S_{jj}. In applying the matrix displacement method the element stiffness matrix $[S]$ is

$$[S] = \begin{bmatrix} S_{ii} & S_{ij} \\ S_{ji} & S_{jj} \end{bmatrix} \quad \text{(in which } S_{ij} = S_{ji}) \tag{16.20.2}$$

In applying the slope deflection method [16.11],

$$\begin{aligned} M_i &= M_{0i} + S_{ii}\theta_i + S_{ij}\theta_j - (S_{ii} + S_{ij})R \\ M_j &= M_{0j} + S_{ji}\theta_i + S_{jj}\theta_j - (S_{ii} + S_{ij})R \end{aligned} \tag{16.20.3}$$

in which M_{0i} and M_{0j} are the fixed-end moments, θ_i and θ_j are the slopes at ends i and j of the flexural element, and R is the clockwise rotation of the element axis.

The column analogy method may be the best way to compute the fixed-end moments, the near-end stiffnesses S_{ii} and S_{jj}, and the cross stiffness S_{ij} or S_{ji} of a flexural element having variable moment of inertia, thus avoiding using the flexibility analysis for pin-end rotations due to applied loads or unit values of end moments. Wang [16.11] has presented a description of this method, as well as the elastic analysis of rigid frames by moment distribution, slope deflection, and matrix displacement methods.

There are tables available for fixed-end moments and flexural stiffness values for slabs having various combinations of column widths, column capitals, and drop panels [16.9, 16.10, 16.131].

Flexural Stiffness of Columns. ACI Commentary-R13.7.4 states that the height of the column is to be measured from middepth of slab above to middepth of slab below, as shown in Fig. 16.20.2. The moment of inertia is to be taken as infinite from the top to the bottom of the slab-beam at the joint (ACI-13.7.4.3). In flat slabs having column capitals, the authors suggest that the $1/I$ value be assumed to vary linearly from zero at the top to its value based on gross cross section at the bottom of the capital.

Flexural Stiffness of Slab-Beam. ACI-13.7.3.3 states that the moment of inertia of slab-beams from center of column to face of column, bracket, or capital shall be assumed equal to that of the slab-beam at face of column, bracket, or capital divided by the quantity $(1 - c_2/L_2)^2$. Although ACI-13.7.3.1 permits the use of gross concrete area for computation of the moment of inertia in gravity load analysis, at the same time ACI-13.3.1.2 requires taking into account the effects of cracking and reinforcement in lateral load analysis. When computers are used, loadings for different gravity load patterns to be combined with lateral load can be listed together in an input

Figure 16.20.2 Basis of calculation of column stiffness.

load matrix; thus, it would be convenient to use only one set of assumptions for flexural stiffness properties, especially in the final analysis after the finished design. ACI Commentary-R13.3.1.2 suggests that it is reasonable to use ACI Formula (9-7), which was developed to account for cracking in the moment of inertia used for deflection computations.

There is not definitive agreement on what constitutes the appropriate assumptions for stiffness, either for gravity load analysis or lateral load analysis. For gravity load analysis, the use of gross section is reasonable because it is the simplest assumption and the results are acceptable. For lateral load analysis, particularly for the unbraced frame where the entire lateral resistance is provided by the flexural stiffness of the slab-beams and columns, the use of gross section for stiffness overemphasizes the resistance to lateral loads. In Chapter 17, devoted to the analysis of two-way floor systems, several analytical models are presented for possible use in lateral load analysis, or in combined gravity and lateral load analysis. For an extended discussion of lateral load analysis, the reader should consult the study by Vanderbilt [16.32]. Perhaps the acceptance of an equivalent beam or reduced beam method of analysis would make possible the use of gross section in lateral load analysis.

Arrangement of Live Load.   When there is a definitely known load pattern, of course, analysis should be made for it (ACI-13.7.6.1). When service live load does not exceed $\frac{3}{4}$ of the service dead load, analysis needs to be made only for full factored dead and live load on all spans (ACI-13.7.6.2). When load patterns in accordance with influence line concepts are used, only $\frac{3}{4}$ of the full factored live load needs to be used (ACI-13.7.6.3); however, factored moments used in design should not be less than those due to full factored dead and live loads on all panels (ACI-13.7.6.4).

Reduction of Negative Moments Obtained at Column Centerlines from Structural Analysis.   Negative moments obtained at interior column centerlines may be reduced to the face of rectilinear or equivalent square (for circular or polygon-shaped supports) supports, but not greater than $0.175L_1$ from the column centerline (ACI-13.7.7). For exterior columns, having capitals or brackets, reduction of negative moments can be made only to a section no greater than halfway between the face of column and edge of the capital or bracket (ACI-13.7.7.2).

Deflections. When the deflection must be calculated for a two-way slab system, the ACI Code (ACI-9.5.3.6) provides little guidance other than that one should take into account "size and shape of panel, conditions of support, and nature of restraints at panel edges." The effective moment of inertia I_e [Eq. (14.4.1)] is required to be used in such calculations. Although a number of techniques have been proposed [16.52–16.71], adaption of the equivalent frame concept seems to have the most promise of being relatively simple to apply and giving reasonable results. This equivalent frame application has been developed by Nilson and Walters [16.54] for essentially uncracked systems and extended by Kripanarayanan and Branson [16.56] for partially cracked load ranges. More recently, Scanlon et al. [16.59, 16.64, 16.66, 16.67, 16.69, and 16.71] have treated the subject in detail.

■ **EXAMPLE 16.20.1**   Assuming the equivalent frame method is to be applied to the two-way slab (with beams) design example described in Section 16.3, obtain the stiffnesses and carry-over factors necessary for the analysis of the equivalent rigid frames A, B, C, and D as shown by the notations in Fig. 16.3.5. Also obtain the fixed-end moments for gravity load. Note that with the information obtained here, the matrix displacement method as described in Chapter 17 can be applied as an alternative (perhaps in preference) to the method of moment distribution.

Solution: (a) Compute flexure properties of slab-beam. The variations in the moment of inertia of the slab-beam in the long and short directions are shown in Fig. 16.20.3. For the long slab-beam, the ratio of moment of inertia between the center and the face of the column to the moments of inertia of the rest of the span is $1.0/(1 - 15/240)^2 = 1.138$; and it is $1.0/(1 - 15/300)^2 = 1.108$ for the short slab-beam (ACI-13.7.3.3). The stiffness K, carry-over factor COF, and fixed-end moment FEM coefficients may be computed by the column analogy method [16.11].

For the long direction, referring to Fig. 16.20.3(a),

$$A = 23.75 + 2(0.879)(0.625) = 23.75 + 1.10 = 24.85$$

$$I = \tfrac{1}{12}(23.75)^3 + 1.10(12.1875)^2 = 1116 + 163 = 1279$$

$$\text{stiffness factor } s_{ii} = L\left(\frac{1}{A} + \frac{Mc}{I}\right)$$

$$s_{ii} = \frac{25}{24.85} + \frac{25(12.5)^2}{1279} = 1.006 + 3.054 = 4.06$$

$$\text{stiffness factor } s_{ij} = L\left(\frac{Mc}{I} - \frac{1}{A}\right)$$

$$s_{ij} = -(1.006 - 3.054) = 2.05$$

$$\text{COF} = \frac{2.05}{4.06} = 0.505$$

Figure 16.20.3 Flexure properties of slab-beam in two-way slab (with beams) design example.

load on analogous column for uniform load ($w = 1.0$)

$$= \tfrac{2}{3}(78.1 - 7.6)(23.75) + 7.6(23.75) + 0.879(7.6)(0.625)$$
$$= 1116 + 180 + 4 = 1300$$

$$\text{FEM coefficient} = \frac{1300}{24.85L_1^2} = \frac{1300}{24.85(625)} = 0.084$$

For the short direction, referring to Fig. 16.20.3(b),

$$A = 18.75 + 2(0.902)(0.625) = 18.75 + 1.13 = 19.88$$
$$I = \tfrac{1}{12}(18.75)^3 + 1.13(9.6875)^2 = 549 + 106 = 655$$
$$s_{ii} = \frac{20}{19.88} + \frac{20(10)^2}{655} = 1.006 + 3.053 = 4.06$$
$$s_{ij} = -(1.006 - 3.053) = 2.05$$
$$\text{COF} = \frac{2.05}{4.06} = 0.505$$

load on analogous column for uniform load ($w = 1.0$)

$$= \tfrac{2}{3}(50 - 6.05)(18.75) + 6.05(18.75) + 0.902(6.05)(0.625)$$
$$= 549 + 113 + 3 = 665$$

$$\text{FEM coefficient} = \frac{665}{19.88L_1^2} = \frac{665}{19.88(400)} = 0.084$$

The flexural stiffness of the slab-beams in frames A, B C, and D are, using the I_{sb} values shown in Fig. 16.20.4,

Frame A, $\qquad K_{sb} = \dfrac{4.06E(66,540)}{300} = 901E$

Frame B, $\qquad K_{sb} = \dfrac{4.06E(57,730)}{300} = 781E$

Frame C, $\qquad K_{sb} = \dfrac{4.06E(39,530)}{240} = 669E$

Frame D, $\qquad K_{sb} = \dfrac{4.06E(33,980)}{240} = 575E$

Note that the moment of inertia values used above are based on the gross cross sections as shown in Fig. 16.20.4, an acceptable procedure for gravity load analysis.

(a) For frames A and B

(b) For frames C and D

Figure 16.20.4 Slab-beam cross sections in two-way slab (with beams) design example.

These stiffness values are likely to be too high for lateral load analysis, since cracking reduces flexural stiffness. ACI-13.3.1.2 states that for lateral load analysis, effects of cracking and reinforcement must be taken into account.

(b) Compute flexure properties of columns. The variations in the moment of inertia of the column section in the long and short directions are shown in Fig. 16.20.5. The stiffness coefficients and carry-over factors may be computed by the column analogy method.

(a) Column section in short direction (for frames C and D)

(b) Column section in long direction (for frames A and B)

Figure 16.20.5 Flexure properties of columns in two-way slab (with beams) design example.

For the short direction, referring to Fig. 16.20.5(a),

$$A = 9.67, \qquad I = \tfrac{1}{12}(9.67)^3 = 75.3$$

$$s_{TT} = \frac{12}{9.67} + \frac{12(6.90)^2}{75.3} = 1.24 + 7.59 = 8.83$$

$$s_{BB} = \frac{12}{9.67} + \frac{12(5.10)^2}{75.3} = 1.24 + 4.15 = 5.39$$

$$s_{TB} = s_{BT} = -\left[\frac{12}{9.67} - \frac{12(6.90)(5.10)}{75.3}\right] = -1.24 + 5.61 = 4.37$$

$$(\mathrm{COF})_{TB} = \frac{4.37}{8.83} = 0.495$$

$$(\mathrm{COF})_{BT} = \frac{4.37}{5.39} = 0.811$$

$$\text{Stiffness at top,} \quad K_{cT} = s_{TT}\frac{EI}{L} = \frac{8.83E(15)^4/12}{144} = 259E$$

$$\text{Stiffness at bottom,} \quad K_{cB} = s_{BB}\frac{EI}{L} = \frac{5.39E(15)^4/12}{144} = 158E$$

For the long direction, referring to Fig. 16.20.5(b),

$$A = 10.00, \qquad I = \tfrac{1}{12}(10)^3 = 83.3$$

$$s_{TT} = \frac{12}{10} + \frac{12(6.73)^2}{83.3} = 1.20 + 6.53 = 7.73$$

$$s_{BB} = \frac{12}{10} + \frac{12(5.27)^2}{83.3} = 1.20 + 4.00 = 5.20$$

$$s_{TB} = s_{BT} = -\left[\frac{12}{10} - \frac{12(6.73)(5.27)}{83.3}\right] = -1.20 + 5.11 = 3.91$$

$$(\mathrm{COF})_{TB} = \frac{3.91}{7.73} = 0.506$$

$$(\mathrm{COF})_{BT} = \frac{3.91}{5.20} = 0.752$$

$$\text{Stiffness at top,} \quad K_{cT} = \frac{7.73E(15)^4/12}{144} = 226E$$

$$\text{Stiffness at bottom,} \quad K_{cB} = \frac{5.20(15)^4/12}{144} = 152E$$

Figure 16.20.6 Torsional constants in two-way slab (with beams) design example (from Example 16.11.1).

(c) Compute torsional stiffnesses of transverse torsional members. The torsional constants C for the transverse members shown in Fig. 16.20.6 are taken from Example 16.11.1. The values for the ratio of I_{sb} to I_s needed to increase the torsional stiffness K_t [ACI Formula (13-6)] for each direction are shown in Fig. 16.20.4.

For frame A, using $I_{sb}/I_s = 12.12$ for 14×21.5 projection below 240×6.5 slab,

$$\text{exterior } K_t = \frac{18E(10,700)}{240(1 - 15/240)^3}(12.12) = 974E(12.12) = 11{,}800E$$

$$\text{interior } K_t = \frac{18E(11,930)}{240(1 - 15/240)^3}(12.12) = 1086E(12.12) = 13{,}200E$$

For frame B, using $I_{sb}/I_s = 19.84$ for 14×21.5 projection below 127×6.5 slab,

$$\text{exterior } K_t = 487E(19.84) = 9660E$$

$$\text{interior } K_t = 543E(19.84) = 10{,}800E$$

For frame C, using $I_{sb}/I_s = 5.75$ for 12×17.5 projection below 300×6.5 slab,

$$\text{exterior } K_t = \frac{18E(19,100)}{300(1 - 15/300)^3}(5.75) = 1340E(5.75) = 7700E$$

$$\text{interior } K_t = \frac{18E(20,700)}{300(1 - 15/300)^3}(5.75) = 1450E(5.75) = 8340E$$

For frame D, using $I_{sb}/I_s = 9.52$ for 12×17.5 projection below 156×6.5 slab,

$$\text{exterior } K_t = 670E(9.52) = 6380E$$

$$\text{interior } K_t = 725E(9.52) = 6900E \quad \blacksquare$$

■ **EXAMPLE 16.20.2**   Assuming the equivalent frame method is to be applied to the flat slab design example described in Section 16.3, obtain the stiffnesses and carry-over factors necessary for the analysis of equivalent rigid frame A in the long direction. Also obtain the fixed-end moments for gravity load.

Solution: **(a)** Compute flexure properties of slab strip. The stiffnesses, carryover factors, and fixed-end moments may be determined by various analysis methods. The column analogy method [16.11] is used in this example. Simmonds and Misic [16.9] have provided design aids to meet the ACI Code assumptions of the "equivalent frame method."

The variation in the moment of inertia along the interior span of the slab strip is shown in Fig. 16.20.7(a). Taking the moment of inertia through the $7\frac{1}{2}$-in. slab as

(a)   Relative moment of inertia

(b)   Analogous column section

(c)   Simple moment diagram, w = 1.0 per unit length

Figure 16.20.7 Flexure properties of slab strip in flat slab design example.

the reference value of 1, the moment of inertia through the drop, where there is a T-section of a 240×7.5 in. flange and an 84×3 in. web, is 1.745. The moment of inertia between the column centerline and the face of the equivalent square column capital is $1.745/(1 - 4.43/20)^2 = 1.745/0.606 = 2.88$.

The variation in the width of the analogous column section is $1/I$, which is shown in Fig. 16.20.7(b). The area of the analogous column section is

$$A = 16.66 + 2(0.573)(1.96) + 2(0.347)(2.21)$$
$$= 16.66 + 2.25 + 1.53 = 20.44$$

The moment of inertia about the midspan, neglecting the moments of inertia of the short segments about their own centroidal axes, is

$$I = \tfrac{1}{12}(16.66)^3 + 2.25(9.31)^2 + 1.53(11.40)^2 = 385 + 195 + 199 = 779$$

$$\text{stiffness factor } s_{ii} = L\left(\frac{1}{A} + \frac{Mc}{I}\right) = 25\left[\frac{1}{20.44} + \frac{12.5(12.5)}{779}\right]$$

$$s_{ii} = 1.22 + 5.01 = 6.23$$

$$s_{ij} = -1.22 + 5.01 = 3.79$$

$$\text{COF} = \frac{3.79}{6.23} = 0.608$$

$$\text{stiffness } K \text{ at end of 20-ft wide slab strip} = \frac{6.23E(7.5)^3/12}{300}(240) = 175E$$

The load on the analogous column is equal to the summation of the product of the width of the analogous column section and the area of the simple beam moment diagram of Fig. 16.20.7(c). Considering the moment areas over the short segments as being trapezoidal, the load on the analogous column is

$$P = \tfrac{2}{3}(78.1 - 43.5)(16.66) + 43.5(16.66) + 2(0.573)(\tfrac{1}{2})(43.5 + 25.2)(1.96)$$
$$+ 2(0.347)(\tfrac{1}{2})(25.2)(2.21)$$
$$= 384 + 725 + 77 + 19 = 1205$$

$$\text{FEM coefficient} = \frac{P}{AL^2} = \frac{1205}{20.44(25)^2} = 0.0943$$

Since the edge column capital is almost equal in size to the equivalent square of the interior column capital, the FEM coefficient, stiffness, and carry-over factor obtained above for the interior span will also be used for the exterior span.

(b) Compute flexure properties of columns. For the interior column, the length is measured between the centerlines of slab thickness, as shown in Fig. 16.20.8(a). The moment of inertia is assumed to be infinite from the top of the slab to the bottom of the drop panel and then the $1/I$ value is taken to vary linearly to the base of the column capital.

(a) Interior column

(b) Exterior column

Figure 16.20.8 Flexure properties of columns in flat slab design example.

The $0.188L$ of the analogous column representing the column capital is divided into four parts, $\Delta L = 0.188L/4$, with $1/I$ of $1/8$, $3/8$, $5/8$, and $7/8$.

$$A = 0.725L + \left(\frac{1}{8} + \frac{3}{8} + \frac{5}{8} + \frac{7}{8}\right)\left(\frac{0.188L}{4}\right) = 0.819L$$

$$\Sigma Ay \text{ from top} = \frac{0.188L}{4}\left[\frac{1}{8}(0.080L) + \frac{3}{8}(0.126L) + \frac{5}{8}(0.174L) + \frac{7}{8}(0.220L)\right]$$
$$+ 0.725L(0.6065L) = 0.4566L^2$$

$$\bar{y} \text{ from top} = \frac{0.4566L^2}{0.819L} = 0.558L$$

$$I = \frac{1}{12}(0.725L)^3 + 0.725L(0.0485L)^2$$
$$+ \frac{0.188L}{4}\left[\frac{1}{8}(0.478L)^2 + \frac{3}{8}(0.432L)^2 + \frac{5}{8}(0.384L)^2 + \frac{7}{8}(0.338L)^2\right]$$
$$= 0.0471L^3$$

The stiffness factors at the top and bottom are

$$s_{TT} = \frac{1}{0.819} + \frac{(0.558)^2}{0.0471} = 1.22 + 6.61 = 7.83$$

$$s_{BB} = \frac{1}{0.819} + \frac{(0.442)^2}{0.0471} = 1.22 + 4.15 = 5.37$$

for which the stiffnesses are

$$K_{cT} = \frac{s_{TT}EI}{L} = \frac{7.83E\pi(9)^4/4}{120} = 336E$$

$$K_{cB} = \frac{s_{BB}EI}{L} = \frac{5.37E\pi(9)^4/4}{120} = 230E$$

The carry-over factors are

$$(\text{COF})_{TB} = \frac{0.558(0.442)/(0.0471) - 1.22}{7.83} = \frac{4.02}{7.83} = 0.513$$

$$(\text{COF})_{BT} = \frac{4.02}{5.37} = 0.749$$

For the exterior column [Fig. 16.20.8(b)],

$$A = 0.742L + \left(\frac{1}{8} + \frac{3}{8} + \frac{5}{8} + \frac{7}{8}\right)\left(\frac{0.171L}{4}\right) = 0.827L$$

$$\Sigma Ay \text{ from top} = \frac{0.171L}{4}\left[\frac{1}{8}(0.077L) + \frac{3}{8}(0.120L) + \frac{5}{8}(0.163L) + \frac{7}{8}(0.206L)\right]$$
$$+ \, 0.742L(0.598L) = 0.4581L^2$$

$$\bar{y} \text{ from top} = \frac{0.4581L^2}{0.827L} = 0.555L$$

$$I = \frac{1}{12}(0.742L)^3 + 0.742L(0.044L)^2$$
$$+ \frac{0.171L}{4}\left[\frac{1}{8}(0.477L)^2 + \frac{3}{8}(0.434L)^2 + \frac{5}{8}(0.391L)^2 + \frac{7}{8}(0.348L)^2\right]$$
$$= 0.0482L^3$$

$$s_{TT} = \frac{1}{0.827} + \frac{(0.554)^2}{0.0482} = 1.21 + 6.37 = 7.58$$

$$s_{BB} = \frac{1}{0.827} + \frac{(0.446)^2}{0.0482} = 1.21 + 4.13 = 5.34$$

$$K_{cT} = \frac{s_{TT}EI}{L} = \frac{7.58E(16)^4/12}{120} = 345E$$

$$K_{cB} = \frac{s_{BB}EI}{L} = \frac{5.34E(16)^4/12}{120} = 243E$$

The carry-over factors are

$$(\text{COF})_{TB} = \frac{0.554(0.446)/0.0482 - 1.21}{7.58} = \frac{3.92}{7.58} = 0.517$$

$$(\text{COF})_{BT} = \frac{3.92}{5.34} = 0.734$$

(c) Compute torsional stiffness of transverse torsional members. From Example 16.11.2,

$$C \text{ (edge beam)} = 18,500 \text{ in.}^4$$
$$C \text{ (interior beam)} = 9800 \text{ in.}^4$$

For the two members, one framing in from each side,

$$K_t \text{ (edge)} = \frac{2(9E)C}{L_2\left(1 - \dfrac{c_2}{L_2}\right)^3} = \frac{2(9E)(18,500)}{240\left(1 - \dfrac{4.5}{20}\right)^3} = 2980E$$

$$K_t \text{ (interior)} = \frac{2(9E)9800}{240\left(1 - \dfrac{4.43}{20}\right)^3} = 1560E \quad \blacksquare$$

SELECTED REFERENCES

General and Historical

16.1.   J. R. Nichols. "Statical Limitations Upon the Steel Requirement in Reinforced Concrete Flat Slab Floors," *Transactions ASCE*, **77**, 1914, 1670–1736.

16.2.   George A. Hool and Nathan C. Johnson. *Concrete Engineers Handbook*. New York: McGraw-Hill, 1918 (pp. 457–486).

16.3.   H. M. Westergaard and W. A. Slater. "Moments and Stresses in Slabs," *ACI Proceedings*, **17**, 1921, 415.

16.4.   H. M. Westergaard. "Formulas for the Design of Rectangular Floor Slabs and Supporting Girders," *ACI Proceedings*, **22**, 1926, 26.

16.5.   Joseph A. Wise. "Design of Reinforced Concrete Slabs," *ACI Proceedings*, **25**, 1929, 712.

16.6.   Joseph DiStasio and M. P. Van Buren. "Slabs Supported on Four Sides," *ACI Journal, Proceedings*, **32**, January–February 1936, 350–364.

16.7.   R. L. Bertin, Joseph DiStasio, and M. P. Van Buren. "Slabs Supported on Four Sides," *ACI Journal, Proceedings*, **41**, June 1945, 537–556.

16.8.   C. P. Siess and N. M. Newmark. "Rational Analysis and Design of Two-Way Concrete Slabs," *ACI Journal, Proceedings*, **45**, December 1948, 273–316.

16.9.   Sidney H. Simmonds and Janko Misic. "Design Factors for the Equivalent Frame Method." *ACI Journal, Proceedings*, **68**, November 1971, 825–831.

16.10.   Alex E. Cardenas, Rolf J. Lenschow, and Mete A. Sozen. "Stiffness of Reinforced Concrete Plates," *Journal of the Structural Division*, ASCE, **98**, ST11 (November 1972), 2587–2603.

16.11.   C. K. Wang. *Intermediate Structural Analysis*. New York: McGraw-Hill, 1983.

Gravity Load—Tests

16.12.   W. L. Gamble, M. A. Sozen, and C. P. Siess. "Measured and Theoretical Bending Moments in Reinforced Concrete Floor Slabs," Civil Engineering Structural Research Series No. 246. Urbana: University of Illinois, June 1962.

16.13.   M. A. Sozen and C. P. Siess. "Investigation of Multi-Panel Reinforced Concrete Floor Slabs," *ACI Journal, Proceedings*, **60**, August 1963, 999–1028.

16.14.   S. A. Guralnick and R. W. LaFraugh. "Laboratory Study of a 45-Foot Square Flat Plate Structure," *ACI Journal, Proceedings*, **60**, September 1963, 1107–1185.

16.15.   David S. Hatcher, Mete A. Sozen, and Chester P. Siess. "Test of a Reinforced Concrete Flat Plate," *Journal of the Structural Division*, ASCE, **91**, ST5 (October 1965), 205–231.

16.16. James O. Jirsa, Mete A. Sozen, and Chester P. Siess. "Test of a Flat Slab Reinforced with Welded Wire Fabric," *Journal of the Structural Division*, ASCE, **92**, ST3 (June 1966), 199–224.

16.17. W. L. Gamble, M. A. Sozen, and C. P. Siess. "Tests of a Two-Way Reinforced Concrete Floor Slab," *Journal of the Structural Division*, ASCE, **95**, ST6 (June 1969), 1073–1096.

16.18. M. Daniel Vanderbilt, Mete A. Sozen, and Chester P. Siess. "Tests of a Modified Reinforced Concrete Two-Way Slab," *Journal of the Structural Division*, ASCE, **95**, ST6 (June 1969), 1097–1116.

16.19. D. S. Hatcher, Mete A. Sozen, and Chester P. Siess. "Test of a Reinforced Concrete Flat Slab," *Journal of the Structural Division*, ASCE, **95**, ST6 (June 1969), 1051–1072.

16.20. E. Ramzy F. Zaghlool, H. A. Rawdon de Paiva, and Peter G. Glockner. "Tests of Reinforced Concrete Flat Plate Floors," *Journal of the Structural Division*, ASCE, **96**, ST3 (March 1970), 487–507.

16.21. Alex E. Cardenas and Paul H. Kaar. "Field Test of a Flat Plate Structure," *ACI Journal, Proceedings*, **68**, January 1971, 50–58.

16.22. Donald D. Magura and W. Gene Corley. "Tests to Destruction of a Multipanel Waffle Slab Structure—1964–1965 New York World's Fair." *ACI Journal, Proceedings*, **68**, September 1971, 699–703.

16.23. A. S. Hall and B. V. Rangan. "Moments in Edge Panels of Flat Plate Floors," *Journal of Structural Engineering*, ASCE, **109**, 11 (November 1983), 2638–2650.

16.24. Abdel Wahid Hago and Prabhakara Bhatt. "Tests on Reinforced Concrete Slabs Designed by Direct Design Procedure," *ACI Journal, Proceedings*, **83**, November–December 1986, 916–924.

16.25. B. Vijaya Rangan and A. S. Hall. "Moment Redistribution in Flat Plate Floors," *ACI Journal, Proceedings*, **81**, November–December 1984, 601–608.

Gravity Load—Analytical Methods

16.26. W. G. Corley, M. A. Sozen, and C. P. Siess. "The Equivalent Frame Analysis for Reinforced Concrete Slabs," Structural Research Series No. 218. Urbana: University of Illinois, Civil Engineering Department, June 1961.

16.27. W. G. Corley and J. O. Jirsa. "Equivalent Frame Analysis for Slab Design," *ACI Journal, Proceedings*, **67**, November 1970, 875–884.

16.28. Maurice P. Van Buren. "Staggered Columns in Flat Plates," *Journal of the Structural Division*, ASCE, **97**, ST6 (June 1971), 1791–1797.

16.29. William L. Gamble. "Moments in Beam Supported Slabs," *ACI Journal, Proceedings*, **69**, March 1972, 149–157.

16.30. Donald J. Fraser. "Equivalent Frame Method for Beam-Slab Structures," *ACI Journal, Proceedings*, **74**, May 1977, 223–228.

16.31. S. K. Sharan, D. Clyde, and D. Turcke. "Equivalent Frame Analysis Improvements for Slab Design," *ACI Journal, Proceedings*, **75**, February 1978, 55–59.

16.32. M. Daniel Vanderbilt. *Equivalent Frame Analysis of Unbraced Reinforced Concrete Buildings for Static Lateral Loads.* Structural Research Report No. 36, Civil Engineering Department, Colorado State University, Fort Collins, June 1981.

16.33. D. J. Fraser. "The Equivalent Frame Method Simplified for Beam and Slab Construction," *Concrete International*, **4**, April 1982, 66–73.

16.34. J. F. Mulcahy and J. M. Rotter. "Moment Rotation Characteristics of Flat Plate and Column Systems," *ACI Journal, Proceedings*, **80**, March–April 1983, 85–92.

16.35. D. J. Fraser. "Simplified Frame Analysis for Flat Plate Construction," *Concrete International*, **6**, September 1984, 32–41.

Pattern Gravity Loading

16.36. J. O. Jirsa, M. A. Sozen, and C. P. Siess. "Pattern Loadings on Reinforced Concrete Floor Slabs," *Journal of the Structural Division*, ASCE, **95**, ST6 (June 1969), 1117–1137.

16.37. Jan C. Jofriet and Gregory M. McNeice. "Pattern Loading on Reinforced Concrete Flat Plates," *ACI Journal, Proceedings*, **68**, December 1971, 968–972.

Lateral Load Analysis on Slab Frames

16.38. David A. Pecknold. "Slab Effective Width for Equivalent Frame Analysis," *ACI Journal, Proceedings*, **72**, April 1975, 135–137.

16.39. Fred Allen and Peter Darvall. "Lateral Load Equivalent Frame," *ACI Journal, Proceedings*, **74**, July 1977, 294–299.

16.40. M. Daniel Vanderbilt. "Equivalent Frame Analysis for Lateral Loads," *Journal of the Structural Division*, ASCE, **105**, ST10 (October 1979), 1981–1998. Disc., **106**, ST7 (July 1980), 1671–1672; **107**, ST1 (January 1981), 245.

16.41. Ziad M. Elias. "Sidesway Analysis of Flat Plate Structures," *ACI Journal, Proceedings*, **76**, March 1979, 421–442.

16.42. Ziad M. Elias and Constantinos Georgiadis. "Flat Slab Analysis Using Equivalent Beams," *ACI Journal, Proceedings*, **76**, October 1979, 1063–1078.

16.43. Ziad M. Elias. "Lateral Stiffness of Flat Plate Structures," *ACI Journal, Proceedings*, **80**, January–February 1983, 50–54.

16.44. Donald J. Fraser. "Elastic Analysis of Laterally Loaded Frames," *Journal of Structural Engineering*, ASCE, **109**, 6 (June 1983), 1479–1489.

16.45. M. Daniel Vanderbilt and W. Gene Corley. "Frame Analysis of Concrete Buildings," *Concrete International*, **5**, December 1983, 33–43.

16.46. I. Paul Lew and Fruma Narov. "Three-Dimensional Equivalent Frame Analysis of Shearwalls," *Concrete International*, **5**, October 1983, 25–30.

16.47. Peter Darvall and Fred Allen. "Lateral Load Effective Width of Flat Plates with Drop Panels," *ACI Journal, Proceedings*, **81**, November–December 1984, 613–617.

16.48. Milija N. Pavlovic and Steven M. Poulton. "On the Computation of Slab Effective Widths," *Journal of Structural Engineering*, ASCE, **111**, 2 (February 1985), 363–377.

16.49. Jack. P. Moehle and John W. Diebold. "Lateral Load Response of Flat-Plate Frame," *Journal of Structural Engineering*, ASCE, **111**, 10 (October 1985), 2149–2164.

16.50. Cheng-Tzu Thomas Hsu. "Lateral Displacement for Unbraced Concrete Frame Buildings," *ACI Journal, Proceedings*, **82**, November–December 1985, 853–862.

16.51. Mary Theresa Cano and Richard E. Klingner. "Comparison of Analysis Procedures for Two-Way Slabs," *ACI Structural Journal*, **85**, November–December 1988, 597–608. Disc. **86**, September–October 1989, 624–626.

Two-Way Slab Deflections

16.52. Mortimer D. Vanderbilt, Mete A. Sozen, and Chester P. Siess. "Deflections of Multiple-Panel Reinforced Concrete Floor Slabs," *Journal of the Structural Division*, ASCE, **91**, ST4 (August 1965), 77–101.

16.53. ACI Committee 435. "State-of-the-Art Report, Deflection of Two-Way Reinforced Concrete Floor Systems," *Deflections of Concrete Structures* (SP-43). Detroit: American Concrete Institute, 1974 (pp. 55–81).

16.54. Arthur H. Nilson and Donald B. Walters, Jr. "Deflection of Two-Way Floor Systems by the Equivalent Frame Method," *ACI Journal, Proceedings*, **72**, May 1975, 210–218.

16.55. B. Vijaya Rangan. "Prediction of Long-Term Deflections of Flat Plates and Slabs," *ACI Journal, Proceedings*, **73**, April 1976, 223–226.

16.56. K. M. Kripanarayanan and D. E. Branson. "Short Time Deflections of Flat Plates, Flat Slabs, and Two-Way Slabs," *ACI Journal, Proceedings*, **73**, December 1976, 686–690.

16.57. P. J. Taylor and J. L. Heiman. "Long-Term Deflection of Reinforced Concrete Flat Slabs and Plates," *ACI Journal, Proceedings*, **74**, November 1977, 556–561.

16.58. B. Vijaya Rangan and Arthur E. McMullen. "A Rational Approach to Control of Slab Deflections," *ACI Journal, Proceedings*, **75**, June 1978, 256–262.

16.59. Andrew Scanlon and David W. Murray. "Practical Calculation of Two-Way Slab Deflections," *Concrete International*, **4**, November 1982, 43–50.

16.60. John A. Sbarounis. "Multistory Flat Plate Buildings: Construction Loads and Immediate Deflections," *Concrete International*, **6**, February 1984, 70–77.

16.61. John A. Sbarounis. "Multistory Flat Plate Buildings: Effect of Construction Loads on Long-Term Deflections," *Concrete International*, **6**, April 1984, 62–70. Disc. **7**, February 1985, 62.

16.62. John A. Sbarounis. "Multistory Flat Plate Buildings: Measured and Computed One-Year Deflections," *Concrete International*, **6**, August 1984, 31–35.

16.63. R. I. Gilbert. "Deflection Control of Slabs Using Allowable Span to Depth Ratios," *ACI Journal, Proceedings*, **82**, January–February 1985, 67–72.

16.64. Cameron J. Graham and Andrew Scanlon. "Deflections of Concrete Slabs Under Construction Loading," *Deflections of Concrete Structures* (SP-86). Detroit: American Concrete Institute, 1985.

16.65. B. Vijaya Rangan. "Estimation of Slab Deflections in Flat Plate Buildings," *ACI Journal, Proceedings*, **83**, March–April 1986, 269–273.

16.66. K. S. Stephen Tam and Andrew Scanlon. "Deflection of Two-Way Slabs Subjected to Restrained Volume Change and Transverse Loads, *ACI Journal, Proceedings*, **83**, September–October 1986, 737–744.

16.67. Cameron J. Graham and Andrew Scanlon. "Long-Time Multipliers for Estimating Two-Way Slab Deflections," *ACI Journal, Proceedings*, **83**, November–December 1986, 899–908.

16.68. N. J. Gardner and H. C. Fu. "Effects of High Construction Loads on the Long-Term Deflections of Flat Slabs," *ACI Structural Journal*, **84**, July–August 1987, 349–360. Disc. **85**, May–June 1988, 359.

16.69. David P. Thompson and Andrew Scanlon. "Minimum Thickness Requirements for Control of Two-Way Slab Deflections," *ACI Structural Journal*, **85**, January–February, 1988, 12–22.

16.70. Amin Ghali. "Prediction of Deflections of Two-Way Floor Systems," *ACI Structural Journal*, **86**, September–October 1989, 551–562.

16.71. Andrew Scanlon and David P. Thompson. "Evaluation of ACI 318 Requirements for Control of Two-Way Slab Deflections," *ACI Structural Journal*, **87**, November–December 1990, 657–661.

Crack Control in Two-Way Slab Systems

16.72. Edward G. Nawy. "Crack Width Control in Welded Fabric Reinforced Centrally Loaded Two-Way Concrete Slabs," *Causes, Mechanism, and Control of Cracking in Concrete* (SP-20). Detroit: American Concrete Institute, 1968 (pp. 211–235).

16.73. Edward G. Nawy and G. S. Orenstein. "Crack Width Control in Reinforced Concrete Two-Way Slabs," *Journal of the Structural Division*, ASCE, **96**, ST3 (March 1970), 701–721.

16.74. Edward G. Nawy and Kenneth W. Blair. "Further Studies on Flexural Crack Control in Structural Slab Systems," *Cracking, Deflection, and Ultimate Load of Concrete Slab Systems* (SP-30). Detroit: American Concrete Institute, 1971, (pp. 1–41).

16.75. Edward G. Nawy. "Crack Control Through Reinforcement Distribution in Two-Way Acting Slabs and Plates," *ACI Journal, Proceedings*, **69**, April 1972, 217–219.

Shear Strength

16.76. Johannes Moe. *Shearing Strength of Reinforced Concrete Slabs and Footings Under Concentrated Loads*, Development Department Bulletin D47. Chicago: Portland Cement Association, April 1961 (130 pp.).

16.77. ACI-ASCE Committee 326. "Report on Shear and Diagonal Tension," *ACI Journal, Proceedings*, **59**, January, February, and March, 1962, 1–30, 277–344, and 352–396.

16.78. Neil M. Hawkins, H. B. Fallsen, and R. C. Hinojosa. "Influence of Column Rectangularity on the Behavior of Flat Plate Structures," *Cracking, Deflection, and Ultimate Load of Concrete Slab Systems* (SP-30). Detroit: American Concrete Institute, 1971 (pp. 127–146).

16.79. Frederic Roll, S. T. H. Zaidi, Gajanan Sabnis, and Kuang Chuang. "Shear Resistance of Perforated Reinforced Concrete Slabs," *Cracking, Deflection, and Ultimate Load of Concrete Slab Systems* (SP-30). Detroit: American Concrete Institute, 1971 (pp. 77–100).

16.80. M. Daniel Vanderbilt. "Shear Strength of Continuous Plates," *Journal of the Structural Division*, ASCE, **98**, ST5 (May 1972), 961–973.

16.81. M. E. Criswell and N. M. Hawkins. "Shear Strength of Slabs: Basic Principles and Their Relation to Current Methods of Analysis," *Shear in Reinforced Concrete*, Vol. 2 (SP-42). Detroit: American Concrete Institute, 1974 (pp. 641–676).

16.82. N. M. Hawkins, M. E. Criswell, and F. Roll. "Shear Strength of Slabs Without Shear Reinforcement," *Shear in Reinforced Concrete*, Vol. 2 (SP-42). Detroit: American Concrete Institute, 1974 (pp. 677–720).

16.83. Neil M. Hawkins, (Chmn.). "The Shear Strength of Reinforced Concrete Members— Slabs, by the Joint ASCE-ACI Task Committee 426 on Shear and Diagonal Tension of the Committee on Masonry and Reinforced Concrete of the Structural Division," *Journal of the Structural Division*, ASCE, **100**, ST8 (August 1974), 1543–1591.

16.84. Brian E. Hewitt and Barrington de V. Batchelor. "Punching Shear Strength of Restrained Slabs," *Journal of the Structural Division*, ASCE, **101**, ST9 (September 1975), 1837–1853.

16.85. Timothy J. Ross and Helmut Krawinkler. "Impulsive Direct Shear Failure in RC Slabs," *Journal of Structural Engineering*, ASCE, **111**, 8 (August 1985), 1661–1677.

16.86. Zdeněk P. Bažant and Zhiping Cao. "Size Effect in Punching Shear Failure of Slabs," *ACI Structural Journal*, **84**, January–February 1987, 44–53.

16.87. Fernando Gonzalez-Vidosa, Michael D. Kotsovos, and Milija N. Pavlovic. "Symmetrical Punching of Reinforced Concrete Slabs: An Analytical Investigation Based on Nonlinear Finite Element Modeling," *ACI Structural Journal*, **85**, May–June 1988, 241–250.

16.88. Ibrahim A. E. M. Shehata and Paul E. Regan. "Punching in RC Slabs," *Journal of Structural Engineering*, ASCE, **115**, 7 (July 1989), 1726–1740.

16.89. Lionello Bortolotti. "Punching Shear Strength in Concrete Slabs," *ACI Structural Journal*, **87**, March–April 1990, 208–219.

16.90. Carl Erik Broms. "Punching of Flat Plates—A Question of Concrete Properties in Biaxial Compression and Size Effect," *ACI Structural Journal*, **87**, May–June 1990, 292–304.

16.91. David I. McLean, Long T. Phan, H. S. Lew, and Richard N. White. "Punching Shear Behavior of Lightweight Concrete Slabs and Shells," *ACI Structural Journal*, **87**, July–August 1990, 386–392.

16.92. John S. Lovrovich and David I. McLean. "Punching Shear Behavior of Slabs with Varying Span–Depth Ratios," *ACI Structural Journal*, **87**, September–October 1990, 507–511.

Shear Reinforcement in Slabs

16.93. W. G. Corley and N. M. Hawkins. "Shearhead Reinforcement for Slabs," *ACI Journal, Proceedings*, **65**, October 1968, 811–824.

16.94. N. M. Hawkins. "Shear Strength of Slabs With Shear Reinforcement," *Shear in Reinforced Concrete*, Vol. 2 (SP-42). Detroit: American Concrete Institute, 1974 (pp. 785–816).

16.95. Paul H. Langohr, Amin Ghali, and Walter H. Dilger. "Special Shear Reinforcement for Concrete Flat Plates," *ACI Journal, Proceedings*, **73**, March 1976, 141–146.

16.96. Amin Ghali, Mahmoud Z. Elmasri, and Walter Dilger, "Punching of Flat Plates Under Static and Dynamic Horizontal Forces," *ACI Journal, Proceedings*, **73**, October 1976, 566–576.

16.97. Walter Dilger, Mahmoud Z. Elmasri, and Amin Ghali. "Flat Plates with Special Shear Reinforcement Subjected to Static Dynamic Moment Transfer," *ACI Journal, Proceedings*, **75**, October 1978, 543–549.

16.98. Frieder Seible, Amin Ghali, and Walter H. Dilger. "Preassembled Shear Reinforcing Units for Flat Plates," *ACI Journal, Proceedings*, **77**, January–February 1980, 28–35.

16.99. Walter H. Dilger and Amin Ghali. "Shear Reinforcement for Concrete Slabs," *Journal of the Structural Division*, ASCE, **107**, ST12 (December 1981), 2403–2420.

16.100. Adel A. Elgabry and Amin Ghali. "Design of Stud-Shear Reinforcement for Slabs," *ACI Structural Journal*, **87**, May–June 1990, 350–361.

Shear–Moment Transfer

16.101. Joseph DiStasio and M. P. Van Buren. "Transfer of Bending Moment between Flat Plate Floor and Column," *ACI Journal, Proceedings*, **57**, September 1960, 299–314.

16.102. N. W. Hanson and J. M. Hanson. "Shear and Moment Transfer Between Concrete Slabs and Columns," *Journal PCA Research and Development Laboratories*, **10**, (1) January 1968, 2–16.

16.103. Neil M. Hawkins and W. Gene Corley. "Transfer of Unbalanced Moment and Shear from Flat Plates to Columns," *Cracking, Deflection, and Ultimate Load of Concrete Slab Systems* (SP-30). Detroit: American Concrete Institute, 1971 (pp. 147–176).

16.104. Adrian E. Long. "Punching Failure of Slabs—Transfer of Moment and Shear," *Journal of the Structural Division*, ASCE, **99**, ST4 (April 1973), 665–685.

16.105. E. Ramzy F. Zaghlool and H. A. Rawdon de Paiva. "Strength Analysis of Corner Column-Slab Connections," *Journal of the Structural Division*, ASCE, **99**, ST1 (January 1973), 53–70.

16.106. E. Ramzy F. Zaghlool and H. A. Rawdon de Paiva. "Tests of Flat-Plate Corner Column-Slab Connections," *Journal of the Structural Division*, ASCE, **99**, ST3 (March 1973), 551–572.

16.107. N. M. Hawkins. "Shear Strength of Slabs With Moments Transferred to Columns," *Shear in Reinforced Concrete*, Vol. 2 (SP-42). Detroit: American Concrete Institute, 1974 (pp. 817–846).

16.108. N. M. Hawkins and W. G. Corley. "Moment Transfer to Columns in Slabs With Shearhead Reinforcement," *Shear in Reinforced Concrete*, Vol. 2 (SP-42). Detroit: American Concrete Institute, 1974 (pp. 847–880).

16.109. Shafiqul Islam and Robert Park. "Tests on Slab-Column Connections with Shear and Unbalanced Flexure," *Journal of the Structural Division*, ASCE, **102**, ST3 (March 1976), 549–568.

16.110. Robert Park and Shafiqul Islam. "Strength of Slab-Column Connections with Shear and Unbalanced Flexure," *Journal of the Structural Division*, ASCE, **102**, ST9 (September 1976), 1879–1901.

16.111. Hans Gesund and Harinatha B. Goli. "Local Flexural Strength of Slabs at Interior Columns," *Journal of the Structural Division*, ASCE, **106**, ST5 (May 1980), 1063–1078.

16.112. V. W. Neth, H. A. R. de Paiva, and A. E. Long. "Behavior of Models of a Reinforced Concrete Flat Plate Edge-Column Connection," *ACI Journal, Proceedings*, **78**, July–August 1981, 269–275.

16.113. S. Unnikrishna Pillai, Wayne Kirk, and Leonard Scavuzzo. "Shear Reinforcement at Slab-Column Connections in a Reinforced Concrete Flat Plate Structure," *ACI Journal, Proceedings*, **79**, January–February 1982, 36–42.

16.114. Harinatha B. Goli and Hans Gesund. "Flexural Strength of Flat Slabs at Exterior Columns," *Journal of the Structural Division*, ASCE, **108**, ST11 (November 1982), 2479–2495.

16.115. B. Vijaya Rangan and A. S. Hall. "Moment and Shear Transfer Between Slab and Edge Column," *ACI Journal, Proceedings*, **80**, May–June 1983, 183–191.

16.116. Denby G. Morrison, Ikuo Hirasawa, and Mete A. Sozen. "Lateral-Load Tests of R/C Slab–Column Connections," *Journal of Structural Engineering*, ASCE, **109**, 11 (November 1983), 2698–2714.

16.117. Denby G. Morrison. "Dynamic Lateral-Load Tests of R/C Column–Slabs," *Journal of Structural Engineering*, ASCE, **111**, 3 (March 1985), 685–698.

16.118. Paul F. Rice and Edward S. Hoffman. "Shear-Moment Transfer," *Concrete International*, **9**, January 1987, 31–35.

16.119. P. R. Walker and P. E. Regan. "Corner Column–Slab Connections in Concrete Flat Plates," *Journal of Structural Engineering*, ASCE, **113**, 4 (April 1987), 704–720.

16.120. Scott D. B. Alexander and Sidney H. Simmonds. "Ultimate Strength of Slab–Column Connections," *ACI Structural Journal*, **84**, May–June 1987, 255–261. Disc. **85**, March–April 1988, 226–232.

16.121. Sidney H. Simmonds and Scott D. B. Alexander. "Truss Model for Edge Column–Slab Connections," *ACI Structural Journal*, **84**, July–August 1987, 296–303. Disc. **85**, May–June 1988, 352–353.

16.122. Ahmad J. Durrani and Hikmat E. Zerbe. "Seismic Resistance of R/C Exterior Connections with Floor Slab," *Journal of Structural Engineering*, ASCE, **113**, 8 (August 1987), 1850–1864.

16.123. Adel A. Elgabry and Amin Ghali. "Tests on Concrete Slab–Column Connections with Stud-Shear Reinforcement Subjected to Shear-Moment Transfer," *ACI Structural Journal*, **84**, September–October 1987, 433–442.

16.124. Jack P. Moehle. "Strength of Slab–Column Edge Connections," *ACI Structural Journal*, **85**, January–February 1988, 89–98. Disc. **85**, November–December 1988, 703–709.

16.125. Jack P. Moehle, Michael E. Kreger, and Roberto Leon. "Background to Recommendations for Design of Reinforced Concrete Slab–Column Connections," *ACI Structural Journal*, **85**, November–December 1988, 636–644.

16.126. ACI-ASCE Committee 352. "Recommendations for Design of Slab–Column Connections in Monolithic Reinforced Concrete Structures," *ACI Structural Journal*, **85**, November–December 1988, 675–696. Disc. **86**, July–August 1989, 496–499.

16.127. Austin Pan and Jack P. Moehle. "Lateral Displacement Ductility of Reinforced Concrete Flat Plates," *ACI Structural Journal*, **86**, May–June 1989, 250–258.

16.128. Neil M. Hawkins, Aibin Bao, and Jun Yamazaki. "Moment Transfer from Concrete Slabs to Columns," *ACI Structural Journal*, **86**, November–December 1989, 705–716.

16.129. B. Vijaya Rangan. "Punching Shear Design in the New Australian Standard for Concrete Structures," *ACI Structural Journal*, **87**, March–April 1990, 140–144.

16.130. Douglas A. Foutch, William L. Gamble, and Harianto Sunidja. "Tests of Post-Tensioned Concrete Slab—Edge Column Connections," *ACI Structural Journal*, **87**, March–April 1990, 167–179.

16.131. ACI Committee 340. *Supplement to Design Handbook Volume 1, Design of Two-Way Slabs, In Accordance with the Strength Design Method of ACI 318-83.* [SP-17(84)(S)]. Detroit: American Concrete Institute, 1985.

PROBLEMS

Two-Way Slabs (with Beams)

16.1 Design the typical interior frame along columns 2-5-7 for the two-way slab system shown. The 13 ft long columns are connected by beams, and no column capitals or drop panels are used. As an initial trial, assume all beams (interior) are 12 × 24 in. overall. Revise beam size as necessary during the design. Determine slab thickness based on ACI-9.5.3, then use the "direct design method"

Problems 16.1 through 16.4

for longitudinal distribution of moments. Show design sketch giving all your decisions, including dimensions, bar sizes, bar lengths, and stirrups for the two spans from column 2 to column 7. The live load is 150 psf, $f'_c = 4000$ psi, and $f_y = 60,000$ psi.

16.2   Design the interior frame of Problem 16.1, except that a 12 in. wall exists at the lower-story level and contains the 24 in. square columns at locations 1, 2, 3, 4, 6, and 10.

16.3   Design the typical interior frame along column lines 4-5-8 for the two-way slab system of Problem 16.1.

16.4   Design the exterior half-frame along column lines 1-2-3 for the two-way slab system of Problem 16.1.

16.5   Design an interior frame in the long direction for a floor system of slabs supported on beams which has 5 panels at 21 ft in one direction and 5 panels at 27 ft in the other direction. The live load is 175 psf and the dead load is 40 psf in addition to the slab weight. Assume that all panels are bounded by beams that are 14 in. wide. Columns 15 in. square and 13 ft long are located at the corners of all panels. Use $f'_c = 4000$ psi, $f_y = 50,000$ psi, and the "direct design method" of the ACI Code.

16.6 Design a floor system of slabs supported on beams which has two panels at 16 ft in one direction and two panels at 21 ft in the other direction. Assume that all panels are bounded by beams 12 in. wide and the columns are 14 in. square and 11 ft long. The live load is 200 psf, and the dead load is 50 psf in addition to the slab weight. Use $f'_c = 3000$ psi, $f_y = 60,000$ psi, and the ACI Code.

16.7 Design a simply supported sidewalk slab for an 18 ft square panel to carry a live load of 250 psf. The panel is supported by beams 12 in. wide on all four sides. There are no walls or columns above the slab. Use $f'_c = 4000$ psi, $f_y = 60,000$ psi, and the ACI Code.

Flat Slabs

16.8 Given the flat slab shown in the accompanying figure. The columns are 24 in. square with columns 1 through 6 existing only below the floor slab, while columns 7 through 9 exist both above and below the floor slab. All columns are 13 ft long center-to-center of floor slabs. The live load is 150 psf, $f'_c = 4000$ psi, and $f_y = 60,000$ psi. Design the flat slab using rectangular column capitals and drop panels.

(a) Determine the slab thickness based on ACI-9.5.3.

(b) Use "direct design method" for longitudinal distribution of moments in interior equivalent frame defined by columns 2, 5, and 7 along its centerline.

(c) Determine transverse distribution and select reinforcement for the column strip (defined by columns 2, 5, and 7) and adjacent half middle strips.

(d) Specify and show details giving lengths and locations of bars.

Problems 16.8, 16.9, and 16.10

16.9   For the flat slab of Problem 16.8, design the interior equivalent frame defined by columns 4, 5, and 8 along its centerline.

16.10   For the flat slab of Problem 16.8, redesign the interior equivalent frame defined by column line 2-5-7, but consider that a 12-in. wall 13 ft high exists along column line 1-2-3 at the story below the slab to be designed.

16.11   Investigate the moment and shear transfer at the exterior support of a flat-slab structure as detailed in the accompanying figure. The exterior support has a 5-ft flat-sided column capital on a 24 in. square column, along with a 7 ft 8 in. width of drop panel that is 11 in. thick. The slab is $8\frac{1}{2}$ in. thick. Assume there is no edge beam or wall at the exterior support location. The factored moment M_u to be transferred is 340 ft-kips and the factored shear V_u is 115 kips. The negative moment reinforcement provided in the column strip is #5 at 10 in. spacing. Use $f'_c = 4000$ psi and $f_y = 60,000$ psi.

Problems 16.11

16.12   Rework through Example 16.10.2, if the service live load is 200 psf instead of 120 psf.

16.13   Rework through Example 16.13.2, if the service live load is 200 psf instead of 120 psf.

16.14   Rework through Example 16.15.1, if the service live load is 200 psf instead of 120 psf.

Flat Plates

16.15   Design a typical interior rigid frame (long direction) of a flat plate floor using the data for the design example of Section 16.3, including the following exception to the example, as assigned by the instructor. Note that the preliminary slab thickness must be determined as part of your design (i.e., the original $5\frac{1}{2}$-in. thickness does not necessarily apply).

(a) Live load 80 psf; $f_y = 60$ ksi; columns 14 in. × 12 in.

(b) Live load 100 psf; $f_y = 60$ ksi; columns 16 in. × 14 in.

(c) Live load 120 psf; $f_y = 60$ ksi; columns 16 in. × 14 in.

(d) Live load 120 psf; $f_y = 60$ ksi; columns 18 in. × 18 in.; panel size 18 ft × 15 ft

16.16 Assuming the slab thickness is $6\frac{1}{2}$ in., investigate the moment and shear transfer in the flat plate design example (see Examples 16.17.1 and 16.18.1) for a service live load of 150 psf instead of 60 psf.

16.17 Rework through Example 16.10.3, if the service live load is 150 psf instead of 60 psf.

16.18 Rework through Example 16.15.2, if the service live load is 150 psf instead of 60 psf.

CHAPTER 17

Equivalent Frame Analysis of Two-Way Floor Systems in Unbraced Frames

17.1 GENERAL INTRODUCTION

The design of two-way floor systems has been presented in Chapter 16, wherein the longitudinal distribution of moments in the equivalent rigid frame* follows the coefficients of the ACI Code, within the limitations set forth in the direct design method. Beyond these limitations, an elastic analysis of the equivalent rigid frame must be made for various gravity load patterns to obtain the longitudinal moment and shear envelopes. For lateral load, however, elastic analysis of the equivalent rigid frame has to be made, whether or not the system is within the limitations of the direct design method for gravity load Note that ACI-13.3.1 states "A slab system may be designed by *any procedure* satisfying conditions of equilibrium and geometric compatibility if shown that the design strength at every section is at least equal to the required strength . . . , and that all serviceability conditions . . . are met."

For gravity load analysis of a typical multibay, multistory frame as shown in Fig. 17.1.1, ACI-13.7.2.5 permits the separate analysis of each floor, with far ends of columns considered fixed in rotation and in sidesway. For lateral load analysis, the implication is that the entire frame should be analyzed as an unbraced frame. The authors believe that the preliminary design, say, of the third floor in Fig. 17.1.1 may be based on results of analysis of a subassembly as shown in Fig. 17.1.2 for both gravity load and lateral load. In this subassembly, the joint rotations at D, E, F, G, H, and K are zero, but sideways at GHK and ABC are permitted. Thus the degree of freedom, in this particular case, is 5 (two sideways plus rotations at A, B, and C); and the analysis can conveniently be made on a microcomputer.

_____

* Note that the concept of "equivalent rigid frame" is applicable in both the direct design method and the equivalent frame method.

Onterie Center, Chicago; provides new concept where columns, spandrel beams, and solid concrete infill panels are clearly expressed in the "diagonalized frame tube;" that is, the combination constitutes the structural system. (Photo by C. G. Salmon.)

Figure 17.1.1 A typical 2-bay, 7-story equivalent rigid frame.

Figure 17.1.2 A subassembly in the 2-bay, 7-story equivalent rigid frame of Fig. 17.1.1.

At this point of presentation, complicating factors such as axial deformation of the columns, reduction of column stiffness due to heavy gravity load, and increase of column moments and shears due to "*P*- delta" effects* are neglected in order that attention may be focused on characteristics of the equivalent rigid frame in a two-way floor system.

— — — —

* In the horizontally deflected position, the gravity load (*P*) times the horizontal deflection (delta) gives the so-called secondary bending moments; hence the term "*P*-delta." This is also treated in Section 15.6.

17.2 ANALYTICAL MODELS FOR ELASTIC FRAME ANALYSIS

The elevation and plan of the subassembly containing a floor with the attached upper and lower columns in a two-way floor system are shown in Figs. 17.2.1(a) and (b). Had the elastic section properties been constant across the width of the equivalent rigid frame, the structure could be modeled simply as a plane frame with two beam

Figure **17.2.1** Analytical model of a two-way floor system.

elements and six column elements, at a degree of freedom of 5. But, since the columns are concentrated along the column line, their restraining effect on the horizontal elements has to be spread in some way across the entire width of the equivalent rigid frame. Furthermore, in the general case, there may be transverse and longitudinal beams through the columns, in addition to the monolithic slab. The most sophisticated model would be to divide the slab into a mesh of finite elements in bending and install "nodes" all over the floor area, with each node having two rotational degrees of freedom and one vertical degree of freedom. (The sidesway degrees of freedom will still need to be included.) This technique has been used in some recent investigations; a summary of these investigations is given by Vanderbilt [16.32].

To model the structure of Figs. 17.2.1(a) and (b) as an elastic frame with straight elements only, it is necessary to treat the slab area as an equivalent beam. One way is to imagine a beam having moment of inertia equal to that of the entire cross section (ACI-Chapter 13, with modifications for cracking and reinforcement) across the full width of the equivalent rigid frame but located at $A'B'C'$ shown in Fig.17.2.1(c) (i.e., offset from the plane defined by the columns). In such a case, the beam ABC located on the column line in Fig. 17.2.1(c) will have no stiffness; but the beam $A'B'C'$ will be connected to the columns by torsion elements AA', BB', and CC'. This is the model implied by the ACI Code and Commentary, and for purpose of identification will be called Model 1.

Investigations by the authors using computer programs written for microcomputers seem to indicate a more rational model, in which the moment of inertia of the cross section in the column strip is placed in beam ABC and that in the middle strip, in beam $A'B'C'$. Again, for purpose for identification, this model will be called Model 2.

It may be noted that the behavior (from tests) of the actual structure and the solution from the analytical model (Models 1 or 2) can be adjusted to correspond through the quantification and classification of the following parameters: (1) moment of inertia (with or without the effect of cracking and reinforcement) and its variation along the length of the beam ABC and $A'B'C'$; (2) moment of inertia and its variation along the length of the columns; and (3) torsional stiffness K_t of the transverse beams as now defined by the ACI Code. As the use of desk-top computers in structural analysis becomes more popular and as more correlations are obtained between analytical solutions and test data, there is little doubt that an appropriate method of gravity and lateral load analysis of unbraced building frames containing two-way floor systems will be achieved.

17.3 TREATMENT OF MODEL 1 AS A PLANE FRAME

Although the limited investigation by the authors indicates that Model 2 (placing column strip stiffness in ABC and middle strip stiffness in $A'B'C'$ of Fig. 17.2.1) gives more rational results than the ACI-implied Model 1 (placing all stiffness within width of equivalent rigid frame in $A'B'C'$, leaving no stiffness in ABC of Fig. 17.2.1), there is an advantage in using Model 1 over Model 2. Model 1 may be "approximately" analyzed as a plane frame having a much smaller degree of freedom than

Degree of freedom = 8
Number of beam elements = 2
Number of column elements = 6
Number of torsion elements = 3

Figure 17.3.1 Model 1 as a space frame. (The term *space frame* is used for simple identification only. In fact, the torsion elements projecting out of the plane of the columns are rigid except in torsional rotations.)

Degree of freedom = 11
Number of beam elements = 2
Number of column elements = 6
Number of torsion elements = 6

(a) As a space frame

Degree of freedom = 5
Number of beam elements = 2
Number of equivalent column elements = 6
(No torsion element)

(b) As a plane frame

Figure 17.3.2 Model 1A: Replacing torsion elements by rotational springs at column ends, distributed to upper and lower columns in proportion to the column flexural stiffnesses (same as K_{ec} model of Vanderbilt [16.32]).

if it is analyzed as a space frame,* as will be shown later. Perhaps it is for this reason that Corley et al. [16.26, 16.27] developed the equivalent column concept, since at that time the moment distribution method was the usual method for structural analysis of plane frames.

As shown in Fig. 17.3.1, when the three torsion elements are treated as such, the degree of freedom of the subassembly (defined as Model 1) is 8.

In Fig. 17.3.2(a), the torsion element at each column is arbitrarily divided into two such elements in parallel, one joined to the lower end of the upper column and the other joined to the upper end of the lower column, with the stiffness of the original element divided proportionally according to the column stiffness at that end.

— — — —

* The term "space frame" is used for simple identification only. In fact, the torsion elements projecting out of the plane of the columns are rigid except in torsional rotation.

Degree of freedom = 9
Number of beam elements = 2
Number of column elements = 6
Number of torsion elements = 4

(a) As a space frame

Degree of freedom = 5
Number of beam elements = 2
Number of column elements = 6
(No torsion element)

(b) As a plane frame

Figure 17.3.3 Model 1B: Replacing torsion elements by rotational springs at slab-beam ends (same as K_{eb} model of Vanderbilt [16.32]).

If the resulting frame, called Model 1A, is still treated as a space frame, its degree of freedom is 11. However, if the six torsion elements and the two beam elements are compressed into the plane of the columns, the torsion elements become six rotational springs set internally at the column ends *before* the column joins the "node," as shown in Fig. 17.3.2(b). The flexural properties of the columns can then be modified owing to the presence of rotational springs at the ends so that the degree of freedom of the plane frame in Fig. 17.3.2(b) is 5.

As an alternative, the torsion element at each interior column may be divided into two such elements in parallel [as shown in Fig. 17.3.3(a)], one joined to the right end of the left beam and the other to the left end of the right beam, with the total torsional stiffness proportioned according to the beam stiffness at that end. The degree of freedom of the resulting frame, called Model 1B, is 9. Again, if the four torsion elements and the two beam elements are compressed into the plane of the columns, they become four rotational springs set internally at the beam ends *before* the beam joins the "node," as shown in Fig. 17.3.3(b). The fixed-end moments and the flexural properties of the beams can then be modified owing to the presence of rotational springs at the ends so that the degree of freedom of the plane frame in Fig. 17.3.3(b) is still 5.

The use of Model 1A by modifying the flexural properties of the columns is the "equivalent column method" of Corley [16.27]; and the use of Model 1B by modifying the fixed-end moments and the flexural properties of the beams is the "equivalent beam method" of Vanderbilt [16.32]. Other models, specifically for flat plate structures using equivalent beams without torsional springs have been used by Pecknold [16.38], Allen and Darvall [16.39, 16.47], and Elias [16.41–16.43].

Whether the results of the equivalent column method [Model 1A of Fig. 17.3.2(b)] are good approximations to those of actually analyzing Model 1 of Fig. 17.3.1 as a space frame can be examined by comparing the slopes θ_{AU}–θ_{BU}–θ_{CU} (subscript U means upper), with the slopes θ_{AL}–θ_{BL}–θ_{CL} (subscript L means lower) of Fig. 17.3.4(a). Likewise, the results of the equivalent beam method [Model 1B of Fig. 17.3.4(b)]

(a) Model 1*A*: Equivalent column method

(b) Model 1*B*: Equivalent beam method

Figure 17.3.4 Plane frame analysis of equivalent rigid frame in a two-way floor system.

may be examined by comparing the slope θ_{BL} (subscript L here means left) with slope θ_{BR} (subscript R means right) of Fig. 17.3.4(b).

In the remainder of this chapter, procedures of analysis for the four models (Model 1, Model 1A, Model 1B, and Model 2) will be described and numerical examples shown.

17.4 REDUCED STIFFNESS MATRIX AND REDUCED FIXED-END MOMENTS FOR A FLEXURAL ELEMENT WITH ROTATIONAL SPRINGS AT ENDS

The presence of a rotational spring at the end of a flexural element is identical to the situation in which the member is attached to a rigid gusset plate at the joint by a semirigid connection [7.1]. Methods of obtaining the reduced fixed-end moments and the reduced stiffness matrix for a member with constant moment of inertia have been treated by Wang [7.1].

For reinforced concrete buildings with two-way floor systems, the beams (whether for the combined column and middle strip as for Model 1, or for the column strip and middle strip separately as for Model 2) and the columns all have variations in moment of inertia along their lengths. Even for such an element with rigid end connection, the element stiffness matrix is no longer

$$[S] = \begin{bmatrix} \dfrac{4EI}{L} & \dfrac{2EI}{L} \\ \dfrac{2EI}{L} & \dfrac{4EI}{L} \end{bmatrix} \qquad (17.4.1)$$

wherein EI is the constant moment of inertia for the entire length L, but it becomes

$$[S] = \begin{bmatrix} S_{ii} & S_{ij} \\ S_{ji} & S_{jj} \end{bmatrix} \qquad (17.4.2)$$

in which $S_{ij} = S_{ji}$ but written as S_{ij} and S_{ji} for theoretical clarification. The best way to show the stiffness of an element is as in Fig. 17.4.1, wherein the "end stiffnesses" S_{ii} and S_{jj} are written at the element ends and the "cross stiffness" $S_{ij} = S_{ji}$ is written near the middle of the element, with lines linking the three items. The end and cross

(a) Flexural element with rigid end connection

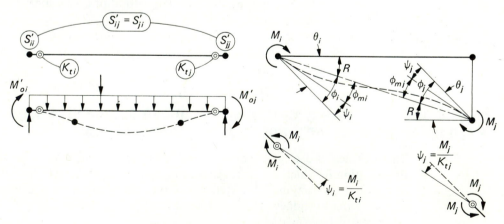

(b) Flexural element with semirigid end connection
(same as with rotational springs at ends)

Figure 17.4.1 Flexural element with rigid and semirigid end connections.

stiffnesses, as well as the fixed-end moments for transverse loads, may best be determined by the column analogy method, as demonstrated by Wang [7.1, Chapter 16].

The presence of rotational springs at the member ends, with stiffnesses K_{ti} and K_{tj}, will cause slip angles ψ_i and ψ_j, both in the *counterclockwise* direction as shown in Fig. 17.4.1(b). By the definition of K_{ti} and K_{tj},

$$\begin{Bmatrix} \psi_i \\ \psi_j \end{Bmatrix} = \begin{bmatrix} 1/K_{ti} & 0 \\ 0 & 1/K_{tj} \end{bmatrix} \begin{Bmatrix} M_i \\ M_j \end{Bmatrix} = [B] \begin{Bmatrix} M_i \\ M_j \end{Bmatrix} \tag{17.4.3}$$

Applying the definition of the regular stiffness matrix $[S]$ to the member itself inside the rotational springs in Fig. 17.4.1(b),

$$\begin{Bmatrix} M_i \\ M_j \end{Bmatrix} = \begin{bmatrix} S_{ii} & S_{ij} \\ S_{ji} & S_{jj} \end{bmatrix} \begin{Bmatrix} \phi_{mi} \\ \phi_{mj} \end{Bmatrix} = [S] \{ \phi_m \} \tag{17.4.4}$$

But, from the geometry shown in Fig. 17.4.1(b),

$$\{ \phi_m \} = \{ \phi \} - \{ \psi \} \tag{17.4.5}$$

Substituting Eq. (17.4.5) in Eq. (17.4.4),

$$\{ M \} = [S] \{ \phi \} - [S] \{ \psi \} \tag{17.4.6}$$

Substituting Eq. (17.4.3) in Eq. (17.4.6),

$$\{ M \} = [S] \{ \phi \} - [S][B] \{ M \}$$

from which

$$[I + SB] \{ M \} = [S] \{ \phi \}$$

and

$$\{ M \} = [I + SB]^{-1} [S] \{ \phi \} \tag{17.4.7}$$

By definition in Fig. 17.4.1(b),

$$\{ M \} = [S'] \{ \phi \} \quad \text{for semirigid end connection} \tag{17.4.8}$$

Comparing Eq. (17.4.8) with Eq. (17.4.7),

$$[S'] = [I + SB]^{-1} [S] \tag{17.4.9}$$

Equation (17.4.9) gives the reduced stiffness matrix of a flexural element with rotational springs at its ends. That the $[S']$ matrix is still symmetric can be proved as follows:

To Prove: $[S'] = [I + SB]^{-1} [S]$ is symmetric.

Proof:
1. Taking the inverse of each side of the above equation, $[S']^{-1} = [S^{-1}][I + SB]$, because in general $[CD]^{-1} = [D^{-1}][C^{-1}]$.
2. Expanding the equation in Step 1, $[S']^{-1} = [S^{-1}I] + [S^{-1} SB] = [S^{-1}] + [B]$.
3. Observing the equation in Step 2, $[S']^{-1}$ is symmetric because $[S^{-1}]$ and $[B]$ are both symmetric.
4. From Step 3, $[S']$ is symmetric, because $[S']^{-1}$ is symmetric.

Next, to obtain formulas for the reduced fixed-end moments under transverse loads, observe first that the rotation R of the member axis is zero in the fixed condition and thus θ's and ϕ's are also zero. There are two ways of reasoning by which the formulas for M'_{0i} and M'_{0j} in Fig. 17.4.1(b) may be obtained. In the first way, the beginning premise is that M'_{0i} and M'_{0j} in Fig. 17.4.1(b) are there to keep $\theta_i = \theta_j = 0$. If the unreduced fixed-end moments M_{0i} and M_{0j} were there instead, then $\theta_{mi} = \theta_{mj} = 0$ (note that when R is zero, θ_{mi} and θ_{mj} are the same as ϕ_{mi} and ϕ_{mj}) but θ_i and θ_j would be, using Eq. (17.4.5),

$$\begin{Bmatrix} \text{unwanted } \theta_i \\ \text{unwanted } \theta_j \end{Bmatrix} = \begin{Bmatrix} \theta_{mi} = 0 \\ \theta_{mj} = 0 \end{Bmatrix} + \begin{Bmatrix} M_{0i}/K_{ti} \\ M_{0j}/K_{tj} \end{Bmatrix} = [B]\begin{Bmatrix} M_{0i} \\ M_{0j} \end{Bmatrix} \qquad \textbf{(17.4.10)}$$

The M'_{0i} and M'_{0j} must then be the sum of the regular fixed-end moments and those to cancel out [thus for the negative sign in Eq. (17.4.11)] the unwanted slopes in Eq. (17.4.10); or

$$\begin{aligned} \{M'_0\} &= \{M_0\} - [S'][B]\{M_0\} \\ &= [I - S'B]\{M_0\} \end{aligned} \qquad \textbf{(17.4.11)}$$

Substituting Eq. (17.4.9) for $[S']$ in Eq. (17.4.11),

$$\{M'_0\} = [I - [I + SB]^{-1}[S][B]]\{M_0\}$$

Premultiplying each side of the above equation by $[I + SB]$,

$$[I + SB]\{M'_0\} = [[I + SB] - [SB]]\{M_0\} = \{M_0\}$$

from which

$$\{M'_0\} = [I + SB]^{-1}\{M_0\} \qquad \textbf{(17.4.12)}$$

The second way of reasoning is more straightforward because the formula for $[S']$ is not required in the process. The beginning premise is the same, in that M'_{0i} and M'_{0j} are there to keep $\theta_i = \theta_j = 0$, which means that θ_{mi} and θ_{mj} (or ϕ_{mi} and ϕ_{mj}) will not be zero but equal to, using Eq. (17.4.5),

$$\{\theta_m\} = \{\theta = 0\} - [B]\{M'_0\} \qquad \textbf{(17.4.13)}$$

The excess of $\{M'_0\}$ over $\{M_0\}$ acting on the member ends between the rotational springs should be consistent with $\{\theta_m\}$ in Eq. (17.4.13) through the regular stiffness matrix $[S]$; or

$$\{M'_0 - M_0\} = -[S][B]\{M'_0\}$$

$$\{M'_0\} + [SB]\{M'_0\} = \{M_0\}$$

from which

$$\{M'_0\} = [I + SB]^{-1}\{M_0\}$$

the same as Eq. (17.4.12).

In a computer program, the formulas for $\{M_0'\}$ and $[S']$, as expressed by Eqs. (17.4.12) and (17.4.9), may be written as

$$
\left.
\begin{aligned}
&1. \quad F_{ti} = \frac{1}{K_{ti}}; \qquad F_{tj} = \frac{1}{K_{tj}} \\[6pt]
&2. \quad T_1 = (S_{ii}S_{jj}) - (S_{ij})^2 \\[6pt]
&3. \quad T_2 = 1 + F_{ti}S_{ii} + F_{tj}S_{jj} + F_{ti}F_{tj}T_1 \\[6pt]
&4. \quad M_{0i}' = [(1 + F_{tj}S_{jj})M_{0i} - (F_{tj}S_{ij})M_{0j}]/T_2 \\[3pt]
&\qquad\quad M_{0j}' = [(1 + F_{ti}S_{ii})M_{0j} - (F_{ti}S_{ij})M_{0i}]/T_2 \\[6pt]
&5. \quad S_{ii}' = (S_{ii} + F_{tj}T_1)/T_2 \\[6pt]
&6. \quad S_{ij}' = S_{ij}/T_2 \\[6pt]
&7. \quad S_{jj}' = (S_{jj} + F_{ti}T_1)/T_2
\end{aligned}
\right\}
\qquad \textbf{(17.4.14)}
$$

The necessary algebraic manipulation to obtain Eqs. (17.4.14) is not shown here but is left as an exercise for the reader.

17.5 THE ACI EQUIVALENT COLUMN

In the 1977 ACI Code, the equivalent column was one having a stiffness obtained by the then ACI Formula (13-6), defined in words as follows: "Flexibility (inverse of stiffness) of an equivalent column shall be taken as the sum of the flexibilities of the actual columns above and below the slab-beam and the flexibility of the attached torsion member ..." This formula was later removed from the Code and placed in Commentary-R13.7.4. The formula is

$$
\frac{1}{K_{ec}} = \frac{1}{\Sigma K_c} + \frac{1}{K_t} \qquad\qquad \textbf{(17.5.1)}
$$

from which

$$
K_{ec} = \frac{\Sigma K_c}{1 + \dfrac{\Sigma K_c}{K_t}} \qquad\qquad \textbf{(17.5.2)}
$$

It can be shown that S_{ii}' of Eq. (17.4.14) is identical to K_{ec} of Eq. (17.5.2), for one column with stiffness K_c. From Eq. (17.4.14), if K_{tj} at the far end is infinity, then

$$
T_2 = 1 + F_{ti}S_{ii}
$$

and

$$
S_{ii}' = \frac{S_{ii}}{T_2} = \frac{S_{ii}}{1 + \dfrac{S_{ii}}{K_{ti}}} \qquad\qquad \textbf{(17.5.3)}
$$

which is identical with Eq. (17.5.2) for $S_{ii} = \Sigma K_c$. So Eq. (17.5.2) is basically correct for one-story buildings in gravity load analysis, as it will be shown later that the equivalent column model (Model 1A) is good for gravity load analysis, although the equivalent beam model (Model 1B) is somewhat better for lateral load analysis. The original idea of Corley for developing the attached transverse torsion element is noteworthy.

17.6 EQUIVALENT COLUMN METHOD (MODEL 1A) vs EQUIVALENT BEAM METHOD (MODEL 1B)

As explained in Section 17.3, the equivalent column method (Model 1A) and the equivalent beam method (Model 1B) are both approximations of the space frame model (Model 1). It turns out that for lateral load analysis both Models 1A and 1B are good approximations with preference for Model 1B; but for gravity load analysis only Model 1A is good. The reason is that for both gravity and lateral loads, the columns are subjected only to end moments causing reverse curvature (that is, they have an inflection point in the midheight region) and the elastic curves in the upper and lower columns are geometrically similar. Hence, the $\theta_{AU}-\theta_{BU}-\theta_{CU}$ are nearly equal to $\theta_{AL}-\theta_{BL}-\theta_{CL}$ in Fig. 17.3.4(a). On the other hand, the moment diagrams for the beams are parabolic for gravity load. Consequently θ_{BL} and θ_{BR} in Fig. 17.3.4(b) are no longer nearly equal. In fact, they are of opposite sign so that there is no continuity of slab-beam at the column line under gravity load.

The authors favor the use of Model 1 on the desk-top computer with little additional programming effort; in fact, there is no need of using the reduced stiffness matrix and the reduced fixed-end moments. If this is the choice, the analyst might just as well go further to use Model 2. One operable version of the computer program is available in the separate supplement to this text. Interested readers may use the source language as a guide for making a more user-friendly version.

■ **EXAMPLE 17.6.1**   By the equivalent column method, analyze the floor subassembly in a building frame with two-way floor systems as detailed in Fig. 17.6.1.

Solution: **(a)** Choice of method. With the popularly available desk-top computers, the matrix displacement method [7.1–7.3] is the method that should be used, not to mention its expandability to analysis of the whole building frame and to consideration of axial deformation, stability functions, and P-delta effects.

(b) Global degree of freedom. The degree of freedom is $NP = 5$, assigned as shown in Fig. 17.3.4(a).

(c) Local stiffness matrix of the beam elements. The formulas shown in Fig. 17.6.2(a) are used to enter the contributions of beams AB and BC to the global stiffness matrix. Input data include

BEAM	NP1	NP2	S_{ii}	S_{ij}	S_{jj}	M_{oi}	M_{oj}
AB	1	2	8	5	10	-100	$+150$
BC	2	3	10	5	8	-120	$+75$

(a) Stiffness values

(b) Loading condition No. 1: FEM in ft–kips as shown
Loading condition No. 2: Lateral loads as shown

Figure 17.6.1 Details of floor assembly.

(d) Local stiffness matrix of the column elements. Equations (17.4.14) are used first to compute the reduced stiffness matrix of each column and then the formulas in Fig. 17.6.2(b) are used to enter the contributions of the six equivalent columns to the global stiffness matrix. Input data include

COLUMN	NP1	NP2	NP3	NP4	S_{ii}	S_{ij}	S_{jj}	H	K_{ti}	K_{tj}
DA	6	1	6	5	5	3	6	10	∞	1.0909
EB	6	2	6	5	5	3	6	10	∞	1.6364
FC	6	3	6	5	5	3	6	10	∞	1.0909
AG	1	6	5	4	5	2.4	4	8	0.9091	∞
BH	2	6	5	4	5	2.4	4	8	1.3636	∞
CK	3	6	5	4	5	2.4	4	8	0.9091	∞

Note that the local degree of freedom is assigned the value $NP + 1 = 6$ wherever there is restraint so the $(NP + 1)$th row and column in the compiled global stiffness matrix are never used. The torsional stiffness at each column intersection is divided into two parts in the ratio of the column stiffnesses; for example, for column DA (in the above table),

P \ X	NP1	NP2
NP1	S_{ii}	S_{ij}
NP2	S_{ji}	S_{jj}

(a) Typical beam element

P \ X	NP1	NP2	NP3	NP4
NP1	S_{ii}	S_{ij}	$+T1$	$-T1$
NP2	S_{ji}	S_{jj}	$+T2$	$-T2$
NP3	$+T1$	$+T2$	$+T3$	$-T3$
NP4	$-T1$	$-T2$	$-T3$	$+T3$

$$T1 = \frac{S_{ii} + S_{ij}}{H} \quad T2 = \frac{S_{jj} + S_{ij}}{H} \quad T3 = \frac{S_{ii} + S_{jj} + 2S_{ij}}{H^2}$$

(b) Typical column element

P \ X	NP1	NP2
NP1	$+K_t$	$-K_t$
NP2	$-K_t$	$+K_t$

(c) Typical torsion element

Figure 17.6.2 Local stiffness matrix of typical beam, column, and torsion elements.

$$K_{tj} = \frac{6}{6 + 5}(2) = 1.0909$$

The modified stiffness values are

COLUMN	S'_{ii}	S'_{ij}	S'_{jj}
DA	3.7308	0.4615	0.9231
EB	3.8214	0.6428	1.2857
FC	3.7308	0.4615	0.9231
AG	0.7692	0.3692	3.0252
BH	1.0714	0.5143	3.0948
CK	0.7692	0.3692	3.0252

(e) The load matrix. The loads that are applied directly at the joints are entered first. Those arising from the fixed-end moments are programmed for internal calculation and compilation. Using loading condition (*LC*) No. 1 for gravity load only, and No. 2 for lateral loading, the resulting load matrix is

$$[P]_{5 \times 2} =$$

P \ LC	1	2
1	+100	0
2	−30	0
3	−75	0
4	0	+36
5	0	+9

(f) Displacements and member end moments. The displacements in terms of kip-ft^2/EI and kip-ft^3/EI are

	GRAVITY LOAD	LATERAL LOAD
θ_A	+11.636	+5.644
θ_B	−2.487	+1.283
θ_C	−6.419	+5.644
Δ_{GHK}	+2.504	+436.69
Δ_{ABC}	+1.384	+266.78

The member end moments in ft-kips are

	GRAVITY LOAD		LATERAL LOAD	
	M_i	M_j	M_i	M_j
Beam *AB*	−19.34	+183.31	+51.57	+41.05
BC	−176.96	+11.21	+41.05	+51.57
Column *DA*	+4.79	+10.55	−109.24	−31.73
AG	+8.79	+3.82	−19.84	−70.01
EB	−2.22	−3.46	−118.27	−49.80
BH	−2.89	−1.78	−32.30	−76.00
FC	−3.54	−6.12	−109.24	−31.73
CK	−5.10	−2.84	−19.84	−70.01

(g) Column end slopes computed from beam end slopes. The θ's in the output are the slopes of beams at the ends of the torsion element. The slopes at the column ends can be computed from

$$\theta_{mi} = \theta_i - \frac{M_i}{K_{ti}}$$

The closeness or divergence of the two upper and lower column slopes indicates the degree of approximation of Model 1A to Model 1. The results are

	INTERSECTION	A	B	C	
Gravity load	θ (upper)	+1.966	−0.370	−0.812	(very good)
	θ (lower)	+1.966	−0.370	−0.812	
Lateral load	θ (upper)	+27.466	+24.973	+27.466	(not too good)
	θ (lower)	+34.728	+31.716	+34.728	

■

■ **EXAMPLE 17.6.2**   Rework Example 17.6.1 except use the equivalent beam method.

Solution: (a) Global degree of freedom. The degree of freedom is $NP = 5$, assigned as shown in Fig. 17.3.4(b).

(b) Local stiffness matrix of the beam elements. Equations (17.4.14) are used first to compute the reduced fixed-end moments and the reduced stiffness matrix of each beam before the formulas in Fig. 17.6.2(a) for the local stiffness matrix are used. Input data include

BEAM	NP1	NP2	S_{ii}	S_{ij}	S_{jj}	K_{ti}	K_{tj}	M_{oi}	M_{oj}
AB	1	2	8	5	10	2	1.5	−100	+150
BC	2	3	10	5	8	1.5	2	−120	+75

The reduced fixed-end moments and the reduced stiffnesses are

BEAM	M'_{oi}	M'_{oj}	S'_{ii}	S'_{ij}	S'_{jj}
AB	−42.22	+33.33	1.4889	0.1667	1.2500
BC	−26.25	+32.50	1.2500	0.1667	1.4889

Note that the carry-over factors are $0.1667/1.4889 = 0.112$ from A to B and $0.1667/1.2500 = 0.133$ from B to A, which means that convergence in moment distribution is extremely fast for gravity load analysis.

(c) Local stiffness matrix of the column elements. The column stiffnesses do not need modification. The input data are the same as in part (d) of the previous example, except there are no rotational springs at the ends.

(d) The load matrix. The load matrix for the gravity and lateral loads, each acting separately, is

$$[P]_{5 \times 2} =$$

P \ LC	1	2
1	+42.22	0
2	−7.08	0
3	−32.50	0
4	0	+36
5	0	+9

(e) Displacements and member end moments. The displacements in terms of kip-ft^2/EI and kip-ft^3/EI are

	GRAVITY LOAD	LATERAL LOAD
θ_A	+3.568	+31.425
θ_B	−0.370	+28.649
θ_C	−2.415	+31.425
Δ_{GHK}	+2.504	+436.20
Δ_{ABC}	+1.383	+249.70

The member end moments in ft-kips are

	GRAVITY LOAD		LATERAL LOAD	
	M_i	M_j	M_i	M_j
Beam AB	−36.97	+33.46	+51.56	+41.05
BC	−27.11	+28.84	+41.05	+51.56
Column DA	+9.60	+20.16	−105.49	−36.18
AG	+16.80	+7.67	−15.38	−73.77
EB	−2.22	−3.46	−113.82	−52.84
BH	−2.89	−1.78	−29.26	−80.44
FC	−8.35	−15.73	−105.49	−36.18
CK	−13.11	−6.69	−15.38	−73.77

Note that the sideways in this Model 1B are almost identical to those in Model 1A. So are the beam and column moments for lateral load. But the beam moments at the ends for gravity load are clearly too small in this Model 1B and should not be relied upon; furthermore, as will be seen below in part (f), slope continuity is *not* satisfied at the junction of the two elastic curves of the beam.

(f) Beam end slopes computed from column end slopes. The θ's in the output are the slopes of the columns at their intersections with the torsion elements. The slopes at the beam ends of the torsion elements can be computed from

$$\theta_{mi} = \theta_i - \frac{M_i}{K_{ti}}$$

The closeness or divergence of the left and right slopes of the beams at the interior support indicates the degree of approximation of Model 1B to Model 1. The results are

	INTERSECTION	A	B	C
Gravity load	θ (left)	(none)	-22.680	-16.836
	θ (right)	$+22.054$	$+17.707$	(none)
			(no good)	
Lateral load	θ (left)	(none)	$+1.283$	$+5.643$
	θ (right)	$+5.643$	$+1.283$	(none)
			(very good)	

The slopes of the slab-beams are $22.680/EI$ counterclockwise on one side and $17.707/EI$ clockwise on the other side of the interior support, indicating that Model 1B should not be used for gravity load. For lateral load analysis, Model 1B gives even better results than Model 1A because there is slope compatibility at the columns as well as at the beam junctions (such as A', B', and C' of Fig. 17.2.1) occurring away from the columns. ■

17.7 TREATMENT OF MODEL 1 AS A SPACE FRAME

The equivalent column method (Model 1A, or K_{ec} model) and the equivalent beam method (Model 1B, or K_{eb} model) are, after all, approximations to the analysis of Model 1 of Fig. 17.3.1 as a space frame. When sideway is ignored in gravity load analysis, Model 1A (not Model 1B) has merit because then the usual moment distribution method can be applied to the subassembly. This was exactly the ingenious idea of Corley et al. [16.26, 16.27] at a time when computers were not conveniently available. With sideways as unknowns in lateral load analysis, the indirect moment distribution method [7.1, 7.2] involving the solution of as many simultaneous equations as there are unknown sideways has to be used. While this approach can itself be programmed on the computer, it is mainly taught, if at all, in schools for the purpose of making students better understand the matrix displacement method.

With an isolated floor assembly as shown in Fig. 17.3.1, the degree of freedom of Model 1 would be larger than that of Models 1A or 1B, by the number of column lines. For an entire tall building frame, say, of 15 stories and 4 column lines, the degree of freedom of Model 1 would be 15 times 4, or 60, larger than that of Models 1A and 1B. In this case, the use of Model 1A for gravity load and Model 1B for lateral load might be justifiable. However, the solution of a large degree-of-freedom problem can be obtained by a proper computer programming technique [7.3, Chapter 12]. In time, it is hoped that Model 1, or Model 2, as advocated in Section 17.8, will be used.

The local stiffness matrix of a typical torsion element is shown in Fig. 17.6.2(c). Here, M_t is taken positive if its rotational vectors act "tension-like" on the ends of the element. Taking the local degree of freedom numbers $NP1$ and $NP2$ at the near and far ends of the element,

$$M_t = K_t(X_2 - X_1) \tag{17.7.1}$$

wherein X_1 and X_2 are the clockwise slopes at the near and far ends of the torsion element, respectively.

■ **EXAMPLE 17.7.1** Rework Examples 17.6.1 and 17.6.2 except use the torsion elements as such in the space frame (Model 1) of Fig. 17.3.1.

Solution: (a) Global degree of freedom. The degree of freedom is $NP = 8$, assigned as shown in Fig. 17.3.1.

(b) Local stiffness matrix of the beam elements. Input data include

BEAM	NP1	NP2	S_{ii}	S_{ij}	S_{jj}	M_{oi}	M_{oj}
AB	4	5	8	5	10	−100	+150
BC	5	6	10	5	8	−120	+75

(c) Local stiffness matrix of the column elements. Input data include

COLUMN	NP1	NP2	NP3	NP4	S_{ii}	S_{ij}	S_{jj}	H
DA	9	1	9	8	5	3	6	10
EB	9	2	9	8	5	3	6	10
FC	9	3	9	8	5	3	6	10
AG	1	9	8	7	5	2.4	4	8
BH	2	9	8	7	5	2.4	4	8
CK	3	9	8	7	5	2.4	4	8

(d) Local stiffness matrix of the torsion elements. Input data include

TORSION ELEMENT	NP1	NP2	K_t
$A'A$	4	1	2
$B'B$	5	2	3
$C'C$	6	3	2

(e) The load matrix. The load matrix for the gravity and lateral loads, each acting separately, is

$$[P]_{8 \times 2} =$$

LC P	1	2
1	0	0
2	0	0
3	0	0
4	+100	0
5	−30	0
6	−75	0
7	0	+36
8	0	+9

(f) Displacements and member end moments from the output. The displacements in terms of kip-ft^2/EI and kip-ft^3/EI are

	GRAVITY LOAD	LATERAL LOAD
θ_A	+1.966	+31.425
θ_B	−0.370	+28.649
θ_C	−0.812	+31.425
θ'_A	+11.636	+5.643
θ'_B	−2.487	+1.283
θ'_C	−6.419	+5.643
Δ_{GHK}	+2.504	+436.20
Δ_{ABC}	+1.383	+249.70

The member end moments in ft-kips are

	GRAVITY LOAD		LATERAL LOAD	
	M_i	M_j	M_i	M_j
Beam $A'B'$	−19.34	+183.31	+51.56	+41.05
$B'C'$	−176.96	+11.21	+41.05	+51.56
Column DA	+4.79	+10.55	−105.49	−36.18
AG	+8.79	+3.82	−15.38	−73.77
EB	−2.22	−3.46	−113.82	−52.84
BH	−2.89	−1.78	−29.26	−80.44
FC	−3.54	−6.12	−105.49	−36.18
CK	−5.10	−2.84	−15.38	−73.77
Torsion $A'A$	−19.34	—	+51.56	—
$B'B$	+6.35	—	+82.10	—
$C'C$	+11.21	—	+51.56	—

The results for the gravity load are exactly the same as those of Model 1A, and the results for the lateral load are exactly the same as those of Model 1B. This exact coincidence is due to the fact that the isolated floor subassembly has all far ends of columns fixed against rotation. If the entire structural frame is analyzed, there will be some rippling effect. However, if the analyst wishes to use an equivalent plane frame, Models 1A and 1B should be used for gravity and lateral load, respectively. ■

17.8 MODEL 2 WITH TORSION ELEMENT CONNECTING COLUMN STRIP TO MIDDLE STRIP

The basic frame model for the structure enclosed between the center lines of adjacent panels of a building with two-way floor systems has been shown in Fig. 17.2.1(c). This basic model contains beams ABC and $A'B'C'$, as well as torsion elements $A'A$, $B'B$, and $C'C$. Model 1 is derived from this basic model by placing all longitudinal flexural stiffness within the width of the equivalent rigid frame in beam $A'B'C'$, with none in beam ABC, in an attempt to analyze the space frame as a plane frame, by, say, the moment distribution method. Consequently, the moments in beam $A'B'C'$ obtained from the analysis have yet to be distributed transversely into the column and middle strips, for the purpose of allocating the reinforcement. By placing all of the longitudinal stiffness away from the plane of the columns (granting that the parameter K_t may be adjusted for doing so), not much unwanted effect is produced in the design moments for gravity load, but the model would certainly underestimate the effectiveness of the actual structure to resist lateral loads applied directly in the plane of the columns.

Considering the current capability of the computer to solve mathematics, one might just as well use the basic model, placing the longitudinal stiffness of the column strip in beam ABC and that of the middle strip in beam $A'B'C'$ (granting again the torsional stiffness K_t might need adjustment to correlate with test data). This model, called Model 2, is believed to be more rational than Model 1. The following example shows the results of a Model 2 analysis.

■ **EXAMPLE 17.8.1** Rework Examples 17.6.1, 17.6.2, and 17.7.1, except place two-thirds of the total longitudinal stiffness and fixed-end moments in the plane of the columns and the remaining one-third at the far ends of the torsion elements (Model 2 analysis).

Solution: The degree of freedom numbers and input data are mostly identical to those in Example 17.7.1 except there are now four beam elements instead of two. The displacements in terms of kip-ft^2/EI and kip-ft^3/EI in the output are

	GRAVITY LOAD	LATERAL LOAD
θ_A	+5.090	+11.058
θ_B	−1.113	+4.964
θ_C	−3.016	+11.058
$\theta_{A'}$	+10.074	+4.777
$\theta_{B'}$	−2.100	−0.107
$\theta_{C'}$	−5.900	+4.777
Δ_{GHK}	+3.073	+230.40
Δ_{ABC}	+1.697	+136.02

The member end moments in ft-kips are

	GRAVITY LOAD		LATERAL LOAD	
	M_i	M_j	M_i	M_j
Beam AB	−43.23	+109.55	+75.52	+69.95
$A'B'$	−9.97	+59.79	+12.56	+7.60
BC	−97.47	+30.21	+69.95	+75.52
$B'C'$	−56.83	+5.77	+7.60	+12.56
Column DA	+13.91	+29.01	−75.64	−56.07
AG	+24.18	+11.12	−32.01	−48.96
EB	−4.70	−8.20	−93.93	−92.94
BH	−6.84	−3.77	−62.48	−63.59
FC	−10.40	−19.62	−75.64	−56.07
CK	−16.35	−8.34	−32.01	−48.96
Torsion $A'A$	−9.97	—	+12.56	—
$B'B$	+2.96	—	+15.21	—
$C'C$	+5.77	—	+12.56	—

Comparing the results of Model 2 with those of Model 1, it is seen that, for gravity load there are higher negative moments at the exterior supports and lower negative moments at the interior support, the column moments are somewhat larger, and the torsion elements transmit less torsional moment. For lateral load, the sidesway is reduced to about 53% of that obtained for Model 1, the column moments will still have to satisfy statics but they are more equalized between the two ends, and the torsional moments are reduced to 20–25% of those from the Model 1 analysis. Although Model 2 is preferred, its use should be deferred, especially for estimates of sidesway, until the torsional stiffness K_t is adjusted by further studies and tests. ■

17.9 CONCLUDING REMARKS

Analysis of concrete frames (including those with two-way floor systems) having their lateral resistance provided only by the stiffnesses of interconnected beams (or slab-beams) and columns should be separated into two parts: (1) analysis for gravity loads only, and (2) analysis for lateral loads plus appropriate simultaneous gravity load. Analytical models that do not allow sidesway should be used for gravity load only, although when the structure or the loading is unsymmetrical a model allowing sidesway should be considered. Simplified assemblies for isolated floors with far ends of columns fixed against rotation but permitted to deflect horizontally may be used for lateral load, although a final review analysis should be made for the whole frame in action.

Gravity Load Analysis.   The plane frame analysis using the equivalent column model (Model 1A, Fig. 17.3.2) with or without the sidesway degree of freedom, seems appropriate as a mathematical model. This model, developed by Corley et al. [16.26, 16.27], has been the one endorsed by the ACI Code since 1971. Both the equivalent frame method and the direct design method as prescribed by ACI-13.3.1.1 are equivalent column models. Vanderbilt [16.32] also indicated that this model is appropriate for gravity load analysis.

Lateral Load Analysis.   For lateral load, ACI-13.3.1.2 states that ". . . analysis of unbraced frames shall take into account effects of cracking and reinforcement on stiffness of frame members." During deliberations of ACI Committee 318 regarding the 1983 ACI Code, strong consideration was given to endorsing the equivalent frame model (Models 1A and 1B, Figs. 17.3.2 and 17.3.3, necessarily with the sidesway degree of freedom included) for lateral load analysis primarily based on the work of Vanderbilt [16.32, 16.40]. According to Vanderbilt [16.32], although there is little difference in the magnitudes of the sidesway obtained by the two equivalent models, the preferred model for lateral load is the equivalent beam model (Model 1B, Fig. 17.3.2), which agrees with the authors' analysis and results presented in this chapter.

Further, the Vanderbilt report [16.32] indicates that for stiffness computations in lateral load analysis, a beam-slab stiffness equal to $\frac{1}{4}$ to $\frac{1}{3}$ of that obtained for gross uncracked concrete may be used in lieu of a calculation taking cracking of the beam-slab into account. This reduction would achieve the result of increasing the computed sidesway. The Model 2 method of analysis is advocated by the authors, wherein the longitudinal stiffnesses of the column strip and the middle strip are placed in beams ABC and $A'B'C'$, respectively, of Fig. 17.2.1(c). Indeed, a major difficulty of any lateral load analysis arises in the stiffness determination; the flexural stiffnesses when Models 1A (Fig. 17.3.2) and 1B (Fig. 17.3.3) are to be used, and additionally an appropriate torsional stiffness when Models 1 (Fig. 17.3.1) or 2 (Fig. 17.2.1) are to be used. For the space frame Models 1A and 1B, the torsional stiffness has been calibrated for the gravity load analysis and then used in Model 1B for the lateral load analysis. With a space frame model (especially the general model of Fig. 17.2.1), use of the proper torsional stiffness is more important, and recommendations based on calibration with actual structures are not available.

SELECTED REFERENCES

The references for lateral load analysis of two-way slab-frame systems are to be found as References 16.38–16.51 of Chapter 16.

CHAPTER 18

Yield Line Theory of Slabs

18.1 INTRODUCTION

Reinforced concrete design methods under the present ACI Code are based on the results of an elastic analysis of the structure as a whole, when subjected to the action of factored loads (ACI-9.2), such as $1.4D + 1.7L$ where D and L refer to service dead and live loads. Actually, the behavior of a statically indeterminate structure is such that after the moment strengths at one or more points have been reached, discontinuities develop in the elastic curve at those points, and the results of an elastic analysis are no longer valid. If there is sufficient ductility, redistribution of bending moments will occur until a sufficient number of sections of discontinuity, commonly called "plastic hinges," form to change the structure into a mechanism, at which time the structure collapses or fails. The term "ultimate load analysis," as opposed to "elastic analysis," relates to the use of the bending moment diagram at the verge of collapse as the basis for design. Other than the provisions for redistribution of moments at the supports of continuous flexural members (ACI-8.4), the present ACI Code has made no allowance for ultimate load analysis. The redistribution as described in ACI-8.4 has been presented and illustrated in Section 10.12.

The design and analysis of two-way slab systems has been treated in Chapters 16 and 17. The factored moments are based on the elastic analysis of an equivalent frame, which has been devised as a simple substitute for the elastic analysis of a three-dimensional system.

The chief concern of this chapter is to develop the yield line theory for two-way slabs. Although not included in the ACI Code, slab analysis by yield line theory may be useful in providing the needed information for understanding the behavior of irregular or single-panel slabs with various boundary conditions.

Flat slabs; Marina City, Chicago. (Courtesy of Portland Cement Association.)

18.2 GENERAL CONCEPT

Although the study of flexural behavior of plates up to the ultimate load may date back to the 1920s [18.1], the fundamental concept of the yield line theory for the ultimate load design of slabs has been expanded considerably by K. W. Johansen [18.2, 18.3]. In this theory the strength of a slab is assumed to be governed by flexure alone; other effects such as shear and deflection are to be separately considered. The steel reinforcement is assumed to be fully yielded along the yield lines at collapse and the bending and twisting moments are assumed to be uniformly distributed along the yield lines.

Yield line theory for one-way slabs is not much different from the limit analysis of continuous beams. On a continuous beam the achievement of flexural strength at one location, say in the negative moment region over a support, does not necessarily constitute reaching the ultimate load on the beam. If the section, having reached its flexural strength, can continue to provide a constant resistance while undergoing further rotation, then the flexural strength may be reached at additional locations. Complete failure theoretically cannot occur until yielding has occurred at several locations (or along several parallel lines in case of one-way slabs) so that a mechanism forms giving a condition of unstable equilibrium.

Consider for example, the one-way slab of finite width shown in Fig. 18.2.1. A uniform loading on the slab will cause uniform maximum negative bending moment along *AB* and *EF* and uniform positive bending moment along *CD*, which is parallel to the supports. When the uniform load is increased until the moments along *AB*, *CD*, and *EF* reach their respective ultimate moment capacities, rotation of the slab segments will occur with the yield lines acting as axes of rotation. Once the ultimate moment capacity is achieved, angle change can occur without additional resisting moment being developed. Thus, under the limiting condition with the slab segments able to rotate with no change in resisting moment, the slab system is geometrically unstable. This condition is known as a "collapse mechanism."

Yield line theory for two-way slabs requires a different treatment from limit analysis of continuous beams, because in this case the yield lines will not in general be parallel to each other but instead form a yield line pattern. The entire slab area will be divided into several segments which can rotate along the yield lines as rigid bodies at the condition of collapse or unstable equilibrium. Some yield line patterns for typical situations are shown in Fig. 18.2.2.

The slab of Fig. 18.2.2(a) has nonparallel supports. At the collapse condition this slab will break into two segments; one segment will have an edge rotating about I and the other will have an edge rotating about II. The positive moment yield line

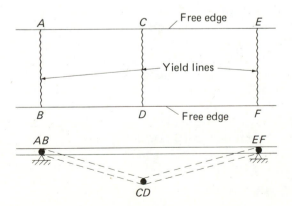

Figure 18.2.1 Collapse mechanism of a one-way slab.

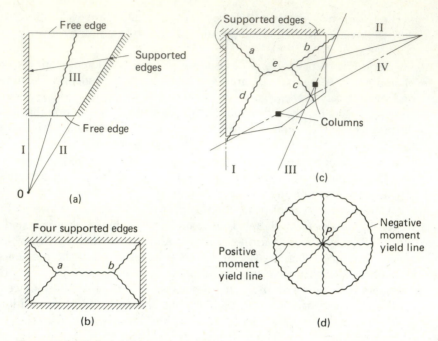

Figure 18.2.2 Typical yield line patterns.

must then intersect lines I and II at their intersection, point 0. The exact position of yield line III will depend on the reinforcement amount and direction, both in the positive and negative moment regions.

For the case of Fig. 18.2.2(b) where a rectangular panel is either simply supported or continuous over four linear supports, the collapse mechanism consists of four slab segments. The exact locations of points a and b will depend on the moment strengths at the supports and the positive moment reinforcement in each direction.

The slab in Fig. 18.2.2(c) is supported along two edges and, in addition, is supported by two isolated columns. The rotational axes for the slab segments at collapse must occur along the supports (lines I and II), and additional rotational axes must pass through the isolated columns. The critical position of the positive moment yield lines a, b, c, d, and e is a function of the reinforcement amount and direction; in the meantime, compatibility of deflection along the yield lines must be maintained during the rigid body rotations of the slab segments.

For a concentrated load at a significant distance from a supported edge, the yield line pattern will be circular as shown in Fig. 18.2.2(d). The circle pattern will be a yield line of negative bending moment, while the radial yield lines are due to positive bending moment. For concentrated loads near a free edge, a fan or partial circular pattern is typical.

18.3 FUNDAMENTAL ASSUMPTIONS

In applying the yield line theory to the ultimate load analysis of reinforced concrete slabs, the following fundamental assumptions are made:

Figure 18.3.1 Typical and idealized M–ϕ relationship for reinforced concrete slab.

1. The steel reinforcement is fully yielded along the yield lines at failure. In the usual case, when the slab reinforcement is well below that in the balanced condition, the moment–curvature relationship [18.4] is as shown in Fig. 18.3.1.
2. The slab deforms plastically at failure and is separated into segments by the yield lines.
3. The bending and twisting moments are uniformly distributed along the yield line and they are the maximum values provided by the moment strengths in two orthogonal directions (for two-way slabs).
4. The elastic deformations are negligible compared with the plastic deformations; thus the slab parts rotate as *plane* segments in the collapse condition.

Assumption No. 3 may be considered to be the yield criterion of orthotropic reinforced concrete slabs. It means that along a yield line as shown in Fig. 18.3.2, the bending moment strength M_{nb} and twisting moment strength M_{nt}, each per unit

Figure 18.3.2 Bending and twisting moments on yield line.

distance along the yield line, are exactly equal to the moment strengths M_{nx} and M_{ny} per unit distance in the y and x directions, respectively. It may be noted that M_{nx} is the strength contributed by the reinforcement in the x direction, and M_{ny} is the strength contributed by the reinforcement in the y direction. Also the sign convention is that the bending moments M_{nx}, M_{ny}, and M_{nb} are positive for tension in the lower portion of the slab and the twisting moment M_{nt} is positive if its vector is directed away from the free body on which it acts.

The bending moment strength M_{nb} and twisting moment strength M_{nt} along the yield line in Fig. 18.3.2 may be expressed in terms of M_{nx} and M_{ny}. Taking equilibrium of moment vectors parallel to the yield line,

$$M_{nb}(L) = M_{nx}(L \sin \theta) \sin \theta + M_{ny}(L \cos \theta) \cos \theta$$

$$M_{nb} = M_{nx} \sin^2 \theta + M_{ny} \cos^2 \theta$$

$$M_{nb} = \frac{M_{nx} + M_{ny}}{2} - \frac{M_{nx} - M_{ny}}{2} \cos 2\theta \qquad \textbf{(18.3.1)}$$

and, taking equilibrium of moment vectors perpendicular to the yield line,

$$M_{nt}(L) = M_{nx}(L \sin \theta) \cos \theta - M_{ny}(L \cos \theta) \sin \theta$$

$$M_{nt} = (M_{nx} - M_{ny}) \sin \theta \cos \theta$$

$$= \frac{(M_{nx} - M_{ny})}{2} \sin 2\theta \qquad \textbf{(18.3.2)}$$

In using Eqs. (18.3.1) and (18.3.2), it is important to note that θ is the counterclockwise angle measured from the positive x axis to the yield line.

18.4 METHODS OF ANALYSIS

There are two methods of yield line analysis of slabs: the virtual work method and the equilibrium method. Based on the same fundamental assumptions, the two methods should give exactly the same results. In either method, a yield line pattern must be first assumed so that a collapse mechanism is produced. For a collapse mechanism, rigid body movements of the slab segments are possible by rotation along the yield lines while maintaining deflection compatibility at the yield lines between slab segments. There may be more than one possible yield line pattern, in which case solutions to all possible yield line patterns must be sought and the one giving the smallest ultimate load would actually happen and thus should be used in design. For instance the failure pattern of the simply supported rectangular slab subjected to uniform load may be that shown either in Fig. 18.4.1(a) or in Fig. 18.4.1(b), depending on the aspect ratio of the rectangular panel and the moment strengths M_{nx} and M_{ny}.

After the yield line pattern has been assumed, the next step is to determine the position of the yield lines, such as defined by the unknown x in Fig. 18.4.1(a) or (b). It is at this point that one may choose to use the virtual work method or the equilibrium method. In the virtual work method, an equation containing the unknown x is established by equating the total positive work done by the ultimate load during

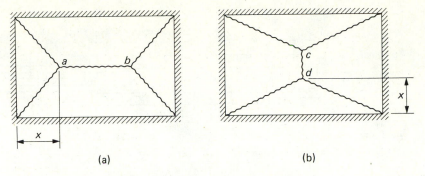

Figure 18.4.1 Yield line patterns of a simply supported rectangular slab.

simultaneous rigid body rotations of the slab segments (while maintaining deflection compatibility) to the total negative work done by the bending and twisting moments on all the yield lines. Then that value of x which gives the smallest ultimate load is found by means of differential calculus. In the equilibrium method, the value of x is obtained by applying the usual equations of statical equilibrium to the slab segments, but the optimal position x is defined by the placement of predetermined nodal forces at the intersection of yield lines. Expressions for the nodal forces in typical situations, once derived, can be conveniently used to avoid the necessity of mathematical differentiation as required in the virtual work method.

In the following sections, yield line analysis for one-way slabs is dealt with first in a manner similar to limit analysis of continuous beams. Then both the virtual work method and the equilibrium method are presented and illustrated for two-way slabs.

18.5 YIELD LINE ANALYSIS OF ONE-WAY SLABS

The continuous slab span shown in Fig. 18.5.1(a) has nominal moment strengths of M_{ni} and M_{nj} provided by top reinforcement at the supports and a nominal strength M_{np} provided by bottom reinforcement within the span. The nominal strengths M_{ni}, M_{nj}, and M_{np} are absolute values of the moment strength per unit slab width. For uniform loading the only possible yield pattern consists of three parallel yield lines as shown in Fig. 18.5.1(a), one each along the left and right supports and one at a distance x from the left support. The moment strength per unit slab width when collapse is imminent should be as shown in Fig. 18.5.1(b). The problem is to determine the collapse load w_u/ϕ per unit slab area in terms of M_{ni}, M_{nj}, and M_{np}, and the span length L. It will be shown that the virtual work and equilibrium methods will give exactly the same results.

Referring to Fig. 18.5.1(c), the rigid body rotations of the slab segments at collapse are measured from the original horizontal positions to those of the dashed lines, where the rotation of the left segment is θ_1 in the clockwise direction and that of the right segment is θ_2 in the counterclockwise direction while maintaining the compatible deflection Δ at the positive yield line. The total positive work done by the uniform

(a) Plan of continuous one-way slab

(b) Moment strength per unit slab width
when collapse condition is imminent

(c) Virtual work method

(d) Equilibrium method

Figure 18.5.1 Yield line analysis of one-way slabs.

load on the left and right segments of unit slab width is

$$\text{total positive work} = (w_u/\phi)x\,\frac{\Delta}{2} + (w_u/\phi)(L - x)\,\frac{\Delta}{2}$$

The total negative work done by M_{ni} and M_{np} on the left segment is $(M_{ni} + M_{np})\theta_1$ and that done by M_{nj} and M_{np} on the right segment is $(M_{nj} + M_{np})\theta_2$, both in absolute quantities. Thus

$$\text{total negative work} = (M_{ni} + M_{np})\theta_1 + (M_{nj} + M_{np})\theta_2$$

in which

$$\theta_1 = \frac{\Delta}{x}, \qquad \theta_2 = \frac{\Delta}{L - x}$$

The principle of virtual work states that the total work done by a force system in equilibrium in going through a virtual rigid body displacement is zero. By means of this principle, one can equate the total positive work to the absolute value of the total negative work; or

$$(w_u/\phi)x\,\frac{\Delta}{2} + (w_u/\phi)(L - x)\,\frac{\Delta}{2} = (M_{ni} + M_{np})\frac{\Delta}{x} + (M_{nj} + M_{np})\frac{\Delta}{(L - x)}$$

Dividing out the compatible deflection Δ from the above equation and solving for w_u/ϕ,

$$\frac{w_u}{\phi} = \frac{2M_{ni}}{Lx} + \frac{2M_{np}}{x(L - x)} + \frac{2M_{nj}}{L(L - x)} \tag{18.5.1}$$

Differentiating Eq. (18.5.1) for w_u with respect to x and setting the derivative to zero, one obtains the quadratic equation,

$$(M_{nj} - M_{ni})x^2 + 2(M_{ni} + M_{np})Lx - (M_{ni} + M_{np})L^2 = 0 \tag{18.5.2}$$

In the virtual work method, then, first the value of x is found by solving the quadratic equation (18.5.2) and the collapse load w_u/ϕ is determined from Eq. (18.5.1).

In the equilibrium method, the value of x is not to be obtained by differential calculus but defined by the magnitude of the nodal forces [V and V shown in Fig. 18.5.1(d)] acting on the slab segments on either side of the positive yield line. In this particular instance, it is known from the elementary theory of bending of beams that the shear at a section of maximum positive bending moment should be zero. From this point on, the unknown values of R_i, R_j, w_u, and x in Fig. 18.5.1(d) are obtained from the four independent equilibrium equations for the entire slab and either of the two slab segments. The left and right reactions on the free bodies of Fig. 18.5.1(d) are found by applying the equilibrium equations to the entire slab; thus

$$R_i = \left[\frac{1}{2}\left(\frac{w_u}{\phi}\right)L - \frac{M_{nj} - M_{ni}}{L}\right], \qquad R_j = \left[\frac{1}{2}\left(\frac{w_u}{\phi}\right)L + \frac{M_{nj} - M_{ni}}{L}\right]$$

Then, summing the vertical forces on the left slab segment,

$$R_i = \frac{1}{2}\left(\frac{w_u}{\phi}\right)L - \frac{M_{nj} - M_{ni}}{L} = \frac{w_u}{\phi}x$$

from which

$$\frac{w_u}{\phi} = \frac{(M_{nj} - M_{ni})}{L(L/2 - x)} \tag{18.5.3}$$

and, taking moments about the positive moment yield line on the left segment,

$$M_{np} = -M_{ni} + \left[\frac{1}{2}\left(\frac{w_u}{\phi}\right)L - \frac{M_{nj} - M_{ni}}{L}\right]x - \frac{1}{2}\left(\frac{w_u}{\phi}\right)x^2 \tag{18.5.4}$$

Substitution of Eq. (18.5.3) into Eq. (18.5.4) results in the same quadratic equation as Eq. (18.5.2). This shows that the two methods, virtual work and equilibrium, give exactly the same results.

In the solution of a numerical problem, the quadratic equation (18.5.2) is first solved for x. Although w_u can then be computed from either Eq. (18.5.1) or Eq. (18.5.3), using Eq. (18.5.1) is far superior to Eq. (18.5.3), because Eq. (18.5.1) contains the sum of three absolute quantities but Eq. (18.5.3) involves the division by a sensitive quantity $(L/2 - x)$.

■ **EXAMPLE 18.5.1**   Given the nominal moment strengths $M_{ni} = 14$ ft-kips, $M_{np} = 16$ ft-kips, and $M_{nj} = 22$ ft-kips per ft width of a 20-ft continuous slab span as shown in Fig. 18.5.2, determine the location x of the positive yield line and the collapse load w_u/ϕ in kips per ft per ft width of slab.

Solution: Using the quadratic equation (18.5.2),

$$(M_{nj} - M_{ni})x^2 + 2(M_{ni} + M_{np})Lx - (M_{ni} + M_{np})L^2 = 0$$

$$(22 - 14)x^2 + 2(14 + 16)(20x) - (14 + 16)(20)^2 = 0$$

$$8x^2 + 1200x - 12,000 = 0$$

$$x^2 + 150x + 75^2 = 1500 + 5625$$

$$x = \sqrt{7125} - 75 = 84.41 - 75 = 9.41 \text{ ft}$$

Substituting the value of $x = 9.41$ ft in Eq. (18.5.1),

$$\frac{w_u}{\phi} = \frac{2M_{ni}}{Lx} + \frac{2M_{np}}{x(L - x)} + \frac{2M_{nj}}{L(L - x)}$$

$$= \frac{2(14)}{20(9.41)} + \frac{2(16)}{9.41(10.59)} + \frac{2(22)}{20(10.59)}$$

$$= 0.149 + 0.321 + 0.208 = 0.678 \text{ kip/ft}$$

Using $w_u/\phi = 0.678$ kip/ft and the end moments of 14 ft-kips and 22 ft-kips, the end reactions, the shear diagram, and the moment diagram are computed and shown in Fig. 18.5.2.   ■

Figure 18.5.2 One-way slab of Example 18.5.1.

18.6 WORK DONE BY YIELD LINE MOMENTS IN RIGID BODY ROTATION OF SLAB SEGMENT

Before taking up the virtual work method of yield line analysis of two-way slabs, it is desirable to derive a general procedure for obtaining the absolute value of the negative work done by the bending and twisting moments acting on the yield line in going through a rigid body rotation of the slab segment. In Fig. 18.6.1(a) are shown the moments acting on the edges of a slab segment having yield line of length L with horizontal and vertical projections of L_x and L_y, respectively. Let this slab undergo a rigid body rotation, whose components are θ_x and θ_y, shown in vector notations in Fig. 18.6.1(b). It can be shown algebraically that the absolute value of the negative work done by the moments $(M_{nb}L)$ and $(M_{nt}L)$ acting on the yield line is equal to the

(a) Total moments on edges   (b) Rigid body rotation

Figure 18.6.1 Work done by yield line moments in rigid body rotation of a slab segment.

positive work done by the moments $(M_{nx}L_y)$ and $(M_{ny}L_x)$ acting on the horizontal and vertical projections of the yield line—for the same rigid body rotation, of course. This is obviously correct because the moments $(M_{nx}L_y)$ and $(M_{ny}L_x)$ on Fig. 18.6.1(a) are the equilibrants of the moments $(M_{nb}L)$ and $(M_{nt}L)$ on the same figure. Certainly then, the negative work done by one set of generalized forces should be equal, numerically, to the positive work done by the alternative set of equilibrating generalized forces.

18.7 NODAL FORCE AT INTERSECTION OF YIELD LINE WITH FREE EDGE

In the equilibrium method of yield line analysis, the position of a yield line is characterized by the insertion of nodal forces at the intersection of a yield line with another yield line, or of a yield line with a free edge. For one-way slabs it has been demonstrated in Section 18.5 that, on the basis of the elementary theory of bending of beams, the nodal force on either side of a positive moment yield line is equal to zero. In this section the expression for the pair of equal and opposite nodal forces V acting on each side of the intersection of a yield line with a free edge in a two-way slab is derived.

Shown in Fig. 18.7.1(a) is a two-way slab with nominal moment strengths of M_{nx} and M_{ny} provided by the bottom reinforcement in the horizontal and vertical directions, respectively. A positive moment yield line is assumed to intersect the free edge at an angle α. The upward nodal force V acts on the left segment and the downward nodal force V acts on the right segment, shown by a dot and a cross for the upward and downward forces in Fig. 18.7.1(a) according to the convention used by Johansen [18.2, 18.3]. For equilibrium the upward and downward nodal forces must be equal in magnitude.

Consider the equilibrium of an infinitesimal slab element shown in either Fig. 18.7.1(b) or (c). By its own definition, a yield line is always at the optimal position. Inasmuch as the edge *ac* in Fig. 18.7.1(b), or the edge *ad* in Fig. 18.7.1(c), is at an

(a) Two-way slab with free edge

(b) Free-body diagram of infinitesimal piece to left of yield line

(c) Free-body diagram of infinitesimal piece to right of yield line

Figure 18.7.1 Nodal force at intersection of yield line with free edge.

angle $\Delta\alpha$ from the yield line, the same yield line moments as those on the yield line *ab* act on edges *ac* or *ad*. By applying the principle of equivalent force systems as indicated by Fig. 18.6.1(a), the free-body diagram shown on the left side of the equal sign in Fig. 18.7.1(b) or (c) is transformed to the equivalent free-body diagram on the right side of the equal sign. Using the equivalent free body and summing the moments about the line *ac* in Fig. 18.7.1(b),

$$M_{ny}(\Delta L_x)\cos{(\alpha - \Delta\alpha)} = V(\Delta L_x)\sin{(\alpha - \Delta\alpha)} \qquad \textbf{(18.7.1)}$$

and summing the moments about the line ad in Fig. 18.7.1(c),

$$M_{ny}(\Delta L_x) \cos (\alpha + \Delta\alpha) = V(\Delta L_x) \sin (\alpha + \Delta\alpha) \tag{18.7.2}$$

Solving for V by using either Eq. (18.7.1) or Eq. (18.7.2),

$$V = M_{ny} \cot \alpha \tag{18.7.3}$$

Note that on the left side of Eq. (18.7.1) or Eq. (18.7.2), $M_{ny}(\Delta L_x)$ represents the net moment along the edges ca and ab, or the edges ba and ad, respectively; while on the right side, the moment of the nodal force V about an axis coincident with ac or ad is involved.

Equation (18.7.3) is used when applying the equilibrium method to a situation where a positive moment yield line intersects a free edge of a slab at an angle other than 90°. This is illustrated by the following example.

■ **EXAMPLE 18.7.1**   The triangular slab ABC shown in Fig. 18.7.2(a) is simply supported along edges AC and BC but has a free edge along AB. The horizontal and vertical reinforcement in the lower face of the slab provides nominal moment strengths $M_{nx} = 8$ ft-kips and $M_{ny} = 10$ ft-kips, each per foot width of slab. Determine the yield line pattern and the collapse uniform load w_u/ϕ.

Solution: (a) Yield line pattern. It has been demonstrated in Fig. 18.2.2(a) that a yield line should pass through the point of intersection of two nonparallel supported

(a) Uniformly loaded triangular slab

(b) The equilibrium method

Figure 18.7.2 Yield line analysis of triangular slab in Example 18.7.1.

edges. In this example, a positive moment yield line CD will break the slab into two segments ACD and BCD, wherein a common compatible deflection Δ of point D can be effected by rigid body rotations of the slab segment ACD about the supported edge AC and of the slab segment BCD about the supported edge BC.

(b) Equilibrium method of finding $BD = x$. The nodal forces V and $-V$ acting on the left and right slab segments of Fig. 18.7.2(b) are computed from Eq. (18.7.3); thus

$$V = M_{ny} \cot \alpha = (+10)\left(\frac{x}{12}\right) = \frac{5}{6} x \text{ kips}$$

Note that the nodal force acts upward in the obtuse angle and it acts downward in the acute angle. The equilibrium of moments about the edge AC of the left segment requires

$$\tfrac{1}{2}(w_u/\phi)(16 - x)(12)\tfrac{1}{3}(DE) = V(DE) + 96 \sin \theta + 10x \cos \theta$$

$$\tfrac{1}{2}(w_u/\phi)(16 - x)(12)(\tfrac{1}{3})(16 - x)(0.6) = \tfrac{5}{6}x(16 - x)(0.6) + 96(0.6) + 10x(0.8)$$

from which

$$\frac{w_u}{\phi} = \frac{576 + 160x - 5x^2}{12(16 - x)^2}$$

Similarly, the equilibrium of moments about the edge BC of the right segment requires

$$\frac{1}{2}(w_u/\phi)(x)(12)\left(\frac{BD}{3}\right) + V(BD) = 96$$

$$\frac{1}{2}\left(\frac{w_u}{\phi}\right)(x)(12)\left(\frac{x}{3}\right) + \frac{5x^2}{6} = 96$$

from which

$$\frac{w_u}{\phi} = \frac{576 - 5x^2}{12x^2}$$

Equating the two expressions for w_u/ϕ,

$$\frac{576 + 160x - 5x^2}{12(16 - x)^2} = \frac{576 - 5x^2}{12x^2}$$

$$x^2 + 14.4x - 115.2 = 0$$

$$x = 5.72 \text{ ft}$$

With the known value of x, the value of w_u/ϕ may be computed,

$$\frac{w_u}{\phi} = 1.048 \text{ kips/sq ft}$$

It can be shown that the same quadratic equation in x may be obtained using the virtual work method by following the procedure as illustrated in Section 18.5.

18.8 NODAL FORCES AT INTERSECTION OF THREE YIELD LINES

In Fig. 18.8.1 are shown three possible yield line patterns for an irregular quadri-lateral slab with four simply supported edges. The optimum positions of the yield lines in each pattern should be such as to give the lowest collapse load. These positions are defined by the locations of points a and b in Fig. 18.8.1(a), of points c and d in Fig. 18.8.1(b), and of point e in Fig. 18.8.1(c). In the equilibrium method the yield lines are characterized by the insertion of predetermined nodal forces. It is the object of this section to derive the expressions for the nodal forces at the intersection of three yield lines, such as at points a, b, c, or d in Fig. 18.8.1. The derivation shown below follows the works of Johansen [18.2, 18.3], and Jones and Wood [18.4, 18.5].

Assume that three yield lines 1–2–3 intersecting at a common point 0 are situated at angles $\phi_1 - \phi_2 - \phi_3$ measured counterclockwise from the positive x axis, as shown in Fig. 18.8.2(a). The nominal moment strengths under the yield lines 1–2–3 are, as shown in Figs. 18.8.2(b) and (c), $M_{nx1} - M_{nx2} - M_{nx3}$ provided by the reinforcement shown horizontally and $M_{ny1} - M_{ny2} - M_{ny3}$ provided by the reinforcement shown vertically, all reinforcement being near the lower face of the slab. The nodal forces V_{12}, V_{23}, and V_{31} are shown by the dots (which mean upward nodal forces) in Fig. 18.8.2(a). Note that for vertical equilibrium at the point of intersection, the sum of V_{12}, V_{23}, and V_{31} must be zero.

Figure 18.8.1 Yield line patterns of a simply supported quadrilateral slab.

Figure 18.8.2 Nodal forces at intersection of three yield lines.

It is important to emphasize at the outset that, by definition, the yield lines 1–2–3 are all situated at their respective optimal positions. The bending and twisting moments on any line passing through the point of intersection and deviating by an infinitesimal angle from a particular yield line should be equal to the bending and twisting moments on that line as provided by the orthogonal moment strengths.

Consider the equilibrium of an infinitesimal slab segment $0AB$ [Fig. 18.8.3(a)], bounded by a differential length $0A$ on yield line 3 and an arbitrary length $0B$ on

(a) Nodal force $V_{1-\Delta 3}$ between yield line 1 and $0A$

(b) Nodal force $V_{2-\Delta 3}$ between yield line 2 and $0A$

(c) Nodal force V_{12} between yield lines 1 and 2

Figure 18.8.3 Determination of nodal force V_{12} between yield lines 1 and 2.

yield line 1. Since BA is at a differential angle $\Delta\alpha$ from $B0$, the bending and twisting moments on both $B0$ and BA are those provided by M_{ny1} and M_{nx1} on their respective horizontal and vertical projections. Likewise the bending and twisting moments on the differential length $0A$ are those provided by M_{ny3} and M_{nx3} on its horizontal and vertical projections. Call the upward nodal force at point 0 inside triangle $B0A$ and bounded by yield line 1 and a different length on yield line 3 by the name $V_{1-\Delta3}$.

Next, write the equation of equilibrium for moments about AB as the axis of rotation for the slab segment $0AB$ in Fig. 18.8.3(a), noting that the moment of the uniform load on $0AB$ about any axis is a differential of the second order and may be neglected.

$$-V_{1-\Delta3}\, 0A \sin(\phi_3 - \phi_1 + \Delta\alpha)$$
$$+ (M_{ny3} - M_{ny1})\, 0A \cos\phi_3 \cos(\phi_1 - \Delta\alpha)$$
$$+ (M_{nx3} - M_{nx1})\, 0A \sin\phi_3 \sin(\phi_1 - \Delta\alpha) = 0$$

Solving the above equation for $V_{1-\Delta3}$ and letting $\Delta\alpha$ approach zero at the limit,

$$V_{1-\Delta3} = \frac{(M_{nx3} - M_{nx1})\sin\phi_3 \sin\phi_1 + (M_{ny3} - M_{ny1})\cos\phi_3 \cos\phi_1}{\sin(\phi_3 - \phi_1)} \qquad \textbf{(18.8.1)}$$

Making a similar analysis for the infinitesimal slab segment $0AC$ [Fig. 18.8.3(b)] bounded by a differential length $0A$ on yield line 3 and an arbitrary length $0C$ on yield line 2 gives the expression for the upward nodal force $V_{2-\Delta3}$ in Fig. 18.8.3(b) as

$$V_{2-\Delta3} = \frac{(M_{nx3} - M_{nx2})\sin\phi_3 \sin\phi_2 + (M_{ny3} - M_{ny2})\cos\phi_3 \cos\phi_2}{\sin(\phi_3 - \phi_2)} \qquad \textbf{(18.8.2)}$$

For vertical equilibrium at the point of intersection, the upward nodal force $V_{\Delta3-1}$ in Fig. 18.8.3(a) in the zone going counterclockwise from the differential length on yield line 3 to yield line 1 is

$$V_{\Delta3-1} = -V_{1-\Delta3} \qquad \textbf{(18.8.3)}$$

and, for the same reason, the upward nodal force V_{12} in Fig. 18.8.3(c) is

$$V_{12} = -V_{2-\Delta3} - V_{\Delta3-1} \qquad \textbf{(18.8.4)}$$

Substitution of Eqs. (18.8.1), (18.8.2), and (18.8.3) into Eq. (18.8.4) gives

$$V_{12} = \frac{(M_{nx3} - M_{nx1})\sin\phi_3 \sin\phi_1 + (M_{ny3} - M_{ny1})\cos\phi_3 \cos\phi_1}{\sin(\phi_3 - \phi_1)}$$
$$- \frac{(M_{nx3} - M_{nx2})\sin\phi_3 \sin\phi_2 + (M_{ny3} - M_{ny2})\cos\phi_3 \cos\phi_2}{\sin(\phi_3 - \phi_2)} \qquad \textbf{(18.8.5)}$$

Replacing each numerical subscript in Eq. (18.8.5) by its successor in the cyclic order of 1–2–3–1 (counterclockwise around the point of intersection) and then once more in the same manner, the following expressions for the upward nodal forces V_{23} and V_{31} as shown in Fig. 18.8.2(a) are obtained.

$$V_{23} = \frac{(M_{nx1} - M_{nx2}) \sin \phi_1 \sin \phi_2 + (M_{ny1} - M_{ny2}) \cos \phi_1 \cos \phi_2}{\sin (\phi_1 - \phi_2)}$$

$$- \frac{(M_{nx1} - M_{nx3}) \sin \phi_1 \sin \phi_3 + (M_{ny1} - M_{ny3}) \cos \phi_1 \cos \phi_3}{\sin (\phi_1 - \phi_3)} \qquad \textbf{(18.8.6)}$$

$$V_{31} = \frac{(M_{nx2} - M_{nx3}) \sin \phi_2 \sin \phi_3 + (M_{ny2} - M_{ny3}) \cos \phi_2 \cos \phi_3}{\sin (\phi_2 - \phi_3)}$$

$$- \frac{(M_{nx2} - M_{nx1}) \sin \phi_2 \sin \phi_1 + (M_{ny2} - M_{ny1}) \cos \phi_2 \cos \phi_1}{\sin (\phi_2 - \phi_1)} \qquad \textbf{(18.8.7)}$$

Equations (18.8.5), (18.8.6), and (18.8.7) are expressions for the upward nodal forces at the intersection of three yield lines.

Nodal Force at Intersection of Yield Line with Free Edge.   The nodal forces at the intersection of a yield line with a free edge, as shown by Fig. 18.8.4, may be obtained by substituting $\phi_1 = 0$, $\phi_2 = \alpha$, $\phi_3 = \pi$, $M_{nx1} = M_{nx3} = M_{ny1} = M_{ny3} = 0$, $M_{nx2} = M_{nx}$, and $M_{ny2} = M_{ny}$ in Eqs. (18.8.5) to (18.8.7); thus

$$V_{12} = -\frac{(-M_{ny2})(-1) \cos \alpha}{\sin (\pi - \alpha)} = -M_{ny} \cot \alpha$$

$$V_{23} = \frac{(-M_{ny2})(+1) \cos \alpha}{\sin (0 - \alpha)} = +M_{ny} \cot \alpha$$

$$V_{31} = \frac{(M_{ny2}) \cos \alpha (-1)}{\sin (\alpha - \pi)} - \frac{(M_{ny2}) \cos \alpha (+1)}{\sin (\alpha - 0)} = 0$$

The above results check with the findings in Section 18.7.

Nodal Forces at Intersection of Three Yield Lines Having Identical M_{nx} and $\bar{M}_{ny}$ Nominal Moment Strengths.   From Eqs. (18.8.5) to (18.8.7) it can be observed that wherever the nominal moment strengths under three intersecting yield lines are identical (that is, $M_{nx1} = M_{nx2} = M_{nx3} = M_{nx}$ and $M_{ny1} = M_{ny2} = M_{ny3} = M_{ny}$) the nodal forces at the intersection are zero. This fact is of great convenience when using the equilibrium method for the yield line analysis of two-way rectangular slabs.

Figure 18.8.4 Nodal forces at intersection of yield line with free edge.

18.9 YIELD LINE ANALYSIS OF RECTANGULAR TWO-WAY SLABS

A typical rectangular two-way slab panel shown in Fig. 18.9.1 has two-way reinforcement within the panel near the bottom face providing positive moment nominal strengths M_{npx} and M_{npy}, and it also has two-way reinforcement along the edges near the top face providing negative moment nominal strengths M_{nnx} and M_{nny}; these strengths are per unit width of slab. The uniform load to give the collapse condition based on the yield line theory may be determined in terms of the sides a and b, and the absolute values of M_{npx}, M_{npy}, M_{nnx}, and M_{nny}.

Yield Line Pattern.   Three possible yield line patterns are shown in Fig. 18.9.2. There is no unknown position in yield line pattern No. 1 of Fig. 18.9.2(a); consequently the nodal forces V need not be predetermined, and their value is dictated by statics alone. The unknowns x and y in yield line patterns Nos. 2 and 3 of Figs. 18.9.2(b) and (c) must be determined by means of differential calculus in the virtual work method; but for the equilibrium method, in this particular case, the nodal forces to define the yield lines are all zero because the moment strengths under a set of three intersecting yield lines are identical.

Analysis for Yield Pattern No. 1.   Assuming a vertical deflection of Δ at the intersection of the diagonal yield lines in Fig. 18.9.3, the deflection at the centroids of the four triangles $A-B-C-D$ is $\Delta/3$. The work done at the collapse condition by the

(a) Dimensions   (b) Top reinforcement   (c) Bottom reinforcement

Figure 18.9.1 A rectangular two-way slab panel.

(a) Yield pattern No. 1   (b) Yield pattern No. 2   (c) Yield pattern No. 3

Figure 18.9.2 Yield line patterns for a rectangular two-way slab panel.

Figure 18.9.3 Analysis for yield pattern No. 1.

uniform load is the product of the total load on the entire panel and $\Delta/3$; thus

$$W = \frac{w_u}{\phi} \, ab\left(\frac{\Delta}{3}\right) \tag{18.9.1}$$

The work done by the yield moments on the boundaries of all four slab segments, referring to Fig. 18.9.3, is

$$W = 2(M_{nny} + M_{npy})(a)\left(\frac{2\Delta}{b}\right) + 2(M_{nnx} + M_{npx})(b)\left(\frac{2\Delta}{a}\right) \tag{18.9.2}$$

Equating Eq. (18.9.1) to Eq. (18.9.2) and solving for w_u,

$$\frac{w_u}{\phi} = 12\left(\frac{M_{nnx} + M_{npx}}{a^2} + \frac{M_{nny} + M_{npy}}{b^2}\right) \tag{18.9.3}$$

Taking moments about the lower edge of slab segment A in Fig. 18.9.3,

$$\frac{1}{2}\left(\frac{w_u}{\phi}\right)a\left(\frac{b}{2}\right)\left(\frac{b}{6}\right) + V\left(\frac{b}{2}\right) = (M_{nny} + M_{npy})(a) \tag{18.9.4}$$

Taking moments about the left edge of slab segment D in Fig. 18.9.3,

$$\frac{1}{2}\left(\frac{w_u}{\phi}\right)b\left(\frac{a}{2}\right)\left(\frac{a}{6}\right) = (M_{nnx} + M_{npx})(b) + V\left(\frac{a}{2}\right) \tag{18.9.5}$$

Eliminating V between Eqs. (18.9.4) and (18.9.5) and solving for w_u/ϕ, the same expression for w_u/ϕ as Eq. (18.9.3) is obtained.

Analysis for Yield Pattern No. 2. Assuming a vertical deflection of Δ at the two points of intersection of the yield lines in Fig. 18.9.4, the work done at the collapse condition by the uniform load on the entire panel is

$$W = 2W_D + 2W_{A1} + 4W_{A2}$$

$$= 2\left[\frac{1}{2}\left(\frac{w_u}{\phi}\right)bx\right]\left(\frac{\Delta}{3}\right) + 2\left(\frac{w_u}{\phi}\right)(a - 2x)\left(\frac{b}{2}\right)\left(\frac{\Delta}{2}\right) + 4\left[\frac{1}{2}\left(\frac{w_u}{\phi}\right)x\frac{b}{2}\right]\left(\frac{\Delta}{3}\right)$$

$$= \frac{w_u}{\phi}\left(\frac{\Delta}{6}\right)(3ab - 2bx) \tag{18.9.6}$$

Figure 18.9.4 Analysis for yield line pattern No. 2.

The work done by the yield moments on the boundaries of all four slab segments is, referring to Fig. 18.9.4,

$$W = 2(M_{nny} + M_{npy})(a)\left(\frac{2\Delta}{b}\right) + 2(M_{nnx} + M_{npx})(b)\left(\frac{\Delta}{x}\right) \qquad \textbf{(18.9.7)}$$

Equating Eq. (18.9.6) to Eq. (18.9.7) and solving for w_u/ϕ,

$$\frac{w_u}{\phi} = \frac{12[b^2(M_{nnx} + M_{npx}) + 2ax(M_{nny} + M_{npy})]}{b^2(3ax - 2x^2)} \qquad \textbf{(18.9.8)}$$

Setting to zero the derivative of Eq. (18.9.8) with respect to x gives the quadratic equation in x,

$$4a(M_{nny} + M_{npy})x^2 + 4b^2(M_{nnx} + M_{npx})x - [3ab^2(M_{nnx} + M_{npx})] = 0 \qquad \textbf{(18.9.9)}$$

Taking moments about the lower edge of slab segment A in Fig. 18.9.4,

$$2\left[\frac{1}{2}\left(\frac{w_u}{\phi}\right)x\frac{b}{2}\right]\left(\frac{b}{6}\right) + \frac{w_u}{\phi}(a - 2x)\left(\frac{b}{2}\right)\left(\frac{b}{4}\right) = (M_{nny} + M_{npy})(a)$$

$$\frac{w_u}{\phi} = \frac{24a(M_{nny} + M_{npy})}{2b^2x + 3b^2(a - 2x)} \qquad \textbf{(18.9.10)}$$

Taking moments about the left edge of slab segment D in Fig. 18.9.4,

$$\frac{1}{2}\left(\frac{w_u}{\phi}\right)bx\left(\frac{x}{3}\right) = (M_{nnx} + M_{npx})(b)$$

$$\frac{w_u}{\phi} = \frac{6(M_{nnx} + M_{npx})}{x^2} \qquad \textbf{(18.9.11)}$$

Equating Eq. (18.9.10) to Eq. (18.9.11) gives the same quadratic equation in x as Eq. (18.9.9).

The condition for $x = a/2$ in Eq. (18.9.9) can be shown to be

$$\frac{M_{nnx} + M_{npx}}{M_{nny} + M_{npy}} = \frac{a^2}{b^2} \qquad \text{for } x = \frac{a}{2} \qquad \textbf{(18.9.12)}$$

which means that if the sum of positive and negative moment reinforcement in the a direction, each per unit width of slab, is equal to (a^2/b^2) times the sum of positive and negative moment reinforcement in the b direction, each per unit width of slab, yield pattern No. 1 prevails.

The condition for $x < a/2$ in Eq. (18.9.9) can be shown to be

$$\frac{M_{nnx} + M_{npx}}{M_{nny} + M_{npy}} < \frac{a^2}{b^2} \qquad \text{for } x < \frac{a}{2} \qquad (18.9.13)$$

which means that in order for yield pattern No. 2 to prevail, the reinforcement in the a direction is less than that for yield pattern No. 1 to control.

Analysis for Yield Pattern No. 3.   By interchanging the subscripts x and y as well as the quantities a and b in Eqs. (18.9.8), (18.9.9), (18.9.10), and (18.9.11), the following equations applicable to yield line pattern No. 3 are obtained. The quadratic equation in y (Fig. 18.9.2) is

$$4b(M_{nnx} + M_{npx})y^2 + 4a^2(M_{nny} + M_{npy})y - [3ba^2(M_{nny} + M_{npy})] = 0 \qquad (18.9.14)$$

Three expressions for w_u/ϕ in terms of y are

$$\frac{w_u}{\phi} = \frac{12[a^2(M_{nny} + M_{npy}) + 2by(M_{nnx} + M_{npx})]}{a^2(3by - 2y^2)} \qquad (18.9.15)$$

$$\frac{w_u}{\phi} = \frac{24b(M_{nnx} + M_{npx})}{2a^2y + 3a^2(b - 2y)} \qquad (18.9.16)$$

$$\frac{w_u}{\phi} = \frac{6(M_{nny} + M_{npy})}{y^2} \qquad (18.9.17)$$

The condition for $y < b/2$ in Eq. (18.9.14) can be shown to be

$$\frac{M_{nnx} + M_{npx}}{M_{nny} + M_{npy}} > \frac{a^2}{b^2} \qquad \text{for } y < \frac{b}{2} \qquad (18.9.18)$$

which means that in order for yield pattern No. 3 to prevail, the reinforcement in the a direction is more than that for yield pattern No. 1 to control.

■ **EXAMPLE 18.9.1**   Determine the controlling yield line pattern and the corresponding collapse condition uniform load for a rectangular two-way slab panel with dimensions as shown in Fig. 18.9.5(a). The slab has reinforcement in the top near the edges and in the bottom within the panel. Obtain solutions for the following three cases:

1. $M_{nnx} + M_{npx} = 6.25$ ft-kips/ft $M_{nny} + M_{npy} = 4$ ft-kips/ft
2. $M_{nnx} + M_{npx} = 2$ ft-kips/ft $M_{nny} + M_{npy} = 4$ ft-kips/ft
3. $M_{nnx} + M_{npx} = 8$ ft-kips/ft $M_{nny} + M_{npy} = 4$ ft-kips/ft

Solution: **(a)** Case 1. The applicable yield line pattern may be determined by comparing the ratio of $(M_{nnx} + M_{npx})$ to $(M_{nny} + M_{npy})$ with the ratio of a^2 to b^2. In this case,

$$\frac{(M_{nnx} + M_{npx})}{(M_{nny} + M_{npy})} = \frac{6.25}{4} = 1.5625, \qquad \frac{a^2}{b^2} = \frac{625}{400} = 1.5625$$

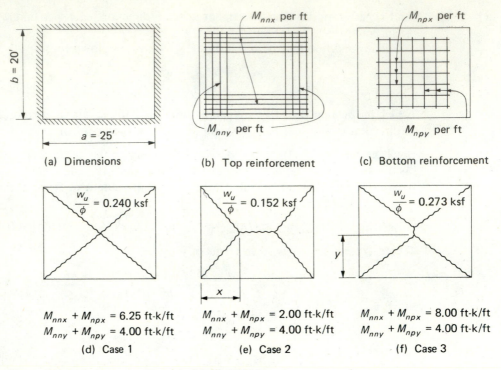

Figure 18.9.5 Rectangular two-way slab of Example 18.9.1.

Since the ratio of reinforcement in direction a to that in direction b per foot slab width is equal to the ratio of a^2 to b^2, the yield pattern is shown in Fig. 18.9.5(d). Then from Eq. (18.9.3),

$$\frac{w_u}{\phi} = 12\left(\frac{M_{nnx} + M_{npx}}{a^2} + \frac{M_{nny} + M_{npy}}{b^2}\right)$$

$$= 12\left(\frac{6.25}{625} + \frac{4}{400}\right) = 0.240 \text{ ksf}$$

(b) Case 2. The ratio of $(M_{nnx} + M_{npx})$ to $(M_{nny} + M_{npy})$ is, in this case,

$$\frac{(M_{nnx} + M_{npx})}{(M_{nny} + M_{npy})} = \frac{2}{4} = 0.5 < \frac{a^2}{b^2} = 1.5625$$

The yield line pattern is as shown in Fig. 18.9.5(e). The quadratic equation (18.9.9) is used to solve for x.

$$4a(M_{nny} + M_{npy})x^2 + 4b^2(M_{nnx} + M_{npx})x - 3ab^2(M_{nnx} + M_{npx}) = 0$$

$$4(25)(4)x^2 + 4(400)(2)x - 3(25)(400)(2) = 0$$

$$x^2 + 8x - 150 = 0$$

$$x = \sqrt{166} - 4 = 8.884 \text{ ft}$$

The same uniform load w_u/ϕ is obtained from Eqs. (18.9.8), (18.9.10), or (18.9.11); the fact that it is so serves as a check on the numerical computation.

$$\frac{w_u}{\phi} = \frac{12[b^2(M_{nnx} + M_{npx}) + 2ax(M_{nny} + M_{npy})]}{b^2(3ax - 2x^2)}$$

$$= \frac{12[400(2) + 2(25)(8.884)(4)]}{400[3(25)(8.884) - 2(8.884)^2]} = 0.152 \text{ ksf}$$

$$\frac{w_u}{\phi} = \frac{24a(M_{nny} + M_{npy})}{2b^2x + 3b^2(a - 2x)}$$

$$= \frac{24(25)(4)}{2(400)(8.884) + 3(400)[25 - 2(8.884)]} = 0.152 \text{ ksf}$$

$$\frac{w_u}{\phi} = \frac{6(M_{nnx} + M_{npx})}{x^2} = \frac{6(2)}{(8.884)^2} = 0.152 \text{ ksf}$$

(c) Case 3. The ratio of $(M_{nnx} + M_{npx})$ to $(M_{nny} + M_{npy})$ is, in this case,

$$\frac{(M_{nnx} + M_{npx})}{(M_{nny} + M_{npy})} = \frac{8}{4} = 2 > \frac{a^2}{b^2} = 1.5625$$

The yield line pattern is as shown in Fig. 18.9.5(f). The quadratic equation (18.9.14) is used to solve for y.

$$4b(M_{nnx} + M_{npx})y^2 + 4a^2(M_{nny} + M_{npy})y - 3ba^2(M_{nny} + M_{npy}) = 0$$

$$4(20)(8)y^2 + 4(625)(4)y - 3(20)(625)(4) = 0$$

$$8y^2 + 125y - 1875 = 0$$

$$y = 9.375 \text{ ft}$$

The same uniform load w_u/ϕ is obtained from Eqs (18.9.15), (18.9.16), or (18.9.17); the fact that it is so serves as a check on the numerical computation.

$$\frac{w_u}{\phi} = \frac{12[a^2(M_{nny} + M_{npy}) + 2by(M_{nnx} + M_{npx})]}{a^2(3by - 2y^2)}$$

$$= \frac{12[625(4) + 2(20)(9.375)(8)]}{625[3(20)(9.375) - 2(9.375)^2]} = 0.273 \text{ ksf}$$

$$\frac{w_u}{\phi} = \frac{24b(M_{nnx} + M_{npx})}{2a^2y + 3a^2(b - 2y)}$$

$$= \frac{24(20)(8)}{2(625)(9.375) + 3(625)[20 - 2(9.375)]} = 0.273 \text{ ksf}$$

$$\frac{w_u}{\phi} = \frac{6(M_{nny} + M_{npy})}{y^2} = \frac{6(4)}{(9.375)^2} = 0.273 \text{ ksf} \quad \blacksquare$$

18.10 CORNER EFFECTS IN RECTANGULAR SLABS

In Section 18.4 on method of yield line analysis, it has been stated that there may be more than one possible yield line pattern, in which case solutions to all possible yield line patterns must be sought and the one giving the smallest collapse load would actually happen and thus should be used in design. Although the three typical yield line patterns for a rectangular two-way slab panel have been shown in Fig. 18.9.2 and their analysis has been completely treated in Section 18.9, it can be demonstrated that the corner yield patterns 4–5–6 shown in Fig. 18.10.1—in one-to-one correspondence to yield patterns 1–2–3 of Fig. 18.9.2—may indeed give a smaller collapse load and therefore control. These corner patterns are complicated to analyze, either by virtual work method or by equilibrium method. For instance, there are three unknowns E–F–G for the yield line positions; then, once the expression for w_u/ϕ is obtained from the virtual work equation as a function of three independent variables, the partial derivative of w_u/ϕ with respect to each of the three unknown variables can be equated to zero. In the equilibrium method, the same set of equations for the positions of points E–F–G may be obtained by inserting the predetermined zero or nonzero nodal forces and applying the moment equation of equilibrium to each of the slab segments.

An analysis [18.6] of a square slab with equal reinforcement in the x and y directions will show that the corner yield pattern No. 4 of Fig. 18.10.2(b) [see also Fig. 18.10.1(a)] results in $w_u/\phi = 22(M_{nn} + M_{np})/a^2$, whereas the regular yield pattern of Fig. 18.10.2(a) indicates $w_u/\phi = 24(M_{nn} + M_{np})/a^2$. M_{nn} and M_{np} are the nominal moment strengths per unit slab width for the negative moment and positive moment regions, respectively, in each direction, and a is the side of the square. Thus the corner pattern is more critical by approximately $(24 - 22)/24 = 8.3\%$. It may be proper then to discount the results of a regular yield pattern analysis as made in Section 8.9 for most rectangular slabs by 8 to 10% for reason of corner effects.

It may be pointed out that the yield line EF in Fig. 18.10.2(b) is a negative moment yield line; thus when there is no negative reinforcement, the moment strength along EF is zero. In this case the crack or yield line EF will not form if the corner A is not held down because the corner would simply lift up. ACI-13.4.6 requires the provision of special reinforcement at exterior corners in both top and bottom of the slab, for a distance in each direction from the corner equal to one-fifth the longer span. The use of negative moment reinforcement near the corner tends to move the

(a) Yield pattern No. 4   (b) Yield pattern No. 5   (c) Yield pattern No. 6

Figure 18.10.1 Corner yield patterns for a rectangular two-way slab panel.

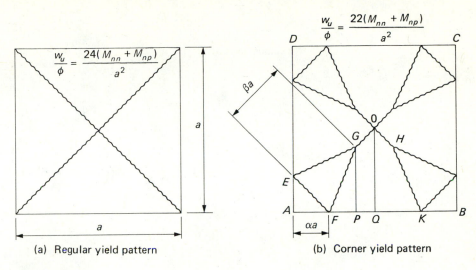

(a) Regular yield pattern   (b) Corner yield pattern

Figure 18.10.2 Square slab panel with equal reinforcement in two directions.

point G farther away from the corner and thus such reinforcement helps to increase the load that will cause a collapse mechanism.

18.11 APPLICATION OF YIELD LINE ANALYSIS TO SPECIAL CASES

The yield line theory of slabs, as has been developed and illustrated in the preceding sections, is particularly suitable for special cases involving irregular shapes or irregular boundary conditions. Prerequisite to the analysis of these cases is the picturing of an applicable yield line pattern. The governing concept here is that rigid body plane rotations of slab segments separated at yield lines are possible under compatible deflection conditions. To this end, the following guides may be provided:

1. Yield lines end at a slab boundary.
2. A yield line (or its prolongation) between two slab segments passes through the intersection of the axes of rotation of the two adjacent slab segments.
3. The axes of rotation lie along lines of supports or pass over column supports.

In addition to those already described, two other yield line patterns to further illustrate the use of the above guides are shown in Fig. 18.11.1.

Special Case.   Shown in Fig. 18.11.2 is a rectangular slab simply supported at three edges and free at the upper edge. The positive moment reinforcement parallel to the a dimension provides a nominal moment strength of M_{nx} per unit of the b distance; and the positive moment reinforcement parallel to the b dimension provides strength M_{ny} per unit of the a distance. Two possible yield patterns are shown in Figs. 18.11.2(c) and (d); the unknown is x in yield pattern No. 1 and it is y in yield pattern No. 2.

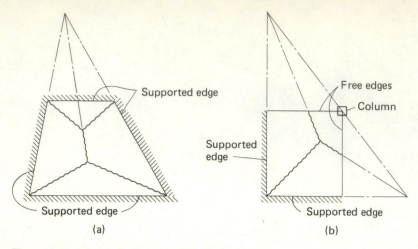

(a)   (b)

Figure 18.11.1 Guides for yield line patterns in special cases.

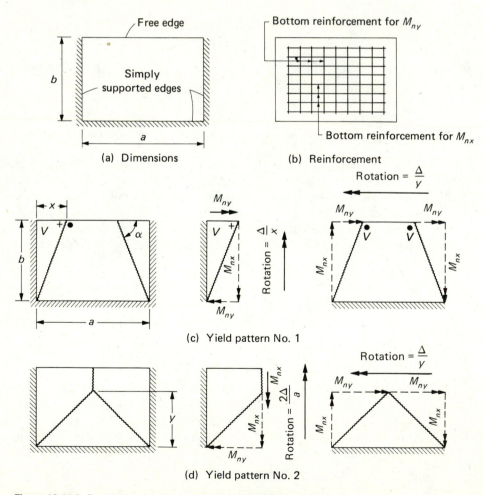

(a) Dimensions   (b) Reinforcement

(c) Yield pattern No. 1

(d) Yield pattern No. 2

Figure 18.11.2 Rectangular slab simply supported at three edges and free at one edge.

For yield pattern No. 1, referring to Fig. 18.11.2(c) and letting Δ be the deflection where the yield line meets the free edge, the virtual work done by the uniform load is

$$W = \frac{1}{2}\left(\frac{w_u}{\phi}\right)bx\left(\frac{\Delta}{3}\right)2 + \frac{1}{2}\left(\frac{w_u}{\phi}\right)xb\left(\frac{\Delta}{3}\right)2 + \frac{w_u}{\phi}(a - 2x)b\,\frac{\Delta}{2}$$

$$= \frac{w_u}{\phi}(\Delta)\left(\frac{b}{6}\right)(3a - 2x)$$

The virtual work done by the yield line moments is

$$W = 2M_{nx}b\left(\frac{\Delta}{x}\right) + 2M_{ny}x\left(\frac{\Delta}{b}\right)$$

Equating the two expressions for W and solving for w_u/ϕ,

$$\frac{w_u}{\phi} = \frac{12(b^2 M_{nx} + x^2 M_{ny})}{b^2 x(3a - 2x)} \tag{18.11.1}$$

Setting to zero the derivative of Eq. (18.11.1) with respect to x,

$$3a\left(\frac{M_{ny}}{M_{nx}}\right)x^2 + 4b^2 x - 3ab^2 = 0 \tag{18.11.2a}$$

In order that the root of Eq. (18.11.2a) be less than $a/2$,

$$\left(\frac{M_{ny}}{M_{nx}}\right) > \frac{4b^2}{3a^2} \tag{18.11.2b}$$

The nodal force V in Fig. 18.11.2(c) may be predetermined by use of Eq. (18.7.3); or

$$V = M_{ny}\cot\alpha = M_{ny}\left(\frac{x}{b}\right)$$

For equilibrium of the triangular segment in Fig. 18.11.2(c),

$$\frac{1}{2}\left(\frac{w_u}{\phi}\right)bx\left(\frac{x}{3}\right) + M_{ny}\left(\frac{x}{b}\right)x = M_{nx}b$$

from which

$$\frac{w_u}{\phi} = \frac{6(b^2 M_{nx} - x^2 M_{ny})}{b^2 x^2} \tag{18.11.3}$$

For equilibrium of the trapezoidal segment in Fig. 18.11.2(c),

$$\frac{1}{2}\left(\frac{w_u}{\phi}\right)xb\left(\frac{b}{3}\right)2 + \frac{w_u}{\phi}(a - 2x)b\left(\frac{b}{2}\right) = 2M_{ny}x + 2M_{ny}\left(\frac{x}{b}\right)b$$

from which

$$\frac{w_u}{\phi} = \frac{24x M_{ny}}{b^2(3a - 4x)} \tag{18.11.4}$$

The same quadratic equation in x as Eq. (18.11.2) is obtained from equating Eq. (18.11.3) to Eq. (18.11.4).

For yield pattern No. 2, referring to Fig. 18.11.2(d) and letting Δ be the deflection at the yield line perpendicular to the free edge, the virtual work done by the uniform load is

$$W = \frac{1}{2}\left(\frac{w_u}{\phi}\right)y\left(\frac{a}{2}\right)\left(\frac{\Delta}{3}\right)2 + \frac{w_u}{\phi}(b-y)\left(\frac{a}{2}\right)\left(\frac{\Delta}{2}\right)2 + \frac{1}{2}\left(\frac{w_u}{\phi}\right)ay\left(\frac{\Delta}{3}\right)$$

$$= \frac{w_u}{\phi}(\Delta)\left(\frac{a}{6}\right)(3b - y)$$

The virtual work done by the yield line moments is

$$W = 2M_{nx}b\left(\frac{2\Delta}{a}\right) + M_{ny}a\left(\frac{\Delta}{y}\right)$$

Equating the two expressions for W and solving for w_u/ϕ,

$$\frac{w_u}{\phi} = \frac{6(4byM_{nx} + a^2M_{ny})}{a^2y(3b - y)} \tag{18.11.5}$$

Setting to zero the derivative of Eq. (18.11.5) with respect to y,

$$4by^2 + 2a^2\left(\frac{M_{ny}}{M_{nx}}\right)y - 3a^2b\left(\frac{M_{ny}}{M_{nx}}\right) = 0 \tag{18.11.6a}$$

In order that the root of Eq. (18.11.6a) be less than b,

$$\left(\frac{M_{ny}}{M_{nx}}\right) < \frac{4b^2}{a^2} \tag{18.11.6b}$$

Since the moment strengths at the three intersecting yield lines in Fig. 18.11.2(d) are identical, the nodal forces are all zero, based on the treatment presented in Section 18.8. For equilibrium of the trapezoidal segment in Fig. 18.11.2(d),

$$\frac{1}{2}\left(\frac{w_u}{\phi}\right)y\left(\frac{a}{2}\right)\frac{a}{6} + \frac{w_u}{\phi}(b-y)\left(\frac{a}{2}\right)\frac{a}{4} = M_{nx}b$$

from which

$$\frac{w_u}{\phi} = \frac{24bM_{nx}}{a^2(3b - 2y)} \tag{18.11.7}$$

For equilibrium of the triangular segment in Fig. 18.11.2(d),

$$\frac{1}{2}\left(\frac{w_u}{\phi}\right)ay\frac{y}{3} = M_{ny}a$$

from which

$$\frac{w_u}{\phi} = \frac{6M_{ny}}{y^2} \tag{18.11.8}$$

The same quadratic equation in y as Eq. (18.11.6) is obtained from equating Eq. (18.11.7) to Eq. (18.11.8).

Thus Eqs. (18.11.1) to (18.11.4) apply to yield pattern No. 1 which cannot happen if (M_{ny}/M_{nx}) is smaller than $4b^2/(3a^2)$, and Eqs. (18.11.5) to (18.11.8) apply to yield

Table 18.11.1 Yield Line Analysis of a Rectangular Slab with Three Supported Edges and One Free Edge (Fig. 18.11.2); $a = 25$ ft, $b = 20$ ft

CASE	1	2	3	4	5	6
M_{ny} (ft-kips/ft)	16	16	16	16	16	16
M_{nx} (ft-kips/ft)	24	18.75	16	8	6.25	4
Yield pattern No. 1						
x (ft), Eq. (18.11.2)	—	12.5	12	9.78	9.01	7.68
w (ksf), Eqs. (18.11.1), (18.11.3), or (18.11.4)	—	0.480	0.427	0.262	0.222	0.167
Yield pattern No. 2						
y (ft), Eq. (18.11.6)	13.22	14.41	15.20	18.75	20	—
w (ksf), Eqs. (18.11.5), (18.11.7), or (18.11.8)	0.549	0.462	0.415	0.273	0.240	—
Yield pattern controlling	No. 2	No. 2	No. 2	No. 1	No. 1	No. 1
Collapse load w_u/ϕ (ksf)	0.549	0.462	0.415	0.262	0.222	0.167

pattern No. 2 which cannot happen if (M_{ny}/M_{nx}) is larger than $4b^2/a^2$. When (M_{ny}/M_{nx}) is between $4b^2/(3a^2)$ and $4b^2/a^2$, analysis for both yield patterns should be made and the one giving the smaller collapse load controls. Although an exact value of (M_{ny}/M_{nx}) in terms of b^2/a^2 (between 1.33 and 4.00) at the transition point where yield pattern No. 1 begins to control over yield pattern No. 2 may be determined, the complication in the algebra does not seem to warrant the effort. Table 18.11.1 shows the results of analysis for six different cases in which the values of (M_{ny}/M_{nx}) become progressively larger while keeping M_{ny} constant.

SELECTED REFERENCES

18.1. A. Ingerslev. "The Strength of Rectangular Slabs," *Journal of Institute of Structural Engineers*, London, **1**, (1) January 1923, 3–14; Disc., 14–19.

18.2. K. W. Johansen. "The Ultimate Strength of Reinforced Concrete Slabs," *Final Report*. Third Congress, International Association for Bridge and Structural Engineering, Liège, Belgium. September 1948 (pp. 565–570).

18.3. K. W. Johansen. *Yield Line Theory*. London: Cement and Concrete Association, 1962.

18.4. L. L. Jones and R. H. Wood. *Yield Line Analysis of Slabs*. New York: Elsevier, 1967.

18.5. R. H. Wood. "Plastic Design of Slabs using Equilibrium Methods," *Flexural Mechanics of Reinforced Concrete*. Proceedings of the International Symposium, Miami, 1964 (pp. 319–336).

18.6. E. Hognestad. "Yield Line Theory for the Ultimate Flexural Strength of Reinforced Concrete Slabs," *ACI Journal, Proceedings*, **49**, March 1953, 637–656.

PROBLEMS

18.1 Assuming a 6 in. slab thickness for the continuous slab span described in Example 18.5.1, the uniform load w_u due to the weight of slab itself may be considered to be $1.4w_D = 1.4(75) = 105$ psf. If the slab supports a transverse wall at 7 ft from the left support line, determine the maximum wall load per transverse ft that will cause a collapse mechanism to occur.

18.2 Solve Example 18.7.1, but use $M_{nx} = 10$ ft-kips and $M_{ny} = 8$ ft-kips, each per foot width of slab.

18.3 For the regular yield pattern solution to Case 1 of Example 18.9.1, investigate the effect of a corner yield pattern in which the negative moment yield line intersects the edges at 5 ft and 4 ft from the corner along the 25 ft and 20 ft edges, respectively.

18.4 Same as Problem 18.3, except for Case 2 of Example 18.9.1.

18.5 Same as Problem 18.3, except for Case 3 of Example 18.9.1.

18.6 Verify the solution for Case 1 in Table 18.11.1.

18.7 Verify the solution for Case 2 in Table 18.11.1.

18.8 Verify the solution for Case 3 in Table 18.11.1.

18.9 Verify the solution for Case 4 in Table 18.11.1.

18.10 Verify the solution for Case 5 in Table 18.11.1.

18.11 Verify the solution for Case 6 in Table 18.11.1.

CHAPTER 19

Torsion

19.1 GENERAL

Reinforced concrete members may be subjected to torsion, frequently in combination with bending and shear. The cantilever member in Fig. 19.1.1(a) is largely subjected to torsion, although some bending and shear also exist due to its own weight. The fixed-ended beam of Fig. 19.1.1(b) is subjected to substantial amounts of bending, shear, and torsion.

Spandrel beams at the edge of a building built integrally with the floor slab are subjected not only to transverse loads but also to a torsional moment per unit length equal to the restraining moment at the exterior end of the slab. Similarly, spandrel girders receive torsional moments from the exterior ends of the floor beams that frame into them.

Torsion on structural systems may be classified into two types: (1) *Statically determinate torsion* (sometimes called "equilibrium torsion"), for which the torsion *can* be determined from statics alone; and (2) *statically indeterminate torsion* (sometimes called "compatibility torsion"), for which the torsion *cannot* be determined

(a)   (b)

Figure 19.1.1 Reinforced concrete members subjected to torsion.

Inclined cracks due to torsion; test by J. P. Klus at the University of Wisconsin, Madison.

from statics and a rotation (twist) is required for deformational compatibility between interconnecting elements, such as a spandrel beam, slab, or column. Both examples in Fig. 19.1.1 are cases of statically determinate torsion.

In cases of statically determinate torsion, as in Figs. 19.1.2(a) and (b), the amount of torsion that the member is required to resist is based on the requirement of statics and is independent of the stiffness of the member. Statically indeterminate torsion, as shown in Figs. 19.1.2(c) and (d), exists in some situations where there would be no torsion if the statical indeterminacy were eliminated. For instance, if the support at A is eliminated in Fig. 19.1.2(c), the torsion is eliminated. Similarly, in Fig. 19.1.2(d), if a flexural hinge is put at B, the torsion is eliminated. For such statically indeterminate torsion situations the amount of torsion in a member depends on the magnitude of the torsional stiffness of the member itself in relation to the stiffnesses of the interconnecting members.

For an excellent overview of the entire subject of torsional phenomena as it affects structures, with specific reference to reinforced concrete design, the reader is referred to Tamberg and Mikluchin [19.3]. An extensive bibliography is also included.

Figure 19.1.2 Comparison of statically determinate torsion (cases *a* and *b*) and statically indeterminate torsion (cases *c* and *d*). (Structures shown in plan view with load *P* at 90° to plane of frame.)

The most extensive treatment of torsion for reinforced and prestressed concrete is the 1983 book by Hsu [19.12]. Collins and Mitchell [19.9, 19.18] have provided a unified rational treatment of the function of reinforcement to resist shear and torsion, and have made design recommendations for reinforced and prestressed concrete. A number of general references are available [19.1–19.18].

In this chapter, brief treatment is given to the computation of torsional stress and torsional rigidity of homogeneous sections, the torsional strength of reinforced concrete sections, the development and background for the ACI Code requirements, design examples, and the effect of torsional stiffness in a continuity analysis.

19.2 TORSIONAL STRESS IN HOMOGENEOUS SECTIONS

A torsional moment T acting on a shaft of homogeneous material as shown in Fig. 19.2.1 causes shear stresses v over the cross section.

Circular Sections.   For a circular section, a plane transverse section before twisting remains plane after twisting. Consequently, the resultant shear stress v at any point is proportional to its distance from the center and is in a direction perpendicular to the radius. Calling h the diameter of the circle [Fig. 19.2.1(b)] and v_t the maximum torsional shear stress at the perimeter,

$$T = \int_A rv\, dA = \int_A r\left(\frac{2v_t}{h}r\right)dA = \frac{2v_t}{h}\int_A r^2\, dA = \frac{2C}{h}v_t$$

in which C, the polar moment of inertia, is

$$C = \int_A r^2\, dA = \int_0^{h/2} r^2 2\pi r\, dr = \left(\frac{2\pi r^4}{4}\right)_0^{h/2} = \frac{\pi h^4}{32}$$

The torsional shear stress v_t becomes

$$v_t = \frac{16T}{\pi h^3} \tag{19.2.1}$$

shear stress
due to torsion v_t

(a)

$dV = v\, dA$

r

$h/2$

(b)

y

x

(c)

Figure 19.2.1 Torsional stress in homogeneous sections.

Table 19.2.1

y/x	1.0	1.2	1.5	2.0	2.5	3	5	∞
α	0.208	0.219	0.231	0.246	0.256	0.267	0.290	0.333

Rectangular Sections.   The torsional shear stress distribution over a rectangular section of dimension x by y cannot be as easily derived as for a circular section. Unlike the circular section where plane transverse sections remain plane after twisting, the noncircular cross section warps under torsion. If plane sections were maintained after twisting, the maximum shear stress would exist at a point farthest from the axis of twist. Such is not the case for rectangular sections. From the mathematical theory of elasticity [9.1, pp. 275–288], it has been found that the maximum torsional shear stress v_t occurs at the midpoint of the long side and parallel to it. The magnitude of v_t is a function of the ratio of y to x (long to short sides) [Fig. 19.2.1(c)]; and

$$v_t = \frac{T}{\alpha x^2 y} \tag{19.2.2}$$

The values for α are given in Table 19.2.1.

T-, L-, and I-Sections.   The torsional shear stress distribution in T-, L-, or I-sections may be approximated by dividing the section into several component rectangles assuming that each component rectangle has a large ratio y/x so that the value for α may be assumed to be $\frac{1}{3}$ [9.1]. The maximum shear stress v_t occurs at the midpoint of the long side y of the rectangle having the greatest thickness x_m and

$$v_t = \frac{Tx_m}{\Sigma \frac{1}{3} x^3 y} \tag{19.2.3}$$

in which x and y are the thickness and side, respectively, of each component rectangle. Since the web of the sections considered is usually thicker than the flange, x_m will usually be the web thickness.

19.3 TORSIONAL STIFFNESS OF HOMOGENEOUS SECTIONS

The torsional stiffness K_t of a member is defined as the ratio of torsional moment T to the angle of twist θ in the length L. The torsional rigidity is usually represented by the symbol GC in which G is the modulus of elasticity in shear and C is the torsion constant. Thus if θ is the total angle of twist in a length L,

$$K_t = \frac{T}{\theta} = \frac{GC}{L} \tag{19.3.1}$$

Circular Sections.   It has been shown in Section 19.2 (Circular Sections) that the torsion constant C of a circular section of diameter h is the polar moment of inertia

$$C = \frac{\pi h^4}{32} \tag{19.3.2}$$

Table 19.3.1

y/x	1.0	1.2	1.5	2.0	2.5	3	4	5
β	0.141	0.166	0.196	0.229	0.249	0.263	0.281	0.291

Rectangular Sections. The torsion constant C of a rectangular section of height y and width x may be expressed [9.1] as

$$C = \beta x^3 y \qquad (19.3.3)$$

in which β is a function of y to x. The values for β are given in Table 19.3.1.

T-, L-, and I-Sections. The torsion constant C of a T-, L-, or I-section may be approximated [9.1] by the expression

$$C = \Sigma \tfrac{1}{3} x^3 y \qquad (19.3.4)$$

in which y and x are the side and thickness of each of the component rectangles into which the section may be divided.

The following is a more exact expression [9.1, p. 278] for the torsion constant, giving values closer to Table 19.3.1 for sections composed of rectangular elements having y/x less than about 10,

$$C = \Sigma \tfrac{1}{3} x^3 y \left(1 - 0.63 \frac{x}{y}\right) \qquad (19.3.5)$$

The reader may note that the ACI Code in its design provisions for torsion (ACI-Chap. 11) makes use of the simpler expression, Eq. (19.3.4). However, in the provisions for slabs as discussed in Chapters 16 and 17 (ACI-13.7.5.3), Eq. (19.3.5) is preferred as it gives a lower estimate of the torsional restraint, and for the uncracked stiffness is more accurate. Although there is some inconsistency in using two different expressions for the torsion constant, the overestimate of stiffness using Eq. (19.3.4) is an acceptable simplification when determining the nominal torsional strength for design. A lower estimate of stiffness using Eq. (19.3.5) is desirable in structural analysis when determining the restraining effect of spandrel members in two-way floor systems.

19.4 EFFECTS OF TORSIONAL STIFFNESS ON COMPATIBILITY TORSION

General Treatment of Torsion on Statically Indeterminate Systems. In order to perform a statically indeterminate structural analysis, it is necessary first to be able to determine the relative stiffnesses of interacting members. The so-called "compatibility torsion" discussed in Section 19.1 is involved. For example, if a spandrel member is *uncracked* and its torsional stiffness GC/L is computed as shown in Section 19.3, the torsional moment that the member will attempt to carry may be very large. As the member cracks, its torsional stiffness reduces drastically, the member will rotate, and the torsional moment carried is likewise reduced.

Postcracking stiffness and torsional moment have been studied by Lampert [19.4] and Collins and Lampert [19.5], who proposed an expression for the torsional rigidity of a cracked section. Since the stiffness is needed *before* the torsional moment can be determined, the cracked section stiffness is not available because it requires knowledge of the steel reinforcement.

Alternatively, Collins and Lampert [19.5] have indicated that in cases of compatibility torsion (where torsional moment depends on torsional stiffness), analysis on the basis of zero torsional stiffness resulted in a design as satisfactory as an analysis using uncracked stiffness. In fact, added steel may increase the torsional moment in the member but have little effect on the twist (rotation). Consequently, it may be more effective to design for a twist (rotation) than for a torque (moment). The purpose of the torsional reinforcement then is to provide ductility and distribute cracks caused by the twist. Such a procedure would be within the spirit of the ACI Code, where ACI-8.6.1 states, "Any reasonable assumptions may be adopted for computing relative flexural and torsional stiffnesses of columns, walls, floors, and roof systems."

Using the zero torsional stiffness assumption, the member resisting torsion (say, a spandrel beam) would be designed for flexure and shear, neglecting torsion; then the torsional stiffness based on cracked section might be computed. The structure may then be analyzed [7.3, Chapter 8] to determine the torsional moments, and the section may be checked according to the ACI Code rules for torsion design.

ACI Code Procedure.   The ACI Code (ACI-11.6.3) provides an optional simple procedure to reduce the design complexity for cases involving compatibility torsion. When a statically indeterminate situation involves torsion, and an internal redistribution of forces can occur as a result of cracking, the factored torsional moment to be used in design is reduced to a minimum value sufficient to provide the necessary rotation capacity (ductility). In other words, if the torsional restraint is omitted in determining the bending moments and shears on the structural elements, the design of those elements may be more conservative than otherwise, but the difficulty of determining the torsional moments has been eliminated. The torsional members must, however, have the ductility to *twist* the necessary amount.

To summarize, the ACI Code provides two options for the design of torsional members when the torsional moment is dependent on the relative stiffness of the interacting members.

1. Estimate the torsional and flexural stiffnesses of all interacting members by making "any reasonable assumptions . . ." (ACI-8.6.1). Determine the moments, shears, and torsional moments by the statically indeterminate analysis using factored loads. Then apply the ACI Code provisions for torsion design.
2. Neglect torsional stiffness in the statically indeterminate structural analysis. Since no torsional moment will then be available, the torsional members must be designed for a strength based on a nominal torsional shear stress under factored load of $4\sqrt{f'_c}$. This procedure is illustrated in Example 19.14.3.

Spandrel Beams and Girders.   Consider the spandrel beam 2*B*4 (Fig. 19.4.1) in the typical slab–beam–girder floor of Example 8.3.1. This beam receives a vertical load

Figure 19.4.1 Floor plan of typical slab—beam—girder construction.

and a torsional moment per unit length from the slab 2S1, which are equal, respectively, to the reaction and restraining moment at the exterior end of slab 2S1. In addition, the beam supports the weight of whatever walls or windows may rest directly on it. Thus the spandrel beam 2B4 is subjected to a torsional moment per unit length in addition to bending and shear. A similar condition exists in the spandrel girder 2G4; however, the torsional moments are applied only at the junction points with the beams.

The torsional moments in the spandrel beams or girders cause torsional shear stresses, which are additive to the bending shear stresses at the *inside* face of the member. The usual approach in design is to provide for the sum of the torsional shear and flexural shear requirements. Since torsional shear stress goes around the member, closed stirrups (hoops) are necessary.

The magnitude of the torsional moment acting uniformly along a spandrel beam (ACI-11.6.3.2) such as 2B4 of Fig. 19.4.1 might be roughly approximated as equal to the restraining moment along the exterior edge of the slab, using a value such as $\frac{1}{24}wL^2$ as given by ACI-8.3.3. Alternatively, the torsional moment may be neglected if the slab is designed assuming there is no restraining moment along the spandrel beam. In such a case, the spandrel beam must be designed for a minimum torsional strength corresponding to that which will provide adequate ductility to twist (see Example 19.14.3).

The design and behavior of spandrel beams has been the subject of many studies [19.84–19.98], the most extensive of which are those of Raths [19.96] and Klein [19.97].

■ **EXAMPLE 19.4.1** Estimate the maximum factored torsional moment T_u in the spandrel beam $2B4$ of Fig. 19.4.1 if the restraining moment at the exterior end of slab panel (4.5-in. slab and a clear span of 11.92 ft) $2S1$ is $M = wL^2/24$. The service live and dead loads are 100 and 56 psf, respectively. Assume an 18×18 in. column and a $13 \times 22\frac{1}{2}$ in. overall size beam. Use $f'_c = 3000$ psi.

Solution: The factored restraining moment M_u along the edge of the stab is approximately

$$w_u = 1.4(56) + 1.7(100) = 249 \text{ psf}$$

$$M_u = \tfrac{1}{24}(0.249)(11.92)^2 = 1.47 \text{ ft-kips/ft width}$$

The torsional moment is largest at the face of the column and decreases nearly linearly to zero at midspan. The factored torsional moment T_u at the face of the column is approximately

$$T_u = (\tfrac{1}{2} \text{ clear span of spandrel})(M_u) = 12.25(1.47) = 18.0 \text{ ft-kips}$$

Note that under the alternate procedure neglecting torsional stiffness, the slab would be designed without restraining moment acting along its exterior edge. The spandrel member would then be designed arbitrarily to have a design strength $\phi T_n = \phi(4\sqrt{f'_c}\tfrac{1}{3}\Sigma x^2 y)$. In this example,

$$\tfrac{1}{3}\Sigma x^2 y = \tfrac{1}{3}[(13)^2(22.5) + (4.5)^2(13.5)] = 1360 \text{ in.}^3$$

considering a slab width equal to $3(4.5) = 13.5$ in. maximum according to ACI-11.6.1.1 as effective to act with the main rectangular portion of the member. Thus, the arbitrary factored moment $T_u = \phi T_n$ to be designed for would be

$$\phi T_n = \phi(4\sqrt{f'_c})\tfrac{1}{3}\Sigma x^2 y = 0.85(4\sqrt{3000})(1360)/12,000 = 21.1 \text{ ft-kips}$$

In this case, the design for ductility using the nominal stress $4\sqrt{f'_c}$ would be the more conservative approach, although simpler. ■

19.5 CRACKING STRENGTH OF PLAIN CONCRETE RECTANGULAR SECTIONS IN TORSION—SKEW BENDING THEORY

According to Eq. (19.2.2), the torsional strength T_n of a plain concrete rectangular section may be expressed

$$T_n = \alpha x^2 y[f_t(\text{max})] \tag{19.5.1}$$

because torsional stress v_t equals maximum principal tensile stress $f_t(\text{max})$ in the situation of pure torsion (same as pure shear). In Eq. (19.5.1), α is an elastic theory coefficient as given in Table 19.2.1, ranging from 0.208 to 1/3 as y/x varies from 1.0 to ∞, and up to 0.50 for plastic theory.

Hsu [19.20] has shown that when $f_t(\text{max})$ is taken at about 5 to $6\sqrt{f'_c}$ (i.e., representing the stress at which concrete cracks in tension), and α at 1/3, Eq. (19.5.1)

Figure 19.5.1 Skew bending of plain concrete rectangular section (according to Hsu [19.20]).

gives the torsional cracking strength T_{cr},

$$T_{cr} = \tfrac{1}{3}x^2y(5 \text{ to } 6\sqrt{f_c'})$$ **(19.5.2)**

Hsu [19.20] has also shown that a torsion failure of a rectangular section does not occur in a spiral form as might be expected from a circular shaft. Instead, a rectangular section in torsion cracks by *bending* about an axis parallel to the wider face of the section and inclined at about 45° to the axis of the beam, as shown in Fig. 19.5.1. This is called the *skew bending theory*.

19.6 STRENGTH OF REINFORCED CONCRETE RECTANGULAR SECTIONS IN TORSION—SKEW BENDING THEORY

Once steel reinforcement, both longitudinal and transverse, is placed in a rectangular section, the behavioral pattern changes from that of plain concrete. The resisting action of transverse reinforcement in the form of closed hoops is similar to that of stirrups resisting flexural shear. Prior to cracking, the reinforcement participates little if at all; but after cracking, the reinforcement carries a large portion of the total torsional moment. The contribution of concrete is only about 40% of the torsional strength of an unreinforced section. The failure mode according to this theory, however, does continue to be one of skew bending such as described in Section 19.5.

The skew bending concept was proposed by Lessig [19.19] and extended by Goode and Helmy [19.24], Collins, Walsh, Archer, and Hall [19.25], and Below, Rangan, and Hall [19.26], all of whom applied it to the case of combined bending and torsion. Hsu [19.22] has applied the concept to the case of torsion alone and has developed the expression that forms the basis for the ACI Code procedure. Hsu [19.21, 19.23], Zia [19.1, 19.2], and Warwaruk [19.10] have provided summaries of the theories relating to rectangular sections in torsion. McMullen and Rangan [19.7] have reviewed the research to clarify the contradictions between the skew bending and the space truss theories (see Section 19.7 for the space truss theory). The following development presents some of the ideas relating to the strength expression developed by Hsu [19.22].

Referring to Fig. 19.6.1, the failure section is assumed to be a plane that is perpendicular to the wider face of the member and inclined at 45° to the axis of the member. The failure plane may be as shown in Fig. 19.6.1, due to twisting moment in the direction indicated. Since a bending mode of failure is assumed, the compression zone is treated as in any beam analysis; that is, it has a depth a over which the compressive stress may be assumed uniform. On the tension side where the concrete cracks and is assumed not to resist tension, the reinforcing hoops have tensile forces P_v in them and the longitudinal bars resist shear across the cracked concrete via dowel action (see Section 5.4). The horizontal and vertical components of this dowel action are designated Q_x and Q_y. As long as the concrete is uncracked and the concrete itself transmits shear, no dowel action exists.

On the compression side [Fig. 19.6.2(a)], the longitudinal bars contribute tensile force P_l; the concrete contributes shear resistance P_s in the failure plane and also compressive resistance P_c normal to the failure plane. The components of the resultant force P are shown in Fig. 19.6.2(b). The hoop reinforcement in compression is neglected because, as has been shown in Section 3.10, the nominal moment strength M_n is not significantly affected by the inclusion of compression reinforcement.

On the tension side, it is noted that no longitudinal force can exist. If such a force were to act, it would have to be counterbalanced by a component of resistance

Figure 19.6.1 Forces acting on skew bending failure section.

(a)

(b)

Figure 19.6.2 Components of resultant force P acting on compression zone of failure plane.

acting oppositely. Since only P_v, Q_x, and Q_y are assumed to be acting, and the resultant force must be directed upward (opposite to P on the compression side), no resultant tension or compression can exist in the longitudinal direction of the tension zone under skew bending.

The reader is reminded that unless the direction of the twisting moment is definitely established by analysis, in usual conditions of providing nominal torsional strength (ACI-11.6.3), a potential failure plane can exist opposite to that in Fig. 19.6.1, having the compression and tension sides interchanged. Thus, the longitudinal forces, stirrup forces, and dowel forces must be resisted on each side of the section.

Strength Attributable to Concrete.   The shear resistance P_s (Fig. 19.6.2) may be expressed as

$$P_s = v_{avg}(y\sqrt{2})a \qquad (19.6.1)$$

where v_{avg} is the average shear stress acting over the compression zone; $y\sqrt{2}$ is the width of the compression zone; and a is the depth of the compression zone.

Alternatively, the shear strength P_s may be considered proportional to the effective area $xy\sqrt{2}$ and to $\sqrt{f_c'}$ [see Eq. (5.5.6) omitting effect of reinforcement]. Thus

$$P_s = k_1 xy\sqrt{f_c'} \qquad (19.6.2)$$

where k_1 is a proportionality constant.

From Fig. 19.6.2(b), one may note that

$$P = \sqrt{2}P_s + P_l \qquad (19.6.3)$$

The first term of Eq. (19.6.3) represents the portion attributable to concrete. Thus the torsional strength T_c attributable to concrete equals the force $\sqrt{2}P_s$ times the moment arm (say, $0.80x$),

$$T_c = \sqrt{2}P_s(\text{arm})$$
$$= \sqrt{2}k_1 xy\sqrt{f_c'}(0.80x)$$
$$= k_2 x^2 y\sqrt{f_c'} \qquad (19.6.4)$$

Equation (19.6.4) is of the same form as Eq. (19.5.2). If $6\sqrt{f'_c}$ is used in Eq. (19.5.2), and 40% of that equation is used in recognition that the concrete strength contribution to the total torsional strength of a *reinforced* section represents only about 40% of the torsional cracking strength of a plain concrete section, then T_c of Eq. (19.6.4) becomes

$$T_c = 0.4[\tfrac{1}{3}x^2y(6\sqrt{f'_c})]$$
$$= 0.8\sqrt{f'_c}\,x^2y \qquad (19.6.5)$$

Strength Attributable to Hoops and Longitudinal Reinforcement. Referring to Figs. 19.6.1 and 19.6.2, the forces P_v, Q_x, and Q_y on the tension side and P_l on the compression side have yet to be considered.

1. The contribution of the closed vertical stirrups (hoops) is

$$P_v = A_t f_y\left(\frac{y_1}{s}\right) \qquad (19.6.6)$$

where y_1/s (see Fig. 19.6.3) is the number of hoops intercepted by the 45° failure plane.

2. The tensile force P_l in the longitudinal bars intercepting the compression concrete zone is

$$P_l = \xi\left(\frac{A_l}{2}\right)f_y \qquad (19.6.7)$$

where ξ is an efficiency factor to account for the longitudinal bars being located at two or more points within the compression zone, and A_l is the total area of all longitudinal bars (assumed to be $A_l/2$ in the compression zone). P_l contributes to the torsional strength since from Eq. (19.6.3) it is part of P.

3. The dowel forces Q_x and Q_y act after the concrete has cracked and these forces may be assumed proportional to the bar cross-sectional area and to the bar lateral displacement, which is proportional to the distance (either $0.5x_2$ or $0.5y_2$) from the center of twist to the bar. Thus

$$Q_x = k_3 A_l y_2$$
$$Q_y = k_3 A_l x_2 \qquad (19.6.8)$$

Figure 19.6.3 Cross-section dimensions.

where k_3 is a proportionality constant. Dowel action in beams subject to torsion has been reviewed by Youssef and Bishara [19.82].

Next let m equal the ratio of the volume of longitudinal bars to the volume of closed hoops such that

$$m = \frac{A_l s}{2A_t(x_1 + y_1)} \qquad (19.6.9)$$

or

$$A_l = A_t\left[\frac{2m(x_1 + y_1)}{s}\right] \qquad (19.6.10)$$

Substitution of Eq. (19.6.10) into Eq. (19.6.7) gives

$$P_l = \xi m\left(1 + \frac{y_1}{x_1}\right)\left(\frac{x_1 A_t f_y}{s}\right) \qquad (19.6.11)$$

Substituting Eq. (19.6.10) into Eq. (19.6.8) gives

$$Q_x = k_3 y_2\left(\frac{2mA_t(x_1 + y_1)}{s}\right)$$

$$= 2\frac{k_3}{f_y}\left(\frac{y_2}{y_1}\right)m\left(1 + \frac{y_1}{x_1}\right)\left(\frac{x_1 y_1 A_t f_y}{s}\right) \qquad (19.6.12)$$

Similarly,

$$Q_y = 2\frac{k_3}{f_y}\left(\frac{x_2}{y_1}\right)m\left(1 + \frac{y_1}{x_1}\right)\left(\frac{x_1 y_1 A_t f_y}{s}\right) \qquad (19.6.13)$$

The torsional resistance from reinforcement then is

$$T_s = P_v\left(\frac{x_1}{2}\right) + P_l\left(\frac{x_2}{2}\right) + 2Q_x\left(\frac{y_2}{2}\right) + 2Q_y\left(\frac{x_2}{2}\right) \qquad (19.6.14)$$

Substitution of Eqs. (19.6.6), (19.6.11), (19.6.12), and (19.6.13) into Eq. (19.6.14) gives

$$T_s = \alpha_t\left(\frac{x_1 y_1 A_t f_y}{s}\right) \qquad (19.6.15)$$

where

$$\alpha_t = \frac{1}{2} + \xi m\left(1 + \frac{y_1}{x_1}\right)\left(\frac{x_2}{2y_1}\right) + 2\frac{k_3}{f_y}m\left(1 + \frac{y_1}{x_1}\right)(x_2^2 + y_2^2)\left(\frac{1}{y_1}\right) \qquad (19.6.16)$$

Assuming $x_2 \approx x_1$ and $y_2 \approx y_1$, the quantity α_t is essentially a function of two parameters, m and y_1/x_1 and might be written as

$$\alpha_t = C_1 + C_2 m + C_3\left(\frac{y_1}{x_1}\right) \qquad (19.6.17)$$

where the constants C_1, C_2, and C_3 may be experimentally determined.

Hsu [19.22] has shown that for equal volume of longitudinal bars to closed hoops (i.e., $m = 1$), α_t may be expressed as

$$\alpha_t = 0.66 + 0.33\left(\frac{y_1}{x_1}\right) \tag{19.6.18}$$

which is used by the ACI Code (ACI-11.6.9.1).

Thus the full nominal strength T_n of rectangular reinforced concrete sections may be written by combining Eqs. (19.6.5) and (19.6.15),

$$T_n = T_c + T_s = 0.8\sqrt{f_c'}\,x^2 y + \alpha_t\left(\frac{x_1 y_1 A_t f_y}{s}\right) \tag{19.6.19}$$

The first term represents the strength attributable to *one rectangle*. When the cross section consists of several rectangular segments, Eq. (19.6.19) becomes

$$T_n = T_c + T_s = 0.8\sqrt{f_c'}\,(\Sigma x^2 y) + \alpha_t\left(\frac{x_1 y_1 A_t f_y}{s}\right) \tag{19.6.20}$$

where the $x^2 y$ is computed for each segment having its short dimension x and long dimension y.

19.7 STRENGTH OF REINFORCED CONCRETE RECTANGULAR SECTIONS IN TORSION—SPACE TRUSS ANALOGY

The space truss analogy approach to the evaluation of torsional capacity was first suggested by Rausch [19.27] and developed by Lampert and Thürlimann [19.4, 19.28]. It was then further explained by Lampert and Collins [19.29], and more recently treated by Müller [19.32] and Rabbat and Collins [19.33]. Recent advances in knowledge of reinforced concrete in torsion seem to lead to the space truss theory as the most rational way to give unified treatment to flexure and torsion. The treatment presented by Collins and Mitchell [19.9] suggests a rational design approach which may eventually supercede the ACI Code approach that is linked to the skew bending theory. The European Concrete Committee (CEB) [19.30] uses the space truss analogy theory as its basis for design.

In determining the torsional strength of reinforced concrete rectangular sections, it has been found that nearly all of the strength is derived from the reinforcement and the concrete that immediately surrounds the steel. Then one may consider a thin-walled box section, as shown in Fig. 19.7.1, as a space truss. The longitudinal bars in the corners contribute tensile forces while the concrete strips between cracks provide compressive resistance. The inclined compressive forces act in a spiral fashion around the box section, giving the compressive forces D_h on the vertical sides and D_b on the horizontal sides.

Consider equilibrium requirements on the section of Fig. 19.7.1. From force equilibrium in the z direction,

$$4Z = 2\frac{\tau t x_2}{\tan \phi_1} + 2\frac{\tau t y_2}{\tan \phi_1} \tag{19.7.1}$$

Figure 19.7.1 Forces on section in torsion according to space truss analogy (from Lampert [19.4]).

where Z is the tensile force in each longitudinal bar in the corner of the section, and τt is the shear flow (force per unit length) in the thin-walled section. Summation of moments of the shear flow forces about the z axis must equal the applied torsional moment,

$$T_n = \tau t y_2 x_2 + \tau t x_2 y_2 \qquad (19.7.2)$$

Next, consider the detail A of Fig. 19.7.1 whose free-body diagram is given in Fig. 19.7.2. Detail A represents the portion of concrete along the z axis of the member

Figure 19.7.2 Free body of detail A (Fig. 19.7.1) on the hollow box section used for the space truss analogy.

tributary to one hoop. Equilibrium in the direction of the hoop tensile force requires

$$A_t f_y = \tau t s \tan \phi_1 \qquad (19.7.3)$$

From Eq. (19.7.3)

$$\tan \phi_1 = \frac{A_t f_y}{\tau t s} \qquad (19.7.4)$$

which when substituted into Eq. (19.7.1) gives

$$2Z = \frac{(\tau t)^2 s}{A_t f_y} (x_2 + y_2) \qquad (19.7.5)$$

Letting the force $Z = (A_l/4) f_y$, Eq. (19.7.5) may be solved for τt,

$$\tau t = \sqrt{\frac{A_l f_y A_t f_y}{2s(x_2 + y_2)}} \qquad (19.7.6)$$

Substitution of Eq. (19.7.6) into Eq. (19.7.2) gives

$$T_n = 2x_2 y_2 \sqrt{\frac{A_l f_y A_t f_y}{2s(x_2 + y_2)}} \qquad (19.7.7)$$

For equal volumes of hoop and longitudinal steel,

$$A_l = \frac{2A_t(x_1 + y_1)}{s} \qquad (19.7.8)$$

Substitution of Eq. (19.7.8) into (19.7.7) gives

$$T_n = 2\left(\frac{A_t f_y}{s}\right) x_2 y_2 \sqrt{\frac{x_1 + y_1}{x_2 + y_2}} \qquad (19.7.9)$$

If $x_1 \approx x_2$ and $y_1 \approx y_2$, Eq. (19.7.9) becomes

$$T_n = 2\left(\frac{A_t f_y}{s}\right) x_2 y_2 \qquad (19.7.10)$$

Comparison of Eqs. (19.7.10) and (19.7.2) will show that

$$\tau t = \frac{A_t f_y}{s}$$

which if substituted into Eq. (19.7.4) will show that $\tan \phi_1 = 1$, or $\phi_1 = 45°$.

One of the features of the more recent treatment of the truss analogy theory, according to the Collins and Mitchell [19.9, 19.31] approach of using inclined compression struts (so-called *compression field action*), is that ϕ_1 is *not* necessarily 45° but will take an optimum angle to permit the greatest strength. Collins and Mitchell [19.9] and the European approach [19.30] accept that $\tan \phi_1$ can be as low as $\frac{3}{5}$ and as high as $\frac{5}{3}$, and need not be 1.0.

One may also note that Eq. (19.7.10) is of the same form as T_s of Eq. (19.6.15), and becomes essentially identical if $\alpha_t = 2$. The simple space truss analogy neglects any torsional strength due to the concrete alone; that is, if either longitudinal *or* hoop steel is not present, this theory gives no torsional capacity.

19.8 STRENGTH OF SECTIONS IN COMBINED BENDING AND TORSION*

In most practical situations, torsion will occur simultaneously with flexure. Although there have been many studies of the interaction between bending and torsion [19.4, 19.9, 19.26, 19.28, 19.34–19.44], there does not seem to be agreement on what is the correct interaction criterion.

Both the skew bending theory [19.19, 19.25, 19.36] and the space truss analogy as developed by Lampert [19.4, 19.28] are in general agreement on the interaction behavior. According to Collins and Lampert [19.29, 19.79] both theories indicate that under positive bending, yielding of bottom reinforcement occurs when equal volumes of longitudinal and transverse steel are used and when equal amounts of longitudinal steel are used in the top and bottom faces (i.e., $A_s' = A_s$). The theories may be approximated by the following, as shown in Fig. 19.8.1,

$$\left(\frac{T_n}{T_{n0}}\right)^2 + \frac{M_n}{M_{n0}} = 1 \tag{19.8.1}$$

where

T_n = nominal torsional strength in the presence of flexure

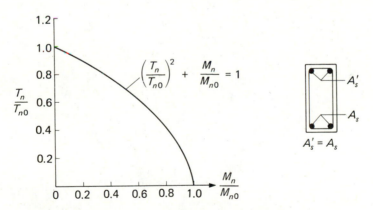

Figure 19.8.1 Bending–torsion interaction diagram for equal tension and compression longitudinal steel (positive moment yield).

— — — —

* The discussion in this section is for the combined action of torsional moment and *positive* bending moment, which if acting alone would cause tension in the bottom steel A_s and compression in the top steel A_s'. In the event that the bending moment is negative, the reader should then interpret A_s' as being the bottom steel and A_s the top steel.

T_{n0} = nominal torsional strength when the member is subjected to torsion alone

M_n = nominal flexural strength in the presence of torsion

M_{n0} = nominal flexural strength when the member is subjected to flexure alone

Unequal Top and Bottom Steel, $A'_s < A_s$. When torsion alone acts, the top and bottom longitudinal steel must resist *equal tensile forces*. When positive bending moment acts with torsion, the bottom steel will yield before the top steel if equal amounts of top and bottom steel are used. However, if there is less top steel than bottom steel, the application of a small amount of positive bending moment will result in an increase in the interactive torsional strength because the compressive force due to flexure tends to counteract the tensile force due to torsion. This mutually beneficial effect reaches its peak (Fig. 19.8.2) when both the top and bottom steel yield simultaneously.

In general, the strength interaction relationship may be stated [19.79], for the case of first yield in bottom reinforcement,

$$\left(\frac{T_n}{T_{n0}}\right)^2 = \frac{1}{r}\left(1 - \frac{M_n}{M_{n0}}\right) \tag{19.8.2}$$

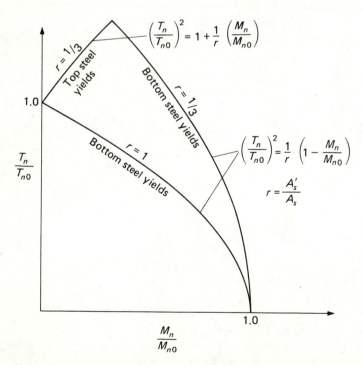

Figure 19.8.2 Bending–torsion interaction relationships (from Collins and Lampert [19.79]).

where $r = (A_s'f_y')/(A_sf_y)$, and f_y' and f_y are the yield stresses of the top and bottom steel, respectively. For the case when the top steel yields first, the expression is

$$\left(\frac{T_n}{T_{n0}}\right)^2 = 1 + \frac{1}{r}\left(\frac{M_n}{M_{n0}}\right) \tag{19.8.3}$$

Equations (19.8.2) and (19.8.3) are shown graphically in Fig. 19.8.2.

Design Implications. In order to examine the implications of Eqs. (19.8.2) and (19.8.3), first consider the bottom steel yield failure mode. Expanding Eq. (19.8.2),

$$rT_n^2 M_{n0} = T_{n0}^2(M_{n0} - M_n) \tag{19.8.4}$$

The strength of a section under flexure alone is

$$M_{n0} = A_sf_y\left(d - \frac{a}{2}\right) \tag{19.8.5}$$

and using Eq. (19.7.7) for the theoretical strength of a member subjected to torsion alone,

$$T_{n0} = 2x_2y_2\sqrt{\frac{A_lf_yA_tf_y}{2s(x_2 + y_2)}} \tag{19.8.6}$$

where x_2 is defined in Fig. 19.6.3. Substitution of Eqs. (19.8.5) and (19.8.6) into Eq. (19.8.4) gives

$$rT_n^2 A_sf_y\left(d - \frac{a}{2}\right) = 4x_2^2y_2^2\left[\frac{A_lf_yA_tf_y}{2s(x_2 + y_2)}\right]\left[A_sf_y\left(d - \frac{a}{2}\right) - M_n\right] \tag{19.8.7}$$

Let $A_l/2$ equal A_s' or A_s, whichever is smaller. Then if hoop steel is designed for torsion alone, with equal volumes of longitudinal and transverse steel, substituting Eq. (19.7.8) for A_l in Eq. (19.8.6) would give

$$T_n = 2\left(\frac{A_tf_y}{s}\right)x_2y_2$$

which gives the amount of hoop steel to be used as

$$\frac{A_tf_y}{s} = \frac{T_n}{2x_2y_2} \tag{19.8.8}$$

Substitution of Eq. (19.8.8) into Eq. (19.8.7) with $r = A_s'/A_s$ and $A_l/2 = A_s'$ gives

$$\frac{A_s'}{A_s}(T_n^2)A_sf_y\left(d - \frac{a}{2}\right) = 4x_2^2y_2^2\left[\frac{A_s'f_y}{(x_2 + y_2)}\right]\left(\frac{T_n}{2x_2y_2}\right)\left[A_sf_y\left(d - \frac{a}{2}\right) - M_n\right]$$

$$T_n\left(d - \frac{a}{2}\right) = \frac{2x_2y_2}{x_2 + y_2}\left[A_sf_y\left(d - \frac{a}{2}\right) - M_n\right] \tag{19.8.9}$$

which upon solving for A_s gives

$$A_s = \frac{T_n(x_2 + y_2)}{2x_2y_2f_y} + \frac{M_n}{f_y(d - a/2)} \tag{19.8.10}$$

Equation (19.8.10) shows that the bottom steel required is the sum of the requirements for flexure and torsion computed separately.

For the top steel yield failure mode, expand Eq. (19.8.3) and let $r = A'_s/A_s$,

$$A'_s T_n^2 = \left[A'_s + A_s \left(\frac{M_n}{M_{n0}} \right) \right] T_{n0}^2 \qquad (19.8.11)$$

Using Eqs. (19.8.5) and (19.8.6) for M_{n0} and T_{n0}, respectively, and letting $A_l/2 = A'_s$,

$$A'_s T_n^2 = \left[A'_s + A_s \frac{M_n}{A_s f_y (d - a/2)} \right] \left[4 x_2^2 y_2^2 \left(\frac{A'_s f_y}{x_2 + y_2} \right) \left(\frac{A_t f_y}{2} \right) \right] \qquad (19.8.12)$$

In the same manner as for the bottom steel yield case, assume the hoop steel is designed for torsion alone. Then substituting Eq. (19.8.8) into Eq. (19.8.12) gives

$$A'_s = \frac{T_n (x_2 + y_2)}{2 x_2 y_2 f_y} - \frac{M_n}{f_y (d - a/2)} \qquad (19.8.13)$$

Thus the top steel requirement is the difference between the flexure and torsion requirements. Usually only minimal longitudinal steel A'_s is needed for combined flexure and torsion.

As will be discussed again in Section 19.12, the ACI Code does not explicitly consider combined bending and torsion. However, the flexural requirement is added to the requirement of longitudinal reinforcement for torsion, so that actually Eq. (19.8.10) is used.

19.9 STRENGTH OF SECTIONS IN COMBINED SHEAR AND TORSION

Rectangular and L-shaped sections have been studied under combined shear and torsion by a few investigators [19.45–19.48]. However, since shear usually accompanies flexure, it is the combination of flexure, shear, and torsion that has received the greatest attention [19.49–19.72]. Generally, though, flexural shear and torsional shear are of significance in those regions where bending moment is low. Thus for design purposes it is most necessary to know the strength interaction relationship between shear and torsion.

Test data have provided a wide range of points on the interaction relationship using torsion and shear coordinates. Because of the unknowns involved, some investigators have proposed a linear interaction equation [19.1, 19.2, 19.54, 19.63] for design purposes. However, a number of studies at the University of Texas [19.45, 19.47, 19.49, 19.53, 19.63] on rectangular, L-shaped, and T-shaped beams have indicated that a quarter-circle interaction relationship is acceptable for members *without* web reinforcement. For members *with* web reinforcement, the interaction is curved but flatter than the quarter circle. The quarter-circle expression is

$$\left(\frac{T_n}{T_{n0}} \right)^2 + \left(\frac{V_n}{V_{n0}} \right)^2 = 1 \qquad (19.9.1)$$

where T_n and V_n are the nominal strengths in torsion and shear, respectively, acting simultaneously; T_{n0} is the nominal strength under torsion alone; and V_{n0} is the nominal strength under shear alone. As will be shown in Section 19.11, Eq. (19.9.1) is used for the ACI Code expression for the strength in combined shear and torsion on sections *without* web reinforcement. However, for web reinforcement the separate requirements for shear and torsion are added together, rather than following Eq. (19.9.1).

19.10 STRENGTH INTERACTION SURFACE FOR COMBINED BENDING, SHEAR, AND TORSION

The effect of the simultaneous application of bending, shear, and torsion may be most easily examined by means of an interaction surface. Such a concept has been used in Section 13.21 for biaxial bending of compression members. Various investigators have proposed interaction surfaces. Some of these surfaces are shown in Figs. 19.10.1

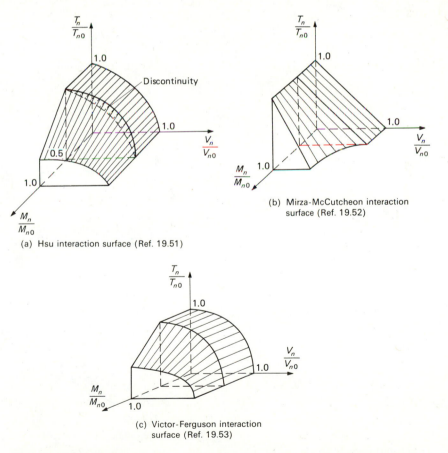

(a) Hsu interaction surface (Ref. 19.51)

(b) Mirza-McCutcheon interaction surface (Ref. 19.52)

(c) Victor-Ferguson interaction surface (Ref. 19.53)

Figure 19.10.1 Interaction surfaces for combined bending, shear, and torsion (for members *without* transverse reinforcement).

and 19.10.2. The work of Collins et al. [19.54] seems to provide the most complete rational approach by correlating the regions of the failure surface with modes of failure as shown in Fig. 19.10.2.

Since most members that are subjected to bending, shear, and torsion will have less longitudinal steel in the compression zone due to flexure alone (top steel in the positive moment zone) than in the tension zone (bottom steel in the positive moment

(a) Interaction surface for beam with weaker top
steel than bottom steel ($A_s' < A_s$)

(b) Mode 1
(bending and torsion)

(c) Mode 2
(low shear–high torsion)

(d) Mode 3 (low bending–high
torsion; weaker top steel)

(e) Mode 4
(high shear–low torsion)

Figure 19.10.2 Interaction surface and failure modes according to Collins et al.
(from Ref. 19.54) (for members *with* transverse reinforcement).

zone), only that case is given in Fig. 19.10.2(a). When equal amounts of top and bottom longitudinal steel are used, the Mode 3 ("negative" yield or yield in top steel of positive moment zone) cannot occur; in such a case the surfaces of Modes 1 and 2 should extend upward until they intersect the T_n/T_{n0} axis.

The following conclusions may be drawn regarding the interaction of bending, shear, and torsion.

1. The interaction between torsion and shear may be represented for most situations (with $A_s' < A_s$) by a quarter circle, and is relatively unaffected by the simultaneous application of bending moment of a magnitude equal to one-third to one-half of the nominal bending strength when shear and torsion are absent.
2. When equal amounts of top and bottom longitudinal steel are used, the quarter-circle shear–torsion interaction still seems acceptable, but there is a reduction in strength when bending moment is also applied.
3. Based on most test results, the straightline shear–torsion interaction, while simple to use, appears overly conservative.

As will be noted in the next section, the use of a quarter-circle shear–torsion interaction, although slightly on the nonconservative side based on a number of experiments, is acceptable by ACI procedures because the ACI shear strength (in the absence of torsion) is taken as that causing an inclined crack. The actual shear strength in many cases is greater, particularly when the shear span to depth ratio is less than about 2.5.

19.11 TORSIONAL STRENGTH OF CONCRETE AND HOOP REINFORCEMENT—ACI CODE

The ACI procedure of design for torsion is based largely on the work of Hsu and the recommendations of ACI Committee 438 [19.73, 19.74]. References are available for special topics such as design for inverted T-girders [19.99, 19.100], ledger beams [19.101], channel-shaped sections [19.102–19.104], and beams having openings [19.105–19.110].

The ACI Code permits neglect of torsion when the factored torsional moment T_u does not exceed about one-quarter of the design cracking strength ϕT_{cr}, where T_{cr} is given by Eq. (19.5.2) when $6\sqrt{f_c'}$ is substituted as $f_t(\text{max})$. Thus, ACI-11.6.1 permits neglect of torsion when

$$T_u \leq \phi(0.5\sqrt{f_c'}\,\Sigma x^2 y) \tag{19.11.1}$$

The strength reduction factor ϕ for torsion is 0.85 (ACI-9.3.2.3). When T_u exceeds the limit of Eq. (19.11.1), design must include the torsion effects as well as shear and flexure.

The general philosophy for designing to accommodate torsional shear is the same as that for flexural shear—that is, to consider that a part of the torsional moment may be carried by the concrete without web reinforcement and the remainder must be carried by closed hoop web reinforcement.

Thus

$$T_n = T_c + T_s \qquad (19.11.2)$$

in which

T_n = total nominal torsional strength

T_c = nominal torsional strength provided by concrete

T_s = nominal torsional strength provided by hoop steel

The nominal strength T_n is given by Eq. (19.6.20) as

$$T_n = 0.8\sqrt{f_c'} \, \Sigma x^2 y + \alpha_t \left(\frac{x_1 y_1 A_t f_y}{s} \right) \qquad [19.6.20]$$

Strength Attributable to Concrete Subject to Torsion Alone.   The first term T_c in Eq. (19.11.2) is the strength attributable to concrete, which, according to the numerator in ACI Formula (11-22), and the first term of Eq. (19.6.20), is

$$T_c = 0.8\sqrt{f_c'}(\Sigma x^2 y) \qquad (19.11.3)$$

Computation of $\Sigma x^2 y$.   In order to compute the torsional property $\Sigma x^2 y$, the cross section is divided into rectangles in such a way that the largest rectangle obtainable is used as one of the components, and the closed hoop reinforcement should then follow its boundaries.

(a)

(b)

Figure 19.11.1 Computation of $\Sigma x^2 y$ for use in torsion design.

Figure 19.11.2 Hollow box sections.

When a monolithic slab is acting as shown in Fig. 19.11.1(a), an effective over-hanging width not exceeding 3 times the flange thickness (ACI-11.6.1.1) is to be used [19.74, 19.80]. In this case the largest rectangle extends from top of flange to bottom of web. However, when the flange is thicker, the main rectangle should extend through-out the entire flange width, as shown in Fig. 19.11.1(b). The smaller rectangles of the configuration should also contain transverse reinforcement, frequently accomplished by using hairpin-type (U-shaped) stirrups anchored within the region of the closed hoop.

The hollow box section of Fig. 19.11.2 may be treated as a solid rectangle (ACI-11.6.1.2) when its wall thickness h is at least $x/4$. When the wall thickness is between $x/10$ and $x/4$, it may still be treated as a solid rectangle, except $\Sigma x^2 y$ must be multiplied by $4h/x$. When the wall thickness is less than $x/10$, the wall stiffness must be considered. The design of thin-walled box girders having h less than $x/10$ should generally be avoided because of the high flexibility and susceptibility to buckling of such walls, and the fact that few experimental results are available to guide the designer.

Strength Attributable to Hoop Reinforcement.   The second terms T_s in Eq. (19.11.2) is the strength attributable to hoop reinforcement, which, according to ACI Formula (11-23), and the second term of Eq. (19.6.20), is

$$T_s = \alpha_t \left(\frac{x_1 y_1 A_t f_y}{s} \right) \tag{19.11.4}$$

where $\alpha_t = 0.66 + 0.33(y_1/x_1)$ but not more than 1.50. For the design of transverse reinforcement, solving Eq. (19.11.4) for A_t/s, and using T_s = required $T_n - T_c$, or $\phi T_s = T_u - \phi T_c$,

$$\frac{A_t}{s} = \frac{T_u - \phi T_c}{\phi \alpha_t x_1 y_1 f_y} \tag{19.11.5}$$

Note that A_t is the area of *one* leg of a closed hoop. The design yield stress shall not exceed 60,000 psi (ACI-11.6.7.4).

Longitudinal Reinforcement.   The development by Hsu [19.22] assumed that the same volume of longitudinal steel would be provided in the distance s as occurs in one hoop of transverse reinforcement. In mathematical terms

$$A_l s = 2A_t(x_1 + y_1) \tag{19.11.6}$$

or

$$A_l = 2\left(\frac{A_t}{s}\right)(x_1 + y_1)$$ **(19.11.7)**

which is ACI Formula (11-24).

Critical Section for Torsion.   According to ACI-11.6.4, the critical section is to be taken at the effective depth d from the face of support, the same as is done for flexural shear alone. The same amount of transverse steel required at the critical section is to be used between the face of support and the critical section.

19.12 PROVISIONS FOR COMBINED TORSION WITH SHEAR OR BENDING—ACI CODE

Combined Shear and Torsion.   As discussed in Section 19.9, the interaction between shear and torsion for members *without* web reinforcement is assumed to be a quarter circle, represented by Eq. (19.9.1).

$$\left(\frac{T_n}{T_{n0}}\right)^2 + \left(\frac{V_n}{V_{n0}}\right)^2 = 1$$ **[19.9.1]**

The shear strength V_n in the presence of torsion T_n is obtained by solving Eq. (19.9.1), after first dividing through by $(V_n/V_{n0})^2$,

$$V_n = \frac{V_{n0}}{\sqrt{1 + \left(\frac{V_{n0}}{T_{n0}}\right)^2 \left(\frac{T_n}{V_n}\right)^2}}$$ **(19.12.1)**

For beams *with* web reinforcement, it is assumed that the quarter-circle interaction relationship is also applicable to the concrete contribution to shear and torsional strengths. Consequently, V_n can be replaced by V_c; also $V_{n0} = 2\sqrt{f'_c}b_w d$, the basic shear strength (ACI-11.3.1.1) of concrete according to the simplified method; and $T_{n0} = 0.8\sqrt{f'_c}(\Sigma x^2 y)$, the basic torsion strength from Eq. (19.11.3). Thus Eq. (19.12.1) becomes

$$V_c = \frac{2\sqrt{f'_c}b_w d}{\sqrt{1 + \left(\frac{V_{n0}}{T_{n0}}\frac{T_n}{V_n}\right)^2}}$$ **(19.12.2)**

Note that

$$\frac{V_{n0}}{T_{n0}} = \frac{2\sqrt{f'_c}b_w d}{0.8\sqrt{f'_c}(\Sigma x^2 y)} = 2.5C_t$$ **(19.12.3)**

where

$$C_t = \frac{b_w d}{\Sigma x^2 y}$$ **(19.12.4)**

Substitution of Eq. (19.12.3) into Eq. (19.12.2) by using $T_n/V_n = T_u/V_u$ gives

$$V_c = \frac{2\sqrt{f_c'}\, b_w d}{\sqrt{1 + \left(2.5 C_t \dfrac{T_u}{V_u}\right)^2}} \qquad (19.12.5)$$

which is ACI Formula (11-5).

The torsion strength T_n in the presence of shear V_n is obtained by solving Eq. (19.9.1), after first dividing through by $(T_n/T_{n0})^2$,

$$T_n = \frac{T_{n0}}{\sqrt{1 + \left(\dfrac{T_{n0}}{V_{n0}}\right)^2 \left(\dfrac{V_n}{T_n}\right)^2}} \qquad (19.12.6)$$

For beams *with* web reinforcement, T_n may be replaced by T_c; also $T_{n0} = 0.8\sqrt{f_c'}(\Sigma x^2 y)$, $V_{n0} = 2\sqrt{f_c'}\, b_w d$ and $V_n/T_n = V_u/T_u$; thus

$$T_c = \frac{0.8\sqrt{f_c'}(\Sigma x^2 y)}{\sqrt{1 + \left(\dfrac{0.4\,V_u}{C_t\,T_u}\right)^2}} \qquad (19.12.7)$$

which is ACI Formula (11-22).

Figure 19.12.1 ACI strength relationship for shear and torsion.

The ACI strength relationship for combined shear and torsion as expressed by Eqs. (19.12.5) and (19.12.7) is shown in Fig. 19.12.1. Since Eqs. (19.12.5) and (19.12.7) represent the strengths attributable to the concrete in members *with* web reinforcement, they underestimate the actual strengths of members *without* web reinforcement. As discussed in Section 19.6, the torsional strength of a concrete member without web reinforcement is considerably greater than the torsional strength T_c representing the concrete contribution to the strength of a member containing web reinforcement.

For the upper limits to strength carried by hoop reinforcement,

$$\max V_s = 8\sqrt{f'_c}\, b_w d$$

as given by ACI-11.5.6.8. Also

$$\max T_s = 4T_c$$

a limit recommended by ACI Committee 438 [19.78] and given by ACI-11.6.9.4.

Combined Bending and Torsion. The ACI Code does not explicitly consider this combination of loadings. For beams *without* web reinforcement, bending was neglected since it can be taken care of by the conservatism inherent in Eqs. (19.12.5) and (19.12.7). For beams *with* web reinforcement, however, the procedure is to design for bending separately and add the longitudinal steel required for flexure to that required for torsion or combined shear and torsion (ACI-11.6.1). From the discussion presented in Section 19.8, it may be apparent that this is an acceptable procedure.

19.13 MINIMUM REQUIREMENTS FOR TORSIONAL REINFORCEMENT—ACI CODE

The minimum requirements for the transverse reinforcement A_t and longitudinal reinforcement A_l are to ensure that there is ductile behavior prior to failure. Hsu [19.23] and Collins [19.76] have found that the strength attributable to concrete when closed hoops *are* present is only about 40% of the torsional capacity of a plain concrete member. Thus, for ductile behavior, a beam *with* web reinforcement should have at least the capacity of one without web reinforcement. In other words, it is desired that

$$T_n \text{ (with hoops)} \geq T_n \text{ (without hoops)} \qquad \textbf{(19.13.1)}$$

or for a rectangular section,

$$\text{Eq. (19.6.19)} \geq T_n \text{ (without hoops)}$$

$$\underbrace{0.8\sqrt{f'_c}(x^2 y)}_{\text{term } A} + \underbrace{\alpha_t\!\left(\frac{A_t}{s}\right)}_{\text{term } B}\underbrace{(x_1 y_1 f_y)}_{\text{term } C} \geq T_n \text{ (without hoops)} \qquad \textbf{(19.13.2)}$$

Thus accepting the 40% value for term A, term B must represent 60% of the capacity of a member containing only a small amount of torsional reinforcement. The condition for the minimum amount of torsional reinforcement then becomes

$$\text{term } B \geq 1.5(\text{term } A)$$

$$\alpha_t \left(\frac{A_t}{s}\right) x_1 y_1 f_y \geq 1.5[0.8\sqrt{f_c'}(x^2 y)] \tag{19.13.3}$$

Solving for A_t gives, for torsion acting alone,

$$A_t \geq \frac{1.2}{\alpha_t}\left(\frac{x^2 ys}{x_1 y_1}\right)\frac{\sqrt{f_c'}}{f_y} \tag{19.13.4}$$

This requirement could reasonably be reduced when transverse reinforcement (stirrups or hoops) is also used to provide for flexural shear. Transverse reinforcement for either purpose improves ductility. Thus when flexural shear requiring transverse reinforcement is present, the minimum transverse reinforcement for both torison and shear $(A_t + A_v/2)$ is

$$A_t + \frac{A_v}{2} \geq \frac{1.2}{\alpha_t}\left(\frac{x^2 ys}{x_1 y_1}\right)\frac{\sqrt{f_c'}}{f_y}\left(\frac{v_{tn}}{v_{tn} + v_n}\right) \tag{19.13.5}$$

where the term $[v_{tn}/(v_{tn} + v_n)]$ represents the assumption that the total shear strength of plain concrete is divided into resistance for torsional shear and flexural shear in the same ratio as the respective values of nominal stresses. Note that as defined in Chapter 5 on shear, A_v is the total area for two legs of a vertical stirrup provided to resist flexural shear.

To simplify Eq. (19.13.5), assume $\alpha_t = 1.2$, $x/x_1 = y/y_1 = 1.2$, and $f_c' = 5000$ psi,

$$A_t + \frac{A_v}{2} \geq \frac{1.2}{1.2}(1.2)(1.2)xs\frac{\sqrt{5000}}{f_y}\left(\frac{v_{tn}}{v_{tn} + v_n}\right) \geq 102\frac{xs}{f_y}\left(\frac{v_{tn}}{v_{tn} + v_n}\right)$$

Rounding the constant off to 100 and multiplying through by 2 gives

$$2A_t + A_v \geq 200\frac{xs}{f_y}\left(\frac{v_{tn}}{v_{tn} + v_n}\right) \tag{19.13.6}$$

For equal volumes of longitudinal and transverse steel,

$$A_l = 2A_t\left(\frac{x_1 + y_1}{s}\right) \tag{19.13.7}$$

Substitution of Eq. (19.13.6) into Eq. (19.13.7) gives

$$A_l \geq \left[200\frac{xs}{f_y}\left(\frac{v_{tn}}{v_{tn} + v_n}\right) - A_v\right]\left(\frac{x_1 + y_1}{s}\right) \tag{19.13.8}$$

ACI Committee 438 [19.78] recommended that the minimum transverse reinforcement for combined shear and torsion should not be required to exceed that for shear alone; that is,

$$2A_t + A_v \geq \frac{50b_w s}{f_y} \tag{19.13.9}$$

as now stated in ACI-11.5.5.5.

Since Eq. (19.13.9) will usually give a lower requirement than Eq. (19.13.6), the recommendation of Committee 438 [19.78] was to increase the required amount of longitudinal reinforcement. There is no complete agreement on whether this is an acceptable procedure [19.76]. Thus the first term of Eq. (19.13.8) was doubled and A_v replaced by $2A_t$,

$$A_l \geq \left[400 \frac{xs}{f_y} \left(\frac{v_{tn}}{v_{tn} + v_n} \right) - 2A_t \right] \left(\frac{x_1 + y_1}{s} \right) \tag{19.13.10}$$

In order to convert stresses into forces to obtain the ACI Code format, one must recognize that torsional *stress* v_{tn} follows from Eq. (19.2.3), where $x_m = x$. Then divide T_u by $\phi \Sigma \frac{1}{3} x^2 y$, and for the shear *stress* v_n divide V_u by $\phi b_w d$. Using Eq. (19.12.4) for C_t, Eq. (19.13.10) becomes

$$A_l \geq \left[400 \frac{xs}{f_y} \left(\frac{T_u}{T_u + V_u/(3C_t)} \right) - 2A_t \right] \left(\frac{x_1 + y_1}{s} \right) \tag{19.13.11}$$

which is ACI Formula (11-25). In the use of Eqs. (19.13.10) or (19.13.11) the value for $2A_t$ is not to be taken less than $50b_w s/f_y$.

Spacing Limitations. The spacing of closed hoops may not exceed $(x_1 + y_1)/4$ nor 12 in., whichever is smaller (ACI-11.6.8.1).

Longitudinal bars must be at least #3 in size and be distributed around the perimeter not farther than 12 in. apart and with one bar in each corner (ACI-11.6.8.2).

These limits (ACI Commentary-R11.6.8.1) are intended "to insure the development of the ultimate torsional strength of the beam, to prevent excessive loss of torsional stiffness after cracking, and to control crack widths."

Termination of Torsion Reinforcement. Closed hoops are permitted to be terminated (ACI-11.6.7.6) at a location $(b_t + d)$ equal to the sum of the width b_t of that part of the cross-section containing closed stirrups resisting torsion plus the effective depth d, beyond the point where the factored torsional moment T_u exceeds ϕT_c. Note that torsion must be included in the design whenever T_u exceeds $\phi(0.5\sqrt{f_c'}\Sigma x^2 y)$, but the amount A_t and A_l of reinforcement required for torsion is based on $\phi T_s = T_u - \phi T_c$.

19.14 EXAMPLES

Several examples are presented to illustrate use of the ACI Code procedures. Table 19.14.1 summarizes most of the ACI provisions on torsion; it is prepared to help the reader in reviewing the computations in the examples.

■ **EXAMPLE 19.14.1** A reinforced concrete spandrel beam has overall dimensions of 10 × 18 in. and is joined integrally with a 6-in. slab (based on Example of Ref. 19.51) as shown in Fig. 19.14.1(a). The section shown is that at the critical location a distance d from the face of support. At this section, the factored loads are negative bending moment $M_u = 75$ ft-kips, shear force $V_u = 9$ kips, and torsional moment $T_u = 8$ ft-kips. Assume that the torsional stiffness was estimated and used in a structural analysis to obtain these design loadings. Check the adequacy of this section, and select the transverse hoop steel required, if any, according to the ACI Code using $f'_c = 4000$ psi and $f_y = 40,000$ psi.

Solution: **(a)** Flexural strength. The required coefficient of resistance R_n is

$$\text{required } R_n = \frac{\text{required } M_n}{bd^2} = \frac{M_u}{\phi bd^2}$$

$$= \frac{75(12,000)}{0.90(10)(16)^2} = 390 \text{ psi}$$

From Fig. 3.8.1, the required reinforcement ratio ρ is

$$\text{required } \rho = 0.011$$

$$\text{required } A_s = 0.011(10)(16) = 1.76 \text{ sq in.}$$

$$\text{provided } A_s = 3(0.79) = 2.37 \text{ sq in.} > 1.76 \qquad \textbf{OK}$$

(a) Given data (from Hsu, [19.51])

(b) Details of torsion reinforcement

Figure 19.14.1 Spandrel beam of Example 19.14.1.

Table 19.14.1 ACI Code Provisions on Torsion

---

1. $T_u = 0$ if computed $T_u \leq \phi(0.5\sqrt{f'_c}\Sigma x^2 y)$ $v_{tn} \leq 1.5\sqrt{f'_c}$ 11.6.1

   $T_u = \phi(4\sqrt{f'_c}\Sigma\frac{1}{3}x^2 y)$, with redistribution $v_{tn} \leq 4\sqrt{f'_c}$ 11.6.3

---

2. $\displaystyle V_c = \frac{2\sqrt{f'_c}\,b_w d}{\sqrt{1 + \left(2.5C_t\dfrac{T_u}{V_n}\right)^2}}$ $\displaystyle v_c = \frac{2\sqrt{f'_c}}{\sqrt{1 + \left(\dfrac{v_{tn}}{1.2v_n}\right)^2}}$ 11.3.1.4; Formula (11-5)

   $T_u > \phi(0.5\sqrt{f'_c}\Sigma x^2 y)$ $\displaystyle v_{tn} = \frac{T_u}{\phi\Sigma\frac{1}{3}x^2 y}$

   $\displaystyle C_t = \frac{b_w d}{\Sigma x^2 y}$ $\displaystyle v_n = \frac{V_u}{\phi b_w d}$

   $N_u = $ axial tension (negative):

   $V_c = 0$; or multiply above V_c by $\left(1 + \dfrac{N_u}{500A_g}\right)$ 11.6.6.2

---

3. $\phi T_n \geq T_u$ $\displaystyle v_{tn} = \frac{T_u}{\phi\Sigma\frac{1}{3}x^2 y}$ 11.6.5; Formula (11-20)

   $T_n = T_c + T_s$ $v_{tn} = v_{tc} + v_{ts}$ 11.6.5; Formula (11-21)

---

4. $\displaystyle T_c = \frac{0.8\sqrt{f'_c}\Sigma x^2 y}{\sqrt{1 + \left(\dfrac{0.4V_u}{C_t T_u}\right)^2}}$ $\displaystyle v_{tc} = \frac{2.4\sqrt{f'_c}}{\sqrt{1 + \left(\dfrac{1.2v_n}{v_{tn}}\right)^2}}$ 11.6.6.1; Formula (11-22)

   $T_u > \phi(0.5\sqrt{f'_c}\Sigma x^2 y)$ $\displaystyle v_{tn} = \frac{T_u}{\phi\Sigma\frac{1}{3}x^2 y}$

   $\displaystyle C_t = \frac{b_w d}{\Sigma x^2 y}$ $\displaystyle v_n = \frac{V_n}{\phi b_w d}$

   $N_u = $ axial tension (negative):

   $T_c = 0$; or multiply above T_c by $\left(1 + \dfrac{N_u}{500A_g}\right)$ 11.6.6.2

---

5. $\displaystyle T_s = \frac{A_t \alpha_t x_1 y_1 f_y}{s}$ $\displaystyle v_{ts} = \frac{A_t \alpha_t x_1 y_1 f_y}{s\Sigma\frac{1}{3}x^2 y}$ 11.6.9.1; Formula (11-23)

   $f_y \leq 60{,}000$ psi $\alpha_t = (0.66 + 0.33 y_1/x_1) < 1.50$ 11.6.7.4; 11.6.9.1

   max $T_s = 4T_c$ max $v_{ts} = 4v_{tc}$ 11.6.9.4

   $\displaystyle A_l = 2A_t\left(\frac{x_1 + y_1}{s}\right)$ 11.6.9.3; Formula (11-24)

   $\displaystyle A_l = \left[\frac{400xs}{f_y}\left(\frac{T_u}{T_u + \dfrac{V_u}{3C_t}} \text{ or } \frac{v_{tn}}{v_{tn} + v_n}\right) - \left(\text{larger of } 2A_t \text{ and } \frac{50b_w s}{f_y}\right)\right]\frac{x_1 + y_1}{s}$ 11.6.9.3; Formula (11-25)

Table 19.14.1   (continued)

6. min $A_v = \dfrac{50 b_w s}{f_y}$	11.5.5.3; Formula (11-14)
for $T_u \le \phi(0.5\sqrt{f_c'}\,\Sigma x^2 y)$ and $V_u > \frac{1}{2}\phi V_c$	
min A_t to satisfy $A_v + 2A_t = \dfrac{50 b_w s}{f_y}$	11.6.9.2; 11.5.5.5 Formula (11-16)
for $T_u > \phi(0.5\sqrt{f_c'}\,\Sigma x^2 y)$ and $V_u > \frac{1}{2}\phi V_c$	
7. max s (closed stirrups) $\le \dfrac{x_1 + y_1}{4} \le 12$ in.	11.6.8.1
torsion reinforcement to terminate at $(d + b_t)$ beyond theoretical point	11.6.7.6
max $s \left(\begin{array}{l}\text{longitudinal bars, } \#3 \text{ or}\\ \text{larger, one in each corner}\end{array}\right) \le 12$ in.	11.6.8.2

Available steel area for longitudinal torsion reinforcement,

$$\text{available } A_s = 2.37 - 1.76 = 0.61 \text{ sq in.}$$

$$\text{available } A_s' = 2(0.60) = 1.20 \text{ sq in.}$$

(b) Compute maximum T_u to neglect torsion. The maximum factored torsional moment T_u that may be neglected (ACI-11.6.1) is

$$\text{limit } T_u = \phi[0.5\sqrt{f_c'}(\Sigma x^2 y)]$$

$$\Sigma x^2 y = (10)^2(18) + (6)^2(18) = 2450 \text{ in.}^3$$

Note that the maximum effective width of slab is 3 times its thickness (ACI-11.6.1.1). Thus,

$$\text{limit } T_u = 0.85[0.5\sqrt{4000}(2450)]\tfrac{1}{12,000} = 5.5 \text{ ft-kips}$$

Since T_u of 8 ft-kips exceeds limit T_u, torsion must be included in design.

(c) Shear reinforcement. The strength attributable to the concrete is given by Eq. (19.12.5) which is ACI Formula (11-5),

$$V_c = \frac{2\sqrt{f_c'}\,b_w d}{\sqrt{1 + \left(2.5 C_t \dfrac{T_u}{V_u}\right)^2}}$$

$$C_t = \frac{b_w d}{\Sigma x^2 y} = \frac{10(16)}{2450} = 0.0653 \text{ per in.}$$

$$V_c = \frac{2\sqrt{4000}(10)(16)/1000}{\sqrt{1 + \left(2.5(0.0653)\dfrac{8(12)}{9}\right)^2}} = 10.1 \text{ kips}$$

$$\phi V_c = 0.85(10.1) = 8.6 \text{ kips}$$

Since $V_u > \phi V_c/2$ (ACI-11.5.5.1), the requirement for transverse reinforcement based on strength is

$$\frac{A_v}{s} = \frac{V_u - \phi V_c}{\phi d f_y} = \frac{9 - 8.6}{0.85(16)(40)} = 0.00074$$

This amount will be combined with the torsion requirement, and the total $A_v + 2A_t$ must be at least $50b_w s/f_y$ (ACI-11.5.5.5).

(d) Transverse torsional reinforcement. From Eq. (19.12.7) which is ACI Formula (11-22),

$$T_c = \frac{0.8\sqrt{f_c'}(\Sigma x^2 y)}{\sqrt{1 + \left(\frac{0.4}{C_t}\frac{V_u}{T_u}\right)^2}}$$

$$T_c = \frac{0.8\sqrt{4000}(2450)/12,000}{\sqrt{1 + \left(\frac{0.4}{0.0653}\frac{9}{8(12)}\right)^2}} = 8.95 \text{ ft-kips}$$

$$\phi T_c = 0.85(8.95) = 7.6 \text{ ft-kips}$$

The requirement of closed hoops for strength is, from Eq. (19.11.5) which is obtained from ACI Formula (11-23),

$$\frac{A_t}{s} = \frac{T_u - \phi T_c}{\phi \alpha_t x_1 y_1 f_y}$$

$$= \frac{(8.0 - 7.6)12}{0.85(1.39)(6.625)(14.625)40} = 0.00105$$

where

$$x_1 = 10 - 2(1.5) - 0.375 = 6.625 \text{ in.}$$

$$y_1 = 18 - 2(1.5) - 0.375 = 14.625 \text{ in.}$$

$$\alpha_t = 0.66 + 0.33\left(\frac{y_1}{x_1}\right) = 0.66 + 0.33\left(\frac{14.625}{6.625}\right) = 1.39 < 1.50 \qquad \textbf{OK}$$

(e) Total transverse reinforcement for combined shear and torsion. The total transverse reinforcement required for strength is

$$\frac{A_v}{s} + \frac{2A_t}{s} = 0.00074 + 0.00210 = 0.00284$$

$$\min\left(\frac{A_v}{s} + \frac{2A_t}{s}\right) = \frac{50b_w}{f_y} = \frac{50(10)}{40,000} = 0.0125 \qquad \text{controls}$$

For #3 closed hoops,

$$\max s = \frac{2(0.11)}{0.0125} = 17.6 \text{ in.}$$

Since this exceeds the spacing limitations of ACI-11.6.8.1,

$$\text{max } s = \frac{x_1 + y_1}{4} = \frac{6.625 + 14.625}{4} = 5.3 \text{ in.}$$

Since this is less than the 12-in. upper limit, the 5.3-in. limit controls.

Use #3 hoops at 5 in. spacing.

(f) Longitudinal torsional reinforcement. According to ACI Formula (11-24) with the strength requirement $2A_t/s = 0.00210$

$$A_l = \frac{2A_t}{s}(x_1 + y_1)$$

$$= 0.00210(6.625 + 14.625) = 0.045 \text{ sq in.}$$

or as a minimum,

$$A_l = \left[\frac{400xs}{f_y}\left(\frac{T_u}{T_u + \dfrac{V_u}{3C_t}}\right) - 2A_t\right]\left(\frac{x_1 + y_1}{s}\right)$$

$$= \left[\frac{400(10)}{40,000}\left(\frac{8(12)}{8(12) + 9/(0.196)}\right) - \frac{2A_t}{s}\right](6.625 + 14.625)$$

$$= \left(0.0676 - \frac{2A_t}{s}\right)(21.25)$$

Since the required $2A_t/s$ of 0.00210 for strength is less than $50b_w/f_y$ of 0.0125, the latter is to be used for $2A_t/s$ in the above equation

$$\text{required } A_l = (0.0676 - 0.0125)(21.25) = 1.17 \text{ sq in.}$$

This is to be distributed around the perimeter of the section at a spacing not to exceed 12 in. In this case, bars must be placed at middepth

$$\frac{A_l}{3} = \frac{1.17}{3} = 0.39 \text{ sq in.}$$

Use 2-#4 bars at middepth.

The longitudinal steel at the top and bottom in excess of that required for flexure is more than adequate for the torsional requirement. Also, #3 transverse reinforcement should be placed in the effective flange portion of the slab and it should be anchored within the main rectangle resisting torsion. Though the ACI Code requires only standard hooks at the location where the hoop closes, Mitchell and Collins [19.81] have recommended the use of 105° hooks (the 105° is the amount of bend from the initial straight bar). Furthermore, when longitudinal steel is to carry torsion at the face of support, the steel should be embedded an amount L_d into the support [19.81]. The reinforced section is shown in Fig. 19.14.1(b). ∎

■ **EXAMPLE 19.14.2**   For the continuous spandrel beam shown in Fig. 19.14.2, design the longitudinal and transverse reinforcement for the factored moment, factored flexural shear, and factored torsional moment, also given in Fig. 19.14.2. Assume that these design loads were obtained after estimating the torsional stiffness and performing a structural analysis. Use $f'_c = 4000$ psi and $f_y = 60,000$ psi.

Solution: **(a)** Flexural strength. The effective depth d is approximately 17.5 in. for one layer of reinforcement. At midspan, neglecting T-beam effect,

$$\text{required } R_n = \frac{M_u}{\phi b d^2} = \frac{83.3(12,000)}{0.90(12)(17.5)^2} = 302 \text{ psi}$$

From Fig. 3.8.1, the required $\rho = 0.0055$ which gives

$$\text{required } A_s = 0.0055(12)(17.5) = 1.15 \text{ sq in.}$$

$$\min A_s = \frac{200}{f_y} bd = \frac{200(12)(17.5)}{60,000} = 0.70 \text{ sq in.}$$

At the support,

$$\text{required } A_s = 1.15(61.7/83.3) = 0.85 \text{ sq in.}$$

The reader may note that less reinforcement than 0.70 sq in. may be used if the amount used is one-third more than required for strength (ACI-10.5.2). Thus the sloping straight line of Figs. 19.14.3(a) and (b) represents approximately $\frac{4}{3}$ of the required A_s in the region between the controlling minimum requirements of ACI-10.5.1 and ACI-12.11.1 (or ACI-12.12.3).

(b) Factored shear V_u and torsion T_u at the critical section. Unless a concentrated load (or girder framing in) exists between the face of support and the distance d therefrom, the maximum shear and torsion for which strength must be provided is at d from the face of support. From Fig. 19.14.2,

$$V_u = 19.4 \text{ kips}$$

$$T_u = 47.7 \text{ ft-kips}$$

(c) Nominal shear strength V_c and nominal torsional moment strength T_c attributable to concrete. Equation (19.12.5), i.e., ACI Formula (11-5), gives

$$V_c = \frac{2\sqrt{f'_c}\, b_w d}{\sqrt{1 + \left(2.5 C_t \dfrac{T_u}{V_u}\right)^2}} \qquad \textbf{[19.12.5]}$$

$$C_t = \frac{b_w d}{\Sigma x^2 y} = \frac{12(17.5)}{(12)^2(20) + (6)^2(18)} = \frac{210}{3528} = 0.0595 \text{ per in.}$$

In computing $\Sigma x^2 y$, the effective width of flange may not exceed 3 times the flange thickness, that is, 18 in. for this beam (ACI-11.6.1.1).

Figure 19.14.2 Spandrel beam including design loadings, for Example 19.14.2 (from Mattock, [19.73]).

$$V_c = \frac{2\sqrt{4000}(210/1000)}{\sqrt{1 + \left(2.5(0.0595)\dfrac{47.7(12,000)}{19.4(1000)}\right)^2}} = 0.44\sqrt{4000}\left(\frac{210}{1000}\right) = 5.9 \text{ kips}$$

Equation (19.12.7), i.e., ACI Formula (11-22), gives

$$T_c = \frac{0.8\sqrt{f_c'}(\Sigma x^2 y)}{\sqrt{1 + \left(\dfrac{0.4}{C_t}\dfrac{V_u}{T_u}\right)^2}} \qquad\qquad [19.12.7]$$

$$T_c = \frac{0.8\sqrt{4000}(3528/12,000)}{\sqrt{1 + \left(\dfrac{0.4}{0.0595}\dfrac{19.4(1000)}{47.7(12,000)}\right)^2}} = 0.78\sqrt{4000}\,\frac{3528}{12,000} = 14.5 \text{ ft-kips}$$

For this example, V_c and T_c are constant along the span because the ratio V_u/T_u is constant along the span.

(d) Maximum nominal strengths permitted. For torsion, ACI-11.6.9.4 gives

$$\max T_s = 4T_c$$

Since $T_n = T_c + T_s$,

$$\max T_n = T_c + 4T_c = 5T_c$$

For this example,

$$\max T_u = \max \phi T_n = 0.85(5)(14.5) = 61.6 \text{ ft-kips} > 47.7 \text{ ft-kips} \qquad\qquad \textbf{OK}$$

For shear, ACI-11.5.4.3 permits

$$\max V_s = 4\sqrt{f_c'}\,b_w d$$

with maximum stirrup spacings not exceeding $d/2$. Beyond $V_s = 4\sqrt{f_c'}\,b_w d$, spacings not exceeding $d/4$ are permitted (ACI-11.5.6.8) to $V_s = 8\sqrt{f_c'}\,b_w d$, the upper limit regardless of stirrup spacing. In this example, for the $d/2$ spacing limit,

$$\max V_s = 4\sqrt{f_c'}\,b_w d = 4\sqrt{4000}(12)(17.5)\tfrac{1}{1000} = 53.1 \text{ kips}$$

which means max V_u for the $d/2$ spacing limit becomes

$$\max V_u = \phi(V_c + \max V_s)$$
$$= \phi(5.9 + 53.1) = 50.1 \text{ kips} > 19.4 \text{ kips} \qquad\qquad \textbf{OK}$$

(e) Transverse reinforcement. For shear, ACI-11.5.6.2 using ACI Formula (11-17) gives

$$\frac{A_v}{s} = \frac{\text{required } \phi V_s}{\phi f_y d} = \frac{V_u - \phi V_c}{\phi f_y d}$$

which at the critical section d from the face of support is

$$\frac{A_v}{s} = \frac{19.4 - 0.85(5.9)}{0.85(60)(17.5)} = 0.0161 \text{ sq in./in.}$$

For torsion, Eq. (19.11.5) from ACI Formula (11-23) requires

$$\frac{A_t}{s} = \frac{\text{required } \phi T_s}{\phi \alpha_t x_1 y_1 f_y} = \frac{T_u - \phi T_c}{\phi \alpha_t x_1 y_1 f_y}$$

which at the critical section d from the face of support is

$$\frac{A_t}{s} = \frac{[47.7 - 0.85(14.5)]12}{0.85(1.30)(8.5)(16.5)60} = 0.0456 \text{ sq in./in.}$$

where

$x_1 = 12 - 2(1.5) - 0.50 = 8.5 \text{ in.}$

$y_1 = 20 - 2(1.5) - 0.50 = 16.5 \text{ in.}$

$\alpha_t = 0.66 + 0.33 y_1/x_1 = 0.66 + 0.33(16.5/8.5) = 1.30 < 1.50$ **OK**

The minimum transverse reinforcement, Eq. (19.13.9) from ACI-11.5.5.5 gives

$$\frac{2A_t}{s} + \frac{A_v}{s} \geq \frac{50 b_w}{f_y} = \frac{50(12)}{60,000} = 0.01$$

For strength, the total requirement is

$$\frac{2A_t}{s} + \frac{A_v}{s} = 2(0.0456) + 0.0161 = 0.107 > 0.01$$

Try #4 closed hoops, $A_v = 2(0.20) = 0.40$ sq in. The maximum spacing of hoops between the face of support and the critical section is

$$\max s = \frac{0.40}{0.107} = 3.74 \text{ in.}$$

The upper limit for spacing of closed hoops anywhere is (ACI-11.6.8.1),

$$\max s = \frac{x_1 + y_1}{4} = \frac{8.5 + 16.5}{4} = 6.25 \text{ in.}$$

and not in any case to exceed 12 in.

(f) Longitudinal reinforcement for torsion at the critical section. Using Eq. (19.11.7) which is ACI Formula (11-24),

$$A_l = \frac{2A_t}{s}(x_1 + y_1) = 0.091(8.5 + 16.5) = 2.27 \text{ sq in.}$$

The minimum value is, from Eq. (19.13.11) which is ACI Formula (11-25),

$$A_l = \left[400 \frac{x}{f_y} \left(\frac{T_u}{T_u + V_u/(3C_t)} \right) - \left(\text{larger of } \frac{2A_t}{s} \text{ and } \frac{50 b_w}{f_y} \right) \right](x_1 + y_1)$$

$$A_l = \left[400 \frac{12}{60,000} \left(\frac{47.7(12)}{47.7(12) + 19.4/[3(0.0595)]} \right) - 0.091 \right](8.5 + 16.5)$$

$= \text{negative}$ (does not control)

Figure 19.14.3 Reinforcement requirements for spandrel beam of Example 19.14.2.

Note that the number 400 in the above equation has psi units, thus f_y must be in psi units.

ACI-11.6.8.2 requires the longitudinal bars to be distributed around the perimeter with a spacing not to exceed 12 in. Since the member depth exceeds 12 in., a layer intermediate between top and bottom is neceesary. Thus use at middepth

$$\frac{A_l}{3} = \frac{2.27}{3} = 0.76 \text{ sq in.}$$

Use 2-#6 bars, one at each side of the member.

(g) Total longitudinal reinforcement for flexure and torsion. At the face of support,

top steel = 0.85 (flexure) + 0.76 (torsion) = 1.61 sq in.

bottom steel = 0.29 (flexure) + 0.76 (torsion) = 1.05 sq in.

At d from the face of support,

top steel = 0.70 (flexure) + 0.76 (torsion) = 1.46 sq in.

bottom steel = 0.29 (flexure) + 0.76 (torsion) = 1.05 sq in.

(h) Transverse steel requirement along the span. Using the same procedure as at the critical section, the requirements are computed as shown in Table 19.14.2. The maximum spacing curve and selected spacings are shown in Fig. 19.14.3(c). Since the factored torsional moment T_u is never less than $\phi(0.5\sqrt{f_c'}\Sigma x^2 y) = 7.9$ ft-kips, closed stirrups are required along the entire span.

(i) Longitudinal steel requirements along the span. The flexural requirements have already been shown in Figs. 19.14.3(a) and (b). The torsional requirements are computed in Table 19.14.3. Because the member depth exceeds 12 in., the longitudinal steel is placed one-third at each of top, bottom, and midheight. The sums of the

Table 19.14.2 Transverse Steel Requirement Along The Span (Example 19.14.2)

LOCATION FROM CENTER OF SUPPORT	V_u (kips)	$\frac{A_v}{s}$ (in.)	T_u (ft-kips)	$\frac{A_t}{s}$ (in.)	$\frac{50b_w}{f_y}$ (in.)	REQUIRED $\frac{2A_t}{s} + \frac{A_v}{s}$ (in.)	MAX s FOR #4 HOOPS (in.)
4	17.7	0.0142	43.5	0.0402	0.010	0.095	4.20
6	15.5	0.0118	38.1	0.0332	0.010	0.078	5.13
8	13.3	0.0092	32.7	0.0262	0.010	0.062	6.25[b]
10	11.1	0.0068	27.2	0.0191	0.010	0.045	6.25[b]
12	8.9	0.0044	21.8	0.0123	0.010	0.029	6.25[b]
15	5.6	0.0006	13.7	0.0017	0.010	0.010[a]	6.25[b]

[a] min $50b_w/f_y$ controls

[b] max $s = (x_1 + y_1)/4$ controls

Table 19.14.3 Longitudinal Steel Requirement for Torsion Along the Span (Example 19.14.2)

LOCATION FROM CENTER OF SUPPORT (ft)	REQUIRED $\dfrac{2A_t{}^*}{s}$ (in.)	ACI FORMULA (11-24)[†] (sq in.)	$\dfrac{400x}{f_y}\left(\dfrac{T_u}{T_u + V_u/3C_t}\right)$	ACI FORMULA (11-25)[‡] (sq in.)	REQUIRED A_l (sq in.)
4	0.0804	2.01	0.067	negative	2.01
6	0.0664	1.66	0.067	negative	1.66
8	0.0524	1.31	0.067	0.38	1.31
10	0.0382	0.96	0.067	0.73	0.96
12	0.0246	0.62	0.067	1.06	1.06
15	0.0034	0.09	0.067	1.43	1.43

* $2A_t/s$ required for torsion, obtained from Table 19.14.2, but need not be less than $50b_w/f_y = 0.010$ for computing longitudinal reinforcement (ACI-11.6.9.3).

[†] $A_l = (2A_t/s)(x_1 + y_1)$

[‡] $\min A_l = \left[\dfrac{400x}{f_y}\left(\dfrac{T_u}{T_u + V_u/3C_t}\right) - \left(\text{larger of } \dfrac{2A_t}{s} \text{ and } \dfrac{50b_w}{f_y}\right)\right](x_1 + y_1)$

Section A-A

Section B-B

Figure 19.14.4 Design details for Example 19.14.2.

flexural requirement and $A_t/3$ are shown as the total requirement for longitudinal steel in Figs. 19.14.3(d) and (e).

In determining the length of longitudinal bars, a conservative interpretation has been made of ACI-11.6.7.6 which requires an extension of $d + b_t$ beyond the theoretical termination point, as in Figs. 19.14.3(d) and (e). The theoretical termination point is taken as that for combined flexure and torsion. It is noted that in some cases ACI-12.10.5 regarding cutting bars in a tension zone may control. The crack control provisions of ACI-10.6.4 must be checked, as well as the deflection if excessive deflection may cause damage. The design details are shown in Fig. 19.14.4. ■

■ **EXAMPLE 19.14.3** Redesign the spandrel beam shown in Fig. 19.14.2 taking the option permitted in ACI-11.6.3 of considering the torsional stiffness to be zero in performing the structural analysis to obtain the design loads. The torsional member is then designed for a nominal torsional stress of $4\sqrt{f_c'}$.

Solution: (a) Flexural strength. Assume that the factored bending moment and shear are not significantly affected by a change in the assumption for torsional stiffness; that is, use Figs. 19.14.2(b) and (c). Note that most structural analysis methods traditionally do not include the effect of torsional stiffness; thus this approach is simple. After completing the structural analysis disregarding torsional stiffness, any member that needs to undergo torsional deformation is then arbitrarily designed to include a torsional stress v_{tn} of $4\sqrt{f_c'}$.

For this example, the requirements for flexure are as in Example 19.14.2 and are shown in Figs. 19.14.3(a) and (b).

(b) Factored shear V_u and torsion T_u forces for which strength must be provided. At d from face of support,

$$V_u = 19.4 \text{ kips} \qquad [\text{see Fig. 19.14.2(c)}]$$

$$T_u = \phi(4\sqrt{f_c'})(\tfrac{1}{3}\Sigma x^2 y) = 0.85(4\sqrt{4000})(1176)\tfrac{1}{12,000} = 21.1 \text{ ft-kips}$$

Note that the 21.1 ft-kips compares with the maximum factored torsional moment of 47.7 ft-kips that was designed for in Example 19.14.2.

(c) Nominal strengths T_c and V_c attributable to the concrete. For torsion, Eq. (19.12.7), which is ACI Formula (11-22), gives at d from the face of support

$$T_c = \frac{0.8\sqrt{f_c'}(\Sigma x^2 y)}{\sqrt{1 + \left(\dfrac{0.4\,V_u}{C_t\,T_u}\right)^2}} = \frac{0.8\sqrt{4000}(3528)/12,000}{\sqrt{1 + \left(\dfrac{0.4}{0.0595}\dfrac{19.4}{(21.1)(12)}\right)^2}} = 13.3 \text{ ft-kips}$$

using $\Sigma x^2 y = 3528$ in.^{3} and $C_t = 0.0595$ per in. from Example 19.14.2, part (c).

For shear, Eq. (19.12.5), which is ACI Formula (11-5), gives at d from the face of support

$$V_c = \frac{2\sqrt{f_c'}\,b_w d}{\sqrt{1 + \left(2.5 C_t \dfrac{T_u}{V_u}\right)^2}} = \frac{2\sqrt{4000}(12)17.5/1000}{\sqrt{1 + \left(2.5(0.0595)\dfrac{21.1(12)}{19.4}\right)^2}} = 12.2 \text{ kips}$$

Though the factored shear force V_u decreases farther from the support, the code-specified factored torsional force T_u, based on a stress of $4\sqrt{f_c'}$ psi used to provide ductility, is taken constant along the length of the beam.

At midspan,

$$V_u = 5.6 \text{ kips}; \qquad T_u = 21.1 \text{ ft-kips}$$

$$T_c = \frac{0.8\sqrt{4000}(3528)/12,000}{\sqrt{1 + \left(\dfrac{0.4}{0.0595}\dfrac{5.6}{21.1(12)}\right)^2}} = 14.8 \text{ ft-kips}$$

$$V_c = \frac{2\sqrt{4000}(12)(17.5)/1000}{\sqrt{1 + \left(2.5(0.0595)\dfrac{21.1(12)}{5.6}\right)^2}} = 3.91 \text{ kips}$$

(d) Transverse reinforcement. For shear, using ACI Formula (11-17) from ACI-11.5.6.2, at the critical section

$$\frac{A_v}{s} = \frac{V_u - \phi V_c}{\phi f_y d} = \frac{19.4 - 0.85(12.2)}{0.85(60)(17.5)} = 0.0101$$

For torsion, using Eq. (19.11.5), which is ACI Formula (11-23), at the critical section

$$\frac{A_t}{s} = \frac{T_u - \phi T_c}{\phi \alpha_t x_1 y_1 f_y} = \frac{[21.1 - 0.85(13.3)]12}{0.85(1.30)(8.5)(16.5)60} = 0.0126$$

where $\alpha_t = 1.30$, $x_1 = 8.5$ in., and $y_1 = 16.5$ in. as computed in Example 19.14.2. The total requirement for strength at d from the face of support is

$$\frac{2A_t}{s} + \frac{A_v}{s} = 2(0.0126) + 0.0101 = 0.0353$$

This exceeds the minimum of $50b_w/f_y$ of 0.01.

The requirement at midspan is

$$\frac{A_v}{s} = \frac{5.6 - 0.85(3.91)}{0.85(60)(17.5)} = 0.0026$$

$$\frac{A_t}{s} = \frac{[21.1 - 0.85(14.8)]12}{0.85(1.30)(8.5)(16.5)60} = 0.0110$$

$$\frac{2A_t}{s} + \frac{A_v}{s} = 2(0.0110) + 0.0026 = 0.0246 > 0.01 \text{ min} \qquad\qquad \textbf{OK}$$

Try #3 hoops, $A_v = 2(0.11) = 0.22$ sq in. The maximum spacing of hoops between the face of support and the critical section is

$$\max s = \frac{0.22}{0.0353} = 6.2 \text{ in.}$$

The upper limit for spacing of closed hoops is (ACI-11.6.8.1),

$$\max s = \frac{x_1 + y_1}{4} = \frac{8.625 + 16.625}{4} = 6.3 \text{ in.}$$

Thus for this beam a practical spacing of 6 in. may be used for the entire beam. Note that the distances x_1 and y_1 have been corrected to agree with the use of #3 hoops.

(e) Longitudinal reinforcement for torsion. Using Eq. (19.11.7), which is ACI Formula (11-24), at the critical section,

$$A_l = \frac{2A_t}{s}(x_1 + y_1) = 2(0.0126)(8.625 + 16.625) = 0.64 \text{ sq in.}$$

where A_t/s has been corrected to agree with the values of x_1 and y_1 using #3 hoops. The minimum value at the critical section is obtained from Eq. (19.13.11), which is ACI Formula (11-25),

$$A_l = \left[\frac{400x}{f_y} \left(\frac{T_u}{T_u + \dfrac{V_u}{3C_t}} \right) - \left(\text{larger of } \frac{2A_t}{s} \text{ and } \frac{50b_w}{f_y} \right) \right](x_1 + y_1)$$

$$= \left[\frac{400(12)}{60,000} \left(\frac{21.1}{21.1 + \dfrac{19.4}{3(0.0595)12}} \right) - 0.0252 \right] (8.625 + 16.625)$$

$$= 0.0307(25.25) = 0.78 \text{ sq in.} \qquad\qquad \text{Controls}$$

Note that the amount A_l will be divided into three equal parts, top of beam, bottom of beam, and midheight in accordance with ACI-11.6.8.2. The longitudinal requirement for torsion is added to the requirement for flexure [Figs. 19.14.3(a) and (b)] and the total requirements for the top and bottom of the beam are shown in Fig. 19.14.5.

(a) Total area of *bottom* reinforcement required for beam

(b) Total area of *top* reinforcement required for beam

Figure 19.14.5 Longitudinal steel requirements for Example 19.14.3.

These requirements can be compared with Figs. 19.14.3(d) and (e) that were obtained when the torsional moments were computed after making stiffness assumptions. Note that in this alternate method, fewer and smaller hoops are required (i.e., #3 @ 6 throughout) and likewise less longitudinal reinforcement is needed. ∎

SELECTED REFERENCES

General

19.1. Paul Zia. "Torsion Theories for Concrete Members," *Torsion of Structural Concrete* (SP-18). Detroit: American Concrete Institute, 1968 (pp. 103–132).

19.2. Paul Zia. "What Do We Know About Torsion in Concrete Members?" *Journal of the Structural Division*, ASCE, **96**, ST6 (June 1970), 1185–1199.

19.3. K. G. Tamberg and P. T. Mikluchin. "Torsional Phenomena Analysis and Concrete Structure Design," *Analysis of Structural Systems for Torsion* (SP-35). Detroit: American Concrete Institute, 1973 (pp. 1–102).

19.4. Paul Lampert. "Postcracking Stiffness of Reinforced Concrete Beams in Torsion and Bending," *Analysis of Structural Systems for Torsion* (SP-35). Detroit: American Concrete Institute, 1973 (pp. 385–433). (Also presented at 1971 Annual Convention, American Concrete Institute, Denver, March 1971.)

19.5. Michael P. Collins and Paul Lampert. "Redistribution of Moments at Cracking—The Key to Simpler Torsion Design?" *Analysis of Structural Systems for Torsion* (SP-35). Detroit: American Concrete Institute, 1973 (pp. 343–383).

19.6. David J. Victor, Narayanan Lakshmanan, and Narayanan Rajagopalan. "Ultimate Torque of Reinforced Concrete Beams," *Journal of the Structural Division*, ASCE, **102**, ST7 (July 1976), 1337–1352.

19.7. Arthur E. McMullen and B. Vijaya Rangan. "Pure Torsion in Rectangular Sections— A Re-Examination," *ACI Journal, Proceedings*, **75**, October 1978, 511–519.

19.8. Madhusudan Chakraborty. "Ultimate Torque of Reinforced Rectangular Beams," *Journal of the Structural Division*, ASCE, **105**, ST3 (March 1979), 653–668.

19.9. Michael P. Collins and Denis Mitchell. "Shear and Torsion Design of Prestressed and Non-Prestressed Concrete Beams," *PCI Journal*, **25**, September/October 1980, 32–100. Disc., **26**, November/December 1981, 96–118.

19.10. J. Warwaruk. "Torsion in Reinforced Concrete," *Significant Developments in Engineering Practice and Research* (SP-72). Detroit: American Concrete Institute, 1981 (pp. 247–277).

19.11. Himat T. Solanki. "Reinforced Concrete Beams in Pure Torsion," *Journal of the Structural Division*, ASCE, **108**, ST12 (December 1981), 2243–2250.

19.12. Thomas T. C. Hsu. *Torsion of Reinforced Concrete*. New York: Van Nostrand Reinhold, 1983.

19.13. Thomas T. C. Hsu and Y. L. Mo. "Softening of Concrete in Torsional Members— Theory and Tests," *ACI Journal, Proceedings*, **82**, May—June 1985, 290–303. Disc. **83**, July–August 1986, 690–693.

19.14. Peter Marti and Keith Kong. "Response of Reinforced Concrete Slab Elements to Torsion," *Journal of Structural Engineering*, ASCE, **113**, 5 (May 1987), 976–993.

19.15. Peter Marti, Peter Leesti, and Waseem U. Khalifa. "Torsion Tests on Reinforced Concrete Slab Elements," *Journal of Structural Engineering*, ASCE, **113**, 5 (May 1987), 994–1010.

19.16. Thomas T. C. Hsu. "Softened Truss Model Theory for Shear and Torsion." *ACI Structural Journal*, **85**, November–December 1988, 624–635.

19.17. Atef H. Bakhsh, Faisal F. Wafa, and Ali A. Akhtaruzzaman. "Torsional Behavior of Plain High-Strength Concrete Beams," *ACI Structural Journal*, **87**, September–October 1990, 583–588.

19.18. Michael P. Collins and Denis Mitchell. *Prestressed Concrete Structures*. Englewood Cliffs, NJ: Prentice Hall, 1991 (Chapter 8).

Skew Bending Concept

19.19. N. N. Lessig. "Determination of Load Carrying Capacity of Reinforced Concrete Element with Rectangular Cross Section Subjected to Flexure with Torsion," Work No. 5. Moscow: Institut Betona i Zhelezobetona (Concrete and Reinforced Concrete Institute), 1959 (pp. 5–28). (Also available as Foreign Literature Study No. 371, PCA Research and Development Labs., Skokie, Illinois.)

19.20. Thomas T. C. Hsu. "Torsion of Structural Concrete—Plain Concrete Rectangular Sections," *Torsion of Structural Concrete* (SP-18). Detroit: American Concrete Institute, 1968 (pp. 203–238). (Also Portland Cement Association Development Department Bulletin D134.)

19.21. Thomas T. C. Hsu. "Torsion of Structural Concrete—A Summary of Pure Torsion," *Torsion of Structural Concrete* (SP-18). Detroit: American Concrete Institute, 1968 (pp. 165–178). (Also Portland Cement Association. Development Department Bulletin D133.)

19.22. Thomas T. C. Hsu. "Ultimate Torque of Reinforced Rectangular Beams," *Journal of the Structural Division*, ASCE, 94, ST2 (February 1968), 485–510. (Also Portland Cement Association Development Department Bulletin D127.)

19.23. Thomas T. C. Hsu. "Torsion of Structural Concrete—Behavior of Reinforced Concrete Rectangular Members," *Torsion of Structural Concrete* (SP-18). Detroit: American Concrete Institute, 1968 (pp. 261–306). (Also Portland Cement Association Development Department Bulletin D135.)

19.24. C. D. Goode and M. A. Helmy. "Ultimate Strength of Reinforced Concrete Beams in Combined Bending and Torsion," *Torsion of Structural Concrete* (SP-18). Detroit: American Concrete Institute, 1968 (pp. 357–377).

19.25. M. P. Collins, P. F. Walsh, F. E. Archer, and A. S. Hall. "Ultimate Strength of Reinforced Concrete Beams Subjected to Combined Torsion and Bending," *Torsion of Structural Concrete* (SP-18). Detroit: American Concrete Institute, 1968 (pp. 379–402).

19.26. Kevin D. Below, B. Vijaya Rangan, and A. Stanley Hall. "Theory for Concrete Beams in Torsion and Bending," *Journal of the Structural Division*, ASCE, ST8 (August 1975), 1645–1660.

Space Truss Analogy

19.27. E. Rausch. *Berechnung des Eisenbetons gegen Verdrehung und Abscheren* (*Design of Reinforced Concrete for Torsion and Shear*). Berlin: Springer-Verlag, 1929.

19.28. Paul Lampert and Bruno Thürlimann. "Ultimate Strength and Design of Reinforced Concrete Beams in Torsion and Bending," *Publications*, International Association for Bridge and Structural Engineering, **31**-1, 1971, 107–131.

19.29. Paul Lampert and Michael P. Collins. "Torsion, Bending and Confusion—An Attempt to Establish the Facts," *ACI Journal, Proceedings*, **69**, August 1972, 500–504.

19.30. CEB-FIP. *Model Code for Concrete Structures, CEB-FIP International Recommendations* (3rd ed.). Paris: Comité Euro-International du Beton, 1978. 348 pp.

19.31. Denis Mitchell and Michael P. Collins. "Diagonal Compression Field Theory—A Rational Model for Structural Concrete in Pure Torsion," *ACI Journal, Proceedings*, **71**, August 1974, 396–408.

19.32. P. Müller. "Failure Mechanisms for Reinforced Concrete Beams in Torsion and Bending," *Publications*, International Association for Bridge and Structural Engineering, 36-II, 1976, 146–163.

19.33. B. G. Rabbat and M. P. Collins. "A Variable Angle Space Truss Model for Structural Concrete Members Subjected to Complex Loading," *Douglas McHenry International Symposium on Concrete and Concrete Structures* (SP-55). Detroit: American Concrete Institute, 1978 (pp. 547–587).

Combined Bending and Torsion

19.34. Hans Gesund, Frederick J. Schuette, George R. Buchanan, and George A. Gray. "Ultimate Strength in Combined Bending and Torsion of Concrete Beams Containing Both Longitudinal and Transverse Reinforcement," *ACI Journal, Proceedings*, **61**, December 1964, 1509–1522.

19.35. John P. Klus and C. K. Wang. "Torsion in Grid Frames," *Torsion of Structural Concrete* (SP-18). Detroit: American Concrete Institute, 1968 (pp. 89–101).

19.36. G. S. Pandit and Joseph Warwaruk. "Reinforced Concrete Beams in Combined Bending and Torsion," *Torsion of Structural Concrete* (SP-18). Detroit: American Concrete Institute, 1968 (pp. 133–163).

19.37. A. A. Gvozdez, N. N. Lessig, and L. K. Rulle. "Research on Reinforced Concrete Beams Under Combined Bending and Torsion in the Soviet Union," *Torsion of Structural Concrete* (SP-18). Detroit: American Concrete Institute, 1968 (pp. 307–336).

19.38. David J. Victor and Phil M. Ferguson. "Reinforced Concrete T-Beams Without Stirrups Under Combined Moment and Torsion," *ACI Journal, Proceedings*, **65**, January 1968, 29–36. Disc., **65**, 560–566.

19.39. V. Ramakrishnan and Y. Ananthanarayana. "Tests to Failure in Bending and Torsion of Reinforced Concrete," *ACI Journal, Proceedings*, **66**, May 1969, 428–431. Disc., 943–944.

19.40. D. W. Kirk and S. D. Lash. "T-Beams Subject to Combined Bending and Torsion," *ACI Journal, Proceedings*, **68**, February 1971, 150–159.

19.41. Hota V. S. GangaRao and Paul Zia. "Rectangular Prestressed Beams in Torsion and Bending," *Journal of the Structural Division*, ASCE, **99**, ST1 (January 1973), 183–198.

19.42. David J. Victor and P. K. Aravindan. "Prestressed and Reinforced Concrete T-Beams Under Combined Bending and Torsion," *ACI Journal, Proceedings*, **75**, October 1978, 526–532.

19.43. P. D. Zararis and G. Gr. Penelis. "Reinforced Concrete T-Beams in Torsion and Bending," *ACI Journal, Proceedings*, **83**, January–February 1986, 145–155.

19.44. Ming B. Leung and William C. Schnobrich. "Reinforced Concrete Beams Subjected to Bending and Torsion," *Journal of Structural Engineering*, ASCE, **113**, 2 (February 1987), 307–321.

Shear and Torsion

19.45. Ugur Ersoy and Phil M. Ferguson. "Behavior and Strength of Concrete L-Beams Under Combined Torsion and Shear," *ACI Journal, Proceedings*, **64**, December 1967, 797–801. Disc., **65**, 477–479.

19.46. John P. Klus. "Ultimate Strength of Reinforced Concrete Beams in Combined Torsion and Shear," *ACI Journal, Proceedings*, **65**, March 1968, 210–215. Disc., 786–791.

19.47. Huey Ming Liao and Phil M. Ferguson. "Combined Torsion in Reinforced Concrete L-Beams with Stirrups," *ACI Journal, Proceedings*, **66**, December 1969, 986–993. Disc., **67**, 475–478.

19.48. Ahmed A. Ewida and Arthur E. McMullen. "Concrete Members under Combined Torsion and Shear," *Journal of the Structural Division*, ASCE, **108**, ST4 (April 1982), 911–928.

Torsion, Shear, and Bending

19.49. Larry E. Farmer and Phil M. Ferguson. "T-Beams Under Combined Bending, Shear and Torsion," *ACI Journal, Preceedings*, **64**, November 1967, 757–766. Disc., **65**, 417–421.

19.50.   E. L. Kemp. "Behavior of Concrete Members Subject to Torsion and to Combined Torsion, Bending, and Shear," *Torsion of Structural Concrete* (SP-18). Detroit: American Concrete Institute, 1968 (pp. 179–201).

19.51.   Thomas T. C. Hsu. "Torsion of Structural Concrete—Interaction Surface for Combined Torsion, Shear, and Bending in Beams Without Stirrups," *ACI Journal, Proceedings*, **65**, January 1968, 51–60, Disc., 566–572.

19.52.   M. S. Mirza and J. O. McCutcheon. Discussion "Torsion of Structural Concrete—Interaction Surface for Combined Torsion, Shear, and Bending in Beams Without Stirrups," *ACI Journal, Proceedings*, **65**, July 1968, 567–570.

19.53.   David J. Victor and Phil M. Ferguson. "Beams Under Distributed Load Creating Moment, Shear, and Torsion," *ACI Journal, Proceedings*, **65**, April 1968, 295–308. Disc., 892–894.

19.54.   Michael P. Collins, Paul F. Walsh, and A. S. Hall. Discussion of "Ultimate Strength of Reinforced Concrete Beams in Combined Torsion and Shear," *ACI Journal, Proceedings*, **65**, September 1968, 786–788.

19.55.   D. L. Osburn, B. Mayoglou, and Alan H. Mattock. "Strength of Reinforced Concrete Beams With Web Reinforcement in Combined Torsion, Shear, and Bending," *ACI Journal, Proceedings*, **66**, January 1969, 31–41. Disc., 593–595.

19.56.   M. S. Mirza and J. O. McCutcheon. "Behavior of Reinforced Concrete Beams Under Combined Bending, Shear, and Torsion," *ACI Journal, Proceedings*, **66**, May 1969, 421–427. Disc., 940–942.

19.57.   Alfred Bishara. "Prestressed Concrete Beams Under Combined Torsion, Bending, and Shear," *ACI Journal, Proceedings*, **66**, July 1969, 525–538. Disc., **67**, 61–63.

19.58.   Arthur E. McMullen and Joseph Warwaruk. "Concrete Beams in Bending, Torsion, and Shear," *Journal of the Structural Division*, ASCE, **96**, ST5 (May 1970), 885–903.

19.59.   Umakanta Behera and Phil M. Ferguson. "Torsion, Shear, and Bending on Stirruped L-Beams," *Journal of the Structural Division*, ASCE, **96**, ST7 (July 1970), 1271–1286.

19.60.   Einar Gausel. "Ultimate Strength of Prestressed I-Beams Under Combined Torsion, Bending, and Shear," *ACI Journal, Proceedings*, **67**, September 1970, 675–678.

19.61.   Priya R. Mukherjee and Joseph Warwaruk. "Torsion, Bending, and Shear in Prestressed Concrete," *Journal of the Structural Division*, ASCE, **97**, ST4 (April 1971), 1063–1079.

19.62.   P. K. Syamal, M. S. Mirza, and D. P. Ray. "Plain and Reinforced Concrete L-Beams Under Combined Flexure, Shear, and Torsion," *ACI Journal, Proceedings*, **68**, November 1971, 848–860.

19.63.   K. S. Rajagopalan and Phil M. Ferguson. "Distributed Loads Creating Combined Torsion, Bending, and Shear on L-Beams with Stirrups," *ACI Journal, Proceedings*, **69**, January 1972, 46–54.

19.64.   K. S. Rajagopalan, Umakanta Behera, and Phil M. Ferguson. "Total Interaction Method for Torsion Design," *Journal of the Structural Division*, ASCE, **98**, ST9 (September 1972), 2097–2117.

19.65. Thomas G. Barton and D. Wayne Kirk. "Concrete T-Beams Subject to Combined Loading," *Journal of the Structural Division*, ASCE, **99**, ST4 (April 1973), 687–700.

19.66. Robert L. Henry and Paul Zia. "Prestressed Beams in Torsion, Bending, and Shear," *Journal of the Structural Division*, ASCE, **100**, ST5 (May 1974), 933–952.

19.67. Lennart Elfgren, Inge Karlsson, and Anders Losberg. "Torsion-Bending-Shear Interaction for Concrete Beams," *Journal of the Structural Division*, ASCE, **100**, ST8 (August 1974), 1657–1676.

19.68. D. Wayne Kirk and David G. McIntosh. "Concrete L-Beams Subject to Combined Torsional Loads," *Journal of the Structural Division*, ASCE, **101**, ST1 (January 1975), 269–282.

19.69. B. Vijaya Rangan and A. S. Hall. "Strength of Rectangular Prestressed Concrete Beams in Combined Torsion, Bending and Shear," *ACI Journal, Proceedings*, **70**, April 1973, 270–278.

19.70. B. Vijaya Rangan and A. S. Hall. "Strength of Prestressed Concrete I-Beams in Combined Torsion and Bending," *ACI Journal, Proceedings*, **75**, November 1978, 612–618.

19.71. K. S. Rajagopalan. "Combined Torsion, Bending, and Shear on L-Beams," *Journal of the Structural Division*, ASCE, **106**, ST12 (December 1980), 2475–2492.

19.72. G. Taylor and J. Warwaruk. "Combined Bending, Torsion, and Shear of Prestressed Concrete Box Girders," *ACI Journal, Proceedings*, **78**, September–October 1981, 335–340.

Design Methods

19.73. Alan H. Mattock. "How to Design for Torsion," *Torsion of Structural Concrete* (SP-18). Detroit: American Concrete Institute, 1968 (pp. 469–495).

19.74. ACI Committee 438. "Tentative Recommendations for the Design of Reinforced Concrete Members to Resist Torsion," *ACI Journal, Proceedings*, **66**, January 1969, 1–8. Disc., **66**, 576–588.

19.75. Thomas T. C. Hsu and E. L. Kemp. "Background and Practical Application of Tentative Design Criteria for Torsion," *ACI Journal, Proceedings*, **66**, January 1969, 12–23. Disc., 591–593.

19.76. Michael P. Collins. Discussion of "Tentative Recommendations for the Design of Reinforced Concrete Members to Resist Torsion," *ACI Journal, Proceedings*, **66**, July 1969, 577–579.

19.77. Umakanta Behera and K. S. Rajagopalan. "Two-piece U-Stirrups in Reinforced Concrete Beams," *ACI Journal, Proceedings*, **66**, July 1969, 522–524.

19.78. ACI Committee 438. Discussion of "Proposed Revision of ACI 318-63: Building Code Requirements for Reinforced Concrete," *ACI Journal, Proceedings*, **67**, September 1970, 686–689.

19.79. Michael P. Collins and Paul Lampert. "Designing for Torsion," *Structural Concrete Symposium.* Toronto: University of Toronto Civil Engineering Department, May 1971 (pp. 38–79).

19.80. David J. Victor. "Effective Flange Width in Torsion," *ACI Journal, Proceedings,* **68**, January 1971, 42–46.

19.81. Denis Mitchell and Michael P. Collins. "Detailing for Torsion," *ACI Journal, Proceedings,* **73**, September 1976, 506–511.

19.82. Mahmoud A. Reda Youssef and Alfred G. Bishara. "Dowel Action in Concrete Beams Subject to Torsion," *Journal of the Structural Division,* ASCE, **106**, ST6 (June 1980), 1263–1277.

19.83. Thomas T. C. Hsu and Y. L. Mo. "Softening of Concrete in Torsional Members— Design Recommendations," *ACI Journal, Proceedings,* **82**, July–August 1985, 443–452. Disc., **83**, July–August 1986, 690–693.

Spandrel Beams

19.84. M. A. Gouda. "Distribution of Torsion and Bending Moments in Connected Beams and Slabs," *ACI Journal, Proceedings,* **56**, February 1960, 757–774. Disc., 1425–1446.

19.85. Robert A. Shoolbred and Eugene P. Holland. "Investigation of Slab Restraint on Torsional Moments in Fixed-Ended Spandrel Girders," *Torsion of Structural Concrete* (SP-18). Detroit: American Concrete Institute, 1968 (pp. 69–88).

19.86. Kolbjorn Saether and N. M. Prachand. "Torsion in Spandrel Beams," *ACI Journal, Proceedings,* **66**, January 1969, 24–30.

19.87. James O. Jirsa, John L. Baumgartner, and Nathan C. Mogbo. "Torsional Strength and Behavior of Spandrel Beams," *ACI Journal, Proceedings,* **66**, November 1969, 926–932. Disc., **67**, 434–435.

19.88. James O. Jirsa. "Torsion in Floor Slab Structures," *Analysis of Structural Systems for Torsion* (SP-35). Detroit: American Concrete Institute, 1973 (pp. 265–292).

19.89. Mario G. Salvadori. "Spandrel-Slab Interaction," *Journal of the Structural Division,* ASCE, **96**, ST1 (January 1970), 89–106.

19.90. Umakanta Behera, K. S. Rajagopalan, and Phil M. Ferguson. "Reinforcement for Torque in Spandrel L-Beams," *Journal of the Structural Division,* ASCE, **96**, ST2 (February 1970), 371–380.

19.91. Ugur Ersoy. "Distribution of Torsional and Bending Moments in Beam-Slab Systems," *Analysis of Structural Systems for Torsion* (SP-35). Detroit: American Concrete Institute, 1973 (pp. 293–324).

19.92. E. L. Kemp and W. J. Wilhelm. "Influence of Spandrel Beam Torsion on Slab Capacity Based on Yield Line Criteria," *Analysis of Structural Systems for Torsion* (SP-35). Detroit: American Concrete Institute, 1973 (pp. 325–341).

19.93. Thomas T. C. Hsu and Kenneth T. Burton. "Design of Reinforced Concrete Spandrel Beams," *Journal of the Structural Division,* ASCE, **100**, ST1 (January 1974), 209–229.

19.94. Thomas T. C. Hsu and Ching-Sheng Hwang. "Torsional Limit Design of Spandrel Beams," *ACI Journal, Proceedings*, **74**, February 1977, 71–79.

19.95. Alfred G. Bishara, Larry Londot, Peter Au, and Majety V. Sastry. "Flexural Rotational Capacity of Spandrel Beams," *Journal of the Structural Division*, ASCE, **105**, ST1 (January 1979), 147–161.

19.96. Charles H. Raths. "Spandrel Beam Behavior and Design," *PCI Journal*, **29**, March–April 1984, 62–131.

19.97. Gary J. Klein. "Design of Spandrel Beams," *PCI Journal*, **31**, September–October 1986, 76–124.

19.98. Y. L. Mo and Thomas T. C. Hsu. "Redistribution of Moments in Spandrel Beams," *ACI Structural Journal*, **88**, January–February 1991, 22–30.

Inverted T-Girders and Ledger Beams

19.99. Sher Ali Mirza and Richard W. Furlong. "Serviceability Behavior and Failure Mechanisms of Concrete Inverted T-Beam Bridge Bentcaps," *ACI Journal, Proceedings*, **80**. July–August 1983, 294–304.

19.100. Sher Ali Mirza and Richard W. Furlong. "Strength Criteria for Concrete Inverted T-Girders," *Journal of Structural Engineering*, ASCE, **109**, 8 (August 1983), 1836–1853.

19.101. Ned M. Cleland and Thomas T. Baber. "Behavior of Precast Reinforced Concrete Ledger Beams," *PCI Journal*, **31**, March–April 1986, 96–117.

Channel-Shaped Sections

19.102. Petar Krpan and Michael P. Collins. "Predicting Torsional Response of Thin-Walled Open RC Members," *Journal of the Structural Division*, ASCE, **107**, ST6 (June 1981), 1107–1127.

19.103. Petar Krpan and Michael P. Collins. "Testing Thin-Walled Open RC Structure in Torsion," *Journal of the Structural Division*, ASCE, **107**, ST6 (June 1981), 1129–1140.

19.104. Ching-Sheng Hwang and Thomas T. C. Hsu. "Mixed Torsion Analysis of Reinforced Concrete Channel Beams—A Fourier Series Approach," *ACI Journal, Proceedings*, **80**, September–October 1983, 377–386.

Beams with Openings

19.105. M. A. Mansur and A. Hasnat. "Concrete Beams with Small Opening under Torsion," *Journal of the Structural Division*, ASCE, **105**, ST11 (November 1979), 2433–2447.

19.106. Mohammad A. Mansur, Seng Kiong Ting, and Seng-Lip Lee. "Ultimate Torque of R/C Beams with Large Openings," *Journal of Structural Engineering*, ASCE, **109**, 8 (August 1983), 1887–1901.

19.107. M. A. Mansur. "Combined Bending and Torsion in Reinforced Concrete Beams with Rectangular Openings," *Concrete International*, **5**, November 1983, 51–58.

19.108. M. A. Mansur and P. Paramasivam. "Reinforced Concrete Beams with Small Opening in Bending and Torsion," *ACI Journal, Proceedings*, **81**, March–April 1984, 180–185.

19.109. Abul Hasnat and Ali A. Akhtaruzzaman. "Beams with Small Rectangular Opening under Torsion, Bending, and Shear," *Journal of Structural Engineering*, ASCE, **113**, 10 (October 1987), 2253–2269.

19.110. W. A. M. Alwis and Mohammad A. Mansur. "Torsional Strength of R/C Beams Containing Rectangular Openings," *Journal of Structural Engineering*, ASCE, **113**, 11 (November 1987), 2248–2258.

Torsion of Prestressed Concrete Beams

19.111. Alfred Bishara. "Prestressed Concrete Beams Under Combined Torsion, Bending, and Shear," *ACI Journal, Proceeding*, **66**, July 1969, 525–538.

19.112. Paul Zia and W. Denis McGee. "Torsion Design of Prestressed Concrete," *PCI Journal*, **19**, March–April 1974, 46–65.

PROBLEMS

All problems are to be worked in accordance with the strength method of the ACI Code, and all loads given are service loads, unless otherwise indicated.

19.1 Determine the reinforcement required on a 12 × 22 in. overall size member to carry a torsional moment of 2 ft-kips dead load and 8 ft-kips live load. Use $f'_c = 4000$ psi and $f_y = 40,000$ psi.

19.2 Determine the reinforcement required for the member in the accompanying figure to carry a torsional moment of 20 ft-kips dead load. Use $f'_c = 4000$ psi and $f_y = 50,000$ psi.

Problem 19.2

19.3   For the beam of the accompanying figure, assume that 4-#9 bars are used in the bottom for the main flexural reinforcement at midspan, with 2-#9 bars extended into the support and properly anchored. Further assume that 2-#7 are used in the top at the supports. What is the nominal torsional strength T_n of the section according to the ACI Code, assuming no simultaneous transverse shear? What is the factored negative bending moment M_u that might be permitted to act at the supports simultaneously when $T_u = \phi T_n$, according to the logic of Section 19.8?

2 – #7

#3 @ 5"

24"

2 – #9

← 14" →

Problem 19.3

19.4   Determine the reinforcement required (including torsion) at midspan and at the supports for the beam of Example 19.4.1 (2B4 of Fig. 19.4.1). Assume the shear is the same as that of a simply supported beam carrying a uniform dead load of 0.65 kips/ft and uniform live load of 0.65 kips/ft. Assume that flexure alone requires longitudinal steel areas of 1.60 sq in. and 2.30 sq in., for positive and negative moment regions, respectively. Use $f'_c = 3000$ psi and $f_y = 40,000$ psi. Show sketches of the cross sections.

19.5   Design the reinforcement to include torsion on the spandrel girder of Example 19.4.2 (2G4 of Fig. 19.4.1). Assume simple beam shears for this span which is continuous at both ends. Assume the reactions to girder 2G4 from beams 2B1 are concentrated loads of 12.3 kips dead load and 15.9 kips live load, and also that the girder has uniform dead load (including its own weight) of 1 kip/ft. Use $f'_c = 3000$ psi and $f_y = 40,000$ psi. Show design sketch.

19.6   Redesign the transverse reinforcement for the beam of Problem 5.1, except consider the total loading to be acting along a line at a distance of 2 in. from the midwidth of the beam. Use the rough approximation that equilibrium torsion is acting and that the torsional moment per unit length equals the uniform loading times 2 in.

19.7   Redesign the transverse reinforcement for the beam of Problem 5.8, except consider the line uniform loading to be acting at 3 in. from the centerline of the beam cross section. Assume equilibrium torsion and use the same approximation as in Problem 19.6.

19.8 Design the reinforcement for the edge beam which is continuous at both ends shown in the accompanying figure. Assume the torsional moment is 76 ft-kips (50% live load) at the face of column and that the torsional moment varies proportionally with the flexural shear (this problem is similar to that of Hsu and Kemp [19.75].

Problem 19.8

19.9 Design the parking garage spandrel beam having a simply supported span of 25 ft center-to-center of bearing (from Reference 19.9). The floor system, consisting of double tees spanning 60 ft, carries a live load of 50 psf. The double tees are supported on the ledge of the spandrel beam. The contribution to R_1 from the dead load of the floor system is 10.7 kips, and from the 50 psf live load is 6.0 kips. Consider the given cross section as a preliminary trial section, with the 8 in. eccentricity of R_1 with respect to R_2 held as constant. The spandrel member will have an attachment to the supporting column, providing lateral support at each vertical support of the spandrel girder. Use $f'_c = 4000$ psi and $f_y = 60,000$ psi and the ACI Code. (Note that the preliminary section is from Reference 19.9 and is for a prestressed concrete member. The final design for a nonprestressed member may require changes from the preliminary given dimensions; those given dimensions may serve as guides for arriving at reasonable proportions.)

(a) Spandrel beam span and loading

(b) Spandrel beam cross section (trial)

Problems 19.9 and 19.10

19.10 Design a spandrel beam of similar L-shape to that of Problem 19.9 for loads R_1 of 5.5 kips dead load and 4.0 kips live load, plus the spandrel girder weight. The span of the spandrel and the locations of the R_1 loads are the same as in Problem 19.9. The eccentricity e is 8 in. (same as for Problem 19.9). Consider the 8 in. used in Problem 19.9 as the minimum thickness; however, the other dimensions should be appropriate for the loads given in this problem. Use $f'_c = 4000$ psi and $f_y = 60{,}000$ psi.

CHAPTER 20

Footings

20.1 PURPOSE OF FOOTINGS

Footings are structural elements that transmit to the soil column loads, wall loads, or lateral loads such as from retained earth. If these loads are to be properly transmitted, footings must be designed to prevent excessive settlement or rotation, to minimize differential settlement, and to provide adequate safety against sliding and overturning.

20.2 BEARING CAPACITY OF SOIL

There must be reliable information on the safe bearing capacity of the soil prior to the design of a footing. It is not within the scope of this text to discuss the details of arriving at the bearing capacity of soil. The allowable bearing capacity of soil is usually determined by the ruling building code, by comparison with existing footings and with related information in the area, by close examination of the soil and study of logs of test borings, by the application of the science of soil mechanics, by load test, or by combinations of the various sources and methods mentioned here.

The following are some of the most common reasons for the many uncertainties concerning soil behavior under a footing.

1. There may be wide variations in soil types, which depend on their geological source, mode of transportation, and sedimentation mechanism.
2. The physical properties and probable behavior under load are unknown and may require extensive testing.
3. Frost action may cause heaving or subsidence.
4. Vibration may cause consolidation of granular material which results in non-uniform settlement.
5. Man-made hazards may exist below the earth surface, such as rock heaps, old sewers, and questionable fill.

Soil pressure is commonly used in design under a working stress philosophy. The soil mechanics/foundations specialist (geotechnical engineer) will establish the

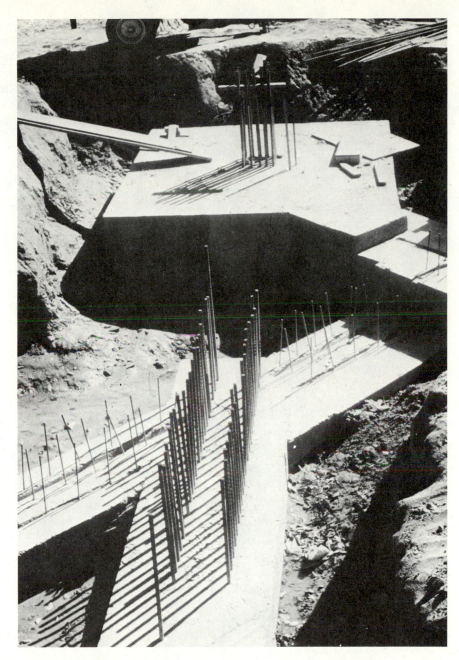

Wall and square spread footings; Kurt F. Wendt Engineering Library, University of Wisconsin, Madison, Wis. (Photo by C. G. Salmon.)

ultimate bearing capacity, applying the appropriate margin for safety, and specify a service load bearing capacity (allowable bearing capacity) to be used in design.

In general, rock is considered the best foundation material; graded sand and gravel are good materials; fine particles of sand and silt are generally questionable; and clay should be studied carefully. The allowable bearing capacity used for design may range from 12,000 psf (575 kN/m^2) or higher for rock to 2000 psf (96 kN/m^2) for soft clay or silty clay. Soils unable to carry 2000 psf generally require piling.

(a) Isolated spread footing

(b) Wall footing

(c) Combined footing, rectangular, $P_A = P_B$

(d) Combined footing, rectangular, $P_B > P_A$

(e) Combined footing, trapezoidal, $P_B > P_A$

(f) Combined footing, trapezoidal, $P_A > P_B$

(g) Combined footing, strap or cantilever

Figure 20.3.1 Types of footings.

20.3 TYPES OF FOOTINGS

Most building footings may be classified as one of the following types (Fig. 20.3.1):

1. Isolated spread footings under individual columns. These may be square, rectangular, or occasionally circular in plan.
2. Wall footings, either flat or stepped, which support bearing walls.
3. Combined footings supporting two or more column loads. These may be continuous with a rectangular or trapezoidal plan or they may be isolated footings joined by a beam. The latter case is referred to as a strap, or cantilever, footing.
4. A mat foundation, which is one large continuous footing supporting all the columns of the structure. This is used when soil conditions are poor but piles are not used.
5. Pile caps, structural elements that tie a group of piles together. These may support bearing walls, isolated columns, or groups of several columns.

20.4 TYPES OF FAILURE OF FOOTINGS

The procedures used for the design of footings in the United States are based primarily on the work of Talbot in 1907 [20.1], Richart in 1946 [20.2], and Moe in 1957–1959 [20.3].

Moe defines the several types of failure that may occur in a slab acted on by concentrated loads. These failure modes are related to the shear span to depth (a/d) ratio, that is, $M_u/V_u d$, and are similar to those described for beams in Section 5.4 (Fig. 5.4.4). The failure mechanisms may be reviewed as follows:

1. Shear-compression failure [Fig. 20.4.1(a)]. Typical with deep sections of short span (low a/d ratios), inclined cracks form that do not cause failure but do extend into the compression zone, thus reducing its size until finally the compression zone fails under the combined compressive and shear stresses.
2. Flexure failure *after* inclined cracks form. Also typical with low a/d ratios, inclined cracks that form first do not cause failure or prevent the development of the nominal moment strength. If embedment of tension steel is adequate, and no failure in the compression zone occurs, the tension steel may reach its yield strength.

(a) Shear-compression failure

(b) Diagonal tension failure

Figure 20.4.1 Shear-related failure mechanisms.

3. Diagonal tension failure [Fig. 20.4.1(b)]. Commonly called *punching shear*, this is typical with medium-span average depth sections (intermediate values of a/d); the slab fails on formation of the inclined cracks around the perimeter of the concentrated load. Test results indicate that the critical section can reasonably be considered at $d/2$ from the periphery of a column or concentrated load.

4. Flexure failure before inclined cracks form. Typical with large values of a/d, no inclined cracks form before flexural strength is reached.

In the design of a footing as well as of a beam, a shear failure should not occur prior to reaching the member flexural strength.

20.5 SHEAR STRENGTH OF FOOTINGS

The shear strength of footings is essentially the shear strength of two-way slab systems, as discussed in Sections 16.15 and 16.18. A summary of the status is presented by ASCE-ACI Task Committee 426 under Chairman N. M. Hawkins [16.83] and confirmed by a 1979 ACI-ASCE Committee 426 Report [20.4]. The six principal variables involved in the strength of slabs without shear reinforcement are (a) the concrete strength, f'_c; (b) the ratio of the side length c of the loaded area to the effective depth d of the slab; (c) the relationship (V/M) between shear and moment near the critical section; (d) the column shape in terms of the ratio β_c of the long side to the short side of the rectangular column; (e) lateral restraints such as may be provided by stiff beams along the boundaries of a slab (no such restraints would normally be acting in the case of footings); and (f) the rate of loading.

As discussed in Section 16.15, ACI-11.12.2.1 gives the nominal *punching* shear strength V_c when no shear reinforcement is used (that is, $V_n = V_c$) as the smallest of

$$V_c = \left(2 + \frac{4}{\beta_c}\right)\sqrt{f'_c}\,b_0 d, \qquad \text{ACI Formula (11-36)} \qquad \textbf{(20.5.1a)*}$$

$$V_c = \left(\frac{\alpha_s}{b_0/d} + 2\right)\sqrt{f'_c}\,b_0 d, \qquad \text{ACI Formula (11-37)} \qquad \textbf{(20.5.1b)*}$$

and

$$V_c = 4\sqrt{f'_c}\,b_0 d, \qquad \text{ACI Formula (11-38)} \qquad \textbf{(20.5.1c)*}$$

— — — —

* For SI, with f'_c in MPa and b_0 and d in mm, ACI 318–89M gives

$$V_c = \frac{1}{6}\left(1 + \frac{2}{\beta_c}\right)\sqrt{f'_c}\,b_0 d \qquad \textbf{(20.5.1a)}$$

$$V_c = \frac{1}{12}\left(\frac{\alpha_s}{b_0/d} + 2\right)\sqrt{f'_c}\,b_0 d \qquad \textbf{(20.5.1b)}$$

$$V_c = \frac{1}{3}\sqrt{f'_c}\,b_0 d \qquad \textbf{(20.5.1c)}$$

where

    b_0 = perimeter of critical section taken at $d/2$ from the loaded area

    d = effective depth of slab

    β_c = ratio of long side to short side of the loaded area

    α_s = 40 for interior columns, 30 for edge columns, and 20 for corner columns

In the application of Eq. (20.5.1b), α_s for an "interior column" applies when the perimeter is four-sided, for an "edge column" when the perimeter is three-sided, and for a "corner column" when the perimeter is two-sided.

For footings in which the bending action is primarily in one direction, the procedure used for beams should be applied as described in Chapter 5. The critical section for such an investigation is to be taken at a distance d from the column face. Figure 20.5.1 summarizes the critical sections to be investigated in regard to shear.

(a) One-way action

(b) Two-way action

Figure 20.5.1 Critical section for shear (inclined cracking) in footings.

20.6 MOMENT STRENGTH OF FOOTINGS AND DEVELOPMENT OF REINFORCEMENT

Research has shown that critical sections for moment and development of reinforcement occur at the face of a reinforced concrete column or wall. The bending in each direction should be considered separately. Richart's tests [20.2] showed that, under service loads, the moment is greater on a strip under the column load than it is out near the corners. However, failure in flexure does not occur until all of the steel has reached yield—that is, after some redistribution of load has taken place. Yield line analyses (see Chapter 18) of footings has been made by Gesund [20.5, 20.6], Jiang [20.7], and Rao and Singh [20.8].

The critical sections for moment and development of reinforcement, however, are to be taken (ACI-15.4.2) at halfway between the middle and the edge of the wall for footings under masonry walls, and at halfway between the face of the column and the edge of the metallic base for footings under steel bases.

20.7 PROPORTIONING FOOTING AREAS FOR EQUAL SETTLEMENT

Differential settlement between footings should be minimized because it may adversely affect the strength of the structure as well as interfere with the fitting of such items as partitions, doors, and ceilings. It is generally assumed that there will be equal settlement if the unit soil pressures due to service loads are equal under all footings. However, the service loads in the columns consist of dead and live loads. The dead load is always there, but in a tall building there is little chance for a column to receive maximum live load from all floors. Therefore most building codes allow a reduction of live loads in columns. Typically, the total live load would have to be carried by columns supporting the roof and one floor. As the number of floors carried by a column increases, the percentage of the total live load that the column must be designed to carry may be reduced from 100% progressively. This percentage might be a minimum of 50% when the number of floors carried by the column is at least eight or nine. The maximum soil pressure under a footing is, then, due to the sum of the dead load in the column, the maximum reduced live load in the column, and the weight of the footing itself.

Soil is a substance that may be relatively elastic such as granular material like sand, or it may be a relatively plastic substance such as clay exhibiting time-dependent deformation under sustained load. Often, for design purposes, equal settlement of footings is presumed when the soil pressure due to sustained loads is the same under all footings. The sustained load may be taken as the sum of the dead load in the column, the weight of the footing itself, and a certain percentage of the maximum reduced live load in the column. In such cases, the relative areas of the footings should be so proportioned that the unit soil pressures under sustained loads would be the same under all footings. This requirement is additional to the requirement that the soil pressure under maximum possible load must not exceed the allowable bearing capacity at each footing.

20.8 INVESTIGATION OF SQUARE SPREAD FOOTINGS

The investigation of square spread footings can best be treated by an illustrative example in which the items considered are (1) soil pressure under the footing, (2) shear (inclined cracking), (3) bending moment, (4) development of reinforcement, and (5) load transfer from column to footing.

■ **EXAMPLE 20.8.1** Check the adequacy of the square footing of Fig. 20.8.1, according to the strength method of the ACI Code. The column axial load is 300 kips dead load and 160 kips live load. The concrete for both the column and the footing has $f'_c = 3000$ psi and $f_y = 40,000$ psi. The allowable soil pressure is 5000 psf. There is a 2-ft earth overburden having a unit weight of 100 pcf.

Solution: **(a)** Soil pressure. The action of soil on the footing is taken to be uniform for isolated footings under concentric loads. The base of the footing must have an area large enough so that the allowable soil pressure will not be exceeded under the action of the column service load, footing weight, and weight of overburden.

Figure 20.8.1 Square spread footing for Example 20.8.1.

$$\begin{aligned}
\text{column load } (300 + 160) &= 460 \text{ kips} \\
\text{footing weight } 10(10)(2.08)(0.150) &= 31 \\
\text{earth } (100 - 4)2(0.100) &= 19 \\
\hline
\text{Total weight on soil} &= \overline{510 \text{ kips}}
\end{aligned}$$

$$\text{soil pressure } p = \frac{510}{(10)(10)} = 5.1 \text{ ksf} \approx 5 \text{ ksf} \qquad \textbf{OK}$$

(b) Shear (inclined cracking) strength. Two possible critical sections must be investigated: one-way action as a beam and two-way action as a slab, as shown in Fig. 20.5.1. The shear to be used is the upward soil pressure less the downward overburden and the footing weight acting outside of the critical section. Since the footing weight and the overburden are usually uniform, the forces acting on the footing may be obtained by using the net upward pressure, which is caused by the column load only. Since the strength method is to be used, the overload factors U must be applied.

Since the action of a square concentrically loaded footing is symmetrical about both axes, the reinforcement in each direction is presumed to do the same work. However, the effective depth cannot be the same for both directions since the bars must cross each other. The average depth d is commonly used except for very shallow footings (say, less than 15 in. deep) where the more conservative value should probably be used. Here the average d is

$$d = 25 - 3 \text{ (i.e., cover)} - 1 \text{ (i.e., bar diameter)} = 21 \text{ in.}$$

For one-way diagonal tension action, the net earth pressure acting upward due to factored loads is

$$p_{\text{net}} = \frac{300(1.4) + 160(1.7)}{100} = 6.92 \text{ ksf}$$

Using the loaded area shown in Fig. 20.8.2(a), the factored shear is

$$V_u = (p_{\text{net}})(\text{effective area}) = 6.92(2.25)10 = 156 \text{ kips}$$

When no shear reinforcement is used, $v_c = 2\sqrt{f'_c}$ unless the more detailed Vd/M procedure is used; thus

$$V_n = V_c = 2\sqrt{f'_c}\, b_w d = 2\sqrt{3000}(120)(21)\tfrac{1}{1000} = 276 \text{ kips}$$

$$[\phi V_n = 0.85(276) = 235 \text{ kips}] > V_u = 156 \text{ kips} \qquad \textbf{OK}$$

For two-way diagonal tension action, the factored shear [Fig. 20.8.2(b)] is

$$V_u = (p_{\text{net}})(\text{effective area})$$
$$= 6.92[100 - 3.75(3.75)] = 595 \text{ kips}$$

When no shear reinforcement is used, the strength using Eqs. (20.5.1) is based on $v_c = 4\sqrt{f'_c}$ when $\beta_c \leq 2$ and $b_0/d \leq 20$ for a four-sided critical section (see ACI-11.12.2); thus

$$b_0/d = 4(3.75)12/21 = 8.6 < 20$$

$$V_n = V_c = 4\sqrt{f'_c}\, b_0 d = 4\sqrt{3000}(4)(3.75)(12)(21)\tfrac{1}{1000} = 828 \text{ kips}$$

$$[\phi V_n = 0.85(828) = 704 \text{ kips}] > V_u = 595 \text{ kips} \qquad \textbf{OK}$$

Figure 20.8.2 Critical sections and loaded area for square spread footing.

Since isolated footings are rarely designed with shear reinforcement, V_c will control the thickness.

(c) Bending moment strength. The critical section and loaded area are as shown in Fig. 20.8.2(a). The factored bending moment is

$$M_u = \frac{p_{net}bl^2}{2} = \frac{6.92(10)(4.0)^2}{2} = 554 \text{ ft-kips}$$

$$\text{required } R_n = \frac{M_u}{\phi bd^2} = \frac{554(12,000)}{0.90(10)(12)(21)^2} = 140 \text{ psi}$$

Referring to Section 3.8,

$$\text{required } \rho = \frac{1}{m}\left(1 - \sqrt{1 - \frac{2mR_n}{f_y}}\right)$$

$$= \frac{1}{15.7}\left(1 - \sqrt{1 - \frac{2(15.7)140}{40,000}}\right) = 0.0036$$

For spread footings, the minimum reinforcement (min $\rho_g = 0.002$) "for structural slabs of uniform thickness" as stated in ACI-10.5.3 applies; thus

$$\text{required } A_s = 0.0036(10)(12)21 = 9.1 \text{ sq in.}$$

$$\text{min required } A_s = 0.002(10)(12)25 = 6.00 \text{ sq in.}$$

$$\text{provided } A_s = 15(0.79) = 11.9 \text{ sq in.}$$

Using 12-#8 would satisfy the requirement of 9.1 sq in.; however, with the low reinforcement ratio used, the 15 bars are not excessive.

(d) Development of reinforcement. The reinforcement must be embedded from the face of the column a distance at least equal to the development length L_d for #8 bars. Referring to Table 6.7.1, the basic development length L_{db} is 23.1 in. Then from Table 6.7.2, the Min $L_d = 21.9$ in. Since the approximate 8 in. center-to-center spacing gives clear spacing exceeding $3d_b$ (i.e., 3 in. for #8 bars) and the 3 in. bottom cover exceeds $2d_b$, Category A applies; $\alpha_{abc} = 1.0$ according to ACI-11.2.3.1(d). In addition, Category D applies for wide lateral spacing where clear spacing exceeds $5d_b$. Thus, $\alpha_{de} = 0.80$, giving "first-level L_d" as

$$\alpha_{abc}\alpha_{de}L_{db} = 1.0(0.80)23.1 = 18.5 \text{ in.} < (\text{Min } L_d = 21.9 \text{ in.})$$

The "first-level L_d" is 21.9 in. No second-level adjustments apply; thus,

$$L_d \text{ (for #8)} = 21.9 \text{ in. (1.8 ft)}$$

Allowing an inch or so cover on the end of the #8 bars, the embedment provided is

$$\text{actual embedment} = 48 - 2 = 46 \text{ in.} > (L_d = 21.9 \text{ in.}) \qquad \textbf{OK}$$

(e) Load transfer from column to footing. ACI-15.8.1 requires all forces acting at the column base to be transferred into the footing. Tensile forces, if any, must be transferred by developed reinforcement, such as bar reinforcement, dowels, or mechanical connectors; however, compressive forces may be transmitted directly by bearing.

The nominal ultimate bearing stress f_b that the base of the column can withstand is $0.85f_c'$ (ACI-10.15.1). The nominal strength P_n in compression based on the *column* concrete strength f_c' is

$$P_n = 0.85f_c'A_g = 0.85(3)(576) = 1470 \text{ kips}$$

$$P_u = 1.4(300) + 1.7(160) = 692 \text{ kips}$$

$$[\phi P_n = 0.70(1470) = 1029 \text{ kips}] > P_u = 692 \text{ kips} \qquad \textbf{OK}$$

Regarding the bearing on the footing concrete, the capacity is increased because the footing area is much larger than the column area, thus permitting a distribution of the concentrated load. The bearing strength for the *footing* concrete is the regular value based on $0.85f_c'$ increased by the multiplier α_b that varies between 1 and 2, as follows:

$$\alpha_b = \sqrt{\frac{A_2}{A_1}} \leq 2$$

where A_1 is the loaded area, which in this example is the 576 sq in. column area; A_2 is the "area of the lower base of the largest frustrum of a pyramid, cone, or tapered wedge contained wholly within the support and having side slopes of 1 vertical to 2 horizontal" (ACI-10.0). In this case, A_2 is the entire footing area, 14,400 sq in.

The bearing stress *on the footing* may control when columns of high-strength concrete rest on footings of low-strength concrete. Since both the column and footing contain the same strength concrete in this example, only the check on bearing in the column was necessary. Bearing in the bottom of the column is important unless the longitudinal reinforcement is developed by extension into the footing or by embedding dowels lapped to the column bars. In this case, the load can be carried without using developed reinforcement.

When the transfer is made by bearing, as in this case, ACI-15.8.2.1 still requires a minimum amount of reinforcement across the joint between the column and footing. Extended longitudinal reinforcement or dowels at least equal to 0.005 times the gross cross-sectional area of the supported member must be provided. When dowels are used, they can be any size #11 and smaller. Though ACI-15.8.2 in 1989 no longer has a size related restriction, ACI-15.8.2.3 relating to development of #14 and #18 column bars indicates dowels larger than #11 bars cannot be used. Dowels comparable in size to the bars being developed should generally be used. The minimum area of developed reinforcement required to go "across interface" of cast-in-place construction is

$$\text{required } A_s = 0.005(576) = 2.88 \text{ sq in.}$$

Using a practical minimum of 4 bars, one in each corner,

$$\text{required } A_s \text{ per bar} = \frac{2.88}{4} = 0.72 \text{ sq in.}$$

Thus the #8 dowels shown are adequate. They must be embedded into the footing a distance equal to the development length L_d in compression for #8 bars.

Using the basic development length formulas of ACI-12.3 for compression bars,

$$L_{db} = \frac{0.02 f_y d_b}{\sqrt{f'_c}} \qquad \qquad \textbf{[6.9.1]}$$

$$= \frac{0.02(40{,}000)1.0}{\sqrt{3000}} = 14.6 \text{ in.}$$

but not less than

$$L_{db} = 0.0003 f_y d_b \qquad \qquad \textbf{[6.9.2]}$$

$$= 0.003(40{,}000)(1.0) = 12.0 \text{ in.}$$

and the final L_d cannot be less than 8 in. Thus

$$L_d \, (\#8) = L_{db} = 14.6 \text{ in.} < 25 \text{ in. available}$$

If the available footing thickness is inadequate, a greater number of smaller diameter bars should be used.

Thus the footing of Fig. 20.8.1 satisfies all ACI requirements. ∎

20.9 DESIGN OF SQUARE SPREAD FOOTINGS

The design of square spread footings involves the determination of the size and depth of the footing and the amount of main reinforcement and dowels so that all of the requirements described in the preceding sections are fulfilled. The design procedures are illustrated by the following example.

■ **EXAMPLE 20.9.1** Design a square spread footing to carry a column dead load of 197 kips and a live load of 160 kips from a 16 in. square tied column containing

#11 bars as the principal column steel. The allowable soil pressure is 4.5 ksf. Consider there is a 2-ft overburden weighing 100 pcf. Use $f'_c = 3000$ psi, $f_y = 40,000$ psi, and the strength method of the ACI Code.

Solution: (a) Estimate the footing weight and determine the plan of the footing. The total weight of the footing, plus any overburden, may be estimated and added to the column load or, as an alternative, the effect of these items in terms of the unit soil pressure may be estimated. In this case, the footing is estimated to be about 2 ft thick, that is, 300 psf, frequently the minimum used by designers. This leaves the net allowable soil pressure that must carry the column load as

$$p_{net} = 4500 - 200 - 300 = 4000 \text{ psf}$$

$$\text{required } A = \frac{357}{4.0} = 89.5 \text{ sq ft}$$

Try 9 ft-6 in. square, $A = 90.3$ sq ft. Note that ACI-15.2.2 requires the base area of a footing to be determined using service loads (unfactored loads) with the allowable soil pressure. This is reasonable since the allowable soil pressure should be determined using principles of soil mechanics and may incorporate varying factors of safety depending on the type of soil and condition of loading.

For the design of the reinforced concrete member, factored loads must be used. Applying overload factors to the column load,

$$P_u = 1.4(197) + 1.7(160) = 548 \text{ kips}$$

$$p_{net} = \frac{548}{90.3} = 6.07 \text{ ksf}$$

The strength reduction factor ϕ will be applied later in the calculations.

(b) Determine depth based on shear (inclined cracking) strength. In most cases the depth necessary for shear without using stirrups controls the footing thickness.

For two-way action [Fig. 20.9.1(a)], assuming a thickness of 24 in.,

$$\text{average } d = 24 - 3(\text{cover}) - 1 \text{ (bar diameter)} \approx 20 \text{ in.}$$

$$V_u = (p_{net})(\text{area}) = 6.07[9.5(9.5) - 3.0(3.0)] = 493 \text{ kips}$$

According to ACI-11.12.2.1 for $\beta_c \leq 2$ and $b_0/d \leq 20$ for a four-sided critical section,

$$b_0/d = 4(16 + 20)/20 = 7.2 < 20 \qquad\qquad \textbf{OK}$$

$$V_c = 4\sqrt{f'_c}\, b_0 d = 4\sqrt{3000}[4(16 + 20)](20)\tfrac{1}{1000} = 631 \text{ kips}$$

$$\phi V_c = 0.85(631) = 536 \text{ kips} > [V_u = 493 \text{ kips}] \qquad\qquad \textbf{OK}$$

No shear reinforcement is required.

For one-way action [Fig. 20.9.1(b)],

$$V_u = 6.07(2.42)(9.5) = 140 \text{ kips}$$

(a) Two-way action　　　　　　　(b) One-way action

Figure 20.9.1 Critical sections for shear in square footing design.

According to ACI-11.12.1.1 and 11.3.1.1,

$$V_c = 2\sqrt{f'_c}\, b_w d = 2\sqrt{3000}(9.5)(12)20\tfrac{1}{1000} = 250 \text{ kips}$$

$$\phi V_c = 0.85(250) = 212 \text{ kips} > [V_u = 140 \text{ kips}] \qquad\qquad \textbf{OK}$$

No shear reinforcement is required. Note that V_c is computed from the simplified procedure which neglects the effect of $\rho V d/M$. The 24-in. thickness is satisfactory for shear.

(c) Check transfer of load at the base of column (ACI-15.8). The compressive design strength ϕP_n based on the nominal ultimate bearing stress $0.85f'_c$ *in the column* is

$$\phi P_n = \phi(0.85f'_c)A_g = 0.70(0.85)(3)(256) = 457 \text{ kips}$$

$$\phi P_n < [P_u = 548 \text{ kips}] \qquad\qquad \textbf{NG}$$

Thus the column load cannot be transferred by bearing alone. It may well be that the minimum dowels required by ACI-15.8.2.1 will be adequate to transfer the excess load.

$$\text{min required } A_s = 0.005(256) = 1.28 \text{ sq in.} \quad \text{(ACI-15.8.2.1)}$$

The excess load to be carried by the dowels is, neglecting the displaced concrete effect,

$$\text{excess } P_u = 548 - 457 = 91 \text{ kips}$$

$$\text{required } A_s = \frac{\text{excess } P_u}{\phi f_y} = \frac{91}{0.70(40)} = 3.25 \text{ sq in.}$$

Logically a stress of $0.85f'_c$ should be subtracted from f_y in the dowels in order to compensate for the displaced concrete; thus more correctly,

$$\text{required } A_s = \frac{\text{excess } P_u}{\phi(f_y - 0.85f'_c)} = 3.47 \text{ sq in.}$$

Use 4-#9 bars as dowels, $A_s = 4.00$ sq. in.

Figure 20.9.2 Dowel anchorage.

The #9 dowels must be developed above and below the junction of column and footing. The development length L_d required in compression according to ACI-12.3 is

$$L_{db} = 0.02 f_y d_b / \sqrt{f'_c}$$
$$= 0.02(40,000)(1.128)/\sqrt{3000} = 16.5 \text{ in.} \qquad \textbf{Controls}$$

but not less than

$$L_{db} = 0.0003 f_y d_b = 0.0003(40,000)(1.128) = 13.5 \text{ in.}$$

and final L_d cannot be less than 8 in. Thus, $L_d = L_{db} = 16.5$ in. The 24-in. thick footing is adequate for straight dowels.

Hooks or bending of bars should not be considered effective in adding to the compressive resistance of bars (ACI-12.5.3). Very often engineers will specify bending of the dowels as shown in Fig. 20.9.2 to prevent their being pushed through the footing during construction and thus reducing the effective embedment distance L_2. In such cases of bending of the dowels, full development of the compressive force is required over the distance L_1.

For this design, if such a bend is used, the available length L_1 is

$$L_1 = 24 - 3 \text{ (cover)} - 2(1) \text{ (footing bars)} - 1.128 \text{ (dowels)} = 17.9 \text{ in.}$$

This exceeds the 16.5 in. required and is therefore acceptable. If it were unacceptable, alternatives would include a thicker footing, a larger number of smaller-sized dowels, or the use of a pedestal.

(d) Design for bending moment strength. The critical section for moment is at the face of the column (Fig. 20.9.3).

$$M_u = \tfrac{1}{2}(6.07)(9.5)(4.08)^2 = 480 \text{ ft-kips}$$

$$\text{required } R_n = \frac{M_u}{\phi b d^2} = \frac{480(12,000)}{0.90(114)(20)^2} = 140 \text{ psi}$$

Referring to Section 3.8,

$$\rho = \frac{1}{m}\left(1 - \sqrt{1 - \frac{2mR_n}{f_y}}\right)$$

Figure 20.9.3 Critical section for bending moment and development of reinforcement.

$$\rho = \frac{1}{15.7}\left[1 - \sqrt{1 - \frac{2(15.7)(140)}{40,000}}\right] = 0.0036$$

required $A_s = \rho bd = 0.0036(114)(20) = 8.21$ sq in.

min required $A_s = 0.002(114)(22) = 5.00$ sq in.   (ACI-10.5.3)

Try 19-#6, $A_s = 8.36$ sq in.; $d = 24 - 3 - 0.75 = 20.3$ in.

$$C = 0.85f'_c ba = 0.85(3)(9.5)(12)a = 291a$$

$$T = A_s f_y = 8.36(40) = 334 \text{ kips}$$

$$C = T; \quad a = 1.15 \text{ in.}$$

$$\phi M_n = 0.90(334)[20.3 - 0.5(1.15)]\tfrac{1}{12} = 494 \text{ ft-kips}$$

$$\phi M_n > [M_u = 480 \text{ ft-kips}] \qquad\qquad \textbf{OK}$$

Use 19-#6 bars ($A_s = 8.36$ sq in.) each way.

(e) Check development of reinforcement. The embedment from the face of the column must be a distance at least equal to the development length L_d for #6 bars. Referring to Table 6.7.1, the basic development length L_{db} is 12.9 in. Then from Table 6.7.2, the Min $L_d = 16.4$ in. The 19-#6 give an approximate 6 in. center-to-center spacing which means clear spacing exceeding about 5.3 in. Since clear spacing exceeds $3d_b$, and the 3 in. bottom cover exceeds $2d_b$, Category A applies; $\alpha_{abc} = 1.0$ according to ACI-11.2.3.1(d). In addition, Category D applies for wide lateral spacing where clear spacing exceeds $5d_b$. Thus, $\alpha_{de} = 0.80$, giving "first-level L_d" as

$$\alpha_{abc}\alpha_{de}L_{db} = 1.0(0.80)12.9 = 10.3 \text{ in.} < (\text{Min } L_d = 16.4 \text{ in.})$$

The "first-level L_d" is 16.4 in. No second-level adjustments apply; thus,

$$L_d \text{ (for #6)} = 16.4 \text{ in.} < 48 \text{ in. available} \qquad\qquad \textbf{OK}$$

Figure 20.9.4 Design sketch for spread footing of Example 20.9.1.

(f) Design sketch. A design sketch as shown in Fig. 20.9.4 is necessary to convey the designer's decision properly.

For practical design of square footings, design aids are available [2.21, 20.9]. ∎

20.10 DESIGN OF RECTANGULAR FOOTINGS

Rectangular footings may be used in locations where space is restricted to prevent the use of a square footing. The procedure for their design is essentially identical with that of square footings, except that one-way shear action and bending moment must be considered in both principal directions.

■ **EXAMPLE 20.10.1**   Design a rectangular spread footing to carry 235 kips dead load and 115 kips live load from an 18-in. square tied column that contains #9 bars. One dimension of the footing is limited to a maximum of 7 ft. The allowable soil pressure is 5500 psf. Neglect the effect of overburden. Use $f'_c = 3000$ psi, $f_y = 40,000$ psi, and the strength method of the ACI Code.

Solution: **(a)** Determine plan of footing. Assume a footing depth of 2 ft, or 300 psf.

$$\text{net soil pressure} = 5500 - 300 = 5200 \text{ psf}$$

$$\text{required } A = \frac{350}{5.2} = 67.3 \text{ sq ft}$$

Space limitation prevents one dimension from exceeding 7 ft; thus

$$\text{length} = \frac{67.3}{7.0} = 9.6 \text{ ft}$$

Try 7 ft × 9 ft 8 in. (area = 67.7 sq ft).

(b) Determine depth required for shear strength. This footing may be long enough for one-way beam action to govern. In such a case, a direct solution for the effective depth d is reasonably practical. The factored column load is

$$P_u = 1.4(235) + 1.7(115) = 525 \text{ kips}$$

The net upward pressure under factored load condition is

$$p_{net} = \frac{525,000}{67.7} = 7750 \text{ psf}$$

Using section A-A in Fig. 20.10.1, and making the nominal shear strength $V_n = V_c$, so that shear reinforcement is not required, which means

$$V_u = \phi V_c = \phi(2\sqrt{f_c'})b_w d$$

$$7750(7.0)(4.08 - d) = 0.85(2\sqrt{f_c'})(7)(d)144$$

$$d = 1.49 \text{ ft (17.9 in.)}$$

Total depth = 17.9 + 3 (cover) + 1 (estimated bar diameter) = 21.9 in. Try 22 in. for total depth. Check this depth for two-way shear action as a slab, using critical section B-B-B-B shown in Fig. 20.10.1 with $d = 18$ in.

$$V_u = 7750[7.0(9.67) - 3.0(3.0)]\tfrac{1}{1000} = 455 \text{ kips}$$

For $\beta_c = 9.67/7.0 = 1.38 < 2.0$, and $b_0/d = 4(18 + 18)/18 = 8 < 20$ for a four-sided critical section,

$$V_c = 4\sqrt{f_c'}b_0 d = 4\sqrt{3000}[4(18 + 18)](18)\tfrac{1}{1000} = 568 \text{ kips}$$

$$\phi V_c = 0.85(568) = 483 \text{ kips} > [V_u = 455 \text{ kips}] \qquad \textbf{OK}$$

(c) Check transfer of load at base of column. Assuming transfer without aid of dowels, the design strength is

$$\phi P_n = \phi(0.85f_c')A_g = 0.70(0.85)(3)(324) = 578 \text{ kips}$$

$$\phi P_n > [P_u = 525 \text{ kips}] \qquad \textbf{OK}$$

Only the minimum dowels required by ACI-15.8.2.1 are needed.

$$\text{min required } A_s = 0.005(324) = 1.62 \text{ sq in.}$$

Use 4-#6 bars as dowels ($A_s = 1.76$ sq in.).

Figure 20.10.1 Rectangular footing for Example 20.10.1.

Minimum embedment length L_1 (Fig. 20.10.1) required for bars in compression,

$$\min L_1 = L_d \,(\#6) = 11.0 \text{ in.} \quad \text{(ACI-12.3)}$$

$$\text{available } L_1 = 22 - 3 - 1.5 - 0.75 = 16.75 \text{ in.} > 11.0 \text{ in.} \qquad \textbf{OK}$$

(d) Design for bending moment strength. The reinforcement in the long direction is distributed uniformly across the 7-ft width, while that in the short direction is concentrated more heavily under the column in a band equal to the footing width and less heavily near the ends. It is prescribed in ACI-15.4.4 that the portion $2/(\beta + 1)$ of the total transverse reinforcement should be placed in the central band (of a width equal to the short side of the footing) wherein β is the ratio of the long side to the short side of the footing. The ratio $2/(\beta + 1)$ may be derived on the basis that the intensity of reinforcement in the central band is twice that of the outer portions.

In the long direction,

$$M_u = 7.75(7.0)\frac{(4.08)^2}{2} = 452 \text{ ft-kips}$$

$$\text{required } R_n = \frac{M_u}{\phi b d^2} = \frac{452(12,000)}{0.90(7)(12)(18.5)^2} = 209 \text{ psi}$$

Since the longitudinal bars are placed below the transverse bars, d has been taken as 18.5 in. in the above calculation.

Using the trial moment arm method, rather than the formula, Eq. (3.8.5), for ρ, assume the moment arm as $0.95d = 17.6$ in. since the value of required R_n is very low.

$$\text{required } A_s = \frac{M_u}{\phi f_y(\text{arm})} = \frac{452(12)}{0.90(40)(17.6)} = 8.55 \text{ sq in.}$$

Check:

$$C = 0.85 f'_c ba = 0.85(3)(84)a = 214a$$

$$T = A_s f_y = 8.55(40) = 342 \text{ kips}$$

$$a = \frac{342}{214} = 1.60 \text{ in.}$$

$$\text{arm} = 18.5 - 0.80 = 17.70 \text{ in.} \approx 17.6 \text{ in. assumed}$$

$$\text{revised required } A_s = 8.55\left(\frac{17.6}{17.7}\right) = 8.5 \text{ sq in.}$$

Use 14-#7, $A_s = 8.4$ sq in. (close enough).

In the short direction,

$$M_u = 7.75(9.67)\frac{(2.75)^2}{2} = 283 \text{ ft-kips}$$

$$\text{assume arm} = 0.95d = 0.95(17.5) = 16.6 \text{ in.}$$

$$\text{required } A_s = \frac{M_u}{\phi f_y(\text{arm})} = \frac{283(12)}{0.90(40)(16.6)} = 5.7 \text{ sq in.}$$

Check:

$$C = 0.85(3)(116)a = 296a$$

$$T = 5.7(40) = 228$$

$$a = \frac{228}{296} = 0.77 \text{ in.}$$

$$\text{arm} = 17.5 - 0.39 = 17.11 \text{ in.}$$

$$\text{revised required } A_s = 5.7\left(\frac{16.6}{17.11}\right) = 5.5 \text{ sq in.}$$

For minimum reinforcement, ACI-10.5.3 requires $\rho_g = 0.002$. For the short direction,

$$\text{min required } A_s = 0.002(116)(22) = 5.1 \text{ sq in.}$$

In this case, the strength requirement controls.

Try 9-#7, $A_s = 5.6$ sq in.

$$\frac{\text{reinforcement in band width}}{\text{total reinforcement}} = \frac{2}{\beta + 1} = \frac{2}{9.67/7.0 + 1} = 0.84$$

Number of bars in the 7-ft band = 9(0.84) = 7.6, say 8. If one bar is placed on each side outside the 7-ft band, a total of 10 bars would be required.

Use 10-#7 bars.

(e) Check development of reinforcement. Referring to Table 6.7.1, the basic development length L_{db} for #7 bars is 17.5 in. Then from Table 6.7.2, the Min L_d = 19.2 in. The 14-#7 in the long direction have an approximate 6 in. center-to-center spacing, which means clear spacing exceeding $3d_b$. In addition, the 3 in. bottom cover exceeds $2d_b$, permitting Category A to apply with α_{abc} = 1.0 according to ACI-11.2.3.1(d). Wide lateral spacing under Category D also applies because clear spacing exceeds $5d_b$. Thus, α_{de} = 0.80, giving "first-level L_d" as

$$\alpha_{abc}\alpha_{de}L_{db} = 1.0(0.80)17.5 = 14.0 \text{ in.} < (\text{Min } L_d = 19.2 \text{ in.})$$

The "first-level L_d" is 19.2 in. No second-level adjustments apply; thus,

$$L_d \text{ (for #7)} = 19.2 \text{ in.}$$

The minimum available embedment is in the short direction and equals 33 in. less an inch or two of cover; this is more than adequate.

The complete design is shown in Fig. 20.10.1. ∎

20.11 DESIGN OF PLAIN AND REINFORCED CONCRETE WALL FOOTINGS

Wall footings carrying direct concentric loads may be of either plain or reinforced concrete. Those that are required to carry moment, such as for the cantilever retaining wall, are treated in Chapter 12. Since the wall footing has bending in only one direction, it may be designed or investigated by considering a typical 12-in. strip along the wall. Many typical walls carry relatively light loads, and the supporting footings are proportioned by using arbitrary minimums. Footings carrying light loads on good soil are often made of plain concrete and may be designed in accordance with *Building Code Requirements for Structural Plain Concrete* [20.10].

■ **EXAMPLE 20.11.1** Determine the adequacy of the plain concrete wall footing of Fig. 20.11.1 to carry a load of 20 kips/linear ft dead load including the wall weight and 8 kips/linear ft live load. Use f'_c = 3000 psi, an allowable soil pressure of 6 ksf, and the ACI Code.

Solution:

$$\text{total service load} = 28 + 6(2.5)(0.145) = 30.2 \text{ kips/ft}$$

$$\text{maximum soil pressure} = \frac{30.2}{6.0} = 5.0 \text{ ksf} < 6 \text{ ksf} \qquad \textbf{OK}$$

The factored bending moment is computed on the critical section at the face of the wall (ACI-15.4.2).

$w = 28$ kips/ft

12"

2' - 6"

6'-0"

Figure 20.11.1 Wall footing for Example 20.11.1.

$$w_u = 20(1.4) + 8(1.7) = 41.6 \text{ kips/ft}$$

net soil pressure under factored load $= 41.6/6 = 6.95$ ksf

$$M_u = 6.95 \frac{(2.5)^2}{2} = 21.7 \text{ ft-kips/ft}$$

For computing the moment of inertia of the section, the bottom 2 in. of concrete placed against the ground are assumed to be of poor quality and required [20.10, Section 6.3.5] to be neglected. Neglecting the bottom 2 in.,

$$I_g = \frac{12(28)^3}{12} = 21,950 \text{ in.}^4$$

For bending of plain concrete, the design strength is based on a linear stress–strain relationship for both tension and compression [20.10, Section 6.3.1] and a conservatively low value of $5\sqrt{f_c'}$ for the modulus of rupture [20.10, Section 6.2.1]. A strength reduction factor of 0.65 is also prescribed [20.10, Section 6.2.2]. Thus,

$$\phi M_n = \phi f_t I_g/(h/2)$$
$$= \frac{0.65(5\sqrt{3000})(21,950)}{14(12,000)} = 23.3 \text{ ft-kips/ft}$$

$$\phi M_n > [M_u = 21.7 \text{ ft-kips/ft}] \qquad\qquad \textbf{OK}$$

Shear strength is adequate since the critical section at a distance d from the face of wall falls near the edge of the footing; ACI-15.5.2, 11.12.1.1, and 11.1.3.1 apply here. ■

■ EXAMPLE 20.11.2   Design a reinforced concrete footing for a 12-in. masonry wall carrying 10 kips/linear ft dead load including the wall weight and 5 kips/ft live load. Use f'_c = 3000 psi, f_y = 40,000 psi, an allowable soil pressure of 4000 psf, and the strength method of the ACI Code.

Solution: Assume the footing depth to be 10 in. at 125 psf. Allowable net soil pressure = 4000 − 125 = 3875 psf. Footing width = 15/3.875 = 3.87 ft. Use 4 ft. It is probable that the thickness will be governed by shear (inclined cracking) which is taken to be critical at a distance d from the face of the wall (Fig. 20.11.2).

Applying the overload factors,

$$w_u = 1.4(10) + 1.7(5) = 22.5 \text{ kips/ft}$$

Net soil pressure under factored load = 22.5/4 = 5.63 ksf. When no shear reinforcement is used, the nominal strength for one-way action, using the simplified procedure, is

$$V_n = V_c = 2\sqrt{f'_c} b_w d$$

which requires that

$$V_u = \phi V_c$$

$$5630(1.5 - d) = 0.85(2\sqrt{3000})(12)(12d)$$

$$2.38d = 1.5 - d$$

$$d = 0.44 \text{ ft (5.3 in.)}$$

$$\text{total thickness} = 5.3 + 3 \text{ (cover)} + 0.5 \text{ (bar radius)} = 8.8 \text{ in.}$$

<u>*Use* 9 in. thickness.</u>

$$\text{check weight} = 113 \text{ psf} \qquad \qquad \textbf{OK}$$

The critical section for bending moment strength on footings under masonry walls occurs halfway between the middle and the edge of the wall (ACI-15.4.2b).

Figure 20.11.2 Wall footing for Example 20.11.2.

$$M_u = \tfrac{1}{2}(5.63)(1.75)^2 = 8.62 \text{ ft-kips/ft}$$

$$\text{required } R_n = \frac{8.62(12{,}000)}{0.90(12)(5.5)^2} = 316 \text{ psi}$$

The steel area may be obtained by formula, Eq. (3.8.5), by trial as in Example 20.9.1, or from Fig. 3.8.1. Using the last,

$$\rho \approx 0.0085$$

$$\text{required } A_s = \rho bd = 0.0085(12)(5.5) = 0.56 \text{ sq in.}$$

Try #5 @ $6\tfrac{1}{2}$ in. ($A_s = 0.57$ sq in./ft). Check strength.

$$C = 0.85f'_c ba = 0.85(3)(12)a = 30.6a$$

$$T = 0.57(40) = 22.8 \text{ kips}$$

$$a = \frac{22.8}{30.6} = 0.75 \text{ in.}$$

$$\phi M_n = 0.90(22.8)[5.5 - 0.5(0.75)]\tfrac{1}{12} = 8.75 \text{ ft-kips} > M_u \qquad \textbf{OK}$$

For development of reinforcement, the basic development length L_{db} and the Min L_d for #5 bars are, from Tables 6.7.1 and 6.7.2, respectively, 9.1 and 13.7 in. The cover and wide spacing indicate $\alpha_{abc} = 1.0$ under Category A, and $\alpha_{de} = 0.80$ under Category D. Thus,

$$\alpha_{abc}\alpha_{de}L_{db} = 1.0(0.80)9.1 = 7.3 \text{ in.} < (\text{Min } L_d = 13.7 \text{ in.})$$

The "first-level L_d" is 13.7 in. No second-level adjustments apply; thus,

$$L_d \text{ (for #5)} = 13.7 \text{ in.}$$

Since 21 in. (the distance from the critical section to the outside edge of footing), less an inch or two for cover, are available for embedment of main reinforcement, this is acceptable.

See Fig. 20.11.2 for the details of the complete design. Some longitudinal reinforcement for shrinkage should probably be provided, say 3-#4 bars. ∎

20.12 COMBINED FOOTINGS

A combined footing is one that usually supports two columns. These may be two interior columns [Fig. 20.12.1(a)] which are so close to each other that isolated footing areas would overlap. If a property line exists at or near the edge of an exterior column, a rectangular [Fig. 20.12.1(b)] or trapezoidal [Fig. 20.12.1(c)] combined footing may be used to support the exterior column and its adjacent interior column. The area of the combined footing may be proportioned for uniform settlement by making its centroid coincide with the resultant of the respective portions of the two column loads that are sustained for long duration. It may be noted that for footings of constant thickness the centroid of the bearing area always coincides with the resultant of the weight of the footing itself.

Figure 20.12.1 Combined footings.

Referring to the frequently occurring situation in Fig. 20.12.1(b), the load P_1 is close to the property line; however, there is adequate space to the right of P_2. Whenever P_2 exceeds P_1, a rectangular combined footing can be used, since it may be made long enough to make the load resultant and the footing centroid coincide. It may be shown that if $\frac{1}{2} < P_2/P_1 < 1$ approximately, a trapezoidal footing could be used. If $P_2/P_1 < \frac{1}{2}$ approximately, then either a strap (Fig. 20.12.2) or a T-shaped spread footing would have to be used.

In the strength computation for a combined footing, maximum loads in columns (full dead load plus reduced live load as discussed in Section 20.7) should be used.

Figure 20.12.2 Cantilever, or strap, combined footing.

Since the resultant of the maximum column loads does not necessarily coincide with that of the sustained column loads producing uniform settlement, under the former loading condition the distribution of the net soil pressure along the footing is not uniform. The deviation from uniformity is usually so small that certain approximate short-cut procedures may be used in determining the shear and moment diagrams in the longitudinal direction.

The ACI Code (see ACI-15.10) does not provide full recommendations for combined footings. Kramrisch [20.11] has provided a detailed treatment of footing design, with particular emphasis on combined footings. However, ACI Committee 336 [20.12] has given design procedures for combined footings and mats, and additional suggestions have been given by Kramrisch and Rogers [20.13], Szava-Kovats [20.14], and Davies and Mayfield [20.15]. When a series of columns is supported by a single footing extending over a large area, the footing is referred to as a *mat*. Mats are treated by Bowles [20.12] as well as committee 336. Basically, transverse steel under each column tends to distribute the column load in the transverse direction. This being considered accomplished, the combined footing itself becomes a beam in the longitudinal direction. Thus it is suggested that the provisions for isolated footings be applied to the transverse direction, and those for beams to the longitudinal direction.

The cantilever or strap footing shown in Fig. 20.12.2 is an alternate design to prevent overturning of an exterior footing placed eccentrically under an exterior column, the edge of which is at or close to a property line. Overturning of the exterior footing is prevented by connecting it with the adjacent interior footing by a strap beam. This strap beam is subjected to a constant shearing force and a linearly varying negative bending moment. Thus it behaves like a cantilever beam; hence the name "cantilever footing."

In the strength computation for a cantilever footing (Fig. 20.12.2), the weight of the strap, the exterior footing, and the interior footing is each assumed to be balanced by the soil pressure caused by it and thus such weight causes no shears and moments in any part of the structure. In the longitudinal direction, the column loads P_{ext} and P_{int} are in equilibrium with the total "net" upward soil pressures R_{ext} and R_{int} under the exterior and interior footings; the upward soil pressure is assumed to be uniform over the entire area of each footing. The exterior footing may be considered as under one-way transverse bending, although some reinforcement in the longitudinal direction is desirable, and the interior footing is under two-way bending as in isolated footings.

Use of a cantilever footing may be justifiable under conditions where the distance between columns is large and a large area of excavation must be avoided. It is usual practice that the bottom surfaces of the exterior footing, the strap, and the interior footing be at the same elevation, but the total thickness of each element may be different, depending on the strength requirements. Certainly it is desirable, unless there are good reasons to the contrary, to make all three elements of constant thickness.

20.13 DESIGN OF COMBINED FOOTINGS

The design of two common types of combined footings will be shown. The first is a rectangular footing, and the second is a strap or cantilever footing.

■ **EXAMPLE 20.13.1** Design a combined footing to support two columns as shown in Fig. 20.13.1(a): $P_A = 350$ kips (40% live load); $P_B = 400$ kips (40% live load); column A is centered 1 ft 3 in. from the property line, and column B is centered 19 ft 3 in. from the property line. Use $f'_c = 3000$ psi, $f_y = 40,000$ psi, maximum soil pressure = 5000 psf, and the strength method of the ACI Code. Assume that the ratio of maximum column loads as given is equal to that of long duration loads in the exterior and interior columns.

Solution: **(a)** Length and width of footing. ACI-15.2.2 indicates base area of footings is to be determined using service loads and allowable soil pressure.

$$\bar{x} \text{ from property line} = \frac{350(1.25) + 400(19.25)}{750} = 10.85 \text{ ft}$$

$$\text{length of footing, } L = 10.85(2) = 21.70 \text{ ft}$$

Use 21 ft-9 in.

Since the design for strength of the footing involves factored loads, there will be an eccentricity no matter how "exact" is the length determination. In the design for shear and bending moment, the soil pressure under factored loads might be taken as linearly varying to account for the eccentric loading, but in the case of a small eccentricity it is probably sufficient to assume a uniform soil pressure as is done in this example.

Assume the footing thickness to be about 2 ft 6 in., or 375 psf.

$$\text{net soil pressure} = 5000 - 375 = 4625 \text{ psf}$$

$$\text{footing width} = \frac{750,000}{4625(21.75)} = 7.46 \text{ ft}$$

Try 7 ft 6 in.

(b) Longitudinal factored shears and factored moments. For gravity loading,

$$\text{column } A, P_u = 1.4(210) + 1.7(140) = 532 \text{ kips}$$

$$\text{column } B, P_u = 1.4(240) + 1.7(160) = 608 \text{ kips}$$

$$\text{net soil pressure under factored load} = \frac{1,140,000}{21.75(7.5)} = 6990 \text{ psf}$$

The factored shear V_u diagram is computed as for a beam and given in Fig. 20.13.1(c). For simplicity, the column loads are taken to be acting along the column centerlines, thus producing the dashed portions within the column widths on the shear diagram. In the computations, the 20-in. diameter column is treated as an equivalent square of side 17.7 in. in accordance with ACI-15.3.

$$\text{net upward uniform pressure} = 7.5(6.99) = 52.4 \text{ kips/ft}$$

$$V_u \text{ at centerline of column } A = +52.4(1.25) = +65.5 \text{ kips}$$
$$+ 65.5 - 532 = -466.5 \text{ kips}$$

Figure 20.13.1 Rectangular combined footing for example 20.13.1.

$$V_u \text{ at centerline of column } B = -52.4(2.5) = -131.0 \text{ kips}$$

$$-131.0 + 608 = +477.0 \text{ kips}$$

$$\text{point of zero shear} = 18\left(\frac{466.5}{466.5 + 477.0}\right)$$

$$= 8.90 \text{ ft from centerline of column } A$$

The factored moment M_u diagram as computed for a beam is given in Fig. 20.13.1(d). Note that the numerical values on the two small end portions of the factored moment diagram are based on assuming all of the column loads to be concentrated at the column centerlines.

$$\text{max } M_u \text{ (computed from left side)} = \frac{52.4(10.15)^2}{2} - 532(8.90)$$

$$= -2035 \text{ ft-kips}$$

$$\text{max } M_u \text{ (computed from right side)} = \frac{52.4(11.60)^2}{2} - 608(9.10)$$

$$= -2006 \text{ ft-kips}$$

The moments as computed from both sides are not exactly the same because a footing length of 21 ft 9 in. is used instead of the computed 20.70 ft and because the distance 8.90 ft to the point of zero shear contains only three significant figures. Use $M_u = 2035$ ft-kips.

(c) Thickness of slab. For moment, the thickness may be based on a desired reinforcement ratio ρ. The maximum permitted by the ACI Code is from Table 3.8.1,

$$\text{max } \rho = 0.75\rho_b = 0.0278$$

For reasonable deflection control, select $\rho = 0.014$, that is, approximately one-half of the maximum permitted. For this value of ρ, using Eq. (3.8.4),

$$R_n = \rho f_y(1 - \tfrac{1}{2}\rho m)$$

$$m = \frac{f_y}{0.85f'_c} = \frac{40}{0.85(3)} = 15.7$$

$$R_n = 0.014(40,000)[1 - 0.5(0.014)(15.7)] = 498 \text{ psi}$$

$$\text{required } d = \sqrt{\frac{M_u}{\phi R_n b}} = \sqrt{\frac{2035(12,000)}{0.90(498)(7.5)(12)}} = 24.6 \text{ in.}$$

The footing is considered a beam in shear computations. One-way action is assumed to control at the distance d from the face of the columns. The shear at a distance d from the face of the 17.7-in. equivalent square column is

$$V_u = 477.0 - \left(\frac{8.85 + d}{12}\right)(52.4) = 438.3 - 4.37d$$

The nominal shear strength when no shear reinforcement is to be used is

$$V_n = V_c = 2\sqrt{f'_c}b_w d$$

unless the more detailed expression involving $\rho V_u d/M_u$ is used. Then,

$$V_u = \phi V_c$$

$$438.3 - 4.37d = 0.85(2\sqrt{3000})(7.5)(12)d/1000$$

$$438.3 - 4.37d = 8.38d$$

$$d = 34.4 \text{ in.}$$

It seems desirable to make the footing deep enough at the desirable reinforcement ratio for the bending moment, but not deep enough to give the extra 10 in. that would be required to eliminate stirrups.

total depth = 24.6 + 2 (cover) + 0.5 (stirrup) + 0.6 (bar radius) = 27.7 in.

Use 28 in. Check the weight, 28(150)/12 = 350 psf.

$$\text{max soil pressure} = \frac{750,000}{21.75(7.5)} + 350 = 4950 \text{ psf} < 5000 \text{ psf} \qquad \textbf{OK}$$

(d) Main longitudinal reinforcement. At the middle of the span,

$$\text{required } R_n = \frac{M_u}{\phi bd^2} = \frac{2035(12,000)}{0.90(7.5)(12)(24.9)^2} = 484 \text{ psi}$$

$$\text{required } A_s = 0.014\left(\frac{484}{498}\right)(7.5)(12)(24.9) = 30.5 \text{ sq in.}$$

Try 20-#11 (approximate $4\frac{1}{2}$ in. spacing); $A_s = 31.2$ sq in.

The anchorage required from the point of maximum moment to the end of the bars is the development length L_d for #11 bars. From Table 6.7.1 the basic development length L_{db} is 45.6 in. and from Table 6.7.2 the Min L_d is 30.9 in. The clear spacing between the 20 bars is

$$\text{clear spacing} = \frac{7.5(12) - 20(1.41)}{20} = 3.09 \text{ in.}$$

assuming the clear distance to the side of the footing is one-half the interior clear spacing. The clear spacing $(2.2d_b)$ exceeds $2d_b$ but not $3d_b$ and the cover (in this case 2.5 in. to the #11 bars) exceeds d_b; therefore Category B (ACI-12.2.3.3) applies and $\alpha_{abc} = 1.4$. The "first-level L_d" is

$$\alpha_{abc}\alpha_{de}L_{db} = 1.4(1.0)45.6 = 63.8 \text{ in.} > (\text{Min } L_d = 30.9 \text{ in.})$$

The "first-level L_d" is 63.8 in. Since the bars have more than 12 in. of concrete cast beneath them, the top bar multiplier, $\alpha_{top} = 1.3$ applies. Thus,

$$L_d (\#11) = \alpha_{top} (63.8) = 1.3(63.8) = 83.0 \text{ in.} \quad (6.9 \text{ ft})$$

If all 20 of the #11 bars are run into the centerlines of the columns, the development requirement is more than adequate, since an anchorage distance of 108 in. is provided on both sides of the point of maximum moment.

Use 20-#11 bars at 18 ft long (spaced approximately at $4\frac{1}{2}$ in.).

Check development length requirement at the points of inflection according to ACI-12.11.3. The Code provision is checked here even though the reinforcement is negative moment reinforcement rather than the positive moment reinforcement as prescribed in ACI-12.11.3. The footing may be visualized for this purpose as an inverted beam with the soil pressure as loading and the columns as supports. The situation is similar to the positive moment requirement in the sense that the required embedment is into the support (column) rather than out into the span as for the negative moment requirement in an ordinary continuous beam. Furthermore, since the inflection points are inside the faces of the columns, the 1.3 factor might be used; however, since the column width (22 in.) is only a small fraction of the footing width (7.5 ft), the authors do *not* recommend using the 1.3 factor.

Since all 20-#11 bars extend through the inflection points,

$$M_n = \frac{2035}{0.90}\left(\frac{31.2}{30.5}\right) = 2300 \text{ ft-kips}$$

$$V_u = 466.5 \text{ kips (at column } A)$$

$$V_u = 477.0 \text{ kips (at column } B)$$

$$L_a = 15 \text{ in. approx (near both columns } A \text{ and } B)$$

In this case the actual distance L_a from the inflection point to the end of the bars is less than the maximum L_a limits of effective depth d or 12 bar diameters; thus the actual distance of 15 in. controls. Since M_u and L_a are the same at both inflection points, the one near the column B having the larger shear V_u controls. At column B inflection point,

$$\frac{M_n}{V_u} + L_a = \frac{2300(12)}{477.0} + 15 = 73 \text{ in.} < L_d = 83 \text{ in.} \qquad \textbf{NG}$$

Since this seems unacceptable, either a rationalization that some increase in strength is available from the support reactions (columns) (i.e., the 1.3 factor) or smaller bars should be used. This is a situation where the 1989 ACI Code increases L_d from 64 to 83 in.

(e) Alternate design of the main longitudinal reinforcement using cutoff. If it seems desirable to cut off some of the tension bars, say 10-#11, and extend the remaining ones into the supports, the general anchorage requirements of ACI-12.10 must be applied. The theoretical point at which the 10-#11 bars are no longer required is indicated in Fig. 20.13.2, corresponding to the moment capacity ϕM_n of 1100 ft-kips. The provision in ACI-12.10.3 requires an extension of at least the effective depth d or 12 bar diameters, whichever is greater, beyond the theoretical cutoff point. In this case the effective depth of 24.8 in. (or 2.1 ft) controls over 12 diameters = 16.9 in. (or 1.4 ft).

In addition, for such a cutoff in the tension bars to be permitted, ACI-12.10.5 must be satisfied. It is assumed here that stirrups will be provided sufficient to satisfy ACI-12.10.5.1. The continuing bars must be embedded L_d beyond the cutoff of the 10-#11 bars according to ACI-12.10.4.

Figure 20.13.2 Bar cutoff alternate for Example 20.13.1.

Determine the development length L_d for the bars being cut as well as for the continuing bars. This length will be shorter than the 83.0 in. computed in part (d) because the spacing of the developing bars will be twice the spacing of the bars were there no cutoff used. Had fewer than one-half the bars been cut, some bars would have remained at the closer spacing and L_d would have remained 83.0 in.; this would have prohibited the cut because the distance from the cut to the maximum moment location would have been less than 83.0 in. With 10-#11 cut, the spacing between those 10 bars will be 2(3.09) = 6.18 in. This exceeds $3d_b$ and since the bars are enclosed with at least minimum stirrups, Category A applies, satisfying ACI-12.2.3.1(a); $\alpha_{abc} = 1.0$. Categories D and E are not applicable; $\alpha_{de} = 1.0$. Thus, the "first level L_d" is,

$$\alpha_{abc}\alpha_{de}L_{db} = 1.0(1.0)45.6 = 45.6 \text{ in.} > (\text{Min } L_d = 30.9 \text{ in.})$$

Then applying the top bar multiplier $\alpha_{\text{top}} = 1.3$ gives,

$$L_d(\#11) = \alpha_{\text{top}}(45.6) = 1.3(45.6) = 59.3 \text{ in.} \ (4.9 \text{ ft})$$

At neither end is adequate distance available (≈ 2 ft from end of bars at column A end of footing to point X_1 and ≈ 3.5 ft from end of bars at column B end of footing to point X_2) to develop ($L_d = 4.9$ ft) the continuing bars through straight embedment. Thus, hooks would be needed at both ends.

When hooks are provided as anchorage beyond an inflection point (as would be the case here at both ends of footing), ACI-12.11.3 need not be satisfied. Referring to Table 6.11.2 and Fig. 6.11.2 for #11 hooked bars,

$$L_{dh} = 20.6 \text{ in.}$$

In Fig. 20.13.2, the development of the #11 hooked bars is shown by the sloping dashed line at each end of the footing starting from the end of the *bar* at 1.5 in. from the side face of footing.

Since the cut bars at X_1 and X_2 are still in the tension zone, ACI-12.10.5 must be satisfied. By inspection of Fig. 20.13.2, ϕM_n exceeds twice M_u, thus, check $V_u/(\phi V_n) \leq 0.75$ for the cuts to be acceptable,

$$\phi V_n = \phi \left(2\sqrt{f_c'} b_w d + \frac{A_v f_y d}{s} \right)$$

$$\phi V_n = 0.85 \left(2\sqrt{3000}(7.5)(12)24.5\tfrac{1}{1000} + \frac{8(0.2)(40)24.5}{12} \right)$$

$$\phi V_n = 205.3 + 111.1 = 316 \text{ kips}$$

The stirrups are designed as multiple loop stirrups (Fig. 20.13.4), in part (h); this check could have been shown there. The check shows the shear strength to be 100% utilized (and even a bit inadequate) when $V_u \approx 340$ kips at the proposed cutoff location. Thus, the cuts cannot be made at the locations X_1 and X_2. The 10-#11 bars must be lengthened to reach the inflection points, in which case they are no longer in the tension zone and ACI-12.10.5 does not apply.

It is clear in this example that cutting bars is impractical when about 4 ft of straight length is saved on 10 bars but the remaining 10 must be hooked.

(f) Longitudinal reinforcement at bottom of footing beyond column centers. The bending moment at the face of column *B* is

$$M_u = \tfrac{1}{2}(52.4)(1.76)^2 = 81.2 \text{ ft-kips}$$

Though certainly not always the case, the moment here appears small enough to require no reinforcement. The strength of the unreinforced section in flexure is computed by using $\phi = 0.65$ according to Section 6.2.2 of *Building Code Requirements for Structural Plain Concrete* [20.10]. Neglecting the bottom 2 in. of thickness,

$$I_g = \tfrac{1}{12}(7.5)(12)(26)^3 = 132{,}000 \text{ in.}^4$$

$$\phi M_n = \phi \left(\frac{f_t I_g}{h/2} \right) = 0.65 \left[\frac{5\sqrt{3000}(132{,}000)}{13(12{,}000)} \right] = 151 \text{ ft-kips}$$

$$\phi M_n > [M_u = 81.2 \text{ ft-kips}] \hspace{2cm} \textbf{OK}$$

No flexural reinforcement in the longitudinal direction is required for strength at the bottom of either cantilever.

(g) Transverse reinforcement. Bending in the transverse direction may be treated in a manner similar to isolated spread footings. The 1940 Joint Committee [20.17] recommended that the transverse reinforcement at each column should be placed uniformly within a band having a width not greater than the width of the column plus twice the effective depth of the footing.

The procedure seems to be reasonable. Certainly the behavior of the footing depends on the overall length-to-width ratio as well as the spacing of the columns. In this design example, the large spacing of the columns and the relatively narrow footing mean that most of the footing between the columns will be subjected only to longitudinal curvature while locally, in the vicinity of the concentrated loads, curvature in both directions will result. Thus the transverse reinforcement is put into bands as shown in Fig. 20.13.3.

$$\text{column } A \text{ band width, } W_A = 1.25 + \frac{8 + 24.8}{12} = 4.0 \text{ ft}$$

$$\text{net factored load pressure in transverse direction} = \frac{532}{7.5} = 71.0 \text{ kips/ft}$$

$$M_u = \tfrac{1}{2}(71.0)(2.83)^2 = 284 \text{ ft-kips}$$

$$d = 28 - 3 \text{ (cover at bottom)} - 0.5 \text{ (bar radius)} = 24.5 \text{ in.}$$

$$\text{required } R_n = \frac{M_u}{\phi b d^2} = \frac{284(12,000)}{0.90(4.0)(12)(24.5)^2} = 132 \text{ psi}$$

From Fig. 3.8.1, $\rho \approx 0.0035$, which exceeds the minimum reinforcement ratio required by ACI-10.5.3 (0.002 from ACI-7.12).

$$\text{required } A_s = 0.0035(4.0)(12)(24.5) = 4.1 \text{ sq in.}$$

Try 7-#7 bars ($A_s = 4.20$ sq in.). Check strength.

$$C = 0.85 f'_c b a = 0.85(3)(4.0)(12)a = 122.4a$$

$$T = 4.20(40) = 168 \text{ kips}$$

$$a = \frac{168}{122.4} = 1.37 \text{ in.}$$

$$\phi M_n = 0.90(168)[24.5 - 0.5(1.37)]\tfrac{1}{12} = 300 \text{ ft-kips} > 284 \text{ ft-kips} \qquad \textbf{OK}$$

Figure 20.13.3 Band width for transverse reinforcement.

Since 2 ft 10 in. is available from face of column, embedment can be provided exceeding the required development length L_d for #7 bars ($L_d = 19.2$ in., Category A, Min L_d controls).

Use 7-#7 bars (approximate 8 in. spacing).

$$\text{column } B \text{ band width } W_B = 2.50 + \frac{8.85 + 24.8}{12} = 5.3 \text{ ft}$$

$$\text{net factored load pressure in transverse direction} = \frac{608}{7.5} = 81.0 \text{ kips/ft}$$

$$M_n = \tfrac{1}{2}(81.0)(3.01)^2 = 367 \text{ ft-kips}$$

$$\text{required } R_n = \frac{M_u}{\phi bd^2} = \frac{367(12,000)}{0.90(5.3)(12)(24.5)^2} = 128 \text{ psi}$$

From Fig. 3.8.1, $\rho \approx 0.0035$, approximately the same as for column A.

$$\text{required } A_s = 0.0035(5.3)(12)(24.5) = 5.5 \text{ sq in.}$$

Try 9-#7 ($A_s = 5.40$ sq in.). Check strength.

$$C = 0.85f'_c ba = 0.85(3)(5.3)(12)a = 162a$$

$$T = 5.40(40) = 216 \text{ kips}$$

$$a = \frac{216}{162} = 1.33 \text{ in.}$$

$$\phi M_n = 0.90(216)[24.5 - 0.5(1.33)]\tfrac{1}{12} = 386 \text{ ft-kips} > 367 \text{ ft-kips} \qquad \textbf{OK}$$

Anchorage of #7 bars is adequate since available embedment of 3.01 ft exceeds L_d of 19.2 in.

Use 9-#7 bars (approximately $7\tfrac{1}{2}$ in. spacing).

(h) Shear reinforcement. The usual approach is to consider the footing as a beam and to provide shear reinforcement on the assumption that the shear (inclined cracking) effect is uniform across the width. This approach seems appropriate in this case with the large distance between columns and the relatively narrow footing width.

The maximum shear to be provided for is at the critical section a distance d from the face of the column. From Fig. 20.13.1, $V_u = 329$ kips. The design shear strength ϕV_c of a beam without shear reinforcement, using the simplified procedure, is

$$\phi V_c = \phi(2\sqrt{f'_c})b_w d$$

$$= 0.85(2\sqrt{3000})(90)(24.8)\tfrac{1}{1000} = 208 \text{ kips}$$

$$\text{required } \phi V_s = V_u - \phi V_c = 329 - 208 = 121 \text{ kips}$$

When shear reinforcement is required, there must be at least the minimum specified by ACI-11.5.5.3, as follows:

$$\text{min } \phi V_s = \phi(50)b_w d = 0.85(50)(90)(24.8)\tfrac{1}{1000} = 95 \text{ kips} > 121 \text{ kips} \qquad \textbf{OK}$$

Thus the 121 kips controls the closest spacing for shear reinforcement. Even though this is a footing, the rules for beams are believed appropriate here. For design of shear reinforcement,

$$\frac{A_v}{s} = \frac{\phi V_s}{\phi f_y d} = \frac{121}{0.85(40)24.8} = 0.144 = \frac{A_s N}{s}$$

In the above expression, A_s is the cross section of the stirrup bar, and N is the number of times the multiple-loop stirrup crosses the neutral axis of the beam. Use a #4 stirrup, with $N = 8$ (see Fig. 20.13.4), spaced at 12 in. This gives

$$\text{provided } \frac{A_v}{s} = \frac{0.20(8)}{12} = 0.133$$

which is about 8% too low.

This may be a situation where computation using the more detailed expression for V_c is justified to verify using the above arrangement.

$$\phi V_c = \phi \left[1.9\sqrt{f_c'} + 2500\,\frac{\rho V_u d}{M_u} \right] b_w d$$

$$\rho = \frac{A_s}{bd} = \frac{20(1.56)}{7.5(12)(24.8)} = 0.0140$$

$$\frac{V_u d}{M_u} = \frac{329,000(24.8)}{990(12,000)} = 0.69 < 1.0 \qquad\qquad \textbf{OK}$$

$$[1.9\sqrt{3000} + 2500(0.0140)(0.69) = 2.34\sqrt{f_c'}] < [3.5\sqrt{f_c'}](\text{max}) \qquad \textbf{OK}$$

$$\phi V_c = 0.85(2.34\sqrt{3000})(90)(24.8)\tfrac{1}{1000} = 243 \text{ kips}$$

$$\phi V_s = V_u - \phi V_c = 329 - 243 = 86 \text{ kips} < [\text{min } \phi V_s = 95 \text{ kips}]$$

The minimum requirement now controls, giving

$$\text{required } \frac{A_v}{s} = \frac{95}{0.85(40)24.8} = 0.113 < 0.133$$

Use #4 stirrups @ 12 in. with $N = 8$, as in Fig. 20.13.4. The 12-in. spacing is the maximum permitted (ACI-11.5.4.3) when v_s does not exceed $4\sqrt{f_c'}$, which corresponds to $\phi V_s = 415$ kips.

The first stirrup is placed at $s/2 = 6$ in. from the face of the column, and the last one should be within $s/2 = 6$ in. of the location where $V_u = \phi V_c/2$, at which shear reinforcement is no longer theoretically required for beams (ACI-11.5.5.1). The stirrup arrangement will be made identical at each column. A few longitudinal bars are placed in the bottom of the footing in order to space and hold in position the stirrups and the transverse reinforcement.

Figure 20.13.4 Multiple-loop stirrup ($N = 8$).

(i) Check shear strength based on two-way action. At the exterior column A, the calculation based on a perimeter at $d/2$ (12.45 in.) from the column face on all four sides would be unrealistic on the side nearest the property line. Instead a three-sided perimeter is used as at column A in Fig. 20.13.3,

$$b_0 = 2(15 + 16 + 12.45) + 22 + 2(12.45) = 133.8 \text{ in.}$$

According to ACI Formula (11-37), $\alpha_s = 30$ for the three-sided perimeter ("edge column"), and when b_0/d does not exceed 15 that equation does not control. In this case,

$$b_0/d = 133.8/24.9 = 5.4 < 15$$

Thus, with b_0/d less than 15 and a loaded area with $\beta_c < 2$, shear strength is based on a nominal stress of $4\sqrt{f_c'}$ according to ACI Formula (11-39). Thus,

$$\phi V_c = 0.85[4\sqrt{3000}(133.8)(24.9)]\tfrac{1}{1000} = 620 \text{ kips}$$

Actually the resisting section for shear is not symmetrical about the column; therefore, the shear distribution is nonuniform and eccentricity of shear should be considered in accordance with ACI-11.12.6.1. In this example, where there is no moment assumed at the base of column, and the moment in the slab at the face of column is small, it may be reasonable to neglect the eccentric effect. Thus, neglecting any eccentricity of shear,

$$V_u = 532 - 6.99 \left[\frac{43.45(46.90)}{144} \right] = 433 \text{ kips}$$

Neglecting any possible contribution of the stirrups used for ϕV_s in one-way shear action,

$$\phi V_n = \phi V_c = 620 \text{ kips} > [V_u = 433 \text{ kips}] \qquad \textbf{OK}$$

A similar calculation for two-way shear action at column B will show the thickness to be adequate without shear reinforcement. Again, it is prudent to consider the failure section (see Fig. 20.13.3) to extend to the end of the footing, rather than merely the circular path at $d/2$ from the column.

(j) Design sketch. The complete design details are given in Fig. 20.13.1. ∎

■ **EXAMPLE 20.13.2** Design a cantilever or strap footing for the situation shown in Fig. 20.13.5, where the property line is at the exterior edge of the exterior column. Column data are given in Table 20.13.1. The distance between column centers is 18 ft. Equal settlement is assumed for DL plus $\frac{1}{2}$LL condition at a uniform pressure of 3.34 ksf. Use $f_c' = 3000$ psi for footing, $f_c' = 3750$ psi for columns, $f_y = 40,000$ psi, and the strength method of the ACI Code.

Solution: **(a)** Size of exterior and interior footings. According to ACI-15.2.2, footing size is always determined using service loads, rather than factored loads. The size of exterior and interior footings is a function of the assumed width of the exterior

Table 20.13.1

COLUMN	SIZE	REINFORCEMENT	LL	DL
Exterior	12 × 12 in.	4-#7	70 kips	55 kips
Interior	14 × 14 in.	8-#8	130 kips	80 kips

Figure 20.13.5 Cantilever, or strap, footing of Example 20.13.2.

footing (which affects the cantilever action and therefore the reactions required of each footing) and the assumed total thickness of each footing (which affects the available net soil bearing capacity). Should these assumed values be revised in the subsequent computation, the sizes of the footings must be revised accordingly. Assume that the total thickness of the entire footing is 24 in. The net uniform soil pressure for both exterior and interior footings is $3.34 - 2(0.150) = 3.04$ ksf. Referring to Fig. 20.13.5 for the dimensions, taking moments about the interior column using the services loads in the equal settlement condition, gives

$$R_{ext} = \frac{(55 + 35)18}{16.25} = 99.7 \text{ kips}$$

and, taking moments about the exterior column gives

$$R_{int} = \frac{(80 + 65)16.25 - (55 + 35)(2.25 - 0.50)}{16.25} = 135.3 \text{ kips}$$

$$\text{area of exterior footing required} = \frac{99.7}{3.04} = 32.8 \text{ sq ft}$$

Using the assumed exterior footing width of 4 ft 6 in.,

$$\text{length of exterior footing} = \frac{32.8}{4.50} = 7.29 \text{ ft} \quad (\text{try 7 ft 4 in.})$$

$$\text{area of interior footing required} = \frac{135.3}{3.04} = 44.5 \text{ sq ft}$$

$$\text{side of square interior footing} = \sqrt{44.5} = 6.67 \text{ ft} \quad (\text{try 6 ft 8 in.})$$

(b) Factored shear and factored moment diagram for strap. Referring to Fig. 20.13.5, applying overload factors U to the maximum load condition gives

$$R_{ext} = \frac{[1.4(55) + 1.7(70)]18}{16.25} = \frac{196(18)}{16.25} = 217 \text{ kips}$$

$$R_{int} = [1.4(80) + 1.7(130)] - \frac{196(1.75)}{16.25} = 333 - 21 = 312 \text{ kips}$$

$$V_u \text{ in strap} = -196 + 217 = +21 \text{ kips}$$

$$M_u \text{ at right end of strap} = 21(3.33) = 70.0 \text{ ft-kips}$$

$$M_u \text{ at left end of strap} = 21(14) = 294 \text{ ft-kips}$$

(c) Design of strap. For the shear requirement, assuming no shear reinforcement is to be used and the simplified expression for strength is used,

$$\phi V_c = \phi(2\sqrt{f_c'})b_w d$$

$$\text{width } b_w \text{ of strap required} = \frac{21,000}{0.85(2\sqrt{3000})21.5} = 10.5 \text{ in.}$$

Assume desirable reinforcement,

$$\rho \approx 0.5\rho_{max} = 0.014$$

for which

$$R_n = 498 \text{ psi} \quad \text{(see Example 20.13.1, part c)}$$

At the junction with the interior footing,

$$\text{width of strap required} = \frac{M_u}{\phi R_n d^2} = \frac{70(12,000)}{0.90(498)(21.5)^2} = 4.1 \text{ in.}$$

At the junction with the exterior footing,

$$\text{width of strap by moment requirement} = \frac{294(12,000)}{0.90(498)(21.5)^2} = 17.1 \text{ in.}$$

Adopt a strap width to vary from 14 in. to 20 in.

$$\text{required } R_n \text{ at wide end} = \frac{M_u}{\phi b d^2} = \frac{294(12,000)}{0.90(20)(21.5)^2} = 424 \text{ psi}$$

$$\text{required } R_n \text{ at narrow end} = \frac{70(12,000)}{0.90(14)(21.5)^2} = 144 \text{ psi}$$

Approximately,

$$\text{required } A_s \text{ at wide end} \approx 0.014\left(\frac{424}{498}\right)(20)(21.5) = 5.2 \text{ sq in.}$$

$$\text{required } A_s \text{ at narrow end} \approx 0.014\left(\frac{144}{498}\right)(14)(21.5) = 1.2 \text{ sq in.}$$

The minimum requirement of ACI-10.5.1 or 10.5.2 applies here because the strap is a beam.

$$\text{min } \rho = \frac{200}{f_y} = 0.005$$

$$\text{min } A_s = 0.005(14)(21.5) = 1.51 \text{ sq in.} \qquad \text{controls}$$

or

$$\text{min } A_s = \tfrac{4}{3}(\text{required } A_s) = \tfrac{4}{3}(1.2) = 1.60 \text{ sq in.}$$

Use 7-#8 ($A_s = 5.53$ sq in.) at the wide end. Extend 3-#8 through the narrow end. Two bars would satisfy min $A_s \geq 1.51$ sq in. but extend three bars to maintain symmetry.

The development length L_d for these #8 bars depends on their clear spacing; in this case, use of Tables 3.9.2, 3.9.3, and 3.9.4 may be simpler than computing the spacing, even though those tables assume stirrups are used. From Table 3.9.4, three #8 require a width of 13.3 in. to satisfy $3d_b$ clear spacing, and 7-#8 require a width of 29.3 in. Thus, at both ends of the strap $3d_b$ is satisfied. The 2 in. clear cover at the top equals $2d_b$. Thus, Category A (see Section 6.7) applies according to ACI-12.2.3.1(d); $\alpha_{abc} = 1.0$. Categories D and E are not applicable; $\alpha_{de} = 1.0$. From Table 6.7.1, the basic development length $L_{db} = 23.1$ in. and from Table 6.7.2, the Min $L_d = 21.9$ in., giving the "first-level L_d" as

$$\alpha_{abc}\alpha_{de}L_{db} = 1.0(1.0)23.1 = 23.1 \text{ in.} > (\text{Min } L_d = 21.9 \text{ in.})$$

Then applying the top bar multiplier $\alpha_{\text{top}} = 1.3$ gives,

$$L_d (\#8) = \alpha_{\text{top}} (23.1) = 1.3(23.1) = 30.0 \text{ in.} \quad (2.6 \text{ ft})$$

The maximum available embedment length into the footing from end of strap is 4 ft 6 in., which is more than adequate.

(d) Investigate one-way shear at the distance d (i.e., 21.5 in.) from the face of the exterior column.

$$V_u = 148 - \frac{21.5}{12}(48.2) = 62 \text{ kips}$$

For no shear reinforcement and the simplified expression for strength,

$$\phi V_c = \phi(2\sqrt{f'_c})b_w d$$
$$= 0.85(2\sqrt{3000})(7.33)(12)(21.5)\tfrac{1}{1000} = 176 \text{ kips} > 62 \text{ kips} \qquad \textbf{OK}$$

(e) Investigate two-way shear action at the critical section $d/2$ from the face of the exterior column. At this location there is an unsymmetrical three-sided perimeter, and there is a bending moment on the section across the 7 ft 4 in. width at the column face. ACI-11.12.6 applies. This shear transfer check is an empirical check of stresses under factored load similar to that for a flat plate at an exterior column. Referring to Fig. 20.13.6, the forces to be considered in analysis of the floor slab and the footing are compared. The floor slab treatment has been discussed in Section 16.18, where the details of the computation are shown. In the floor slab the direction of the shear is the opposite of that in the strap footing; however, the moment direction is the same. Thus, the maximum factored shear stress occurs at the interior side of the critical section for the floor slab, but at the exterior edge of the footing slab.

At the face of the exterior column, the factored moment* is

$$M_u = 74 \text{ ft-kips (at face of column)}$$

ACI-11.12.6.1 requires the following fraction of unbalanced moment to be transferred by eccentricity of shear:

$$1 - \frac{1}{1 + \frac{2}{3}\sqrt{\frac{b_1}{b_2}}} = 1 - \frac{1}{1 + \frac{2}{3}\sqrt{\frac{12 + 21.5/2}{12 + 21.5}}} = 0.355$$

Thus,

$$M_{uv} = 0.355M_u = 0.355(74) = 26.3 \text{ ft-kips}$$

----

* This longitudinal moment is taken from the factored moment diagram of Fig. 20.13.5(b). Where the strap joins the exterior footing there is a sudden discontinuity which may cause an indeterminate redistribution of longitudinal and transverse moments.

(a) Floor slab   (b) Cantilever (strap) footing

Figure 20.13.6 Comparison of forces to be transferred between floor slab and column to those between exterior footing and column in cantilever (strap) footing.

Using the symbols of Fig. 16.18.1, for this example $b_1 = 22.75$ in. and $b_2 = 33.5$ in. The factored shear that must be carried is

$$V_u = \frac{217}{4.5(7.33)}\left[4.5(7.33) - \frac{22.75(33.5)}{144}\right] = 182 \text{ kips}$$

The section properties for the three-sided critical section for shear transfer, as shown in Fig. 20.13.6(b), are

$$A = 21.5[2(22.75) + 33.5] = 1699 \text{ sq in.}$$

$$x_2 = \frac{2(22.75)(11.375)}{2(22.75) + 33.5} = 6.55 \text{ in.}$$

$$J_c = 21.5\left[\frac{2(22.75)^3}{3} - (45.5 + 33.5)(6.55)^2\right] + \frac{22.75(21.5)^3}{6}$$

$$= 133{,}600 \text{ in.}^4$$

At the exterior side of column,

$$v_{u1} = \frac{182{,}000}{1699} + \frac{26.3(12{,}000)(22.75 - 6.55)}{133{,}600} = 107 + 38 = 145 \text{ psi}$$

At the interior side of column,

$$v_{u2} = \frac{182{,}000}{1699} - \frac{26.3(12{,}000)(6.55)}{133{,}600} = 107 - 15 = 92 \text{ psi}$$

For β_c not exceeding 2.0, and $b_0/d = [3(12) + 4(10.75)]/21.5 = 79/21.5 = 3.7$ which is less than 15 for a three-sided critical section, the factored load stress is limited to $\phi(4\sqrt{f'_c})$ as a maximum; for this example, 186 psi (ACI-11.12.2.1). The computed maximum stress of 145 psi is below 186 psi, and therefore is acceptable.

(f) Investigate development of reinforcement at the inflection point (ACI-12.11.3). The Code provision is checked here, even though the reinforcement is negative moment reinforcement rather than the positive moment reinforcement as prescribed in ACI-12.11.3 because the footing may be visualized as an inverted beam as discussed in Example 20.13.1, part (d).

Tension reinforcement is 3-#8 at the narrow end of the strap near the inflection point.

$$C = 0.85f'_c ba = 0.85(3)(14)a = 35.7a$$

$$T = 3(0.79)40 = 95 \text{ kips}$$

$$a = \frac{95}{35.7} = 2.66$$

$$M_n = 95(21.5 - 1.33)\tfrac{1}{12} = 160 \text{ ft-kips}$$

$$V_u = 85 \text{ kips}$$

The #8 bars are proposed to be terminated at the interior side of the interior column, giving about 2.5 ft of embedment from the inflection point.

$$L_a = 2.5 \text{ ft,} \qquad \text{or } d = 1.79 \text{ ft,} \qquad \text{or } 12d_b = 1.0 \text{ ft}$$

$$\frac{M_n}{V_u} + L_a = \left(\frac{160}{85} + 1.79\right)12 = 44 \text{ in.} > [L_d = 30.0 \text{ in.}] \qquad \textbf{OK}$$

(g) Investigate development of reinforcement in the top of the exterior column footing. Tension reinforcement is 7-#8.

$$C = 0.85f'_c ba = 0.85(3)(7.33)(12)a = 224a$$

$$T = 7(0.79)40 = 221 \text{ kips}$$

$$a = \frac{221}{224} = 0.99 \text{ in.}$$

$$M_n = 221[21.5 - 0.5(0.99)]\tfrac{1}{12} = 387 \text{ ft-kips}$$

$$V_u = 148 \text{ kips}$$

$$1.30\frac{M_n}{V_u} + L_a = 1.30\frac{387(12)}{148} + 4 = 45 \text{ in.} > [L_d = 30.0 \text{ in.}] \qquad \textbf{OK}$$

Note that ACI-12.11.3 allows a 30% increase to reflect the ends of the reinforcement compressed by a reaction; also L_a is the embedment beyond the *centerline of the support*, which is 4 in. in this case.

It is observed that the assumption of zero earth pressure under the strap has been made; hence in construction the region below the strap should be disturbed and the strap should be formed on the bottom. Furthermore, liberal anchorage lengths (perhaps even hooks) should be provided into the exterior footing [as shown in Fig. 20.13.5(a)] to accommodate fully the tensile force in the top of the footing across to the exterior column. Often some of the steel extending into the exterior footing is flared to distribute more effectively the load from the 20-in. width to the 88-in. width.

(h) Design of exterior footing. In the transverse direction,

$$d = 24 - 3 \text{ (i.e., cover)} - 0.5 \text{ (i.e., est. bar radius)} = 20.5 \text{ in.}$$

$$M_u \text{ at edge of strap} = \left(\frac{217}{7.33}\right)\frac{(2.83)^2}{2} = 118 \text{ ft-kips}$$

$$\text{required } R_n = \frac{M_u}{\phi b d^2} = \frac{118(12,000)}{0.90(54)(20.5)^2} = 70 \text{ psi}$$

From Fig. 3.8.1, required $\rho = 0.002 < [\min \rho_g = 0.002 \text{ (ACI-10.5.3)}]$,

$$\text{required } A_s = 0.002(54)(24) = 2.59 \text{ sq in.}$$

The beam shear (one-way action) at d from the face of the column in the 7.33-ft direction of the footing is

$$V_u = \frac{217}{7.33}(1.12) = 33 \text{ kips}$$

$$\phi V_c = \phi(2\sqrt{f_c'})b_w d$$
$$= 0.85(2\sqrt{3000})(54)(20.5)\tfrac{1}{1000} = 103 \text{ kips} > 33 \text{ kips} \qquad \textbf{OK}$$

Use 6-#6 bars ($A_s = 2.64$ sq in.). Since bars are to extend the full 7 ft 4 in. length (less minimum cover) of the exterior footing, development of reinforcement for #6 bars is automatically provided; that is,

$$\text{available embedment} = 36 \text{ in.} > [L_d (\#6) = 16.4 \text{ in.}] \qquad \textbf{OK}$$

(i) Design of interior footing.

$$\text{net soil pressure under overload} = \frac{312}{(6.67)^2} = 7.0 \text{ ksf}$$

$$M_u = 7.0(6.67)\frac{(2.75)^2}{2} = 177 \text{ ft-kips}$$

$$\text{required } R_n = \frac{M_u}{\phi b d^2} = \frac{177(12,000)}{0.90(80)(20)^2} = 74 \text{ psi}$$

Check two-way action for shear. When $\beta_c < 2$ and $b_0/d < 20$ for a four-sided critical section, the nominal strength V_c is based on $4\sqrt{f_c'}$. Thus,

$$\phi V_c = \phi 4 \sqrt{f_c'}\, b_0 d = 0.85(4\sqrt{3000})(4)(34)20 \tfrac{1}{1000} = 506 \text{ kips}$$

$$V_u = 7.0[(6.67)^2 - (2.83)^2] = 225 \text{ kips} < 506 \text{ kips} \qquad \textbf{OK}$$

From Fig. 3.8.1, required $\rho \approx 0.0025$,

$$\text{required } A_s = 0.0025(80)(20) = 4.0 \text{ sq in.}$$

Use 9-#6 bars each way ($A_s = 3.96$ sq in.). The 2 ft 9 in. from face of column to edge of footing provides adequate embedment to develop the #6 bars ($L_d = 16.4$ in.).

(j) Design sketch. The final details of the design are shown in Fig. 20.13.5(a). Note that the moment capacity ϕM_n diagram in Fig. 20.13.5(b) for the strap portion is approximate; the #8 bars are extended as far as possible toward the narrow end as the width narrows from 20 in. to 14 in. ∎

20.14 PILE FOOTINGS

The principles and methods to be used in the design of pile caps are little different from those of spread footings. The following, however, may be noted.

1. Computations for moments and shears may be based on the assumption that the reaction from any pile is concentrated at the center of the pile (ACI-15.2.3).
2. In computing the external shear on any section through a footing supported on piles, the portion of the pile reaction to be assumed as producing shear on the section shall be based on straightline interpolation between full value when the pile center is at one-half the pile diameter outside the section and zero value when the pile center is at one-half the pile diameter inside the section (ACI-15.5.3).
3. In reinforced concrete pile footings, the thickness above the reinforcement at the edge shall not be less than 12 in. (ACI-15.7).

Note that pile caps frequently must be designed for shear considering the member as a deep beam. In other words, when piles are located inside the critical sections d (for one-way action) or $d/2$ (for two-way action) from the face of column, the shear cannot be neglected. ACI Commentary-R15.5.3 suggests using ACI-11.8 for designing such pile caps. The problem requires the most attention when large loads are carried by a few piles, such as the two-pile cap having 100-ton piles. There is no agreement about the proper procedure to use. For guidance, the reader is referred to the *CRSI Handbook* [2.23], Rice and Hoffman [10.2, 20.18], and Gogate and Sabnis [20.19].

SELECTED REFERENCES

20.1.   A. N. Talbot. *Reinforced Concrete Wall Footings and Column Footings.* Urbana, Illinois: Eng. Experiment Station Bulletin No. 67, University of Illinois, March 1913.

20.2. Frank E. Richart. "Reinforced Concrete Wall and Column Footings," *ACI Journal, Proceedings*, **45**, October 1948, 97–127; November 1948, 237–260.

20.3. Johannes Moe. *Shearing Strength of Reinforced Concrete Slabs and Footings Under Concentrated Loads* (Bulletin No. D47). Chicago: Portland Cement Association Research and Development Laboratories, April 1961.

20.4. ACI-ASCE Committee 426. *Suggested Revisions to Shear Provisions for Building Codes.* Detroit, American Concrete Institute, 1979. 82 pp.

20.5. Hans Gesund. "Flexural Limit Analysis of Concentrically Loaded Column Footings," *ACI Journal, Proceedings*, **80**, May–June 1983, 223–228.

20.6. Hans Gesund. "Flexural Limit Design of Column Footings," *Journal of Structural Engineering*, ASCE, **111**, 11 (November 1985), 2273–2287.

20.7. Da Hua Jiang. "Flexural Strength of Square Spread Footings," *Journal of Structural Engineering*, ASCE, **109**, 8 (August 1983), 1812–1819. Disc. **112**, 8 (August 1986), 1925–1926.

20.8. Kanakapura S. Subba Rao and Shashikant Singh. "Lower-Bound Collapse Load of Square Footing," *Journal of Structural Engineering*, ASCE, **113**, 8 (August 1987), 1875–1979.

20.9. Richard W. Furlong. "Design Aids for Square Footings," *ACI Journal, Proceedings*, **62**, March 1965, 363–371.

20.10. ACI Committee 318, *Building Code Requirements for Structural Plain Concrete* (ACI 318.1). Detroit: American Concrete Institute, 1989.

20.11. Fritz Kramrisch. "Footings," *Handbook of Concrete Engineering* (Mark Fintel, ed.). New York; Van Nostrand Reinhold Company, 1974 (Chapter 5, pp. 111–140).

20.12. ACI Committee 336. "Suggested Design Procedures for Combined Footings and Mats," *ACI Structural Journal*, **85**, May–June 1988, 304–324. Disc., **86**, January–February 1989, 111–116.

20.13. Fritz Kramrisch and Paul Rogers, "Simplified Design of Combined Footings," *Journal of Soil Mechanics and Foundations Division*, ASCE, **87**, SM5 (October 1961), 19–44.

20.14. Leslie J. Szava-Kovats. "Design of Combined Footings Using Support Reaction and Moment Influence Lines of Continuous Beam on Elastic Supports," *ACI Journal, Proceedings*, **64**, June 1967, 312–319.

20.15. Gwynne Davies and Brian Mayfield. "Choosing Plan Dimensions for an Eccentrically Loaded Footing Slab," *ACI Journal, Proceedings*, **69**, May 1972, 285–290.

20.16. Joseph E. Bowles. "Mat Design," *ACI Journal, Proceedings*, **83**, November–December 1986, 1010–1017. Disc., **84**, September–October 1987, 449.

20.17. *Report of the Joint Committee of Standard Specifications for Concrete and Reinforced Concrete.* Detroit: American Concrete Institute, 1940.

20.18. Paul F. Rice and Edward S. Hoffman. "Pile Caps—Theory, Code, and Practice Gaps," CRSI Professional Members' Structural Bulletin No. 2, February 1978.

20.19.   Anand B. Gogate and Gajanan M. Sabnis. "Design of Thick Pile Caps," *ACI Journal, Proceedings,* **77,** January–February 1980, 18–22.

PROBLEMS

All problems* are to be done in accordance with the strength method of the ACI Code, and all loads given are *service* loads, unless otherwise indicated.

20.1   Design a square spread footing to support a 14 in. square tied column carrying a dead load of 120 kips and a live load of 90 kips. The column reinforcement consists of #8 bars. Use $f'_c = 4000$ psi for the column, $f'_c = 3000$ psi for the footing, and $f_y = 60,000$ psi. Use a 6 ft 9 in. square footing. (Column, 360 mm square; DL = 530 kN; LL = 400 kN; column bars, #25M; $f'_c = 30$ MPa (column); $f'_c = 20$ MPa (footing); $f_y = 400$ MPa; footing, 2 m square.)

20.2   Design a square spread footing to support a 20 in. square tied column carrying a dead load of 400 kips and a live load of 264 kips. The column reinforcement consists of #11 bars. Use $f'_c = 3000$ psi for both the column and footing, $f_y = 60,000$ psi, and allowable soil pressure = 5 ksf. Include a design sketch (Column, 500 mm square; DL = 1800 kN; #40M bars; $f'_c = 20$ MPa; $f_y = 400$ MPa; soil pressure = 240 kN/m².)

20.3   Investigate the transfer of load from column to footing for the two conditions of the accompanying figure. The column is 20 in. square ($f'_c = 4000$ psi) containing 12-#10 bars ($f_y = 40,000$ psi) spirally reinforced. The footing is 10 ft square and 28 in. thick ($f'_c = 3000$ psi) and is adequately reinforced. Use the provisions of ACI-15.8 and Section 6.2 of Reference 20.10 on plain concrete.

(a) Without pedestal   (b) With pedestal

Problem 20.3

20.4   Design a spread footing to carry a load from an 18 × 32 in. tied column. The dead load and live load are each 230 kips. Because of the closeness of the

* Many problems may be solved as problems stated in Inch-Pound units, or as problems in SI units using quantities in parenthesis at the end of the statement. The SI conversions are approximate to avoid implying higher precision for the given information in metric units than that for Inch-Pound units.

property line (see the accompanying figure), the footing cannot exceed 8 ft perpendicular to that line. Use $f'_c = 3000$ psi, $f_y = 60,000$ psi, and allowable soil pressure = 5 ksf. (460 mm × 810 mm column; DL = LL = 1000 kN; $f'_c = 20$ MPa; $f_y = 400$ MPa; allowable soil pressure = 240 kN/m².)

Problem 20.4

20.5 Design a plain concrete footing to carry a long 12 in. concrete block wall which must transmit 6 kips/ft (60% live load) to the footing. Use $f'_c = 3000$ psi and allowable soil pressure = 4 ksf. (300 mm wall; loading = 90 kN/m; $f'_c = 20$ MPa; soil $p = 190$ kN/m².)

20.6 Design a reinforced concrete footing to carry a 12 in. concrete wall to carry 20 kips/ft (60% live load). Use $f'_c = 3000$ psi, allowable soil pressure = 3 ksf and $f_y = 40,000$ psi. (300 mm wall; 300 kN/m; $f'_c = 20$ MPa; soil $p = 140$ kN/m²; $f_y = 300$ MPa.)

20.7 Redesign for the conditions of Example 20.13.2 a rectangular combined footing.

20.8 Design a rectangular combined footing for the situation shown in the accompanying figure. Column data are in the following tabulation:

COLUMN	SIZE	REINFORCEMENT	LL	DL
Exterior	12 in. square	#9 bars	45 kips	120 kips
Interior	12 in. square	#9 bars	90 kips	155 kips

Equal settlement is taken for DL plus LL conditions at a uniform soil pressure of 3 ksf. Use $f'_c = 4000$ psi and $f_y = 60,000$ psi.

Problem 20.8

20.9 Design a rectangular combined footing for the conditions of the accompanying figure. Assume that the system is within a building basement and is to have a 4 in. concrete slab over the footing. Use $f'_c = 3000$ psi, $f_y = 60,000$ psi, and allowable soil pressure for the given loads = 5 ksf. ($f'_c = 20$ MPa; $f_y = 400$ MPa; soil $p = 240$ kN/m^2.)

Property line
18" × 18" column (460 × 460 mm)
$P = 200^k$ (50% live load)
(900 kN)
20" diameter column (500 mm)
$P = 450^k$ (40% live load)
(2000 kN)
9"
(230 mm)
16'–0"
(4.9 m)

Problem 20.9

20.10 Rework Problem 20.8 using a strap, or cantilever, footing.

20.11 Rework Problem 20.9, except assume that the exterior column load is 450 kips and the interior column load is 250 kips with 50% and 40% live load, respectively. In this assignment, however, only the plan size and the shear and moment diagrams are required. (Exterior column, 2000 kN; interior column, 100 kN.)

20.12 Rework Problem 20.8, except assume that, in addition, the space to the right of the interior column is restricted to 1 ft 6 in. In this assignment, however, only the plan size and the shear and moment diagrams are required.

21

Introduction to Prestressed Concrete

21.1 PRESTRESS

Prestress means a stress that acts even though no dead or live load is acting. The principle of prestressing has been used for centuries. For example, wooden barrels may be made by tightening metal bands or ropes around barrel staves. The tensile stress in the bands causes a compression between the staves, thus making the barrel tight. In the making of early wheels, the wooden spokes and rim were first held together by a hot metal tire. Upon cooling, the tensile stress due to shrinkage in the metal would then compress the wooden rim and spokes together. In bolted joints, the bolt is pretensioned by tightening, which in turn precompresses the elements being joined.

The primary application of prestressing on a large scale today is in concrete construction. In general, prestress involves the imposition of stresses opposite in sign to those which are caused by the subsequent application of service loads. For example, prestressing wires placed eccentrically in a simple beam, as shown in Fig. 21.1.1, produce in the concrete an axial compression as well as a negative bending moment. Thus it is possible to keep the entire section in compression when service loads are added. This is a great advantage since concrete is weak in tension but strong in compression. Of course, steel is used to impose the prestress though less is required for prestressed concrete than in ordinary reinforced concrete. In general, it may be said that prestress provides a means for the most efficient use of material—that is, steel in tension and concrete in compression.

Prestressed concrete T-girders erected on precast columns for parking garage. (Photo by C. G. Salmon.)

(a) Effect of service loads on beam

(b) Effect of prestress with eccentricity

(c) Combined effects of service load and prestress

Figure 21.1.1 Opposite effects of service load and prestress on simply supported beam.

21.2 HISTORICAL BACKGROUND

The general concepts of prestressed concrete were first formulated in the period 1885–1890 by C. F. W. Doehring in Germany and P. H. Jackson in the United States. These early applications were handicapped by the low steel strengths obtainable. Steel stressed to low tension levels will not precompress concrete adequately to maintain its compression after shrinkage and creep take place.

The theory of prestressed concrete was first propounded by J. Mandl of Germany in 1896. It was further advanced by M. Koenen of Germany in 1907 (the first recognition of losses in prestress force from elastic shortening of concrete), and by G. R. Steiner in the United States in 1908. Steiner recognized shrinkage losses and suggested retensioning after shrinkage had occurred.

In practical uses, R. E. Dill of the United States in 1928 produced prestressed planks and fence posts. Circular prestressing of storage tanks began about 1935, but no significant linear prestressing (beams, slabs, planks, etc.) was done until about 1950. The Walnut Lane Bridge in Philadelphia, built in 1949–1950, was the first major use of linear prestressing in the United States.

In Europe, however, linear prestressing began about 1928 and advanced rapidly with the work of F. Dischinger, E. Freyssinet, E. Hoyer, G. Magnel, Y. Guyon, P. Abeles, and F. Leonhardt. With the publication of Magnel's work on the loss of stress in work-hardened steels in 1944, the basic theory of prestressed concrete was sufficiently complete for successful economical applications. In the United States, T. Y. Lin has been a leading proponent and practitioner.

The use of prestress is now widespread in nearly every type of simple structural element, as well as in many statically indeterminate structures. Zollman [21.1] has presented his interesting "Reflections" on the beginnings of prestressed concrete in America. Methods of inducing prestress are ingenious and unique. Many textbooks are available that describe the methods and applicable theories [21.2–21.9].

21.3 ADVANTAGES AND DISADVANTAGES OF PRESTRESSED CONCRETE CONSTRUCTION

The original concept of prestressed concrete was that it was crack-free under service loads. Especially when a structure is exposed to the weather, elimination of cracks prevents corrosion. Also a crack-free prestressed member has greater stiffness under service loads because its entire section is effective.

Prestressed concrete in several respects is more predictable than ordinary reinforced concrete. It permits accommodation of both shrinkage and creep reasonably well. High-strength concrete may be more efficiently utilized by merely adjusting the prestress force.

Precompression of the concrete reduces the tendency for inclined cracking, and the use of curved tendons provides a vertical component to aid in carrying the shear. Shear strength is more consistent than in ordinary reinforced concrete.

Other features of prestressed concrete are its high ability to absorb energy (impact resistance), its high fatigue resistance due especially to the low steel stress variation resulting from the high initial pretension, and its high live load capacity arising from the ability of the prestressing tendon to support the dead load. Use of prestressed concrete also permits partial testing of both steel and concrete through application of prestress.

Some of the disadvantages of prestressed concrete construction are as follows: (1) the stronger materials used have a higher unit cost; (2) more complicated form-work may be necessary; (3) end anchorages and bearing plates are often required; (4) labor costs are greater; and (5) more conditions must be checked in design and closer control of every phase of construction is required.

Short-span members and single-unit applications of any kind are likely to be uneconomical in prestressed concrete. However, economy is usually achieved when units can be standardized and the same unit repeated many times. For many situations, the desirability of achieving a certain advantage is sufficient to justify a higher initial cost.

Currently, prestressing need not create a crack-free structure at service load. In fact, prestressing to various levels of stress can provide the whole range of results, from "fully" prestressed, when at service load tension does not occur, to nonpre-stressed as discussed in preceding chapters. The level of prestressing can be used that will accomplish the desired crack control or stiffness objective. So-called "partial" prestressing has become common in construction. In general, this chapter focuses on prestressed concrete where the section is *uncracked* at service load.

21.4 PRETENSIONED AND POSTTENSIONED BEAM BEHAVIOR

Since discussion in this chapter is limited to beams, it is well to consider at this stage the behavior of such a member as it relates to the method of inducing the prestress. The most commonly used procedure is to put a specified tensile force into the wires by stretching them between two anchorages prior to placing the concrete. The con-crete is then placed and the wires become bonded to the concrete throughout their length. After the concrete has cured, the wires are cut at the anchorages. The imme-diate shortening of the wires transfers through bond (i.e., interaction between the steel and the surrounding concrete) a compressive stress to the concrete. Such a process is called *pretensioning*.

The behavior associated with *pretensioning* will be described step by step. In addition, the terminology and allowable values will be given in accordance with Chapter 18 of the ACI Code.

Step 1, as shown schematically in Fig. 21.4.1(a), is to stretch the wires between two anchorages in the casting yard sufficiently to introduce a tensile stress f_{si} into the wires, which according to ACI-18.5.1 may not exceed 94% of the specified yield strength f_{py}, but not greater than the lesser of (1) 80% of the specified tensile strength f_{pu}, or (2) the maximum value recommended by the manufacturer of prestressing tendons or anchorages. The quality-controlled concrete is then placed in the forms and frequently is steam cured. Concrete strength must be adequately developed by

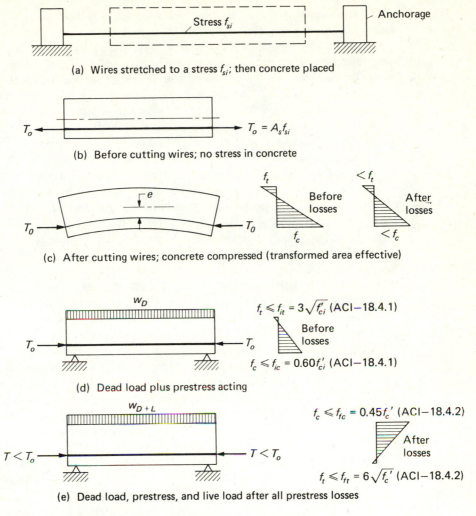

(a) Wires stretched to a stress f_{si}; then concrete placed

(b) Before cutting wires; no stress in concrete

(c) After cutting wires; concrete compressed (transformed area effective)

(d) Dead load plus prestress acting

(e) Dead load, prestress, and live load after all prestress losses

Figure 21.4.1 Stages of behavior up to service load—*pretensioned* beam.

the time the compression is to be introduced; thus high early-strength cement is usually used. Generally, the concrete strength f'_{ci} at transfer is specified by designers to be 4000 to 4500 psi.

Step 2 is to cut the wires. Acting through bond, the force T_0 in the wires acts as a compressive force on the entire effective (transformed) section. The stress in the concrete goes from zero [Fig. 21.4.1(b)] before the wires are cut to that shown in Fig. 21.4.1(c) after they have been cut. Once the prestress has been introduced, certain losses of prestress begin to occur. Loss of prestress may arise from slip at the anchorage, elastic shortening of the concrete member, creep and shrinkage of the concrete, relaxation of steel stress, and frictional losses due to intended or unintended curvature in the tendons. It is true that some small portion of such losses may occur prior to the transfer of stress to the concrete; however, it is practical and conservative to assume that the losses occur after the transfer.

Dead load of the flexural member will, of course, be acting simultaneously with the prestressing force once the wires have been cut and the transfer of stress is accomplished. Dead load combined with prestress is shown in Fig. 21.4.1(d), where the most critical stress situation occurs immediately after transfer and before most losses have taken place. Limiting values (ACI-18.4.1) for this temporary situation are a tensile stress at the top of the beam of $3\sqrt{f'_{ci}}$ (approximately 40% of the cracking strength) and a compressive stress at the bottom equal to 60% of the concrete strength f'_{ci} which has been developed at the time of transfer. One reason for holding the temporary tensile stress to such a low value is to prevent any possibility of an upward buckling of the beam resulting from sudden cracking at the top. Frequently no reinforcement (nonprestressed) exists to restrain such cracking. At the *end* of a simply supported member, the temporary tensile stress in the concrete is permitted (ACI-18.4.1c) to reach $6\sqrt{f'_{ci}}$, a conservative value for the modulus of rupture (see Section 1.8).

Step 3 is the service condition of dead load, live load, and prestress where, after losses, ACI-18.4.2 permits a net tensile stress at the bottom not to exceed $6\sqrt{f'_c}$, along with compressive stress at the top not to exceed $0.45f'_c$. Since tendons are usually placed near the bottom surface, there will be little danger of sudden cracking and buckling.

Previous to the 1971 ACI Code, no tensile stress was permitted if the member was exposed to freezing temperatures or to a corrosive environment. ACI Commentary-R18.4.2(b) suggests using reduced tensile stress to eliminate possible cracking under an adverse corrosive environment, but does not suggest specific reduced values. Corrosive atmosphere exists when chemical attack may occur from such sources as seawater, corrosive industrial atmosphere, or sewer gas. The ACI Code limit of $6\sqrt{f'_c}$ implies cover according to ACI-7.7.3.1; with a corrosive environment the cover should be increased above the minimum.

Further, the tensile stress in the precompressed zone may exceed $6\sqrt{f'_c}$ to a maximum of $12\sqrt{f'_c}$, when special calculations show that the deflection requirements of ACI-9.5.4 and cover requirements of ACI-7.7.3.2 are satisfied.

The alternative to pretensioning is *posttensioning*. In a posttensioned beam, the concrete is first cast either with a hollow tube enclosed or with the unstressed tendons coated with grease or mastic to prevent bond with the concrete, as shown in Fig. 21.4.2. An end plate or anchorage is placed against each end of the member; then, once the concrete is sufficiently cured, the wires are pulled by jacking against the end plates. Tensile stress in the tendons immediately after anchorage is not to exceed 70% of specified tensile strength f_{pu} [ACI-18.5.1(c)]. During the tensioning process, elastic

Figure 21.4.2 Posttensioned member.

shortening occurs, frictional losses take place, and the dead load moment becomes partially active due to the induced curvature. Thus the jacking force must account for these losses. Losses that occur after tensioning will thus be less in this case then in pretensioning. For posttensioning, the stresses induced and the allowable values at the different stages are essentially the same as those described in detail for pretensioning. However, since the tendons are not bonded to the concrete, the posttension force acts only on the plain concrete, at least until reaching the situation of Fig. 21.4.1(d). Prior to the imposition of live load, the tendons are usually grouted (space in the ducts is filled). If such grouting is properly done, the live load stresses may be computed on the transformed section, the same as for pretensioning.

21.5 SERVICE LOAD STRESSES ON FLEXURAL MEMBERS—TENDONS HAVING VARYING AMOUNTS OF ECCENTRICITY

In order to demonstrate some of the attributes of prestressed concrete, the first example is one with the prestressing elements placed at the centroid of the section, giving a uniform precompression of the concrete.

■ **EXAMPLE 21.5.1** For the section shown in Fig. 21.5.1 assume that the member is pretensioned by 2.30 sq in. of steel wire having a maximum acceptable initial tensile stress of 175,000 psi. The prestress wires are centered at the centroid of the section. The concrete has $f'_c = 5000$ psi ($n = 7$), and it is to be assumed that the concrete has attained a strength of $f'_{ci} = 4000$ psi at the time of transfer. Determine: (a) the stresses due to prestress immediately after transfer; (b) the temporary stresses when the member is used on a 40-ft simple span; and (c) the service live load moment capacity according to the ACI Code, allowing for a 20% loss of prestress due to creep, shrinkage, and other sources.

Figure 21.5.1 Section and stresses for Example 21.5.1.

Solution: (a) Stress due to prestress immediately after transfer.

$$T_0 = f_{si}A_s = 175(2.30) = 402.5 \text{ kips}$$

This force T_0 acts as a compressive force on the transformed section immediately after cutting the wires. Thus

$$f_c = \frac{T_0}{A_c + nA_s} = \frac{T_0}{A_g + (n-1)A_s} = \frac{402,500}{600 + 6(2.3)} = \frac{402,500}{614} = 656 \text{ psi}$$

The decrease in steel stress is

$$\Delta f_s = n f_c = 7(656) = 4590 \text{ psi}$$

Thus elastic shortening may be considered to have caused a loss of tensile stress, so that the remaining tensile stress in the wires is $175,000 - 4590 = 170,400$ psi. The loss in this case is 2.6%, but it could be as high as 5%.

It should further be noted that, although it may be theoretically correct to use the transformed section, in ordinary practice it is common and sufficiently accurate in most cases to use the gross section. Since the prestressing force is applied at the centroid of the gross section in the present problem, f_c is uniform over the entire section, or

$$f_c = \frac{402,500}{600} = 671 \text{ psi}$$

which is little different from the value of 656 psi determined above by using the transformed section.

(b) Temporary stress—prestress plus dead load.

$$w_D = \frac{20(30)}{144}(150) = 625 \text{ plf}$$

$$M_D = \tfrac{1}{8}(0.625)(40)^2 = 125 \text{ ft-kips}$$

Using the approximate method with gross moment of inertia I_g and neglecting the transformed area of reinforcement,

$$I_g = \tfrac{1}{12}(20)(30)^3 = 45,000 \text{ in.}^4$$

$$f(\text{initial prestress} + \text{DL}) = -\frac{402,500}{600} \mp \frac{125(12,000)(15)}{45,000}$$

$$= -671 \mp 500$$

$$= -1171 \text{ psi} \quad (\text{compression, top})$$

$$= -171 \text{ psi} \quad (\text{compression, bottom})$$

These stresses are acceptable based on temporary stress restrictions immediately after transfer and before losses,

$$f_c(\text{max}) = 0.6 f'_{ci} = 2400 \text{ psi}$$

$$f_t(\text{max}) = 3\sqrt{f'_{ci}} = 190 \text{ psi}$$

(c) Service live load moment capacity.

$$f(\text{prestress} - \text{losses} + \text{DL}) = -0.8(671) \mp 500 = -537 \mp 500$$

$$= -1037 \text{ psi} \quad \text{(compression, top)}$$

$$= -37 \text{ psi} \quad \text{(compression, bottom)}$$

Based on stress at service load, after allowance for all prestress losses,

$$f_c(\text{max}) = 0.45f_c' = 2250 \text{ psi}$$

$$f_t(\text{max}) = 6\sqrt{f_c'} = 423 \text{ psi}$$

The stress available for live load may then be computed.

$$f(\text{prestress} - \text{losses} + \text{DL} + \text{LL}) = -1037 + f_{LL} = -2250 \text{ psi} \quad \text{(top)}$$

$$= -37 + f_{LL} = +423 \text{ psi} \quad \text{(bottom)}$$

$$f_{LL}(\text{max})(\text{top}) = -1213 \text{ psi}$$

$$f_{LL}(\text{max})(\text{bottom}) = +460 \text{ psi} \qquad \textbf{(Controls)}$$

Thus

$$\frac{M_L(15)}{45,000} = 460 \text{ psi}$$

$$M_L = \frac{45,000(460)}{15(12,000)} = 115 \text{ ft-kips}$$

The wide divergence between the maximum acceptable live load stresses of 1213 psi compression and 460 psi tension indicates the need for an unsymmetrical section or some arrangement to equalize them better. A study of Fig. 21.5.1 will show that the most economical arrangement would be for the initial prestress variation to offset the pattern of final stress under full dead plus live load. ■

■ **EXAMPLE 21.5.2** Repeat the solution of Example 21.5.1, except locate the tendons 5 in. from the bottom of the section (Fig. 21.5.2).

Solution: **(a)** Temporary stress (prestress + DL immediately after transfer). Using properties of the gross section as for the preceding example, the prestressing force $T_0 = 402.5$ kips applied with an eccentricity of 10 in. gives

$$f(\text{initial prestress} + \text{DL}) = -671 \pm \frac{402,500(10)(15)}{45,000} \mp 500$$

$$= -671 \pm 1342 \mp 500$$

$$= +171 \text{ psi} \quad \text{(tension, top)}$$

$$= -1513 \text{ psi} \quad \text{(compression, bottom)}$$

It is to be noted that the temporary tensile stress of 171 psi at the top is nearly equal to the allowable value of $3\sqrt{f_{ci}'} = 190$ psi for such stress. Any significantly greater eccentricity, therefore, would require a reduction in the prestressing force.

Figure 21.5.2 Section and stresses for Example 21.5.2.

(b) Final stress (prestress + DL), after allowance for 20% prestress loss, is

$$f(\text{prestress} + \text{DL} - \text{losses}) = +171 - 0.2(-671 + 1342)$$
$$= +37 \text{ psi} \quad (\text{top})$$
$$= -1513 - 0.2(-671 - 1342)$$
$$= -1110 \text{ psi} \quad (\text{bottom})$$

(c) Service live load moment capacity (see Example 21.5.1 for allowables). Based on final dead load plus live load conditions,

$$+37 + f_{LL} = -2250 \text{ psi} \quad (\text{top}), \qquad f_{LL} = -2287 \text{ psi}$$
$$-1110 + f_{LL} = +423 \text{ psi} \quad (\text{bottom}), \qquad f_{LL} = +1533 \text{ psi}$$

Since the neutral axis for live load resistance is assumed to be at middepth, $f_{LL} = +1533$ psi controls.

$$M_L = \frac{45,000(1533)}{15(12,000)} = 383 \text{ ft-kips}$$

Thus, increasing the eccentricity of the prestressing force increases the live load capacity until the limit is reached when the temporary stress at transfer reaches its maximum permissible value, either at the top or at the bottom of the section. ■

It is to be noted that the magnitude of the prestress over the concrete section is constant for the entire span when the tendons are straight, whereas the magnitude of dead and live load stresses is a maximum at only one point. For straight tendons, the complete stress situation near the supports on simple spans approaches that of Fig. 21.5.2(b), less losses, because the superimposed dead and live load stresses vanish. Because of this difficulty, tendons frequently are placed so as to have an eccentricity that varies from zero at points of low external bending moment to a maximum in the region of high external bending moment. This variation in eccentricity may be accomplished in pretensioning by holding down the stressed tendons at midspan, or at other locations such as at the one-third points. In posttensioning, the ducts or greased tendons are simply draped (held at the ends and permitted to take a natural deflected shape) such that desired eccentricities at the ends and at midspan are achieved; the points in between will lie on a curved path.

21.6 THREE BASIC CONCEPTS OF PRESTRESSED CONCRETE

When considering the stresses in prestressed concrete under service load conditions, there are three general patterns of thought that may be applied.

Homogeneous Beam Concept. The homogeneous beam concept is used in Section 21.5 wherein the prestressing effectively eliminates cracking and the combined stress formula, $P/A \pm Mc/I$, may be used to investigate the section. Two examples appear in Section 21.5.

Internal Force Concept. The internal force approach uses the equilibrium of internal forces; steel takes the tension and concrete takes the compression as shown in Fig. 21.6.1. This approach is analogous to the internal-couple method used for nonprestressed reinforced concrete. At service load in reinforced concrete, the *points of action* of the forces C and T ($C = T$) are *independent* of the magnitude of applied bending moment, depending only on the cross-sectional dimensions and the modulus of elasticity ratio n; thus the magnitude of the forces is directly proportional to the applied bending moment. In prestressed concrete, the *magnitude* of internal forces is *independent* of applied bending moment, depending only on the prestress and the percentage of losses. In this case the location of the force C must vary with the applied loading. The approach may be summarized by the following steps:

1. A known prestress force put into the steel defines T.
2. An applied moment M is put on the beam.
3. For equilibrium, the moment arm = M/T and $C = T$.
4. Knowing the magnitude and point of action of the force C, the stress in the concrete may be computed as

$$f = \frac{C}{A} \pm \frac{Cey}{I} \tag{21.6.1}$$

■ **EXAMPLE 21.6.1** Apply the dead load moment of 125 ft-kips and live load moment of 383 ft-kips (as computed in Example 21.5.2) to the rectangular beam of Fig. 21.5.2, using the internal force concept. Determine the service load stresses at transfer and under final conditions.

Solution: (a) At transfer, the prestress force T_0 is 402.5 kips.

$$C = T_0 = 402.5 \text{ kips}$$

Figure 21.6.1 Internal force concept of prestressing.

When the applied moment is 125 ft-kips, the moment arm of the internal forces must be

$$\text{arm} = \frac{M_D}{C} = \frac{125(12)}{402.5} = 3.73 \text{ in.}$$

This means the compressive force C is eccentric to the middepth by an amount,

$$e = 15 - 5 \text{ (to steel)} - 3.73 \text{ (arm)}$$
$$= 6.27 \text{ in.} \quad \text{(below middepth)}$$

$$f = -\frac{402.5}{600} \pm \frac{402.5(6.27)15}{45,000}$$

$$f(\text{top}) = -671 + 842 = +171 \text{ psi} \quad \text{(tension)}$$

$$f(\text{bottom}) = -671 - 842 = -1513 \text{ psi} \quad \text{(compression)}$$

exactly the same as in Example 21.5.2, Fig. 21.5.2(d).

(b) At the final condition, the prestress force T_e is $0.8(402.5) = 322$ kips after losses.

$$C = T_e = 322 \text{ kips}$$

$$\text{arm} = \frac{M_D + M_L}{C} = \frac{(125 + 383)12}{322} = 18.95 \text{ in.}$$

$$e = 15 - 5 - 18.95 = -8.95 \text{ in.} \quad \text{(above middepth)}$$

$$f = -\frac{322}{600} \mp \frac{322(8.95)15}{45,000}$$

$$f(\text{top}) = -537 - 960 = -1497 \text{ psi} \quad \text{(compression)}$$

$$f(\text{bottom}) = -537 + 960 = +423 \text{ psi} \quad \text{(tension)}$$

exactly as shown in Fig. 21.5.2(e). ■

Load Balancing Concept. The load balancing approach visualizes prestressing primarily as a process of balancing loads on the member. The prestressing tendons are placed so that the eccentricity of the prestressing force varies in the same manner as the moments from applied loads, which if exactly done would result in zero flexural stress. Only the axial stress P/A (P is the horizontal component of force in tendon) would act. Refer to Fig. 21.6.2(a) showing the parabolically draped prestressing tendon. Figure 21.6.2(b) shows the free body of forces acting on the concrete due to prestress alone. The prestressing effect may be considered as an upward uniform load if the tendon is parabolically draped. The maximum prestress moment of Te_{max} at midspan can be equated to an equivalent uniformly loaded beam moment, $w_p L^2/8$; thus

$$w_p = \frac{8 Te_{\text{max}}}{L^2} = \text{equivalent uniform load (acting upward)} \qquad \textbf{(21.6.2)}$$

Let

$$w_{\text{net}} = w \text{ (actual downward load)} - w_p$$

(a) Member having parabolically draped tendon

(b) Forces acting on concrete from prestress alone

Figure 21.6.2 Load balancing concept of prestressing.

then

$$M_{net} = \frac{w_{net}L^2}{8} \qquad\qquad \textbf{(21.6.3)}$$

and

$$f = -\frac{C}{A} \mp \frac{M_{net}y}{I} \qquad\qquad \textbf{(21.6.4)}$$

If the tendons are not parabolically draped, the actual net moment (applied load moment minus prestress moment) may be used for M_{net} in Eq. (21.6.4).

■ **EXAMPLE 21.6.2**   Compute the stresses on the beam of Fig. 21.5.2 using dead load and live load moments of 125 and 383 ft-kips, respectively, using the load balancing concept.

Solution: The maximum moment due to prestress at transfer is

$$M_{prestress} = 402.5\left(\frac{10}{12}\right) = 335 \text{ ft-kips}$$

$$M_{net} = M_D - M_{prestress} = +125 - 335 = -210 \text{ ft-kips}   \text{(negative bending)}$$

$$f = -\frac{402.5}{600} \mp \frac{(-210)(12)15}{45,000}$$

$$f(\text{top}) = -671 + 842 = +171 \text{ psi}   \text{(tension)}$$

$$f(\text{bottom}) = -671 - 842 = -1513 \text{ psi}   \text{(compression)}$$

At the final condition,

$$M_{net} = 125 + 383 - 322(10)/12 = 240 \text{ ft-kips} \quad \text{(positive bending)}$$

$$f = -\frac{322}{600} \mp \frac{240(12)(15)}{45,000}$$

$$f(\text{top}) = -537 - 960 = -1497 \text{ psi} \quad \text{(compression)}$$

$$f(\text{bottom} = -537 + 960 = +423 \text{ psi} \quad \text{(tension)}$$

The results agree with those previously obtained. ■

21.7 LOSS OF PRESTRESS

The amount of prestress actually existing in a prestressed concrete member is not easily measured. The total force in the tendons at the time of prestressing is all that may conveniently be determined. Various losses reduce the prestress to that lower value available to resist the load. The difference between the final available prestress and the initial value is referred to as the loss of prestress.

In practice the initial prestress is usually determined by a pressure gage on the jack and may be verified by a direct measurement of the tendon elongation. In the pretensioned members, the uniformity of initial prestress may be verified at several points. In certain posttensioning procedures, the prestressing force will diminish due to friction at points remote from the jacking source. Initial prestress is, however, generally known with good accuracy.

Elastic Shortening. The loss of prestress due to elastic shortening can be easily determined. For example, let T_0 be the prestressing force that is applied at the centroid of the concrete section in a pretensioned member. If T_f is the final tensile force in the tendons just after elastic shortening has occurred, the strain (unit shortening) in the concrete may be expressed as

$$\epsilon_c = \frac{f_c}{E_c} = \frac{T_f}{A_c E_c} \tag{21.7.1}$$

where $A_c = A_g - A_s$. The change in strain in the tendons as a result of losses is

$$\Delta\epsilon_s = \frac{T_0 - T_f}{A_s E_s} \tag{21.7.2}$$

Equating the expressions for ϵ_c and $\Delta\epsilon_s$ gives

$$\frac{T_0}{T_f} = \frac{A_c + nA_s}{A_c} = \frac{A_T}{A_c} \tag{21.7.3}$$

The loss of prestress Δf_s is

$$\Delta f_s = \frac{T_0 - T_f}{A_s} = \frac{nT_f}{A_c} = \frac{nT_0}{A_T} \qquad (21.7.4)$$

As a practical matter, the loss in prestress Δf_s, regardless of whether or not the prestressing force is applied at the centroid of the gross section, may be taken approximately as

$$\Delta f_s = \frac{nT_0}{A_g} \qquad (21.7.5)$$

More correctly, the loss in prestress due to elastic shortening and bending of the section should be obtained as n times the computed compressive stress in the concrete adjacent to the tendons.

In the posttensioning case, the tendons are not stretched simultaneously. Further, the elastic shortening occurs gradually during the tensioning operation. The various methods of accounting for these gradual losses are adequately described elsewhere [21.2–21.8].

■ **EXAMPLE 21.7.1**   Determine the percent loss of prestress due to elastic shortening and bending in the pretensioned member of Fig. 21.5.2. $f'_c = 5000$ psi with $n = 7$; $f_{si} = 175,000$ psi.

Solution: (a) Loss due to elastic shortening, neglecting bending. "Exact" method [see also part (a) of Example 21.5.1],

$$\Delta f_s = \frac{nT_0}{A_T} = \frac{7(402.5)}{614} = 4.59 \text{ ksi}$$

$$A_T = 20(30) + (7-1)2.30 = 614 \text{ sq in.}$$

$$\text{percent loss} = \frac{4.59}{175} = 2.62\%$$

Approximate method,

$$\Delta f_s = \frac{nT_0}{A_g} = \frac{7(402.5)}{600} = 4.70 \text{ ksi}$$

$$\text{percent loss} = \frac{4.70}{175} = 2.69\%$$

There is no significant difference in the two results. Three percent is a typical value for loss due to elastic shortening in a pretensioned beam, whereas in a posttensioned beam it would be on the order of $1\frac{1}{2}\%$ average.

(b) Loss including bending due to dead load of 125 ft-kips. By the approximate method (using gross section instead of transformed section), the stress in *the concrete* *adjacent to the tendons* is (using the internal-force concept),

$$\text{arm} = \frac{M_D}{T_0} = \frac{125(12)}{402.5} = 3.72 \text{ in.}$$

$$C = T_0 = 402.5 \text{ kips}$$

$$f_c = -\frac{C}{A_g} - \frac{C(15 - 5 - 3.72)10}{I_g}$$

$$= -\frac{402.5}{600} - \frac{402.5(6.28)10}{45,000}$$

$$= -671 - 562 = -1233 \text{ psi}$$

This is the stress in concrete 5 in. from the bottom of the beam. The *change* in steel stress is nf_c.

$$\Delta f_s = nf_c = 7(1.233) = 8.6 \text{ ksi}$$

$$\text{percent loss} = \frac{8.6}{175} = 4.9\% \quad \blacksquare$$

Creep in Concrete.   Creep is the time-dependent deformation that occurs in concrete under stress, and has already been discussed in Section 1.10 (Chapter 1) and Section 14.6 (Chapter 14). The strain due to creep will vary with the magnitude of stress and in general may be assumed to vary with the elastic strain from about 100% in humid atmosphere to about 300% in very dry atmosphere.

In Chapter 14, the creep coefficient C_t is defined as

$$C_t = \frac{\text{creep strain, } \epsilon_{cp}}{\text{initial elastic strain, } \epsilon_i} \qquad \textbf{(21.7.6)}$$

The elastic strain in the concrete at the centroid of the section is (f_c = stress at centroid)

$$\epsilon_i = \frac{f_c}{E_c}$$

$$\epsilon_{cp} = C_t \epsilon_i = C_t \left(\frac{f_c}{E_c} \right) \qquad \textbf{(21.7.7)}$$

The strain in the concrete due to creep equals the decrease in strain in the steel; thus

$$\Delta \epsilon_s = \epsilon_{cp} = C_t \left(\frac{f_c}{E_c} \right) \qquad \textbf{(21.7.8)}$$

also

$$\Delta \epsilon_s = \frac{\Delta f_s}{E_s} \qquad \textbf{(21.7.9)}$$

Then, equating Eqs. (21.7.8) and (21.7.9),

$$\Delta f_s = C_t nf_c \qquad \textbf{(21.7.10)}$$

The coefficient C_t may be determined using the general expressions given in Section 14.6 (Chapter 14), or as suggested by Zia, Preston, Scott, and Workman [21.10], use $C_t = 2.0$ for pretensioned members and $C_t = 1.6$ for posttensioned members. The stress f_c in Eq. (21.7.10) is recommended [21.10] to be taken as $(f_{cir} - f_{cds})$, where f_{cir} (the subscript r means residual) is the net compressive stress in the concrete at the center of gravity of tendons immediately after the prestress has been applied to the concrete, and f_{cds} (the subscript s means superimposed) is the stress in concrete at the center of gravity of tendons due to all superimposed permanent dead loads that are applied to the member after it has been prestressed. For example, f_{cir} would correspond to the $f_c = 1233$ psi computed in Example 21.7.1(b), because the prestress and the beam dead load act simultaneously when the prestress is applied. Typical values for percentage loss of prestress due to creep are from 5 to 6%. Pretensioned beams will exhibit more creep than posttensioned beams because the prestress is imposed when the concrete is at an earlier age; age at loading is a major factor in determining the magnitude of creep.

Shrinkage in Concrete. Shrinkage is the volume change in concrete that occurs with time, as discussed in Section 1.10 (Chapter 1) and Section 14.7 (Chapter 14). The loss of prestress due to shrinkage may be expressed as

$$\Delta f_s = \epsilon_{sh} E_s \qquad\qquad (21.7.11)$$

where ϵ_{sh} is the shrinkage strain in concrete (see Section 14.7). In Chapter 14 are given general expressions for evaluating shrinkage strain. Zia et al. [21.10] recommend computing ϵ_{sh} by starting with a value of 550×10^{-6} in./in. as the basic ultimate shrinkage strain, multiplying by $(1 - 0.06V/S)$ to correct for the volume V to surface S ratio, and then multiplying by $(1.5 - 0.015H)$ to correct for the relative humidity H. For posttensioned members, an additional reduction factor is used to account for the time between the end of moist curing and the application of prestress.

Relaxation of Steel Stress. Relaxation is taken to mean the loss of stress in steel under nearly constant strain at constant temperature. Loss due to relaxation varies widely for different steels, and such loss should be provided for in accordance with test data furnished by the steel manufacturers. This loss is generally assumed to be in the range of 2 to 3% of the initial steel stress for stress-relieved strands and about 1% for low-relaxation strands (see Section 1.12). Zia et al. [21.10] have provided a formula for this computation. The percentage loss of prestress relating to relaxation varies with the type of tendon and the ratio of initial prestress to tensile strength of tendon.

Friction Losses in Posttensioned Members. There will be frictional losses, which are generally small, in the jacking equipment as well as friction between the tendons and the surrounding material (either duct or actual concrete member), due to intended or unintended curvature in the tendons. The friction between tendons and surrounding material is not small and may be considered as partly a length effect and partly a curvature effect.

(a) Element of curved tendon

(b) Force polygon

Figure 21.7.1 Friction losses in posttensioned member.

Referring to Fig. 21.7.1, let dx be a segment of a curved tendon. Assume that the tendon is being jacked from the left end by the force P_s which results in a force P_x at some distance to the right; these forces define the limits for the tension t. The full angle enclosed within the arc is α.

For equilibrium of the entire segment dx, refer to Fig. 21.7.1(b); the normal force dN is

$$dN = t\left(\frac{d\alpha}{2}\right) + \left(t + \frac{dt}{d\alpha}\,d\alpha\right)\frac{d\alpha}{2} \qquad (21.7.12)$$

and neglecting infinitesimals of higher order,

$$dN = 2t\left(\frac{d\alpha}{2}\right) = t\,d\alpha \qquad (21.7.13)$$

The friction force developed along the length dx is

$$\mu\,dN = \mu t\,d\alpha \qquad (21.7.14)$$

Summation of forces along the tendon gives

$$t - \mu t\,d\alpha - (t + dt) = 0$$

$$\frac{dt}{t} = -\mu\,d\alpha \qquad (21.7.15)$$

Integrating to obtain the total effect over the entire curved portion included within the angle α,

$$\int_{P_s}^{P_x} \frac{dt}{t} = \int_0^\alpha -\mu\,d\alpha \qquad\qquad \textbf{(21.7.16)}$$

$$\log_e P_x - \log_e P_s = -\mu\alpha \qquad\qquad \textbf{(21.7.17)}$$

$$\frac{P_x}{P_s} = e^{-\mu\alpha} \qquad\qquad \textbf{(21.7.18)}$$

Replacing the friction force term $\mu\alpha$ with the following expression, which contains a friction part due to curvature and a length effect (wobble effect), $\mu\alpha + KL_x$, thus

$$\frac{P_x}{P_s} = e^{-(\mu\alpha + KL_x)} \qquad\qquad \textbf{(21.7.19)}$$

or

$$P_s = P_x e^{(\mu\alpha + KL_x)} \qquad\qquad \textbf{(21.7.20)}$$

which is ACI Formula (18-1) in ACI-18.6.2.1. Note that $\alpha = L/R$, the length L of the curve divided by R, the radius of curvature. L_x is the length of prestressing tendon element from jacking end to any point x.

As an approximation, when $P_s - P_x$ is small (such as not more than 15 to 20% of the jacking force P_s) the friction force may be assumed to be constant. If the friction force is assumed to be proportional to the force P_x, then (see Fig. 21.7.2)

$$\mu N = \mu P_x \alpha \qquad\qquad \textbf{(21.7.21)}$$

and assuming the wobble effect KL_x is also proportional to P_x, equilibrium requires

$$P_s = P_x + P_x(\mu\alpha + KL_x)$$

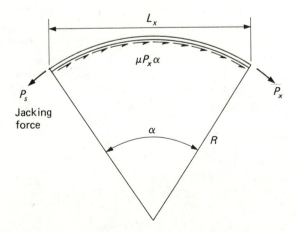

Figure 21.7.2 Approximate procedure for friction losses.

Table 21.7.1 Average Percentages of Loss of Prestress [21.6]

	PRETENSIONING (PERCENT)	POSTTENSIONING (PERCENT)
Elastic shortening and bending of concrete member	4	1
Creep of concrete	6	5
Shrinkage of concrete	7	6
Relaxation (creep) in steel	8	8
Totals	25	20

or

$$P_s = P_x(1 + \mu\alpha + KL_x) \tag{21.7.22}$$

which is ACI Formula (18-2), permitted for use when $(\mu\alpha + KL_x)$ does not exceed 0.3.

Practical Design Consideration—Total Losses. The total loss in prestress may be expressed in unit strains, total strains, unit stresses, or in percentage of initial prestress. Although it is difficult to generalize the amount of prestress loss, Lin and Burns [21.6] have suggested that for average steel and concrete properties and for average curing conditions the values in Table 21.7.1 may be taken as representative.

Zia et al. [21.10] have indicated that the upper limit for total loss in steel stress (not including friction loss) for stress-relieved strand in normal-weight concrete may be assumed as 50,000 psi.

More detailed treatment of losses of prestress is to be found in the textbooks by Collins and Mitchell [21.2], Libby [21.3], Nawy [21.4], Naaman [21.5], Lin and Burns [21.6], and Nilson [21.7], as well as in the PCI Committee Report [21.11], and Zia et al. [21.10].

■ **EXAMPLE 21.7.2** The posttensioned beam of Fig. 21.7.3 containing a cable of 72 parallel wires, $A_s = 3.60$ sq in., is to be tensioned 2 wires at a time. The jacking stress is to be measured by a pressure gage. The wires are to be stressed from one end of the member to a value f_1 to overcome frictional loss, then released to a value f_2, so that immediately after anchoring an initial prestress of 144 ksi is obtained. Compute f_1 and f_2, as well as the final design stress after all losses, according to the ACI Code. Assumptions are as follows:

(a) Coefficient of friction $\mu = 0.50$
(b) Wobble coefficient, $K = 0.0008$ (per ft)
(c) Deformation at anchorages and slip of wires = 0.06 in.
(d) Shrinkage strain, $\epsilon_{sh} = 0.0002$
(e) Steel relaxation = 5% of initial prestress (144 ksi)
(f) $E_s = 29,000$ ksi; $E_c = 5000$ ksi

Figure 21.7.3 Posttensioned beam of Example 21.7.2.

Solution: (a) Loss due to friction. Using the "exact" method, Eq. (21.7.20),

$$P_s = P_x e^{(\mu\alpha + KL_x)}$$

$$\mu\alpha = 0.5\left(\frac{45}{203.125}\right) = 0.1108$$

$$KL_x = 0.0008(45) \qquad = \frac{0.0360}{0.1468}$$

$$P_s = P_x e^{(0.1468)} \qquad = 1.158P.$$

The above expression can also be used in terms of unit stress f; in this case if f_x is desired to be 144 ksi, then the initial stress f_1 to overcome frictional loss is

$$f_1 = 1.158(144) = 167 \text{ ksi}$$

Using the approximate expression, Eq. (21.7.22),

$$P_s = P_x(1 + 0.1468) = 1.1468P_x$$

or

$$f_1 = 1.1468(144) = 165 \text{ ksi}$$

(b) Anchorage slip. For tensioning from one end,

$$\epsilon_s = \frac{0.06}{45(12)} = 0.00011$$

$$\Delta f_s = \epsilon_s E_s = 0.00011(29,000) = 3.2 \text{ ksi}$$

To allow for anchorage slip, tension to $f_1 = 167$ ksi then release to $f_2 = 144 + 3.2 = 147.2$ ksi. The minimum stress $f_{si} = 144$ ksi will then exist at both ends.

(c) Elastic shortening due to posttensioning two wires at a time. The wires tensioned first will have the greatest loss, having some additional loss as each succeeding pair of wires is tensioned. The paired tensioned last will have zero loss. Thus Δf_s in the first pair, given by Eq. (21.7.5), is

$$\Delta f_s = \frac{n T_0}{A_g}$$

$$n = \frac{29,000}{5000} = 5.8, \qquad \text{say 6 to nearest whole number}$$

$$T_0 = \frac{35}{36}(3.60)(144) = 504 \text{ kips}$$

$$\Delta f_s \text{ (first pair)} = \frac{6(504)}{18(36)} = 4.66 \text{ ksi}$$

$$\Delta f_s \text{ (last pair)} = 0 \text{ ksi}$$

$$\text{average } \Delta f_s = 2.33 \text{ ksi} \ (1.6\% \text{ of } 144 \text{ ksi})$$

(d) Creep loss. Using the procedure recommended by Zia et al. [21.10], f_c in Eq. (21.7.10) is computed as $(f_{cir} - f_{cds})$. The *net* concrete stress at midspan at the centroid of the steel tendons is

$$f_{cir} = \frac{T_0}{A_g} + \frac{T_0 e^2}{I_g} - \frac{M_D e}{I_g}$$

where

$$T_0 = 3.60(144) = 518.4 \text{ kips}$$

$$M_D = \frac{1}{8} \frac{18(36)}{144}(0.15)(45)^2 = 170.9 \text{ ft-kips}$$

Then,

$$f_{cir} = \frac{518.4}{18(36)} + \frac{518.4(12)12}{18(36)^3/12} - \frac{170.9(12)12}{18(36)^3/12}$$

$$= 0.80 + 1.07 - 0.35 = 1.52 \text{ ksi} \quad \text{(at the midspan)}$$

In this case, if there is no additional superimposed permanent dead load, f_{cds} is zero. At the support where there is zero eccentricity of the tendons and zero dead load moment,

$$f_{cir} = \frac{T_0}{A_g} = 0.80 \text{ ksi}$$

The average value of $(f_{cir} - f_{cds})$ should be used as f_c in Eq. (21.7.10) for posttensioned beams,

$$f_c = \frac{1.52 + 0.80}{2} = 1.16 \text{ ksi}$$

Note again that in this example f_{cds} is taken as zero.

$$\epsilon_c = \text{elastic strain} = \frac{f_c}{E_c} = \frac{1.16}{5000} = 0.000232 \text{ in./in.}$$

and for a posttensioned beam using $C_t = 1.6$,

$$\epsilon_{cp} = C_t\epsilon_c = 1.6(0.000232) = 0.000371 \text{ in./in.}$$

$$\Delta f_s = \epsilon_{cp}E_s = 0.000371(29{,}000) = 10.8 \text{ ksi} \quad (7.5\%)$$

(e) Shrinkage loss. Using the procedure of Zia et al. [21.10], the basic shrinkage strain is

$$\text{basic } \epsilon_{sh} = 550 \times 10^{-6} \text{ in./in.}$$

The correction factor (CF) for volume/surface (V/S) ratio is

$$(CF)_{V/S} = 1 - 0.06\frac{V}{S} = 1 - 0.06(6) = 0.64$$

where

$$\frac{V}{S} = \frac{18(36)}{2(18) + 2(36)} = 6$$

The correction factor (CF) for humidity H is

$$(CF)_h = 1.5 - 0.015H = 1.5 - 0.015(70) = 0.45$$

for 70% relative humidity. The adjusted shrinkage strain then becomes

$$\epsilon_{sh} = (\text{basic } \epsilon_{sh})(CF)_{V/S}(CF)_h = (550 \times 10^{-6})(0.64)(0.45)$$
$$= 0.000158 \text{ in./in.}$$

$$\Delta f_s = \epsilon_{sh}E_s = 0.000158(29{,}000) = 4.58 \text{ ksi} \quad (3.2\%)$$

(f) Relaxation in steel.

$$\Delta f_s = 0.05(144) = 7.2 \text{ ksi} \quad (5.0\%)$$

(g) Total losses

	LOSS OF STRESS	
	ksi	PERCENT
Elastic shortening	2.33	1.6
Creep in concrete	10.8	7.5
Shrinkage	4.6	3.2
Relaxation in steel	7.2	5.0
	Total = 24.9 ksi	17.3%

The final design prestress under dead load plus live load, after losses, is

$$f_{se} = 144 - 24.9 = 119.1 \text{ ksi} \quad ■$$

21.8 STRENGTH OF FLEXURAL MEMBERS—ACI CODE

Design must include consideration of all significant load stages. Primarily, these stages are (1) initial stage, including the period before and during prestressing, as well as the transfer of prestress to the concrete; (2) intermediate stage during transportation and erection; (3) final stage under service load, after losses; and (4) overload stage, where cracking and ultimate strength are important. Though the actual number of conditions to be investigated varies with the situation, ordinarily the initial service condition involves beam dead load plus prestress *before* losses; the final service condition involves full dead plus live load *after* losses; and the strength must be adequate for possible overload.

The initial and final service load stages have been considered in Section 21.5. A beam may be properly prestressed to carry service loads with little deflection and generally without cracking but, because of a small moment arm to the centroid of the steel, may have a nominal moment strength M_n that gives an inadequate margin of safety. On the other hand, if only the strength M_n were considered, service loads might cause excessive cracking, camber (upward deflection), or deflection.

Balanced Strain Condition. As first described in Section 3.5, the balanced strain condition occurs when the concrete strain ϵ_c at the extreme compression fiber is 0.003 at the instant the steel reaches its yield strain $\epsilon_y = f_y/E_s$. In prestressed concrete members, the steel used does not exhibit the well-defined yield point that occurs with ordinary deformed bars, so that the concept of balanced failure is nebulous. The balanced condition was used with ordinary deformed bars as a frame of reference to ensure a ductile failure mode. ACI-18.8.1 uses the criterion $\omega_p = \rho_p f_{ps}/f_c' = 0.36\beta_1$ for prestressed concrete members to represent the equivalent of $0.75\rho_b$ used for ordinary reinforced concrete. The term ρ_p refers to the reinforcement ratio A_{ps}/bd for the prestressing steel, and f_{ps} is the stress in the prestressed reinforcement when the nominal moment strength M_n is reached.

Prior to 1983, 0.3 was used instead of $0.36\beta_1$; for $f_c' = 5000$ psi, $\beta_1 = 0.80$ giving $0.36\beta_1 = 0.288$. The recent use of concrete strengths above 5000 psi has required a lower limit on ω_p for satisfactory ductility. The criterion $\rho_p f_{ps} \leq 0.36\beta_1$ essentially limits the neutral axis location to be not farther than $0.423d$ from the extreme compression fiber when the strength M_n is reached, which is comparable to the limit of $0.75x_b$ as defined in Chapter 3. (This is further explained by Example 21.8.2.)

The constant 0.423 may be explained as follows. Starting with the ACI Code limitation,

$$\frac{\rho_p f_{ps}}{f_c'} \leq 0.36\beta_1$$

Then, multiplying both sides by $f_c' bd$ gives

$$\rho_p f_{ps} bd \leq 0.36\beta_1 f_c' bd$$

The left-hand side of this inequality is $T = A_{ps} f_{ps} = \rho_p f_{ps} bd$. The right-hand side is the compressive force $C = 0.85 f_c' b\beta_1 x$. When $x = 0.423d$, the two sides of the above inequality are identical.

The following discussion in regard to under- and overreinforcement is restricted to pretensioned construction and posttensioned construction in which the tendons are grouted. The nominal moment strength M_n of posttensioned members in which tendons are ungrouted is, in general, less than that of a bonded beam.

Most of the following development applies for members containing only prestressed reinforcement (so-called *fully* prestressed members); thus, any nonprestressed steel effect is omitted. In 1983, changes were made in ACI Code expressions to include nonprestressed steel effects to provide for *partially* prestressed members.

Underreinforced Beams.   For cases where $\rho_p f_{ps}/f_c' \leq 0.36\beta_1$, the nominal moment strength may be determined for rectangular sections, in accordance with the principles of Chapter 3 (Section 3.3), by

$$M_n = T\left(d - \frac{a}{2}\right) \tag{21.8.1}$$

where $T = A_{ps}f_{ps}$, $C = 0.85f_c'ba$, and from $C = T$,

$$a = \frac{A_{ps}f_{ps}}{0.85f_c'b} = \frac{\rho_p b d f_{ps}}{0.85f_c'b} = \frac{\rho_p f_{ps}}{0.85f_c'}d \tag{21.8.2}$$

and A_{ps}, ρ_p, and f_{ps} refer to the area, reinforcement ratio, and tensile stress for the prestressed reinforcement. Note that f_{ps} is the average stress in the prestressing steel *when nominal moment strength M_n is reached*; it is used instead of f_y because the steel usually exhibits no well-defined yield point (see Fig. 21.8.1).

Substitution of Eq. (21.8.2) into Eq. (21.8.1) gives

$$M_n = A_{ps}f_{ps}\left(d - \frac{\rho_p f_{ps}}{1.7f_c'}d\right) \tag{21.8.3}$$

Thus for any concrete cross section, the nominal moment strength depends on ρ_p and f_{ps}.

The actual stress in the prestressed reinforcement when the nominal strength is reached may not be easily determined, particularly when the specific stress–strain curve for the steel used is not available. Referring to Fig. 21.8.1, for low percentages of steel and therefore higher stress f_{ps}, the strain may be nearly 0.05; whereas for higher percentages of reinforcement, the strain may be closer to 0.01 (approximately corresponding to yield stress). Thus f_{ps} is not the same for all beams.

Since test results have indicated good agreement [21.12] with Eq. (21.8.3) up to $\rho_p f_{pu}/f_c' = 0.3$, f_{ps} has been approximated as $0.85f_{pu}$ (representing the sharp change in slope of the stress–strain curve) when $\rho_p f_{pu}/f_c' = 0.3$ and it can be assumed to vary linearly from the aforementioned value to $f_{ps} = f_{pu}$ when $\rho_p f_{pu}/f_c' = 0$. Thus, in lieu of more accurate stress–strain data, the steel stress f_{ps} in members with bonded *stress-relieved tendons* can be taken as

$$f_{ps} = f_{pu}\left(1 - 0.5\rho_p \frac{f_{pu}}{f_c'}\right) \tag{21.8.4}$$

Figure 21.8.1 Typical stress–strain curve for high-tensile steel (stress-relieved) wire.

when f_c' does not exceed 5000 psi. Note that f_{pu} is the tensile strength of the pre-stressed reinforcement (see Fig. 21.8.1).

With the increasing use of *low-relaxation prestressing strands* where the yield strength f_{py} equals $0.90f_{pu}$, instead of *stress-relieved tendons* where f_{py} equals $0.85f_{pu}$, and the increasing use of f_c' greater than 5000 psi, where β_1 is less than 0.80, there was a need to improve the estimate of f_{ps} previously based on Eq. (21.8.4). Thus, Eq. (21.8.4) was modified in 1983 to become

$$f_{ps} = f_{pu}\left(1 - \frac{\gamma_p}{\beta_1}\frac{\rho_p f_{pu}}{f_c'}\right) \tag{21.8.5}$$

which is the present ACI Formula (18-3) omitting the terms for nonprestressed rein-forcement. Note the values of γ_p/β_1 in Table 21.8.1. The value used for f_{ps} from Eq. (21.8.5) will be *lower than previously* for higher strength concrete with stress-relieved tendons; however, with low-relaxation strands the values used for f_{ps} will be higher than previously.

Table 21.8.1 Values for γ_p/β_1 for Eq. (21.8.5)

	CONCRETE STRENGTH, f_c' (psi)			
γ_p	5000	6000	7000	8000
0.40 (for $f_{py}/f_{pu} \geq 0.85$)	0.50	0.53	0.57	0.62
0.28 (for $f_{py}/f_{pu} \geq 0.90$)	0.35	0.37	0.40	0.43

Overreinforced Beams. For situations in which $\rho_p f_{pu}/f_c' > 0.36\beta_1$ (i.e., equivalent to $x > 0.423d$), corresponding roughly to ρ exceeding $0.75\rho_b$ for nonprestressed concrete beams, the higher prestressed reinforcement ratio is permitted but the strength M_n must be evaluated using no more than the amount corresponding to $\rho_p f_{pu}/f_c' = 0.36\beta_1$. When $\rho_p f_{pu}/f_c' > 0.36\beta_1$ for a member containing only prestressed reinforcement, ACI-18.8.2 states "design moment strength shall not exceed the moment strength based on the compression portion of the moment couple."

■ **EXAMPLE 21.8.1** Determine the nominal moment strength M_n of the pretensioned bonded section investigated in Example 21.5.2, and evaluate whether or not the strength is adequate for the given service loads. The concrete has $f_c' = 5000$ psi and the stress-relieved prestressing strand has $f_{pu} = 250,000$ psi. Assume 20% prestress losses and an average stress–strain relationship for the steel, as given in Fig. 21.8.1.

Solution: The approximate stress when nominal strength is reached may be taken, according to ACI-18.7.2, or Eq. (21.8.5), as

$$f_{ps} = f_{pu}\left(1 - \frac{\gamma_p}{\beta_1}\rho_p\frac{f_{pu}}{f_c'}\right)$$

$$\rho_p = \frac{A_{ps}}{bd} = \frac{2.30}{20(25)} = 0.0046$$

For $f_c' = 5000$ psi, $\beta_1 = 0.80$; and for stress-relieved strand $f_{py}/f_{pu} \geq 0.85$, $\gamma_p = 0.40$. Thus,

$$f_{ps} = 250\left[1 - \frac{0.40}{0.80}(0.0046)\frac{250}{5}\right] = 250(1 - 0.115) = 221 \text{ ksi}$$

Check $\rho_p f_{ps}/f_c'$ against $0.36\beta_1$.

$$\rho_p\frac{f_{ps}}{f_c'} = 0.0046\left(\frac{221}{5}\right) = 0.203 < [0.36\beta_1 = 0.288] \qquad \textbf{OK}$$

From Fig. 21.8.2(c),

$$C_u = 0.85f_c'ba = 0.85(5)(20)a = 85a$$

$$T_u = A_{ps}f_{ps} = 2.30(221) = 508 \text{ kips}$$

$$C_u = T_u$$

$$a = \frac{508}{85} = 5.98 \text{ in.}$$

$$x = \frac{a}{\beta_1} = \frac{5.98}{0.8} = 7.47 \text{ in.}$$

Figure 21.8.2 Nominal strength M_n of section for Example 21.8.1.

The additional strain ϵ_{s2} due to the superimposed nominal moment M_n is

$$\epsilon_{s2} = 0.003 \frac{17.53}{7.47} = 0.00704$$

which, when added to the strain due to prestress after losses, gives

$$\epsilon_s = \epsilon_{s1} + \epsilon_{s2} = \frac{140}{29,000} + 0.00704 = 0.00483 + 0.00704 = 0.0119$$

$$f_{si} \text{ (initial)} = 175,000 \text{ psi}$$

$$\Delta f_s \text{ (losses)} = 0.2(175,000) = 35,000 \text{ psi}$$

The strain ϵ_s of 0.0119 corresponds approximately to that for a stress f_{ps} of 225 ksi, using a typical stress–strain relationship for a steel with $f_{pu} = 250$ ksi, such as in Fig. 21.8.1. The value of f_{ps} agrees closely with the starting assumption of 221 ksi based on Eq. (21.8.4).

Thus, the nominal moment strength is

$$M_n = T_u \left(d - \frac{a}{2} \right) = \left[508 \left(25 - \frac{5.98}{2} \right) \right] \frac{1}{12} = 932 \text{ ft-kips}$$

The factored service load moment M_u based on the service loads of Example 21.5.2 is

$$M_u = 1.4(125) + 1.7(383) = 826 \text{ ft-kips}$$

$$\phi M_n = 0.90(932) = 839 \text{ ft-kips} > (M_u = 826 \text{ ft·kips}) \quad \textbf{OK} \quad \blacksquare$$

■ **EXAMPLE 21.8.2**   For a typical prestressed concrete beam, using high tensile strength steel wire having a stress–strain curve as in Fig. 21.8.1, show that the balanced strain condition at nominal strength M_n may be approximated by $\rho_p = 0.48\beta_1 f'_c / f_{ps}$.

Solution: (a) Consider that crushing strain for concrete is 0.003 as prescribed by the ACI Code. Referring to Fig. 21.8.1, the proportional limit is about $0.85f_{pu}$, above which value strain increases more rapidly than stress. Thus the approximate balanced strain condition occurs when $\epsilon_c = 0.003$ and $\Delta\epsilon_s$ (change in steel strain due to external loading) equals the strain ϵ_{ps} at $f_{ps} = 0.85f_{pu}$ less the initial prestress strain ϵ_i. For the balanced strain condition,

$$C_b = 0.85f'_c ba_b$$

$$T_b = A_{psb}f_{ps} = \rho_{pb}bdf_{ps}$$

$$C_b = T_b$$

$$a_b = \left(\frac{d}{0.85}\right)\rho_{pb}\frac{f_{ps}}{f'_c}$$

$$a_b = \beta_1 x_b$$

$$\frac{x_b}{d} = \frac{1}{0.85\beta_1}\rho_{pb}\frac{f_{ps}}{f'_c}$$

or

$$\rho_{pb} = 0.85\beta_1\frac{f'_c}{f_{ps}}\left(\frac{x_b}{d}\right)$$

where ρ_{pb} is the prestressed reinforcement ratio at the balanced strain condition.

At $f_{ps} = 0.85f_{pu}$, $\epsilon_{ps} = 0.0075$ from Fig. 21.8.1. Also, if the initial tension in the prestressing steel is $0.60f_{pu}$, then the initial tensile strain in the steel (representing a compressive strain in concrete) is

$$\epsilon_i = \frac{0.60f_{pu}}{E_s} = \frac{150}{30,000} \approx 0.005$$

Thus

$$\Delta\epsilon_s = \epsilon_{ps} - \epsilon_i = 0.0075 - 0.005 = 0.0025$$

Then, from Fig. 21.8.3,

$$\frac{x_b}{d} = \frac{0.003}{0.003 + 0.0025} = 0.545$$

$\epsilon_c = 0.003$

$\dfrac{x}{d} = 0.545$

d

$\Delta\epsilon_s = \epsilon_{ps} - \epsilon_i$

Figure 21.8.3 Strain diagram for the approximate balanced strain condition due to superimposed loads.

Substituting into the ρ_{pb} expression gives

$$\rho_{pb} = 0.85\beta_1 \frac{f'_c}{f_{ps}} (0.545) = 0.46\beta_1 \frac{f'_c}{f_{ps}}$$

(b) Suppose that $\epsilon_c = 0.0035$ instead of 0.003. There will be no change in $\Delta\epsilon_s$ if it is still assumed $f_{ps} = 0.85f_{pu}$,

$$\frac{x_b}{d} = \frac{0.0035}{0.0035 + 0.0025} = 0.584$$

$$\rho_{pb} = 0.85\beta_1 \frac{f'_c}{f_{ps}} (0.584) = 0.50\beta_1 \frac{f'_c}{f_{ps}}$$

In effect the values of ρ_{pb} represent the "balanced" amount of reinforcement, which if exceeded would have the concrete reach crushing strain while the steel is still in the elastic range. Probably $0.50\beta_1 f'_c/f_{ps}$ represents a reasonable upper bound and $0.46\beta_1 f'_c/f_{ps}$ represents a typical value for the balanced condition. Thus the ACI Code implied use of $0.48\beta_1 f'_c/f_{ps}$ as the balanced reinforcement ratio seems reasonable; consequently, $0.36\beta_1 f'_c/f_{ps}$ reasonably represents 0.75 of the balanced reinforcement ratio. ∎

21.9 CRACKING MOMENT

One of the features of prestressed concrete is that under service load it is usually crack free. To be sure that an adequate reserve exists against cracking, ACI-18.8.3 requires the total amount of prestressed and nonprestressed reinforcement to be adequate to develop a factored load in flexure at least 1.2 times the cracking load calculated using a modulus of rupture equal to $7.5\sqrt{f'_c}$ for normal-weight concrete (see ACI-9.5.2.3 for lightweight concrete). The basic cracking moment requirement is a means of ensuring that cracking will occur *before* flexural strength is reached, and by a large enough margin so that significant deflection will occur to warn that the strength M_n is being approached. The typical member will have a fairly large margin between cracking strength and flexural strength but the designer must be certain by checking it.

The ACI Code waives the above requirement when the flexural member has ϕV_n and ϕM_n at least twice V_u and M_u, respectively.

∎ **EXAMPLE 21.9.1** Compute the cracking moment M_{cr} and check its acceptability according to the ACI Code, for the beam of Example 21.8.1 (Fig. 21.8.2). Use $f'_c = 5000$ psi and assume the effective prestress, after losses, is 140 ksi.

Solution: **(a)** Use the homogeneous beam concept (Section 21.6). The effective prestress force is

$$T_e = A_{ps}f_{se} = 2.30(140) = 322 \text{ kips}$$

Using the following equation to find the cracking moment,

$$-\left(\begin{array}{c}\text{axial} \\ \text{prestress}\end{array}\right) - \left(\begin{array}{c}\text{moment} \\ \text{prestress}\end{array}\right) + \left(\begin{array}{c}\text{external} \\ \text{loads}\end{array}\right) = \left(\begin{array}{c}\text{cracking} \\ \text{stress}\end{array}\right)$$

$$-\frac{T_e}{A_g} - \frac{T_e e y_t}{I_g} + \frac{M_{cr} y_t}{I_g} = f_r = 7.5\sqrt{5000} = 530 \text{ psi}$$

$$-\frac{322}{600} - \frac{322(10)15}{45,000} + \frac{M_{cr}(15)}{45,000} = 0.530$$

$$M_{cr} = \left[\frac{0.530(45,000)}{15} + 321.5(10) + \frac{322(45,000)}{600(15)}\right]\frac{1}{12}$$

$$= 133 + 268 + 134 = 535 \text{ ft-kips}$$

(b) Use the internal force concept (Section 21.6). Consider that the cracking moment M_{cr} is comprised of two parts,

$$M_{cr} = M_1 + M_2$$

where M_1 is the superimposed moment necessary to give zero stress in the precompressed tension zone (see Fig. 21.9.1), and M_2 is the additional moment to cause cracking assuming that zero stress exists at the tension face when M_2 is applied (see Fig. 21.9.1).

$$C = T_e = 322 \text{ kips}$$

$$M_1 = T_e(e + e_1) = 322\left(\frac{10 + 5}{12}\right) = 403 \text{ ft-kips}$$

$$M_2 = \frac{f_r I_g}{y_t} = \frac{0.530(45,000)}{15(12)} = 133 \text{ ft-kips}$$

$$M_{cr} = M_1 + M_2 = 403 + 133 = 536 \text{ ft-kips}$$

(c) Check ACI-18.8.3. From Example 21.8.1, the nominal moment strength M_n provided is 932 ft-kips. Since ACI-18.8.3 requires reinforcement "to develop a

Figure 21.9.1 Computation of cracking moment.

factored load" exceeding 1.2 M_{cr},

$$\phi M_n > 1.2 M_{cr}$$

$$0.90(932) > 1.2(536)$$

$$840 \text{ ft-kips} > 640 \text{ ft-kips} \qquad \textbf{OK}$$

Thus cracking will occur soon enough before reaching nominal moment strength so that large deflection will give a warning of impending failure.

In addition it is noted that the overload factor against cracking at midspan is

$$\text{FS against cracking} = \frac{536}{125 + 383} = 1.05$$

If more reserve against cracking is desired, the tensile stress permitted under final service conditions should be reduced below $6\sqrt{f'_c}$. Cracking at service load may or may not be detrimental. Nonprestressed beams usually crack under service load. ∎

21.10 SHEAR STRENGTH OF MEMBERS WITHOUT SHEAR REINFORCEMENT

In general, the ideas presented in Chapter 5 regarding shear strength for nonprestressed beams are also applicable to prestressed beams. An excellent summary of background for the ACI Code expressions for shear strength of prestressed concrete beams is given by MacGregor and Hanson [5.48]. More information on shear strength of prestressed concrete is available in the ASCE-ACI Committee 426 Reports [5.3, 5.11]. Collins and Mitchell [19.9] have presented an excellent unified treatment of shear and torsion behavior for both prestressed and nonprestressed concrete beams, along with proposed design rules.

It is known, however, that a prestressed concrete beam generally performs better under high shear conditions than does an ordinary reinforced concrete beam. Consider Eq. (5.3.1) as derived in Chapter 5 for maximum principal stress,

$$f_t(\text{max}) = \frac{f_t}{2} + \sqrt{\left(\frac{f_t}{2}\right)^2 + v^2} \qquad \textbf{[5.3.1]}$$

In nonprestressed concrete the normal stress f_t is a tensile stress on one side of the neutral axis. When the concrete is prestressed, f_t is a compressive stress throughout the member. Replacing f_t with $-f_t$ it is apparent that the magnitude of the principal tensile stress decreases,

$$f_t(\text{max}) = \frac{-f_t}{2} + \sqrt{\left(\frac{f_t}{2}\right)^2 + v^2} \qquad \textbf{(21.10.1)}$$

The angle α that the maximum principal tensile stress makes with the beam axis is greater for a prestressed concrete beam than for an ordinary reinforced concrete beam, as shown in Fig. 21.10.1. Inclined cracking will be less likely to occur and, if it occurs, will be more nearly horizontal than in nonprestressed members.

Two types of inclined cracks are possible in prestressed concrete beams: (1) the flexure-shear crack that occurs in a beam previously cracked due to flexure; and (2) the web-shear crack that occurs in the thin web of a previously uncracked beam (see

(a) Principal stress — reinforced concrete beam

(b) Principal stress — prestressed concrete beam

Figure 21.10.1 Comparison of directions of principal tensile stress.

Fig. 5.4.1). Whereas only the flexure-shear crack is common in nonprestressed beams, both types represent potential cracks in prestressed concrete beams.

Flexure-Shear Cracking Strength. The flexure-shear crack arises from high principal stress near the interior extremity of a flexural crack. Experimental studies have shown that the shear corresponding to the flexure-shear crack is that which causes flexural cracking at approximately a distance $d/2$ from the load point, in the direction of decreasing bending moment [21.13] (see Fig. 21.10.2). The 1963 ACI Code formula for flexure-shear cracking strength was the experimentally determined linear

Figure 21.10.2 Flexure-shear type of inclined crack.

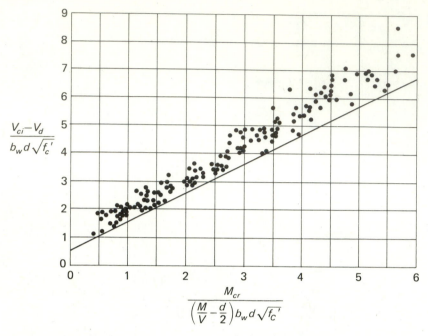

Figure 21.10.3 Comparison of the shear corresponding to flexure-shear cracking in prestressed concrete beams with the ratio of the flexural cracking moment to the shear span (both axes nondimensionalized by dividing by $b_w d\sqrt{f'_c}$) (from Sozen and Hawkins [21.13]).

relationship of Fig. 21.10.3. Because the relationship between moment and shear (M/V) was not the same for both dead and live loads in the tests, the effects of dead load were not included in the linear relationship.

The ordinate of the plot (see Fig. 21.10.3) is thus in terms of $V_{ci} - V_d$, where V_d is the service dead load shear force and V_{ci} is the total nominal shear strength. The abscissa involves the net cracking moment M_{cr} due to service loads. The net cracking stress used for computing M_{cr} equals the modulus of rupture (assumed conservatively at $6\sqrt{f'_c}$ psi), plus the compressive stress f_{pe} provided by the prestressing force (after losses) occurring at the extreme fiber at which tensile stresses are caused by the applied loads, less the tensile stress f_d due to service dead load. Thus

$$M_{cr} = \frac{I}{y_t}(6\sqrt{f'_c} + f_{pe} - f_d) \qquad (21.10.2)^*$$

which is ACI Formula (11-12).

— — — —

* For SI, ACI 318–89M, for f'_c, f_{pe}, and f_d in MPa, M_{cr} in N·mm, I in mm⁴, and y_t in mm, gives

$$M_{cr} = \frac{I}{y_t}\left(\frac{1}{2}\sqrt{f'_c} + f_{pe} - f_d\right) \qquad (21.10.2)$$

Figure 21.10.4 Shear–moment relationship.

The abscissa (Fig. 21.10.3) involves $M_{cr}/(M/V - d/2)$, which represents the applied load shear at $d/2$ from the cross section being investigated. A moment M and a shear V act on section *A-A* of Fig. 21.10.4. Assume that M and V give rise to a moment crack at $d/2$ from section *A-A*. From the basic shear–moment relationship, the change in moment between two points equals the area under the shear diagram between those points. Thus referring to Fig. 21.10.4,

$$M - M_{cr} = \frac{V + V_{cr}}{2}\left(\frac{d}{2}\right) \tag{21.10.3}$$

Since the difference between V and V_{cr} will usually be small, assume $V_{cr} = V$; thus

$$M - M_{cr} = V\left(\frac{d}{2}\right) \tag{21.10.4}$$

Solving for M_{cr},

$$M_{cr} = M - V\left(\frac{d}{2}\right)$$
$$= V\left(\frac{M}{V} - \frac{d}{2}\right) \tag{21.10.5}$$

Finally, the shear on the section investigated becomes

$$V = \frac{M_{cr}}{M/V - d/2} \tag{21.10.6}$$

Thus the linear relationship for the flexure-shear cracking strength resulting from high principal stress in the vicinity of a flexural crack is, according to Fig. 21.10.3,

$$\frac{V_{ci} - V_d}{b_u d \sqrt{f_c'}} = 0.6 + \frac{M_{cr}}{(M/V - d/2)b_w d \sqrt{f_c'}} \tag{21.10.7}$$

Since 1971, ACI-11.4.2 has simplified the expression by eliminating the subtracted $d/2$ term.

Thus the nominal flexure-shear cracking strength V_{ci}, ACI Formula (11-11), is

$$V_{ci} = 0.6\sqrt{f'_c}\,b_w d + \frac{V_i M_{cr}}{M_{max}} + V_d \geq 1.7\sqrt{f'_c}\,b_w d \qquad \textbf{(21.10.8)*}$$

In Eq. (21.10.8), M_{max} replaces M of Eq. (21.10.7) and represents the maximum moment that can occur at the section under consideration, due to externally applied factored loads (i.e., applied loads other than beam weight and prestress, unless the prestress causes an external reaction). V_i replaces V of Eq. (21.10.7) and represents the shear force at the section due to the factored loading that caused maximum moment. In other words, one uses the moment *envelope* values for M_{max} along with the corresponding shears, rather than the shear envelope values which would be larger. Of course, where partial span loadings are not considered, the full-span loading gives V_i and M_{max} for each point along the span. When full-span uniform loading is used in a simply supported span, $V_i = w(L - 2x)/2$ and $M_{max} = wx(L - x)/2$, and

$$\frac{V_i}{M_{max}} = \frac{L - 2x}{x(L - x)} \qquad \textbf{(21.10.9)}$$

In effect, Eq. (21.10.8) gives the flexure-shear cracking strength as the sum of (1) the shear due to actual dead load (beam weight) without overload factor, (2) the superimposed factored live load which is sufficient to cause flexural cracking, and (3) the additional load ($0.6\sqrt{f'_c}\,b_w d$) which will cause the flexural crack to initiate the inclined flexure-shear crack.

Web-Shear Cracking Strength.   The web-shear crack arises in a beam previously uncracked due to flexure. Such a crack is typical near the support of a thin-webbed section (see Fig. 5.4.1) as a result of high principal tensile stress.

For this development the beam may reasonably be considered as homogeneous. Thus Eq. (21.10.1) may be directly applied,

$$f_t(\text{max}) = \frac{-f_t}{2} + \sqrt{\left(\frac{f_t}{2}\right)^2 + v^2} \qquad \textbf{[21.10.1]}$$

Since tests have demonstrated that cracks of this type usually originate near the centroid of the section, the stresses refer to that location. Solving Eq. (21.10.1) for v,

---

* For SI, ACI 318–89M, for f'_c in MPa, b_w and d in mm, and V in N, gives

$$V_{ci} = \frac{1}{20}\sqrt{f'_c}\,b_w d + \frac{V_i M_{cr}}{M_{max}} + V_d \geq \frac{1}{7}\sqrt{f'_c}\,b_w d \qquad \textbf{(21.10.8)}$$

$$\left[f_t(\text{max}) + \frac{f_t}{2}\right]^2 = \left(\frac{f_t}{2}\right)^2 + v^2$$

$$[f_t(\text{max})]^2 + f_t(\text{max})f_t + \left(\frac{f_t}{2}\right)^2 = \left(\frac{f_t}{2}\right)^2 + v^2$$

$$v = f_t(\text{max})\sqrt{1 + \frac{f_t}{f_t(\text{max})}} \qquad (21.10.10)$$

where f_t is the compressive stress at the level of the centroid, and $f_t(\text{max})$ = principal tensile stress $\leq$ tensile strength of concrete if no cracks are to form.

Since Eq. (21.10.10) should agree with the criteria for shear strength of non-prestressed beams, $f_t(\text{max})$ is taken as $3.5\sqrt{f'_c}$. The equation then becomes

$$v = 3.5\sqrt{f'_c}\sqrt{1 + \frac{f_t}{3.5\sqrt{f'_c}}} \qquad (21.10.11)$$

A plot of Eq. (21.10.11) is given in Fig. 21.10.5, illustrating that this equation may be approximated by a straight line,

$$v = 3.5\sqrt{f'_c} + 0.3f_t \qquad (21.10.12)$$

or, in ACI Code terminology, multiplying by $b_w d$ to give nominal shear strength,

$$V_{cw} = (3.5\sqrt{f'_c} + 0.3f_{pc})b_w d \qquad (21.10.13)$$

where f_{pc} is defined as the compressive stress (psi) in the concrete, after losses, at the centroid of the section resisting the applied loads, or if the centroid lies within the flange on a T-section, f_{pc} is the stress at the junction of the flange and web.

Figure 21.10.5 Comparison of theoretical maximum shear stress with straightline approximation.

If prestressing tendons are draped, a vertical component V_p arises which assists in carrying the shear. Therefore ACI-11.4.2 gives ACI Formula (11-13) as

$$V_{cw} = (3.5\sqrt{f'_c} + 0.3f_{pc})b_w d + V_p \qquad (21.10.14)*$$

Alternatively, the web-shear cracking strength may be determined (ACI-11.4.2.2) using the principal stress equation, Eq. (21.10.10), where the principal stress $f_t(\text{max})$ is limited to $4\sqrt{f'_c}$. This would give Eq. (21.10.11) with coefficients 4 instead of 3.5. Thus using f_{pc} for f_t, Eq. (21.10.11) multiplied by $b_w d$ with V_p added becomes

$$V_{cw} = 4\sqrt{f'_c}\sqrt{1 + \frac{f_{pc}}{4\sqrt{f'_c}}}\,b_w d + V_p \qquad (21.10.15)*$$

which is an acceptable alternate to using Eq. (21.10.14).

The nominal shear strength V_n at which inclined cracking is imminent is given by the lesser of V_{ci} and V_{cw}, determined from ACI Formulas (11-11) and (11-13) for normal-weight concrete.

When applying Eqs. (21.10.8) and (21.10.14) or (21.10.15) for V_{ci} and V_{cw}, the effective depth d is to be taken as the distance from the extreme compression fiber to the centroid of the prestressing tendons, or as 80% of the overall depth of the member, whichever is greater.

In computing the prestress effect f_{pc}, full prestress after losses may be used only when the prestressing tendons are embedded a distance exceeding their required development length from the section being investigated. The critical section for maximum shear is generally at $h/2$ from the face of support (ACI-11.1.3.2), so that the region between the face of support and $h/2$ therefrom must be designed for the shear at the critical section.

ACI Code Simplified Alternative for Shear Strength. When the member has an effective prestress f_{se} at least equal to 40% of the tensile strength f_{pu} of the flexural reinforcement, the nominal shear strength may be taken as

$$V_c = \left[0.6\sqrt{f'_c} + 700\frac{V_u d}{M_u}\right]b_w d \qquad (21.10.16)*$$

which is ACI Formula (11-10). This equation may be considered a linear gradient between that using the minimum nominal unit stress $v_c = 2\sqrt{f'_c}$ and an upper bound of $v_c(\text{max}) = 5\sqrt{f'_c}$. Supporting data for this alternate relationship are shown in Fig. 21.10.6 from MacGregor and Hanson [5.48].

----

* For SI, ACI 318–89M, for f'_c and f_{pc} in MPa, b_w and d in mm, and V in N, gives

$$V_{cw} = 0.3(\sqrt{f'_c} + f_{pc})b_w d + V_p \qquad (21.10.14)$$

$$V_{cw} = \frac{1}{3}\sqrt{f'_c}\sqrt{1 + \frac{3f_{pc}}{\sqrt{f'_c}}}\,b_w d + V_p \qquad (21.10.15)$$

$$V_c = \left[\frac{1}{20}\sqrt{f'_c} + 5\frac{V_u d}{M_u}\right]b_w d \qquad (21.10.16)$$

Figure 21.10.6 Alternate equation for computing v_c for prestressed beams (from MacGregor and Hanson [5.48]).

When applying Eq. (21.10.16) from ACI-11.4.1, V_u is the maximum shear due to factored loading at the section and M_u is the simultaneously occurring moment; $V_u d/M_u$ is also limited to a maximum value of 1.0; and d is the actual effective depth to the centroid of prestressed reinforcement. Equation (21.10.16) represents essentially the flexure-shear cracking strength expressed in a manner similar to that for non-prestressed reinforced concrete. Since the percentage of reinforcement is low in prestressed concrete, a constant has been used instead of the variable ρ.

■ **EXAMPLE 21.10.1** Determine the nominal shear strength $V_n = V_c$ at a section 4 ft from the supports of the 40 ft simple span shown in Fig. 21.10.7. The effective prestress force T_e (after losses) is 322 kips. The beam is to support a service live load of 1.38 kips/ft in addition to the beam weight of 0.625 kip/ft. The concrete has $f'_c = 5000$ psi.

Solution: **(a)** Simplified alternate procedure, using ACI Formula (11-10), Eq. (21.10.16).

$$V_c = \left(0.6\sqrt{f'_c} + 700\,\frac{V_u d}{M_u}\right)b_w d$$

Figure 21.10.7 Section for Example 21.10.1.

In the application of the above formula, the symbols V_u and M_u indicate use of the shear and moment including overload factors. Assume partial span loading is not considered,

$$w_u = 0.625(1.4) + 1.38(1.7) = 0.87 + 2.34 = 3.21 \text{ kips/ft}$$

$$V_u = 3.21(20 - 4) = 51.4 \text{ kips}$$

$$M_u = \tfrac{1}{2}(3.21)(4)(36) = 231 \text{ ft-kips}$$

$$\frac{V_u d}{M_u} = \frac{51.4(30 - 11)}{231(12)} = 0.35$$

where d *in this term* is the actual d, 19 in.

The normal unit stress capacity is

$$v_c = 0.6\sqrt{5000} + 700(0.35) = 42 + 245 = 287 \text{ psi}$$

$$v_c \text{ (upper limit)} = 5\sqrt{f_c'} = 354 \text{ psi} > 287 \text{ psi} \qquad \textbf{OK}$$

$$v_c \text{ (lower limit)} = 2\sqrt{f_c'} = 141 \text{ psi} < 287 \text{ psi} \qquad \textbf{OK}$$

Thus at this section the prestressed member has a contribution v_c to the shear strength attributable to the concrete of 287 psi. The nominal shear strength V_c at 4 ft from the support would be

$$V_c = v_c b_w d = 0.287(20)(24) = 138 \text{ kips}$$

where d is taken as the larger of $0.8h$ [i.e., $0.8(30) = 24$ in.] and actual d (i.e., 19 in.).

(b) Flexure-shear cracking strength using the more exact procedure, ACI Formula (11-11), Eq. (21.10.8).

$$V_{ci} = 0.6\sqrt{f_c'}\, b_w d + \frac{V_i M_{cr}}{M_{max}} + V_d$$

Using Eq. (21.10.2),

$$M_{cr} = \frac{I}{y_t}(6\sqrt{f_c'} + f_{pe} - f_d)$$

$$I = \tfrac{1}{12}(20)(30)^3 = 45{,}000 \text{ in.}^4 \quad \text{(neglecting steel)}$$

$$y_t = 15 \text{ in.}$$

f_{pe} = compressive stress due to prestressing force

$$= \frac{322,000}{20(30)} + \frac{322,000(4)(15)}{45,000} = 537 + 429 = 966 \text{ psi}$$

$e = 4$ in. at 4 ft from center of support

$$f_d = \text{dead load stress} = \frac{My}{I} = \frac{625(4)(36)(12)(15)}{2(45,000)} = 180 \text{ psi}$$

$$M_{cr} = \frac{45,000}{15(12,000)} (6\sqrt{5000} + 966 - 180)$$

$$= 0.25(424 + 966 - 180) = 0.25(1210) = 302 \text{ ft-kips}$$

M_{max} = maximum moment due to externally applied factored loads, except beam weight at section being investigated

$$= \tfrac{1}{2}(1.38)(4)(36)(1.7) = 169 \text{ ft-kips}$$

V_i = corresponding shear due to factored load at section investigated

$$= 1.38(20 - 4)(1.7) = 37.5 \text{ kips}$$

$$\frac{V_i M_{cr}}{M_{max}} = \frac{37.5(302)}{169} = 67.0 \text{ kips}$$

$$V_d = \text{dead load shear} = 0.625(20 - 4) = 10.0 \text{ kips}$$

$$d = 30 - 11 = 19 \text{ in.} \quad \text{or} \quad 0.80h = 0.8(30) = \underline{24 \text{ in.}} \quad \text{(use larger value)}$$

$$V_{ci} = 0.6\sqrt{5000}(20)(24)\tfrac{1}{1000} + 67.0 + 10.0 \doteq 97.4 \text{ kips}$$

The flexure-shear cracking strength is not to be taken less than

$$\min V_{ci} = 1.7\sqrt{f_c'}\, b_w d$$

$$= 1.7\sqrt{5000}(20)(24)\tfrac{1}{1000} = 57.7 \text{ kips} < 97.4 \text{ kips} \qquad \textbf{OK}$$

(c) Web-shear cracking strength using the more exact procedure, ACI Formula (11-13), Eq. (21.10.14).

$$V_{cw} = [3.5\sqrt{f_c'} + 0.3f_{pc}]b_w d + V_p$$

f_{pc} = compressive stress in concrete at centroid of section resisting live load, due to all applied loads (zero due to all except T_e/A_c in this case)

$$= \frac{322,000}{20(30)} = 537 \text{ psi}$$

$$d = 19 \text{ in.} \quad \text{or} \quad 0.80(30) = \underline{24 \text{ in.}} \quad \text{(use larger value)}$$

$$V_{cw} = [3.5\sqrt{5000} + 0.3(537)](20)(24)\tfrac{1}{1000} + \tfrac{1}{12}(322) = 223 \text{ kips}$$

Since $V_{ci} < V_{cw}$, $V_c = V_{ci} = 97.4$ kips at 4 ft from the support for the member without shear reinforcement. The simplified alternate equation gives a nominal shear strength

42% higher than obtained from the more exact procedure; 138 kips compared to 97.4 kips. ∎

21.11 SHEAR REINFORCEMENT FOR PRESTRESSED CONCRETE BEAMS

The computations for shear reinforcement are essentially the same as for nonprestressed concrete as developed in Chapter 5. The total nominal shear strength V_n may be represented as the sum of the amount V_c attributable to the concrete and the amount V_s attributable to the shear reinforcement; thus Eq. (5.8.1) still applies,

$$V_n = V_c + V_s \qquad \text{[5.8.1]}$$

The nominal strength V_c attributable to the concrete may be determined by (1) Eq. (21.10.16) when the effective prestress is at least 40% of the tensile strength of the steel, or (2) the *smaller* of Eqs. (21.10.8) and (21.10.14) for V_{ci} and V_{cw}. The second and more accurate method using V_{ci} and V_{cw} may be used whatever the magnitude of prestress.

For the shear reinforcement contribution, Eq. (5.11.7) is used,

$$V_s = \frac{A_v f_y d}{s} \qquad \text{[5.11.7]}$$

where

A_v = effective area of shear reinforcement at any section

s = spacing of the shear reinforcement

Just as for nonprestressed reinforced concrete beams, prestressed concrete beams must have at least a minimum amount of shear reinforcement whenever $V_u > \phi V_c / 2$ (ACI-11.5.5.1). However, this requirement may be waived if tests are made showing that the required flexural and shear strengths can be developed when shear reinforcement is omitted. Since prestressed concrete members are usually of standardized shapes with many identical or similar members manufactured, manufacturers frequently have made tests demonstrating that no shear reinforcement is required.

When minimum shear reinforcement is required, ACI-11.5.5.4 requires using the *lesser* of

(1) ACI Formula (11-14), as follows,

$$A_v = 50 \frac{b_w s}{f_y} \qquad \text{(21.11.1)}$$

which is also used for nonprestressed concrete, and

(2) where the effective prestress force equals at least 40% of the tensile strength f_{pu} of the steel, ACI Formula (11-15), as follows,

$$A_v = \frac{A_{ps}}{80} \left(\frac{f_{pu}}{f_y} \right) \left(\frac{s}{d} \right) \sqrt{\frac{d}{b_w}} \qquad \text{(21.11.2)}$$

where

A_{ps} = area of prestressed reinforcement

f_{pu} = tensile strength of prestressed reinforcement

f_y = specified yield strength of shear reinforcement

s = stirrup spacing, which may not exceed 0.75 of the depth of the member, or 24 in., whichever is smaller (ACI-11.5.4.1)

d = distance from the extreme compressive fiber to the centroid of the prestressed reinforcement (but need not be taken less than $0.8h$)

■ **EXAMPLE 21.11.1** Determine the maximum spacing of #3 U stirrups at a point 4 ft from the support on the beam of Example 21.10.1 (Fig. 21.10.7). The prestressing steel has an area of 2.30 sq in. and a tensile strength $f_{pu} = 250,000$ psi, the yield strength of the stirrup reinforcement $f_y = 40,000$ psi, and the concrete has $f'_c = 5000$ psi. The service live load is 1.38 kips/ft, and the dead load is 0.625 kip/ft.

Solution: The inclined cracking strength $V_c = V_{ci}$ as determined in Example 21.10.1, is

$$V_{ci} \text{ (controls)} = 97.4 \text{ kips} \quad \text{(more exact procedure)}$$

At 4 ft from centerline of support,

$$V_u = [1.4(0.625) + 1.7(1.38)](20 - 4) = 3.21(20 - 4) = 51.4 \text{ kips}$$

$$V_u = 51.4 \text{ kips} > \frac{\phi V_c}{2} = \frac{0.85(97.4)}{2} = 41 \text{ kips}$$

Thus a minimum amount of shear reinforcement must be provided, unless tests are made to justify its omission. Thus using Eq. (21.11.1)

$$\frac{A_v}{s} = 50 \frac{b_w}{f_y} = 50 \left(\frac{20}{40,000} \right) = 0.025$$

or, alternatively, using the more elaborate formula, Eq. (21.11.2),

$$\frac{A_v}{s} = \frac{A_{ps}}{80} \left(\frac{f_{pu}}{f_y} \right) \left(\frac{1}{d} \right) \sqrt{\frac{d}{b_w}}$$

$$= \frac{2.30}{80} \left(\frac{250}{40} \right) \left(\frac{1}{24} \right) \sqrt{\frac{24}{20}} = 0.0082$$

Using the smaller of 0.025 and 0.0082, the minimum A_v/s (for #3 U stirrups) is

$$s = \frac{2(0.11)}{0.0082} = 30.1 \text{ in.}$$

but, according to ACI-11.5.4.1, the spacing may not exceed $0.75h = 0.75(30) = 22.5$ in. or 24 in., whichever is smaller. Thus #3 stirrups could be placed no farther apart than 22.5 in. unless tests are performed. ■

21.12 DEVELOPMENT OF REINFORCEMENT

Following the basic concepts established for nonprestressed reinforced concrete in Chapter 6, development of reinforcement must also be considered in prestressed members. The prestressing force must be transferred into the concrete by bond (interaction between concrete and steel strand) in the pretensioned beam, and the length required to accomplish this is called the "transfer length." This, of course, occurs in end regions and, in effect, anchors the tendons.

The mechanism of transfer is summarized by Zia and Mostafa [21.14]. The high stress in the pretensioned tendon must, on cutting of the tendon, be transferred to the concrete so that equilibrium is achieved. The situation in the transfer zone is quite different from that in the anchorage zones of nonprestressed concrete.

In nonprestressed concrete, the bars are stressed after being cast in the concrete. As they are stressed they decrease slightly in diameter and tend to draw away (at least create a slight tensile stress in the direction transverse to the bars) from the surrounding concrete. Thus a reasonably large length is necessary to transfer a force $A_s f_y$ in the steel to the surrounding concrete.

In prestressed concrete, the wires or strands are pretensioned to a high level of stress, thus initially reducing the wire diameter prior to the placing of the concrete. Once the concrete is cured and the wires are cut at the ends, the wires tend to shorten and correspondingly increase in diameter. Thus a compression between the wires and concrete is produced. The stress in the cut wire must increase from zero at the free end to the prestress value at a certain distance from the free end. This distance is known as the "transfer length." During the accomplishment of the transfer, probably in the first few inches from the free end, it is solely the friction that is developed as the slipping wires compress against the concrete. Farther from the end, adhesive bond, that is, slip resistance, certainly contributes to the transfer.

The transfer length L_t typically is about 50 diameters for a strand, and about 110 to 120 diameters for a single wire [21.14]. These values assume a clean strand (or wire) surface, a gradual release of the prestressing force to the concrete, and a steel stress of 140 to 150 ksi after transfer. For strands with a slightly rusted surface, the transfer distance is certainly less, and for a sudden release of stress such as can occur by burning the strand, the transfer length may easily be 20% greater. The transfer length does not seem to be affected by variation in concrete strength [21.15].

The importance of the transfer length depends on the member under consideration. It seems to be of little importance except on short members and in situations where flexural cracking may occur in the transfer zone.

Development Length for Prestressing Strand. Exactly as for nonprestressed reinforcement, the tension in the prestressing steel necessary to achieve nominal flexural strength M_n must be developed by embedment or end anchorage or a combination thereof (ACI-12.1). The purpose is to prevent general slip prior to achieving the necessary moment strength M_n; thus the steel stress must increase from the effective prestress value f_{se} to the value f_{ps} used in computing the moment strength. The net anchorage length available to accommodate a change in the tensile force is the development length L_d to the end of the strand from the section in question, less the transfer length L_t.

The following empirical relationship was established by Hanson and Kaar [21.15],

$$(f_{ps} - f_{se}) \text{ in ksi} = \frac{L_d - L_t}{d_b}$$ **(21.12.1)**

where d_b is the nominal strand diameter.

The calculated stress f_{ps} in the prestressed reinforcement occurring when the nominal strength M_n is reached must not cause the reinforcement to slip relative to the surrounding concrete; thus

$$f_{ps}(\text{ksi}) \leq f_{se}(\text{ksi}) + \frac{L_d - L_t}{d_b}$$ **(21.12.2)**

Next, it is necessary to obtain an expression for L_t/d_b. Using the concept used in deriving Eq. (6.2.1), that is, the tensile force $A_s f_{se}$ in the strand must be transmitted to the surrounding concrete by a "stress" u_s acting around the perimeter πd_b over the embedment length L_t,

$$A_s f_{se} = u_s \pi d_b L_t$$

$$\left(\frac{\pi d_b^2}{4}\right) f_{se} = u_s \pi d_b L_t$$ **(21.12.3)**

Since the actual strand (three- or seven-wire) properties differ from those based on the nominal diameter, a correction must be applied. The true circumference is taken as $\frac{4}{3}\pi d_b$; and the true area A_s as $0.725\pi d_b^2/4$. Thus

$$0.725\left(\frac{\pi d_b^2}{4}\right) f_{se} = u_s\left(\frac{4}{3}\right)\pi d_b L_t$$ **(21.12.4)**

$$\frac{L_t}{d_b} = \frac{2.175 f_{se}}{16 u_s}$$ **(21.12.5)**

If the average bond stress u_s in the transfer zone is taken as 400 psi for clean strands [21.15],

$$\frac{L_t}{d_b} = \frac{2.175}{16(0.4)} f_{se} = \frac{1}{2.94} f_{se} \quad \text{say}, \frac{1}{3} f_{se}$$ **(21.12.6)**

where f_{se} is measured in ksi.

Substituting Eq. (21.12.6) in Eq. (21.12.2) gives

$$f_{ps} \leq f_{se} + \frac{L_d}{d_b} - \frac{1}{3} f_{se}$$

and solving for L_d, the necessary development length to the end of the strand,

$$L_d = (f_{ps} - \tfrac{2}{3} f_{se}) d_b$$ **(21.12.7)**

which is the formula given in ACI-12.9.1 for the three- or seven-wire pretensioning strands. Note that the expression in parenthesis uses stresses f_{ps} and f_{se} in ksi units, but the resulting expression ($f_{ps} - \frac{2}{3}f_{se}$) is treated as a dimensionless constant. Investigation may be restricted to those cross sections nearest each end of the member that are required to develop their nominal strengths M_n.

■ **EXAMPLE 21.12.1**   Investigate the development of reinforcement for the beam of Example 21.10.1 (Fig. 21.10.7). Assume $f_{ps} = 221$ ksi as computed in Example 21.8.1. The effective prestress, after losses, may be assumed to be 0.8(175) = 140 ksi.

Solution: Since only at midspan is the maximum moment strength M_n required, the total embedment from midspan must exceed the development length L_d,

$$L_d = \left[221 - \tfrac{2}{3}(140)\right]d_b$$

$$L_d = (221 - 93.3)(0.5) = 64 \text{ in.} < 240 \text{ in.} \qquad \textbf{OK}$$

It is noted that f_{ps} used in this equation may be taken as that value required for the necessary strength at a section. In this case required nominal strength M_u/ϕ of 918 ft-kips was only slightly less than the provided M_n of 932 ft-kips. However, if the requirement was much below that provided, f_{ps} could have been taken as less than 221 ksi even at midspan. Development of reinforcement usually is not a problem on simply supported prestressed concrete beams.   ■

21.13 PROPORTIONING OF CROSS SECTIONS FOR FLEXURE WHEN NO TENSION IS PERMITTED

In order to give further insight into the variables involved, the following discussion is presented on proportioning the cross section. All cross sections used here will consist of rectangular flanges and web, though in practical design 90° junctions between flanges and web are avoided because of forming difficulties.

In this brief treatment, the only case treated is when no tension is permitted, either at the initial condition (at transfer of prestress before losses) or at the final condition (full dead and live load after losses). Thus the desired stress distributions are shown in Fig. 21.13.1.

It can be shown that on any cross section having an axis of symmetry, there are two points on the axis of symmetry, known as the "kern points," that are located at distances k_b and k_t from the centroid, as shown in Fig. 21.13.1. Each kern point represents the farthest distance from the centroid at which a resultant force can act without inducing a stress of opposite sign at the extreme fiber in the opposite direction from the centroid. This means that the stress distribution varies from zero at the top to maximum at the bottom, when the resultant force acts on the lower kern point; or it varies from zero at the bottom to maximum at the top when the resultant force acts on the upper kern point. Referring to Fig. 21.13.1, and using Eq. (21.6.1) with the internal force concept

$$f = \frac{C}{A} - \frac{Ck_b y_t}{I} = 0 \qquad \textbf{(21.13.1a)}$$

Figure 21.13.1 Stress distributions desired when no tension is permitted—*small* girder moment.

and

$$f = \frac{C}{A} - \frac{Ck_t y_b}{I} = 0 \qquad \textbf{(21.13.1b)}$$

Solving Eqs. (21.13.1ab) for k_b and k_t,

$$k_b = \frac{I}{A y_t} = \frac{r^2}{y_t} \qquad \textbf{(21.13.2a)}$$

and

$$k_t = \frac{I}{A y_b} = \frac{r^2}{y_b} \qquad \textbf{(21.13.2b)}$$

in which r is the radius of gyration of the cross section.

The reader may recall from Section 21.6 that as the load applied to a prestressed concrete beam changes, the *position* of the internal force C must also change; that is, the moment arm (see Fig. 21.13.1) measured from the position of the prestressing steel must increase as the load increases. Thus when the moment M_D due to dead load of the girder is acting at the initial condition (immediately after transfer of prestress), the moment arm of the internal couple is $(arm)_D$; and when moment due to dead plus live load is acting, the moment arm is $(arm)_{D+L}$.

In order to achieve the triangular distribution at the initial condition, the prestressing steel must have its centroid below the lower kern point by the amount $(arm)_D = M_D/T_0$ (see Fig. 21.13.1). If the girder moment M_D is small, it may be possible and practical to position the steel at $e = k_b + (arm)_D$ below the centroid of the section. If the moment M_D is large, the distance $[k_b + (arm)_D]$ may extend so far below the centroid of the section as to be too close to the bottom for proper cover or even outside the section. Thus for larger girder weight (large M_D), the optimum condition of triangular stress distribution at the initial condition may be impossible to achieve.

Preliminary Design for Small Girder Moment.   When the girder moment M_D is small (say M_D representing 0.2 or less of the total, $M_D + M_L$), the $(arm)_D$ will be small and C can probably be located at the lower kern point; that is, the steel will be located at $e = k_b + M_D/T_0$ from the centroid of the section. Observing Fig. 21.13.1 again,

$$(arm)_{D+L} - (arm)_D = k_t + k_b$$

$$\frac{M_D + M_L}{T_e} - \frac{M_D}{T_0} = k_t + k_b$$

If M_D is small, M_D/T_0 in the above equation may be assumed to be M_D/T_e, then

$$\frac{M_L}{T_e} \approx k_t + k_b$$

or

$$T_e \approx \frac{M_L}{k_t + k_b} \qquad\qquad \textbf{(21.13.3)}$$

The maximum stress in Fig. 21.13.1 may be obtained from the stress at the centroid (which is T/A_c without bending stress) by the linear relationship. Normally the ratio of the allowable stresses f_{ic} at the initial condition to f_{fc} at the final condition is approximately the same as the ratio of the initial prestress force T_0 to the final prestress force T_e. This means that the economical section should have $y_t \approx y_b$, or $k_t \approx k_b$, that is, be symmetrical. For a rectangular section $k_t = k_b = h/6$, thus $k_t + k_b = h/3$. For the I-shaped section that is more efficient to resist bending, $k_t + k_b > h/3$, say about $h/2$. Thus Eq. (21.13.3) may be approximated as

$$T_e \approx \frac{M_L}{0.5h} \qquad\qquad \textbf{(21.13.4)}$$

If the section is symmetrical, the stress at the centroid will be half of the maximum; thus the required area A_c of the section would be, for the initial condition at transfer,

$$\text{required } A_c = \frac{T_0}{0.50f_{ic}} \qquad\qquad \textbf{(21.13.5)}$$

and for the final condition of dead plus live load after losses,

$$\text{required } A_c = \frac{T_e}{0.50f_{fc}} \qquad\qquad \textbf{(21.13.6)}$$

For the preliminary selection of cross section in case of small girder moment M_D, the following steps may be followed:

1. Select the overall depth h of the section. This is somewhat arbitrary but in the absence of other limitations, the guidelines of Lin and Burns [21.6] may

be followed:
(a) Use 70% of the depth that would be used for nonprestressed reinforced concrete construction.
(b) For *slabs* having light loading, $h \geq 1/55$ of the span L.
(c) For *slabs* having heavy loading, $h \geq L/35$.
(d) On bridges, h ranges between $L/15$ and $L/25$.
(e) As a rule of thumb, h (inches) for *beams* can be approximated by the empirical formula, $1.5\sqrt{M(\text{ft-kips})}$ to $2.0\sqrt{M}$.

2. Compute the approximate T_e from Eq. (21.13.4).
3. Determine the approximate A_c from Eq. (21.13.6).
4. Proportion a symmetrical I-shaped section.
5. With this preliminary section, compute the section properties and locate the desired distance e of the steel centroid from the section centroid (CG),

$$e = k_b + (\text{arm})_D = k_b + \frac{M_D}{T_0} \qquad \textbf{(21.13.7)}$$

where $T_0 = T_e/\eta$ and η is the proportion of initial prestress remaining after losses.

6. If the steel can be located at the desired e, then T_e is more correctly determined,

$$T_e = \frac{M_D + M_L}{(\text{arm})_{D+L}} = \frac{M_D + M_L}{e + k_t} \qquad \textbf{(21.13.8)}$$

Then $T_0 = T_e/\eta$ and a new value of e is established; the iterative process is repeated until the desired accuracy is obtained.

7. Equations (21.13.5) and (21.13.6) are then used to determine the required A_c. When the equations give significantly different requirements, the section may be changed to become somewhat unsymmetrical with respect to the centroidal axis. In such a case the average stress is not half of the maximum. In general, from Fig. 21.13.1,

$$\text{required } A_c = \frac{T_0 h}{f_{ic} y_t} \qquad \textbf{(21.13.9)}$$

$$\text{required } A_c = \frac{T_e h}{f_{fc} y_b} \qquad \textbf{(21.13.10)}$$

The minimum area A_c is obtained when Eqs. (21.13.9) and (21.13.10) give the same result.

Preliminary Design for Large Girder Moment.   When the girder moment M_D exceeds about 0.2 to 0.3 of the total moment $M_D + M_L$, the $(\text{arm})_D$ will be too large to permit the steel distance e to be at $k_b + M_D/T_0$ from the centroid of the section. Thus the initial stress distribution at transfer cannot be triangular but instead will be trapezoidal (see Fig. 21.13.2). The final condition, which can still give a triangular stress

Figure 21.13.2 Stress distributions when no tension is permitted—*large* girder moment.

distribution, will probably govern. Thus

$$T_e = \frac{M_D + M_L}{e + k_t} \tag{21.13.11}$$

When the final condition controls, more of the area A_c should be located at the top where the highest stress occurs. Thus an unsymmetrical section is indicated—for instance, a T-shaped section. As an approximation, Lin and Burns [21.6] suggest $(e + k_t) \approx 0.65h$ for use in Eq. (21.13.11). Generally $e + k_t$ will vary from $0.3h$ to $0.8h$, with the average about $0.65h$. For preliminary design, Eq. (21.13.11) then becomes

$$T_e \approx \frac{M_D + M_L}{0.65h} \tag{21.13.12}$$

The required area A_c would then be

$$\text{required } A_c = \frac{T_e}{\text{avg} f_c} = \frac{M_D + M_L}{0.65h(\text{avg} f_c)} \tag{21.13.13}$$

For the unsymmetrical section, y_b/h in Eq. (21.13.10) will be greater than 0.5, say 0.6. Equation (21.13.13) for design would then be

$$\text{required } A_c = \frac{M_D + M_L}{0.65h(0.6f_{fc})} = \frac{2.6(M_D + M_L)}{hf_{fc}} \tag{21.13.14}$$

For the preliminary selection of cross section in case of large girder moment, the following steps may be followed:

1. When $M_D/(M_D + M_L) > 0.2$ to 0.3, use Eq. (21.13.14) to estimate A_c after establishing the overall depth according to Step 1 for small girder moment.
2. Proportion an unsymmetrical section; a T-section may be a practical choice.

3. With the preliminary section, compute the section properties and establish the distance e from the centroid of the section to the centroid of the pre-stressing steel. For large girder moment, one will find

$$e < k_b + \frac{M_D}{T_0}$$

If e can exceed $k_b + M_D/T_0$, then the procedure for the small girder moment case is to be followed.

4. With the steel located, T_e can be determined using Eq. (21.13.8). From that, $T_0 = T_e/\eta$.

5. The required area A_c based on the final condition is then determined using Eq. (21.13.10).

6. Check the required area based on the initial condition. Referring to Fig. 21.13.2, a trapezoidal stress distribution should occur.

The maximum compressive stress f is, using Eq. (21.6.1) with the internal force concept,

$$f = \frac{C}{A} + \frac{Cey}{I} \tag{21.13.15}$$

Since $C = T_0$, $A = A_c$, $I = A_c r^2$, $e = e'$ (Fig. 21.13.2), and $y = y_b$,

$$f = \frac{T_0}{A_c} + \frac{T_0 e' y_b}{A_c r^2} \le f_{ic}$$

$$= \frac{T_0}{A_c}\left[1 + \frac{e - M_D/T_0}{k_t}\right] \le f_{ic} \tag{21.13.16}$$

The required area A_c based on the initial condition is

$$\text{required } A_c = \frac{T_0}{f_{ic}}\left[1 + \frac{e - M_D/T_0}{k_t}\right] \tag{21.13.17}$$

Again the minimum area A_c will be obtained when Eqs. (21.13.10) and (21.13.17) give the same result.

■ **EXAMPLE 21.13.1**   Design a cross section for a 30-in. deep girder whose girder moment $M_D = 45$ ft-kips. The live load moment to be carried is 300 ft-kips. The initial prestress $f_{si} = 175{,}000$ psi and assume 20% losses. The allowable service load stresses are $f_{it} = 0$, $f_{ic} = 2400$ psi, $f_{ft} = 0$, and $f_{fc} = 2250$ psi. Omit checks of flexural strength, cracking moment, shear strength, and development of reinforcement.

Solution: (a) Preliminary design. The girder moment as a percent of the total moment is

$$\frac{M_D}{M_D + M_L} = \frac{45}{45 + 300} = 0.13 < 0.2$$

Approach as a small girder moment design. Using Eqs. (21.13.4) and (21.13.6),

$$T_e \approx \frac{M_L}{0.5h} = \frac{300(12)}{0.5(30)} = 240 \text{ kips}$$

$$T_0 = \frac{T_e}{\eta} = \frac{240}{0.8} = 300 \text{ kips}$$

$$\text{required } A_c = \frac{T_0}{0.5f_{ic}} = \frac{300}{0.5(2.40)} = 250 \text{ sq in.}$$

Since $f_{fc}/f_{ic} = 2.25/2.4 = 0.94$ is greater than $T_e/T_0 = 0.8$, the equation based on the initial condition is controlling if the section is symmetrical. Though a slightly unsymmetrical section (with the centroid below the middepth in this case) would give the section of minimum area, the procedure is illustrated using a symmetrical one.

Try a 30-in. deep section with flanges 5×17 in. and a 4-in. thick web, $A_c = 250$ sq in. [Fig. 21.13.3(a)].

(b) Determine the section properties.

	AREA, A_c (sq in.)	I (in.^{4})
17×30 rectangle	510	38,250
13×20 sides	-260	$-8,667$
	250	29,583

$$r^2 = \frac{I}{A_c} = 118.3 \text{ sq in.}, \qquad k_t = k_b = \frac{r^2}{y_b} = \frac{r^2}{y_t} = \frac{118.3}{15} = 7.89 \text{ in.}$$

(c) Locate centroid of prestressed reinforcement. Using approximate T_e, find approximate T_0.

$$T_0 \approx \frac{T_e}{\eta} = \frac{249}{0.8} = 300 \text{ kips}$$

$$\text{desired } e = k_b + \frac{M_D}{T_0} = 7.89 + \frac{45(12)}{300} = 9.69 \text{ in.}$$

The available distance is 15 in., less appropriate cover. Thus $e = 9.69$ in. would be acceptable. Recalculate T_e using Eq. (21.13.8),

$$T_e \approx \frac{M_D + M_L}{e + k_t} = \frac{(45 + 300)12}{9.69 + 7.89} = 235 \text{ kips}$$

$$\text{revised } T_0 = \frac{235}{0.8} = 294 \text{ kips}$$

$$\text{revised } e = 7.89 + \frac{45(12)}{294} = 9.73 \text{ in.}$$

This is in close agreement with the previous value of 9.69 in. If the first estimate had been farther off, additional iterations may have been needed.

(d) Check whether the area A_c is adequate. Using Eqs. (21.13.9) and (21.13.10),

$$\text{required } A_c \text{ (initial condition)} = \frac{T_0 h}{f_{ic} y_t} = \frac{294(30)}{2.4(15)} = 245 \text{ sq in.}$$

$$\text{required } A_c \text{ (final condition)} = \frac{T_e h}{f_{fc} y_b} = \frac{235(30)}{2.25(15)} = 209 \text{ sq in.}$$

The section is adequate.

Use 5 × 17 in. flanges with a 4 in. web ($A_c = 250$ sq in.). If made slightly unsymmetrical, the area could be reduced to somewhere between 209 and 245 sq in. A final check of initial and final stresses should be made as done in Section 21.6, but the illustration is omitted here. ■

■ **EXAMPLE 21.13.2** Redesign the cross section of Example 21.13.1 for $M_D = 200$ ft-kips and $M_L = 145$ ft-kips. Note that the total $M_D + M_L = 345$ ft-kips is the same as in Example 12.13.1.

Solution: **(a)** Preliminary design. The girder moment as a percent of the total moment is

$$\frac{M_D}{M_D + M_L} = \frac{200}{345} = 0.58 > 0.2 \text{ to } 0.3$$

Approach as a large girder moment design. Using Eq. (21.13.14),

$$\text{required } A_c = \frac{2.6(M_D + M_L)}{h f_{fc}} = \frac{2.6(345)12}{30(2.25)} = 159 \text{ sq in.}$$

An unsymmetrical shape is to be selected; try flanges 5 × 14 in. and 5 × 8 in. with a 4-in. web having $A_c = 190$ sq in. [Fig. 21.13.3(b)]. The larger the ratio of M_D to

(a) Example 21.13.1   (b) Example 21.13.2

Figure 21.13.3 Sections for design examples.

$(M_D + M_L)$ the larger may be the numerator coefficient (2.6 used above); thus an area somewhat larger than indicated by the formula was used.

(b) Determine section properties. Referring to Fig. 21.13.3(b), first locate the centroid of the area measured from the top.

	AREA, A_c (sq in.)	ARM, y (in.)	A_cy (in.^{3})	I (in.^{4})
5 × 10 top flange projection	50	2.5	125	313
I_0				104
5 × 4 bottom flange projection	20	27.5	550	15,125
I_0				42
4 × 30 web (full depth)	120	15	1800	27,000
I_0				9,000
	190		2475	51,584

$$y_t = \bar{y} = \frac{\Sigma Ay}{\Sigma A} = \frac{2475}{190} = 13.03 \text{ in.}$$

$$I_0 = I - A\bar{y}^2 = 51,584 - 190(13.03)^2 = 19,300 \text{ in}^4$$

$$r^2 = \frac{I_0}{A_c} = 101.7 \text{ sq in.}$$

$$k_t = \frac{r^2}{y_b} = \frac{101.7}{16.97} = 5.99 \text{ in.}$$

$$k_b = \frac{r^2}{y_t} = \frac{101.7}{13.03} = 7.81 \text{ in.}$$

(c) Locate centroid of prestressed reinforcement. Assume that with adequate cover the steel may be centered 4 in. from the bottom of the section. Then

$$e = y_b - 4 = 16.97 - 4 = 12.97 \text{ in.}$$

and using Eq. (21.13.8),

$$T_e = \frac{M_D + M_L}{e + k_t} = \frac{345(12)}{12.97 + 5.99} = 218 \text{ kips}$$

then

$$T_0 = \frac{218}{0.8} = 273 \text{ kips}$$

Check

$$k_b + \frac{M_D}{T_0} = 7.81 + \frac{200(12)}{273} = 7.81 + 8.79 = 16.60 \text{ in.} > e$$

This shows the tendons cannot be located far enough from the centroid of the section to give a triangular stress distribution at the initial condition.

(d) Check whether the area A_c is adequate. Using Eqs. (21.13.10) and (21.13.17),

$$\text{required } A_c \text{ (final condition)} = \frac{T_e h}{f_{fc} y_b} = \frac{218(30)}{2.25(16.97)} = 171 \text{ sq in.}$$

$$\text{required } A_c \text{ (initial condition)} = \frac{T_0}{f_{ic}}\left[1 + \frac{e - M_D/T_0}{k_t}\right]$$

$$= \frac{273}{2.4}\left[1 + \frac{12.97 - 8.79}{5.99}\right] = 193 \text{ sq in.}$$

This is close enough but shows that the initial condition governs in this case, the reason being that area has been shifted from the bottom of the section to the top where it is needed for the final condition. The minimum area section for this case would be slightly more symmetrical than the chosen one.

A final check of stresses (initial and final conditions) should be made (see Section 21.6) to verify the result.

The general line of reasoning presented here may also be used when tension is permitted at initial or final conditions, or both. Lin and Burns [21.6] provide a detailed treatment of design of sections when tension is permitted. ■

21.14 ADDITIONAL TOPICS

Many other topics have been omitted from this introductory treatment of prestressed concrete. Topics such as the practical design approaches, use of I-shaped and non-symmetrical sections, prestressing of continuous members, stresses in end blocks, partial prestressing, deflections, composite construction, and other specific applications, are adequately and extensively treated in recent textbooks devoted entirely to the subject [21.2–21.7].

SELECTED REFERENCES

21.1. Charles C. Zollman. "Reflections on the Beginnings of Prestressed Concrete in America," *PCI Journal*, Part 1, **23**, May/June 1978, 22–48, Part 2, **23**, July/August 1978, 29–63; Part 9, **25**, January/February 1980, 124–145, March/April 1980, 94–117, May/June 1980, 123–152.

21.2. Michael P. Collins and Denis Mitchell. *Prestressed Concrete Structures*. Englewood Cliffs, NJ: Prentice-Hall, 1991, 766 pp.

21.3. James R. Libby. *Modern Prestressed Concrete, Design Principles and Construction Methods* (4th ed.). Princeton, New Jersey: Van Nostrand Reinhold, 1990, 872 pp.

21.4. Edward G. Nawy. *Prestressed Concrete, A Fundamental Approach*. Englewood Cliffs, New Jersey: Prentice-Hall, 1989, 739 pp.

21.5. Antoine E. Naaman. *Prestressed Concrete Analysis and Design—Fundamentals*. New York: McGraw-Hill, 1982, 670 pp.

21.6. T. Y. Lin and Ned H. Burns. *Design of Prestressed Concrete Structures* (3d ed.). New York: Wiley, 1981, 646 pp.

21.7. Arthur H. Nilson. *Design of Prestressed Concrete*. New York: Wiley, 1978, 526 pp.

21.8. Yves Guyon. *Prestressed Concrete*, Vols. 1 and 2. New York: Wiley, 1960.

21.9. Gustave Magnel. *Prestressed Concrete* (2d ed.). London: Concrete Publications, 1950.

21.10. Paul Zia, H, Kent Preston, Norman L. Scott, and Edwin B. Workman. "Estimating Prestress Losses," *Concrete International*, **1**, June 1979, 32–38.

21.11. PCI Committee on Prestress Losses. "Recommendations for Estimating Prestress Loss," *PCI Journal*, **20**, July–August 1975, 43–75.

21.12. J. R. Janney, E. Hognestad, and D. McHenry. "Ultimate Flexural Strength of Prestressed and Conventionally Reinforced Concrete Beams," *ACI Journal, Proceedings*, **52**, January 1956, 601–620.

21.13. M. A. Sozen and N. M. Hawkins. Discussion of "Shear and Diagonal Tension Report," Report of ACI-ASCE Committee 326, *ACI Journal, Proceedings*, **59**, September 1962, 1341–1347.

21.14. Paul Zia and Talat Mostafa. "Development Length of Prestressing Strands," *PCI Journal*, **22**, September/October 1977, 54–65.

21.15. Norman W. Hanson and Paul H. Kaar. "Flexural Bond Tests of Pretensioned Prestressed Beams," *ACI Journal, Proceedings*, **55**, January 1955, 783–802.

PROBLEMS

All problems are to be worked in accordance with the ACI Code using the allowable concrete stresses in ACI-18.4.1 and ACI-18.4.2, unless otherwise indicated.

21.1 The rectangular beam of the accompanying figure contains pretensioned steel with an initial tensile stress of 160 ksi ($f_{pu} = 250$ ksi; low-relaxation strand having $f_{py} = 0.90f_{pu}$). The concrete has $f'_{ci} = f'_c = 5000$ psi ($n = 7$). The beam is on a simple span of 35 ft.

(a) Determine the concrete stresses at top and bottom, and the steel stress, at transfer immediately after the wires are cut at the ends.

(b) Recompute the stresses in (a) immediately after a 20% loss in prestress. What is the maximum service live load that can be superimposed on the beam? Consider only the section of maximum bending moment, and *omit* consideration of flexural strength M_n.

Problem 21.1

21.2 Based on the midspan cross section of the figure for Problem 21.1, investigate whether or not it is possible to increase the live load moment capacity by either or both of the following:

(a) Increase the initial prestress above 160 ksi.

(b) Decrease the eccentricity.

Assume there is a 20% loss of initial prestress. Determine the maximum service live load capacity possible by adjusting the prestress or the eccentricity or both but still not violating the ACI Code limitations. Omit consideration of flexural strength M_n.

21.3 The rectangular section of the accompanying figure has been pretensioned by a force of 300 kips after all losses. If $f'_{ci} = f'_c = 5000$ psi, what uniformly distributed live load may be safely carried on a 40-ft simple span? Omit consideration of flexural strength M_n.

Problem 21.3

21.4 For the live load determined in Problem 21.1, determine the maximum permissible eccentricity of the prestressing tendons at the $\frac{1}{4}$ point of the span to satisfy service load allowable stress limits. Omit consideration of flexural strength M_n.

21.5 For the live load determined in Problem 21.3, determine the maximum permissible eccentricity of the prestressing tendons at the $\frac{1}{8}$ point of the span to satisfy service load allowable stress limits. Omit consideration of flexural strength M_n.

21.6   A straight pretensioned member 35 ft long is 18 in. square in cross section. It is concentrically prestressed with 2.24 sq in. of high tensile strength steel wire. The wires are stressed originally to 145 ksi and are anchored to end bulkheads. Calculate the loss and percent of loss of prestress in the wires due to elastic shortening of the concrete at transfer using both the approximate and the "exact" methods. Use $f'_c = 6000$ psi ($n = 6.5$).

21.7   An 18-in. square concrete member is posttensioned by four cables each with an area of 0.56 sq in. These cables are stressed one after another, each to a stress of 145 ksi. Without taking any account of the eccentricity of the cables, compute the loss and percent of loss of prestress in each cable due to the elastic shortening of the concrete. Compute the average loss of prestress. Assume $n = 6.5$.

21.8   The symmetrical double cantilever beam shown is to be prestressed by a single cable *ABCDE*. The cable consists of 12 wires, each of 0.20 in. diameter, and is to be prestressed simultaneously from both ends of the member. It is desired that the minimum stress in the cable immediately after stressing and before any creep or shrinkage losses take place be 145 ksi. The cable is such that the friction constant $\mu = 0.50$ and the wobble effect $K = 0.0010$. Determine the steel stress at the jack, the percentage of friction losses, and the extension that will be required at each jack. Solve by the following methods: (a) Neglect variation in tension throughout the length, using ACI Formula (18-2), $P_s = P_x(1 + \mu\alpha + KL_x)$; (b) Neglect variation in tension at every point along the length of the curve but consider variation from segment to segment, using ACI Formula (18-2); and (c) Use "exact" expression, ACI Formula (18-1).

Problem 21.8

21.9   Determine the nominal moment strength M_n and the cracking moment M_{cr} for the pretensioned bonded section of the figure for Problem 21.1. The concrete has $f'_c = 5000$ psi and the steel has $f_{pu} = 250$ ksi (stress-relieved strand). Assume the average stress–strain curve of Fig. 21.8.1 is to be used for the steel.

21.10 Assuming no special tests are to be made, determine the number and spacing for #3 U stirrups for the beam of Problem 21.1 if a service load of 0.48 kip/ft is acting. Use the alternate procedure for V_c of ACI-11.4.1, as well as the more exact procedure of AC-11.4.2.

21.11 Assuming no special tests are to be made, determine the number and spacing for #3 U stirrups for the beam of Problem 21.3 if the maximum service load computed in that problem is acting. Use and compare both procedures of ACI-11.4.

21.12 Design of section for *small* dead load moment.

(a) Make a preliminary design (use rectangular flanges and a web) for a section of a prestressed beam to resist a total bending moment of 960 ft-kips assuming that the moment due to the girder weight is 70 ft-kips. The overall depth of the section is to be 42 in. and the effective prestress f_{se} in the steel is 136 ksi. In selecting a section assume the minimum thickness of components (flanges or web) is 5 in.

(b) Make the final design for the preliminary section you selected for part (a) revising as you find necessary, with the objective of obtaining a *minimum* cross-sectional area. The selected cross-sectional area should not exceed 460 sq in.; however, the "best" design will be presumed to be the one having the least cross-sectional area.

(c) Make a check of stresses at initial (transfer) and final conditions. Use the following control stresses:

$$f_{ic} = 2400 \text{ psi} \qquad f_{fc} = 2250 \text{ psi} \qquad f_{si} = 160 \text{ ksi}$$
$$f_{it} = 0 \text{ psi} \qquad f_{ft} = 0 \text{ psi} \qquad f_{se} = 136 \text{ ksi}$$

21.13 Design of section for *large* dead load moment. The data are the same as Problem 21.12, except that the moment due to the girder weight is 650 ft-kips, instead of 70 ft-kips; the minimum thickness for components (flanges or web) is 4 in.; and the cross-sectional area should not exceed 340 sq in.

22

Composite Construction

22.1 INTRODUCTION

Composite construction, as defined herein, is the use of a cast-in-place concrete slab placed upon and interconnected to a prefabricated beam so that the combined beam and slab will act together as a unit. The prefabricated beam may be a rolled or built-up steel shape, a precast reinforced concrete beam, a prestressed concrete beam, a timber beam, or even light-gage steel decking. The interconnection to obtain the single unit action is by combinations of mechanical shear connectors, friction, and shear keys.

In the early 1900s, a type of composite construction was used where a steel I-shaped section was fully encased in concrete placed integrally with the slab. The use of encased beams is still permitted [2.18] but such use is rare. The composite beam and slab construction presently used began to appear in the 1930s. Since about 1940, nearly all usage has been with a slab attached to one flange of a prefabricated beam by means of mechanical connectors. This type of composite construction has been widespread in bridge design since the early 1950s and in buildings since about 1960. Present design methods are the result of extensive research into composite section behavior [22.1–22.15].

Throughout this chapter, emphasis is on the slab composite with precast reinforced concrete and prestressed concrete, as covered by the recommendations of the Joint ASCE-ACI Committee on Composite Construction [22.1], and by the ACI Code. The slab composite with a steel beam is covered by the AISC Specification [2.18], and detailed treatment is provided elsewhere [15.26, 22.14, 22.15, 22.17, 22.18].

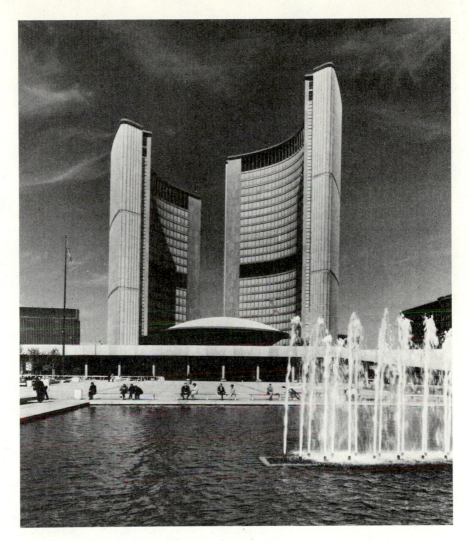

City Hall, Toronto, Ontario, Canada. (Courtesy of Portland Cement Association.)

22.2 COMPOSITE ACTION

Consider a concrete slab atop the flange of a steel or precast concrete beam as shown in Fig. 22.2.1. First, if the system of slab and beam is not acting composite, only friction will provide interaction; thus little of the longitudinal action is carried by the slab. The static system with friction neglected is shown in Fig. 22.2.2(a), wherein the slab and the beam each carry separately a portion of the load. When the non-composite system deforms under vertical load, the lower surface of the slab is in tension and elongates while the upper surface of the beam is in compression and

Figure 22.2.1 Concrete slab and prefabricated beam.

shortens. Thus a discontinuity will occur at the plane of contact. Since friction is neglected, only vertical internal forces act between the slab and beam.

When a system acts compositely [Fig. 22.2.2(b)], no relative slip occurs between the slab and the beam. Horizontal forces (shears) are developed which would shorten the lower surface of the slab and elongate the upper surface of the beam. Thus the discontinuity at the contact surface may be eliminated when sufficiently large horizontal shear resistance can develop. It is noted that the deflection of the composite system will be significantly less than that of the noncomposite system.

In an actual beam-slab system, the degree of composite action may vary over a wide range. For instance, a steel beam used with a concrete slab without mechanical shear connectors will generally develop little composite action, while friction between a concrete beam and a slab may develop nearly the full composite action. Present methods entirely neglect friction (bond) between a steel beam and the concrete slab (unless beam is encased) but consider such bond under certain conditions when the beam is made of precast concrete.

(a) Deflected noncomposite beam   (b) Deflected composite beam

Figure 22.2.2 Comparison of deflected beams with and without composite action.

22.3 ADVANTAGES AND DISADVANTAGES OF COMPOSITE CONSTRUCTION

The significant feature of a composite system is a stiffer and stronger structure than can be obtained from the same beam and slab acting noncompositely. In general, the advantages over noncomposite construction are (1) smaller and shallower beams may be used, (2) longer spans are possible without encountering deflection problems, (3) the toughness (impact capacity or energy absorption) is greatly increased, and (4) the overload capacity is substantially greater.

Some of the factors that tend to weigh against this construction are (1) the cost of the connectors which offsets some of the saving in beam material, (2) the cost of placing the mechanical shear connectors, particularly on nonencased steel beams where they are required without exception, and (3) the erection and construction difficulties encountered when the projecting connectors impede or prevent workmen from walking on the beams.

Most indications are, however, in favor of designing for composite interaction wherever a cast-in-place slab is used.

22.4 EFFECTIVE SLAB WIDTH

A slab acting compositely with a beam behaves the same as in an ordinary reinforced concrete T-section, as discussed in Chapter 9. Referring to Fig. 22.4.1, the variables that control the effective slab width are (1) the ratio of slab thickness to total beam depth, t/h, (2) the ratio of beam span to beam width, L/b_w, (3) the ratio of beam span to beam spacing, L/b_0, (4) the type of loading, and (5) Poisson's ratio. Just as for the T-section of Chapter 9, here also the effective width b_E is to be taken in accordance with ACI-8.10 as the smallest of the following for interior beams: (1) one-fourth of the beam span length, $L/4$, (2) center-to-center spacing of beams, i.e., b_0, and (3) beam web width b_w plus 16 times the slab thickness t, i.e., $b_w + 16t$. Full details for interior, exterior, and isolated beams are given in Section 9.3. When the prefabricated beam is of structural steel [Fig. 22.4.1(b)], instead of precast concrete, the AISC *LRFD Specification* [2.18, LRFD-I3.1] uses for an interior beam the smaller of $L/4$ and the center-to-center spacing b_0, thus eliminating the slab thickness t as a specification variable.

Figure 22.4.1 Variables controlling effective slab width.

22.5 COMPUTATION OF SECTION PROPERTIES

The elastic section properties (area, moment of inertia, section modulus) of the composite section are needed for computation of actual working stresses and deflections under service loads. For such section properties, the transformed section concept (see Section 4.5) is used to convert all areas of the composite section into an equivalent homogeneous member. When a steel beam is used, the concrete slab is converted into equivalent steel by using a slab width equal to b_E/n, where $n = E_s/E_c$, the ratio of the modulus of elasticity of the steel beam to that of the concrete slab. When the prefabricated beam is either reinforced or prestressed concrete, the 28-day compressive strength f'_c is frequently different for the beam and slab; thus E_c is different. In that case the slab may be converted into equivalent beam material by using a slab width of

$$\text{equivalent } b_E = \frac{b_E E_c(\text{slab})}{E_c(\text{beam})} = \frac{b_E n_{\text{beam}}}{n_{\text{slab}}} \tag{22.5.1}$$

■ **EXAMPLE 22.5.1** Compute the elastic section modulus values for the composite steel–concrete section of Fig. 22.5.1. The W21 × 57 steel section has a depth of 21.06 in., flange width of 6.555 in., moment of inertia about its middepth of 1170 in.[4], and an area of 16.7 sq in. The yield strength of steel is 36,000 psi. The slab is of concrete with $f'_c = 3000$ psi ($n = 9$).

Solution: The properties for the composite section are shown in Table 22.5.1. The distance y is measured from the centroid (axis x-x of Fig. 22.5.1) of the steel section.

Table 22.5.1 Properties of Composite Section in Example 22.5.1

	EFFECTIVE AREA, A	ARM y	Ay	Ay^2	I_0
Slab	60.0	13.53	811.8	10,984	180
W section	16.7	0	0	0	1170
Totals	76.7		811.8	10,984	1350

$$I_x = I_0 + Ay^2 = 1350 + 10,984 = 12,334 \text{ in.}^4$$

$$\bar{y} = \frac{811.8}{76.7} = 10.58 \text{ in.}$$

$$I = 12,334 - 76.7(10.58)^2 = 3748 \text{ in.}^4$$

$$y_t = 10.53 + 6.00 - 10.58 = 5.95 \text{ in.}$$

$$y_b = 10.53 + 10.58 = 21.11 \text{ in.}$$

Figure 22.5.1 Steel–concrete composite section for Example 22.5.1.

$$S_t = \frac{I}{y_t} = \frac{3748}{5.95} = 630 \text{ in.}^3$$

$$S_b = \frac{I}{y_b} = \frac{3748}{21.11} = 177 \text{ in.}^3$$

It is to be noted that the neutral axis of the composite section falls slightly in the concrete slab (0.05 in.). Usually the concrete on the tension side is entirely neglected, but here no correction is made since the amount of concrete in tension is negligible. For cases where the tension concrete is to be considered inactive, the neutral axis under service load is located as for ordinary beams (see Chapter 4). ■

■ **EXAMPLE 22.5.2** Compute the elastic section modulus values for the composite precast concrete beam and concrete slab system of Fig. 22.5.2. The slab concrete has $f'_c = 3000$ psi ($n = 9$), while the precast beam has $f'_c = 6000$ psi ($n = 6.5$).

Solution: The cross section may be converted into a homogeneous beam of material having the same E_c as concrete with $f'_c = 6000$ psi. Using Eq. (22.5.1), the

Figure 22.5.2 Precast beam and cast-in-place slab composite section for Example 22.5.2.

transformation factor η for the slab is either

$$\eta = \frac{E_c(\text{slab})}{E_c(\text{beam})} = \frac{57,000\sqrt{3000}}{57,000\sqrt{6000}} = 0.707$$

or

$$\eta = \frac{n_{\text{beam}}}{n_{\text{slab}}} = \frac{6.5}{9} = 0.72$$

In this case, $\eta = 0.72$ is used.

The conversion into equivalent beam concrete ($f'_c = 6000$ psi) is not actually necessary except in cases either where the precast beam is prestressed and therefore its gross section is fully effective or where the concrete area in the web and above the neutral axis is significant enough to be considered. In this example, the computations of Table 22.5.2 have neglected compression in the web.

Table 22.5.2 Properties of Composite Section in Example 22.5.2

	EFFECTIVE AREA, A	ARM y	Ay	Ay²	I_0
Slab	216	2.0	432	864	288
Steel in precast beam	26	24.0	624	14,976	—
Totals	242		1056	15,840	288

$$I_{\text{top}} = I_0 + Ay^2 = 288 + 15,840 = 16,128 \text{ in.}^4$$

$$\bar{y} = y_t = \frac{1056}{242} = 4.36 \text{ in.}$$

$$I_{cr} = 16,128 - 242(4.36)^2 = 11,520 \text{ in.}^4$$

$$y_b = 24.0 - 4.36 = 19.64 \text{ in.}$$

$$S_t = \frac{I_{cr}}{y_t} = \frac{11,520}{4.36} = 2640 \text{ in.}^3$$

$$S_b = \frac{I_{cr}}{y_b} = \frac{11,520}{19.64} = 587 \text{ in.}^3. \quad \blacksquare$$

22.6 WORKING STRESSES WITH AND WITHOUT SHORING

When no temporary falsework or shoring is used to prevent deflection of the precast member while the slab is being placed and cured, the precast member alone must support its own weight plus the weight of the freshly placed slab. The composite section then resists the live load and any additional superimposed dead load, together with long-time effects from creep and shrinkage. On the other hand, if temporary supports are used to carry the precast beam and the slab concrete until such concrete has achieved about 75% of its 28-day compressive strength f'_c, then the composite section will carry the entire load. Thus working stresses under service load may be computed as follows:

Without shoring,

$$f = \frac{M_D}{S_p} + \frac{M_L}{S_c} \qquad\qquad (22.6.1)$$

where M_D is the moment due to dead load produced prior to the time at which the cast-in-place concrete attains 75% of its specified 28-day compressive strength, M_L is the moment due to live load and superimposed dead load, S_p is the effective section modulus of the precast or prefabricated beam, and S_c is the effective section modulus of the composite section.

With shoring,

$$f = \frac{M_D + M_L}{S_c} \qquad\qquad (22.6.2)$$

■ **EXAMPLE 22.6.1**   For the section of Example 22.5.2, compute the stresses due to a service dead load moment of 70 ft-kips and a service live load moment of 105 ft-kips. Reinforcement has $f_y = 60,000$ psi. Consider the case (a) without shoring and (b) with shoring.

Solution: (a) Without shoring. Since the precast beam must carry the dead load prior to curing of the slab, its neutral axis location is required; thus

$$\tfrac{1}{2}(14)x^2 = 26(20 - x)$$

$$x = 6.96 \text{ in.}$$

$$\text{arm} = 20 - \frac{x}{3} = 20 - 2.25 = 17.68 \text{ in.}$$

$$f \text{ (tension, steel)} = \frac{M_D}{A_s(\text{arm})} = \frac{70(12)}{4.0(17.68)} = 11.9 \text{ ksi}$$

$$f \text{ (compression, concrete)} = \frac{11.9}{6.5}\left(\frac{6.96}{13.04}\right) = 0.98 \text{ ksi}$$

The additional stresses due to the live load acting on the composite section are

$$f \text{ (tension, steel)} = \frac{nM_L}{S_b} = \frac{6.5(105)(12)}{587} = 14.0 \text{ ksi}$$

$$f \text{ (compression, } n = 6.5 \text{ concrete)} = \frac{M_L}{S_t} = \frac{105(12)}{2640} = 0.48 \text{ ksi}$$

$$f \text{ (actual compression, } n = 9 \text{ concrete)} = 0.48(6.5)/9 = 0.34 \text{ ksi}$$

Figure 22.6.1 Service load stresses in precast concrete–concrete composite section with and without shoring.

The maximum stress in the reinforcement is $11.9 + 14.0 = 25.9$ ksi, which exceeds the ACI allowable value of 24 ksi. The maximum stress in the concrete at the top fiber of the precast beam is

$$f = 0.98 + \frac{105(12)(0.36)}{11{,}520} = 0.98 + 0.04 = 1.02 \text{ ksi}$$

The stress distribution without shoring is given in Fig. 22.6.1(b).

(b) With shoring.

$$f \text{ (tension, steel)} = \frac{n(M_D + M_L)}{S_b} = \frac{6.5(175)(12)}{587} = 23.3 \text{ ksi}$$

$$f \text{ (compression, } n = 6.5 \text{ concrete)} = \frac{M_D + M_L}{S_t} = \frac{175(12)}{2640} = 0.79 \text{ ksi}$$

$$f \text{ (actual compression, } n = 9 \text{ concrete)} = 0.79(6.5)/9 = 0.57 \text{ ksi}$$

The corresponding stress distribution on a homogeneous section where $n = 6.5$ is given in Fig. 22.6.1(c).

Thus it would appear that the reinforcement is overstressed in the system without shoring, whereas it is within the allowable value when shoring is used. ∎

22.7 STRENGTH OF COMPOSITE SECTIONS

Moment strength M_n of a composite section entirely of reinforced concrete is computed as explained for T-sections in Section 9.4. When the composite section includes a steel shape or a prestressed concrete beam, the basic principles are only slightly modified. Since the moment strength M_n is unrelated to the sequence of loading and the relative amounts of live and dead load, it is independent of whether or not shoring is used.

■ **EXAMPLE 22.7.1** Determine the nominal moment strength M_n of the steel–concrete composite section of Example 22.5.1, using basic statics. The steel section has $f_y = 36,000$ psi and the concrete has $f'_c = 3000$ psi.

Solution: Determine whether the depth a of the rectangular stress block lies above or below the bottom of the slab. If depth a is at the bottom of the slab

$$C_{max} = 0.85f'_c b_E a = 0.85(3)(90)(0.85)(6) = 1170 \text{ kips}$$

$$T_{max} = A_s f_y = 16.7(36) = 601 \text{ kips}$$

It will be observed from Fig. 22.7.1(c) that the value of T_{max} as computed above is an overestimate if the depth a lies at the bottom of the slab. However, by comparing C_{max} to T_{max} it is also obvious that the depth a is less than 6 in. Thus, as in an ordinary T-section where the depth a is less than the flange thickness,

$$C = 0.85f'_c ab = 0.85(3)(a)(90) = 230a$$

$$T = A_s f_y = 16.7(36) = 601 \text{ kips}$$

$$a = \frac{601}{230} = 2.62 \text{ in.}$$

$$x = \frac{a}{\beta_1} = \frac{2.62}{0.85} = 3.08 \text{ in.}$$

For this case, the strain at the top of the steel section is

$$\epsilon'_s = \frac{0.003}{3.08}(6.0 - 3.08) \approx 0.0028 > \left[\epsilon_y = \frac{36}{29,000} = 0.00124\right]$$

which means the strain on the entire steel section is at least yield strain ϵ_y. Thus

$$\text{arm} = \frac{d}{2} + t - \frac{a}{2} = \frac{21.06}{2} + 6.0 - \frac{2.62}{2} = 15.22 \text{ in.}$$

$$M_n = T(\text{arm}) = 601(15.22)\tfrac{1}{12} = 762 \text{ ft-kips}$$

Thus the nominal moment strength M_n is 762 ft-kips whether or not shoring is used. ■

Figure 22.7.1 Section for Example 22.7.1.

Figure 22.7.2 Section for Example 22.7.2.

■ **EXAMPLE 22.7.2**   Determine the nominal moment strength M_n of the concrete composite section of Fig. 22.7.2. The slab has $f'_c = 3000$ psi while the precast beam has $f'_c = 6000$ psi, and the reinforcement has $f_y = 60,000$ psi.

Solution:

$$C_{\max} = 0.85f'_c b_E t = 0.85(3)(75)(4) = 765 \text{ kips}$$

$$T_{\max} = A_s f_y = 4(60) = 240 \text{ kips}$$

Since $C_{\max} > T_{\max}$, $a < t$. Thus

$$a = \frac{240}{0.85(3)(75)} = 1.25 \text{ in.}$$

$$M_n = 240[24.0 - 0.5(1.25)]\tfrac{1}{12} = 468 \text{ ft-kips}$$

This most typical situation with $a < t$ is treated exactly as in Chapter 9. If $a > t$, some contribution from the 6000 psi concrete would be included in the compressive force. ■

22.8 SHEAR CONNECTION

Working Stress Concept.   As discussed in Section 22.2, in order for the slab to act together with the prefabricated beam, the horizontal shear forces must be developed between the slab and beam. Consider the uniformly loaded beam of Fig. 22.8.1 along with the shear stress distribution across a typical section of the beam. It is the shear stress v_1 that must be developed by the connection between slab and beam. Under the service load (working stress) condition, it is seen from Fig. 22.8.1 that the shear stress v_1 varies from zero at midspan to a maximum at the support. Consider an elemental slice of the beam, as in Fig. 22.8.2. The shear force per unit distance along the span is $dC/dz = v_1 b_E = VQ/I$. Thus if a given connector has an allowable capacity of q kips, the maximum spacing s to provide the required capacity is

$$s \leq \frac{q}{VQ/I} \qquad\qquad\qquad \textbf{(22.8.1)}$$

Figure 22.8.1 Shear stress distribution across a steel–concrete composite section.

where V is the total shear at the section, Q is the statical moment of the effective slab area above the neutral axis with respect to the neutral axis of the composite section, and I is the moment of inertia of the transformed composite section neglecting the area of concrete in tension.

Strength Concept. If one uses the strength concept, the shear connectors share equally in carrying the total compressive force developed in the concrete slab as the nominal moment strength M_n is approached. This means, referring to Fig. 22.8.1(a), that shear connection is required to transfer the compressive force developed at mid-span to the prefabricated beam in the distance $L/2$, since no compressive force can exist in the concrete slab at the end of the span where zero moment exists. The compressive force at strength M_n to be accommodated could not exceed that which the concrete can carry,

$$C_{\max} = 0.85 f_c' b_E t \quad \text{(upper bound)} \qquad (22.8.2)$$

or, if the tensile force below the bottom of the slab at strength M_n is less than $C_{\max}$,

$$T_{\max} = A_s f_y \qquad (22.8.3)$$

Figure 22.8.2 Force required of shear connectors—working stress method.

(a) Straight stud   (b) L-connector   (c) Flexible   (d) Spiral
                                                          channel

Figure 22.8.3 Shear connectors for concrete slab steel beam construction.

Thus, for individual connectors each having a strength q_{ult} when failure is imminent, the total number of connectors N required between the points of maximum and zero bending moment is

$$N = \frac{C_{max}}{q_{ult}} \quad \text{or} \quad \frac{T_{max}}{q_{ult}} \tag{22.8.4}$$

whichever is smaller.

Connection of Slab to Beam.   It can also be noted that the connection and the beam must provide the same nominal moment strength M_n. Under working loads, however, the beam resists dead load and live load, but the connection may need to resist only the load coming on after the slab has acquired its strength (primarily the live load). If the connection is designed to carry only the live load, a higher factor of safety should be used. Approximately the same result is achieved if the connection is designed to carry dead load as well as live load with the usual safety provisions.

Several types of connectors are as follows:

1. Stud shear connector, straight [Fig. 22.8.3(a)] and L-shaped [Fig. 22.8.3(b)], welded to the steel beam in concrete–steel construction.
2. Flexible channel shear connector [Fig. 22.8.3(c)], welded to the steel beam in concrete–steel construction.
3. Spiral shear connector [Fig. 22.8.3(d)], welded to the steel beam in concrete–steel construction.
4. Reinforcing bar stirrups from the precast beam, fully anchored into the slab.
5. Friction, or bond, in combination with vertical ties, for slab on precast reinforced or prestressed concrete beam. This type of shear connection is adequate for most of these cases. While friction, or bond, alone may be sufficient, at least a minimum amount of vertical ties must be used (ACI-17.6) unless the contact surfaces are subject to a nominal stress $V_u/(\phi b_v d)$ not greater than 80 psi, *and* "clean, free of laitance, and intentionally roughened." Minimum ties are to provide a sort of clamping action to prevent buckling of the concrete slab, which would suddenly break the bond.
6. Shear keys, for all cases of concrete–to–concrete composite action where friction, or bond, is inadequate. Since these keys are acting very nearly in pure

shear rather than in diagonal tension, determination of the strength of such a key according to the general principles in Chapter 5 will be unduly conservative. The shear-friction concept of ACI-11.7 should be used for the design of the keys, as discussed in Section 5.16.

22.9 DEFLECTIONS

Since, in general, one of the advantages of using composite construction is to obtain shallower members, the calculation of deflection is important. For deflections arising from live load, or dead load in shored construction, properties of the composite section may be used in accordance with the principles of Chapter 14. For the time-dependent effect, the work of Branson [22.13, 22.16–22.18] may be used. Specifically, for a slab on precast reinforced or prestressed concrete the provisions of ACI-9.5.5 should be followed. For a slab on a steel beam, the AISC [2.18] suggests using the ordinary value of $n = E_s/E_c$ (short-time loading) when computing either immediate live load or sustained load deflections, while the Joint ACI-ASCE Committee [22.1] recommended using $2n$ for sustained load deflections.

22.10 SLAB ON PRECAST REINFORCED CONCRETE BEAM—STRENGTH DESIGN

Essentially, the principles of Chapter 3 are used, with full realization that no distinction is made between shored and unshored members. In addition, ACI-17.2 simply states general requirements, including ACI-17.2.7 which requires composite members to meet the deflection control requirements of ACI-9.5.5. This means that whenever excessive deflection may cause damage, deflections must be computed.

To assure the composite action, the horizontal shear must be transferred across the contact surface. For ordinary design, ACI-17.5.2 uses a horizontal shear nominal strength V_{nh} computed as

$$V_{nh} = v_{nh}b_v d_c \qquad (22.10.1)$$

where

v_{nh} = nominal unit stress capable of being transmitted on contact surface (values as obtained from ACI-17.5.2 are described below)

b_v = width of cross section being investigated for horizontal shear

d_c = distance from extreme compression fiber to centroid of tension reinforcement, for the entire composite section

The maximum values of v_{nh} from ACI-17.5.2 are as follows:

1. When the contact surface is intentionally roughened, clean and free of laitance, with no vertical ties used, max v_{nh} = 80 psi.
2. When the contact surface is clean but *not* intentionally roughened, and when vertical ties having a minimum area of $A_v = 50b_w s/f_y$ are spaced not more than 4 times the slab thickness (i.e., least dimension of the supported element), nor 24 in., max v_{nh} = 80 psi.

3. When the contact surface is intentionally roughened,* clean, free of laitance, and minimum ties as in (2) are used, max $v_{nh} = 350$ psi.

4. When the nominal stress v_{nh} exceeding 350 psi is desired, design for horizontal shear must be made using the shear-friction provisions of ACI-11.7, as explained in Section 5.16.

Thus the design requirement of ACI-17.5.2 may be stated as

$$\phi V_{nh} \geq V_u \tag{22.10.2}$$

where

V_u = total shear force at section due to factored loads

ϕ = 0.85 for shear (ACI-9.3)

V_{nh} = nominal strength computed according to Eq. (22.10.1)

As an alternative to the above procedure of ACI-17.5.2, ACI-17.5.3 provides that the actual compressive or tensile force in any segment may be computed, and then provision is made to transfer that force as horizontal shear to the supporting element. Since the term "any segment" could mean anything from an elemental segment of span to the full distance between the maximum moment point and a point of contraflexure, the ACI statement would seem to suggest any procedure from Eq. (22.8.1), relating to an elemental segment, to Eq. (22.8.4), for a longer finite length of span. In other words, the total horizontal shear to be transferred must be accommodated by some rational process.

For the lower levels of horizontal shear stress covered under ACI-17.5.2.1 through 17.5.2.3 (categories 1 through 3 above), the shear variation is already taken into account when the stress $v_u = V_u/\phi b_v d_c$ is computed. Recent (1991) proposals being considered for the ACI Code have indicated that the slip (≈ 0.02 in.) at a concrete–to–concrete interface when maximum horizontal shear resistance is achieved, is much less than the corresponding slip (0.2 to 0.4 in.) at a concrete–to–steel interface when headed steel stud shear connectors are used. Thus, significant redistribution of horizontal shear cannot occur along a concrete–to–concrete interface. This means that ties should be distributed to correspond with the shear along the member; that is, use Eq. (22.8.1). Because of the longer distribution of horizontal shear in the concrete–to–steel interface, the strength approach using Eq. (22.8.4) is appropriate for that situation. However, when ties are designed for strength as horizontal shear connectors under ACI-17.5.2.4 and ACI-17.5.3, the authors recommend they be distributed along the member in accordance with Eq. (22.8.1). The strength q per tie (or set of ties) can be determined as for shear-friction from ACI Formula (11-26).

For precast prestressed concrete beams, the minimum tie area may be taken as ACI Formula (11-15), Eq. (21.11.2), if the effective prestress force is at least equal to 40% of the tensile strength of the flexural reinforcement.

— — — —

* ACI–17.5.2.3 indicates that the interface must have a full amplitude of $\frac{1}{4}$ in. of roughness to satisfy the requirement of intentional roughness (based on Reference 22.4) to use $v_{nh} = 350$ psi.

■ **EXAMPLE 22.10.1**   Design a composite slab on a simply supported precast reinforced concrete beam span of 24 ft. The spacing of beams is 8 ft center-to-center.
The cast-in-place slab is 4 in. thick, and the live load to be carried is 200 psf. Use
f'_c (slab) = 3000 psi, f'_c (precast beam) = 4000 psi, f_y = 40,000 psi, and the strength
method of the ACI Code.

Solution: (a) Loads and solution procedure. As a preliminary to the actual solution, it is to be noted that the use of precast members speeds construction and the
use of composite action reduces the required depth of the beam.

To design a composite beam without using temporary shoring, the precast beam
is first designed to carry its own weight plus the weight of freshly placed concrete.
Of course, the noncomposite system must also carry temporary load due to work-
ers, equipment, runways, and impact, plus the dead weight of forms. The loads are:

Loads on the noncomposite precast beam,

$$
\begin{aligned}
\text{4-in. slab, } (4/12)(0.15)(8) &= 0.4 \text{ kip/ft} \\
\text{estimated beam weight} &= 0.2 \text{ kip/ft} \\
\hline
\text{dead load} &= 0.6 \text{ kip/ft} \\
\text{temporary load, } 0.050(8) &= 0.4 \text{ kip/ft}
\end{aligned}
$$

Load on the composite section,

$$\text{live load, } 0.200(8) = 1.6 \text{ kips/ft}$$

Temporary live and dead construction loads frequently are not included in the design of the precast noncomposite section, but rather the overload that may occur is
accepted as a short-duration reduction in the factor of safety.

When deflection is to be investigated, service load moments are needed; so they
could be computed first, and then overload factors are applied. For permanent loads
on the noncomposite section,

$$M_D = \tfrac{1}{8}(0.6)(24)^2 = 43.2 \text{ ft-kips}$$

For the live load on the composite section,

$$M_L = \tfrac{1}{8}(1.6)(24)^2 = 115 \text{ ft-kips}$$

Several factors may control the size of the precast beam: (1) dead load moment
requirement for a rectangular precast beam, (2) total load moment requirement acting
on the T-shaped composite section, and (3) total load shear on the T-shaped section.
Items (2) and (3) are treated similarly to the procedure discussed in Sections 9.6 and
10.2 for T-sections.

(b) Moment on precast noncomposite section. Assume a desirable reinforcement
ratio ρ about one-half the maximum permitted, say 0.018 (see Table 3.6.1 for ACI

Code maximum). Then the desired R_n is

$$R_n = \rho f_y(1 - \tfrac{1}{2}\rho m)$$
$$= 0.018(40,000)[1 - 0.5(0.018)(11.8)] = 644 \text{ psi}$$

$$m = \frac{f_y}{0.85f_c'} = \frac{40,000}{0.85(4000)} = 11.8$$

$$M_u = 1.4(43.2) = 60.5 \text{ ft-kips}$$

$$\text{required } bd^2 = \frac{M_u}{\phi R_n} = \frac{60.5(12,000)}{0.90(644)} = 1250 \text{ in.}^3$$

$$\min h = \frac{L}{16}(0.8) = \frac{24(12)}{16}(0.8) = 14.4 \text{ in.}$$

The precast beam must be at least 14.4 in. deep (ACI-Table 9.5a) unless deflection is computed even if the member is not supporting or attached to construction likely to be damaged by excessive deflection. If $h = 15$ in., $d \approx 12.5$ in.; then

$$\text{required } b = \frac{1250}{(12.5)^2} = 8 \text{ in.}$$

$$\text{required } A_s \approx 0.018(8)(12.5) = 1.8 \text{ sq in.}$$

There is no problem to fit the steel required for the beam *before* the live load is applied. However, the greater reinforcement requirement for the *total* load will probably make it desirable to use a width exceeding 8 in.

(c) Determine reinforcement for composite section.

$$M_u = 1.4M_D + 1.7M_L = 1.4(43.2) + 1.7(115) = 256 \text{ ft-kips}$$

$$\text{required } M_n = \frac{M_u}{\phi} = \frac{256}{0.90} = 284 \text{ ft-kips}$$

$$\text{effective width } b_E = \tfrac{1}{4}(24)(12) \quad \text{or } 12 + 16(4) \quad \text{or } 8(12)$$
$$= 72 \text{ in.} \qquad \text{or } 76 \text{ in.} \qquad \text{or } 96 \text{ in.}$$
$$= 72 \text{ in.}$$

Assuming the neutral axis to be in the slab when nominal strength M_n is reached,

$$C = 0.85f_c'b_Ea = 0.85(3)(72)a = 184a$$

$$T = f_yA_s = 40A_s$$

$$C = T$$

$$A_s = \frac{184a}{40} = 4.60a$$

If a is typically somewhat less than $t/2$, say 1 to 2 in., two layers of steel will be required even if the beam width is increased. Try $b = 12$ in. and $h = 15$ in. for precast beam (Fig. 22.10.1). Then $d_c = 15 + 4 - (\approx 3.5) = 15.5$ in.

Figure 22.10.1 Section for Example 22.10.1.

$$\text{required } M_n = C(d_c - 0.5a)$$

$$284(12) = 184a(15.5 - 0.5a)$$

$$a^2 - 31a = -37.04$$

$$a = 1.24 \text{ in.} < 4.0 \text{ in.} \qquad \textbf{OK}$$

$$\text{required } A_s = 4.60a = 4.60(1.24) = 5.70 \text{ sq in.}$$

Try 6-#9 bars in two layers ($A_s = 6.00$ sq in.).
A check by basic statics may be made as follows:

$$T = 40(6.0) = 240 \text{ kips}$$

$$a = \frac{240}{184} = 1.30 \text{ in.}$$

$$M_n = 240(15.5 - 0.65)\tfrac{1}{12} = 297 \text{ ft-kips} > 284 \text{ ft-kips} \qquad \textbf{OK}$$

(d) Investigate construction loads.

$$M_{\text{temp}} = \tfrac{1}{8}(0.6 + 0.4)(24)^2 = 72 \text{ ft-kips} \quad \text{(service load)}$$

For the precast section,

$$C = 0.85f'_c b_w a = 0.85(4)(12)a = 40.8a$$

$$T = f_y A_s = 40(6.0) = 240 \text{ kips}$$

$$a = \frac{240}{40.8} = 5.88 \text{ in.}$$

$$M_n = 240(11.5 - 2.94)\tfrac{1}{12} = 171 \text{ ft-kips}$$

Though no special safety requirements are given for temporary loads, it may be reasonable to use the dead load factor 1.4; thus

$$M_u = 1.4(72) = 101 \text{ ft-kips}$$

$$\phi M_n = 0.90(171) = 154 \text{ ft-kips} > (M_u = 101 \text{ ft-kips}) \qquad \textbf{OK}$$

Use 6-#9 in two layers. If deflection is important, it should be checked.

(e) Investigate shear transfer.

$$V_u = \tfrac{1}{2}\big[1.4(0.6) + 1.7(1.6)\big]24 = 42.7 \text{ kips}$$

$$v_{nh} = \frac{V_{nh}}{b_v d_c} = \frac{V_u}{\phi b_v d_c} = \frac{42,700}{0.85(12)(15.5)} = 270 \text{ psi}$$

Since $v_{nh} > 80$ psi, friction alone may not be relied on to transmit horizontal shear. However, since $v_{nh} < 350$ psi, the surface is to be intentionally roughened and minimum ties will be used. For #3 ties,

$$\max s = \frac{A_v f_y}{50 b_w} = \frac{0.22(40,000)}{50(12)} = 14.6 \text{ in.} \qquad \textbf{(Controls)}$$

or

$$\max s = 4t = 4(4) = 16 \text{ in.}$$

but not greater than 24 in. in any case.

Use #3 ties @ 12 in. In general, these should be stirrups which project out at the top of the precast beam. This projecting portion of the stirrup is then cast into the slab as shown in Fig. 22.10.1. Even when stirrups in excess of the minimum percentage are required for shear, all of these should be extended into the slab to serve as ties. ∎

22.11 SLAB ON STEEL BEAM

Design requirements for the slab on steel beam are covered by AISC *Load and Resistance Factor Design Specification* [2.18]. Nominal strength M_n is determined recognizing there is no distinction between shored and unshored construction. A satisfactory section must have adequate composite section strength to carry the total factored load. When the system is unshored, the steel section alone must have adequate strength to carry its self load, the slab dead load, and any construction loads.

To ensure the composite action, mechanical connectors are required for all cases except where beams are totally encased. Connectors are designed for strength, applying Eqs. (22.8.2) through (22.8.4), and may be spaced uniformly between sections of maximum and zero moment.

Examples of design are not presented because steel section properties are necessary, and this subject is treated in detail by Salmon and Johnson [15.26].

22.12 COMPOSITE COLUMNS

The composite column was first discussed in Chapter 13 where the two major types of such columns are shown in Fig. 13.2.1. The general approach to the short column is the same as for regular reinforced concrete columns described in Chapter 13. The specific ACI Code rules for both short and long composite columns are in ACI-10.14. The work of Furlong [22.19–22.22] provides the basis for the ACI Code design of steel-encased concrete columns, with supporting data from the work of Roderick and Rogers [22.23] and Knowles and Park [22.24].

Basically, every composite column, whether concrete-encased steel sections or steel-encased concrete, must be specifically designed to have shear transfer between concrete and steel. The so-called combination column, where concrete merely fills a pipe column, does not fit this category.

ACI-10.14.3 requires that any axial load strength assigned to be carried by concrete must be transferred to concrete by "members or brackets in direct bearing" on the concrete. Connectors such as lugs, plates, or reinforcing bars welded to the structural shape before the concrete is cast are suitable to transfer by direct bearing the force in the concrete. If the force assigned to the concrete is

$$C_c = 0.85f'_c A_c \tag{22.12.1}$$

and the connectors each have a capacity q_{ult}, the total number N of connectors required is

$$N = \frac{C_c}{q_{\text{ult}}} \tag{22.12.2}$$

Another modification for composite column design is a modified expression for radius of gyration given by ACI-10.14.5

$$r = \sqrt{\frac{0.2E_c I_g + E_s I_t}{0.2E_c A_g + E_s A_t}} \tag{22.12.3}$$

where I_t and A_t represent the moment of inertia and area, respectively, of structural steel or tubing in a composite section.

Further, in computing the moment magnification factor (see Chapter 15) the effective EI may not exceed

$$\max EI = \frac{0.2E_c I_g}{1 + \beta_d} + E_s I_t \tag{22.12.4}$$

Though Eqs. (22.12.3) and (22.12.4) are mentioned above for completeness, their use relates directly to length effects on columns dealt with in Chapter 15. Symbols not defined herein are standard ACI symbols and are used throughout Chapter 15.

SELECTED REFERENCES

22.1.   ACI-ASCE Committee 333. "Tentative Recommendations for Design of Composite Beams and Girders for Buildings," *ACI Journal, Proceedings,* **57**, December 1960, 609–628; also *Journal of the Structural Division,* ASCE, **86**, ST12 (December 1960), 73–92.

22.2.   Ivan M. Viest. "Review of Research on Composite Steel-Concrete Beams," *Journal of the Structural Division,* ASCE, **86**, ST6 (June 1960), 1–21.

22.3.   B. Grossfield and C. Birnstiel. "Tests of T-Beams with Precast Webs and Cast-in-Place Flanges," *ACI Journal, Proceedings*, **59**, June 1962, 843–851.

22.4.   J. C. Saemann and George W. Washa. "Horizontal Shear Connections Between Precast Beams and Cast-in-Place Slabs," *ACI Journal, Proceedings*, **61**, November 1964, 1383–1409.

22.5.   Paul H. Kaar, Ladislav B. Kriz, and Eivind Hognestad. "Precast-Prestressed Concrete Bridges: 1. Pilot Tests of Continuous Girders," *Journal of the PCA Research and Development Laboratories*, **2**, 2 (May 1960), 21–37. (Also PCA Development Department, *Bulletin D34.*)

22.6.   Norman W. Hanson. "Precast-Prestressed Concrete Bridges: 2. Horizontal Shear Connections," *Journal of the PCA Research and Development Laboratories*, **2**, 2 (May 1960), 38–58. (Also PCA Development Department, *Bulletin D35.*)

22.7.   Peter R. Barnard. "A Series of Tests on Simply Supported Composite Beams," *ACI Journal, Proceedings*, **62**, April 1965, 443–456. Disc., 1629–1631.

22.8.   William R. Spillers. "On Composite Beams," *Journal of the Structural Division*, ASCE, **91**, ST4 (August 1965), 17–21.

22.9.   John C. Badoux and C. L. Hulsbos. "Horizontal Shear Connection in Composite Concrete Beams Under Repeated Loads," *ACI Journal, Proceedings*, **64**, December 1967, 811–819.

22.10.   Alan H. Mattock and Sterling B. Johnston. "Behavior Under Load of Composite Box-Girder Bridges," *Journal of the Structural Division*, ASCE, **94**, ST10 (October 1968), 2351–2370.

22.11.   R. Paul Johnson. "Research on Steel-Concrete Composite Beams," *Journal of the Structural Division*, ASCE, **96**, ST3 (March 1970), 445–459.

22.12.   R. Paul Johnson. "Longitudinal Shear Strength of Composite Beams," *ACI Journal, Proceedings*, **67**, June 1970, 464–466.

22.13.   Dan E. Branson. "Time-Dependent Effects in Composite Concrete Beams," *ACI Journal, Proceedings*, **61**, February 1964, 213–230. Disc., 1207–1209.

22.14.   Charles G. Salmon and James M. Fisher. "Composite Steel-Concrete Construction," *Handbook of Composite Construction Engineering* (ed. by Gajanan Sabnis). New York: Van Nostrand Reinhold, 1979 (Chapter 2).

22.15.   John P. Cook. *Composite Construction Methods*. New York: Wiley, 1977.

22.16.   Dan E. Branson. *Deformation of Concrete Structures*. New York: McGraw-Hill, Inc., 1977 (pp. 226–249).

22.17.   Dan E. Branson. "Reinforced Concrete Composite Flexural Members," *Handbook of Composite Construction Engineering* (ed. by Gajanan Sabnis). New York: Van Nostrand Reinhold, 1979 (Chapter 4).

22.18.   Dan E. Branson. "Prestressed Concrete Composite Flexural Members," *Handbook of Composite Construction Engineering* (ed. by Gajanan Sabnis). New York: Van Nostrand Reinhold, 1979 (Chapter 5).

22.19. Richard W. Furlong. "Strength of Steel-Encased Concrete Beam-Columns," *Journal of the Structural Division*, ASCE, **93**, ST10 (October 1967), 113–124.

22.20. Richard W. Furlong. "Design of Steel-Encased Concrete Beam-Columns," *Journal of the Structural Division*, ASCE, **94**, ST1 (January 1968), 267–281.

22.21. Richard W. Furlong. "Design Tables for Composite Columns," Preprint 1531, ASCE Annual Meeting, October 18–22, 1971, St. Louis, Mo.

22.22. Richard W. Furlong. "Steel-Concrete Composite Columns," *Handbook of Composite Construction Engineering* (ed. by Gajanan Sabnis). New York: Van Nostrand Reinhold, 1979 (Chapter 6).

22.23. J. W. Roderick and D. F. Rogers. "Load-Carrying Capacity of Simple Composite Columns," *Journal of the Structural Division*, ASCE, **95**, ST2 (February 1969), 209–228.

22.24. Robert B. Knowles and Robert Park. "Axial Load Design for Concrete Filled Steel Tubes," *Journal of the Structural Division*, ASCE, **96**, ST10 (October 1970), 2125–2153.

PROBLEMS

All problems are to be worked in accordance with the strength method of the ACI Code unless otherwise indicated.

22.1 Determine the nominal moment strength M_n for live load plus superimposed dead load applied to the composite section for the section of the accompanying figure. The concrete has $f'_c = 4500$ psi ($n = 7.5$), and the steel is A36 with $f_y = 36,000$ psi.

W 14 × 22
$I = 199$ in.^{4}
13.74"
0.335"   4 $\frac{1}{2}$"
Beam span = 30 ft
Beam spacing = 10 ft
0.23"
5.0"

Problem 22.1

22.2 Determine the service live load moment capacity available for superimposed load on the composite slab on the precast beam of the accompanying figure. Use f'_c (slab) = 3000 psi ($n = 9$), f'_c (beam) = 6000 psi ($n = 6.5$), and $f_y = 40,000$ psi.

Problem 22.2

22.3 For the composite beam of the accompanying figure, determine the service live load moment capacity. Use f'_c (slab) = 3000 psi ($n = 9$), f'_c (beam) = 4500 psi ($n = 7.5$), and $f_y = 50,000$ psi.

Problem 22.3

22.4 For the beam of Problem 22.2, determine what is necessary to provide proper shear transfer for maximum live load.

22.5 For the beam of Problem 22.3, determine what is necessary to provide proper shear transfer for maximum live load.

22.6 Determine the depth and reinforcement required for a 12 in. wide by 20 in. deep precast reinforced concrete beam on a span of 36 ft. The supported slab is 4 in. thick, and the beam spacing is 8 ft. The live load is 125 psf. Use f'_c (slab) = 3500 psi ($n = 8.5$), f'_c (beam) = 5000 psi ($n = 7$), and $f_y = 60,000$ psi. Assume that the construction is to be made without shoring.

22.7 Investigate the economics of the composite concrete–concrete beam for the data of Problem 22.6, except use a different beam spacing (4 ft, 6 ft, 7 ft, 9 ft, 10 ft, as assigned by instructor). As a consequence of different beam spacing, the required slab thickness may change. Consider slab concrete at $100.00/cu yd (including forms) and steel at $0.30/lb.

22.8 Repeat Example 22.10.1, except use a simply supported span of 22 ft, beam spacing of 7 ft 6 in., with a 5 in. thick cast-in-place slab. The live load is 250 psf, and $f_y = 60,000$ psi. Other data are the same as the example.

Index

LIST OF SYMBOLS

R_n = coefficient of resistance for strength design, $M_n/(bd^2)$

s = spacing of shear reinforcement measured along the axis of the member

     = pitch of spiral reinforcement in a column measured center-to-center of spiral bar (Fig. 13.9.1)

s_2 = vertical spacing of longitudinal shear reinforcement (Section 5.14)

S_{ii}, S_{jj} = stiffness at ends i and j, respectively, of a flexural member ij; S_{ii} is moment per unit angle at end i; S_{jj} is moment per unit angle at end j (Section 16.20 and Chapter 17)

S'_{ii} = $(S_{ii} + F_{tj}T_1)/T_2$ (Section 17.4); $S_{ii}/(1 + S_{ii}/K_{ti})$ (Section 17.5)

S_{ij}, S_{ji} = cross stiffness, moment at end i caused by unit angle at end j, and moment at end j caused by unit angle at end i, respectively, of member ij (Section 16.20 and Chapter 17)

S'_{ij} = S_{ij}/T_2 (Section 17.4)

S'_{jj} = $(S_{jj} + F_{ti}T_1)/T_2$ (Section 17.4)

S_p = effective section modulus of precast or prefabricated beam

t = thickness of slab (Chapters 9, 16, and 17)

     = time; days after loading (Section 14.6)

t_a = age at loading, days (Section 14.6)

T = tensile force; torsional force (Chapter 19)

T_0 = force in concrete due to prestress before losses have occurred (Chapter 21)

T_1 = tensile force in steel to balance C_c (Section 3.12)

     = $S_{ii}S_{jj} - (S_{ij})^2$ (Section 17.4)

T_2 = tensile force in steel to balance C_s (Section 3.12)

     = $1 + F_{ti}S_{ii} + F_{tj}S_{jj} + F_{ti}F_{tj}T_1$ (Section 17.4); $1 + F_{ti}S_{ii}$ (Section 17.5)

T_b = tensile force in steel for the balanced strain condition

T_c = nominal torsional strength attributable to concrete

T_e = effect prestress force after all prestress losses (Chapter 21)

T_f = final tensile force in the tendons after elastic shortening has occurred (Chapter 21)

T_n, T_{n0} = nominal torsional strength; nominal torsional strength when member is subject to torsion alone (Chapter 19)

T_s = nominal torsional strength attributable to steel reinforcement, $\alpha_t x_1 y_1 A_t f_y/s$ (Chapter 19)

T_u = factored torsional moment

     = force in prestressed reinforcement when nominal strength M_n is reached, $A_{ps}f_{ps}$ (Chapter 21)

U = ACI Code factors for safety related to overload; factored load, factored moment, factored shear, factored axial force (Section 2.6)

v_c = nominal shear stress attributable to concrete, $V_c/(b_w d)$

v_n = nominal shear stress, $V_u/(\phi b_w d)$

v_{nh} = nominal ultimate unit stress capable of being transmitted on contact surface (values from ACI-17.5.4) (Chapter 22)

v_s = nominal shear stress attributable to the reinforcement, $V_s/(b_w d)$

v_t = torsional shear stress (Chapter 18)

v_{tc} = nominal torsional stress, $T_c/(\Sigma x^2 y)$ (Chapter 19)

v_{tn} = nominal torsional stress, $T_u/(\phi \frac{1}{3}\Sigma x^2 y)$ (Chapter 19)

v_{ts} = nominal torsional stress, $T_s/(\Sigma x^2 y)$ (Chapter 19)

V_c = nominal shear strength attributable to concrete

V_{ci} = nominal flexure-shear cracking strength (Chapter 21)

V_{cw} = nominal web-shear cracking strength (Chapter 21)

$V_n; V_{n0}$ = nominal shear strength, $V_c + V_s$; nominal shear strength in the absence of torsion (Chapter 19)

V_s = nominal shear strength attributable to shear reinforcement

V_u, V_{ux}, V_{uy} = factored shear force; in the x direction; in the y direction (Fig. 11.2.1)

w = crack width at tension face of a section (Section 4.11)

w_c = density of concrete (for computing E_c)

w_D = service dead load per unit length or area

w_L = service live load per unit length or area

w_u = factored load per unit length or area

x = distance from compression face of section to neutral axis

     = M_n/V_u (Section 6.15)

     = short dimension of a rectangular element for use in computing torsional properties (Chapters 16 and 19)

x_1 = narrow dimension of core of a section subject to torsion (Fig. 19.6.3)

x_2 = narrow dimension center-to-center of longitudinal bars of a section subject to torsion (Fig. 19.6.3)

x_b = neutral axis distance for the balanced strain condition

y = long dimension of a rectangular element for use in computing torsional properties (Chapters 16 and 19)

y_1 = long dimension of core of a section subject to torsion (Fig. 19.6.3)

y_2 = long dimension center-to-center of longitudinal bars of a section subject to torsion (Fig. 19.6.3)

y_t = distance from neutral axis to extreme fiber of concrete in tension

z = factor of ACI Code relating to crack width, $f_s \sqrt[3]{d_c A}$ (Section 4.11)

LIST OF SYMBOLS

α = angle of inclination of shear reinforcement measured from axis of member

= $E_{cb}I_b/E_{cs}I_s$; ratio of flexural stiffness of beam section to the flexural stiffness of a width of slab bounded laterally by the centerline of adjacent panel, if any, on each side of the beam (Section 16.4)

= constant used to determine the load contour curve for biaxial bending at a constant compression strength P_n (Section 13.21)

= $PL^2/(\pi^2 EI)$ (Chapter 15)

α_{abc} = bar development multiplier (Section 6.7) relating to lateral spacing, cover, and confinement provisions of ACI-12.2.3.1 to 12.2.3.3

α_1, α_2 = α in the direction of L_1 and L_2, respectively (Section 16.12)

α_c = $\Sigma K_c/\Sigma(K_s + K_b)$, ratio of flexural stiffness of the columns above and below the slab to the combined flexural stiffness of the slab and beams at a joint taken in the direction moments are being determined (Sections 16.9 and 16.10)

α_{cd} = bar development multiplier (Section 6.7) relating to wide lateral spacing or confinement by closely spaced ties or spirals provisions of ACI-12.2.3.4 to 12.2.3.5

α_{epx} = bar development multiplier (Section 6.7) of ACI-12.2.4.3 relating to epoxy-coated bars

α_{exs} = bar development multiplier (Section 6.7) of ACI-12.2.5 relating to having excess reinforcement over that required

α_f = angle between shear-friction reinforcement and the shear plane

α_{lw} = bar development multiplier (Section 6.7) of ACI-12.2.4.2 relating to lightweight aggregate concrete

α_m = average α for all beams along the edges of a panel (Section 16.5)

α_{max} = angle measured from the vertical plane to the principal plane

α_{min} = minimum α_c to avoid using the positive moment multiplier (Chapter 16, Table 16.9.1)

α_s = parameter for two-way shear action; 40 for interior columns, 30 for edge columns, and 20 for corner columns (Sections 16.15 and 20.5)

α_t = dimensionless factor relating to torsional strength, $0.66 + 0.33y_1/x_1$ (Chapter 19)

α_{top} = bar development multiplier (Section 6.7) of ACI-12.2.4.1 relating to top bars (bars having more than 12 in. cast beneath them in one lift)

β = ratio of long side to short side of a rectangular spread footing (Chapter 20)

= M_{nx}/M_{0x} or M_{ny}/M_{0y} (Section 13.21)

= proportion of slab panel side a represented by corner yield pattern (Fig. 18.10.2)

β_a = ratio of service dead load per unit area to service live load per unit area (Section 16.9)

= deflection coefficient that depends on the degree of fixity at supports, the variation in moment of inertia along the span, and the distribution of the loading (Section 4.10 and Chapter 14)

β_b = ratio of area of longitudinal bars cut off to the total area of bars at the section (Section 6.12)

β_1 = ratio a/x; depth of rectangular stress distribution to the depth to the neutral axis

β_c = ratio of long side to short side of a concentrated load or reaction area

β_d = proportion of factored axial load that is sustained (Chapter 15)

β_t = $E_{cb}C/(2E_{cs}I_s)$ (Section 16.12)

β_w = $E_c f_r = 50.8w$ (Section 14.11)

γ = ratio $(d - d')/h$ (Fig. 13.16.4)

= ratio of distance from heel of footing to front face of wall to entire footing length L (Fig. 12.4.1)

= dimensionless constant relating to beam–column joint; 1.4 if joint is confined laterally by members framing in; 1.0 if joint is not adequately confined (Chapter 11)

= I_e/I_g (Section 14.11)

γ_D = I_e/I_g for $M_{max} = M_D$ (Section 14.11)

γ_{D+L} = I_e/I_g for $M_{max} = M_{D+L}$

γ_p = dimensionless constant used in predicting f_{ps} (Section 21.8)

δ = magnification factor

δ_b = magnification factor δ for a braced frame compression member (Section 15.9), $C_m/(1 - P_u/\phi P_c)$

δ_s = magnification factor δ for an unbraced frame compression member to account for P–Δ effect (Section 15.10)

= positive moment multiplier to account for pattern loading effect (Section 16.9)

Δ_{1u}, Δ_{2u} = lateral deflection under factored loads; first-order and second-order analysis, respectively

Δ = deflection

Δ_1 = additional secondary deflection on a braced member subject to transverse loading

Δ_{cp+sh} = creep and shrinkage deflections combined

Δ_{cp} = deflection due to creep alone (Section 14.6)

Δ_D = dead load deflection

Δ_{D+L} = dead plus live load deflection

Δ_i = instantaneous or short time deflection

Δ_L = live load deflection

$(\Delta_i)_D$ = instantaneous deflection due to dead load

$(\Delta_i)_{D+L}$ = instantaneous deflection due to dead plus live load

$(\Delta_i)_L$ = instantaneous deflection due to live load

Δ_m = total midspan deflection